Frontispiece. *Littorina (Neritrema) sitkana* and *L. (Littorina) scutulata* among barnacles on shore at Seldovia, Alaska (x1.6).

SYSTEMATICS AND EVOLUTION OF *LITTORINA*

David G. Reid

The Ray Society

Made and printed in Great Britain by
Henry Ling Limited
The Dorset Press, Dorchester, Dorset

Set in Photina

ISBN 0 903874 26 1

Sold by The Ray Society,
Registered Charity Number 208082
c/o Intercept Ltd., P.O. Box 716
Andover, Hampshire, SP10 1YG

CONTENTS

PREFACE

Snails of the genus *Littorina*, periwinkles to the English-speaking world, are small, drab, familiar, and are common on all the chill northern seashores. Yet, throughout the history of the study of molluscs, they have held an unusual fascination. What are the secrets of this appeal?

To the early shell collectors the subtle variety of these snails was a challenge to the emerging science of taxonomy. So it has remained, with the study of anatomy, genetics and molecular biology making successive contributions, resolving old controversies while revealing new ones. Even today, the name *Littorina* is synonymous with taxonomic difficulty. The first aim of this book is, therefore, to review the history of the taxonomic study of the genus and to produce the first comprehensive systematic monograph for more than a century. To achieve this, I have tried to gather the diverse literature, from historical taxonomic accounts to modern studies of population genetics. I have examined more than 4000 samples from the major museums of the world and have made extensive collections in the field. From this material, I have evaluated and described the most taxonomically useful morphological characters of the 19 living species. This will not be the final word on the classification of these species, and I have tried to highlight the areas of controversy to which future research will contribute.

Variation of form may pose problems for taxonomy, but it is the raw material of evolution. Indeed, the study of evolution is the study of variation, of its origin, its maintenance and its function. It has long been recognized that molluscan shells are particularly suitable subjects for such investigation. Of all marine gastropods, some species of *Littorina* are among the most variable in this respect, for their shells may be polymorphic in colour, diverse in sculpture, and strikingly different in shape from one shore to the next. My second aim has been to describe and illustrate this marvellous variation in the shells of *Littorina*, and to summarize the numerous studies that have examined its causes. Intraspecific variation exists at every scale, between geographical regions, among habitats, from one microhabitat to another on the same shore, and between individuals, male or female, juvenile or adult. Most previous studies have been limited to one or a few species, or to a small geographical area. By taking a broader approach, I have tried to show that there are some common patterns in the initially bewildering diversity of this variation. The causes of such patterns include the reduced gene flow within those species which lack widespread dispersal of pelagic eggs and larvae, the importance of wave action and of crab predation as forces of natural selection acting on those shell characters under genetic control, and the possible ecophenotypic effects of growth rate and salinity.

Where they occur, littorinids are generally abundant and may play a dominant role in ecological interactions on northern rocky shores. Species of *Littorina* exhibit a remarkable range of reproductive strategies, from the broadcasting of pelagic egg capsules to the brooding of developing embryos. They are hardy and amenable subjects for physiological and biochemical experiments. Many are known to harbour a range of parasites and to be preyed upon by crabs, fish and birds. They have been used as monitoring organisms in pollution assessment, some are edible, and one is commercially collected. For all these reasons, *Littorina* species have been intensively studied. Much of this work is outside the scope of this book. Nevertheless, a common theme running through many of these investigations is, or should be, a concern with evolutionary relationships. My third aim has therefore been to reconstruct the phylogeny of the living *Littorina* species, as a framework for their comparative biology. Evidence has come from the cladistic analysis of their morphological characters, from published allozyme studies, from recent analyses of DNA sequences and from the first survey of their known fossil record. While the phylogeny remains hypothetical, it receives strong support from the large measure of agreement revealed by these diverse approaches.

The science of genetics has provided a further dimension to the study of variation and here, too, this group has been a focus for research. Genetic studies of allozymes, and of the DNA itself, have been used to infer patterns of gene flow and of reproductive isolation. As a result, species of *Littorina* have emerged as model systems for research on genetic differentiation, including the microevolutionary processes leading to speciation. At the macroevolutionary level, however, *Littorina* species have been relatively neglected. Accordingly, my final aim has been to redress this, by considering the speciation processes that have produced their diversity, the biogeographic history that has resulted in their distribution, and the adaptations that have shaped their morphology.

It is my hope that, by providing a taxonomic and phylogenetic foundation, this book will contribute to the wider recognition of *Littorina* as a molluscan group of unparalleled opportunity for studies of evolution and comparative biology.

ACKNOWLEDGEMENTS

The idea for this work was first suggested by John Peake and John Taylor of The Natural History Museum, London, and without this it would not have been attempted. Throughout the course of my study I have greatly appreciated the help, encouragement and good fellowship of my closest colleagues at The Natural History Museum, in particular John Taylor, Peter Mordan and Kathie Way. I also record my gratitude to successive Keepers of Zoology at this museum, John Peake and Colin Curds, for their continued support. Among colleagues at other institutions, I owe much to the late Joe Houbrick (National Museum of Natural History, Washington, D.C.), to Alexander Golikov (Zoological Institute, St Petersburg) and to Anders Warén (Swedish Museum of Natural History, Stockholm), for their invaluable advice and inspiration. I have benefited greatly from the use of material and literature collected by the late Joe Rosewater (National Museum of Natural History, Washington, D.C.), who devoted much of his career to the study of the Littorinidae.

I have visited many other institutions and thank all the curators who have permitted me to examine the collections in their charge. I appreciate the patience and care of those who have sent loans of material from other museums: Elana Benamy (Academy of Natural Sciences of Philadelphia), Warren Blow (National Museum of Natural History, Washington, D.C.), Philippe Bouchet (Muséum National d'Histoire Naturelle, Paris), Jean DeMouthe (California Academy of Sciences), Mary Garback (Academy of Natural Sciences of Philadelphia), Raye Germon (National Museum of Natural History, Washington, D.C.), Shelley Greenhouse (National Museum of Natural History, Washington, D.C.), David Heppell (National Museum of Scotland), Sveinn Jakobsson (Icelandic Museum of Natural History), A. W. Janssen (Nationaal Natuurhistorisch Museum, Leiden), Michael Kellogg (California Academy of Sciences), Dick Kilburn (Natal Museum), Elizabeth Kools (California Academy of Sciences), Agnès Lauriat-Rage (Muséum National d'Histoire Naturelle, Paris), Jim McLean (Los Angeles County Museum of Natural History), Tom Meijer (Rijks Geologische Dienst, Haarlem), G. Pinna (Museo Civico di Storia Naturale di Milano), David Robinson (Academy of Natural Sciences of Philadelphia), Tom Schiøtte (Zoological Museum, Copenhagen), Sadako Takeuti (Saito Ho-on Kai Museum of Natural History, Sendai), Jane Topping (Canadian Museum of Nature, Ottawa) and Anders Warén (Swedish Museum of Natural History, Stockholm).

Many others have also very kindly collected and donated specimens for this study, or have lent material from their private collections: Sylvia Behrens Yamada, Elizabeth Boulding, Jim Carlton, Juliet Dyson, Ilídio Félix-Alves, Yuri Gladenkov, Jim Goedert, Andrei Granovitch, Tadashige Habe, Steve Hawkins, Keith Hiscock, Dietrich Kadolsky, Dan Levinson, Peter Long, Margaret & Jack Lukin, Louie Marincovich, Julia Nunn, Tomowo Ozawa, G. Høpner Petersen, Cesare Sacchi, Sergei Sergievsky, Tom Shirley, Shelagh Smith, Son Min Ho, Antony Sutcliffe, Andrei Tatarenkov, John Taylor, Satoru Uozumi, Domenico Voltolina, Thierry Warmoes, Tom Warwick, Kathie Way, Pam & Ron Woolford and Nadezhda Zaslavskaya.

During my own field trips, many friends and colleagues have provided assistance and hospitality: Kenji Aketa (Hokkaido), Luis Burnay (Portugal), Serge Gofas (France), John Grahame (U. K.), Steve Jeffcoat (U. K.), Alan Kohn (Washington), David McGrath (Ireland), Kinzo Matsubayashi (Kyushu), Akihiko Matsukuma (Kyushu), Peter van Marion (Norway), Aevar Petersen (Iceland), Emilio Rolán (Spain), Robin Seeley (Maine), Jon-Arne Sneli (Norway) and Kunihiko Takada (Hokkaido). Collecting trips were funded mostly by The Natural History Museum, and one in part by the Percy Sladen Fund of the Linnean Society of London.

At all stages of this work I have benefited greatly from the advice of colleagues. For help in locating and interpreting fossil material, I am indebted to Jim Goedert, Marji Johns, Dietrich Kadolsky, Tom Meijer, Noel Morris, Liz Nesbitt, Richard Preece, LouElla Saul, Geerat Vermeij and the late Phil Cambridge. The sequencing of mitochondrial DNA, which has contributed to a greater understanding of the phylogeny of *Littorina* species, was carried out at The Natural History Museum by Elaine Rumbak (supported by Natural Environment Research Council grant no. GR3/7854A, to Richard Thomas and DGR); the resulting data were analysed by Richard Thomas, who has given me much advice in this field. John Taylor helped in many ways, and in particular with the interpretation of shell microstructure. Ian Tittley kindly identified the marine algae. For locating obscure literature references I am grateful to Alan Kabat, and to Dietrich Kadolsky and David Heppell for advice on the taxonomic literature and for discussions on points of nomenclature. Kerstin & Bo Johannesson and John Grahame have most generously made available their manuscripts in the process of publication.

For their excellent photographic work I thank members of The Natural History Museum Photographic Unit, Phil Crabb, Pat Hart, Nick Hayes and Harry Taylor. Fine histological preparations were made by David Cooper. Assistance with electron microscopy was kindly provided by Sue Barnes, Chris Jones and Louisa Jones. I am most grateful to Stephen Ridgway for patient help with computer programs. For translation of Russian texts I am indebted to Zara Frenkiel, and of those in German to Klaus Sattler.

I record my most sincere thanks to those who have performed the onerous task of reading this book in manuscript form, Kerstin Johannesson, Dietrich Kadolsky, Alan Kohn and John Taylor; they made many suggestions for improvement, but those errors that remain are of course entirely my own. In addition, I am especially grateful to Elizabeth Platts, editor of this volume on behalf of the Ray Society, for her invaluable help and advice during its preparation for publication.

Chapter 1

INTRODUCTION

Littorina is one of the most thoroughly studied of all marine gastropod genera. The intensity and scope of research on this group can be seen from the references listed at the end of this book, of which over 1300 are concerned with *Littorina* species. For the past five years the *Zoological Record* has listed an average of 43 references to members of this genus annually. Research on littorinids has been the subject of three recent international symposia in a continuing series (Johannesson, Raffaelli & Hannaford Ellis, 1990; Grahame, Mill & Reid, 1992; Fioroni & Reid, 1994; Mill & McQuaid, 1995).

The scientific study of *Littorina* species began with early descriptions of shells and faunistic accounts. New species proliferated, and were based only on descriptions of shells. Several nineteenth century conchologists included monographs of *Littorina* in their great iconographies (Philippi, 1845–1850; Küster, 1853, 1856; Reeve, 1857–1858; Weinkauff, 1878, 1882; Tryon, 1887), although the concept of the genus was then wider than it is today, so that it included many littorinids now classified in other genera. With the emergence of evolutionary ideas late in the nineteenth century, species concepts became broader and the number of recognized species declined. The attentions of some conchologists turned to the description of the extreme intraspecific variation observed in some *Littorina* species (Dautzenberg & Fischer, 1912, 1915), and the first studies of shell morphometrics were made (Bumpus, 1898; Colman, 1932; Knudsen, 1949).

With the development of the new discipline of ecology, *Littorina* species figured largely in descriptive accounts of northern-temperate seashores (e.g. Ricketts & Calvin, 1939; Lewis, 1964; T. A. & A. Stephenson, 1972; Foster, de Vogelaere, Oliver, Pearse & Harrold, 1991; Mathieson, Penniman & Harris, 1991), and in numerous studies of their patterns of zonation and community structure (e.g. Colman, 1933; Dexter, 1947; Evans, 1947; Moyse & Nelson-Smith, 1963; Underwood, 1973). Indeed *Littorina* species are so characteristic of the upper tidal levels on rocky shores that, in their universal classification of littoral zones, T. A. & A. Stephenson (1949) defined the 'supralittoral fringe' as the zone dominated by *Littorina*. Detailed studies of the autecology of the more common species were made, such as that of *L. littorea* by Moore (1937, 1940) and of *L. saxatilis* by Berry (1961). *Littorina* species were found to be suitable subjects for the study of animal behaviour and in particular of movements in relation to zonation (e.g. Barkman, 1955; Newell, 1958; Underwood, 1972; Gendron, 1977; Petraitis, 1982), of activity rhythms (e.g. Petpiroon & Morgan, 1983) and feeding responses (e.g. Bakker, 1959; Watson & Norton, 1985a, 1987). They have also been widely used in physiological and biochemical studies, for example of temperature tolerance (e.g. Evans, 1948; Murphy & Johnson, 1980), osmoregulation (e.g. Rumsey, 1973; Berger, 1986a; Taylor & Andrews, 1988), respiration (e.g. Newell & Pye, 1971; Russell-Hunter & McMahon, 1974; McMahon & Russell-Hunter, 1977; Shumway, Lesser & Crisp, 1993; McMahon, Russell-Hunter & Aldridge, 1995), nutrition (e.g. Emerson & Duerr, 1967; Williams, 1970; Grahame, 1973; Merdsoy & Farley, 1973) and excretion (e.g. Duerr, 1968; Heil & Eichelberg, 1983; Taylor & Andrews, 1987). *Littorina* species are frequently infected by digenean trematodes, and there is a large literature on these parasites and their effects on snail populations (e.g. Stunkard, 1930, 1957; James, 1969; Robson & Williams, 1971; Lauckner, 1987; Granovitch, 1992).

Anatomical investigation of *Littorina* species was at first mainly concerned with the reproductive system (Linke, 1933), larvae and embryology (Delsman, 1914; Thorson, 1946), but in 1962 Fretter & Graham described *L. littorea* in detail, as a standard example of prosobranch anatomy. Soon afterwards, studies of anatomy began to make an impact on the taxonomy of the group. In Europe an additional three or four species were recognized from their reproductive structures (Sacchi & Rastelli, 1966; Heller, 1975a; Sacchi, 1975; Hannaford Ellis, 1979), and additional species were also found in the northern Pacific (Habe, 1979; Murray, 1979; Reid, Zaslavskaya & Sergievsky, 1991; Reid & Golikov, 1991). These unexpected discoveries, in a genus that had been considered taxonomically well known, coincided with great advances in the fields of molecular genetics, community ecology, and the study of variation and natural selection. As a result, the last quarter of the twentieth century has seen an explosion of interest in this group.

In the field of ecology, there have been studies of the population structure and dynamics of the newly distinguished species (Emson & Faller-Fritsch, 1976; Faller-Fritsch, 1977; Raffaelli, 1978a; Raffaelli & Hughes, 1978; Hughes, 1980; Atkinson & Newbury, 1984). Within the genus *Littorina* there is a remarkable diversity of types of development, including pelagic and encapsulated larval development and ovoviviparity, which provided the opportunity to test theories of reproductive strategy (Hughes & Roberts, 1980; Hart & Begon, 1982; Hughes, 1995). Another popular area of research has been the interaction between *Littorina* species and their foodplants, leading to the discovery of the important role of grazing in the community organization of temperate rocky shores (e.g. Lubchenco, 1980, 1982; Bertness, 1984; Petraitis, 1987; Janke, 1990; Williams, 1990a; Vadas, 1992). Competitive interactions among grazers have also been studied (Behrens Yamada & Mansour, 1987).

At first the extreme variability of shell form and coloration in *Littorina* species defied explanation (Fischer-Piette & Gaillard, 1971), but improved resolution of the taxonomy of the genus led to renewed investigation of this problem. Early observations described populations having patulous shells (with enlarged apertures and low spires) in wave-exposed habitats, and populations with larger, thick-walled shells in sheltered habitats. These trends were explained in terms of a genotypic response to natural selection imposed by wave action and crab predators (e.g. Heller, 1976; Raffaelli, 1979a; Elner & Raffaelli, 1980; Janson & Sundberg, 1983; Johannesson, 1986). This has been supported by measurements of survivorship in the field and estimates of the heritability of shell shape (Newkirk & Doyle, 1975; Janson, 1982a, b; Boulding & Hay, 1993; Boulding & van Alstyne, 1993; B. & K. Johannesson, 1996), although some direct ecophenotypic effects upon shell form have also been reported (Kemp & Bertness, 1984; Boulding, Buckland-Nicks & van Alstyne, 1993). That natural populations can indeed show a rapid and adaptive morphological response to the arrival of a new predator was shown by Seeley's (1986) study of *L. obtusata* and the crab *Carcinus*. Variation in colour is also striking in several *Littorina* species, and correlations between the frequencies of colour morphs and the visual properties of the background suggest that this variation is sometimes under selection (e.g. Heller, 1975b; Smith, 1976; Byers, 1990; Reid, 1993). In one case experiments with fish predators have shown that visual selection is indeed possible (Reimchen, 1979), although pleiotropic effects unconnected with visual appearance have also been implicated (e.g. Sergievsky, 1992).

The development of electrophoretic techniques for the investigation of enzyme variation has profoundly increased the understanding of interspecific differentiation and intraspecific variation among *Littorina* species. Allozyme studies have in many cases supported the recognition of new species that were first identified on anatomical grounds (e.g. Ward & Warwick, 1980; Mastro, Chow & Hedgecock, 1982; Ward & Janson, 1985; Mill & Grahame, 1988; Knight & Ward, 1991). However, in other instances distinctive shell varieties have been shown to be genetically so similar that they are regarded as 'ecotypes' of variable species (e.g. Moyse, Thorpe & Al-Hamadani, 1982; Janson & Ward, 1984, 1985; K. & B. Johannesson, 1990). An important difference has been revealed in the genetic structure of *Littorina* species according to their life histories; those with pelagic veliger larvae are genetically uniform over large geographical distances, as a result of high gene flow caused by larval dispersal; in contrast, those which lack a dispersive larval stage often show considerable differentiation between local populations, even over a scale of a few metres (e.g. E. M. Berger, 1973; Janson & Ward, 1984; Janson, 1987a; Johannesson, 1992; Tatarenkov & Johannesson, 1994). The localized genetic differentiation seen in the members of the latter group has important evolutionary implications, raising the possibility of sympatric speciation (Boulding, 1990; Johannesson & Sundberg, 1992; K. Johannesson, B. Johannesson & Rolán-Alvarez, 1993). Recently, studies of DNA have also been used to investigate interspecific and intraspecific differentiation (Crossland, Coates, Grahame & Mill, 1993, 1995).

Another new technique, the cladistic analysis of morphological characters, has contributed to an understanding of the macroevolution of *Littorina*. As a result, the genus has been redefined in a more restricted sense than before, as a monophyletic group, and relationships with other genera in the family Littorinidae have been investigated (Reid, 1989a). The same method has been used to reconstruct phylogenetic relationships among the species of *Littorina* (Reid, 1990b), and the result is in good agreement with phylogenetic analyses of allozyme frequencies (Ward, 1990; Backeljau & Warmoes, 1992; Zaslavskaya, Sergievsky & Tatarenkov, 1992a) and of mitochondrial DNA sequences (Rumbak, Reid & Thomas, 1994; Reid, Rumbak & Thomas, 1996). In combination with information from the sparse fossil record, and from the present geographical distributions of *Littorina* species, these phylogenetic studies have permitted analysis of the historical biogeography of the group, showing that the modern radiation of species in the northern Atlantic is the product of the late Pliocene trans-Arctic migration of ancestral species from the northern Pacific (Reid, 1990c).

Interest in *Littorina* has also been of a more applied nature. *Littorina littorea* is an edible species, formerly widely collected for food throughout its range in Europe and North America, and is still the basis of a commercial fishery in parts of Britain and France (e.g. Jeffreys, 1865; Nobre, 1886; Wright, 1936; Navaz y Sanz, 1942; Bequaert, 1943; Cole, 1956). Since they are such common and widespread molluscs of the intertidal zone, *Littorina* species have been widely used for monitoring pollution by heavy metals, organic compounds and radionuclides, and in laboratory studies of the physiological and cellular effects of such pollutants (e.g. Ireland & Wootton, 1977; Straughan & Hadley, 1978; Bryan, Langston, Hummerstone, Burt & Ho, 1983; Crothers, 1983; Livingstone, 1988; Mason, 1988; Brough & White, 1990; Cajaraville, Marigomez & Angulo, 1990; Marigomez & Ireland, 1990; McKay & Fox, 1991; Bjerregaard & Depledge, 1994). In common with some other gastropods, *Littorina* species also show development of male characteristics in the female reproductive system in response to the antifouling compound tributyltin (Fioroni, Deutsch, Stroben & Oehlmann, 1992; Bauer, Fioroni, Ide, Liebe, Oehlmann, Stroben & Watermann, 1995).

This brief review can give only an indication of the broad scope and great volume of research on *Littorina* species. Some aspects of this work have been reviewed in detail by others, including anatomy (Fretter & Graham, 1962, 1994; Reid, 1989a, 1990b), ecology (Clay, 1967; Fretter & Graham, 1980, 1994; Raffaelli, 1982; Faller-Fritsch & Emson, 1985), predators (Pettitt, 1975), grazing (Norton, Hawkins, Manley, Williams & Watson, 1990) and biochemical genetics (Ward, 1990). A bibliography of the literature on littorinids has also appeared (Pettitt, 1974a, b, 1979).

The present work is arranged in three main parts. First there are accounts of the morphological characters, of both shell and anatomy, that have been used in the taxonomy of *Littorina* (Chapter 3). Taking a broad view of variation within the genus as a whole, the known biological functions of these characters are reviewed, as well as the causes and significance of their intraspecific variation. This provides a basis for the major part of the work, a systematic monograph of all 19 recognized Recent and four fossil species of *Littorina* (plus six fossil species of uncertain status) (Chapter 4). For each one there is an extensive synonymy, listing of type specimens, and a review of its taxonomic history; for the most variable and well-studied species this is complex, and their histories are often illustrative of broader trends in malacological systematics. There follows a brief diagnosis with the main features for identification, and then detailed descriptions of anatomy and shell. Intraspecific variation has long confused and preoccupied workers on *Littorina*; here, in contrast to almost all previous studies of the group, a broad geographical approach is taken, describing variation throughout the range. The numerous recent studies of local and small-scale variation in shell form are reviewed, and comparisons are drawn between species in order to demonstrate common trends. A brief summary of habitat, zonation on the shore, and diet is given, as well as the geographical distribution, which is now known in detail for the majority of species. Fossil records of *Littorina* are sparse and scattered, and for the first time the fossil history of the whole group is considered. For each species a review is given of the contribution made by recent advances in biochemical genetics to an understanding of patterns of variation and microevolution. In the third part the macroevolutionary history of this entire monophyletic group is examined in detail (Chapters 5 and 6). Evidence is gathered from cladistic analysis of morphological characters, from the fossil record, allozymes and DNA analyses, to reconstruct a consensus phylogeny for the genus. This is integrated with distributional data to consider historical biogeography and speciation mechanisms, and phylogenetic correlations between morphological and ecological traits are used to reveal likely adaptive characters. As a result, the evolutionary radiation of the genus *Littorina* is now one of the best understood among marine molluscs.

Chapter 2

MATERIAL AND METHODS

Synonymies and types

Nineteen Recent and four fossil species of *Littorina* are here recognized as valid, and described in the Systematic Descriptions (Chapter 4). In addition, a further six fossil species of uncertain status are discussed. In the synonymies the attempt has been made to include all new names of specific, subspecific or infrasubspecific rank (including emendations and *nomina nuda*), which total 327. All new combinations have also been included. Intentional emendations of generic and specific names have been listed individually, but misspellings have been entered under the correct spelling, and noted. Otherwise, the synonymies are comprehensive, but not exhaustive; criteria used in their compilation have included major taxonomic monographs and faunistic works, good illustrations, widely used identification guides, anatomical works (excluding those relating to organ systems not discussed herein, unless there is a systematic component), genetic and biochemical studies that have drawn systematic conclusions, and descriptions of shell form or coloration and its variation. Ecological and physiological studies, faunal lists and atlases have not been listed in the synonymies, unless they qualify on other grounds.

Frequently, authors have included several species under a single name. This may be clear from given figures, descriptions or distributions, or from references to other published descriptions or figures. In such cases, an entry has been made in the synonymy of each of the included species, with the note 'in part', and a reference to the others. Some species cannot always be recognized unequivocally from descriptions or figures, for example members of the *Littorina scutulata* group, *L. sitkana* group, *L. obtusata* group and *L. saxatilis* group. Where it is unclear, but likely, that more than one species was included, the entry is followed by '*s.l.*' (*sensu lato*) and appears only in the synonymy of the species after which the group is informally named. A query preceding an entry in a synonymy indicates uncertainty about the species involved.

The synonymies are given in condensed form, following each name of specific or lower rank through its various combinations, in chronological order of first usage of each combination. For a new name, the authority follows the name without intervening punctuation, and type locality, location of types and notes are given in parentheses. New names (including unjustified emendations) are regarded as available, unless indicated as *nomina nuda*, misspellings, monstrosities, hybrids or infrasubspecific names, in which case they are unavailable (International Commission on Zoological Nomenclature, 1985: arts 1b, 10–20). If a new name is a junior primary or secondary homonym, the senior homonym follows the word 'not'. Subsequent users of the name in the same combination follow a full stop. Subsequent users of a name in a new combination are distinguished by a long dash between the name and the author; if the usage is a misidentification, this is indicated in parentheses by the word 'not' followed by the original form of the incorrectly used name. An equality sign precedes the recognized valid name (if different from the original) of homonyms and misused taxa. Figures referred to are of shells unless otherwise indicated.

All available type specimens have been examined, as indicated by the word 'seen'. A few have not been located, or are considered lost, as noted. Lectotypes have been designated only as required for valid species names, or where necessary to resolve taxonomic confusion, as for example where syntypes are clearly or potentially not all conspecific. Similarly, neotypes have been designated only for valid species names, or in order to remove uncertainty about the identities of taxa for which types are lost. The holotype, lectotype, neotype or, in a few instances, a paratype or paralectotype, of each valid species is illustrated. One new species is described, and one replacement name introduced.

Many new names in this conchologically variable genus were originally introduced as 'varieties', 'forms' or other infrasubspecific categories, and these require special comment. Although names that are definitely infrasubspecific have been included in the synonymies, they are excluded from the species group and from nomenclatural consideration (ICZN, 1985: art. 45e); types are therefore not designated, although original localities and the whereabouts of original and figured specimens are mentioned, if known. Obviously infrasubspecific names include those first introduced as an addition to a trinomen (ICZN, 1985: art. 45 f). Names introduced as 'varieties' or 'forms' can be interpreted to be of either subspecific or infrasubspecific rank. If published after 1960 they are infrasubspecific; if published before 1961 they are subspecific, 'however, if the content of the work reveals that infrasubspecific rank is meant, the name is infrasubspecific unless, prior to 1985, it has been treated as an available name and either adopted as the name of a species or subspecies or treated as a senior homonym, in which cases the name is deemed to be subspecific from the date of its establishment' (ICZN, 1985: art. 45g). Therefore, varieties and forms introduced before 1961 are generally treated as available names, with type specimens. Exceptions are the 'var.', 'var. ex col.', 'var. ex sculpt.', 'var. hybr.' and 'form' categories of Middendorff (1849, 1851), Schrenck (1867), Daniel (1883), Dautzenberg and co-workers (Dautzenberg, 1887, 1893a, b; Dautzenberg & Durouchoux, 1900, 1906, 1913; Dautzenberg & Fischer, 1912, 1915, 1925), Pallary (in Seurat, 1924) and Schlesch (1930); all these are regarded as infrasubspecific, as is evident from the content of the works. These unavailable names do not enter into homonymy. Nevertheless, homonyms are indicated for these names, in case some may in future be deemed to be available.

It should be noted that the names used in the text for 'geographical varieties' are intentionally infrasubspecific, with no nomenclatural standing.

Material

This monograph is based upon the examination of all material in the following institutions: The Natural History Museum (London), National Museum of Scotland (Edinburgh), National Museum of Ireland (Dublin), Muséum National d'Histoire Naturelle (Paris), Museum für Naturkunde (Berlin), Zoological Museum (Copenhagen), Naturhistoriska Riksmuseet (Stockholm), Zoological Museum (University of Oslo), Trondheim University Museum, Icelandic Museum of Natural History (Reykjavík), Zoological Museum (University of Moscow), Zoological Institute (St Petersburg), National Museum of Natural History (Smithsonian Institution, Washington, D.C.) and National Science Museum (Tokyo). Additional material and type specimens have been borrowed from other institutions, as noted. Field observations and collections have been made throughout the British Isles, in France, Spain, Portugal, Morocco, Italy, Tunisia, Sweden, Norway, Iceland, Maine, New Brunswick, California, Washington, Alaska and Japan. Personally collected material is deposited in The Natural History Museum (London). All figured shells and anatomical material are from the collection of The Natural History Museum (London), except as noted.

The numbers of lots, excluding fossils, of Recent species examined (4258 in total), and of specimens dissected, are listed in Table 1. Fossil material is much less well represented in collections, and numbers of specimens seen are mentioned in the Systematic Descriptions.

Anatomical characters

Preservation. Living specimens were relaxed in a solution of 7.5% magnesium chloride (7.5 g hydrated crystals in 100 ml distilled water) for 2 to 10 hours as necessary. Fixation was done in 10% buffered seawater formalin (1 part 40% aqueous solution of formaldehyde diluted with 9 parts seawater and buffered with excess borax) for at least one week, before rinsing in water and storage in 80% ethanol.

Head-foot. Pigmentation of the body was unaffected by preservation, and was described from preserved specimens. The spiral coiling of the operculum was described by the opercular ratio (diameter of spiral part divided by maximum length, Fig. 2C).

Penis (Fig. 1A). Penes were drawn by camera lucida from relaxed living specimens, from material preserved as described above (which did not differ significantly in shape from relaxed living specimens) or, where noted, from museum material of unknown preparation. In fact the method of preservation, including whether or not the animals were relaxed before fixation, appeared to have no appreciable effect upon the final penial shape of preserved animals. The penial filament was defined as the portion between the tip and the distal mamilliform penial gland, or (in those species lacking mamilliform glands) as the smooth distal portion lacking annular wrinkles. Immature individuals, and those parasitized by trematodes (with digenean cercariae visible in the digestive gland and gonad), were not

Table 1. Numbers of lots, excluding fossils, of Recent species of *Littorina* examined, with numbers of individual penes and pallial oviducts dissected (histological preparations in parentheses), samples of sperm and spawn (or embryos, for *L. saxatilis*), and numbers of radulae examined with scanning electron microscope.

Species	Lots	Penes	Sperm	Oviducts	Spawn or embryos	Radulae
L. striata	29	10	7	8 (2)	1	7
L. keenae	61	15 (1)	3	11 (1)	—	4
L. scutulata	54	39	3	39	—	6
L. plena	63	61	6	45	—	6
L. squalida	95	11	2	17	—	6
L. littorea	693	19	3	15 (1)	—	6
L. kasatka	12	19	2	21 (2)	—	11
L. brevicula	136	28	6	31	—	6
L. mandshurica	55	28	6	16	—	6
L. sitkana	250	73	2	156	2	10
L. horikawai	26	7	1	2	—	4
L. subrotundata	73	44	4	82	2	20
L. aleutica	27	22	3	22	1	8
L. natica	20	36	—	24	2	11
L. obtusata	636	90 (4)	21	72 (3)	2	16
L. fabalis	424	78 (1)	14	60 (1)	1	15
L. compressa	267	92	4	77	—	14
L. arcana	104	242	5	298	—	9
L. saxatilis	1233	389	19	802 (1)	54	26

used. In the anatomical drawings the dotted flask-shaped outlines within the mamilliform penial glands represent the muscular reservoirs containing their secretion, which are often visible by transparency; this convention also serves to indicate the number of mamilliform glands when these are not visible in lateral view. It should be noted that reduction in size, and even loss, of the penis occurs in many species outside the breeding season.

Sperm. Sperm were extracted by fine glass pipette from the seminal vesicle of relaxed, living specimens. Samples were fixed in a 1% solution of formalin in seawater, and examined under a coverslip with a compound microscope, at high magnification (x1000), unstained and with standard bright-field illumination. Lengths of euspermatozoa and maximum diameters of paraspermatozoa were measured with a calibrated eyepiece graticule at least five times in each sample. In some cases sperm were examined from whole animals, previously fixed and stored in seawater formalin; although more difficult to observe, because of agglutination of euspermatozoa, the sizes and shapes of paraspermatozoa were unchanged, as shown by comparison with samples freshly prepared (as described above) from some of the same individuals. Storage in 80% ethanol resulted in a shrinkage of about 20% in the paraspermatozoa.

Oviduct (Fig. 1B). Preserved animals were used for dissection. The pallial oviduct can be seen on the right side of a female following removal of the shell. In most cases the spiral form of the egg groove was visible without dissection, as were the positions of the albumen and capsule glands. The capsule gland (including the covering gland, p. 17) was distinguished by its opaque white or cream colour, often with a reddish or brown translucent portion, contrasting with the somewhat translucent whitish albumen and jelly glands. The arrangement of spiral loops of the egg groove was sometimes obscured in those individuals in the process of spawning, when the glands are enlarged and swollen, and was often most clearly visible in slightly immature or spent animals. Immature individuals with a small oviduct, and those with digenean parasites in the digestive gland and gonad, were not used. Dissection was necessary to trace the extent of the renal oviduct and of the copulatory bursa. The seminal receptacle was visible as an opalescent bulb. To confirm the interpretation of the structure of the oviduct, serial histological sections were prepared of several species (Table 1). These sections were stained with Masson's trichrome, which distinguishes the blue albumen gland and bluish translucent capsule gland from the red opaque capsule gland (Reid, 1986a). A standard shading convention is adopted in all figured oviducts: sparse stipple, albumen gland; intermediate stipple, opaque capsule gland and covering gland; dense stipple, seminal receptacle; vertical hatching, translucent capsule

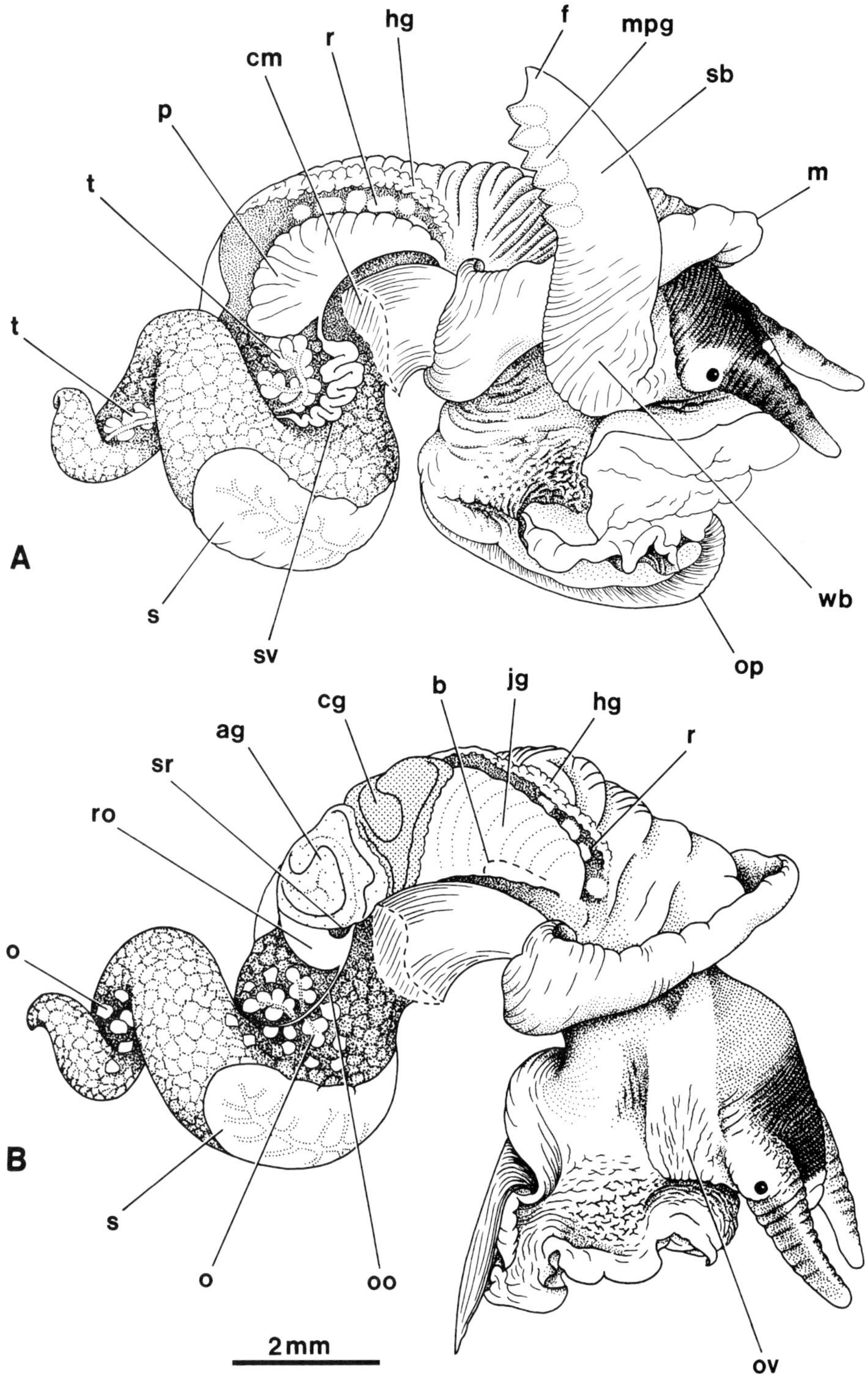

Figure 1. Mature male (**A**) and female (**B**) of a representative *Littorina* species (*L. (Neritrema) compressa*, Helford River, Cornwall, England), showing principal organs visible in animals removed from their shells. Specimens are relaxed, intact and undissected, and are shown as in life, except that the mantle edge has been recurved to show the head-foot, and in the male the penis, which normally lies folded back within the mantle cavity, has been lifted outside the mantle. Abbreviations: ag, albumen gland of pallial oviduct; b, copulatory bursa (visible by dissection); cg, opaque capsule gland (including covering gland) of pallial oviduct; cm, insertion of columellar muscle on shell; f, filament of penis; hg, hypobranchial gland; jg, jelly gland of pallial oviduct; m, mantle edge (recurved); mpg, mamilliform penial gland (visible by transparency); o, ovary (ramifying in digestive gland); oo, ovarian oviduct; op, operculum; ov, ovipositor; p, prostate gland; r, rectum; ro, renal oviduct; s, stomach; sb, smooth region of penial base, with opaque subepithelial glandular tissue; sr, seminal receptacle; sv, seminal vesicle; t, testis (ramifying in digestive gland); wb, wrinkled region of penial base.

gland; solid black, lumen; thickest line, route of egg groove (visible by transparency); dashed line, extent of copulatory bursa (usually only visible by sectioning or dissection); dotted lines, septa of jelly gland (or, in *L. saxatilis*, brood pouch) visible by tranparency.

Spawn. *Littorina striata* produced egg capsules while held temporarily in small containers of sea water in the laboratory. Benthic spawn masses of *L. sitkana*, *L. obtusata* and *L. fabalis* were collected in the field, and those of *L. subrotundata*, *L. aleutica* and *L. natica* were available in museum collections. For most other species, details of spawn were available from the literature.

Radula and foregut. Entire radulae were dissected from the buccal mass and uncoiled; relative radular length was measured as total radular length divided by shell height. Only radulae from adult animals were used, unless otherwise noted. Radulae were cleaned of tissue and mucus by soaking in a bleaching solution (1% w/v sodium hypochlorite, 8% w/v sodium chloride) for up to 5 minutes, followed by gentle cleaning with a fine paintbrush and thorough rinsing in three changes of distilled water. Comparison with radulae cleaned only mechanically confirmed that the chemical cleaning did not affect the appearance of the teeth. Heating was avoided, since this caused the cusps of the teeth to become recurved. Each cleaned radula was mounted damp on a thin layer of polyvinyl acetate glue on a glass coverslip, and the outer marginal teeth quickly folded outwards with fine needles before the radula had dried. The specimens were coated with gold and palladium before examination in a scanning electron microscope. Teeth from the central part of the radula were examined, where they were fully formed, but unworn. Scanning electron micrographs were taken in each of three orientations: in standard flat view from vertically above the radula (to show shapes of teeth), at an angle of 45° from the front end of the radula (to show shapes of tooth cusps), and at an angle of 45° from the side of the radula (to show relief). For each radula examined, the length to width ratio of one rachidian tooth was measured (total length of tooth divided by maximum basal width, in standard flat orientation).

Salivary glands were examined by dissection and easily traced by their pinkish colour, lying beneath or (if enlarged) surrounding the mid-oesophagus.

Shell characters

Shell height (H) was measured as the maximum dimension parallel to the axis of coiling, and shell breadth (B) as the maximum dimension perpendicular to this when the shell was held in standard apertural view (Fig. 2A, B). The length of the aperture (LA) was measured as the distance from the junction of the outer lip with the body whorl to the furthest point of the anterior lip, in the plane of the aperture. The width of the columella (C) was measured in *L. obtusata* and *L. fabalis*, as the maximum width from pillar to inner lip, including any pseudo-umbilical area (Fig. 2B). Measurements were made with vernier calipers, or for small specimens with the calibrated eyepiece graticule of a binocular microscope, to 0.1 mm. Only adult shells (with a thickened, slowly growing apertural margin) were measured. Three simple ratios were used as indices of shell shape: the height to breadth ratio (H/B), the relative spire height (SH=H/LA), and the relative columellar width (C/LA).

Details of shell sculpture and protoconch are illustrated by scanning electron micrographs. The protoconch is the larval shell formed before metamorphosis; in species with planktotrophic development it is terminated by a sinusigeral rib, whereas in those with nonplanktotrophic development and intracapsular metamorphosis it is terminated by a straight growth line, after which the spiral sculpture of the teleoconch begins. The number of whorls and diameter of the protoconch were measured as shown in Fig. 2D. Intact protoconchs are rarely found, except in the smallest juveniles. The finest spiral striations of the teleoconch are termed microstriae. Coarser sculptural elements are termed ribs and grooves, and the strongest ribs are described as carinae. The ribs and grooves are often too irregular and variable in strength to permit precise, repeatable counts of their numbers.

Details of shell mineralogy were taken from Taylor & Reid (1990), and for additional Recent species the same techniques of acetate peels and scanning electron microscopy of ground and etched surfaces were employed. For rare fossil specimens which could not be sectioned, shell layers and crystal structure could sometimes be seen with a high-powered binocular microscope, and sometimes with the scanning electron microscope at broken shell edges, provided that recrystallization had not occurred.

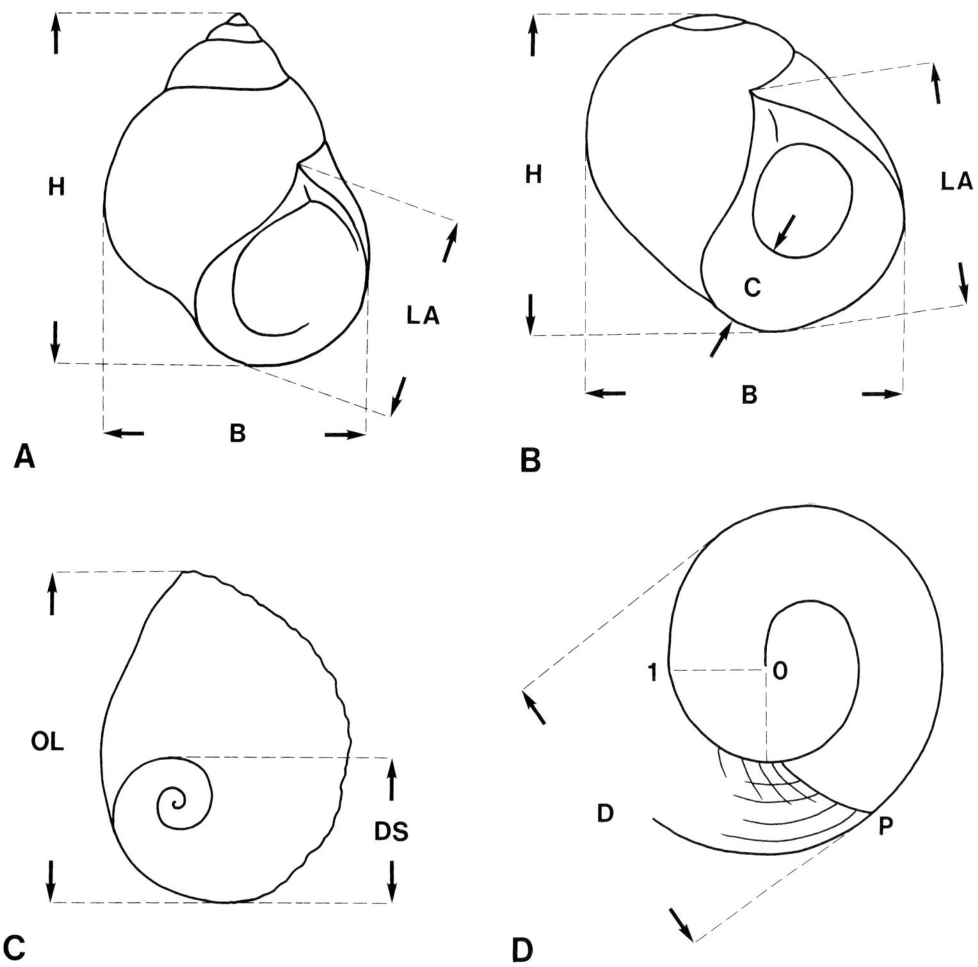

Figure 2. Measurements of shells (**A,** *Littorina littorea*; **B,** *L. fabalis*), operculum (**C,** *L. littorea*) and protoconch (**D,** *L. horikawai*). Abbreviations: B, shell breadth; C, width of columella; D, diameter of protoconch; DS, diameter of opercular spiral; H, shell height; LA, length of aperture; OL, total length of operculum; P, end of protoconch (1.7 whorls in this example) and origin of teleoconch sculpture; 0, origin of suture of protoconch; 1, end of first whorl of protoconch.

Distribution

Information on habitat, zonation and diet was summarized from the literature, and from personal observation in the field. For the description of zonation on the shore, the system adopted is that of biologically-defined zones, as developed by T. A. & A. Stephenson (1949) and modified by Lewis (1964) (see Russell, 1991, for review of alternative systems). The littoral zone is divided into the upper littoral fringe (from the upper limit of barnacles to the top of the belt of littorinids and the black lichen *Verrucaria*), and the lower eulittoral zone (from the upper limit of barnacles down to the upper limit of the kelp *Laminaria*), and below this the sublittoral zone. For fossil records a combination of examined specimens and published records was used, as indicated. Maps of the modern geographical distributions were compiled almost entirely from material examined (Table 1). For most species, shells alone were adequate for unequivocal identification, so records based on dry or anatomical material are not distinguished on the maps, and both appear as circles. For those species requiring anatomical examination for unequivocal identification (*L. scutulata, L. plena, L. arcana, L. saxatilis*), probable records based on identification of shells alone are indicated as triangles. Occasionally, reliable published records have been included on the maps, as asterisks, where they extend the geographical range. Full lists of material examined are too voluminous for inclusion, but have been deposited in The Natural History Museum (London).

Comparisons of species

Descriptions are provided of the characters that are most useful for discrimination between closely related, or morphologically similar, species. These are tabulated for groups of species among which confusion is most likely. In cases of comparison between pairs of species, the 'percentage accuracy' listed for various characters reflects the percentage of the combined samples (see Table 1 for sample sizes) of the two species that would have been misidentified using that character alone (actual identification having been established using those characters which are taken as diagnostic for each species, i.e. 100% accurate).

Phylogenetic analysis

Details of the cladistic analysis of morphological characters are given in Chapter 5. The sequence of the species in the Systematic Descriptions (Chapter 4) follows, as far as possible, a phylogenetic arrangement (compare with Fig. 119).

Abbreviations

Throughout the text, only the generic name *Littorina* is abbreviated as '*L.*'. DGR indicates personal collection by the author, of material deposited in The Natural History Museum (London). ICZN is the International Code of Zoological Nomenclature. The following abbreviations are used for institutions:

BMNH	The Natural History Museum, London
CAS	California Academy of Sciences, San Francisco
DMNH	Delaware Museum of Natural History, Greenville
IM	Ipswich Museum, Suffolk
IMNH	Icelandic Museum of Natural History, Reykjavík
IRSNB	Institut Royal des Sciences Naturelles de Belgique, Brussels
LACM	Los Angeles County Museum of Natural History
LSL	Linnean Society of London
MCZ	Museum of Comparative Zoology, Harvard University
MCSNM	Museo Civico di Storia Naturale di Milano
MNHNP	Muséum National d'Histoire Naturelle, Paris
NCM	Norwich Castle Museum, Norfolk
NMC	Canadian Museum of Nature, Ottawa
NMI	National Museum of Ireland, Dublin
NMS	National Museum of Scotland, Edinburgh
NNML	Nationaal Natuurhistorisch Museum, Leiden
NSMT	National Science Museum, Tokyo
PMM	Palaeontological Museum, Moscow
RAMM	Royal Albert Memorial Museum, Exeter, Devon
RGD	Rijks Geologische Dienst, Haarlem
SHKM	Saito Ho-on Kai Museum of Natural History, Sendai
SM	Wood End Museum of Natural History, Scarborough, Yorkshire
SMC	Sedgwick Museum, Cambridge University
SMNH	Naturhistoriska Riksmuseet, Stockholm
TM	Trondheim University Museum
UCLA	University of California, Los Angeles
UCMP	Museum of Paleontology, University of California, Berkeley
USNM	National Museum of Natural History, Smithsonian Institution, Washington, D.C.
UWP	Burke Museum, University of Washington, Seattle
ZISP	Zoological Institute, St Petersburg
ZMB	Museum für Naturkunde, Berlin
ZMC	Cambridge University Museum of Zoology
ZMO	Zoological Museum, University of Oslo
ZMK	Zoological Museum, University of Copenhagen

Excluded species

Before the generic revisions of the Littorinidae by Bandel & Kadolsky (1982) and Reid (1986a, 1989a), many unrelated littorinids were included in the genus *Littorina*, because of the superficial similarity of their smooth or spirally striated shells (e.g. Rosewater, 1970). The genus is now defined in a narrow, phylogenetic sense as a monophyletic group almost entirely restricted to the Northern Hemisphere (p. 39). Within this region a number of species which have in the past been familiar as members of *Littorina* have now been reassigned to other genera. These include *Melarhaphe neritoides* (Linnaeus, 1758), *Nodilittorina punctata* (Gmelin, 1791), *Nodilittorina ziczac* (Gmelin, 1791), *Littoraria irrorata* (Say, 1822) and *Littoraria pintado pullata* (Carpenter, 1864).

The genus *Mainwaringia* Nevill, 1885, is an enigmatic group consisting of two species from Southeast Asia and India, *M. leithii* (E. A. Smith, 1876) and *M. rhizophila* Reid, 1986 (Reid, 1986b). Although well-known anatomically, its relationships have in the past been unclear. In a cladistic analysis of the morphological characters of all littorinid subgenera, *Mainwaringia* appeared near the base of the Littorininae in the most parsimonious tree, although subjective weighting of certain characters suggested the possibility of inclusion within the clade *Littorina*, in a more derived position (Reid, 1989a). Possible synapomorphies with *Littorina* include the paraspermatozoa without rod pieces, the form of the egg groove in the pallial oviduct, the shape of the egg capsules, and some aspects of radular tooth shape. However, other characters that are plesiomorphic within the Littorininae, supporting its basal position in the subfamily, include the absence of a longitudinal division of the foot, the closure of the penial vas deferens, the aragonitic shell, and salivary glands partly anterior to the circumoesophageal nerve ring. Some of these apparently plesiomorphic features could, however, be explained as adaptations to its unusual tropical and muddy habitat (Reid, 1990b; Taylor & Reid, 1990). Furthermore, its geographical distribution could be accounted for by the biogeographical model of Reid (1990c), which proposed a close correspondence between phylogenetic position and geographical distribution in the genus *Littorina*. For these reasons, Reid (1990b) tentatively included the two *Mainwaringia* species within *Littorina*. However, the results of phylogenetic analysis of mitochondrial DNA sequences have recently shown that *Mainwaringia* is certainly not a member of the clade *Littorina*, and must be excluded from this genus (Reid, Rumbak & Thomas, 1996).

Some other taxa that are here excluded from *Littorina*, or regarded as indeterminate, are listed below:

Turbo aureus Brown, 1827=*Homalopoma sanguineum* (Linnaeus, 1758), according to Forbes & Hanley (1852: 259).
Littorina ? globosa Jeffreys, 1884. Indeterminate fossil.
Littorina headleyi Bell, 1919. New name for *Littorina globosa* Jeffreys, 1884.
Nerita littoralis Linnaeus, 1758=*Theodoxus fluviatilis* (Linnaeus, 1758), see p. 202.
Algamorda parva Wang, 1982. Not a *Littorina* species, see p. 179.
Littorina pedroana Conrad, 1855. A *Lacuna* species, see p. 80.
Nerita rapula 'Chiereghini' Nardo, 1847=*Cyclope neritea* (Linnaeus, 1758), according to Brusina (1870).
Littorina rufa 'Récluz' Reeve, 1857. Types not in BMNH. Indeterminate.

Chapter 3

MORPHOLOGY

In this chapter the morphological characters that are most useful for the taxonomic discrimination and phylogenetic analysis of *Littorina* species are described and evaluated, using the traditional, although artificial, division into features of the soft anatomy and of the adult shell (teleoconch). Information about the functional significance of the characters is included, but discussions of their evolution and adaptation are deferred until Chapter 6, where they are examined in a phylogenetic context. Comparisons with other genera of the Littorinidae are limited, having been discussed in detail elsewhere (Reid, 1989a). Common patterns of variation in *Littorina* species are emphasized, so that the Systematic Descriptions (Chapter 4) can be understood in the context of variation within the genus as a whole. Intraspecific variation has been studied mainly in characters of the shell, which are most accessible and easily measured. Indeed, the extreme intraspecific variability in shell form and coloration shown by some species of *Littorina* is one of the most notable features of the genus. It has also been the source of most of the taxonomic confusion that has surrounded this group. Only since species have been defined using stable anatomical characters, and their coherence confirmed by molecular data, has it been possible to understand and explain this variability. Anatomical characters will therefore be considered first.

Anatomical characters

Head-foot and operculum

Most *Littorina* species show intraspecific variation in the degree to which black pigmentation is developed on the head, tentacles and sides of the foot. In *L. saxatilis s.l.* James (1968a) found that newly-emerged juveniles were unpigmented, and reported gradual darkening of the pattern throughout life. However, darkening with age, at least among larger snails, was not confirmed by Davis (1973) in *L. saxatilis s.s.*, who instead described wide variation in pigmentation within populations, as seems generally the case throughout the genus. In *L. obtusata* and *L. fabalis*, which both exhibit striking colour polymorphism of the shell, there is a partial correlation between the colour of the shell and the degree of pigmentation of the head-foot, yellow shells usually containing unpigmented animals, and brown or olive shells usually black animals, but this is not invariably the case and much variation is known (e.g. Sacchi, 1974; Smith, 1976). Parasitism by digenean trematodes may cause the sole of the foot to become orange or brown, as a result of the release of carotenoid pigments from the damaged digestive gland (Willey & Gross, 1957; Huxham, Raffaelli & Pike, 1993), but is not known to affect the distribution of black pigment on the rest of the body. Differences in black pigmentation have sometimes been used to characterize local populations or ecotypes (James, 1968a; Davis, 1973; Rolán & Templado, 1987). Despite this variability, pigmentation of the head-foot, particularly of the head and tentacles, has occasionally been used in identification, for example in the discrimination of *L. littorea* from *L. saxatilis s.l.* (James, 1968b), and of *L. sitkana* from *L. subrotundata* (Boulding, Buckland-Nicks & van Alstyne, 1993). Pigmentation of the ovipositor (Goodwin & Fish, 1977) and of the penis (Boulding *et al.*, 1993) have also been emphasized, but in the present study are shown to be subject to much intraspecific variation. Only in the case of the conchologically similar pair, *L. scutulata* and *L. plena*, has pigmentation of the head and tentacles been found to be a reliable and useful means of identification.

The shape of the operculum, as measured by the opercular ratio, is recorded in the Systematic Descriptions. While there are apparently some differences between species, this has not been used for identification. In *Nodilittorina* Bandel & Kadolsky (1982) found a partial correlation between opercular shape and tidal zonation level, and suggested that a more tightly coiled operculum functioned both to close the more circular aperture of high-level species, and to provide an increased thickness of opercular material for reduction of evaporative water loss. No such correlation has been found, either with tidal level or latitude, among *Littorina* species.

Male reproductive system

The form of the penis has been recognized as one of the most useful taxonomic characters at the specific level in the Littorinidae since the early comparative studies of Abbott (1954b) and Rosewater (1970). In *Littorina* it has often provided one of the first pieces of evidence for the separation of conchologically similar species, for example *L. fabalis* and *L. obtusata* (Sacchi & Rastelli, 1966), *L. compressa* and *L. saxatilis* (Heller, 1975a; Sacchi, 1975), *L. arcana* and *L. saxatilis* (Hannaford Ellis, 1979), *L. plena* and *L. scutulata* (Murray, 1979), and both *L. kasatka* and *L. natica* from other species similar to *L. sitkana* (Reid, Zaslavskaya & Sergievsky, 1991; Reid & Golikov, 1991). The importance of this character is emphasized throughout the Systematic Descriptions (Chapter 4).

The anatomy of the male reproductive system in the Littorinidae has been reviewed by Reid (1986a, 1989a), and that of *Littorina* described in detail by Linke (1933) and Fretter & Graham (1962, 1994). The tubules of the testis ramify in the digestive gland and connect to a coiled seminal vesicle, and thence to the prostate gland, which is a deep groove between two flaps of glandular tissue in the pallial wall on the right side of the animal (Fig. 1A). The penis is a large, muscular appendage situated on the right side, just below and behind the head, with an open, ciliated, groove (the penial vas deferens) along its antero-dorsal edge, through which sperm are conducted. Three types of glandular structures have been distinguished on the penis of *Littorina* (Reid, 1989a). Most obvious are the mamilliform penial glands, visible as prominent papillae along the postero-ventral edge and adaxial surface of the wrinkled penial base. Each of these mamilliform glands consists of several clusters of subepithelial glandular cells, which discharge through cytoplasmic extensions into a single large reservoir, lined by a tall epithelium and surrounded by a muscular coat (often partly visible externally, by transparency, as the hyaline core of the papilla) that effects discharge of the secretion through the tip of the papilla (Linke, 1933; Buckland-Nicks & Worthen, 1992). A simpler type of gland consists of single cells or groups of cells beneath the surface epithelium of the penis, which discharge to the surface by cytoplasmic extensions. These simple, subepithelial glands are concentrated particularly around the penial sperm groove, in both the base and filament of the penis, and the same glandular material extends into the lateral papilla or projection on the penis of *L. scutulata* and *L. plena*. They are sometimes visible externally, especially in living animals, as swollen and opaque regions on the penis. Both the mamilliform glands and the subepithelial glands produce a neutral, viscous secretion, which is believed to secure the penis in position in the mantle cavity of the female during copulation (Linke, 1933; Reid, 1986a, 1989a). Since the papillae of the mamilliform glands are retractile, it has been suggested that they might function by suctorial adhesion (Buckland-Nicks & Worthen, 1992). A third type of glandular cell are the goblet cells, particularly abundant in the columnar epithelium of the penial filament, which produce an acidic, mucous secretion, perhaps serving for lubrication during copulation (Reid, 1989a). During copulation the penial filament is believed to be inserted into the copulatory bursa of the female, into which sperm are transferred.

The penis, absent in small juveniles, develops gradually until sexual maturity, but a number of factors can affect its appearance, even after maturity is attained. In species with a restricted breeding season, there may be a regression in size or even disappearance of the penis outside this period. In a few cases the entire penis has been observed to be shed (Grahame, 1969; Le Breton, 1970, 1971; Pettitt, 1973a; Daguzan, 1976b), although whether this is a normal process or a response to environmental stress is disputed (Deutsch & Fioroni, 1992). Whatever the process, seasonal regression is common in *Littorina*, having been observed or inferred in ten species. In those populations, or species, in which breeding is continuous throughout the year, seasonal regression does not occur (e.g. Sacchi, 1969a; Sacchi & Torelli, 1973; Daguzan, 1976b; Goodwin, 1978). Parasitism by digenean trematodes can also affect the penis, sometimes causing stunting (e.g. Williams & Brailsford, 1990). The presence of a penis is normally diagnostic of males, but development of a small penis has occasionally been recorded as an abnormality in females; such pseudohermaphroditism is sometimes a response to the anti-fouling pollutant tributyltin (TBT) (Fioroni, Oehlmann & Stroben, 1990; Fioroni, Deutsch, Stroben & Oehlmann, 1992; Bauer, Fioroni, Ide, Liebe, Oehlmann, Stroben & Watermann, 1995). This abnormality has so far been found only in *L. littorea*, *L. kasatka* and *L. saxatilis*, and then very rarely.

Although the number of mamilliform glands has been widely used as a taxonomic character, it is subject to considerable intraspecific variation. In species with a small number (1–7) of glands, they may sometimes be entirely absent (e.g. *L. keenae*, *L. brevicula*), and in *L. kasatka* the absence of glands appears to be a characteristic of some entire populations. Where glands are more numerous (as, for example in *L. saxatilis*, on which most studies of their variability have been carried out), their number is generally correlated with body size (as measured by shell height) (Heller, 1975a; Raffaelli, 1979a; but note that both authors probably confused *L. arcana* and *L. saxatilis*; Janson, 1982a). There may also be differences between populations or ecotypes in the numbers of glands (James, 1968a; Davis, 1973; Reid, 1993) and more precisely in the regression of gland number on shell size (Janson, 1982a). Some authors have attempted to correlate this with exposure to wave action; both Davis (1973) and James (1968a) found glands to be more numerous in populations from wave-exposed sites, although the converse was found by Janson (1982a). Where the number of glands is high, they are likely to be arranged in more than one row (Davis, 1973; Hannaford Ellis, 1979; Raffaelli, 1979a). As a result of this variability, the taxonomic use of penial characters has to be carefully evaluated, and in general it is differences in overall shape (for example the relative lengths of filament and base, and shape of the filament) that distinguish species, rather than numbers of glands.

Since penial form is so diverse, and often diagnostic of species of *Littorina*, it has been suggested that it may play a role in species recognition, or even function as an isolating mechanism preventing interspecific mating (Heller, 1975a). Since the number of mamilliform glands is so variable, this may be less important in species recognition than overall shape (Raffaelli, 1979a). The suggestion of a role in species recognition is in agreement with some behavioural observations. Males often display rather indiscriminate mating behaviour, and in some cases appear unable to discern the sex or even the species of another individual without mounting the shell and attempting copulation (Gibson, 1964; Saur, 1990). Attempted copulations between males can therefore be seen in the field, although such matings are of short duration and sperm are probably not transferred (Linke, 1934; Gibson, 1964; Raffaelli, 1977; Saur, 1990; Johannesson, Rolán-Alvarez & Ekendahl, 1995). One study has, however, found evidence that males of *L. littorea* are able to distinguish between the mucous trails of males and females (Erlandsson & Kostylev, 1995). Males may also exert some selectivity after mounting has occurred, for example preferring large to small females (Erlandsson & Johannesson, 1994). Despite numerous observations of attempted interspecific copulations both in the field (Thompson, 1852; Sauvage, 1873; Raffaelli, 1977; Saur, 1990) and in the laboratory (Warwick, Knight & Ward, 1990), successful hybridization has been reported in only one combination of species, as mentioned below. It is not known exactly how interspecific hybridization is prevented, but some interspecific matings are apparently brief (Saur, 1990). If hybrids are inviable or less fit, it is likely to be advantageous to both sexes to avoid interspecific sperm transfer. It is known that females can repel males that are attempting copulation, by rasping the penis with the radula (Saur, 1990), raising the possibility that if penis shape is a species recognition character, females may be able to prevent copulation by males that are not conspecific. Alternatively, or additionally, males may be able to recognize conspecific females after insertion of the penis into the bursa, as suggested by the fact that the size and position of the bursa are often diagnostic of closely related species (p. 18). There is not, however, an obvious morphological correspondence of a 'lock and key' type between the shape of the penial filament (at least at rest) and that of the conspecific bursa (p. 19). Interestingly, the only known example of interspecific hybridization in *Littorina* has been achieved in the laboratory between male *L. saxatilis* and female *L. arcana* (Warwick *et al.*, 1990), and this pair of sympatric, probable sister-species, show only partial differentiation of penial form. Hannaford Ellis (1979) showed significant differences in both penial shape and number of mamilliform glands between these two species, but with some overlap between them. No natural hybrids have yet been found in this case, and prezygotic reproductive isolation may be enhanced by other behavioural mechanisms (Warwick *et al.*, 1990) or by segregation in different microhabitats (Hannaford Ellis, 1985).

Classically, biological species have been defined in terms of 'isolating mechanisms' that prevent interbreeding, and which arise accidentally during allopatric speciation, but can be enhanced by selection (reinforcement) if incipient species are reunited (Mayr, 1963; p. 373). This idea leads to the prediction that isolating mechanisms should be more pronounced in sister-species pairs that are sympatric or partly so. Among sympatric, closely-related, sister-taxon pairs of *Littorina*, penes are diagnostically distinct in three cases (*L. scutulata* and *L. plena*; *L. obtusata* and *L. fabalis*; *L. compressa* and *L. saxatilis*+*arcana*) and show small overlap in two cases (*L. brevicula* and *L. mandshurica*; *L. saxatilis* and *L. arcana*). Only in the case of three species in the *L. sitkana* group (*L. sitkana*, *L. subrotundata* and *L. natica*), which are not, however, immediate sister-species (p. 361), are there no obvious qualitative differences in the penes of sympatric *Littorina* species. Among allopatric sister-species, penes are entirely distinct in one case (*L. littorea* and *L. squalida*) and partly so in two pairs (*L. sitkana* and *L. horikawai*; *L. aleutica* and *L. natica*). The sample size is small, but the prediction is not obviously fulfilled. (There is, however, a possible bias here, since differences in penial form contribute largely to the initial recognition of species as distinct; nevertheless, genetic data have so far supported the morphological definitions of *Littorina* species; p. 375). There is a suggestion that hybridization between *L. arcana* and *L. saxatilis* in the laboratory is more successful with individuals from allopatric than sympatric populations (Warwick *et al.*, 1990), but no attempt has yet been made to connect this with possible character displacement in penial form (which could provide evidence for reinforcement), nor has character displacement in penial form been noticed in the areas of sympatry of partially allopatric species. An alternative view of 'isolating mechanisms' is that they appear during the speciation process, as a result of selection, because they serve to promote successful interbreeding between individuals, and constitute the 'specific mate recognition system' (SMRS) (Paterson, 1985). They may act as isolating mechanisms too, but that was not their primary function. Penial form is more appropriately regarded as part of the SMRS of *Littorina*, because this explains the marked differences between allopatric species.

Spermatozoa

The spermatozoa of *Littorina* are dimorphic, consisting of euspermatozoa and paraspermatozoa, of which only the former appear to be involved in fertilization (review by Reid, 1989a). The euspermatozoa (or eupyrene sperm) are filiform, and details of their structure are only visible by electron microscopy; no significant differences in ultrastructure have been reported among eight species of *Littorina* (Buckland-Nicks, 1973; Buckland-Nicks & Chia, 1977; Paviour, Mill & Grahame, 1989). In the present study, light microscopy has been used to measure the lengths of euspermatozoa in 17 species. In all but *L. striata* and *L. keenae* these range from 38 to 58 μm; within this group the ranges of length of some species do not overlap, but more intensive sampling would be required to assess the significance of this. In *L. striata* and *L. keenae* the euspermatozoa are significantly larger, from 81 to 91 μm in length. Females store euspermatozoa for up to one year in the seminal receptacle, and copulate repeatedly with different males, so sperm competition is likely (Erlandsson & Johannesson, 1994). Furthermore, interspecific matings have been observed (p. 14), but whether post-copulatory sperm choice by females occurs, or could contribute to reproductive isolation between species, is unknown.

There is some uncertainty about the homology of the paraspermatozoa of littorinids with those of other prosobranchs, and for this reason they were formerly referred to as 'paraspermatic nurse cells' (Reid, 1989a). However, details of their development from the germinal epithelium in the testis suggest that they are closely similar, and the term 'paraspermatozoa' is appropriate (Healy & Jamieson, 1993; 'apyrene sperm' of Buckland-Nicks & Chia, 1986). The paraspermatozoa are round or oval cells, ranging in diameter from 8 to 21 μm, and in the tubules of the testis a bunch of euspermatozoa become attached by their acrosomal tips to each paraspermatozoon, forming a composite structure called a spermatozeugma. With the light microscope, paraspermatozoa can be seen to be packed with large rounded vesicles; in addition, a refractile body of similar size can sometimes be seen, which is the degenerating nucleus, although in many species this seems to disappear before the paraspermatozoa reach maturity. The ultrastructure of the paraspermatozoa of *Littorina* has been described by Buckland-Nicks & Chia (1977) and Paviour *et al.* (1989). Their function appears to be largely a mechanical one, the formation of spermatozeugmata preventing premature dispersal of euspermatozoa from the open genital ducts of the male, and once within the copulatory bursa they are said to degenerate (review by Reid, 1989a). In some genera of Littorinidae, notably *Littoraria* (Reid, 1986a) and *Nodilittorina*, the paraspermatozoa are more elaborate, containing additional rod-shaped bodies (Healy & Jamieson, 1993). In these two genera the form of the paraspermatozoa is remarkably diverse, and can sometimes be used to distinguish between species. However, in *Littorina* they are quite uniform, with only minor differences among species (Fig. 3). These include small differences in the size of the vesicles (particularly large in *L. fabalis*, for example), and a tendency for the nucleus to persist in some species (e.g. *L. striata*, *L. plena*, *L. brevicula*, *L. subrotundata*).

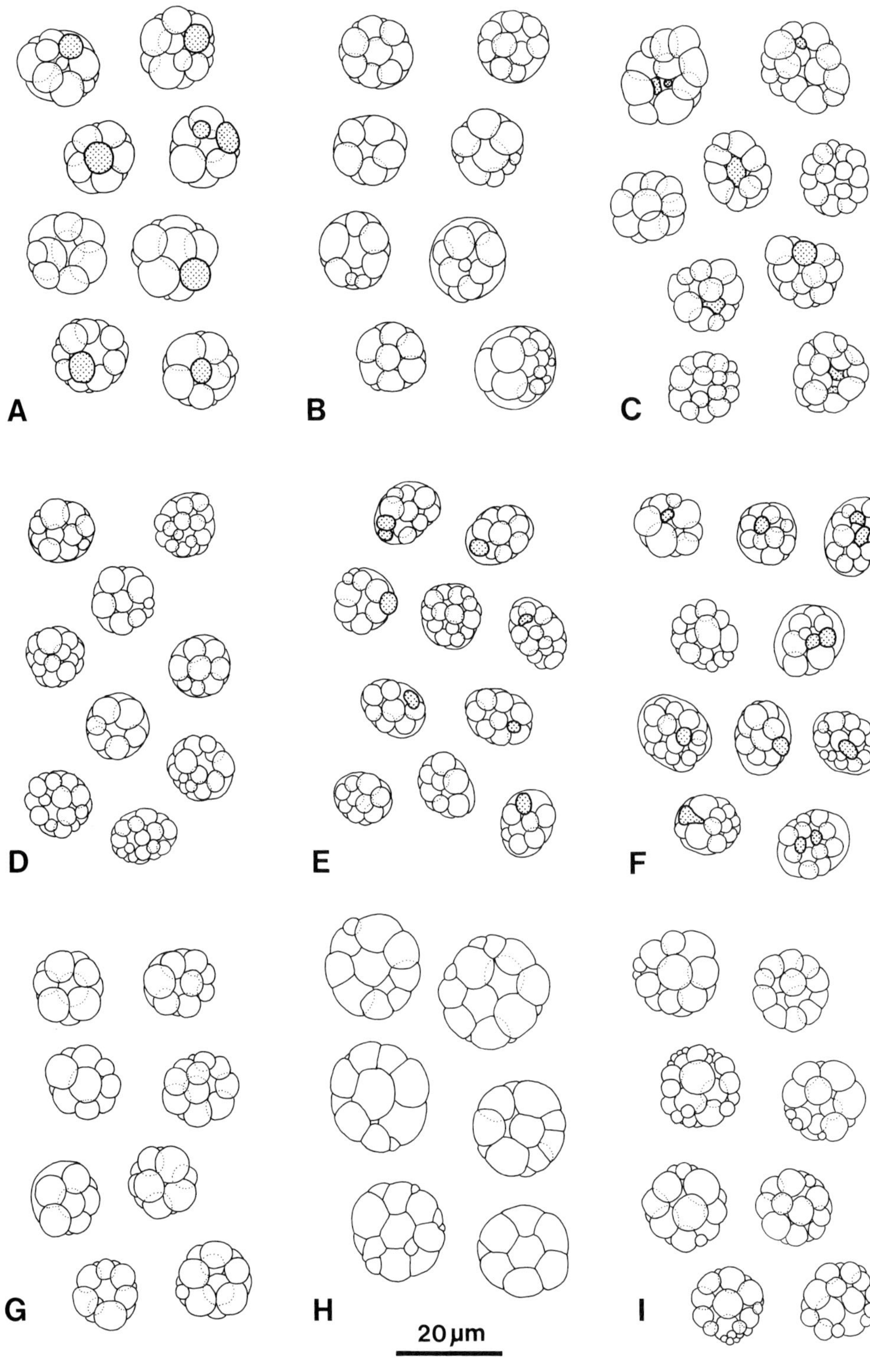

Figure 3. Paraspermatozoa of *Littorina* species, seen by light microscopy of freshly fixed (**A, B, D, G–I**) or formalin-preserved (**C, E, F**) material. **A,** *L. (Liralittorina) striata*, Caloura, São Miguel, Açores; **B,** *L. (Planilittorina) keenae*, Shell Beach, south of San Luis Obispo, California; **C,** *L. (Littorina) plena*, Seldovia, Alaska; **D,** *L. (Littorina) littorea*, Appledore Island, Isles of Shoals, Maine; **E,** *L. (Littorina) brevicula*, Akkeshi, Hokkaido, Japan; **F,** *L. (Neritrema) subrotundata*, Seldovia, Alaska; **G,** *L. (Neritrema) obtusata*, Orchard Bay, Isle of Wight, England; **H,** *L. (Neritrema) fabalis*, Orchard Bay, Isle of Wight, England; **I,** *L. (Neritrema) saxatilis*, Orchard Bay, Isle of Wight, England. Key: stipple, nucleus. Each sample from one individual; attached euspermatozoa not shown.

Female reproductive system

The form of the littorinid oviduct has been reviewed in detail by Reid (1986a, 1989a), and that of *Littorina* itself has been described by Hannaford Ellis (1979; *L. arcana*) and at the ultrastructural level by Buckland-Nicks & Chia (1990; *L. sitkana*); these reports form the basis of the following description. There have also been accounts of the gross anatomy and histology of the oviduct of *Littorina* species by Linke (1933), Fretter & Graham (1962) and Fretter (1980), but these failed to describe correctly the spiral arrangement of the egg groove and used a different terminology for some of the glandular components (see Reid, 1986a, for discussion).

The female reproductive tract of *Littorina* consists of the ovary, which ramifies in the digestive gland, a narrow proximal oviduct of ovarian origin, a glandular renal section of the oviduct, and finally a large distal part formed from the mantle wall (Fig. 1B). This last portion of the tract, the pallial oviduct, has a complex structure concerned with its functions of reception, transport and storage of sperm, production of albumen and other materials encapsulating the eggs, and in one species with the brooding of embryos. The basic structure of the pallial oviduct is that of a laterally flattened tube, receiving the renal oviduct at the posterior end and opening anteriorly into the mantle cavity. This has been illustrated in diagrammatic form by Reid (1986a) in the genus *Littoraria*, and in *Littorina* it is essentially similar. Within the slit-like lumen of the tube, the ventral ciliated channel is the sperm groove, and the dorsal ciliated channel is the egg groove. Two sperm sacs are associated with the pallial oviduct, the copulatory bursa which generally separates from the sperm groove near its anterior end, and the seminal receptacle which separates from the posterior end, at the junction between the renal and pallial portions of the oviduct. During copulation, sperm are deposited into the bursa and subsequently transported along the sperm groove to the seminal receptacle, where long-term storage takes place.

Several glandular components are differentiated along the length of the dorsal egg groove, which coat the eggs with successive layers as they are moved forwards through the length of the groove. At the posterior end the first gland that the eggs encounter is the albumen gland, followed by the covering gland, the capsule gland and finally the jelly gland, the last of which produces the voluminous gelatinous matrix in which the encapsulated eggs of some species are embedded. At the levels of gross anatomy and histology, however, only albumen, capsule (with opaque proximal and translucent distal parts) and jelly glands can usually be identified (see Material and Methods, p. 6); the covering gland cannot be distinguished from the adjacent opaque capsule gland, and the two are therefore not differentiated in the anatomical drawings of the Systematic Descriptions (Chapter 4). The position and function of the covering gland have only recently been established (Buckland-Nicks & Chia, 1990), and it was not mentioned by Reid (1989a). Histological examination shows that the secretory cells of the albumen and capsule glands are subepithelial, arranged in closely-packed lobules beneath the basement membrane of the ciliated epithelium that lines the egg groove, and that their secretions are discharged through cytoplasmic extensions passing between the epithelial cells. In the jelly gland, however, the large secretory cells usually lie within the epithelium, although with some subepithelial development in *L. striata* and *L. keenae* (Reid, 1989a). In those species in which the jelly gland is well-developed (*L. keenae* and *L. (Neritrema)* species) its walls are thrown into closely-packed, fleshy septa. Uniquely in *L. saxatilis* the jelly gland has become modified to form a brood pouch with thin, septate walls, containing the developing embryos. The ultrastructure of the glandular cells of the oviduct, and their relationships to the layers of the egg capsule, have been described in detail by Buckland-Nicks & Chia (1990).

The most unusual feature of the littorinid pallial oviduct is that the dorsal egg groove does not follow a straight pathway from its posterior end to its anterior opening, but is thrown into a series of spirally-wound loops (although the ventral sperm channel of the common lumen of the pallial oviduct runs approximately straight). This complex path presumably serves to lengthen the passage of the eggs through the secretory glands of the oviduct, increasing the rate of egg production, and the several possible configurations in *Littorina* are shown diagrammatically in Figure 113. The spiral form of the pallial oviduct can be understood from the illustrations of its early ontogenetic stages in *L. arcana* (Fig. 97A–E), showing progressive development of the looped pathway of the egg groove (compare with diagram of pathway of ova along egg groove in Fig. 113I). The internal structure of the pallial oviduct (as seen by gross dissection and histological examination) is shown by serial sections of those of *L. keenae* (Fig. 13A–F; cf. diagram of Fig. 113C) and *L. littorea* (Fig. 33G–M; cf. diagram of Fig. 113D).

Within species there is some individual variation in the appearance of the pallial oviduct. This structure gradually develops in juveniles until sexual maturity is attained. The spiral loops of the egg groove develop at an early stage (Fig. 97A–E), so that the adult pattern of loops is visible long before maturity, but only close to maturity does the bursa achieve its final proportions and do the glandular tissues become visibly differentiated. While spawning is taking place the glands are enlarged and swollen, but although the pathway of the egg groove may then be difficult to trace without sectioning, the arrangement of spiral loops through which it passes remains unchanged. Outside the breeding season the pallial oviduct is reduced in size and often darker in colour, but does not disappear, and again the spiral pattern remains unchanged. There is nevertheless some individual variation in the precise details of the spiral loops of the egg groove, as illustrated in the figures of the Systematic Descriptions. Geographical variation in this pattern has been found in *L. compressa*. Parasitism by digenean trematodes may result in shrinkage, or complete disappearance, of the pallial oviduct (e.g. Robson & Williams, 1971). Abnormalities of the pallial oviduct may also result from exposure to high levels of the pollutant tributyltin (TBT) and include a ventral split and eventual transformation into a prostate gland (Bauer *et al.*, 1995).

Details of the pallial oviduct were first used as taxonomic characters in *Littorina* by Sacchi (1975) in his separation of the oviparous *L. compressa* from the ovoviviparous *L. saxatilis*, and then by Hannaford Ellis (1978a, 1979) in her description of the new species *L. arcana*. Since then, the pallial oviduct has been shown to be one of the most important morphological characters of the family Littorinidae, and is used for the phylogenetic definition of generic groups (Reid, 1986a, 1989a, 1990b). The details of the pattern of loops of the egg groove, relative sizes of its component glands, and size of the copulatory bursa, have also been shown to be of importance in the discrimination of closely-related *Littorina* species (Reid, 1990a; Reid, Zaslavskaya & Sergievsky, 1991; Reid & Golikov, 1991) and are therefore emphasized throughout the Systematic Descriptions. The precise configuration of the pathway of the egg groove can be a useful taxonomic character (for example among species of *L. (Neritrema)* in the northern Pacific), but its functional significance remains obscure. It may perhaps be a fortuitous consequence of the lengthening of the egg groove. It does not appear to be related to the shape of the egg capsules, which are moulded from the opposing lips of the capsule gland on either side of the egg groove (Buckland-Nicks & Chia, 1990).

The relative sizes of the component glands of the pallial oviduct, however, are both useful taxonomic characters and of clear functional significance. They are directly related to the type of spawn produced, which is in turn correlated with the type of larval development (p. 19), and examination of the oviduct therefore permits prediction of these characteristics of species in which they are unknown. In species with pelagic eggs (and planktotrophic larvae) the capsule gland is at least as large, and usually much larger, than the albumen gland, and the jelly gland is small, not swollen, and seldom visibly septate (*L. striata* and all members of the subgenus *Littorina*). In contrast, those species which produce benthic gelatinous egg masses (within which larvae undergo intracapsular metamorphosis) show albumen, capsule and jelly glands of approximately equal size, and the jelly gland is large, swollen and conspicuously septate (all members of the subgenus *Neritrema*, except *L. saxatilis*). *Littorina keenae* is unique in the genus, for its oviduct appears somewhat intermediate between the two groups just described; it has a capsule gland much larger than the albumen gland, but also a swollen and septate jelly gland; this species is known to reproduce by pelagic egg capsules, but uniquely these are released in gelatinous masses (Schmitt, 1979), explaining the enlarged jelly gland. *Littorina saxatilis* is also unique in the genus; here the capsule gland is reduced to an inconspicuous loop much smaller than the albumen gland (indeed the capsule gland may have been lost, and the remaining loop may be only the covering gland), while the jelly gland is modified to form a voluminous, septate, but thin-walled brood pouch, filled with developing embryos, indicating that it is ovoviviparous. It has therefore been possible to predict the type of spawn and larval development that can be expected in *L. kasatka* and *L. horikawai*.

As discussed earlier (p. 14), the shape of the penial filament is perhaps a species recognition character used by the animals themselves to promote successful interbreeding, rather than a classical isolating mechanism. Since the penis is inserted into the copulatory bursa during successful mating, it is interesting to consider whether the bursa may serve a similar function. If the size of the bursa were an isolating mechanism, selected for the purpose of preventing interspecific mating, it would be predicted that differences should be more pronounced between sister species that are sympatric or partly so, as a result of reinforcement. Among sympatric, closely-related, sister-taxon pairs of *Littorina*, the bursa is distinctly different in four cases (*L. scutulata* and *L. plena*; *L. brevicula* and *L. mandshurica*; *L. obtusata* and *L. fabalis*; *L. arcana* and *L. saxatilis*), to which may be added one more with only slight overlap (*L. compressa* and *L. arcana+saxatilis*). There is only one example of a sympatric pair without distinctive sizes of bursa (*L. sitkana+horikawai* and *L. subrotundata*), and this is not a pair of immediate sister-species (indeed they may be less closely related than morphological similarity might suggest, since there is no strong molecular evidence for their relationship; p. 360). Among allopatric sister-taxa there are no obvious differences in the bursae in two cases (*L. sitkana* and *L. horikawai*; *L. aleutica* and *L. natica*), whereas in a third case (*L. squalida* and *L. littorea*) the difference is distinct (in this case, however, there is some disagreement between molecular, geological and palaeontological evidence, and a stage of previous sympatry cannot be ruled out; p. 90). The sample size is small, but the prediction does appear to be fulfilled, with two somewhat doubtful exceptions. There is therefore some weak evidence that the size of the bursa may have diverged in sympatric sister species in order to prevent interspecific hybridization, although this does not preclude an initial role in species recognition. It must also be pointed out that there is not necessarily a morphological correspondence, of a 'lock and key' type, between the size of the penial filament (at least as observed at rest; during copulation the size of the filament may be very different, e.g. Bingham, 1972, in *Littoraria irrorata*) and that of the conspecific bursa. In four cases members of a pair of sister-species differ strikingly in both size of the penial filament and in size of the bursa; in two of these the longer filament is found in the species with the longer bursa (*L. squalida* and *L. littorea*; *L. brevicula* and *L. mandshurica*), but in the remaining two the longer filament occurs in the species with the shorter bursa (*L. scutulata* and *L. plena*; *L. obtusata* and *L. fabalis*).

Other characters of the female reproductive anatomy have also been used in taxonomy; these include the size of a ciliated field in the pallial wall adjacent to the rectum (Hannaford Ellis, 1979), and the coloration of the ovipositor (Goodwin & Fish, 1977), but are not of general application.

Spawn, larval development and protoconch

Littorina species produce two principal types of spawn, pelagic or benthic. Pelagic spawn is found in all species belonging to the subgenera *Liralittorina*, *Planilittorina* and *Littorina*. It consists of transparent, variously-sculptured capsules, ranging in diameter between 240 and 1100 μm. The diameter of the ovum is from 70 to 120 μm in seven species, but 120 to 160 μm in *L. littorea* (with a single report as high as 205 μm). Ova are contained singly or in groups of up to 41 within each capsule, depending upon the species (Table 15). The form of the capsules has been recorded in eight of the nine species in this group (that of *L. kasatka* is unknown), and in each case is distinctive, and apparently diagnostic of the species. Capsule form, size and number of contained ova have been used for the taxonomic discrimination of *L. scutulata* and *L. plena* (Murray, 1979; Mastro, Chow & Hedgecock, 1982). Unfortunately there is little information on possible intraspecific variation in the spawn; in the well-known species *L. littorea*, the diameter varies from 310 to 1100 μm, the diameter of the ovum from 120 to 205 μm, and the number of ova per capsule from 1 to 9 (p. 101), although the shape of the capsule appears to be constant. Some variation in shape is indicated in the several published figures of the capsule of *L. brevicula* (Fig. 44P–R). The possible functional significance of capsule shape is unknown, but it has been suggested that large capsules (which carry several ova) may be less vulnerable to filter-feeding predators (Reid, 1990b; p. 401). Egg capsules are usually shed singly or in strings of ephemeral mucus, but in *L. keenae* they are released in coiled gelatinous masses up to 15 mm long, which disintegrate, liberating the small individual capsules, after 24 hours in seawater (Schmitt, 1979). It has been speculated that this serves to protect the eggs from desiccation or abrasion (Schmitt, 1979), or again from filter-feeding predators (Reid, 1990b), or may permit rapid release of spawn. Pelagic spawn is highly unusual among gastropods and has been indicated as a significant adaptation which permitted the Littorininae to colonize the littoral fringe, where benthic egg masses would be susceptible to desiccation (Reid, 1989a).

Benthic spawn is found in a further nine species, all of which are members of the subgenus *Neritrema*. Individual spawn masses are deposited on rock or macrophytes, in damp microhabitats, and measure up to 17 or 37 mm in length; communal spawning, producing much larger masses, is common in *L. sitkana*. Within the gelatinous matrix, ova are enclosed singly in simple biconvex, oval or spherical capsules, 520 to 1000 μm in diameter, and measurements of the diameters of individual ova range from 175 to 285 μm (Table 15). Once again, little is known about possible intraspecific variation. In *L. sitkana* the diameter of the capsule varies between 550 and 1000 μm, perhaps with a geographical component (p. 150). Differences between similar species have been documented, but are less striking than in species with pelagic capsules. For example, in contrast to *L. sitkana*, the spawn of *L. subrotundata* is not laid communally, and the two halves of the capsule wall are uniformly thin rather than of unequal thickness (Boulding *et al.*, 1993); the spawn masses of *L. obtusata* and *L. fabalis* differ slightly in outline, and in size of the ova (Goodwin & Fish, 1977). The histochemistry of the egg capsule of *L. obtusata* has been studied by Goodwin (1979) and the ultrastructure of that of *L. sitkana* by Buckland-Nicks & Chia (1990). One member of this subgenus, *L. (Neritrema) saxatilis*, is unique in being ovoviviparous. The ova (180–211 μm in diameter) are unencapsulated, surrounded only by the egg covering (Losse & Greven, 1993), and are retained within the brood pouch (the modified jelly gland) of the pallial oviduct.

A fundamental distinction can be made between two types of larval development in gastropods, planktotrophic and nonplanktotrophic (e.g. Jablonski & Lutz, 1980, 1983). In the former the veliger larva feeds in the plankton, whereas in the latter the larva reaches metamorphosis without intake of planktonic food. In *Littorina* all nonplanktotrophic species undergo intracapsular metamorphosis (so-called 'direct' development), hatching as miniature crawling snails; no species are known to hatch as a swimming, but non-feeding, veliger larva (lecithotrophic development). There is an exact correlation between spawn type and development; pelagic capsules hatch into planktotrophic veligers, while benthic egg masses contain nonplanktotrophic embryos. The ovoviviparous *L. saxatilis* also undergoes nonplanktotrophic development and intracapsular metamorphosis. The adaptive significance of the type of development, and its profound evolutionary and biogeographical significance, are considered in Chapter 6.

The type of larval development is reflected in the form of the protoconch, the larval shell formed before metamorphosis (Reid, 1989a). This is rarely preserved in adult shells, which suffer erosion and damage in the littoral zone, and is best seen in newly-settled, or newly-hatched, juveniles. In planktotrophic species, the protoconch measures between 240 and 390 μm in diameter (Table 15), and consists of 2.2 to 3 whorls. Details of its form are known only in *L. striata* (Fig. 9E) and *L. littorea* (Fish & Fish, 1977); in the former the surface is sculptured by a pattern of short, oblique, wrinkles arranged in spiral rows, and in the latter by minute tubercles in spiral bands, while in both the protoconch is terminated by a marked sinusigeral rib (representing the apertural 'beak' between the velar lobes of the larva). In contrast, the protoconch of nonplanktotrophic species is generally much larger, from 390 to 600 μm in diameter (but of a greater range, 350 to 800 μm, in *L. saxatilis*; Table 15), and consists of only 1.4 to 1.7 whorls (up to 2.1 in *L. saxatilis*) (Figs 60B, 104E–J); the surface is smooth or bears rows of minute tubercles, and the line terminating the protoconch (indicating the point at which hatching occurred) is straight and parallel to the subsequent growth lines of the teleoconch, with no sinusigeral flexure.

Radula and foregut

The radula of littorinids is of the taenioglossate type, that is with seven teeth in each V-shaped row: the central rachidian, and on each side a lateral, inner marginal and outer marginal tooth. The radula is relatively long, extending back from the buccal mass and coiling in a spiral over the mid-oesophagus. Traditionally, the form of the radular teeth has been regarded as both a useful taxonomic and phylogenetic character in gastropod systematics, and consequently the radulae of *Littorina* species have frequently been illustrated. However, there have been few comparative studies or assessments of intraspecific variability. Some slight differences have sometimes been reported between similar species, as in the case of *L. scutulata* and *L. plena* (Mastro *et al.*, 1982), and *L. obtusata* and *L. fabalis* (Goodwin & Fish, 1977; Watson & Norton, 1987), but other authors have found radulae of closely-related *Littorina* species to be insufficiently differentiated to be useful as taxonomic characters (Heller, 1975a; Raffaelli, 1979a; Reid & Golikov, 1991).

In the Systematic Descriptions relative radular lengths (radular length divided by shell height) of adults have been recorded. These range between 0.9 and 3.5 in most species, but the upper limit exceeds 5.0 in *L. brevicula*, *L. arcana* and *L. saxatilis*, and in *L. striata* the radula may be up to 11.8 times as long as the shell. Variability in this character may be even greater than these figures suggest. In *L. saxatilis s.l.* the relative radular length increases from about 1 in embryos to 4 or more at a shell height of about 5 mm, and thereafter declines to between 2 and 3 (James, in Fischer-Piette, Gaillard &

James, 1964; James, 1968a; Seshappa, 1976). Even more striking variation was found by Kizaki (1987) in a population of *L. brevicula*; in April the relative radular length was about 1, but this increased to 5 or 6 by October, as a consequence of a cessation of feeding activity (and of radular wear) during the summer months, while radular production continued. Furthermore, there was a small difference between the sexes in this seasonal pattern. In adults of *L. littorea*, *L. obtusata* and *L. saxatilis*, new radular teeth are produced at the rate of 5–6 rows per day (Isarankura & Runham, 1968). Among the Littorinidae as a whole, the longest radulae occur in species which graze on hard substrates, suggesting a correlation with the rate of tooth wear (Reid, 1989a). However, it is clear that ontogenetic change and patterns of feeding activity can have profound effects on relative radular length, so that without detailed study it is unlikely to be a useful taxonomic character.

There is also ontogenetic variation in the form of the radular teeth. In juveniles of *L. saxatilis* and *L. arcana* (less than about 5 mm in shell height) the cusps of all teeth are relatively longer and more pointed, and the outer marginal bears 7–8 cusps (4–6 in adults) (Seshappa, 1976; Raffaelli, 1979a) (Fig. 104D). There is apparently an allometric relationship with size (as measured by shell height), since radulae of this 'juvenile' form are found not only in juveniles, but also in small adults, for example of the barnacle ecotype (Figs 98H, 105E). The ontogeny of the radula requires further study, but this allometric relationship appears to be common in the genus, having also been observed in *L. littorea*, *L. fabalis* and *L. compressa*. This may also explain the relatively elongate cusps of the small species *L. kasatka* and *L. horikawai*. There have been few studies of intraspecific variation in shapes and numbers of tooth cusps among adults, but these include descriptive accounts of *L. littorea*, *L. obtusata*, *L. fabalis* and *L. saxatilis* (Johansen, 1901; Goodwin & Fish, 1977; Rolán-Alvarez, Rolán & Johannesson, 1996). The descriptions and illustrations in the Systematic Descriptions represent the first account of intraspecific variation in the genus as a whole. From these it can be seen that radular characters differ principally between groups of species, and are seldom useful for taxonomic discrimination of closely-related taxa. Intraspecific variation is especially pronounced in *L. scutulata*, *L. kasatka*, *L. horikawai*, *L. subrotundata* and *L. saxatilis*, but in most cases this cannot yet be correlated with geography or habitat. In contrast to the intraspecific variability in the shells, that of the radula does not appear to be noticeably greater in species with nonplanktotrophic development (subgenus *Neritrema*). It is notable that variation in the shapes of the principal cusps of the five central teeth in each row are closely correlated, for example if an individual has unusually small and pointed cusps, this will be true of all the five central teeth.

The functional significance of the form of the radular teeth is still not well understood. Several authors have classified radulae into functional groups, by correlating tooth form with diet. Steneck & Watling (1982), for example, contrasted the taenioglossate radula with other categories and suggested that it functions as a rake-like tool. Rosewater (1980) classified littorinid radulae into four functional types, based on the form of the rachidian tooth, cusp shapes and habitat (as an indication of diet); three of these occur in *Littorina*: rhomboidal (with a broad rachidian and a diet of macroalgae), moderate and pick (which together form a continuum; with long, sharp cusps and a diet of microalgae on rocks in the littoral fringe). Other workers have made more detailed comparisons of the diets and radulae of a few species. On the Pacific coast of North America *L. keenae* has a radula with fine, chisel-shaped cusps; its diet includes only epilithic and endolithic microalgae, diatoms and lichens (Behrens Yamada, 1992), and it is unable to consume macroalgae (Dahl, 1964). The cusps of *L. sitkana*, on the other hand, are short and pointed, apparently more suitable for puncturing the epidermis of the macroalgae which form part of its diet, while it is unable to utilize the sparse epilithic and endolithic microalgal flora in the littoral fringe (Sacchi & Voltolina, 1987; Voltolina & Sacchi, 1990; Behrens Yamada, 1992). The radula of *L. scutulata* is somewhat intermediate between these two extremes; its cusps are not as long as those of *L. keenae*, but it is likewise able to feed on superficially bare rock surfaces (abraded rock particles are found in the guts of both species), while it can in addition consume macroalgae (Voltolina & Sacchi, 1990; Behrens Yamada, 1992).

The form and function of the littorinid radula have been studied in most detail in *L. littorea* and *L. obtusata* (Steneck & Watling, 1982; Norton, Hawkins, Manley, Williams & Watson, 1990). During the feeding stroke the radula is protruded and scraped over the substrate with the marginal teeth splayed outwards, leaving a characteristic grazing trace of scratch marks; as the radula is withdrawn, these teeth fold inwards, gathering up the food particles (Ankel, 1936a, 1938; Fretter & Graham, 1962; Petraitis & Sayigh, 1987; Hawkins, Watson, Hill, Harding, Kyriades, Hutchinson & Norton, 1989). The feeding traces of these two species are similar, but the scratches made by *L. obtusata* are broader and deeper. The preferred food of *L. littorea* is thin, foliose algae such as *Ulva* (although diatom films and macroalgal sporelings are also consumed), and the short, pointed tooth cusps (like those of *L. sitkana*) appear to function by tearing the algal thallus. Supporting this is the finding that the edibility of several macroalgae to *L. littorea* is inversely correlated with their puncture strength, but unaffected by their abrasion resistance (Watson & Norton, 1985a, b). In contrast, the cusps of *L. obtusata* are broad and blunt, and used to excavate the thallus of fucoid macroalgae; in this case the edibility of algae is negatively correlated with abrasion resistance, and not with puncture strength (Watson & Norton, 1985b, 1987). The radula of *L. fabalis* is very similar to that of *L. obtusata*, but its diet includes a larger component of epiphytic diatoms, rasped from the surface of fucoids (Watson & Norton, 1987). It is, however, difficult to predict the 'toughness' of an alga from its growth form, or from mechanical tests, because resistance of the alga depends on both its own cellular properties and on the shape and hardness of the radula grazing it, so that actual radulae should preferably be used in tests of algal resistance (Padilla, 1985, 1989). For example, the severity of tooth wear in *L. littorea* has been found to be correlated with cellular structure of the alga grazed, rather than with mechanical measures of puncture or scratch resistance (Bertness, Yund & Brown, 1983).

In the present study the adult radulae of *Littorina* species have been divided into three types (for the purpose of cladistic analysis, p. 349), based on the morphology of the principal cusps of the five central teeth. These may be long and narrow (e.g. *L. keenae*), short and pointed (e.g. *L. littorea*) or short and blunt (e.g. *L. obtusata*). There is an obvious correspondence with the observations and functional inferences of previous authors described above, suggesting that the three types are functional groups specialized for diets of epilithic and endolithic microalgae; foliose, filamentous algae and diatoms; and macroalgae and their epiphytes, respectively. Correlations between diet and radular form throughout the genus, and possible adaptive interpretations, will later be explored in a phylogenetic context (p. 405). The outer marginal teeth have been classified into two types (Reid, 1989a, 1990b), one with a pronounced axial rib from the neck to the base of the tooth, a prominent flange on its outer side, and long, finger-like cusps (e.g. *L. keenae*), and the other with a poorly-developed basal rib and outer flange, and shorter, less numerous, blunt or rounded cusps (e.g. *L. littorea*). The outer marginal teeth are splayed and move furthest across the substrate during the feeding stroke, and are therefore believed to play a major role in the collecting of microalgae by a brushing action; where microalgae form a large component of the diet, this action may perhaps be performed most effectively by the finer cusps of the first type.

Classification of radular types is made difficult not only by the considerable variation in adult radular form found within some species, but also by the ontogenetic changes in the radula. Among juveniles and small adults, the three groups of cusp shapes of the five central teeth are less apparent, because the cusps are all narrow and elongated. They therefore resemble the 'long and narrow' category of the adult radula. The cusps of the outer marginal are finger-like and sometimes more numerous than in the adult, resembling those of adult *L. keenae*. There are no records of the diet of juvenile *Littorina* species, but it is likely that only relatively large individuals have sufficiently strong buccal musculature for penetrating and excavating algal thalli (Watson & Norton, 1987). The juvenile form of the radula is therefore probably related to the greater importance of microalgae in the diet of juveniles, and resembles the adult radular form of species that exploit a microalgal food source throughout their lives.

A single character of the foregut has also been included in the Systematic Descriptions and phylogenetic analysis. In all *Littorina* species except *L. striata* and *L. keenae* the salivary glands are much enlarged, extending up the sides of the mid-oesophagus and sometimes even enclosing the radular sac above it (Reid, 1989a, 1990b). The functional significance is unknown.

Shell characters

The causes of intraspecific variation

In order to understand the range of shell variation, and to evaluate its taxonomic significance, it is first necessary to consider the intrinsic and extrinsic factors that contribute to the determination of shell form. These are, however, still poorly understood. It has long been recognized that there are some predictable associations between the shell form of a population and the habitat in which it occurs. Differences in form between contrasting habitats have often been explained as adaptive, in other words as the result of natural selection, by such agents as predators or wave action. Implicit in such explanations is the assumption that the observable differences have at least a partly genetic basis, although this has been demonstrated in relatively few cases. In the most well-studied species, *L. saxatilis*, a genetic component has been shown by significant mother-offspring regressions and correlations of shell shape (Newkirk & Doyle, 1975; J. E. Smith, 1981; Grahame & Mill, 1993). Differences in growth rate are also at least partly determined genetically, as shown by reciprocal transplants in the field, and by laboratory rearing (Janson, 1982b; Behrens Yamada, 1987). Although this species and others can be reared in the laboratory (Warwick, 1982), there has as yet been only a single study of genetic correlations between shapes of siblings reared under different conditions (Boulding & Hay, 1993, in *L. subrotundata*). Juveniles reared together in the laboratory are said to retain the distinctive shell characters of their parent populations (Janson, 1982a; K. & B. Johannesson & Rolán-Alvarez, 1993), as confirmed by detailed morphometric analyses (B. & K. Johannesson, 1996), but no controlled laboratory crosses have been reported. Shell coloration too has generally been assumed to have a genetic basis when adaptive explanations for variation have been proposed, but again the experimental evidence is slight, restricted to brief references to preliminary breeding studies, of which details have not been published (Reimchen, 1979, 1989, in *L. fabalis*; Atkinson & Warwick, 1983, in *L. saxatilis*).

A genetic component of interpopulation differences in shell shape or coloration is a necessary, but not sufficient, condition for explaining these differences as the result of natural selection. Correlations between shell form and the presumed selective agent can at best provide only circumstantial evidence, and experimental studies are necessary to verify that morphological traits do indeed confer the expected differences in fitness. To predict the evolutionary outcome of changes in selection pressure, it is also necessary to demonstrate significant additive genetic variance in the genetic component, and this has only been done by Boulding & Hay (1993). Other explanations for genetic differences among populations may not require the direct action of natural selection; genotypes may be selectively neutral and subject to the random processes of drift and founder effects, or may be influenced by pleiotropy or linkage with other characters unrelated to the visible differences between phenotypes. These possibilities have been considered mainly in relation to shell colour, as discussed later (p. 32).

It is, however, becoming clear that phenotypic differences are not solely the result of genetic differentiation, but that shell form shows a degree of phenotypic plasticity under the influence of microenvironmental factors. This ecophenotypic variation has so far received relatively little study in *Littorina* species, although it is well known in intertidal muricids and in several freshwater pulmonates. One of the most important influences appears to be the rate of growth, which is in turn affected by such environmental factors as food availability and quality, feeding time, and densities of conspecifics and competing species. However, the effects are complex, and may differ between species. In *L. littorea* it has been found that at lower densities individuals grew faster and produced shells with thinner walls, more rounded apertures, greater internal volume and a more globose, low-spired shape (Kemp & Bertness, 1984). *Littorina subrotundata* likewise grew more rapidly when reared at lower density and produced a larger, more thin-walled shell, but in contrast the form was more tall-spired than at a slower growth rate (Boulding & Hay, 1993). Spiral sculpture was more likely to develop at low growth rates in *L. sitkana* cultured in the laboratory (Boulding *et al.*, 1993). However, by laboratory rearing of morphologically distinct populations of *L. saxatilis*, B. & K. Johannesson (1996) found little evidence of phenotypic plasticity. In *Nucella*, as in *L. subrotundata*, a slow growth rate is correlated with the production of a thicker shell (Palmer, 1981), but there are conflicting reports of the connection between growth rate and shape (Spight, 1973; Brown & Quinn, 1988). The production of thin-walled shells during rapid growth is perhaps explained by a limitation on the rate at which shell material can be deposited (Palmer, 1981; Kemp & Bertness, 1984), while the influence of growth rate upon shape may possibly be a consequence of geometrical constraints (Vermeij, 1980). Surprisingly, predators may exert direct ecophenotypic effects upon shell shape. Species of *Nucella* respond to the water-borne scent of crabs (especially when these are fed on conspecific snails) by producing thicker-walled shells with more prominent apertural teeth, which are more resistant to attack by crabs. This is achieved partly by modification of their behaviour which reduces feeding activity, thus influencing shell shape through the resulting reduction of growth rate. In addition there is apparently a direct effect, since starved snails also modify their shell shape in the presence of crabs (Appleton & Palmer, 1988; Palmer, 1990). It is not known whether *Littorina* species also modify shell growth in this way, but the possibility remains, since for example *L. littorea* alters its behaviour in response to the juice of crushed conspecifics (Hadlock, 1980). Turbulence and wave action may be another source of ecophenotypic effects, as in the enlargement of the area of the pedal sole in response to wave action shown by *Nucella* (Etter, 1988a). There is some evidence of similar phenotypic plasticity in the freshwater pulmonate *Lymnaea*, in which flowing water induces both a larger foot and shell aperture (Lam & Calow, 1988). Other non-genetic influences on shell form are less well known and have yet to be studied experimentally. Gastropods from water of low salinity often show thin-walled shells, perhaps because of the increased energetic cost of calcification; at the intraspecific level, this may involve phenotypic plasticity. Calcification is also energetically more costly at low temperatures (Graus, 1974; Clarke, 1983). Overall growth rate is slower at low temperatures, providing another mechanism for an ecophenotypic effect of temperature on shell form (Vermeij, 1980). Parasitic infection too may affect the rate of growth, and in some cases is believed to cause gigantism in *Littorina* (James, 1969). Damage to the growing edge of a shell may also affect its final form, resulting in premature rounding of the aperture (van Marion, 1981). It is important to recognize that while morphological differences that are the result of ecophenotypic responses may be functional (as in the heavier and more resistant shells produced in the presence of crabs, or larger feet and apertures which improve resistance to dislodgement by waves), they cannot be described as adaptive, since they are not themselves the result of natural selection. On the other hand, the ability to show phenotypic plasticity, particularly if demonstrably functional, very probably is adaptive (to demonstrate this would require a phylogenetic comparison, see Adaptation, p. 394).

Allometry, change in shape with increasing size, has a complex relationship with growth rate and adult size, and may partly explain the phenotypic plasticity of shell shape. One common pattern in *Littorina* species is that the apical angle of the shell first increases (flaring), remains constant and then decreases (doming) as maximum size is approached, in parallel with the acceleration and decline in growth rate (Vermeij, 1980). Where there are differences in size between populations in contrasting microhabitats, allometric relationships may contribute directly to observed differences in shape, and must be taken into account in morphometric analyses (Sundberg, 1988). However, allometric relationships are not invariant and may themselves by subject to ecophenotypic or genetic influences; in the latter case they may be adaptive and under selection (Palmer, 1990).

Potentially, both ecophenotypic and allometric effects may contribute to observed intraspecific variation in all *Littorina* species, irrespective of factors such as population size, type of development and dispersal. However, these factors will influence to a great extent the effect of natural selection upon genetic differentiation. It is perhaps the most striking feature of shell variation in *Littorina* that those species with planktotrophic development show less intraspecific variation, in both space and time, than those without a dispersive, pelagic larval phase. The planktotrophic species are widely dispersed by their pelagic eggs and veliger larvae, ensuring high levels of gene flow throughout large populations, within which any small-scale differences in selection are swamped by recruitment from a wider area. In contrast, the species with nonplanktotrophic development (those with development to the crawling stage within either benthic egg masses or a brood pouch, classified together as the subgenus *Neritrema*) are divided into localized populations, relatively isolated from each other by low levels of gene flow, permitting genetic differentiation through either microenvironmental differences in selection or random genetic drift. This connection between genetic structure and larval development in *Littorina* species has been documented in numerous studies of allozyme variation (e.g. Janson & Ward, 1984; Janson, 1987a; Johannesson, 1992; review by Ward, 1990). Since the range of microhabitats occupied by most nonplanktotrophic species is not obviously greater than that of the less variable planktotrophic ones, the observation of their greater intraspecific variability does itself support the importance of the genetic control of shell form in this genus. The evolutionary and biogeographic significance of larval development and dispersal will be considered in Chapter 6.

From a taxonomic point of view, the more intraspecifically uniform shells of planktotrophic species provide better characters for identification than the variable shells of nonplanktotrophic species. Of the nine Recent planktotrophic species, only two (the *L. scutulata* group) cannot always be readily separated by their shells, whereas the ten Recent *Neritrema* species fall into three groups (*L. sitkana*, *L. obtusata* and *L. saxatilis* groups) within which intraspecific variability is often too great for identification using shells alone. As a heuristic means of classifying the range of shell types in *Littorina* species, variation will be discussed here, and in the Systematic Descriptions (Chapter 4), under the headings of geographical, ecotypic, sculptural, and ontogenetic variation, as well as sexual dimorphism, monstrosities and colour variation. These categories are in fact different sources of variation that are superimposed. Although shell variation may at first seem bewildering in its diversity, a number of common patterns can be discerned.

Geographical variation

In the Systematic Descriptions the more conspicuous patterns of geographical variation have been described by designating 'geographical varieties' with Latin names. This is convenient, because some of these names are already in use in the literature (e.g. *L. palliata*, *L. groenlandica*). Where the most familiar geographical variety corresponds to the type of the species, it has been named *typica*. Other names have been adopted from taxa described by early authors, while a few are new. It must, however, be emphasized that these names are informal and have no nomenclatural status. In the past, such recognizable geographical entities have been described as subspecies. However, none of the cases involves obvious geographical isolation on an island, or other disjunction of a range. The term 'subspecies' carries genetic and evolutionary implications, and its use would not seem warranted when little is yet known about patterns of genetic differentiation, gene flow and the possibility of interbreeding on a geographical scale. Furthermore, it is considered undesirable to add further to the already complicated nomenclature of these species. For these reasons, the neutral term 'geographical variety' has been preferred. It is used in a purely descriptive sense, to describe a recognizable geographical type, with no implication that there is a common basis for this scale of morphological variation in different species. Conspicuous geographical variation can be seen in most species with a sufficiently extensive geographical range, including both planktotrophic and nonplanktotrophic taxa, although it is most pronounced in the latter. Several possible causes of geographical variation can be considered, but these are often confounded and poorly understood. In several cases (*L. plena*, *L. brevicula*, *L. sitkana*) there are no sharp boundaries between the geographical varieties, consistent with explanations involving either gradual genetic clines or geographical gradients in ecophenotypic effects. In the other examples (the Atlantic *Neritrema* species) there are rather sharp boundaries between geographical varieties which can be correlated with a likely cause.

One common geographical trend (although not sufficiently conspicuous to distinguish named varieties) is the occurrence of individuals of the largest size in the northern parts of the range. This is not the case in every species, but is seen in *L. squalida*, *L. littorea*, *L. sitkana*, *L. subrotundata*, *L. obtusata* and *L. fabalis*. This latitudinal trend occurs in many invertebrates, and a connection with low rates of cell division and enlarged cell size at low temperatures has been proposed (Ray, 1960). Another possibility is that since algal productivity may be high in nutrient-rich northern seas, and reproductive seasons shorter, larger size may be a reflection of more favourable conditions for somatic growth (Vermeij, 1978). For these intertidal snails, which feed only when submerged or damp, feeding time might be reduced in warmer climates. It is unclear whether there might also be a connection between temperature, growth rate and shell shape. In *L. littorea* the shells from the southern parts of the range are more globular (Dautzenberg & Fischer, 1912) and thinner-walled (Vermeij, 1982), which are both characteristics of shells of this phenotypically plastic species that are produced at high growth rates (Kemp & Bertness, 1984).

The energetics of calcification at low temperatures may also influence shell form on a geographical scale. Gastropods from high latitudes are typically thin-walled, and of a globular shape (which represents the most efficient use of shell material for enclosing the body), which has been explained by the higher energetic cost of calcification in cold water, in which calcium carbonate is more soluble (Graus, 1974; Clarke, 1983). At a given temperature there may also be a maximum rate at which shell deposition can occur, thereby constraining body growth (Palmer, 1981). Specimens of *Littorina* from high latitudes are indeed often thin-walled, but where trends in shape are apparent they are in fact more tall-spired. Energetic considerations might explain why, in the absence of opposing selection, shells should be as thin as possible, but on present evidence appear unlikely to contribute to geographical trends in shell form.

Vermeij (1973) has described interspecific latitudinal gradients in shell form in members of the Littorinidae in high-intertidal habitats; these include an increase in relative spire height (and corresponding reduction in aperture size) and in development of nodulose sculpture towards the tropics. These trends were interpreted as adaptive, reducing the area of contact between foot and the hot substrate, and maximizing the surface area of the shell for re-radiation of heat, relative to its absorption from direct sunlight. Small size will also increase the surface to volume ratio, making convective heat loss more favourable. Adaptation for temperature regulation is one possible explanation for the typically small, tall-spired shells of *L. saxatilis* found in the southern parts of its range.

It has long been known that shells of *L. obtusata* from subarctic regions are characteristically taller-spired and thinner-walled (geog. var. *palliata*) than southern shells. The explanation was first suggested by Seeley (1986) after her observation that a change from northern to southern phenotype had occurred at sites in New England following the northward spread of the predaceous crab *Carcinus maenas* during the twentieth century. She showed that the heavy, globose southern variety was more resistant to crab predation, and interpreted the morphological change as the result of a genotypic response to intense selection. Since the morphological response was so rapid, some have doubted this genetic mechanism, proposing instead a direct ecophenotypic effect of the crabs (Palmer, 1990). Nevertheless, in a related species, *L. subrotundata*, it has been confirmed that there is sufficient additive genetic variance in this character to respond rapidly to strong selection (Boulding & Hay, 1993). It is notable that the planktotrophic species, *L. littorea*, did not show a measurable morphological response to the arrival of *Carcinus* in New England, which supports the view that low gene flow and local adaptation were required for the rapid phenotypic response in *L. obtusata* (Vermeij, 1982). Based on close correlations between the occurrence of the crab-resistant variety of *L. obtusata* and the distribution of *Carcinus* in Iceland and Norway, it appears that Seeley's explanation is a general one. Furthermore, the four other Atlantic species of the subgenus *Neritrema* also display a closely similar pattern of geographical variation, with tall-spired, thin-walled varieties in Canada, Iceland and northern Norway where *Carcinus* is absent. In the Pacific Ocean there is probably a similar gradient of decreasing intensity of crushing predation at higher latitudes, which may account for some geographical trends in *L. sitkana* and *L. subrotundata* (Vermeij, 1992b; Boulding *et al.*, 1993), but the pattern is much less clear and consistent than in the Atlantic.

On a smaller geographical scale, less conspicuous clines in shell shape have been reported around the coast of Britain in *L. saxatilis*. The regional variations can sometimes be correlated with the presence of a closely related sympatric species, *L. arcana*, and may then be a case of character displacement (Grahame & Mill, 1989, 1992; Grahame, Mill & Brown, 1990). The significance of other regional variations in *L. saxatilis* is not yet understood (Mill & Grahame, 1995).

Ecotypic variation

The 'ecotype' (or 'ecomorph') concept was first introduced into discussions of intraspecific variation in *Littorina* species by Fretter & Graham (1980) and Janson & Ward (1985). The term 'ecotype' is taken from Turesson (1922), who applied it to ecological forms of plants, defining it as an 'ecological unit to cover the product arising as a result of the genotypical response of an ecospecies to a particular habitat'. This term is appropriate for similar phenotypes that are found in similar habitats, and of which the phenotypic characters are at least partially determined genetically. It does not, however, imply that these characters are the result of phylogenetic descent, nor that their genetic basis is homologous. Rather, their similarity is likely to be due to convergence under similar regimes of natural selection. The correlation between shell form and microhabitat is often very striking among *Littorina* species, and may occur on such a small scale, for example within transects on a single shore, that ecotypes have sometimes been mistaken for distinct, sympatric species. This has given rise to much of the taxonomic confusion in the genus, which has only been resolved with the advent of genetic studies. The taxonomic history and debate about the specific status of the ecotypes of each species are reviewed in the Systematic Descriptions (Chapter 4). Evidence for the conspecificity of superficially distinct ecotypes includes anatomical similarity (e.g. Reid, 1993), intergradation of shell morphology in intermediate or transitional microhabitats (e.g. Janson & Sundberg, 1983; Janson & Ward, 1985; B. & K. Johannesson, 1990; K. Johannesson, B. Johannesson & Rolán-Alvarez, 1993), similarity of allozyme frequencies (e.g. Moyse, Thorpe & Al-Hamadani, 1982; Janson & Ward, 1985), inferred patterns of gene flow (e.g. Janson & Ward, 1984; K. & B. Johannesson, 1990; Johannesson *et al.*, 1993; Rolán-Alvarez, Rolán & Johannesson, 1996) and successful hybridization (e.g. Ward, Warwick & Knight, 1986; Warwick, Knight & Ward, 1990). Of critical importance is the inference that gene flow between different ecotypes on the same shore is higher than between the same ecotypes on geographically separated shores, which is the test by which ecotypes are distinguished from 'sibling' species (K. & B. Johannesson, 1990).

Ecotypic variation is particularly well developed in those *Littorina* species with nonplanktotrophic development (subgenus *Neritrema*). For example, on shores of strong wave-exposure the shells of *L. saxatilis* are more thin-walled, low-spired and have relatively larger apertures than those from sheltered shores (e.g. Heller, 1976; Raffaelli, 1979a; Janson, 1982a, 1983b), but shells of the planktotrophic *L. littorea* are seldom significantly different in shape or thickness on shores of contrasting wave-exposure (e.g. Currey & Hughes, 1982; Janson, 1987a; Ash, 1989; Crothers, 1992). This comparison of species differing in type of development and dispersal suggests that localized adaptation, permitted by low gene flow, is an important component of ecotypic differentiation. As discussed earlier (p. 23), the partially genetic determination of ecotypic differences has been confirmed by some studies of *L. saxatilis* and *L. subrotundata*, although the possible contribution of ecophenotypic effects should not be discounted, and requires further study. The combination of low vagility of adult snails and steep selection gradients appears able to maintain striking ecotypic differences even over distances as small as 1 to 10 m in *L. saxatilis* (e.g. Janson, 1983b; Janson & Sundberg, 1983; B. & K. Johannesson, 1990; Johannesson *et al.*, 1993). In at least one case adjacent ecotypes of this species display a degree of assortative mating, so that differentiation is further promoted by partial genetic isolation (Johannesson, Rolán-Alvarez & Ekendahl, 1995). Ecotypic differentiation of this kind may be the first step in a process of sympatric speciation (p. 381).

In the Systematic Descriptions intraspecific variation between microhabitats has been described by reference to four ecotypes, termed moderate, wave-exposed, barnacle and sheltered/brackish ecotypes. This is essentially a subjective and arbitrary division of a continuum of shell forms, for where the characteristic microhabitats of these ecotypes adjoin or merge into each other, intermediate forms occur (e.g. Janson & Sundberg, 1983; Janson & Ward, 1985; Johannesson *et al.*, 1993; Reid, 1993). This recognition of four principal ecotypes is based on taxonomic tradition and on recent ecological and genetic work on *L. saxatilis*, the species in which the basis of ecotypic variation has been most thoroughly investigated. Thus the moderate ecotype corresponds with *L. rudis* of nineteenth century authors and of Heller (1975a), and with the S morph of Janson (1982a, b). The wave-exposed ecotype was included under the name *L. patula* by early authors and Heller (1975a), and named the E morph by Janson (1982a, b). The barnacle ecotype has for long been distinguished as a separate species, *L. neglecta* (e.g. Heller, 1975; Fish & Sharp, 1985; Caley, Grahame & Mill, 1995; Grahame, Mill, Hull & Caley, 1995), although considered an ecotype by many recent authors (K. & B. Johannesson, 1990; Reid, 1993; Crossland, Coates, Grahame & Mill, 1995). The sheltered/brackish ecotype too has until recently been widely considered a distinct species, *L. tenebrosa* (e.g. Fretter & Graham, 1980), although it has now been shown to be an ecotype (Janson & Ward, 1985). Debate about whether additional shell forms are sufficiently distinctive to be recognized as ecotypes, or even as named varieties, may continue (Barnes, 1993), but is simply a matter of personal preference. The traditional system adopted here is convenient for description, and is particularly useful for emphasizing the analogous ecotypic variation which occurs in other *Neritrema* species less well known than *L. saxatilis*. Not all the *Neritrema* species display the complete range of ecotypes, for few occupy such a wide range of habitats as *L. saxatilis*. Nor does the classification of shores, for example as 'moderate' or 'wave-exposed', correspond exactly in different species, because species occupy different microhabitats (tidal levels, crevices, etc.). The analogous ecotypic variation of some *Neritrema* species is illustrated in Figure 4. In the brief descriptions that follow, the ecotypes are characterized mainly by shape, size and thickness, with less emphasis on the more variable attributes of sculpture and colour. It must be emphasized that recognition of ecotypes of a species is relative to the moderate ecotype of that species alone; there are close parallels of trends in relative variation among species, but shell shapes do not correspond exactly.

The moderate ecotype is the most typical and abundant form of most species, occurring on shores that are neither strongly wave-exposed nor extremely sheltered. Shells are thick-walled, of moderate spire height and often of large size in comparison with the morphological range of the particular species. In geographical regions where crab predators are common, it is believed that they exert a strong selective influence on shell form (although, as mentioned previously, a direct ecophenotypic component is also a possibility; Palmer, 1990). Laboratory trials have shown that shells of this ecotype are more resistant to crushing by crabs than are those of the thinner-walled, wave-exposed ecotype (Elner & Raffaelli, 1980; Johannesson, 1986; Boulding & van Alstyne, 1993), so that adaptation is assumed. A smaller or narrower aperture may also be adaptive, protecting the vulnerable aperture from insertion of large crab chelae (Johannesson, 1986). Adaptive strategies are, however, not always comparable; for example, *L. fabalis* achieves some protection against crabs by maturation and

Figure 4. Ecotypic variation in *Littorina (Neritrema)* species. **Column 1** (left), moderate ecotype (large; thick-walled; moderate spire); **column 2**, wave-exposed ecotype (usually smaller; thinner-walled; low spire); **column 3**, barnacle ecotype (very small; thin-walled; low spire; often black and/or white); **column 4** (right), sheltered/brackish ecotype (small; thin-walled; tall spire). **Row 1** (top), *Littorina saxatilis*; **row 2**, *L. compressa* (which lacks a sheltered/brackish ecotype); **row 3**, *L. subrotundata*; **row 4** (bottom), *L. obtusata* (which lacks a barnacle ecotype). Scale bar = 5 mm.

thickening of the shell at small size, whereas *L. obtusata* may avoid predation by growing rapidly to large size, so that size gradients vary with wave exposure in opposite directions in these two species (Reimchen, 1982).

In contrast, on wave-exposed shores crabs are not usually common, and the main selective force is dislodgement by waves. In comparison with the moderate ecotype, shells of the wave-exposed ecotype are thin-walled, usually smaller, and with a lower spire and correspondingly enlarged and more circular aperture. Shells of this form show higher survival on wave-exposed shores than transplanted moderate ecotypes (Janson, 1983b; Boulding & van Alstyne, 1993), and the shape may be adaptive because it reduces shell weight (Heller, 1976), drag (Denny, 1988) and accommodates a larger foot for adherence (J. E. Smith, 1981; Grahame & Mill, 1986). However, once again an ecophenotypic component cannot be discounted (Etter, 1988a).

On wave-exposed shores there may be selection for small size (or early maturation), to permit animals to shelter in rock crevices and other refuges, and there is therefore a correspondence between shell size and that of available refuges (Emson & Faller-Fritsch, 1976; Raffaelli & Hughes, 1978). The barnacle ecotype appears to be an extreme example of this trend, for it is found only on wave-exposed shores, sheltering not only among abundant barnacles, but also in mussel beds and algal turf. Shells are less than about 5 mm in maximum height, globular in shape, with relatively enlarged and rounded apertures, and are often smooth. The coloration is frequently black, white, grey, or banded or tessellated in these colours, or a polymorphism of these types, which might suggest visual selection (Hannaford Ellis, 1984; Fish & Sharp, 1985; Reid, 1993).

The fourth ecotype is characteristic of extremely sheltered inlets, lagoons, seagrass beds, salt marshes and estuaries, where conditions are often brackish. It is named the sheltered/brackish ecotype, because these two environmental conditions are frequently correlated and, since measurements of salinity are seldom available, it cannot be certain which variable is the more significant. Shells from these habitats are generally tall-spired, thin-walled, smooth, of small size and often yellowish with brown bands or tessellations. There are several possible selective or ecophenotypic influences that might contribute to this characteristic form. The tall spire and small aperture have been interpreted as an adaptive response to severe desiccation stress on sheltered shores with little wave splash (Newkirk & Doyle, 1975), while the coloration may suggest visual selection (Knight, Hughes & Ward, 1987) or a pleiotropic connection with tolerance of high temperature or low salinity (Sacchi, 1974). Unlike the other ecotypes, this one has a parallel in the variation of planktotrophic species, for both *L. littorea* and *L. squalida* develop small, thin, tall-spired shells in brackish and lagoonal situations. This suggests that direct ecophenotypic effects may play a part; low salinity might induce small, thin-walled shells, for example. Another possibility is that algal and detrital food may be abundant in productive habitats such as salt marshes, leading to rapid growth, which does induce tall-spired, thin-walled shells in some species such as *L. subrotundata* (Boulding & Hay, 1993), although not in others such as *L. littorea* (Kemp & Bertness, 1984).

This overview of ecotypic variation is necessarily a simplification, for the interrelationships between genetic adaptation, ecophenotypic plasticity and allometry are undoubtedly complex and likely to differ between shores, making generalizations difficult. Particular traits may be subject to a variety of influences; shell size (as measured by shell height), for example, is under both genetic and ecophenotypic control, and natural selection may be related to wave action and crevice size, crab predators and demographic factors such as size-specific mortality and fecundity. Sometimes the selective forces may conflict, as for example where crabs and strong wave action occurring together may select for reduced and enlarged aperture size respectively (Johannesson, 1986). Furthermore, although shell traits are often discussed separately, they may not be independent. Thus geometrical considerations preclude certain shell forms; a low spire is necessarily accompanied by an enlarged aperture, for example, because otherwise an umbilicus would appear, mechanically weakening the shell. Allometric relationships may also constrain the range of shell form (Vermeij, 1980; Sundberg, 1988), as may genetic correlations between shell traits (Boulding & Hay, 1993).

Sculptural variation

Variation in the development of ribs and grooves is common in most species of *Littorina*, although once again most pronounced among the nonplanktotrophic *Neritrema* species. Just two species of the genus (*L. striata* and *L. aleutica*) sometimes develop nodulose shells. As in the case of ecotypic variation, the first line of investigation has been to seek correlations between microhabitats or environmental variables and shell sculpture. For example, in *L. saxatilis* populations from wave-exposed shores are often strongly sculptured, and those on sheltered shores and in salt marshes weakly so, or entirely smooth (e.g. Fischer-Piette & Gaillard, 1960; Fischer-Piette, Gaillard & James, 1964; Davis, 1973; Heller, 1975a), although the correlation is not close, and there are many exceptions. In *L. striata* the more nodulose forms are smaller and occur on more sheltered shores and at higher tidal levels (Fischer, 1946, 1953, 1967; de Wolf, van Dongen, Brito & Backeljau, 1994; Britton, 1995). A large-scale geographical pattern is seen in *L. brevicula*, *L. sitkana* and *L. aleutica*.

Several possible functions of shell sculpture have been suggested. Thickened ribs of shell material might strengthen the shell, so that strong sculpture could be expected in environments where calcification is costly or where a heavy shell is disadvantageous. However, Brandwood (1985) found no evidence that ridging increased shell strength in *L. saxatilis*, and pointed out that the absence of sculpture in low-salinity environments argues against this explanation. An alternative suggestion is that ridged shells are more difficult to dislodge from the tight crevices in which they shelter on wave-exposed shores. Purely hydrodynamic considerations cannot explain a preponderance of sculptured shells on wave-washed shores, for surface irregularities increase drag (Denny, 1988). At high intertidal levels on tropical shores, surface sculpture may serve in temperature regulation, by increasing heat loss through convection and re-radiation (Vermeij, 1973).

A genetic basis of shell sculpture has yet to be thoroughly investigated in *Littorina* species, but has been suggested by laboratory rearing in *Nodilittorina hawaiiensis* (Struhsaker, 1968, as *Littorina picta*), and by Mendelian inheritance in *Nucella* (Palmer, 1985). One study has shown that sculptural differences persist in laboratory-reared progeny of two contrasting ecotypes of *L. saxatilis* (Johannesson *et al.*, 1993). If there is generally a significant genetic component of shell sculpture in *Littorina* species, the differences between habitats described above may indeed be adaptive. However, there are also suggestions that there may be a strong ecophenotypic component, for in two of the most variable species, *L. saxatilis* (Gaillard, 1965) and *L. sitkana* (Boulding *et al.*, 1993), there are reports that development of sculpture in the progeny of strongly ribbed parents is much reduced when reared under favourable conditions of growth in the laboratory. A connection between strong sculpture and slow growth might explain the presence of ribbed *L. saxatilis* on exposed shores (if, for example, feeding time were restricted by wave action), and smooth ones in salt marshes (where food may be abundant). It might also account for the predominance of large, smooth shells of *L. sitkana* in the more productive northwestern part of its range in the Pacific. In *L. striata* too an ecophenotypic effect is likely, for growth may be slow at the highest levels of the littoral fringe where sculptured forms predominate.

Shell sculpture is therefore too variable to be a useful taxonomic character in many cases. Nevertheless, details of the form of the ribs may serve to separate some otherwise similar species, for example *L. saxatilis* and *L. compressa*. Spiral microstriae are a finer degree of surface ornament than ribs and grooves, and cover the surface of most species. The presence or strength of microstriae also sometimes distinguish closely-related species, for example the pairs *L. obtusata* and *L. fabalis*, and *L. sitkana* and *L. subrotundata*.

Ontogenetic variation

One aspect of ontogenetic variation in shell form is allometry, the change in shape with increasing size. This can be seen from simple linear regressions of shell measurements on shell height, if they are non-linear or do not pass through the origin (e.g. Janson, 1982a), from changing ratios of shell dimensions with increasing shell height (e.g. van Marion, 1981), and in logarithmic plots of regressions, from which the allometric constant k (in the equation $y=bx^k$, where y and x are shell dimensions, the latter generally shell height, the usual measure of size) can be found (e.g. Kemp & Bertness, 1984; Lowell, Fletcher, Grahame & Mill, 1994). Allometric relationships may themselves alter during the life of an individual; for example, in *Littorina* species the universal pattern in the relation between shell breadth and shell height is of positive ($k>1$) followed by negative ($k<1$) allometry. This is clearly visible as an initial flaring of the spire (i.e. increase in the apical angle of the shell) in the first few whorls of the teleoconch, followed by constancy and then final doming of the spire (i.e. decrease in the apical angle), although the duration of each phase varies both within and between species. This pattern of growth is common in gastropods (Vermeij, 1980). Allometric relationships introduce complications in intraspecific comparisons of shell shapes, and the perception of allometry depends upon the variable used as a measure of size, which may be height, body weight or a multidimensional factor (Sundberg, 1988; Palmer, 1990).

From the constancy of allometric relationships in some gastropod species (Gould, 1966), the idea has arisen that these relationships are determined genetically, and subject to adaptive modification. However, it is becoming clear that they can also be directly influenced by environmental factors, including presence of crabs (Palmer, 1990, in *Nucella*) and factors which modify growth rate (Kemp & Bertness, 1984, in *L. littorea*; Boulding & Hay, 1993, in *L. subrotundata*). Differences in allometric relationships among populations in different microhabitats may be functional (Palmer, 1990) and, if genetically determined, could be considered adaptive. A possible example in *Littorina* is the narrowing of the final whorl in populations of the moderate ecotype of *L. saxatilis* that are believed to be exposed to heavy crab predation; this shell form permits the combination of a relatively low spire and constricted aperture, which if growth were isometric could only be achieved by coiling around an umbilicus.

Another aspect of ontogenetic change in shell form is the thickening of the shell that takes place as maximum size is approached, giving rise to positive allometry of thickness with size, which may well be adaptive in relation to crushing predation (Lowell *et al.*, 1994). Thickening is especially obvious at the outer lip of the aperture, and in the Systematic Descriptions specimens with a thickened, non-growing, apertural lip are taken to be of 'adult size'. Thickening also proceeds over all internal surfaces of the shell (Taylor & Reid, 1990). Nevertheless, in most *Littorina* species shell growth is apparently indeterminate. In *L. saxatilis* Moreteau (1976) showed that thickening of the apertural lip indicated only a temporary cessation of growth, and that phases of growth and non-growth alternated throughout life, with the latter predominating with age. In sub-Arctic regions prolonged interruptions to growth occur during the winter, and are marked on the shell of *L. saxatilis* and *L. obtusata* as conspicuous growth rings (Stroud & Knudsen, 1982; Sergievsky, Granovitch & Mikhailova, 1991). Probably in all species, shocks caused by temperature and slight damage result in minor interruptions to growth, visible externally as growth lines (Ekaratne & Crisp, 1984). In two species, *L. obtusata* and *L. fabalis*, growth is more nearly determinate, for at least in their southern geographical varieties there is a rather abrupt change in shape as adult size is reached, involving descent of the suture (junction with the preceding whorl) towards the base, as well as thickening of the lip.

Sexual dimorphism

Statistically-significant dimorphism in shell height has been found in ten of the eleven species of *Littorina* in which it has been investigated, and in all these cases it is the females that are slightly larger. The larger size of females is probably due to their faster growth (Moore, 1937; Daguzan, 1977; Chow, 1987b; B. & K. Johannesson, 1996), although greater longevity has also been claimed (Pelseneer, 1926; Daguzan, 1977). In *L. littorea* males mate preferentially with larger (and therefore more fecund) females, and may thus exert sexual selection for size (Erlandsson & Johannesson, 1994). Whether there is ever sexual selection for large male size, either by female choice or by competition between males for mates (so far observed only in *L. keenae*, Gibson, 1964), is not known (Erlandsson & Johannesson, 1994). The opposite possibility, selecting for smaller males (e.g. Rolán-Alvarez, Zapata & Alvarez, 1995, in *L. fabalis*), might be imposed in environments where the weight of the male, attached to the female's shell during copulation, significantly increases the chance of dislodgement of the pair from the substrate, by waves or wind. Mechanical considerations which necessitate a fit between the genitalia of a mating pair may, however, constrain the degree of possible dimorphism. This may account for correlations between the shell sizes of mating pairs that have been observed in some, but not all studies (Gibson, 1964; Raffaelli, 1977; Erlandsson & Johannesson, 1994), although size-based mate choice, or ecological size-segregation, are also possible explanations (Johannesson, Rolán-Alvarez & Ekendahl, 1995). The degree of dimorphism may vary among populations, and it may be absent in some populations of species known to be dimorphic elsewhere, as in *L. obtusata* (e.g. Sacchi, 1967b, 1968b, 1969b), *L. saxatilis* (e.g. Hughes, 1979; van Marion, 1981; Sacchi, 1981; Grahame & Mill, 1989) and *L. littorea* (e.g. Moore, 1937; Daguzan, 1977; Saur, 1990). In *L. fabalis* dimorphism is more pronounced on sheltered shores (Goodwin & Fish, 1977; Williams, 1994), and in *L. saxatilis* in populations of small adult size (Gallardo & Götting, 1985), but no explanations have been suggested for these observations.

Sexual dimorphism in shape has been reported less frequently, in only six species. It is most pronounced in *L. squalida* (pers. obs.) and *L. fabalis* (e.g. Sacchi, 1968a, b, 1969b; Gallardo & Götting, 1985). In both these species the males are slightly more patulous, with a lower spire and larger aperture. This shell form resembles wave-exposed ecotypes, which are adapted for adherence; conceivably, this shape might serve the same function in males, permitting a larger foot size for adherence to the female during copulation. In *L. saxatilis* there are some conflicting accounts of the differences between the sexes (e.g. Daguzan, 1977; Sacchi & Torelli, 1979; Sacchi, 1981), but some have found females to be broader, especially across the penultimate whorl, which might serve to accommodate the large brood pouch (Jeffreys, 1865; Grahame & Mill, 1992).

Monstrosities

Malformations of the adult shell, including keeled and elongated forms, have been recorded in several of the more common species of *Littorina*. Frequently, these are clearly the result of damage to the apertural margin during growth, usually by breakage, which may often be caused by unsuccessful predation attempts by crabs. Even if breakage does not result in an obvious malformation, shell shape may still be affected in a more subtle way; for example, van Marion (1981) found that following a breakage the regrown shell of *L. saxatilis* had a smaller apical angle and more rounded aperture. Interruption to normal coiling may also be the result of settlement of barnacles or algae near the growing aperture, but epibiotic growths are surprisingly rare in *Littorina* species, even at lower tidal levels on the shore (Sacchi, 1970, 1975; Wahl & Sönnichsen, 1992). Most reports of extreme malformations of the shell refer to *L. littorea*, in which they include reversed coiling, sharp keels, turreting of the whorls, double apertures, pearl formation, and even natural survival of animals without shells at all (p. 107). However, these frequent reports are probably only a reflection of the appeal of this large species to shell collectors, and of its commercial gathering for food in large quantities.

The monstrosities described above may be curious, but are of no taxonomic or evolutionary importance. Of greater interest are the extreme forms of *L. obtusata* that occur at frequencies of less than 10% in some populations from northern Norway and Iceland. These shells may be carinate or unusually tall-spired, but the cause is unknown. There is no obvious connection with damage to the apertural or mantle edge, epizoic encrustation, or parasitism by Digenea, and nor does low temperature seem likely to responsible. Their geographical restriction does suggest the likelihood of a genetic basis, since examples are not known in apparently similar environments elsewhere. It is speculated that these shells simply represent extremes of the range of shell variation, in populations in which natural selection on shell form is relaxed (p. 216). The only comparable and likewise mysterious case concerns the famous monstrosities of *L. littorea* from the Late Pliocene Norwich Crag formation of East Anglia, where similar greatly elongated and carinate shells occur in a single shell bed over a small area (a maximum distance of about 5 km). These are so remarkable that they have received six specific and 14 varietal names, but a continuous range of intermediates connects them with the normal shells with which they are found. The absence of any similar forms among living populations of this well-known species suggests that an ecophenotypic effect (such as low temperature or low salinity) is unlikely, and again the possibility of increased genetic variability under localized conditions of relaxed selection is suggested (p. 108).

Colour variation

All living species of *Littorina* show some variation in shell colour, but the term 'colour polymorphism' is restricted to those in which discrete, non-intergrading, morphs can be distinguished. In *L. keenae*, *L. scutulata* and *L. plena* a range of colour patterns can be found within some populations, but the variation is not clearly discontinuous, and these species are therefore not described as polymorphic. Those species known to be polymorphic include *L. littorea*, *L. squalida*, *L. brevicula* and all members of the nonplanktotrophic subgenus *Neritrema*, with the exception of *L. natica* (Fig. 5). These are polymorphic within each species as a whole, and most strikingly so in *Neritrema*, but polymorphism within populations is infrequent in all except *L. obtusata* and *L. fabalis*, in which it is the rule. It is not yet clear whether the polymorphism is homologous throughout the genus, for the range of colour differs between species, and the genetic and structural bases of shell colour are poorly understood. Phylogenetic analysis does support the view that the polymorphism is homologous throughout

Figure 5. Variation in shell colour within populations of *Littorina* species (x1.6). **Top row,** *Littorina (Littorina) scutulata*, Shell Beach, near Pismo Beach, south of San Luis Obispo, California; **row 2,** *L. (Littorina) brevicula*, Muroran, Hokkaido, Japan; **row 3,** *L. (Littorina) littorea*, Lyme Regis, Dorset, England; **row 4,** *L. (Neritrema) sitkana*, Neah Bay, Washington; **row 5,** *L. (Neritrema) obtusata*, Rubh' Aird-mhicheil, South Uist, Outer Hebrides, Scotland; **row 6,** *L. (Neritrema) compressa*, Finavarra, Co. Clare, Ireland; **row 7,** *L. (Neritrema) saxatilis*, Höfdakaupstadur, Iceland. Each sample taken from a single shore.

(p. 399). Detailed studies of colour polymorphism in *Littorina* have been almost entirely restricted to the European species *L. obtusata*, *L. fabalis*, *L. compressa*, *L. arcana* and *L. saxatilis*. Because of its variability, shell colour is usually of little use as an identification character, although exceptions include the similar pair *L. scutulata* and *L. plena*, which can often be distinguished by colour pattern. Particular colours may be rare in certain species (the tessellated morph in *L. compressa*, the olive morph in *L. fabalis*, for example), which may also aid identification.

Several systems of classification have been used to describe this polymorphism. Early authors developed a formal system in which each recognizable 'var. ex colore' was given a Latin name, as applied to *L. littorea* (Dautzenberg & Fischer, 1912, 1925), *L. obtusata s.l.* (Dautzenberg & Fischer, 1915) and *L. saxatilis* (Middendorff, 1849; Dautzenberg & Fischer, 1912). With some modification, combining those varieties which intergrade, this system has worked satisfactorily in population genetic studies of *L. obtusata* and *L. fabalis*, in which the principal morphs are quite discrete (e.g. Sacchi, 1974; Smith, 1976; Reimchen, 1979; Ekendahl, 1995). However, in *L. saxatilis s.l.* the distinctions between colour morphs are less clear, and subjective description of morphs led to an unwieldy system of no less than 25 varieties (Fischer-Piette & Gaillard, 1971). Accordingly, Pettitt (1973) devised a classification based on an imaginary system of genetic control, with six 'loci' describing the traits ground colour, band number, band colour, tessellation, interrupted lines and lines in grooves, each with a number of 'alleles' according to colour or presence/absence. This system was more objective, and has been adopted in some studies of the *L. saxatilis* group (e.g. Raffaelli, 1979b; Atkinson & Warwick, 1983). It has also been modified for use in *L. obtusata* (Sergievsky, 1983a, 1984, 1992). The structural basis of shell pigmentation has unfortunately received little attention, but a study of the distribution of pigments in the shell layers of *L. obtusata* has raised the possibility that a more meaningful biochemical and genetic classification might be achieved (Mikhailova & Naumov, 1990).

All recent studies of colour polymorphism in *Littorina* have been based on the assumption that it is genetically determined, although there is little direct evidence that this is so. Nevertheless, the indirect evidence is persuasive; this includes the localized and unpredictable occurrence of rare morphs, suggestive of drift and founder effects (e.g. Fischer-Piette & Gaillard, 1968), the restricted geographical distribution of some morphs, the discrete colour variation within populations (Smith, 1976), and in a few cases the presence of morphs in predicted Hardy-Weinberg proportions (Kalabushkin, 1976; Kartavtsev, 1976). Direct evidence of heritability of colour has reportedly been obtained in *L. obtusata* and *L. fabalis* (Reimchen, 1974, quoted in Smith, 1976; Reimchen, 1979, 1989) and in *L. saxatilis* (Atkinson & Warwick, 1983), but no details have been published. There have been no suggestions of ecophenotypic effects, for example of diet, on shell coloration in *Littorina* species, although the possibility has been mentioned in the genus *Littoraria* (Reid, 1986a).

The assumption of a genetic basis thus appears to be justified, and the maintenance and adaptive significance of colour polymorphism can be considered. Cook & Garbett (1992) approached these evolutionary processes as a hierarchy of questions. First, it must be established whether the colour variation is the result of neutral mutation or is under selection. Such is the bewildering variety of shell colours in *L. saxatilis s.l.*, sometimes within single populations, that it has been suggested that the variation might be selectively neutral (Fischer-Piette, Gaillard & James, 1963; Fischer-Piette & Gaillard, 1968, 1971). Nevertheless, this can be eliminated as a general explanation, since numerous examples are now known of correlations between colour frequencies and various environmental variables, which include wave-exposure, salinity, and colour and visual diversity of the background. Such correlations may be suggestive, but cannot in themselves demonstrate selective processes. Secondly, it is recognized that selection based upon the phenotype is not the only nonrandom process able to maintain polymorphism. Another theoretical possibility is assortative mating; there is no evidence of this among the colour morphs of *L. obtusata* (Sergievsky, 1983b), but it does occur among certain ecotypes of *L. saxatilis* (Johannesson, Rolán-Alvarez & Ekendahl, 1995), although in this case phenotypic differences involve not only colour, but also shape and size. Heterosis is an additional possibility, although until the genetic basis of shell colour is understood, this cannot be evaluated. The third question considers the nature of selection on the phenotype, which may be related to its visual appearance, or to non-visual properties, which are connected with colour only by pleiotropy or linkage.

Some authors have interpreted correlations of colour frequencies with variables of the physical environment as evidence for the importance of non-visual effects. For example, in Britain and southern Europe the olive morph of *L. obtusata* predominates in sheltered habitats, and the reticulated brown morph in those exposed to waves (e.g. Sacchi, 1967b, 1974; Smith, 1976; Goodwin & Fish, 1977). This led to comparisons of the desiccation resistance (Sacchi, 1969a), thermal tolerance (Sacchi, 1972a) and behaviour (Sacchi, 1969a, 1974) of the principal morphs, but although small differences were found, these could not be related to the observed frequencies in the field. However, in the more extreme environment of the White Sea, the different physiological properties of morphs of this species can be related to their distributions, and include tolerance of low salinity, high temperatures and resistance to digenean infection (review by Sergievsky, 1992). Selection of a non-visual kind can evidently be significant in some circumstances. It has also been invoked to explain field distributions of the morphs of *L. saxatilis* (e.g. Fischer-Piette & Gaillard, 1971; Sacchi, 1974, 1979; Raffaelli, 1979b), but without experimental evidence.

Nevertheless, many authors have found correlations between the frequencies of colour morphs and the visual properties of the habitat, and have interpreted these as evidence for visual predation. Some of the best examples are of associations of morphs in the *L. saxatilis* group with rock and barnacle substrates of similar colour (Davis, 1973; Heller, 1975b; Matthews & Montgomery, 1978; Byers, 1990; Dytham, Grahame & Mill, 1990; Reid, 1993), although these may not apply throughout the range of the species (Fischer-Piette & Gaillard, 1968; Naylor & Begon, 1982; Raffaelli, 1979b). Similarly, the three common morphs of *L. obtusata* have each been found to be most common on the fucoid alga on which they appear most cryptic (Smith, 1976). Unfortunately, the only direct experimental evidence of visual predation, and of resulting selection for crypsis, is in the case of *L. fabalis* preyed upon by a blenny (Reimchen, 1979). Other predators in the intertidal zone, with the necessary acuity and colour vision, might include birds and possibly crabs (Pettitt, 1975). Visual selection may also be a result of climatic factors; white shells of *L. saxatilis* reach significantly lower temperatures under strong illumination than dark morphs, although whether this affects fitness in the field is not yet known (Sergievsky & Granovitch, 1989; Sergievsky, 1992). The muricid *Nucella* occupies lower intertidal levels than *L. saxatilis*, and may have a lower tolerance of heat or desiccation, but here climatic selection is an important cause of variation in shell colour between shores (Etter, 1988b).

Directional selection, whether visual or not, may explain large-scale differences between populations, but if polymorphism is to be maintained within a population by selection, there must be a balancing or frequency-dependent component (review by Reid, 1987). One of the ways in which this might be achieved is by disruptive selection in a coarse-grained environment, with a balance between selection and migration of genotypes. The grain-size of the environment is assessed in relation to the dispersal ability of the species, and so in planktotrophic species the size of patches would have to be very large to maintain polymorphism in this way. However, in the nonplanktotrophic *L. obtusata*, without a pelagic dispersal phase, both spatial and temporal variation between the fucoid weeds on which it lives might provide a basis for disruptive selection (Smith, 1976). In a fine-grained environment a stable polymorphism can still be maintained under certain conditions (Endler, 1978, 1981). For example, morphs might be equally cryptic, random samples of the background pattern; this could apply to black, white and banded morphs of barnacle ecotypes. Some morphs might mimic certain background elements, as has been claimed for the resemblance of *L. obtusata* to fucoid bladders, and polymorphism could be maintained by a kind of frequency-dependence. Another possibility for maintenance in a fine-grained environment is if morphs choose the background on which they are most cryptic, and there is some limited experimental evidence for this in *L. fabalis* (Ekendahl, 1995). A final possibility, independent of environmental grain-size, is apostatic selection, a type of frequency-dependent selection that is caused by the tendency of predators to specialize on common types, while overlooking rare morphs. This might explain, for example, the low but persistent frequencies of red morphs in populations of *L. littorea*, when these do not resemble any elements of the background. Some indirect evidence for frequency-dependent selection has been claimed from the correlation between frequencies of the same morph in populations of *L. saxatilis* and *L. arcana*, which differs significantly from a value of 1 (Atkinson & Warwick, 1983). The possible interactions between crypsis, apostasy and frequency-dependent predation are complex, and the several processes outlined are not mutually exclusive.

In summary, the significance of shell colour and the processes responsible for its variation within species and polymorphism within populations, are complex. Furthermore, they may differ not only between species, but even within the habitat or geographical range of a single species. Interpretations are frequently speculative, and based only on correlations observed in the field. The genetic basis and selective processes are still poorly understood, and there is clearly a need for further experimental work.

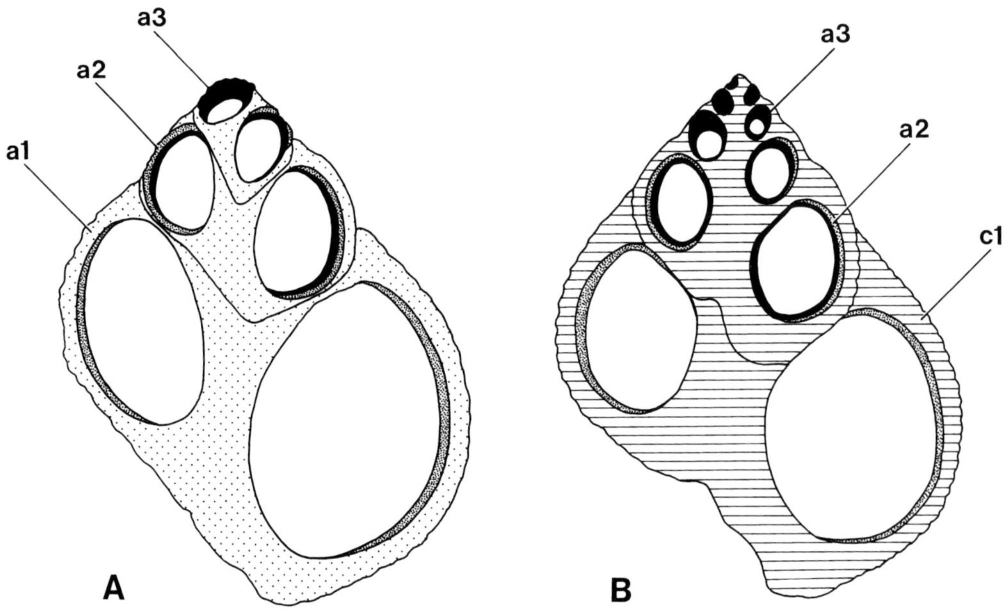

Figure 6. Diagrammatic longitudinal sections of shells of *Littorina (Liralittorina) striata* (**A**) and *L. (Littorina) littorea* (**B**), showing shell layers. Abbreviations: a1, crossed-lamellar aragonite layer 1 (primary lamellae parallel to apertural margin); a2, crossed-lamellar aragonite layer 2 (primary lamellae normal to apertural margin); a3, crossed-lamellar aragonite layer 3 (spire infill); c1, irregular-prismatic calcite.

Shell microstructure and mineralogy

In most gastropod shells calcium carbonate is present in the mineral polymorph aragonite, whereas calcite is of much more restricted occurrence (Bøggild, 1930). In two species of *Littorina, L. striata* and *L. keenae*, the shell is entirely aragonitic, whereas in all remaining species there is an outer layer of calcite. The mineralogy and microstructure of the shell of *Littorina* species has been described in detail by Taylor & Reid (1990), from which the following account has been taken. In the aragonitic species the shell consists of three layers (Fig. 6A). Each has a crossed-lamellar microstructure (as in layer a2 of Fig. 7A, B); in the outermost, thickest layer the long axes of the primary lamellae run parallel to the apertural margin, but in the thinner second layer they run normal to the margin. The third, innermost layer is secreted high in the spire, and shows variable orientation of the lamellae. In the remaining species the outermost layer is replaced by a similarly thick layer of calcite (Fig. 6B), with a distinctive irregular-prismatic structure (layer c1 of Fig. 7B). At the apertural margin shell material is laid down in micro-growth bands corresponding to tidal cycles, which are visible using certain preparation techniques (Ekaratne & Crisp, 1982).

The presence of the calcite layer, which occupies most of the thickness of the shell, is an important phylogenetic character, but is not used for identification of living species of *Littorina*, for which many other taxonomic characters are available. For fossil species, however, shell microstructure can be extremely useful. In many respects, the shells of *Littorina* do not differ consistently from those of the two related genera *Nodilittorina* and *Littoraria*. However, since neither of these other genera are known to possess calcite layers, its presence in a fossil is strong evidence for inclusion in *Littorina*. Recrystallization, in which the entire shell material is replaced by coarsely crystalline calcite, may have occurred in some fossils, and all trace of the original shell microstructure is then lost. However, in many fossils of calcitic *Littorina* species the contrast between a thin, inner, chalky-white layer and a thick, outer, translucent one can still be seen, even with a binocular microscope.

Taylor & Reid (1990) pointed out that, in the Littorinidae as a whole, groups with calcitic shells are found only in temperate and higher latitudes. They suggested that intertidal shelled animals are particularly exposed to the danger of shell dissolution, because of accumulation of carbon dioxide during the night in inshore seawater. Dissolution of calcium carbonate is most likely to occur in cold water, in which it is more soluble. Since calcite is less soluble than aragonite, they argued that shells with an outer calcite layer should be more resistant to dissolution.

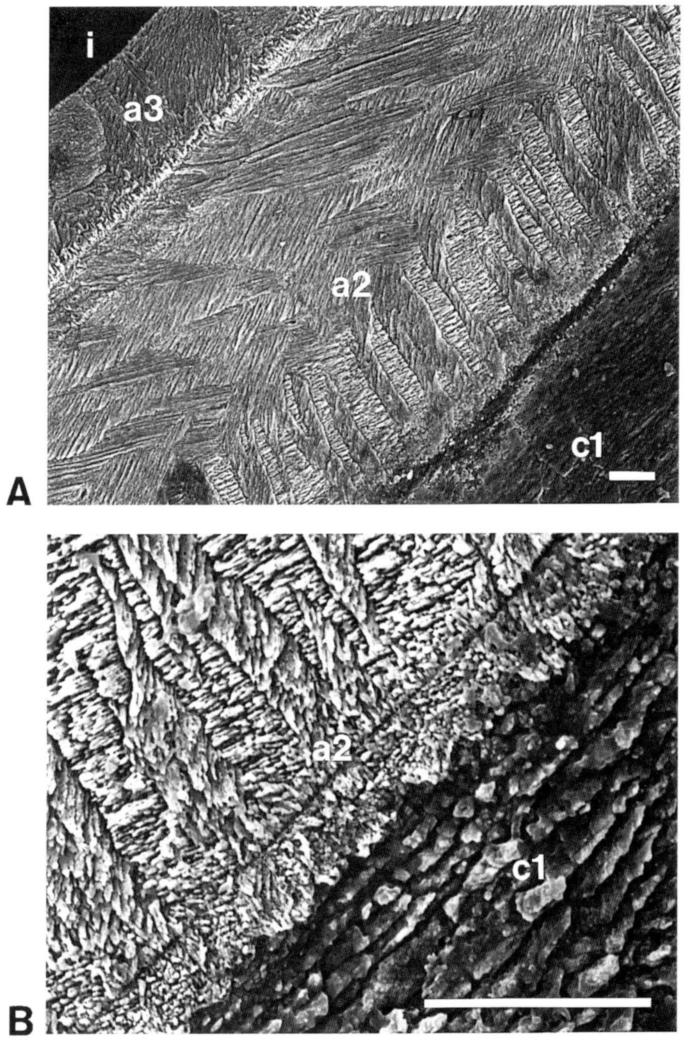

Figure 7. Shell microstructure, seen by scanning electron microscopy of ground and etched fractured surface. **A,** *Littorina (Neritrema) subrotundata*, showing three shell layers. **B,** *Littorina (Neritrema) aleutica*, showing high-power detail of junction between aragonite and calcite layers. Abbreviations: a2, crossed-lamellar aragonite layer 2 (primary lamellae normal to apertural margin); a3, crossed-lamellar aragonite layer 3 (spire infill); c1, irregular-prismatic calcite (outermost); i, interior of shell. Scale bars = 10 μm.

Chapter 4

SYSTEMATIC DESCRIPTIONS

Genus *Littorina* Férussac, 1822

Synonymy

Turbinarius Froriep, in Duméril, 1806: 165 (type species by monotypy *Turbo littoreus* Linnaeus, 1758).

Littorina Férussac, 1822: XI, XXXIV (type species *Turbo littoreus* Linnaeus, 1758, by subsequent designation (Anton, 1838, as *Litorina*, but see ICZN, 1985: art. 67h); see ICZN Opinion 1159 (Melville, 1980); Official List of Generic Names in Zoology no. 2106; proposed as subgenus of *Paludina*; subsequent misspellings include *Lithorina* Hartmann, 1840; *Litormia* Mørch, 1849; *Littorrina* Gabb, 1861; *Litonia* Weinkauff, 1878; *Litorinna* Dall, 1884; *Littornia* Jenkins & Grocock, 1891; *Litornia* Preston, 1917).

Litorina Menke, 1828: 24–25 (unjustified emendation of *Littorina* Férussac; type species *Turbo littoreus* Linnaeus, 1758; see ICZN, 1985: art. 67h).

Bacalia 'Gray' H. & A. Adams, 1854: 312 (from Gray, 1840: 147 (*nomen nudum*); in synonymy of *Littorina*; type species *Turbo littoreus* Linnaeus, 1758, by subsequent designation (Winckworth, 1922); unavailable name (ICZN, 1985: art. 11e)).

Cricostoma 'Klein' Mørch, 1857: 80 (from Klein, 1753, non-binominal; in synonymy of *Littorina*; unavailable name (ICZN, 1985: art. 11e)).

Isonema 'Hall' Tryon, 1883: 240 (in synonymy of *Littorina*; type species *Turbo littoreus* Linnaeus, 1758, by subsequent designation (Bequaert, 1943, but attributed to Provancher, 1891); unavailable name (ICZN, 1985: art. 11e); not *Isonema* Meek & Worthen, 1866, although Tryon may have intended this taxon).

Algaroda Dall, 1918: 137 (type species *Littorina litorea* Linnaeus (i.e. *Turbo littoreus* Linnaeus, 1758) by original designation; proposed as section of *Littorina*).

Ezolittorina Habe, 1958a: 9 (type species *Littorina squalida* Broderip & Sowerby, 1829, by monotypy).

Brevilittorina Golikov, Dolgolenko, Maximovich & Scarlato, 1990: 295 (*nomen nudum*; included species *B. brevicula*, *B. mandshurica*).

Taxonomic history

Linnaeus (1758) described four species in the first section ('Neritoidei aperturae margine columnari') of his genus *Turbo*; these, in order, were *T. obtusatus*, *T. neritoides*, *T. littoreus* and *T. muricatus*. It could, with some justification, therefore be argued that he considered these littorinid species to be typical of *Turbo*, and indeed Swainson (1840) took *T. littoreus* as the type species of *Turbo*. However, the type species of this genus had previously been designated as *T. petholatus* Linnaeus, 1758, by Montfort (1810).

In 1806 Froriep published a German translation of Duméril's *Zoologie Analytique* in which, by citing included species for the first time, he made available many of Duméril's generic names. One of these was *Turbinarius*, with *Turbo littoreus* as the only named species. Were priority to be rigorously enforced, *Turbinarius* should replace its junior objective synonym *Littorina*. Nevertheless, since the name *Turbinarius* seems to have been entirely overlooked by subsequent authors, the maintenance of current usage of *Littorina* is desirable (strictly, this requires suppression of the earlier name by use of the ICZN plenary powers; ICZN, 1985: art. 79).

The generic name *Littorina* was first introduced by Férussac (1822) as a subgenus of *Paludina*, but in such a manner that there have been differing interpretations as to its validity and much debate about subsequent designations of a type species. Férussac (1822: IX–X) listed each of the four Linnaean species mentioned above, as well as *Turbo afer* Gmelin, 1791, as '*Paludina*. Marine' but, in naming the new subgenus that was to contain them, he used only the French vernacular 'littorine' (p. XI). On p. XXXIV he did use the Latin *Littorina*, but did not describe the new subgenus, nor refer to included species. The majority of subsequent authors have considered this as a valid introduction of the generic name. Subsequent designation of *Turbo littoreus* Linnaeus, 1758, as the type species has in the past often been credited to Blainville (1828), and in this choice he was followed by many authors (e.g. Anton, 1838; Deshayes & Milne-Edwards, 1843; Herrmannsen, 1846; Gray, 1847; H. & A. Adams, 1854; Reeve, 1857; Tryon, 1887; Cossmann, 1915; Winckworth, 1922, 1932; Bequaert, 1943; Rosewater, 1970, 1981; Bandel & Kadolsky, 1982; Reid, 1989a). Nevertheless, Blainville (1828) had referred only to the vernacular name 'Littorine', and his designation of a type species could therefore be considered invalid. The next designation was that of Rang (1829), who specified *L. littoralis* as the type species. This, however, is definitely invalid, since this species (which must be assumed to be *Nerita littoralis* Linnaeus, 1758) was not among those included in *Littorina* on its original publication (Férussac, 1822), although it has been argued that *Turbo littoreus* may have been intended by Rang (Herrmannsen, 1846; Kadolsky, 1970). Notwithstanding, Rang's designation was accepted by some authors (Dall, 1909, 1918; Thiele, 1929; Wenz, 1938; Nordsieck, 1968, 1982; Golikov & Kusakin, 1978; Palmer, 1989). All of these assumed that the two Linnaean species *Nerita littoralis* and *Turbo obtusatus* are synonymous, whereas in fact it is here argued (and fixed by designation of a neotype) that the former is the species better known as *Theodoxus fluviatilis* (p. 202). Yet another interpretation of the correct type species of *Littorina* is based on the view that the generic name was not validly introduced by Férussac (1822), since the Latin *Littorina* was provided with neither description nor included species. This leads to the conclusion (Kennard, 1942; Winckworth, 1951) that the genus was validly introduced by Payraudeau (1826), from which the type species can be established by monotypy as *Littorina basterotii* Payraudeau, 1826 (=*Turbo neritoides* Linnaeus, 1758). Fortunately, this confusion has been resolved by a ruling of the ICZN (Melville, 1980: Opinion 1159), which established that *Littorina* was made available by Férussac (1822) and set aside all designations of type species prior to that of *Turbo littoreus* Linnaeus, 1758, by Anton (1838) (for detailed discussion of this case see Kadolsky, 1970; Heppell, 1970, 1971; Cernohorsky, 1971; Holthuis, 1971; Melville, 1980).

In practice, these differing views on the type species of *Littorina* have had little effect on the inclusion of species within the genus (since most authors have adopted a definition of the genus sufficiently broad to include all three proposed type species), although they have affected the nomenclature of its constituent subgenera. The spelling of the name *Littorina* was altered to *Litorina* by Menke (1828), *littus* and *litus* being alternative classical spellings. This is regarded as an intentional (although unjustified) emendation, since the same word was spelled in this way twice in the original publication (as *Litorina litorea*; ICZN, 1985: art. 33b); *Litorina* is an objective synonym of *Littorina* Férussac, and is available with its own author and date (ICZN, 1985: art. 33b). It should be noted that although species-group names spelled with a single or double consonant, such as '*litorea*' and '*littorea*', are deemed identical variant spellings (ICZN, 1985, art. 58), in the genus group a one-letter difference is sufficient to prevent homonymy (ICZN, 1985, art. 56b). This emendation was followed mainly by the German school (e.g. Anton, 1838; Philippi, 1845–1850; Küster, 1853; Weinkauff, 1878, 1883). In either form, the generic name passed into general use between 1830 and 1840, replacing the earlier use of *Turbo* (see Synonymies of species). Since then, the species here assigned to *Littorina* have, with very few exceptions, been placed together in this genus. However, until recently, most authors have employed the name in a wider sense. In the early monographs, for example, not only *Littorina sensu stricto*, but also those species now assigned to *Melarhaphe*, *Cenchritis*, *Tectarius*, *Littoraria* and *Nodilittorina* (see Reid, 1989a, for species lists) were all included (Deshayes & Milne Edwards, 1843; Philippi, 1845–1850; Reeve, 1857; Weinkauff, 1878, 1883). Following Gray (1847b) and H. & A. Adams (1854) the trochiform, spinose and nodulose shells were removed from the genus (usually as *Tectarius*, *Risella* and others). Those species remaining in *Littorina* were characterized simply by their superficially similar shells, which were smooth or with spiral ribs and lacked nodulose sculpture (Tryon, 1883, 1887; Thiele, 1929; Bequaert, 1943; Rosewater, 1970, 1981). A few authors also removed some of the species now classified as *Littoraria* (Cossmann, 1915; Wenz, 1938). Gradually, it was realized that shell sculpture alone was a poor guide to generic classification (Abbott, 1954b; Rosewater, 1970, 1981), but not until 1982 did Bandel & Kadolsky attempt the first diagnoses of the genera of the Littorinidae, based on a range of characters including not only the shell, but also the operculum, penis, spawn and radula. They thus removed species of *Nodilittorina*, *Littoraria* and *Melarhaphe*, establishing a concept of *Littorina* similar to that used here. This was supported by the comparative anatomical work of Reid (1986a), who identified the two consecutive spirals of the albumen and capsule glands in the pallial oviduct as a possible synapomorphy of the genus. There followed a more comprehensive cladistic analysis of the anatomical characters of all known subgenera of the Littorinidae (Reid, 1989a), which established the monophyly of *Littorina* and found an additional synapomorphy in the absence of rod-pieces in the paraspermatozoa. A list of known species was also attempted and included the problematical *L. striata*. This study was used as the basis for a phylogenetic analysis at the species level (Reid, 1990b), in which there was some doubt about the possible inclusion of the two species of *Mainwaringia* (here excluded, pp. 11, 359). Although the monophyly of the genus has not been unequivocally upheld by analyses of mitochondrial DNA sequences, there is as yet no strong reason to reject it (Rumbak, Reid & Thomas, 1994; Reid, Rumbak & Thomas, 1996; p. 359).

Many authors have attempted subgeneric grouping of the species here assigned to *Littorina*. If the type species of *Littorina* was taken as *L. littorea*, those species differing from it by smooth, globose shells (resembling *L. obtusata*) were distinguished either as *Neritoides* (Brown, 1827; H. & A. Adams, 1854; Tryon, 1883, 1887; Dautzenberg & Durouchoux, 1900; Kobelt, 1907; Winckworth, 1922, 1932; Dall, 1923) or *Neritrema* (Récluz, 1869; Bequaert, 1943; Smith, 1982). If, on the other hand, either *L. obtusata* or *'L.' neritoides* were regarded as the type species of *Littorina*, then *L. littorea* and *L. squalida* were placed in the subgenus *Algaroda* (Dall, 1918, 1921; Thiele, 1929; Wenz, 1938; Kennard, 1942; Winckworth, 1951; Nordsieck, 1968, 1982; Golikov & Kusakin, 1978). This division was supported by the different spawn of these two groups (Winckworth, 1922). An emphasis on spawn and development also led to the separation of the supposedly ovoviviparous species in the subgenus *Littorivaga* (Dall, 1918, 1921; Winckworth, 1922, 1932; Thiele, 1929; Wenz, 1938; Bequaert, 1943; Nordsieck, 1968, 1982; Smith, 1982). Some authors have promoted these groups to generic status, because of the supposed significance of their different modes of reproduction (Winckworth, 1922; Habe, 1951, 1958a; Palmer, 1989). The genus *Ezolittorina* of Habe (1958a) was established for *L. squalida*, because of its large egg capsules containing many ova, but did not gain acceptance outside the Japanese literature. The cladistic analysis of anatomical characters resulted in the transfer of *L. striata* to *Littorina*, in the subgenus *Liralittorina*, and the recognition of the new subgenus *Planilittorina*, for *L. keenae*, while *Neritrema* was accepted as a monophyletic group for all the nonplanktotrophic species (Reid, 1989a, 1990b; p. 366). Further details are given in the Taxonomic Histories of the subgenera.

Diagnosis

Shell: spiral grooves usually present, but sometimes smooth; occasionally nodulose; protoconch of planktotrophic or nonplanktotrophic type; aragonitic, crossed-lamellar, with outer layer of irregular-prismatic calcite in most species. Male: anterior vas deferens and penial vas deferens open; penis not bifurcate; scattered simple penial glands present; one or more mamilliform penial glands of uniform size usually, but not invariably, present. Sperm: rod-pieces absent from paraspermatozoa. Female: egg groove coiled in two or three successive spiral loops through albumen and capsule glands of pallial oviduct. Spawn and development: either pelagic capsules, benthic gelatinous masses, or ovoviviparous with brooding in pallial oviduct. Alimentary system: salivary glands posterior to nerve ring, usually enlarged. (Diagnosis after Reid, 1989a).

Subgenus *Liralittorina* Rosewater, 1981

Synonymy

Liralittorina Rosewater, 1981: 36 (type species *Littorina striata* King & Broderip, 1832, by original designation; proposed as subgenus of *Nodilittorina*).

Taxonomic history

The generic classification of the only known species of this subgenus, *L. striata*, has long been problematical, because of certain resemblances in both shell and anatomical characters to the three genera *Littorina*, *Nodilittorina* and *Melarhaphe* (see species account following). *Liralittorina* was introduced as a subgenus of *Nodilittorina* by Rosewater (1981), because *L. striata* differed from typical species of *Nodilittorina* in its simple penis, lacking the bifurcation, glandular disc and single mamilliform gland. Since phylogenetic analyses of morphology (Reid, 1989a, 1990b; p. 346), allozymes (Backeljau & Warmoes, 1992; p. 356) and DNA (Rumbak, Reid & Thomas, 1994; Reid, Rumbak & Thomas, 1996; p. 358) have supported its basal position within *Littorina* and rather distant relationship with its congeners, the subgenus has been retained.

Diagnosis

As for the only known species, *L. striata* (p. 42).

Littorina (Liralittorina) striata King & Broderip, 1832

(Figs 3A, 6A, 8–11)

Synonymy

Littorina striata King & Broderip, 1832: 345 (Port Praya [Cape Verde Islands]; holotype BMNH 1968323, seen, Fig. 10A; not a secondary homonym of *Turbo striatus* Schumacher, 1817 (not da Costa, 1778)=*Littoraria scabra* (L.), see Reid, 1986a: 95). Reeve, 1857: sp. 40; pl. 8, fig. 40; pl. 14, fig. 73. Drouët, 1858: 26. Dautzenberg, 1890: 157. Locard, 1897: 492. Nobre, 1930: 55. Nobre, 1937: 39. Marche-Marchad, 1956: 42 (in part, from distribution includes *Nodilittorina miliaris helenae* (E. A. Smith)). Duffus & Johnston, 1969: 32. Bandel, 1974: 107; figs 18–21, 54, 55 (not 38, 39 as in text; radula). Pérez Sánchez & Moreno Batet, 1991: 177. Poppe & Goto, 1991: 93–94; fig. 28; pl. 11, fig. 22. Backeljau & Warmoes, 1992: 9–24. Backeljau, de Wolf, van Dongen & Brito, 1994: 239–240.

Litorina striata—Philippi, 1847, **2**: 199–200; *Litorina* pl. 4, fig. 9. Dunker, 1853: 10–11; pl. 2, figs 49–51 (in part, includes *Nodilittorina granosa* (Philippi)). Weinkauff, 1882: 50–51; pl. 6, figs 9, 12. Weinkauff, 1883: 218. Watson, 1886: 575–576 (in part, includes *Nodilittorina granosa*). Watson, 1897: 291.

Littorina (Melarhaphe) striata—H. & A. Adams, 1854, **1**: 314 (as *Melaraphe*). Tryon, 1887: 247; pl. 44, figs 58, 59 (as *Melaraphe*). Dautzenberg, 1889: 44 (as *Melaraphe*). Dautzenberg, 1927: 116–117 (as *Melaraphe*; in part, includes *Nodilittorina granosa*). Fischer, 1967: 12–15. Nordsieck, 1968: 41; pl. 6, fig. 23.31 (as *Melaraphe*). Sacchi, 1974: 90–91 (as *Melaraphe*). Nordsieck & Garcia-Talavera, 1979: 56–57; pl. 11, fig. 5 (as *Melaraphe*). Nordsieck, 1982: 61–62; pl. 26, figs 23.31, 23.31a (as *Melaraphe*).

Melarhaphe striata—Fischer, 1943: 5–6 (as *Melaraphe*; in part, from distribution includes *Nodilittorina miliaris helenae*). Fischer, 1946: 286, 289 (as *Melaraphe*; in part, from distribution includes *Nodilittorina miliaris helenae*). Fischer, 1953: 140 (as *Melaraphe*).

Nodilittorina (Granulilittorina) striata—Rosewater, 1970: 424 (generic assignment with a query).

Nodilittorina (Liralittorina) striata—Rosewater, 1981: 36–38; pl. 1N, O; pl. 2K; pl. 3E (operculum); pl. 4F (radula); pl. 4M (penis); pl. 6A, C, E, G, H. Bandel & Kadolsky, 1982: 3, 10.

Littorina (Liralittorina) striata—Reid, 1989a: 100; pl. 2, fig. 2k; fig. 7l (penis); fig. 8c (paraspermatozoa); fig. 10k (oviduct); fig. 11p (spawn); fig. 14n (radula). Reid, 1990b: 1–19; fig. 1a; fig. 2b (oviduct); fig. 3a (spawn); fig. 4a (radula). Taylor & Reid, 1990: 199–215; fig. 25 (mineralogy).

Nodilittorina striata—Buys, 1992: 108, pl. 4, fig. 2.

Littorina arenica 'Nuttall' Jay, 1839: 73 (Fayal [Faial, Azores]; *nomen nudum*; presumably this is *L. striata*, the only species common in the Azores).

Litorina arenica 'Nuttall' Dunker, 1845: 166 (Loanda [Angola; in error]; insula S. Vincente [São Vicente, Cape Verde Islands]; Fayal [Faial, Azores]; here restricted to Faial; no types in ZMB).

Littorina (Melarhaphe) striata var. *arenica*—Dautzenberg, 1927: 117 (as *Melaraphe*).

Littorina canariensis d'Orbigny, 1839: 78–79; pl. 6, figs 8–10 (Santa-Cruz de Ténériffe [Canary Islands]; 13 syntypes BMNH 1854.9.28.78, seen). Burnay & Monteiro, 1977: 24; fig. 12. von Cosel, 1982: 48.

Littorina affinis d'Orbigny, 1839: 79; pl. 6, figs 11–13 (Santa Cruz de Ténériffe [Canary Islands]; 13 syntypes BMNH 1854.9.28.79, seen). Locard, 1897: 492.

Litorina affinis—Philippi, 1846a, **2**: 146; *Litorina* pl. 2, fig. 19. Dunker, 1853: 11; pl. 2, figs 28, 29. Weinkauff, 1882: 99–100; pl. 14, fig. 9. Weinkauff, 1883: 218.

Tectarius affinis—H. & A. Adams, 1854, **1**: 315.

Littorina (Melarhaphe) striata var. *affinis*—Dautzenberg, 1927: 117 (as *Melaraphe*; in part, includes *Nodilittorina granosa*). Nordsieck, 1982: 62; pl. 26, fig. 23.31a (as *Melaraphe*; form *affinis*).

Melarhaphe striata var. *affinis*—Fischer, 1946: 289.

Littorina striata var. *affinis*—Duffus & Johnston, 1969: 32.

Nodilittorina affinis—Burnay & Monteiro, 1977: 24–25; fig. 13.

Littorina planaxis G. B. Sowerby, 1844: 153 (Tertiary deposit; St Jago, Cape de Verde Archipelago; type probably lost (R. J. Cleevely, quoted by Rosewater, 1978)).

Litorina globosa Dunker, 1845: 165–166 (Loanda [Angola], in error, here corrected to Tenerife, Canary Islands; no types in ZMB; lectotype (here designated) Dunker, 1853: pl. 4, fig. 9). Philippi, 1848, **3**: 64; *Litorina* pl. 7, fig. 10. Dunker, 1853: 9–10; pl. 4, figs 9, 10. Weinkauff, 1882: 82; pl. 11, fig. 4.

Littorina (Melarhaphe) striata var. *globosa*—Dautzenberg, 1927: 116–117 (as *Melaraphe*).

Tectarium miliare—Lamy, 1907: 149 (not *Littorina miliaris* Quoy & Gaimard, 1833; in part, includes *Nodilittorina miliaris* and *N. granosa*).

Taxonomic history

Littorina canariensis and *L. arenica* were quickly synonymized with *L. striata* (Philippi, 1847). However, because of the variability in shape and sculpture shown by *L. striata*, many nineteenth century conchologists continued to recognize as separate species, in addition to *L. striata*, the lower-spired *L. globosa* and the nodulose *L. affinis* (Philippi, 1846a–48; Dunker, 1853; Weinkauff, 1882). Weinkauff (1883) reduced *L. globosa* to a variety of *L. striata* and Watson (1886) considered *L. affinis* to be the young of *L. striata*, while Tryon (1887) included all four names as synonyms of *L. striata*. Some continued in use as varietal names (e.g. Dautzenberg, 1927: vars *globosa, affinis, arenica*). Despite its queried inclusion in the synonymy of *L. striata* given by Fischer (1967), *Tectarium turbinoides* Locard, 1897, is not a littorinid.

There has been some confusion with other nodulose littorinids from the tropical Atlantic that are members of the genus *Nodilittorina*. Both Dunker (1853) and Dautzenberg (1927) included *N. granosa* in the synonymy of *L. striata*. Jeffreys (1872) recorded *L. striata* from St Helena, but this was presumably a misidentification of *N. miliaris helenae*, the only littorinid known from the island (Smith, 1890; Rosewater, 1981). This error has been perpetuated in the records of *L. striata* from St Helena given by Fischer (1943, 1946) and Marche-Marchad (1956).

A greater number of generic assignments have been used for this than for any other species of *Littorina*. Characters of the shell led to its inclusion in *Melarhaphe* (as a subgenus of *Littorina*, e.g. H. & A. Adams, 1854; Tryon, 1887; Dautzenberg, 1927; Fischer, 1967; Sacchi, 1974; Nordsieck, 1982; or as a genus, Fischer, 1943, 1946, 1953). However, because of its nodulose sculpture, the *affinis* form was included in *Tectarius* (H. & A. Adams, 1854) or *Nodilittorina* (Rosewater, 1970; Burnay & Monteiro, 1977). Rosewater (1981) created a new subgenus, *Nodilittorina (Liralittorina)*, for this species on anatomical grounds, and was followed by Bandel & Kadolsky (1982). As a result of cladistic analysis of anatomical characters, it has recently been returned to *Littorina*, as the basal member of the clade, in the monotypic subgenus *Liralittorina* (Reid, 1989a, 1990b). This grouping has received some support from the analysis of allozyme frequencies (Backeljau & Warmoes, 1992) and from mitochondrial DNA sequence data (Rumbak, Reid & Thomas, 1994; Reid, Rumbak & Thomas, 1996). Generic placement and phylogenetic relationships are examined in detail later (p. 48).

Diagnosis

Mature shell height 5.0–19.4 mm. Shell turbinate to conical; parietal area of body whorl eroded. Mineralogy crossed-lamellar aragonite only. Sculpture of spiral ribs and microstriae, often eroded; two peripheral ribs and one near suture sometimes nodulose, especially on smaller shells. Colour greyish with fine brown flecks; aperture blackish brown with cream basal band. Penis simple, tapering, lacking mamilliform glands. Pallial oviduct of two simple loops of albumen then capsule gland; jelly gland not enlarged; bursa at posterior end of straight section. Spawn of small cupola-shaped pelagic capsules sculptured by concentric rings, containing single eggs; development planktotrophic. Distribution: Azores, Madeira, Canary Islands, Cape Verde Islands.

Anatomical description

Head-foot (Fig. 8C, D). Head grey to black; tentacles with fine transverse black bands, sometimes an unpigmented stripe across eye and tentacle base; sides of foot speckled pale or dark grey. Opercular ratio 0.45–0.67.

Penis (Fig. 8H–M). Base finely wrinkled, slightly pigmented; mamilliform glands absent; subepithelial glandular tissue around sperm groove in base. Filament tapering, 30–50% total length, sperm groove

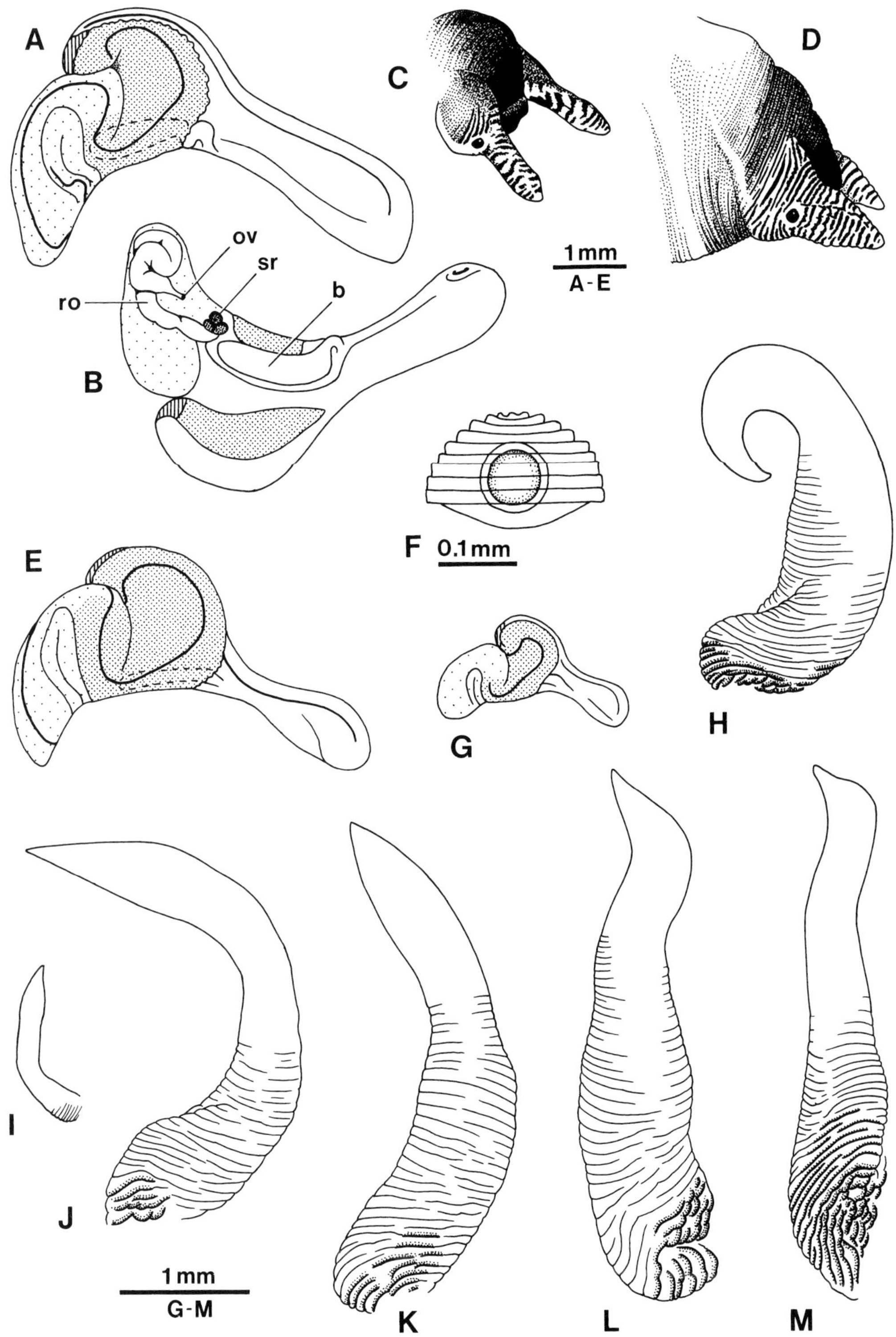

Figure 8. *Littorina (Liralittorina) striata*: **A, B, E, G,** pallial oviducts; **A, B,** lateral and medial views of the same pallial oviduct; **C, D,** head pigmentation; **F,** pelagic egg capsule; **H–M,** penes. **A–C, F, H, J,** Los Gigantes, Tenerife, Canary Islands; **D, E, M,** Caloura, São Miguel, Azores; **G, I,** Monte de Leste, Sal Island, Cape Verde Islands (shells of nodulose form); **K, L,** Canical, Madeira. Shell heights: **A, B,** 12.0 mm; **C,** 8.7 mm; **D, E,** 10.7 mm; **G,** 5.5 mm; **H,** 9.0 mm; **I,** 5.0 mm; **J,** 8.7 mm; **K,** 9.1 mm; **L,** 9.4 mm; **M,** 8.3 mm. For shading conventions of pallial oviducts see p. 6. Abbreviations: b, copulatory bursa; ov, junction with ovarian oviduct; ro, renal oviduct; sr, seminal receptacle.

ending terminally. Rosewater (1981) reported that the sperm groove was open, and indeed in preserved specimens the tips of the groove can be pulled apart with forceps. However, in living animals the groove is closed by the adherence of the epithelia lining it. This degree of closure is less complete than in other cases of 'superficial closure' in littorinids (Reid, 1989a), but the groove does appear to be functionally closed.

Sperm (Fig. 3A). Euspermatozoa 81–89 μm. Paraspermatozoa 12–17 μm diameter, spherical; vesicles large and distinct; spherical nucleus persists.

Oviduct (Fig. 8A, B, E). Loop of albumen gland followed by loop of capsule gland; capsule gland with distal reddish translucent portion; straight section (jelly gland) with small amount of subepithelial glandular tissue beneath secretory epithelium, not septate. Bursa separates at posterior end of straight section. Seminal receptacle lobulate.

Spawn and development (Figs 8F, 9E). Pelagic, cupola-shaped capsules, 240 μm diameter, sculptured by about 9 concentric rings, containing single eggs 70 μm diameter. Development planktotrophic. Protoconch 260 μm diameter, 3 whorls, sculptured by undulating spiral ribs and diagonal wrinkles. Breeding season: summer (in BMNH collections from Canaries, Azores and Madeira, females have been found to be immature in samples taken in April, May and November, but mature in June, July and September; field-collected animals spawned in the laboratory in September).

Radula and foregut (Fig. 9G–J). Relative radular length 2.84–11.8. Rachidian: length/width 1.52–2.00; single very long, blunt or rounded cusp, one pointed basal cusp on each side. Lateral: four cusps, central two long, bluntly rounded, others pointed. Inner marginal: four cusps, third from inside long, bluntly rounded, others pointed. Outer marginal: 7–10 long, pointed cusps; narrow neck, projecting flange on either side of base. Salivary glands not enlarged.

Shell variation (Figs 9A–D, F; 10)

Geographical variation. In this species there is little obvious differentiation on a geographical scale between island groups. A trend of decreasing aperture size, and therefore taller spire, towards the southern parts of the range has been found by de Wolf (pers. comm.).

Sculptural variation. Superficially, the nodulose shell often known as 'var. *affinis* ' (Fig. 10B, C, F) looks quite distinct from typical *L. striata* with only spiral sculpture or slightly undulating spiral ribs giving a granulose appearance (Fig. 10A, D, E, G–K), and many early authors retained *affinis* as a separate species (see Synonymy, p. 41). In the present study no anatomical or radular differences have been found between nodulose and typical forms. In large series from a single locality the whole range of intermediates can often be found (Dautzenberg, 1927; Fischer, 1967; pers. obs.), and there is no consistent geographical trend in shell sculpture. Allozyme studies have found no differentiation between sympatric samples of smooth and nodulose shells (de Wolf, van Dongen, Brito & Backeljau, 1994), and they exhibit similar heat coma temperatures (Vedel & Depledge, 1995). The variation is therefore intraspecific.

There is some controversy about the description and interpretation of this sculptural polymorphism. Watson (1886) was the first to place *L. affinis* in the synonymy of *L. striata*, observing that since the early whorls were always more or less nodulose, *L. affinis* was simply the young of *L. striata*. Rosewater (1981) also described nodules on the spire whorls. Similarly, Britton (1995) stated that most individuals show 'sequentially dimorphic' shell ornamentation, those smaller than about 5 mm producing a nodulose shell, which is replaced (or 'covered over') on subsequent whorls by spiral striae. These reports are difficult to reconcile with the present study. Examination of large series of shells has shown that the degree of nodulation does not normally change appreciably during the growth of an individual; only occasionally do larger shells have more pronounced granules on the spire whorls. Instead, nodulose shells are found to attain smaller size (up to 9.8 mm) than striate or merely granulose shells. Large and old shells often show some erosion of the apical whorls, but by examination of small and well-preserved shells (Fig. 9A–D, F) it can be verified that sculpture is usually constant. Certainly the nodules at and above the periphery of small shells are not 'covered

Figure 9. *Littorina (Liralittorina) striata*: **A–D**, juvenile shells; **E**, protoconch; **F**, detail of sculpture; **G**, **H**, two views of radula (shell H=11.8 mm) at 45° (**G**) and flat (**H**); **I**, **J**, two views of radula (shell H=11.5 mm) at 45° (**I**) and side view (**J**). **A**, **D–F**, La Restinga, El Hierro, Canary Islands; **B**, **C**, Monte de Leste, Sal Island, Cape Verde Islands; **G**, **H**, Porto Moniz, Madeira; **I**, **J**, Los Gigantes, Tenerife, Canary Islands. Scale bars: **A–D**=1 mm; **E**=100 µm; **F**=500 µm; **G**, **H**=50 µm; **I**=50 µm; **J**=50 µm.

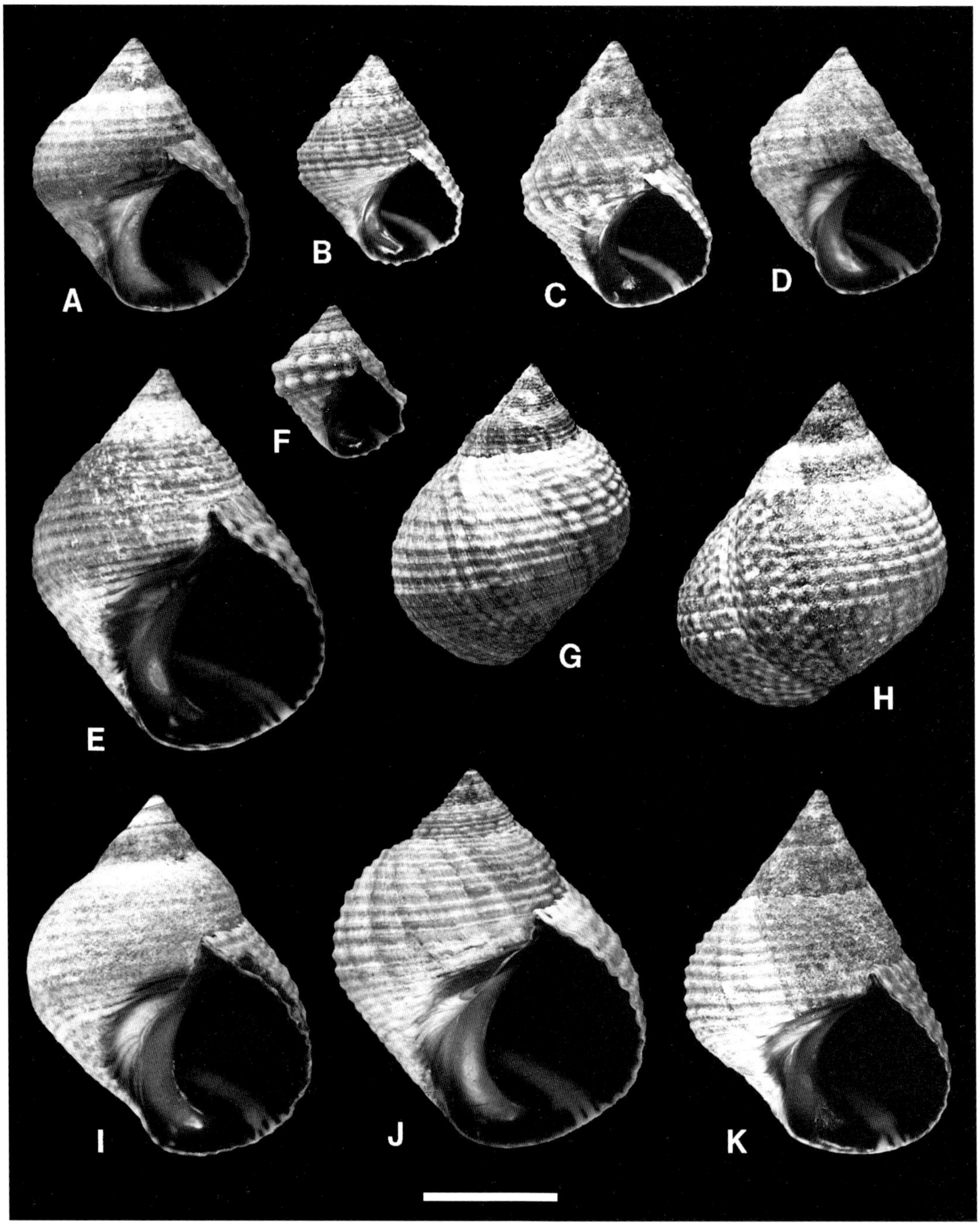

Figure 10. *Littorina (Liralittorina) striata*. **A,** *Littorina striata* King & Broderip, 1832; holotype, BMNH 1968323; Port Praya, Cape Verde Islands. **B, C, H, K,** Lanzarote, Canary Islands. **D, I,** Feteira, Faial, Azores. **E, G,** Los Gigantes, Tenerife, Canary Islands. **F,** Monte de Leste, Sal Island, Cape Verde Islands. **J,** Porto Moniz, Madeira. Scale bar = 5 mm.

over' by subsequent growth; they remain visible on the spire unless obscured by erosion. This suggests an alternative interpretation of Britton's (1995) observation that striate shells occurred in all size classes of mixed populations, in which no nodulose shells were larger than 7 mm.

The functional significance of shell form in littorinids has been discussed, in relation to temperature control and conservation of water, by Vermeij (1973). Where several species of littorinid occur together on tropical shores, the species at higher tidal levels tend to be more strongly sculptured and often nodulose; this is believed to increase the surface area for heat loss by convection and re-radiation. In *L. striata* several authors have noted that shell sculpture is connected with tidal level, substrate and aspect of the shore. Fischer (1946, 1953, 1967) observed that nodulose shells occurred at higher tidal levels than smooth forms. In a detailed study of two sites on the island of São Miguel in the Azores, Britton (1995) found that nodulose shells were common on a black basaltic shore, where they did indeed predominate in the littoral fringe, whereas in the upper eulittoral zone larger striate shells were in the majority; in contrast, on a shore with paler rocks nodulose shells were scarce at all levels. Others have found nodulose shells to be more common on sheltered shores, and smooth forms on wave-exposed coasts (de Wolf *et al.*, 1994). The small, nodulose shells thus predominate on dark rocks heated by the sun, at the highest tidal levels, and on shores with little wave-splash, and are therefore found in just those situations where heat and desiccation stress are likely to be most severe. A small, nodulose shell is likely to dissipate heat more efficiently than a large, striate one in these microhabitats, although this has yet to be confirmed experimentally.

Although the sculptural polymorphism is likely to be of functional significance, it is not known how it is determined, or how the distribution of sculptural types on the shore is achieved. Britton (1995) believed that most individuals passed through a nodulose stage, and that the distribution was achieved by differential migration. This is not, however, consistent with his observation of a similar size distribution at both upper and lower levels on the shore, nor does it explain the scarcity of nodulose forms on the shore with paler and cooler rocks. If, as suggested here, nodulose individuals do not generally become striate with growth, other explanations can be sought. Because of the planktonic dispersal of this species, it seems unlikely that the sculptural polymorphism can be genetically determined. Samples of nodulose and striate shells show no detectable allozyme differentiation (de Wolf *et al.*, 1994), and levels of interpopulation genetic variation are low (Backeljau, Breugelmans, Brito, de Bruyn, de Wolf & Timmermans, 1995). In the face of extensive gene flow, a genetically-based sculptural polymorphism could only be maintained by differential settlement or, after settlement, by strong selection or migration. From reports of heat coma temperatures (Vedel & Depledge, 1995) and measurements of evaporative water loss, Britton (1995) did not believe that mortality from heat and desiccation were likely to be sufficient to cause the differential distribution of sculptural types.

Another, more likely, explanation is that an ecophenotypic effect is involved. As in some other *Littorina* species (p. 29), sculpture could be a function of growth rate. Individuals settling higher on the shore in the littoral fringe, or anywhere where algal food is less abundant and opportunities for feeding while wetted by spray are more limited, should grow more slowly and attain smaller size, and as a result may develop nodules. Alternatively, there might be a direct effect of temperature on the pattern of shell growth. Dislodgement and migration will result in some mixing of the forms within shores, and may also lead to some cases of change in sculpture during growth. These suggestions remain to be investigated, but are in agreement with the facts on development of sculpture, distribution of sculptural types on the shore, and the inferred high levels of gene flow. A similar, and even more striking, sculptural polymorphism occurs in *Nodilittorina australis* (Gray, 1826), in which nodulose shells are likewise smaller than striate ones, and change in sculpture on a single shell is unusual (see Rosewater, 1970, although he interpreted this as two species with rare hybrids; Reid, 1989a).

Sexual dimorphism. Females have been found to be significantly larger (de Wolf, pers. comm.).

Distribution

Habitat. *Littorina striata* is common on bare rocks in the upper eulittoral and littoral fringe; its lower limit is among the uppermost barnacles and it extends up into the splash zone, a little below the highest levels occupied by *Melarhaphe neritoides* (Fischer, 1946, 1967; Bandel, 1974; Hawkins, Burnay, Neto, Tristão da Cunha & Frias Martins, 1990; Morton, 1990a). Mean densities of up to 469 per m^2 have been recorded on São Miguel in the Azores, and there is a tendency for smaller individuals to occur higher on the shore (Britton, 1995). It may be found in crevices and high-level pools (Fischer, 1967; Bandel, 1974), and is most abundant on wave-exposed coasts (Fischer, 1967; Hawkins *et al.*, 1990). Bullock (1995) recorded several small juveniles from midlittoral coralline algal turf in the Azores, but the identification is questionable, since the figured shell is not this species. Only one study of the feeding behaviour of this species has been found (Granado & Caballero, 1991), but this employed ground samples of low-tide algae suspended in agar to assess palatability, and so can have little relevance to the natural diet. In view of the habitat of this species, it presumably grazes on epilithic and endolithic microalgae and perhaps encrusting lichens.

Range (Fig. 11). This species is limited to the oceanic high islands of the warm temperate and subtropical northeastern Atlantic: the Azores, Madeira, the Canary Islands and the Cape Verde Islands. Records from St Helena by Jeffreys (1872) and later authors (Fischer, 1943, 1946; Marche-Marchad, 1956) are almost certainly based on misidentification of the only littorinid reliably reported from the island, *Nodilittorina miliaris helenae*. Contrary to the note by Rosewater (1981), Lamy's (1907) record of *Tectarium miliare* from São Tomé refers to *N. granosa*, although *L. striata* was included in the synonymy. There are a small number of records from the African mainland. The early ones from 'Guinea inferior' (Philippi, 1847) and 'Loanda' (Dunker, 1845, 1853; Weinkauff, 1882) are apparently erroneous, but Rosewater (1981) recorded a single specimen in a mixed lot of littorinids from Port Marshall, Liberia, in the collection of ZMK, so perhaps occasional specimens may occur on the mainland. The African coast nearer to the Cape Verde and Canary Islands is not well represented in museum collections.

Fossil record. Two fossil records have been traced. One is that of Sowerby (1844), who described his species *Littorina planaxis* 'from a tertiary deposit, beneath a great basaltic stream, at St Jago in the Cape de Verde Archipelago'. This specimen is now lost, but it can be determined from the original description of a flattened columellar lip, and the author's comparison with the Recent species. The other is a Pleistocene record from San Catalina on Gran Canaria (Rothpletz & Simonelli, 1890, as *L. affinis*).

Discussion

Shell variation and genetics. Investigations of the population genetics of this species have shown that allozyme frequencies are rather uniform on local and geographical scales: on a single island in the Azores (de Wolf *et al.*, 1994), among islands in the group (Backeljau *et al.*, 1995) and between the Azores, Canaries and Madeira (de Wolf, pers. comm.). This implies a high degree of gene flow, consistent with the presumably wide dispersal of the pelagic eggs and planktotrophic larvae. Such widely-dispersed species are often morphologically uniform also (Janson, 1987a; Johannesson, 1992), so the rather variable shell of *L. striata* is unexpected. However, there is little obvious morphological differentiation even on a geographical scale between the island groups on which it occurs and, as discussed earlier (p. 47), the remarkable variation in shell sculpture may be an ecophenotypic effect. Nodulose sculpture may perhaps develop if growth is slow, but this suggestion requires experimental investigation. It has been reported that esterases show an unusually high degree of variation between individuals within populations (Backeljau *et al.*, 1995).

Phylogenetic relationships. The generic assignment and phylogenetic relationships of this species are controversial. H. & A. Adams (1854) included *L. striata* in the subgenus *Melarhaphe*, with other mainly tropical species distinguished by relatively thin, acuminate, spirally striated and colourful shells (now largely classified as *Littoraria* and *Nodilittorina*). This classification was followed by many authors who considered only shell characters (e.g. Tryon, 1887; Dautzenberg, 1927; Fischer, 1967; Nordsieck, 1982). Species with nodulose sculpture were excluded from *Littorina*, and consequently the *affinis* form was placed in *Tectarius* by H. & A. Adams (1854). With increasing emphasis on characters of the penis, radula and egg capsules in the classification of the Littorinidae, the use of *Melarhaphe* was restricted, and *L. striata* was tentatively reclassified in *Nodilittorina (Granulilittorina)* (Rosewater, 1970), a decision which still evidently owed much to its nodulose or granulose shell sculpture. This generic assignment was reiterated by Rosewater (1981), citing both radular morphology and the tendency towards nodulation on the shell. However, because of its simple penis, lacking the conspicuous basal glands typical of other *Nodilittorina* species, he placed it in a new, monotypic subgenus, *Liralittorina*. He

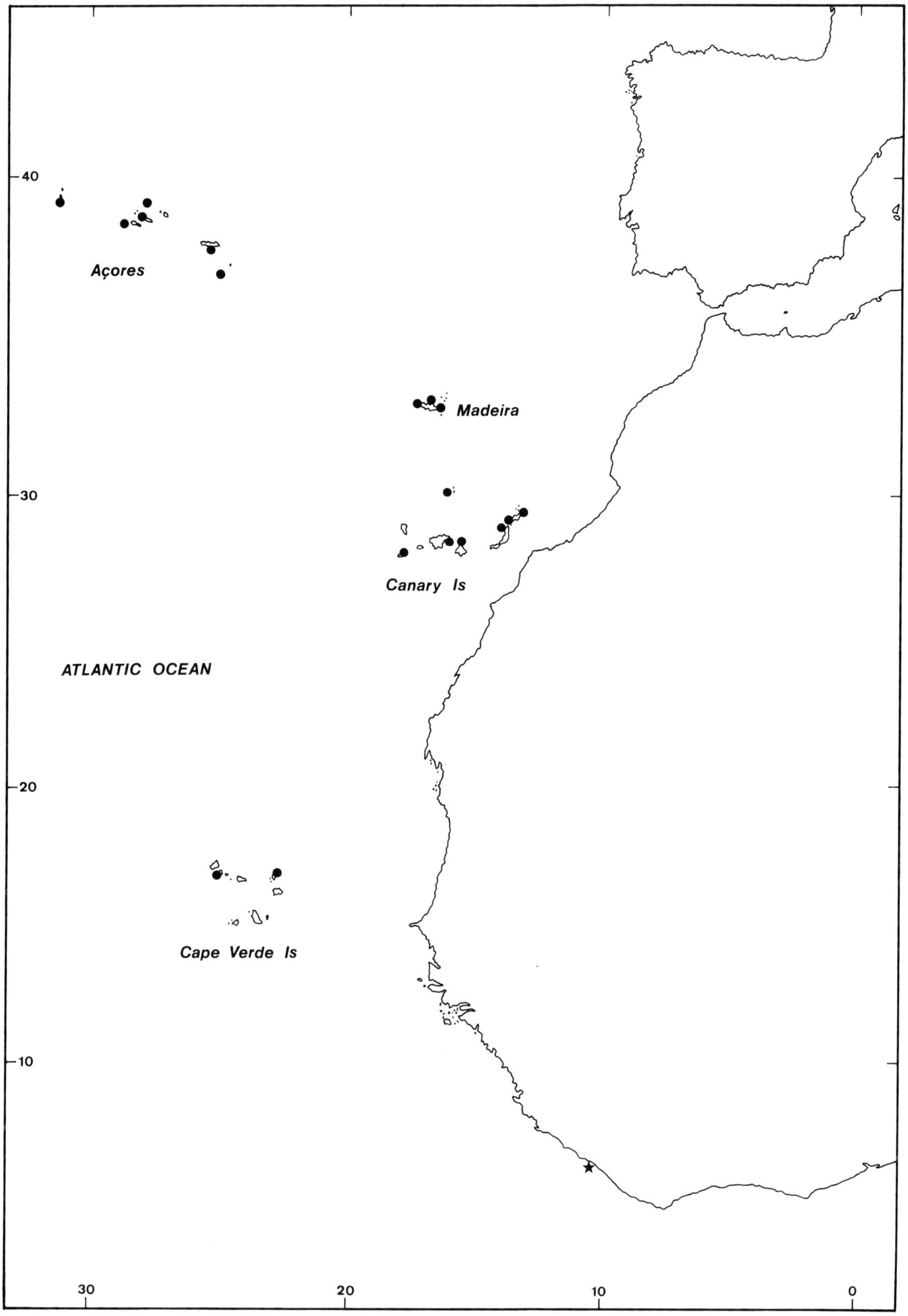

Figure 11. *Littorina (Liralittorina) striata*: distribution map. Key: circles, specimens examined; asterisk, record from literature (Rosewater, 1981).

also noted the superficial similarity of the penis to that of *Melarhaphe neritoides*, but excluded *L. striata* from *Melarhaphe* because its penial sperm groove was thought to be open, and it lacked the basal cusps on the rachidian tooth that are present in *M. neritoides*. Bandel & Kadolsky (1982) followed this classification of *L. striata* in their revision of western Atlantic *Nodilittorina* species.

More recently, *L. striata* has been returned to *Littorina*. In a cladistic analysis of the morphological characters of all recognized subgenera of the family, Reid (1989a) showed that although *L. striata* resembles *Nodilittorina* species in its nodulose and entirely aragonitic shell, its radula, and cupola-shaped egg capsule, these are plesiomorphic characters. More significant are the paraspermatozoa lacking rod-pieces, and the two consecutive loops of the spiral egg groove in the pallial oviduct; these are two apomorphic characters shared with all species of *Littorina*. This species was therefore considered to be the basal member of the clade *Littorina* (Reid, 1990b; p. 351), and the subgenus *Liralittorina* was retained. This classification has since received some support from the phylogenetic analysis of allozyme data (Backeljau & Warmoes, 1992; p. 356), although conflicting results can be obtained by the use of different methods of tree construction (Backeljau, de Wolf, van Dongen & Brito, 1994). Phylogenetic analyses of mitochondrial DNA sequence data have also given some support to the position of *L. striata* as the basal branch of the clade *Littorina* (Rumbak *et al.*, 1994; Reid *et al.*, 1996; p. 358). Interestingly, the ecology of this species also resembles that of many *Nodilittorina* species, which likewise occur on bare rocks in the littoral fringe, in warm regions and on oceanic high islands (p. 394).

Similar species. Over most of its range *L. striata* is sympatric with *Melarhaphe neritoides*. The latter is distinguished as a smaller species, usually less than 7 mm in height, which is almost smooth (lacking nodules, spiral ribs and microstriae) and has indistinct sutures. In *M. neritoides* the cephalic tentacles are black with a longitudinal unpigmented stripe (Reid, 1989a: fig. 5c), the penis is similar to that of *L. striata*, but the form of the pallial oviduct is quite different, without spiral looping of the capsule gland (Reid, 1989a: fig. 9l). *Littorina saxatilis* also occurs in the Azores and Canary Islands, but is rare; in these localities it has a smooth and often tessellated shell; there are mamilliform glands on the penis and the females are ovoviviparous.

In the case of strongly-sculptured forms of *L. striata*, confusion is much more likely with nodulose species of *Nodilittorina*, although these are apparently never sympatric with *L. striata* (unless *L. striata* does rarely occur on the African coast). *Nodilittorina granosa* occurs only on the West African coast and adjacent islands, and its shell has about six equal rows of nodules, generally in axial alignment, above the periphery (Rosewater, 1981: pl. 1, figs H, I). *Nodilittorina miliaris* (Quoy & Gaimard) is a highly variable species, here considered to include the typical subspecies from Ascension Island (Rosewater, 1981: pl. 1, figs J, K; Bandel & Kadolsky, 1982: figs 48, 49), the subspecies *helenae* from St Helena (Rosewater, 1981: pl. 1, figs L, M; Bandel & Kadolsky, 1982: fig. 51) and the subspecies *vermeiji* Bandel & Kadolsky (Bandel & Kadolsky, 1982: figs 46, 47). Typically, *N. miliaris* has three to four rows of nodules above the periphery and these are not usually axially aligned; there may also be secondary sculpture of narrow cords in the areas between the nodulose ribs, so that the spiral sculpture appears irregular. The most nodulose forms of *L. striata* usually develop only three rows of nodules (two at the periphery and one near the suture). However, some examples of *N. miliaris* from St Helena are almost without nodules; in these cases the less regular spiral ribs aid their separation from *L. striata*. Anatomical characters are more reliable. The two *Nodilittorina* species have a penis with both a single mamilliform gland and a glandular disc on a bifurcation of the base, and in the female there is but a single spiral loop in the egg groove of the pallial oviduct, where it passes through the albumen gland.

Subgenus *Planilittorina* Reid, 1989

Synonymy

Planilittorina Reid, 1989a: 100–101 (type species *Littorina keenae* Rosewater, 1978, by original designation; proposed as subgenus of *Littorina*).

Taxonomic history

This subgenus was established for *L. keenae* by Reid (1989a), because anatomical comparisons and cladistic analysis of morphological characters suggested that it was not closely related to other recognized subgenera of *Littorina*. It differed from the more plesiomorphic *Liralittorina* in the presence of two mamilliform penial glands, an enlarged jelly gland in the pallial oviduct, and biconvex egg capsules released in a pelagic gelatinous mass. By comparison with the more derived subgenus *Littorina*, it lacked the apomorphic characters of an outer calcite shell layer, three consecutive spiral loops of the egg groove in the pallial oviduct, and enlarged salivary glands. The only known living species is *L. keenae*, but the fossil *L. sookensis* is here assigned to this subgenus for the first time.

Diagnosis

Shell: turbinate to high-turbinate; smooth or with spiral grooves; parietal area of body whorl eroded and flattened; mineralogy of crossed-lamellar aragonite only. Anatomy: as for the only known living species, *L. keenae* (see below).

Littorina (Planilittorina) sookensis Clark & Arnold, 1923

(Fig. 12)

Synonymy

Littorina sookensis Clark & Arnold, 1923: 165; pl. 37, figs 4a, 4b, 5a, 5b (in the sea cliffs east of the mouth of Kirby Creek, 6 miles west of Sooke, Vancouver Island [type locality from original description; on label: 12 miles west of Sooke], Upper Oligocene or Lower Miocene, CAS loc. 231; holotype CAS Geology 231.02, seen, Fig. 12A, D; paratype CAS Geology 61805.43, seen, Fig. 12B, C). Weaver, 1942: 329; pl. 67, figs 23, 24, 29. Reid, 1990c: 37–39.

Diagnosis

Shell height up to 23 mm. Shell turbinate; sutures not impressed; body whorl with angulation at shoulder. Mineralogy probably entirely aragonitic. Sculpture of 13–21 spiral ribs on last whorl. Columella broad, hollowed; parietal area of body whorl eroded. Distribution: Upper Oligocene Sooke formation of Vancouver Island, Canada.

Shell variation (Fig. 12)

Only two specimens of this species have been seen, holotype and paratype. They differ in shape, the holotype having a relatively taller spire (H/B=1.28, cf. 0.99 for paratype). Both show a distinctive angulation or keel at the shoulder. The aperture is also similar in the two, with a broad, hollowed columella, and a strongly eroded crescentic area on the parietal wall adjacent to the aperture. The shells are somewhat worn, so that details of the sculpture are unclear, but in the holotype there are about 21 spiral ribs on the last whorl, with marked grooves between them, and in the smaller paratype only about 13 ribs. On the columellar pillar of the holotype crossed-lamellar structure is visible with a binocular microscope, suggesting that the shell is not recrystallized. No differentiation of an outer shell layer can be seen, so the shell may be entirely aragonitic, although sectioning would be required for confirmation.

Distribution

Fossil record. The only records that have been found are those of the original authors, from the Sooke formation near Sooke on Vancouver Island (Clark & Arnold, 1923). Although they described only two specimens, they mentioned four localities, so additional specimens may exist. They placed the Sooke formation in the Upper Oligocene or Lower Miocene, and since then there has been continuing debate about its correlation. Some studies of the foraminifera have suggested ages as young as Lower Pliocene, but subsequent work has indicated a Lower Miocene age (Jeletzky, 1973). Molluscan correlations have generally supported a Lower Miocene (Jeletzky, 1973) or Upper Oligocene age (Durham, 1944; Armentrout, 1975), and the latter is apparently now accepted (Muller, Cameron & Northcote, 1981; Irving & Massey, 1990). The molluscan assemblage of the Sooke formation

Figure 12. *Littorina (Planilittorina) sookensis*. **A, D,** *Littorina sookensis* Clark & Arnold, 1923; holotype, CAS Geology 231.02; E of mouth of Kirby Creek, 6 miles W of Sooke, Vancouver Island, Canada; Sooke formation, Upper Oligocene. **B, C,** paratype, CAS Geology 61805.43; between mouths of Muir and Kirby Creeks, W of Otter Point, Sooke, Vancouver Island, Canada; Sooke formation. Scale bar=10 mm.

represents a strand-line facies, and suggests a temperate climate, similar to that prevailing between Vancouver Island and northern California at the present time (Clark & Arnold, 1923).

Discussion

Phylogenetic relationships. If the outer calcite layer is indeed absent, *L. sookensis* lacks this synapomorphy of the subgenera *Littorina* and *Neritrema*. This layer is also absent in the two Recent species at the base of the *Littorina* clade, *L. striata* from the Atlantic islands and *L. keenae* from the northeastern Pacific. These two also share with *L. sookensis* a crescentic, eroded, parietal area, apparently a symplesiomorphy, since this is present in most members of the sister-genus, *Nodilittorina*. As noted by Clark & Arnold (1923), *L. sookensis* appears most closely related to *L. keenae*, with which it shares a similar wide, hollowed columella and uniformly rounded anterior margin of the aperture. It has therefore been placed in the same subgenus, *Planilittorina*. The maximum size of *L. keenae* is similar (22.1 mm), but its sculpture is weaker, and it does not develop a keeled shoulder. The climatic conditions inferred for the Sooke formation are similar to those over the range of *L. keenae*, from southern Oregon to Baja California.

This species is the oldest certain species of *Littorina*, although earlier fossils are doubtfully assigned to the genus (pp. 340–344).

Littorina (Planilittorina) keenae Rosewater, 1978

(Figs 3B, 13–16)

Synonymy

Littorina planaxis 'Nuttall' Jay, 1839: 73 (Upper California; *nomen nudum*).
Litorina planaxis 'Nuttall' Philippi, 1847, **2**: 201; *Litorina* pl. 4, fig. 16 (California superior; lectotype (Rosewater, 1978: fig. 2; Fig. 15A herein)+3 paralectotypes BMNH 1990004, seen; when examined by Rosewater, another lot (BMNH 1912.4.12.30–34) was mixed with the syntypic series, these have now been separated; lectotype is specimen figured by Philippi, 1847; not *Littorina planaxis* G. B. Sowerby, 1844=*L. striata*). Carpenter, 1864: 655. Weinkauff, 1882: 49; pl. 6, figs 5, 8. Weinkauff, 1883: 214.
Littorina planaxis—Reeve, 1857: sp. 38; pl. 8, fig. 38a, b. Cooper, 1860: 374. Keep, 1888: 68; fig. 53. Keep, 1904: 216; fig. 233. Packard, 1918: 320; pl. 35, figs 4, 8. Keep & Baily, 1935: 198; fig. 179. Abbott, 1954a: 134; pl. 20, fig. a. Abbott, 1974: 67; text-fig. 553. Murray, 1979: 469–474; fig. 1C (spawn). Schmitt, 1979: 359–366; fig. 1 (spawn).
Littorina (Melarhaphe) planaxis—H. & A. Adams, 1854, **1**: 314 (as *Melaraphe*). Tryon, 1887: 248; pl. 43, fig. 55; pl. 44, fig. 57 (as *Melaraphe*; in part, other figs are *Littoraria intermedia* (Philippi) and *Nodilittorina hawaiiensis* Rosewater & Kadolsky).
Littorina (Littorivaga) planaxis—Dall, 1921: 153. Oldroyd, 1927: 61–62. Burch, 1945: 12, 37.
Littorina (Littorina) planaxis—Rosewater, 1970: 423.
Littorina patula Gould, 1849: 83 (San Francisco; holotype USNM 5636, seen, Rosewater, 1978: figs 3, 4; not *Littorina patula* Thorpe, 1844). Gould, 1852: 197–198; Atlas (1861) pl. 14, figs 237, 237a.
Littorina keenae Rosewater, 1978: 124; figs 1–6 (new name for *Litorina planaxis* Philippi, 1847). Rosewater, 1980: 5–8; figs 5, 6 (radula). Ricketts, Calvin, Hedgpeth & Phillips, 1985: 19–20, 32; fig. 8; fig. 9 (radula). Kozloff, 1987: 209; fig. 12.25.
Littorina (Planilittorina) keenae—Reid, 1989a: 100–101; pl. 2, fig. 2l; fig. 7m (penis); fig. 10l (oviduct); fig. 11r (spawn); fig. 14o (radula). Reid, 1990b: 1–19; fig. 1b; fig. 2c (oviduct); fig. 3c (spawn). Taylor & Reid, 1990: 199–215; figs 3, 4 (mineralogy).

Taxonomic history

The history of this species is relatively straightforward. Although introduced by Jay (1839) as a *nomen nudum*, the name *L. planaxis* was validated by Philippi (1847). Gould (1849) described *L. patula*, noting in 1852 that it might be the same as *L. planaxis*, which he believed had not been described. This species was long familiar by the name *L. planaxis*, but Rosewater (1978) renamed it *L. keenae*, because *Litorina planaxis* Philippi, 1847, was a junior homonym of *Littorina planaxis* G. B. Sowerby, 1844 (the difference in spelling of the generic names is disregarded, ICZN, 1985: art. 57e).

This species was made the type of the monotypic subgenus *Planilittorina* by Reid (1989a), in recognition of its unique oviduct and spawn type, and relatively basal position within the genus (Reid, 1990b).

Diagnosis

Mature shell height 5.9–22.1 mm. Shell high-turbinate; parietal area of body whorl eroded away and thus flattened. Mineralogy of crossed-lamellar aragonite only. Sculpture smooth or with fine spiral lines, often eroded; prominent axial growth lines. Colour grey-brown, marbled or flecked with greyish white; aperture brown with cream basal band. Penis with 1–3 (usually 2) mamilliform glands; tapering filament, 30–50% total length. Pallial oviduct of loop of albumen gland, then multispiral capsule gland; jelly gland enlarged, septate; bursa at posterior end of straight section. Spawn of biconvex pelagic capsules with lateral rim, containing single eggs; development planktotrophic. Distribution: California and Baja California.

Anatomical description

Head-foot (Fig. 13H, I). Head and tentacles black; unpigmented stripe across eye, usually extending across tentacle base; sides of foot mottled black. Opercular ratio 0.38–0.47.

Penis (Fig. 13H, K–R). Base wrinkled, slightly pigmented in most specimens; usually with two mamilliform penial glands, basal one shorter, broader, surrounded by part of subepithelial glandular tissue that extends up into penial filament, around sperm groove; second, distal, mamilliform gland occasionally absent (Fig. 13R), or two may rarely be present in this position (Fig. 13O). Filament tapering, 30–50% total length. Sperm groove open, ending terminally.

Sperm (Fig. 3A). Euspermatozoa 84–91 µm. Paraspermatozoa 12.5–15 µm diameter, spherical; vesicles large, distinct, no nucleus visible.

Oviduct (Fig. 13A–G). Loop of albumen gland followed by large, multispiral loop of capsule gland with apical reddish translucent portion; straight section an enlarged, septate jelly gland with secretory epithelium. Bursa originates posterior to jelly gland, incompletely separated until beneath albumen gland (see sections in Fig. 13C–F). Seminal receptacle simple.

Spawn and development (Fig. 13J). Pelagic, biconvex capsules 300–400 µm diameter with peripheral flange; containing single (rarely 2–3) ova 89–90 µm diameter (Murray, 1979; Schmitt, 1979; Abbott & Haderlie, 1980, quoting Imai, 1964, give a smaller mean ovum diameter of 77 µm). Eggs are released from female in an elongate, coiled, gelatinous mass 5–15 mm long, within which capsules are embedded in a helical arrangement; egg mass is not attached to substrate, and distintegrates after about 24 h in seawater, releasing individual capsules (Schmitt, 1979). Shell width of veliger at hatching 137 µm (Murray, 1979). Development planktotrophic. Protoconch 240 µm diameter, about 3 whorls, sculpture not preserved in available specimens. Spawning season reported as March to July at both Bodega Bay, northern California, and at Santa Catalina Island, southern California (Schmitt, 1979), and as March to September at Bodega Bay (Chow, 1987), but a few snails can be found copulating and spawning at all times of year (Schmitt, 1979; Abbott & Haderlie, 1980). Development planktotrophic (Murray, 1979).

Radula and foregut (Fig. 14C–H). Relative radular length 3.49–4.69. Rachidian: length/width 1.07–1.75; single long, bluntly rounded cusp, one small pointed cusp on each side. Lateral: four cusps, central two larger, third from inside longest and blunt, others pointed. Inner marginal: four cusps, third from inside long and bluntly rounded, other pointed. Outer marginal: 7–9 long, pointed cusps; narrow neck, projecting flange either side of base. Salivary glands not enlarged.

Shell variation (Figs 14A, B; 15)

The intensity of shell patterning shows some variability, and occasionally more or less distinct bands are present (Fig. 15D) although, since the range appears to be continuous, the species is not described as polymorphic. In general the shell characters of this species are rather uniform throughout its range (Fig. 15). North (1954) observed that shells were smaller in more exposed areas, perhaps because larger snails were removed by strong wave action. According to Frank (1979), the maximum sizes occur at the extremes of the geographical range. There is no evidence of sexual dimorphism in shell size (Gibson, 1964; Chow, 1987b). Shells sometimes show conspicuous growth lines (Fig. 15M), which may represent annual interruptions to growth in this long-lived species (up to 14 years, Behrens Yamada, 1992).

Distribution

Habitat. This is the most common and characteristic mollusc of superficially bare rocks in the upper eulittoral and littoral fringe on the open coasts of California (Hewatt, 1937; Bock & Johnson, 1967; Ricketts, Calvin, Hedgpeth & Phillips, 1985; Chow, 1987; Foster, de Vogelaere, Oliver, Pearse & Harrold, 1991; Littler, Littler, Murray & Seapy, 1991). It usually clusters in crevices when dry, and can sometimes be found in stagnant high-level pools. Most of the larger snails inhabit the splash zone, up to 7 m above high water on wave-exposed shores (North, 1954; Ricketts *et al.*, 1985), but may move downwards during neap-tide periods (Abbott & Haderlie, 1980). Small and medium-sized snails occur at lower levels, in the zone of *Balanus glandula* and the red alga *Endocladia muricata* (Glynn, 1965), but even here the animals prefer a rock substrate (Dahl, 1964). The lower limit of the vertical distribution of *L. keenae* may be determined by wave action, predators and perhaps also by parasitic infection (Behrens Yamada, 1992). The vertical range of *L. keenae* is higher than that of *L. scutulata*

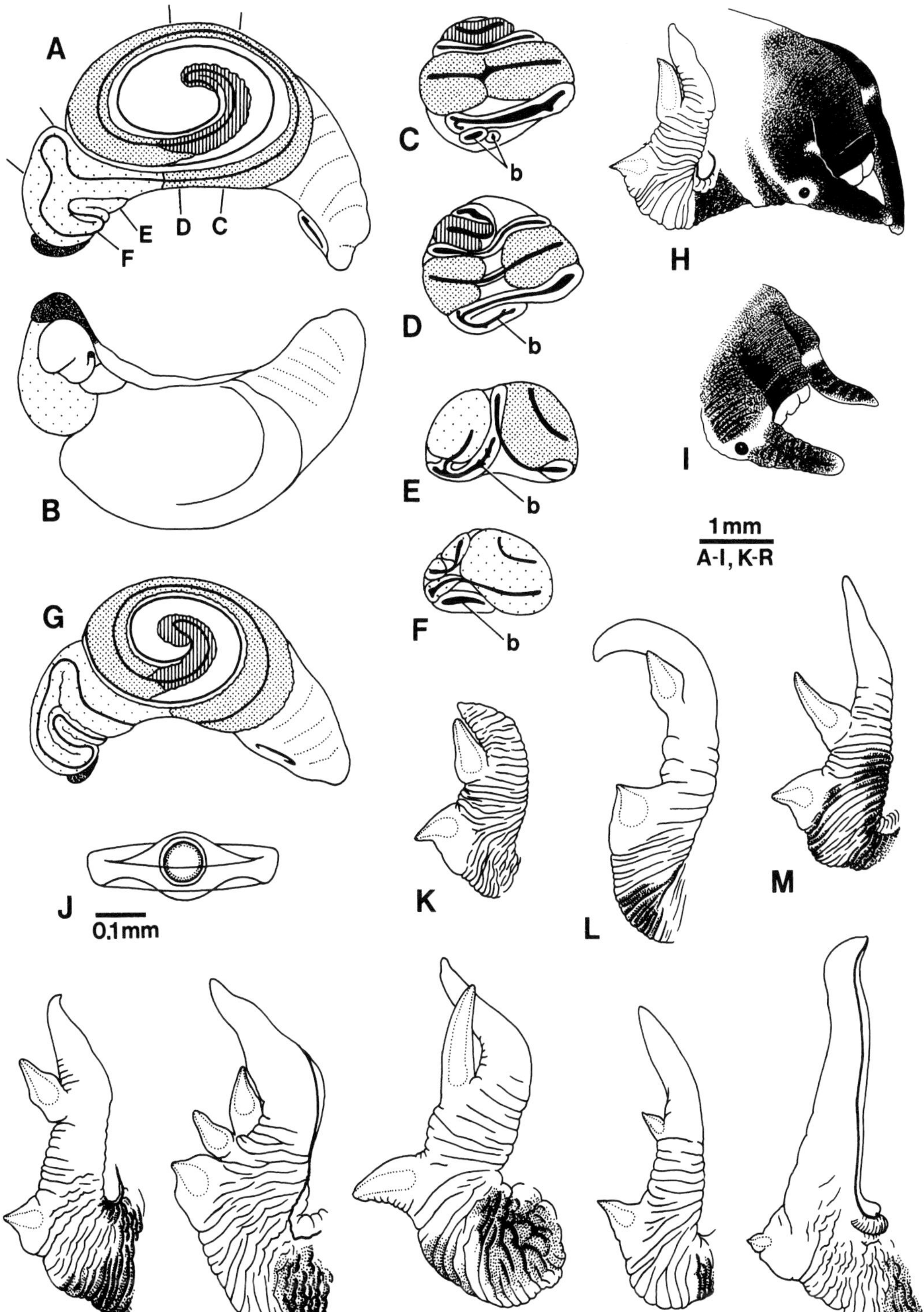

Figure 13. *Littorina (Planilittorina) keenae*: **A–G**, pallial oviducts; **A–F**, lateral view, medial view and transverse serial sections (taken at levels indicated in **A**, viewed from anterior end, right lateral side of animal towards top of page) of same oviduct; **H**, **I**, head pigmentation; **J**, pelagic egg capsule (after Schmitt, 1979); **H**, **K–R**, penes. **A–F**, **I**, **K**, **N–R**, Shell Beach, near Pismo Beach, California; **G**, **H**, **M**, Palos Verdes Peninsula, California; **J**, Bodega Bay, California; **L**, Point Lobos, Monterey, California. Shell heights: **A–F**, 15.0 mm; **G**, 11.3 mm; **H**, 11.0 mm; **I**, 8.5 mm; **K**, 10.1 mm; **L**, 12.0 mm; **M**, 12.5 mm; **N**, 15.7 mm; **O**, 18.7 mm; **P**, 18.8 mm; **Q**, 9.1 mm; **R**, 18.7 mm. For shading conventions of pallial oviducts see p. 6. Abbreviation: b, copulatory bursa.

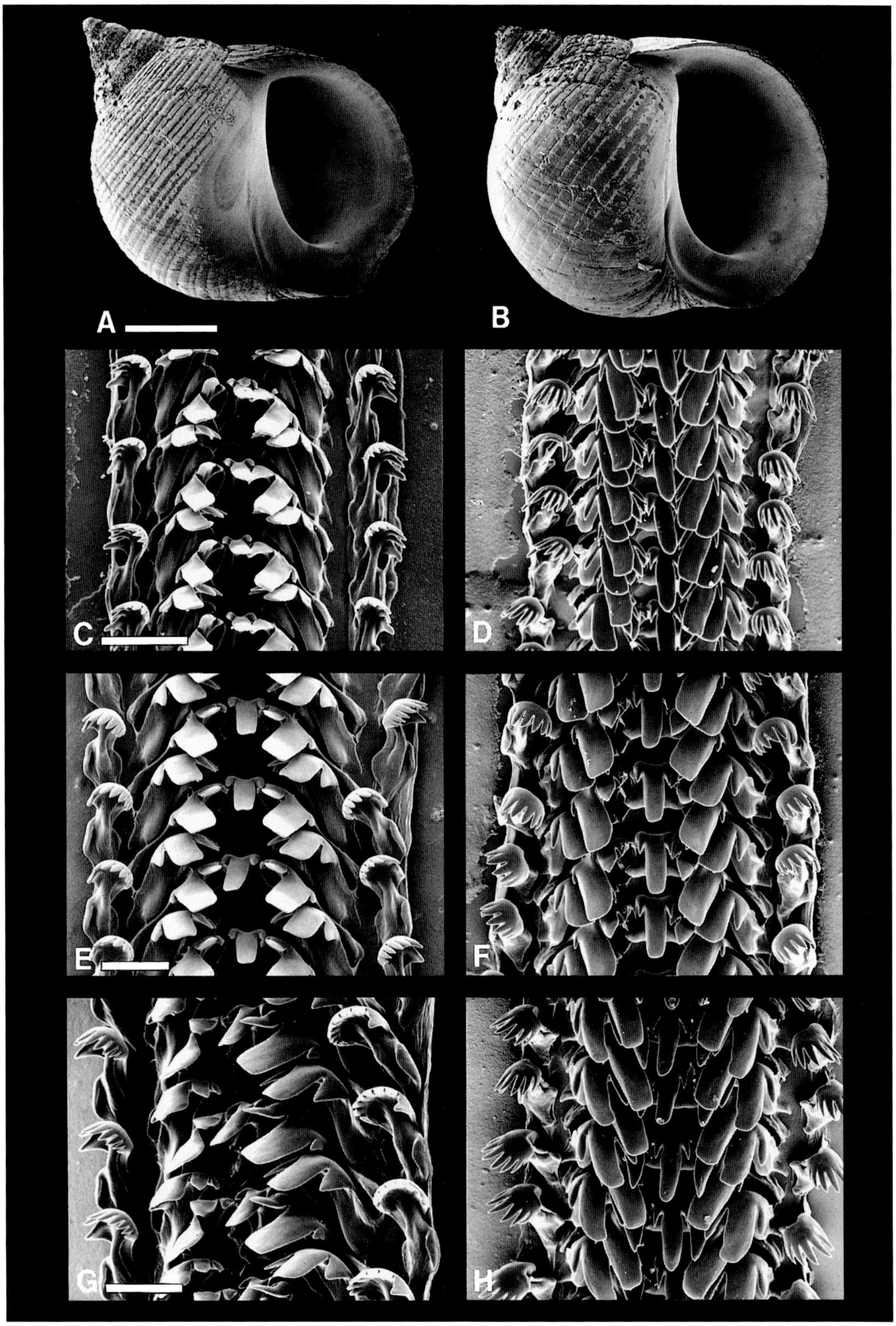

Figure 14. *Littorina (Planilittorina) keenae*: **A, B,** juvenile shells; **C, D,** two views of radula (shell H=8.0 mm), flat (**C**) and at 45° (**D**); **E–G,** three views of radula (shell H=15.0 mm), flat (**E**), at 45° (**F**) and side view (**G**); **H**, radula (shell H=11.3 mm), at 45°. **A, B, H,** Palos Verdes Peninsula, California; **C, D,** Ensenada, Mexico; **E–G,** Shell Beach, near Pismo Beach, California. Scale bars: **A, B**=1 mm; **C, D**=50 µm; **E, F**=50 µm; **G, H**=50 µm.

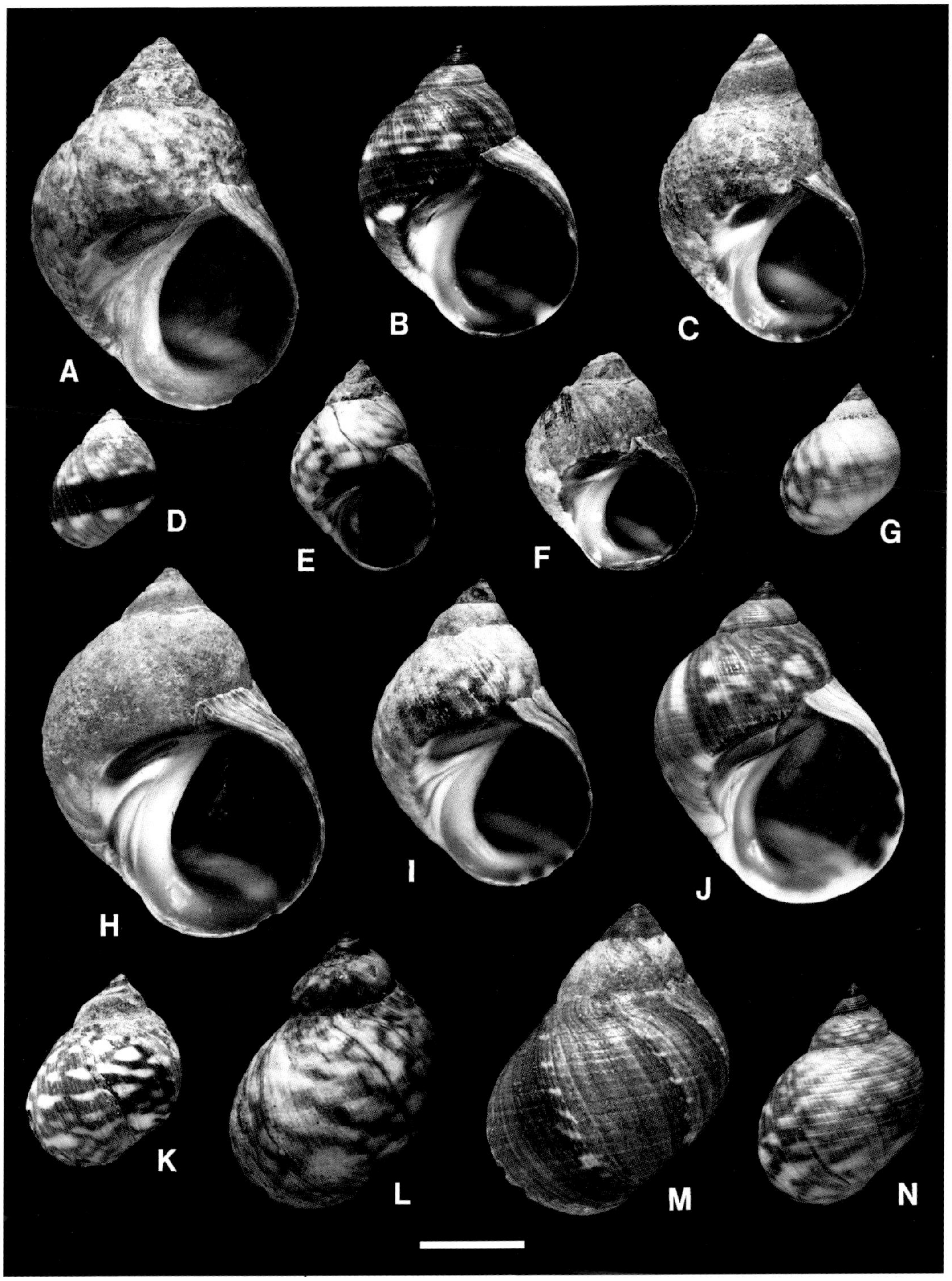

Figure 15. *Littorina (Planilittorina) keenae.* **A,** *Litorina planaxis* Philippi, 1847; lectotype, BMNH 1990004; California superior. **B, E, J, K, N,** Ensenada, Mexico. **C, D, F, G,** Palos Verdes Peninsula, California. **H,** Shell Beach, near Pismo Beach, California. **I,** Ocean Beach, San Diego, California. **L, M,** 6 miles S of Carmel, California. Scale bar=5 mm.

and *L. plena*, with which it is frequently found, but there is some overlap with these species; possible competition between them has been discussed by Behrens Yamada (1992).

The diet consists mainly of black lichens (Behrens Yamada, 1992) and also includes a range of epilithic and endolithic green, blue-green and red microalgae, scraped from the surface together with rock particles (Foster, 1964; Glynn, 1965; Behrens Yamada, 1992). Macroalgae are eaten only when fragmented or small enough to be consumed entire (Dahl, 1964).

Range (Fig. 16). The range over which *L. keenae* has previously been recorded is from Charleston, Oregon (Keen & Doty, 1942; Behrens Yamada, 1977a) south to 24.5°N on the Pacific coast of Baja California (Behrens Yamada, 1992). This range can now be extended southwards to Cabo San Lucas at the tip of Baja California on the basis of a single record (USNM 780668). However, it was not found on a recent visit to this locality (pers. obs., 1994), so occurrence there may be rare or sporadic. Further south is the record from Socorro Island given by Dall (1921); this was presumably based on the sole collection from this locality in the USNM (no. 207284), made before 1898 by C. B. Turrill. This record is considered unreliable, because samples of *L. plena* from the same locality by the same collector are of doubtful provenance (see remarks on *L. plena*, p. 79), and because collections from Socorro Island in LACM and CAS do not include *L. keenae*. The record from Cocos Island (Keen, 1971) seems unlikely and requires confirmation. At the northern extreme of the range there have been two records from the vicinity of Puget Sound, Washington. One was again due to Dall (1921), presumably based on lot USNM 3928 from Puget Sound. Behrens Yamada (1977a) stated that these shells were from San Francisco, but did not explain why. The second was by Oldroyd (1924), who reported a personal collection from Brown Island, San Juan County; the location of these specimens is unknown, but Behrens Yamada (1977a) noted that specimens under this name in the collection of the Friday Harbor Laboratories were worn *L. sitkana*. Other modern authors have agreed that *L. keenae* is not found in Washington (Urban, 1962; Abbott & Haderlie, 1980). To investigate the causes of range limitation in this species, Behrens Yamada (1977b) transplanted individuals to San Juan Island, Washington. Since they not only survived for five years, but also produced viable spawn, she concluded that either larval mortality or southerly currents prevent northward dispersion. Although found on continental shores, it prefers open coasts of oceanic aspect and occurs on the Channel Islands and Isla de Guadalupe; its distribution is therefore of the oceanic type (p. 369).

Fossil record. Pleistocene occurrences in California have been listed by Cooper (1888), Arnold (1903), Grant & Gale (1931) and Woodring, Bramlette & Kew (1946). There are also records of this species from the Upper Miocene San Pablo formation (Gabb, 1869; Arnold, 1906; Weaver, 1909), but these are probably misidentifications of *L. remondii* (p. 61). Woodring *et al.* (1946) noted that fossil shells from many localities in the Palos Verdes Hills, California, reached a maximum size of only 9.6 mm, while Recent shells from the same region were larger, up to 20.8 mm.

Discussion

Spawn type. The chief peculiarity of this species is its spawn type. No other littorinid is known to produce a free gelatinous egg mass from which pelagic capsules are released, although others with pelagic capsules may extrude them in ephemeral mucous strings (e.g. *Littoraria nebulosa* (Lamarck); Bandel, 1974). Schmitt (1979) speculated that this type of egg mass may protect the capsules from desiccation or abrasion when initially released on the shore, or may enhance the likelihood of eggs being swept into the sea by a strong wave. It may also be significant that the gelatinous mass is much larger than an individual capsule, thus reducing the risk of predation by filter feeders such as mussels and barnacles which are abundant on Californian shores (Reid, 1990b). An additional or alternative suggestion is that this system allows egg release to take place more rapidly than would be possible if capsules were to be spawned individually, which may be important for a species found at high tidal levels where contact with wave splash is brief. This unusual spawn is reflected in the structure of the pallial oviduct, which shows a unique combination of well-developed capsule gland (indicating pelagic capsules) and enlarged, septate jelly gland (indicating production of a gelatinous mass around the capsules).

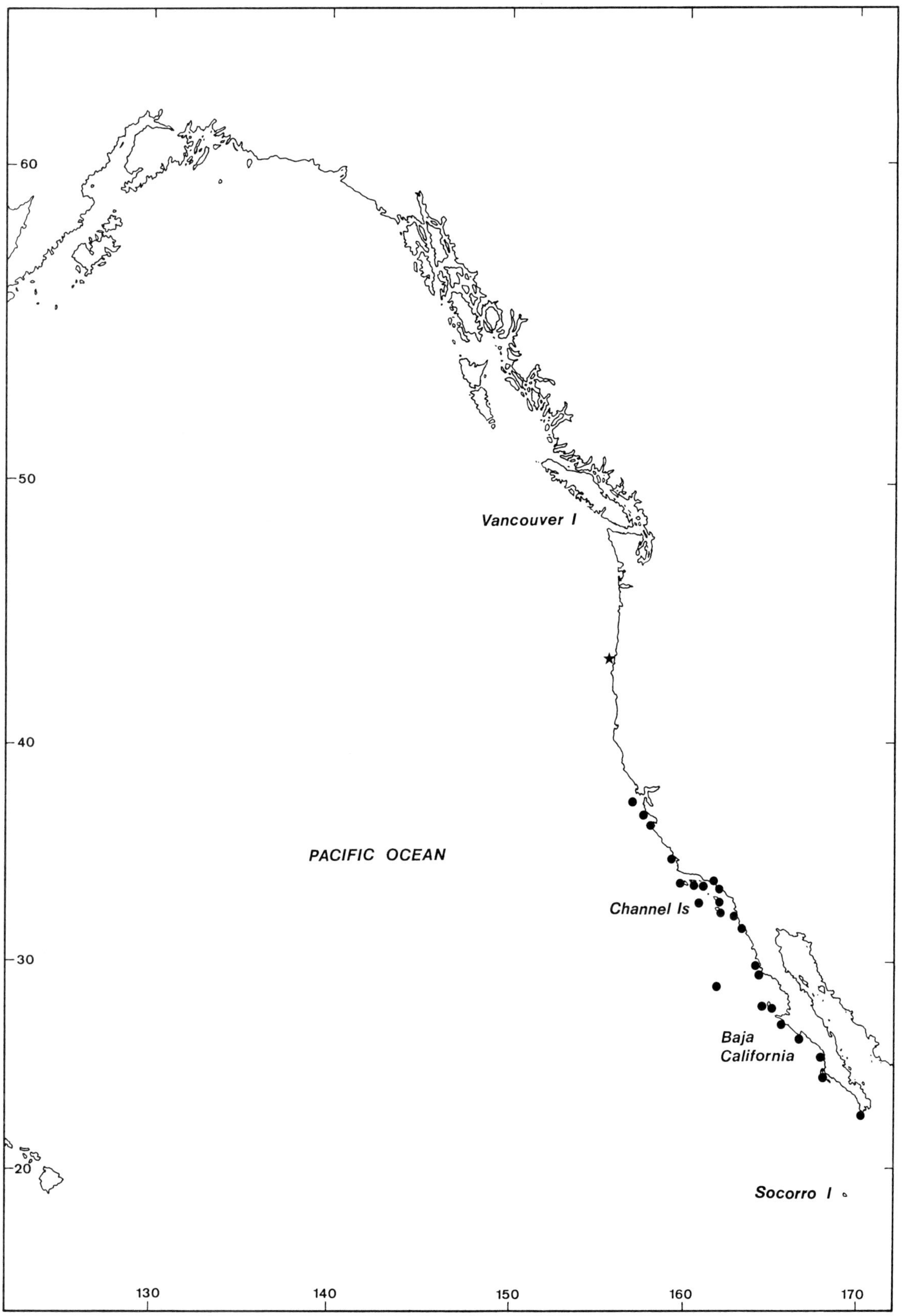

Figure 16. *Littorina (Planilittorina) keenae*: distribution map. Key: circles, specimens examined; asterisk, record from literature (Behrens Yamada, 1977a).

Intraspecific variation. This species is morphologically rather uniform throughout its range. This is consistent with a low degree of genetic differentiation between populations and probable high gene flow, as a consequence of its pelagic dispersal (Correa, 1994).

Phylogenetic relationships. In phylogenetic analyses of morphological characters *L. keenae* appears in a relatively basal position on the cladogram, between *L. striata* and all other *Littorina* species (Reid, 1989a, 1990b; p. 351). It shares the synapomorphy of biconvex egg capsules with all other species except *L. striata*, but lacks the synapomorphies of an outer calcite layer of the shell, and enlarged salivary glands. This position is strongly supported by analysis of mitochondrial DNA sequences (Rumbak, Reid & Thomas, 1994; Reid, Rumbak & Thomas, 1996; p. 358). Shell characters suggest that *L. sookensis* from the Upper Oligocene of Vancouver Island may be closely related to *L. keenae* (p. 52).

Similar species. Only *L. plena* and *L. scutulata* are sympatric with *L. keenae*, and both have a narrower and more elongate shell that is usually more conspicuously tessellated than in *L. keenae*. The eroded area of the parietal region adjacent to the columella, typical of *L. keenae*, is absent in the other two, and indeed is only shared with *L. striata* among other Recent members of the genus; this feature is, however, frequent in species of the sister-genus *Nodilittorina*.

Subgenus *Littorina* Férussac, 1822

Synonymy

See Synonymy of genus *Littorina* (p. 38).

Taxonomic history

In the past some authors have accepted *Turbo obtusatus* as the type species of *Littorina* (p. 39). Those that regarded the closely-related pair *L. littorea* and *L. squalida* as subgenerically distinct had then to follow Dall (1918, 1921) in adopting his sectional name *Algaroda*, with *L. littorea* as its type (Thiele, 1929; Wenz, 1938; Nordsieck, 1968, 1982; Golikov & Kusakin, 1978; Palmer, 1989, at generic rank). Dall himself subsequently changed his mind and used the subgenus *Littorina* for this group (Dall, 1923). Some have argued for *Turbo neritoides* (=*Melarhaphe neritoides*) as the type of *Littorina* and have likewise then used *Algaroda* as a subgenus for *L. littorea* (Kennard, 1942; Winckworth, 1951). However, most authors have accepted *Turbo littoreus* as the type species of *Littorina*, and this was confirmed by Opinion 1159 of the ICZN (Melville, 1980); *Algaroda* thus becomes a junior objective synonym of *Littorina*.

While a broad concept of the genus was current, those authors employing subgeneric categories usually included in the nominal subgenus a variety of species (now assigned to *Littorina sensu stricto*, *Nodilittorina* and *Littoraria*) that remained after separation of more well-defined subgenera such as *Neritoides* (=*Neritrema*) and *Melarhaphe* (e.g. H. & A. Adams, 1854; Tryon, 1883, 1887). Some employed newly available information on spawn characteristics to imply that the nominal subgenus should contain only species with pelagic eggs and larvae (Winckworth, 1922, 1932; Bequaert, 1943; also probably intended by Dall, 1909, 1918, for *Algaroda*). Winckworth (1922) even suggested that spawn characteristics were of sufficient importance to distinguish *Littorina*, *Neritoides* and *Littorivaga* as genera. Habe (1958a) also emphasized spawn characteristics, using the generic name *Littorina* only for species with small, asymmetrically biconvex egg capsules (*L. littorea*, *L. brevicula*) and creating a new genus, *Ezolittorina*, for *L. squalida*, because of its larger, biconvex capsules containing many eggs. The subgenus *Littorina* as used by Rosewater (1970, 1981) was almost equivalent to the genus *Littorina* used here. The new generic name *Brevilittorina* (*nomen nudum*) was used by Golikov, Dolgolenko, Maximovich & Scarlato (1990), without description, but presumably to recognize the close relationship between the included species, *L. brevicula* and *L. mandshurica*. After a cladistic analysis of a range of morphological characters, Reid (1990b) used the subgenus *Littorina* for all the planktotrophic species except *L. striata* and *L. keenae*, but noted that this was a paraphyletic group. Subsequent phylogenetic analyses of mitochondrial DNA sequences have shown that the constituent species are quite closely related, but have not clearly resolved their relationships (Rumbak, Reid & Thomas, 1994; Reid, Rumbak & Thomas, 1996; p. 358).

Diagnosis

Shell: turbinate or conical; usually with spiral grooves, rarely smooth; protoconch of planktotrophic type; aragonitic, with thick outer layer of irregular-prismatic calcite. Male: 0–100 mamilliform glands of equal size. Female: jelly gland not enlarged; egg groove coiled in three successive spiral loops. Spawn: pelagic, biconvex capsules, commonly containing more than 1 ovum. Radula: outer marginal teeth with or without outer basal projection, cusps long or short; cusps of five central teeth long and rectangular or short and pointed. Alimentary system: salivary glands enlarged.

Littorina (Littorina) remondii Gabb, 1866

(Fig. 17)

Synonymy

Littorina remondii Gabb, 1866: 14; pl. 2, figs 23, 23a (Pliocene of Kirker's Pass, California [Contra Costa Co.; San Pablo formation, Upper Miocene, Clark, 1915]; lectotype (Keen & Bentson, 1944) ANSP 4334, seen, Fig. 17A; 3 paralectotypes ANSP 77549, seen, Fig. 17B, C). Gabb, 1869: 80. Clark, 1915: 403, 405, 422, 425, 484; pl. 65, fig. 13. Stewart, 1927: 346; pl. 32, fig. 5. Grant & Gale, 1931: 781 (as *remondi*). Woodring, Stewart & Richards, 1940: 84–85. Keen & Bentson, 1944: 167. Adegoke, 1969: 160.

Litorina remondii—Cooper, 1888: 246. Arnold, 1906: 24 (as *remondi*). Weaver, 1909: 264 (as *remondi*).

Littorinopsis (Touzinia) remondii—Cossmann, 1915: 63 (as *remondi*).

Littorina planaxis—Gabb, 1869: 80 (probably applies to *L. remondii*; in part, includes *Litorina planaxis* Philippi, 1847=*Littorina keenae*).

Litorina planaxis—Arnold, 1906: 24 (probably applies to *L. remondii*; not Philippi, 1847). Weaver, 1909: 264 (probably applies to *L. remondii*; not Philippi, 1847).

Littorina pittsburgensis Clark, 1915: 405, 422, 484–485; pl. 65, fig. 22 (upper beds of San Pablo group in the region of Kirker Pass, University of California locality 1482, California [Contra Costa Co.], Miocene; lectotype, here designated, UCMP 11649, seen, Fig. 17H). Woodring *et al.*, 1940: 84–85.

Nematurella pittsburgensis—Keen & Bentson, 1944: 167 (see Hanna, 1923).

Taxonomic history

The original author (Gabb, 1869) listed both his new species and *L. planaxis* (=*L. keenae*) from the type locality, and both Arnold (1906) and Weaver (1909) also used the name *L. planaxis* for shells from the same locality; although these specimens have not been seen, it is considered likely that they were lower-spired examples of *L. remondii*, which appears to be a variable species. Clark (1915) described *L. pittsburgensis*, from the upper San Pablo beds, distinguishing it from *L. remondii* in the lower beds in the same group, citing the more rounded whorls and lack of sculpture in the former, but the lectotype is worn, and development of the peripheral keel is variable in *L. remondii*. Hanna (1923) made the ill-informed suggestion that *L. pittsburgensis* should be placed in *Nematurella* (Hydrobiidae), because of the apparent posterior sulcus of the aperture.

Diagnosis

Shell height up to 14 mm (estimated). Shell high-turbinate to conical; sutures slightly channelled or indistinct; spire whorls flat-sided or slightly rounded; last whorl usually sharply angled at periphery. Mineralogy of thick external calcite and thin internal aragonite layers. Sculpture of up to 17 fine and irregular incised spiral lines on last whorl; peripheral rib and adjacent basal rib may develop as rounded carinae; sometimes up to 4 less distinct carinae above periphery. Columella moderately narrow, anterior part of pillar angled away from axis of coiling. Distribution: Upper Miocene of San Pablo formation, central California.

Shell variation (Fig. 17)

Only eleven specimens have been seen of this species, and further material is required to assess its variability. From the available material, shell shape is moderately variable (H/B=1.21–1.51). Most specimens show traces of fine, spiral ribs and grooves, but these are usually somewhat irregularly

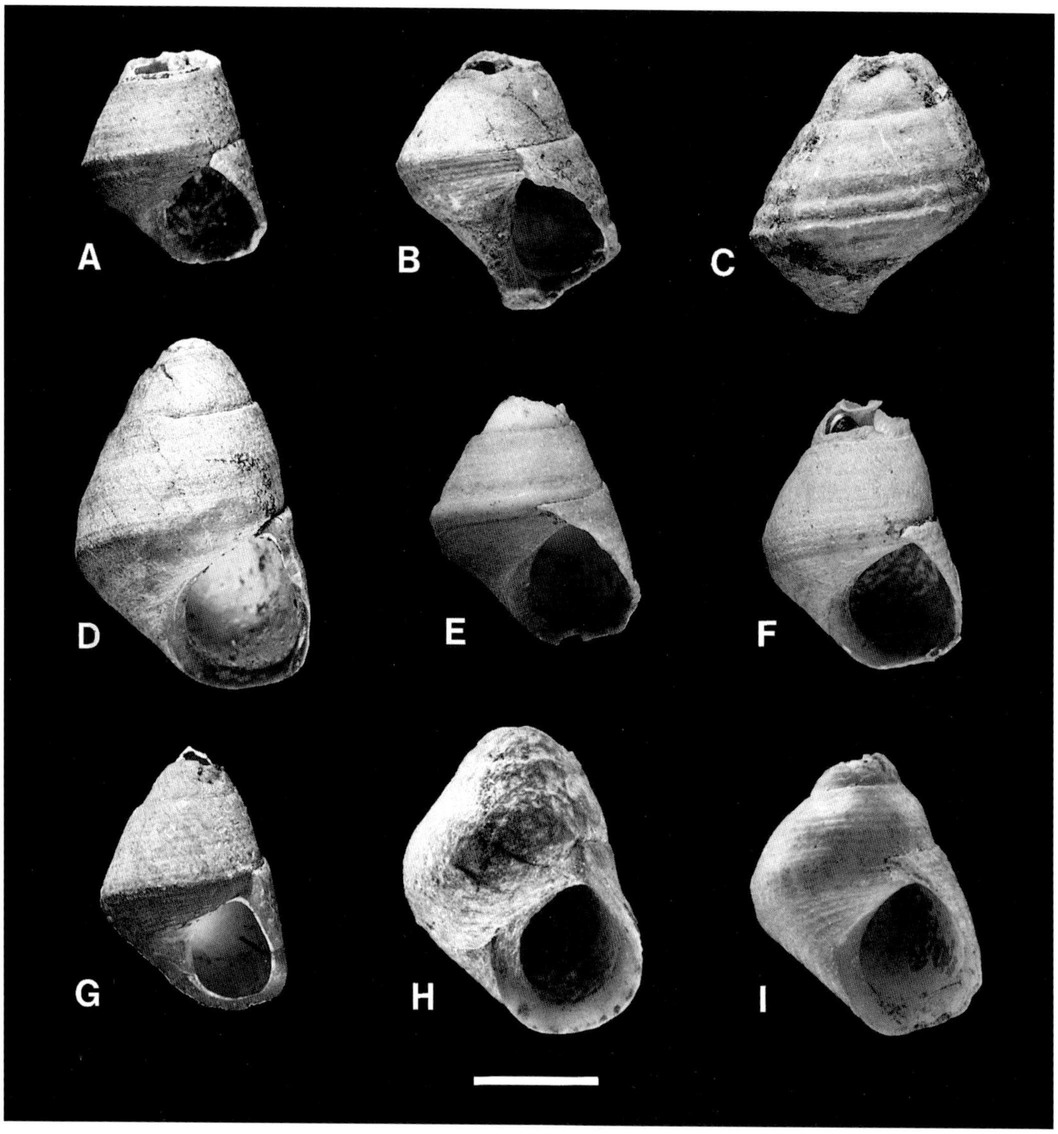

Figure 17. *Littorina (Littorina) remondii.* **A–C,** *Littorina remondii* Gabb, 1866; lectotype (**A**), ANSP 4334; paralectotypes (**B, C**), ANSP 77549; Kirker's Pass, Contra Costa Co., California; San Pablo formation, Upper Miocene. **D–G,** Kirker's Pass, Contra Costa Co., California; San Pablo formation, Upper Miocene; CAS 61191.01. **H,** *Littorina pittsburgensis* Clark, 1915; lectotype, UCMP 11649; Kirker's Pass, Contra Costa Co., California; San Pablo formation, Upper Miocene. **I,** bank of Kirker's Creek, Kirker Pass–Mount Diablo, California; San Pablo formation, Upper Miocene; UCMP 11610. Scale bar=5 mm.

spaced and often indistinct. There is a variable tendency towards carination at the periphery. In some examples the periphery is only indistinctly angled (Fig. 17H), whereas in others there may be a strong peripheral keel, marked by an enlarged peripheral rib (Fig. 17E). In several specimens a second large, but less prominent, rib is present on the base, adjacent to the periphery (Fig. 17A). The base has a rather flattened profile, often with a slightly concave middle part. In one of the paralectotypes of *L. remondii* (Fig. 17C) there are four rounded ribs on the body whorl above the peripheral rib, becoming less distinct towards the suture. Where the peripheral carina is well developed, the attaching suture appears slightly channelled. A constant feature is the relatively narrow and characteristically angled columella. In the holotype of *L. pittsburgensis* (Fig. 17H) the outer lip of the aperture is relatively thick, and meets the body whorl at an acute angle, giving the impression of a 'posterior sinus'; it was for this reason that Hanna (1923) advocated transfer of the species from *Littorina* to *Nematurella*. In fact, this development of the aperture is not uncommon in other *Littorina* species, e.g. *L. squalida* (Fig. 28I) and *L. mandshurica* (Fig. 51K). The two layers of the shell, corresponding to the thin inner aragonite and outer calcite layers, can be seen clearly at the broken edge of the aperture of the shell in Figure 17G.

Variable carination as seen in *L. remondii* is unusual in the subgenus *Littorina*, although somewhat similar variation is present in *L. squalida* and *L. mandshurica*. In *L. littorea* carination is known only in the monstrosities from the Pliocene Norwich Crag. The possibility has therefore been considered that *L. remondii* might be an abnormal form, perhaps of *L. petricola*, a Pliocene fossil from California. However, the carinations are not as extreme, nor as variable in position, as in the abnormal examples of *L. littorea*. Furthermore, the differences in whorl profile, spiral sculpture and columellar shape appear to distinguish *L. remondii* consistently from *L. petricola* (see below), whereas these characters are not consistently different in the monstrous and typical forms of *L. littorea*.

Distribution

Fossil record. This species has been reliably recorded only from the upper and lower beds of the Upper Miocene San Pablo formation at the type locality, Kirker's Pass, Contra Costa County, California. Adegoke (1969) identified some small (3–4 mm) shells from the Temblor formation (Middle to Late Miocene) of Kings County, California, as *L. remondii*, but some of these have been examined (UCMP 36806, 36807) and found to be too worn for determination.

Discussion

Phylogenetic relationships. This is an enigmatic species, known from only one locality, and is without obvious close relatives among living species. The combination of an outer calcite layer and conical profile indicate placement in the subgenus *Littorina*. However, the distinctive carination of *L. remondii* is not exactly reproduced in any Recent *Littorina* species. The closest appears to be *L. plena*, in which columellar form can be similar (Fig. 24J), tall and straight-sided forms can occur (Fig. 24F) and in which one or two peripheral ribs can be slightly more prominent than the rest (Fig. 24F, I). However, the incised spiral sculpture of *L. plena* is usually more regular than that seen in *L. remondii*, and the peripheral ribs are never developed into carinae. *Littorina remondii* is tentatively suggested to be related to the pair of Recent sister-species, *L. plena* and *L. scutula*. Analyses of mitochondrial DNA have estimated the age of divergence between *L. scutulata* and *L. plena* as about 12 Ma (p. 365) which, pre-dating *L. remondii*, implies that *L. remondii* cannot have been directly ancestral to either. Some of the specimens of *L. remondii* with more rounded whorls and stronger incised sculpture (Fig. 17F, I) do show weak resemblance to *L. squalida* and *L. petricola*, and a connection with that lineage is another possibility.

The rather tall, straight-sided spire and keeled periphery are characters associated with some Recent species of *Nodilittorina*, a genus typical of tropical to warm-temperate climates, that at present extends as far north as the Gulf of California and the Pacific coast of southern Baja California. During the Upper Miocene the climate of central California was warmer than today, and Panamic molluscs extended further north than at present (Addicott, 1970). However, Recent species of *Nodilittorina* generally show sculpture of regularly spaced incised spiral grooves, a long and straight columellar pillar, and often an eroded parietal area adjacent to the aperture, and these characters are not shared with *L. remondii*. Furthermore, the outer calcite layer of the shell excludes *L. remondii* from *Nodilittorina*.

Similar species. There has been some confusion in the literature with *L. petricola*, also from California, but recorded only from Pliocene deposits (Arnold & Hannibal, 1913; Woodring, Stewart & Richards, 1940; p. 92). However, the two appear to be distinct. The whorls of *L. remondii* are flatter and usually show at least one carina at the periphery. The whorls of *L. petricola* are more rounded and, even in the tall-spired forms from King's County (Fig. 31B), the periphery is not sharply angled or carinate. More significantly, the columella of *L. remondii* is distinctive, being narrower and angled to the right of the axis of coiling in apertural view; in comparison, that of *L. petricola* is wider and the axis of coiling passes through the pillar.

***Littorina (Littorina) scutulata* Gould, 1849**

(Figs 5, 18–21)

Synonymy

Littorina scutulata Gould, 1849: 83 (Puget Sound [Washington State]; lectotype (Johnson, 1964) USNM 5640, seen, Fig. 20A, this is probably specimen figured by Gould (1861); paralectotype USNM 612308, seen). Gould,

1852: 200; atlas (1861) pl. 14, figs 241, 241a. Cooper, 1860: 374 (*s.l.*). Keep, 1888: 68; fig. 52 (*s.l.*). Keep, 1904: 216; fig. 232 (*s.l.*). Abbott, 1954a: 134; pl. 20, fig. c (*s.l.*). Harger, 1972: fig. 6 (spawn). Buckland-Nicks, Chia & Behrens, 1973: 359–365; figs 3, 18–23 (spawn, embryo, larva). Abbott, 1974: 67–68; text-fig. 554 (in part, includes reference to *L. plena* Gould, includes *Littoraria pintado pullata* (Carpenter)). Murray, 1979: 469–474; fig. 1B (spawn); fig. 2A (penis). Murray, 1982: 233–238. Mastro, Chow & Hedgecock, 1982: 239–246; fig. 3 (penis, spawn); fig. 4 (radula). Buckland-Nicks & Chia, 1984: 311–313; figs 2, 3 (oviduct). Ricketts, Calvin, Hedgpeth & Phillips, 1985: 21, 32–34; fig. 11 (*s.l.*). Chow, 1987a: 359–366. Sacchi & Voltolina, 1987: 209–234; fig. 2d; fig. 3b (oviduct); fig. 3c (penis); fig. 5 (radula). Strathmann, 1987: 238; fig. 11.6 (spawn).

Litorina scutulata—Carpenter, 1864: 656 (in part, includes reference to *L. plena*). Weinkauff, 1882: 85–86; pl. 11, figs 13–16 (in part, includes reference to *L. plena*). Weinkauff, 1883: 222 (in part, includes reference to *L. plena*).

Littorina (Melarhaphe) scutulata—Tryon, 1887: 250; pl. 45, figs 98–100, 2, 3 (in part, another fig. is *L. plena*, includes reference to *Littoraria pintado pullata*; as *Melaraphe*). Dall, 1921: 153 (*s.l.*). Oldroyd, 1927: 62 (*s.l.*). Keep & Baily, 1935: 199; fig. 183 (*s.l.*) (as *Melaraphe*). Burch, 1945: 12, 37 (*s.l.*).

Littorina (Littorina) scutulata—Rosewater, 1970: 423 (*s.l.*). Reid, 1989a: 101; fig. 11n (spawn). Reid, 1990b: 1–19; fig. 1c; fig. 2d (oviduct); fig. 3g (spawn); fig. 4b (radula). Taylor & Reid, 1990: 199–215 (mineralogy).

Littorina lepida Gould, 1849: 83 (Puget Sound; lectotype (Mastro *et al.*, 1982) USNM 5637, seen, Gould, 1861: pl. 14, figs 238, 238a, Fig. 20B herein; paralectotype USNM 677095, seen; 4 paralectotypes MCZ 169222, not seen). Gould, 1852: 199; atlas (1861) pl. 14, figs 238, 238a–c.

Littorina (Melarhaphe) pullata—Burch, 1945: 12 (in part, includes *Littoraria pintado pullata*; not *Litorina pullata* Carpenter, 1864).

Note: In the synonymy '*s.l.*' (*sensu lato*) following a reference indicates that the author's concept of *L. scutulata* probably included *L. plena*, but without specific reference to, or a figure of, the latter. In the text, '*L. scutulata s.l.*' indicates that both species may have been included, in contrast to '*L. scutulata s.s.*' (*sensu stricto*) where identification is not in doubt.

Taxonomic history

Gould (1849) described three closely allied species of *Littorina*, *L. scutulata*, *L. lepida* and *L. plena*. *Littorina lepida* was said to differ from *L. scutulata* by its more prominent spotted pattern, its reddish aperture with anterior white band, and stronger spiral sculpture, but both were collected from Puget Sound. *Littorina plena*, from San Francisco, was described as having a globose last whorl with more ovate shape and with smaller aperture than *L. scutulata*, and again an anterior white band in the aperture was mentioned.

Carpenter (1864) synonymized *L. lepida* and *L. plena* under *L. scutulata*, and subsequent authors also recognized only a single species. However, in 1979 Murray found dimorphic egg capsules and penes within *L. scutulata* and concluded that this taxon was a complex of two species of similar shell morphology. He identified his '*L. scutulata* type II' as *L. scutulata s.s.*, since only penes and spawn of this type were found at the type locality, Puget Sound. Later, he separated the two species using discriminant analysis of shell measurements (with 4% probability of misclassification) and also noted that *L. scutulata* less frequently shows a pale basal band (Murray, 1982). Use of the discriminant functions to classify type specimens confirmed the association of *L. scutulata s.s.* with the penial and spawn characters of 'type II', while 'type I' was identified as *L. plena* (with *L. lepida* as a synonym). Allozyme electrophoresis was used by Mastro, Chow & Hedgecock (1982) to confirm the separation of *L. plena* from *L. scutulata*; these authors also examined the resoaked animals of the two paralectotypes of *L. plena* and found their penial morphology to correspond with Murray's 'type I', although they could not reliably assign *L. lepida* to either species.

In the present study additional characters have been found for discrimination of *L. scutulata* and *L. plena* (shell colour pattern, pigmentation of head-foot, oviduct structure, radula; p. 71, Table 2). On the basis of its coarsely tessellated colour pattern, *L. lepida* is here synonymized with *L. scutulata*, contradicting the result of the discriminant analysis by Murray (1982).

Several authors have either included *Litorina pullata* Carpenter, 1864, as a synonym of *Littorina scutulata* (Tryon, 1887; Abbott, 1974), or have used the name for a lined colour form of *L. scutulata* (Burch, 1945). However, *pullata* is now recognized as an eastern Pacific subspecies of *Littoraria pintado* (Wood, 1828) (see Reid, 1986a: 64).

Diagnosis

Mature shell height 4.7–18.0 mm. Shell conical. Mineralogy of external irregular-prismatic calcite and internal crossed-lamellar aragonite. Sculpture smooth or with fine spiral lines. Colour brown, olive or black, coarsely tessellated, spotted or banded with white to grey, most strongly patterned at suture and periphery; pale basal band usually absent. Head black, tentacles with transverse black bands. Penis with tapering filament, small bifurcation at tip; mamilliform glands absent, sperm groove ends subterminally. Pallial oviduct of two consecutive loops of albumen gland followed by large spiral loop of capsule gland; straight section relatively short, jelly gland not enlarged or septate; bursa large, extending from halfway along straight section back to seminal receptacle. Spawn of large biconvex pelagic capsules sculptured by two flanged rims of different diameters, containing 3–10 eggs; development planktotrophic. Distribution: Baja California to southeastern Alaska.

Anatomical description

Head-foot (Fig. 18A–E). Head black; tentacles with an unpigmented stripe across eye, usually extending across tentacle base, predominant pattern of transverse black bands and flecks, only rarely a pale or interrupted longitudinal streak or almost all black; sides of foot black or mottled dark grey. Opercular ratio 0.45–0.60.

Penis (Fig. 18N-R). Base wrinkled, slightly pigmented; mamilliform penial glands absent. Filament tapering with terminal bifurcation, not clearly separated from base but approximately 40–50% total length. Sperm groove open, ending on smaller of the two projections at tip, therefore subterminally. Subepithelial glandular tissue (reddish in living specimens) around sperm groove in filament and extending into larger terminal projection. Sacchi & Voltolina (1987) reported a reduction in size of the penis during winter in some males.

Sperm. Euspermatozoa 38–44 µm. Paraspermatozoa 12.5–16 µm diameter, spherical to slightly ovoid; vesicles small, indistinct or absent; no nucleus visible.

Oviduct (Fig. 18F–K). Two consecutive loops of albumen gland followed by large spiral loop of capsule gland with distal reddish translucent portion; straight section relatively short, jelly gland not enlarged or clearly septate (although indistinct internal wrinkles sometimes visible externally, Fig. 18H, K). Bursa large, originating halfway along straight section, extending back to level of seminal receptacle. Seminal receptacle simple.

Spawn and development (Fig. 18L, M). Pelagic, biconvex capsules sculptured by two flanged rims of unequal diameters; maximum diameter 700–1000 µm; containing 1–14 (usually 3–10) eggs 105 µm diameter (Buckland-Nicks, Chia & Behrens, 1973; Murray, 1979; Mastro *et al.*, 1982). Behrens Yamada (1989) recorded 1–10 eggs per capsule at San Juan Island, Washington, and up to 36 eggs per capsule in Oregon, but it is not clear that capsules of *L. plena* were excluded. Both Abbott & Haderlie (1980) and Buckland-Nicks *et al.* (1973) quoted a student report by Imai (1964) which claimed that a spawn mass is produced containing singly encapsulated eggs, but this was presumably based on a misidentification of *L. keenae*. Development planktotrophic. Size of hatching veligers given as 160 µm shell length (Buckland-Nicks *et al.*, 1973) and 155 µm shell width (Murray, 1979). Protoconch about 240 µm diameter in poorly preserved available specimens; Buckland-Nicks *et al.* (1973) recorded larval shells 300–360 µm in length just before metamorphosis. Spawning season May to September in Puget Sound (Buckland-Nicks *et al.*, 1973); March to October around southern Vancouver Island (Behrens Yamada, 1989); sporadically from March to September in Bodega Bay, California (Chow, 1987b). In contrast, breeding was reported 'at all seasons except summer in California' by Abbott & Haderlie (1980), but this was perhaps quoted from Glynn (1965), whose attached, transparent capsules with shelled embryos were clearly not of this species.

Radula and foregut (Fig. 19B–H). Relative radular length 1.22–2.31. Rachidian: length/width 1.00–1.33; central cusp varies from bluntly rounded to long; one pointed cusp and one outer denticle on each side. Lateral: four cusps, third from inside largest and blunt, others pointed. Inner marginal:

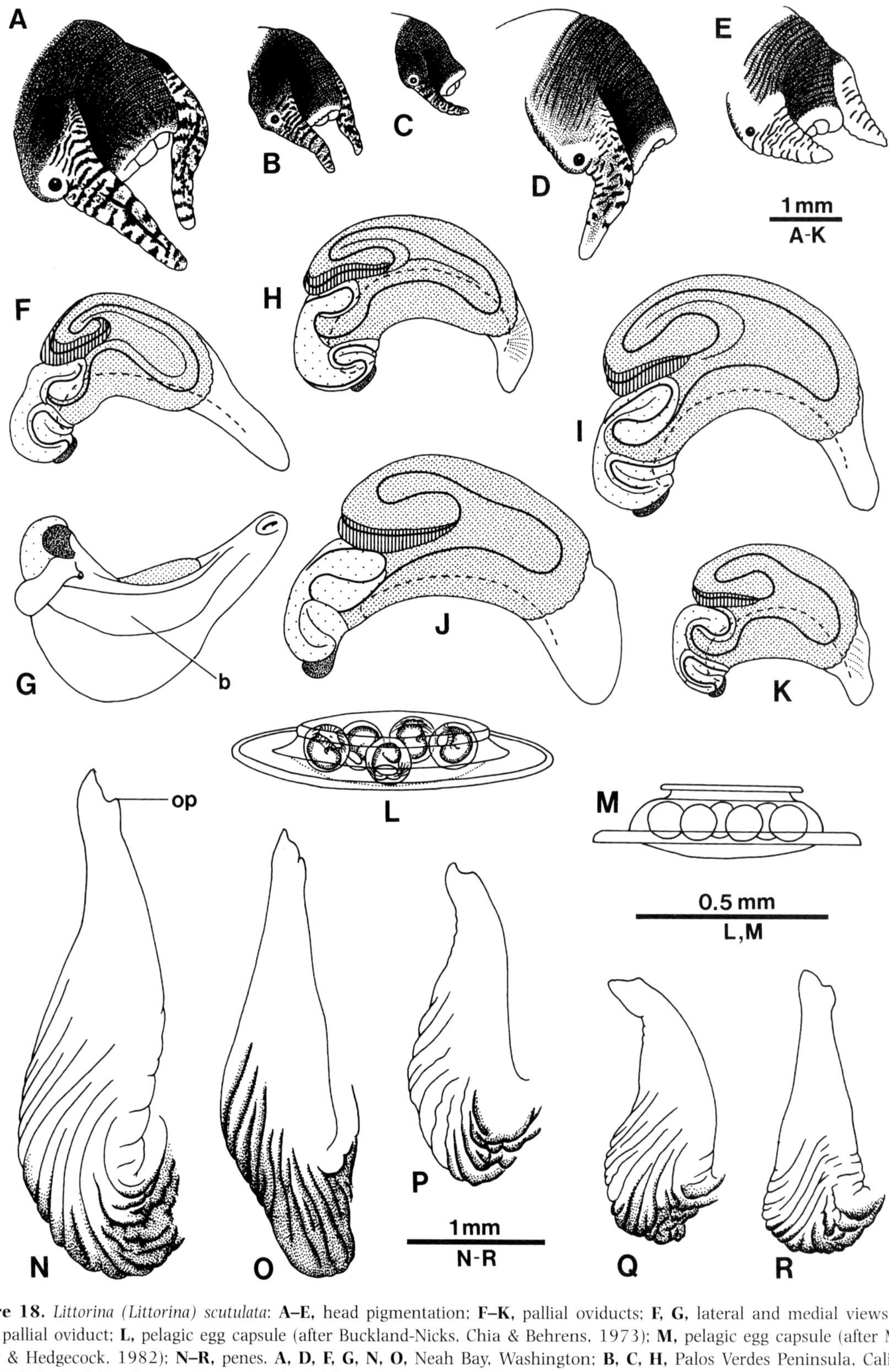

Figure 18. *Littorina (Littorina) scutulata*: **A–E,** head pigmentation; **F–K,** pallial oviducts; **F, G,** lateral and medial views of the same pallial oviduct; **L,** pelagic egg capsule (after Buckland-Nicks, Chia & Behrens, 1973); **M,** pelagic egg capsule (after Mastro, Chow & Hedgecock, 1982); **N–R,** penes. **A, D, F, G, N, O,** Neah Bay, Washington; **B, C, H,** Palos Verdes Peninsula, California; **E, I, J, P, Q,** Shell Beach, near Pismo Beach, California; **K, R,** Ensenada, Mexico; **L,** San Juan Island, Washington. Shell heights: **A,** 11.5 mm; **B,** 6.2 mm; **C,** 5.0 mm; **D,** 12.2 mm; **E,** 10.0 mm; **F, G,** 11.1 mm; **H,** 9.7 mm; **I,** 12.0 mm; **J,** 14.4 mm; **K,** 9.5 mm; **N,** 12.0 mm; **O,** 12.2 mm; **P,** 9.4 mm; **Q,** 9.4 mm; **R,** 8.3 mm. For shading conventions of pallial oviducts see p. 6. Abbreviations: b, copulatory bursa; op, opening of penial sperm groove.

four cusps, third from inside largest and blunt, others pointed. Outer marginal: 7–9 moderate or long pointed cusps; narrow neck, projecting flange either side of base. Salivary glands enlarged.

Shell variation (Figs 5, 19A, 20)

Shell shape is rather uniform in this species, which does not show conspicuous ecotypic (Sacchi & Voltolina, 1987) or geographical variation. Chow (1987b) reported that females attained significantly larger size than males at Bodega Bay, as a result of faster growth. However, in southern Vancouver Island Sacchi & Voltolina (1987) found that females were only slightly (and not usually significantly) larger than males. Animals from pools have been said to be larger than those from exposed rocks (North, 1954), and larger size has been correlated with physico-chemical conditions in individual pools by Humphrey & Macy (1931). Figures 5 and 20 show the range of colour variation, from entirely black, through the more common tessellated forms, to spotted and occasional banded individuals. These colour forms are mixed in many populations, but since the variation appears to be continuous, it is not described as a polymorphism. Chow (1987a) reported stronger tessellation in a sheltered harbour than on a nearby exposed coast at Bodega Bay.

Distribution

Habitat. The vertical range of this species extends from the middle eulittoral into the bare areas of the littoral fringe, on both wave-exposed and protected rocky shores (Chow, 1987b; Sacchi & Voltolina, 1987; Voltolina & Sacchi, 1990). It is most abundant in the high eulittoral, frequently on macroscopic red, brown and green algae, in pools and in barnacle beds (Hewatt, 1937; T. A. & A. Stephenson, 1961a, b; Dahl, 1964; Bock & Johnson, 1967; Ricketts, Calvin, Hedgpeth & Phillips, 1985; Foster, de Vogelaere, Oliver, Pearse & Harrold, 1991; Behrens Yamada, 1992). It appears to be infrequent or absent in the most sheltered habitats, such as salt marshes. Juveniles are frequent in barnacle tests, crevices and algal holdfasts (Dahl, 1964; Glynn, 1965; Bock & Johnson, 1967; Jensen, 1981). The distribution of juveniles has been reported as above that of adults by Bock & Johnson (1967), but Chow (1975) found them both above and below adults, and recorded an increase in mean size in an upshore direction. Adults make vertical movements with the rise and fall of the tide (Glynn, 1965). In the southern part of its range (south of Charleston, Oregon) *L. scutulata* is often sympatric with *L. keenae* and, although their distributions partly overlap, *L. scutulata* is more common at lower levels and in less exposed situations (Dahl, 1964). Competition with *L. keenae* for scarce algal food may limit the upshore distribution of *L. scutulata*, while predation may impose the low-shore limit (Behrens Yamada, 1992). To the north it is sympatric with *L. sitkana*, but extends higher into the littoral fringe, is more tolerant of exposure to waves, and is less dependent on sheltered and humid microclimates (Sacchi & Voltolina, 1987; Behrens Yamada, 1992).

Littorina scutulata is an opportunistic omnivore, grazing on epilithic and endolithic microalgae and lichens on rocks in the littoral fringe and, at lower levels on the shore, on diatom films, on a wide range of red, brown and green macroalgae and on plant litter (Castenholz, 1961; Dahl, 1964; Foster, 1964; Nicotri, 1977; Jensen, 1981; D'Antonio, 1985; Sacchi & Voltolina, 1987; Voltolina & Sacchi, 1990). Microalgae and diatoms appear to be the major components of the diet.

No detailed comparisons with the habitat and diet of the broadly sympatric and morphologically similar species *L. plena* have yet been made. Many of the references to the ecology of *L. scutulata* before discrimination of the two species by Murray (1979, 1982) must have included both. However, where both species have been explicitly recognized, no differences in habits or zonation have previously been recorded (Chow, 1987b, 1989; Sacchi & Voltolina, 1987; Behrens Yamada, 1992). In this study, personal observation and data with museum collections suggest that, while the two species are frequently found together, *L. scutulata* tends to predominate in the more exposed localities, whereas in very sheltered situations and in salt marshes *L. plena* may occur alone.

Range (Fig. 21). Following Dall (1921), the range of *L. scutulata* has generally been quoted as from Kodiak Island in Alaska to Bahía de Tortuga in Baja California, and Socorro Island. However, in the light of the recognition of *L. plena* as a separate species (Murray, 1979, 1982), all previous records must be reexamined. So far, only Murray (1979) and Mastro *et al.* (1982) have commented on the distribution of *L. scutulata s.s.*, recording it from San Diego to Vancouver Island. In the present study

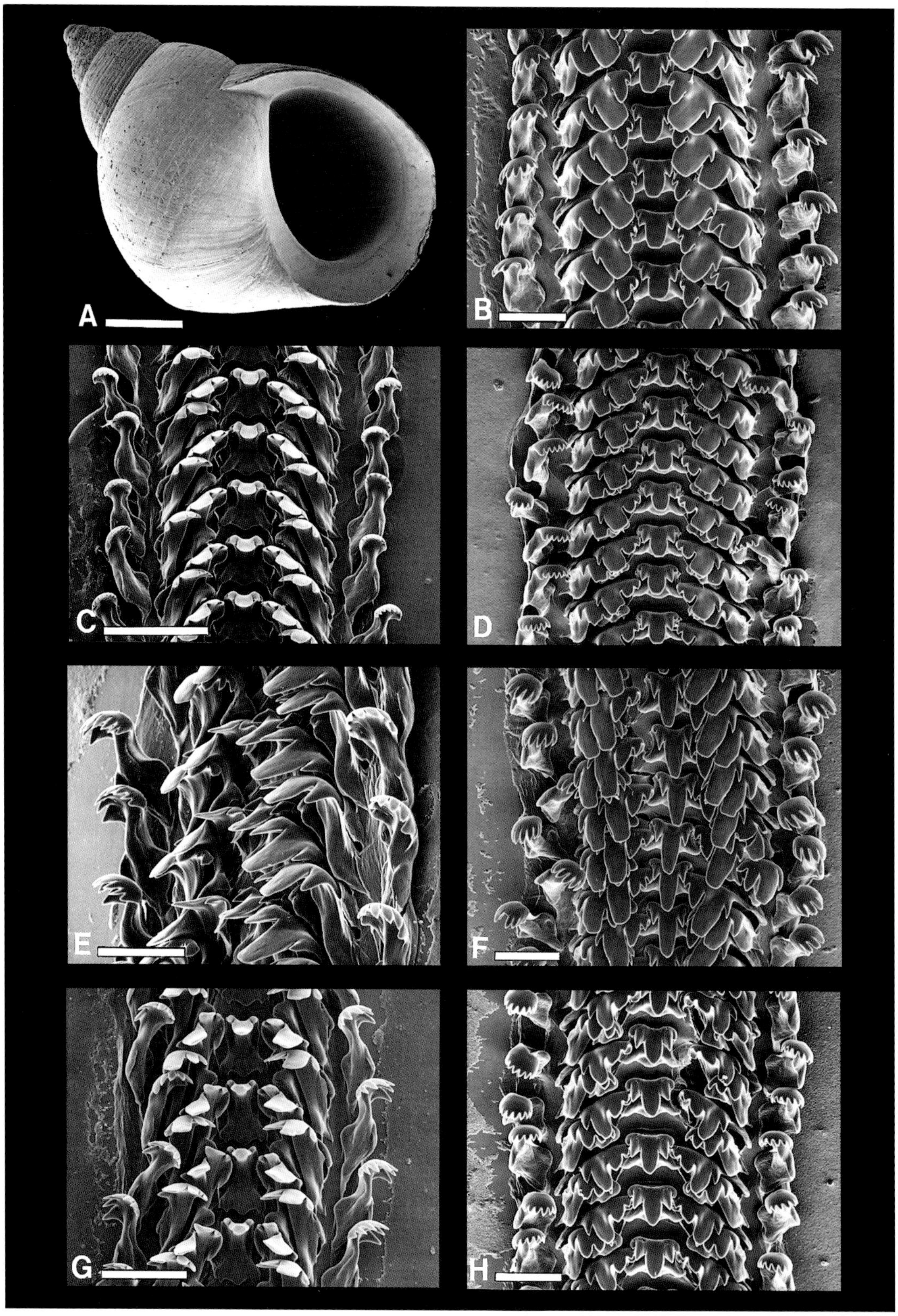

Figure 19. *Littorina (Littorina) scutulata*: **A,** juvenile shell; **B,** radula (shell H=12.2 mm), at 45°; **C, D,** two views of radula (shell H=11.5 mm), flat (**C**) and at 45° (**D**); **E, F,** two views of radula (shell H=12.0 mm), side view (**E**) and at 45° (**F**); **G,** radula (shell H=6.2 mm), flat; **H,** radula (shell H=9.2 mm), at 45°. **A, G,** Palos Verdes Peninsula, California; **B–D,** Neah Bay, Washington; **E, F,** Shell Beach, near Pismo Beach, California; **H,** Monterey, California. Scale bars: **A**=1 mm; **B**=50 µm; **C, D**=100 µm; **E**=50 µm; **F**=50 µm; **G**=50 µm; **H**=50 µm.

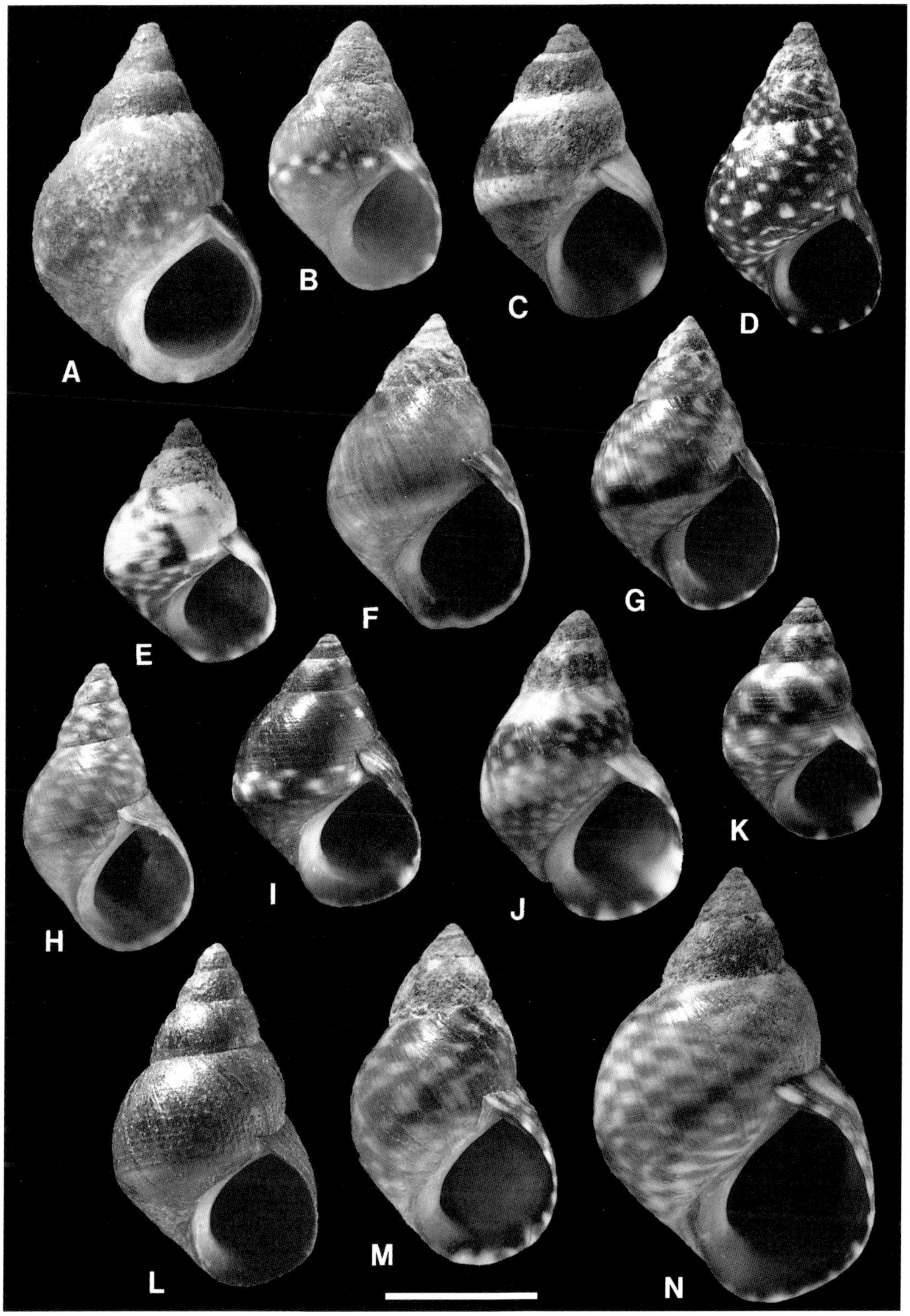

Figure 20. *Littorina (Littorina) scutulata*. **A,** *Littorina scutulata* Gould, 1849; lectotype, USNM 5640; Puget Sound, Washington. **B,** *Littorina lepida* Gould, 1849; lectotype, USNM 5637; Puget Sound, Washington. **C,** Point Lobos, Monterey, California. **D, L,** Westport, Washington. **E,** Three Tree Point, Puget Sound, Washington. **F, G, J, M, N,** Shell Beach, near Pismo Beach, California. **H,** Rompiente Point, Baja California, Mexico; LACM 71–162. **I, K,** Palos Verdes Peninsula, California. Scale bar=5 mm.

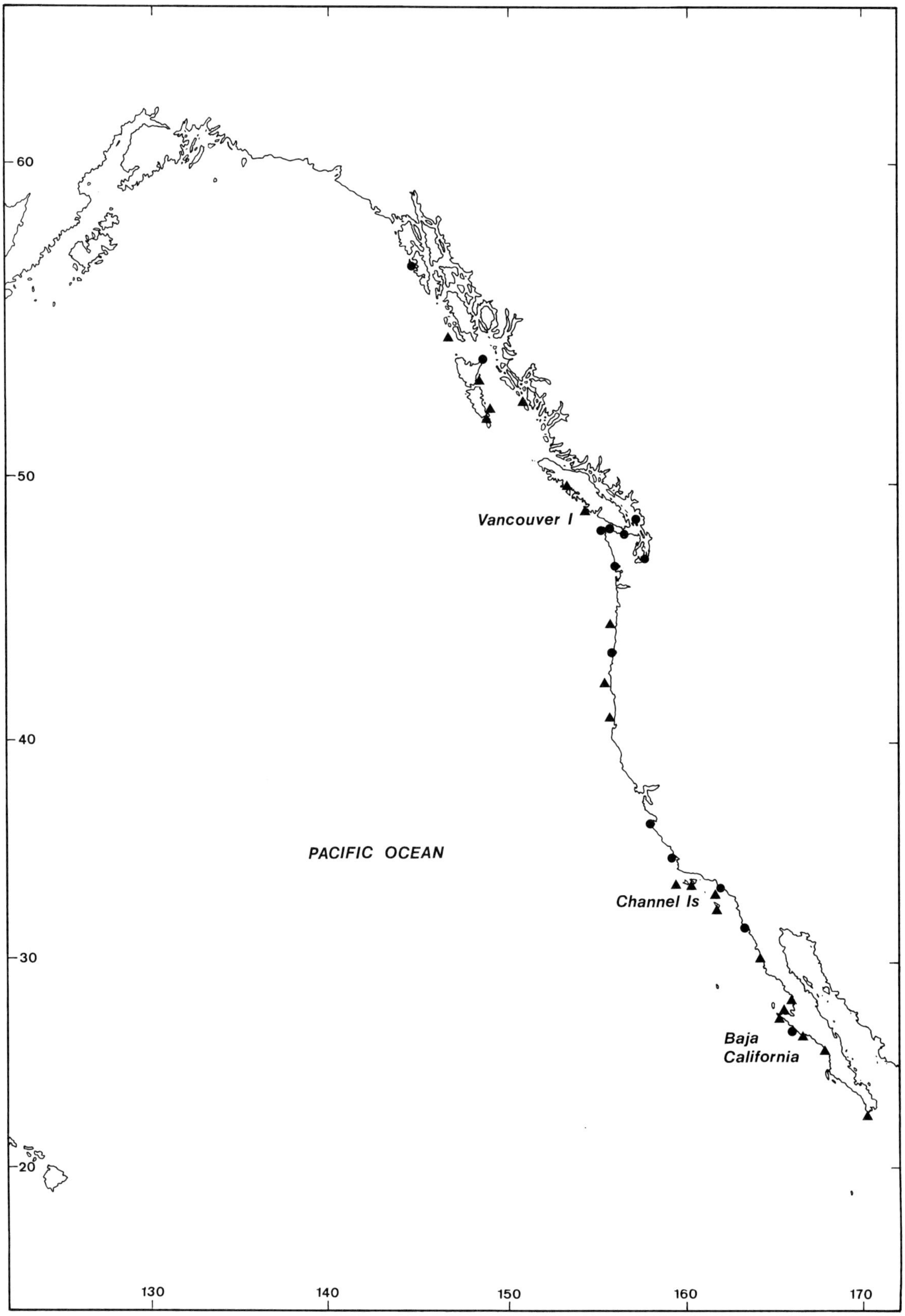

Figure 21. *Littorina (Littorina) scutulata*: distribution map. Key: circles, anatomical material examined; triangles, probable determination based on examination of shells only.

preserved material has confirmed the occurrence of this species from Sitka, Alaska (57°N; USNM 749836), south to Punta San Hipolito, Baja California (27°N; LACM 79–23), and a dry sample (USNM 780670) extends the range southwards to Cabo San Lucas. A recent visit to Cabo San Lucas (pers. obs., 1994) failed to find this species, so it may be of rare or sporadic occurrence at this extremity of the range. The species is also present on the Channel Islands of California. All the collections from north of Sitka that have been examined are of *L. plena*, as are the samples from Socorro Island recorded by Dall (1921) (although, as discussed in the remarks on *L. plena*, this locality is considered unreliable).

Fossil record. Two lots from the Pliocene San Diego Formation in San Diego have been seen (LACM 323 and 309) which closely resemble *L. scutulata s.l.* The shells are elongate, with no spiral sculpture except for an indistinct rib at the periphery, and measure up to 5.7 mm in height. There are a number of Pleistocene records of *L. scutulata s.l.* from California (Cooper, 1888; Arnold, 1903; Woodring, Bramlette & Kew, 1946; Valentine & Meade, 1960).

Discussion

Intraspecific variation. The relative uniformity of shell characters in this species, and lack of obvious geographical or ecotypic variation, are in agreement with the supposed widespread dispersal (and hence gene flow) in a species with pelagic eggs and planktotrophic larvae (see Janson, 1987a; Johannesson, 1992; for other littorinid examples). However, two studies of allozyme variation among populations of *L. scutulata s.s.* have found surprising levels of differentiation, comparable with those in some nonplanktotrophic species (Mastro *et al.*, 1982, recalculated by Ward, 1990; Correa, 1994).

Phylogenetic relationships. Cladistic analyses of morphological characters suggest that *L. scutulata* and *L. plena* are closely related, sharing the synapomorphies of double-rimmed egg capsules with many ova, and penes lacking mamilliform glands, but with a lateral glandular projection (Reid, 1990b; p. 351). Shell morphology and colour pattern are also extremely similar. The genetic study of Mastro *et al.* (1982) compared the two species and failed to find any diagnostic alleles among ten enzymes; the mean genetic identity (Nei, 1972) was 0.707. Nevertheless, phylogenetic analyses of mitochondrial DNA sequence data have lent only limited support to this sister-species relationship; they suggest that the divergence between them may be of long standing, with estimates of its age falling between 2.48 and 18.63 Ma (mean 12.81 Ma) (Reid, Rumbak & Thomas, 1996; p. 356). The Upper Miocene fossil *L. remondii* (p. 61) may be related to *L. scutulata* and *L. plena*.

Similar species. Littorina scutulata and *L. plena* are remarkably similar in shell morphology, to the extent that following their first description by Gould (1849) they were universally regarded as synonymous. Their separate status was revealed during a study of reproduction and development by Murray (1979), who observed differences in egg capsule morphology which have been confirmed by subsequent work (Mastro *et al.*, 1982; Strathmann, 1987). *Littorina scutulata* produces capsules 700 to 1000 μm in diameter, sculptured by two rims of unequal diameter, and containing 1 to 14 ova, whereas in *L. plena* the capsules measure 1100 μm in diameter, are sculptured by two rims of almost equal diameter, and contain 4 to 41 ova. Murray (1979) also found a difference in the penis, that of *L. scutulata* having a subterminal opening of the sperm groove and a small terminal projection, in contrast to that of *L. plena* with a terminal opening and a large projection near the base of the long filament. These characters were constant over a wide geographical range from Vancouver Island to southern California, and they have been used as diagnostic of the species in all subsequent studies (Murray, 1982; Mastro *et al.*, 1982; Chow, 1987a). No other diagnostic characters have hitherto been reported, although subtle differences in shell shape can be detected by multivariate statistical techniques (Murray, 1982; Chow, 1987a) and there are significant differences in allele frequencies (Mastro *et al.*, 1982).

Clearly there is a need for additional characters for the rapid, accurate identification of those individuals that are not mature males or spawning females, and characters that can be used in the field would greatly facilitate ecological studies. The present work has confirmed the consistent form of the penis over the geographical range of each species (Figs 18N–R, 22M–R). In females differences have been found in the pallial oviduct; in *L. scutulata* the copulatory bursa is long, extending back to the level of the seminal receptacle, and the straight distal section of the oviduct is relatively shorter in relation to the spiral glandular part, and not obviously septate (Fig. 18F–K); in *L. plena* the bursa is shorter, extending only to the midpoint of the spiral capsule gland, the straight section is relatively longer and its internal septation is visible from the outside (Fig. 22A, B, J–L). All mature males and females can therefore be identified by simple dissection.

A much more rapid, though not completely reliable, means of identification is provided by the pigmentation of the cephalic tentacles. In *L. scutulata* the predominant pattern is of transverse black bands and flecks, sometimes with a faint or broken longitudinal streak (Fig. 18A–E). In *L. plena* there is usually a broad, continuous, longitudinal black stripe, with less pronounced transverse lines and flecks, or the tentacles are sometimes entirely black (Fig. 22C–I). During the present study 122 specimens of *L. scutulata* from 13 sites, and 131 of *L. plena* from 12 sites, have been dissected. In 3% of *L. scutulata* and 14% of *L. plena* the tentacles were entirely black, and in less than 1% of *L. scutulata* there was an uninterrupted longitudinal black line. Therefore, using the above criteria of tentacle pattern, 4% of *L. scutulata* and none of *L. plena* were misclassified.

Radulae were examined by Mastro *et al.* (1982), who detected a small, but significant, difference in the length/width ratio of the rachidian tooth (wider in *L. scutulata*). In the present study relative radular length was greater in *L. scutulata*, the tooth cusps were less sharply pointed, and there were 7 to 9, rather than 6 to 7, cusps on the outer marginal teeth. Differences do appear to exist in the radulae, but are not likely to be useful for identification purposes.

Shell characters were at first believed to show too much overlap and variability to permit separation of *L. scutulata* and *L. plena* (Murray, 1979). However, it was soon noticed that a pale band on the shell base, visible also within the aperture, is more frequent in *L. plena*, permitting identification with a chance of error of between 13 and 20% (Murray, 1982; Mastro *et al.*, 1982; Chow, 1987a). Opinions differ on the significance of shell tessellation; Murray (1982) noted tessellation only in *L. plena*, whereas Mastro *et al.* (1982) and Chow (1987a) found more prominent tessellation in *L. scutulata*. From the present study of samples covering the whole geographical range, it is clear that it is not the prominence of tessellation, but the scale of the pattern that is significant. In *L. scutulata* the tessellation is coarse, and often most striking in a band at the periphery and another at the suture; on the final revolution of a point on the periphery of the shell there are 8 to 14 pale spots. In *L. plena* the tessellation is finer and less conspicuous, with usually 17 to 20 spots on the final revolution of the line corresponding to the more or less well-developed peripheral keel. Either species may be uniformly dark in colour, but this is more frequent in *L. plena*. The difference in tessellation is striking when Figures 20 and 24 are compared; of the patterned shells illustrated, only 24G and 24L would be misidentified using this criterion. Taking finely tessellated and uniformly dark shells to be *L. plena*, accuracy is estimated at about 90%.

Shell size, shape and sculpture are more variable, and particularly so in *L. plena* (Fig. 24). Previous studies agree that *L. scutulata* is on average the larger (mean height of 11.6 mm, compared with 8.6 mm for *L. plena*, Murray, 1982; maximum size for *L. scutulata* 17 mm, for *L. plena* 11 mm, Mastro *et al.*, 1982). In the present study the species have more often than not been found to occur sympatrically, and at all such localities *L. scutulata* attains larger size. However, where *L. plena* occurs alone (as, for example, in some sheltered bays in the vicinity of San Francisco) it can equal or even slightly exceed the maximum size of 18.0 mm recorded for *L. scutulata*. Differences in shell shape have been investigated by multivariate analyses. Murray (1982) used three linear shell measurements and the number of whorls to derive discriminant functions which classified the two species from a range of localities with an error rate of only 4%. Chow (1987a) repeated this analysis for samples from a single exposed locality near Bodega Bay and achieved error rates of 11% for *L. scutulata* and 5% for *L. plena*, with functions incorporating shell length, width, number of whorls, presence of basal band and degree of shell tessellation. The same functions performed less successfully with samples from a sheltered harbour nearby, demonstrating that, because of intraspecific variability, discriminant functions derived for local populations cannot be applied uncritically elsewhere. Both these studies found a slightly greater apical angle in *L. plena*, but from Figures 20 and 24 it is clear that this is not consistent. At some localities, however, shape differences may be obvious; in Washington and further north *L. plena* is often the narrower and more tall-spired, whereas in California and further south the reverse is sometimes the case. Spiral lines on the shell may be faint or obsolete in both species, but strong spiral grooves and a pronounced peripheral keel are only ever developed in examples of *L. plena*.

Differences in distribution can also now be given. There is broad overlap in geographical range, but *L. scutulata* is the more southern species (23 to 57°N for *L. scutulata*; 30 to 61°N for *L. plena*). The two

Table 2. Comparison of *Littorina scutulata* and *L. plena*.

Character	*L. scutulata*	*L. plena*	% accuracy	References
Cephalic tentacles	Fig. 18A–E. Transverse black bands and flecks	Fig. 22C–I. Broad, unbroken black stripe with some transverse flecks, or all black	96–100	This study
Penis	Fig. 18N–R	Fig. 22M–R	100	Murray, 1979; this study
Projection of filament	Small, terminal	Large, basal		
Length of filament	40-50% total length of base	70-80% total length of base		
Opening of sperm groove	Subterminal	Terminal		
Pallial oviduct	Fig. 18F–K	Fig. 22A, B, J–L	100	This study
Copulatory bursa	Long, extends to seminal receptacle	Shorter, extends to mid-point of capsule gland		
Straight section	Relatively shorter, not septate	Relatively longer, septate		
Spawn	Fig. 18L, M	Fig. 22S, T	100	Murray, 1979; Mastro *et al.*, 1982; Strathmann, 1987
Capsule diameter	700–1000 μm	1100 μm		
Shape	2 rims of unequal diameter	2 rims of equal diameter		
Number of ova	1–14	4–41		
Radula	Fig. 19B–H	Fig. 23C-H		This study
Relative length	1.22–2.31	1.10–1.41		
Largest tooth cusps	Blunt	Pointed		
Outer marginal tooth	7–9 cusps	6–7 cusps		
Shell	Figs 19A, 20	Figs 23A, B; 24		
Pale basal band	Usually absent	Usually present	80-87	Murray, 1979; Chow, 1987a
Tessellation	Coarse	Fine	90	This study
Mean (maximum) height	11.6 (18.0) mm	8.6 (18.7) mm		Murray, 1982; this study
Discriminant functions			89-96	Murray, 1982; Chow, 1987a
Sculpture	Absent to faint	Absent to strong incised lines		This study
Distribution				This study
Latitude	Fig. 21. 23–57°N	Fig. 25. 30–61°N		
Habitat	Exposed and sheltered rocks	Moderately exposed and sheltered rocks: salt marshes		

are frequently sympatric on rocky shores of varying exposure, but in general *L. scutulata* seems to predominate on exposed coasts and *L. plena* in sheltered bays and salt marshes. Only *L. scutulata* has been recorded from the Channel Islands of California, suggesting that its distribution may be the more oceanic in character. No comparisons of zonation or diet have yet been made but, with the aid of new characters for identification, these may now be possible.

The principal differences between *L. scutulata* and *L. plena* are summarized in Table 2. For field studies it is recommended that samples should first be identified in the laboratory using the highly reliable and easily observed anatomical characters (penis, pallial oviduct, cephalic tentacles) in order to assess local variation in shell coloration, shape and sculpture. With care, it is then possible to identify the great majority of specimens in the field using shell characters alone.

Littorina (Littorina) plena Gould, 1849

(Figs 3C, 22–25)

Synonymy

Littorina plena Gould, 1849: 84 (San Francisco; lectotype (Murray, 1982) USNM 5635, seen, Fig. 24A, this is probably specimen figured by Gould, 1861; 2 paralectotypes (with resoaked animals) USNM 677096, seen).

Gould, 1852: 201–202; atlas (1861) pl. 14, fig. 236a, b. Murray, 1982: 233–238 (in part, includes *L. lepida* Gould=*L. scutulata*). Mastro, Chow & Hedgecock, 1982: 239–246; fig. 3 (penis, spawn); fig. 4 (radula). Chow, 1987a: 359–366. Sacchi & Voltolina, 1987: 209–234; fig. 2c; fig. 3a (penis). Strathmann, 1987: 238; fig. 11.5 (spawn).

Litorina scutulata var. *plena*—Carpenter, 1864: 656.

Littorina (Littorina) plena—Reid, 1989a: 101; pl. 2, fig. 2m. Reid, 1990b: 1–19; fig. 1d; fig. 3i (spawn). Taylor & Reid, 1990: 199–215.

Littorina (Melarhaphe) scutulata—Tryon, 1887: 250; pl. 45, fig. 1 (in part, other figs are *L. scutulata* Gould; as *Melaraphe*).

Littorina scutulata—Packard, 1918: 320–321; pl. 35, figs 2, 3 (not Gould, 1849).

Littorina scutulata 'type I'—Murray, 1979: 469–474; fig. 1A (spawn); fig. 2B (penis).

Note: See Synonymy of *L. scutulata* (p. 63) for references to authors who used the name *L. scutulata* in a broad sense that probably included *L. plena*.

Taxonomic history

As described earlier (p. 64), *L. plena* was synonymized with *L. scutulata* by Carpenter (1864), and has only recently been reestablished as a recognized species (Murray, 1979, 1982; Mastro, Chow & Hedgecock, 1982).

Diagnosis

Mature shell height 3–18.7 mm. Shell ovate-conical to high conical. Mineralogy of external irregular-prismatic calcite and internal crossed-lamellar aragonite. Sculpture usually with incised spiral lines. Colour brown, olive or black, often with fine grey tessellation or mottling; pale basal band usually present. Head black, tentacles usually with broad longitudinal black band. Penis bifurcate, with lateral glandular projection on long, tapering filament; mamilliform glands absent; sperm groove ends terminally. Pallial oviduct of two consecutive loops of albumen gland followed by large spiral loop of capsule gland; straight section with septate jelly gland; bursa extends from anterior end of straight section back to midpoint of capsule gland. Spawn of large pelagic capsules sculptured by two flanged rims of same diameter, containing 17–32 eggs; development planktotrophic. Distribution: California to Gulf of Alaska.

Anatomical description

Head-foot (Fig. 22C–I). Head black; tentacles with an unpigmented stripe across eye, usually extending across tentacle base, predominant pattern of a broad continuous longitudinal black stripe, usually with less pronounced transverse black lines and flecks, tentacles occasionally all black; sides of foot grey to black. Opercular ratio 0.41–0.56.

Penis (Fig. 22M–R). Base wrinkled, slightly pigmented; mamilliform penial glands absent. Filament long, tapering, often coiled, approximately 70–80% total length; with bifurcation near its base, producing a lateral papilla. Superficially, this projection resembles a mamilliform gland, but it lacks the duct and reservoir of the latter. Sperm groove open, ending terminally. Subepithelial glandular tissue (reddish in living specimens) around sperm groove in filament and extending into lateral papilla.

Sperm (Fig. 3C). Euspermatozoa 38–54 μm. Paraspermatozoa 10–20 μm diameter, spherical to ovoid; vesicles moderate to large, usually distinct; nucleus sometimes visible.

Oviduct (Fig. 22A, B, J–L). Two consecutive loops of albumen gland followed by large spiral loop of capsule gland with distal reddish translucent portion; straight section with septate (but not swollen) jelly gland. Bursa extends from near anterior end of straight section back to midpoint of capsule gland. Seminal receptacle simple.

Spawn and development (Fig. 22S, T). Pelagic, biconcave capsules with two flanged rims of equal diameter; 1100 μm diameter; containing 4–41 (usually 17–32) eggs 96 μm diameter (Murray, 1979;

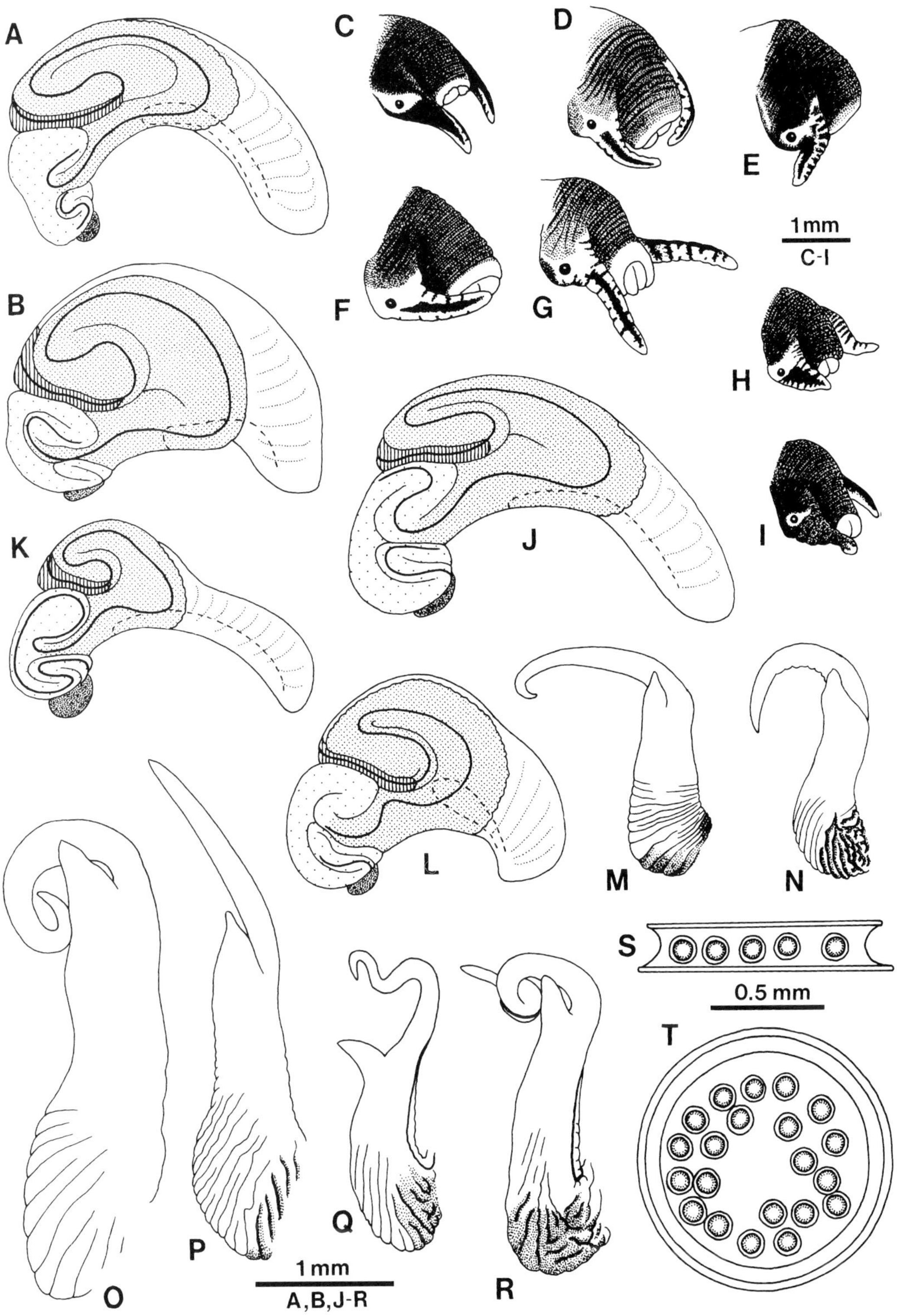

Figure 22. *Littorina (Littorina) plena*: **A, B, J–L,** pallial oviducts; **C–I,** head pigmentation; **M–R,** penes; **S, T,** two views of pelagic egg capsule (after Strathmann, 1987). **A, H, I, N,** Shell Beach, near Pismo Beach, California; **B, C, M,** Ensenada, Mexico; **D, K,** Colin's Cove, San Juan Island, Washington; **E, F, R,** Neah Bay, Washington; **G, J,** Tongue Point, near Port Angeles, Washington; **L, Q,** Seldovia, Alaska; **O,** Tomales Bay, California (LACM 49–15); **P,** Berkeley, California (LACM 67757). Shell heights: **A,** 7.0 mm; **B,** 7.2 mm; **C,** 8.0 mm; **D,** 5.4 mm; **E,** 9.8 mm; **F,** 8.5 mm; **G,** 7.1 mm; **H,** 7.0 mm; **I,** 7.0 mm; **J,** 7.1 mm; **K,** 5.4 mm; **L,** 6.0 mm; **M,** 6.2 mm; **N,** 5.5 mm; **O,** 12.5 mm; **P,** 8.9 mm; **Q,** 6.6 mm; **R,** 7.3 mm. For shading conventions of pallial oviducts see p. 6.

Mastro *et al.*, 1982; Strathmann, 1987). Veligers hatch at shell width of 169 μm; development planktotrophic (Murray, 1979). Protoconch approximately 300 μm diameter, but not well preserved in available specimens. Spawning season includes April to August in central and northern California (Murray, 1979); sporadic spawning reported from February to October at Bodega Bay, California (Chow, 1987b).

Radula and foregut (Fig. 23C–H). Relative radular length 1.10–1.41. Rachidian: length/width 1.00–1.24; central cusp pointed, sometimes slightly rounded; one pointed cusp and one outer denticle on each side. Lateral: four cusps, third from inside largest and bluntly rounded to pointed, others pointed. Inner marginal: four cusps, third from inside largest and bluntly rounded to pointed, others pointed. Outer marginal: 6–7 long pointed cusps; narrow neck, projecting flange either side of base. Salivary glands enlarged.

Shell variation (Figs 23A, B; 24)

Geographical variation. Ecotypic variation in shell shape has not been observed in *L. plena*, but on a geographical scale the variation in shape is striking. Shells from the northern part of the range are more tall-spired and can usefully be designated as geog. var. *alaskana*.

> Geog. var. *typica* (Fig. 24A–C, E, G–P): H up to 18.7 mm; H/B=1.38–1.65; SH=1.57–1.94. Shell smooth to spirally striated; periphery rounded or with slight keel. Range: Baja California to San Juan Archipelago, Washington.
>
> Geog. var. *alaskana* (Fig. 24D, F): H up to 11.6 mm; H/B=1.71–2.00; SH=1.94–2.14. Shell sculptured with moderate to strong incised lines; periphery usually slightly keeled. Range: Strait of Juan de Fuca to Alaska.

There is no sharp distinction between the two varieties, nor a sudden geographical transition between them. In northern Washington the populations include taller forms (Fig. 24D); samples from British Columbia are mostly tall-spired, while those from Alaska seem to be exclusively of the taller form (Fig. 24F).

Sculptural variation. Shell sculpture ranges from almost smooth to incised spiral lines; this feature is often variable even at a single locality (e.g. shells from Ensenada, Baja California, in Fig. 24B, J, M, and from Westport, Washington, in Fig. 24C, I, K, L).

Sexual dimorphism. Chow (1987b) recorded significant sexual dimorphism in this species, females being larger as a result of higher growth rates.

Colour variation. Colour varies from black through brown to olive, usually with a finely spotted, tessellated or zigzag pattern. Available samples appear to show continuous variation, and therefore the species is not described as polymorphic. A noticeable aspect of ecotypic variation in this species is that specimens from sheltered localities show more pronounced tessellated patterning on the shell (e.g. Fig. 24C, I, K, L, all from a salt marsh, and Fig. 24D from a sheltered cobble shore); a similar effect has been noted in *L. scutulata* (Chow, 1987a). Figure 24G illustrates an unusual low-spired form with strong brown and white colour patterning; there are no habitat details for this specimen, which was found in a lot together with more typical shells.

Distribution

Habitat. Littorina plena occurs most commonly in beds of barnacles and mussels on rocky shores, but also extends just into the littoral fringe (Chow, 1987b, 1989). It can be found in tufts of red algae low in the eulittoral zone, and among *Fucus* and *Ulva* at higher levels. Juveniles are abundant among barnacles, and larger snails occur upshore (Chow, 1989; Geller, 1991). It tolerates a range of conditions of wave-exposure, from moderately exposed coasts to sheltered bays and harbours. It is frequent on pilings, stones and tufts of *Fucus* in the runnels of salt marshes.

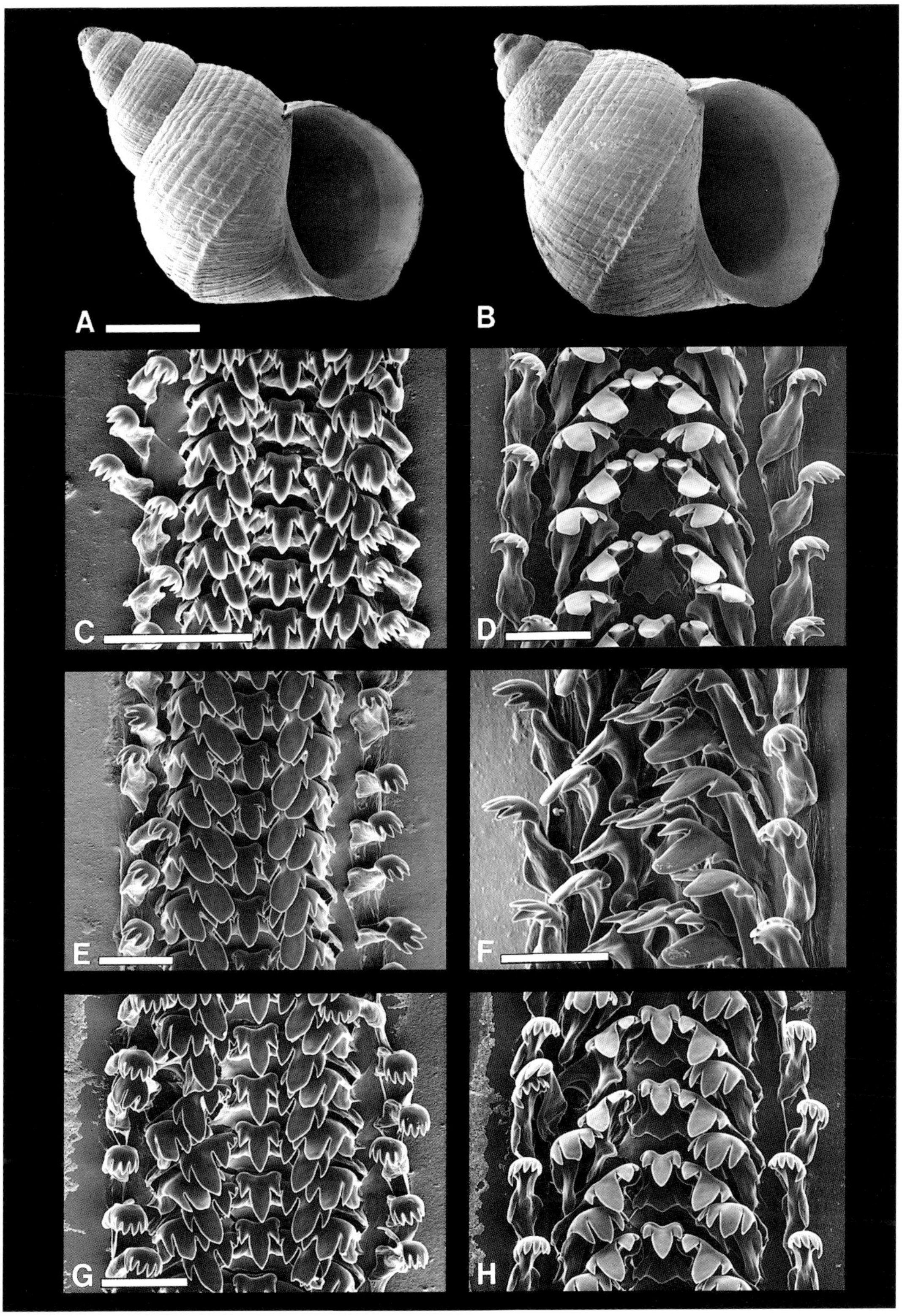

Figure 23. *Littorina (Littorina) plena*: **A, B,** juvenile shells; **C,** radula, at 45° (shell H=5.2 mm); **D–F,** three views of radula (shell H=8.5 mm), flat (**D**), at 45° (**E**), side view (**F**); **G, H,** two views of radula (shell H=8.5 mm), at 45° (**G**), flat (**H**). **A, B,** Westport, Washington; **C,** Seldovia, Alaska; **D–F,** Neah Bay, Washington; **G, H,** Ensenada, Mexico. Scale bars: **A, B**=1 mm; **C**=50 μm; **D**=50 μm; **E**=50 μm; **F**=50 μm; **G, H**=50 μm.

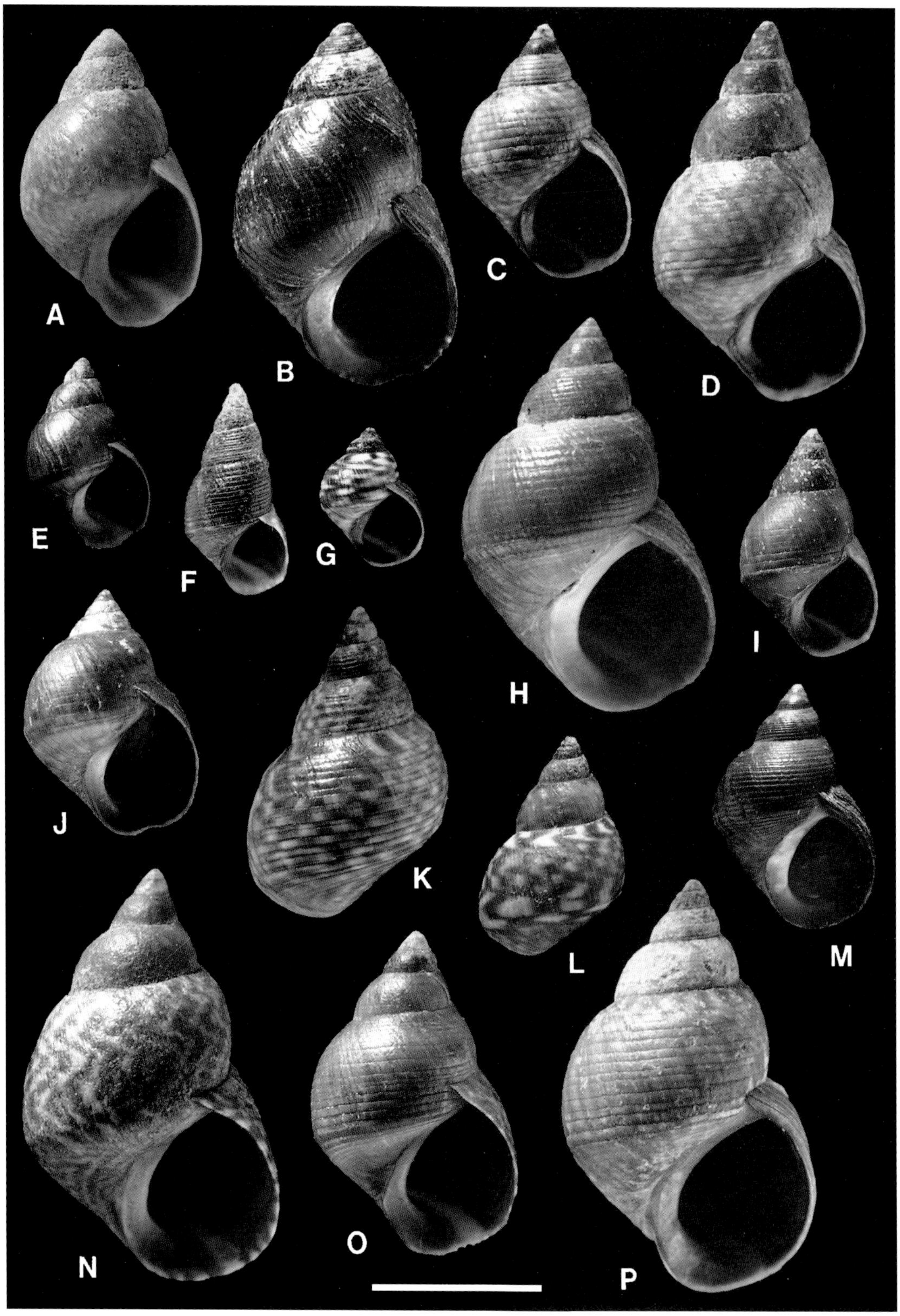

Figure 24. *Littorina (Littorina) plena*: **A–C, E, G–P,** geog. var. *typica*; **D, F,** geog. var. *alaskana*. **A,** *Littorina plena* Gould, 1849; lectotype, USNM 5635; San Francisco. **B, J, M,** Ensenada, Mexico. **C, I, K, L,** Westport, Washington. **D,** Neah Bay, Washington. **E,** Shell Beach, near Pismo Beach, California. **F,** Seldovia, Kenai Peninsula, Alaska. **G,** Bakers Beach, San Francisco, California; USNM 803493. **H,** San Francisco, California. **N,** 'Socorro Island' (unreliable locality); USNM 207285. **O, P,** Oyster Cove, Tomales Bay, California; LACM 48–39. Scale bar=5 mm.

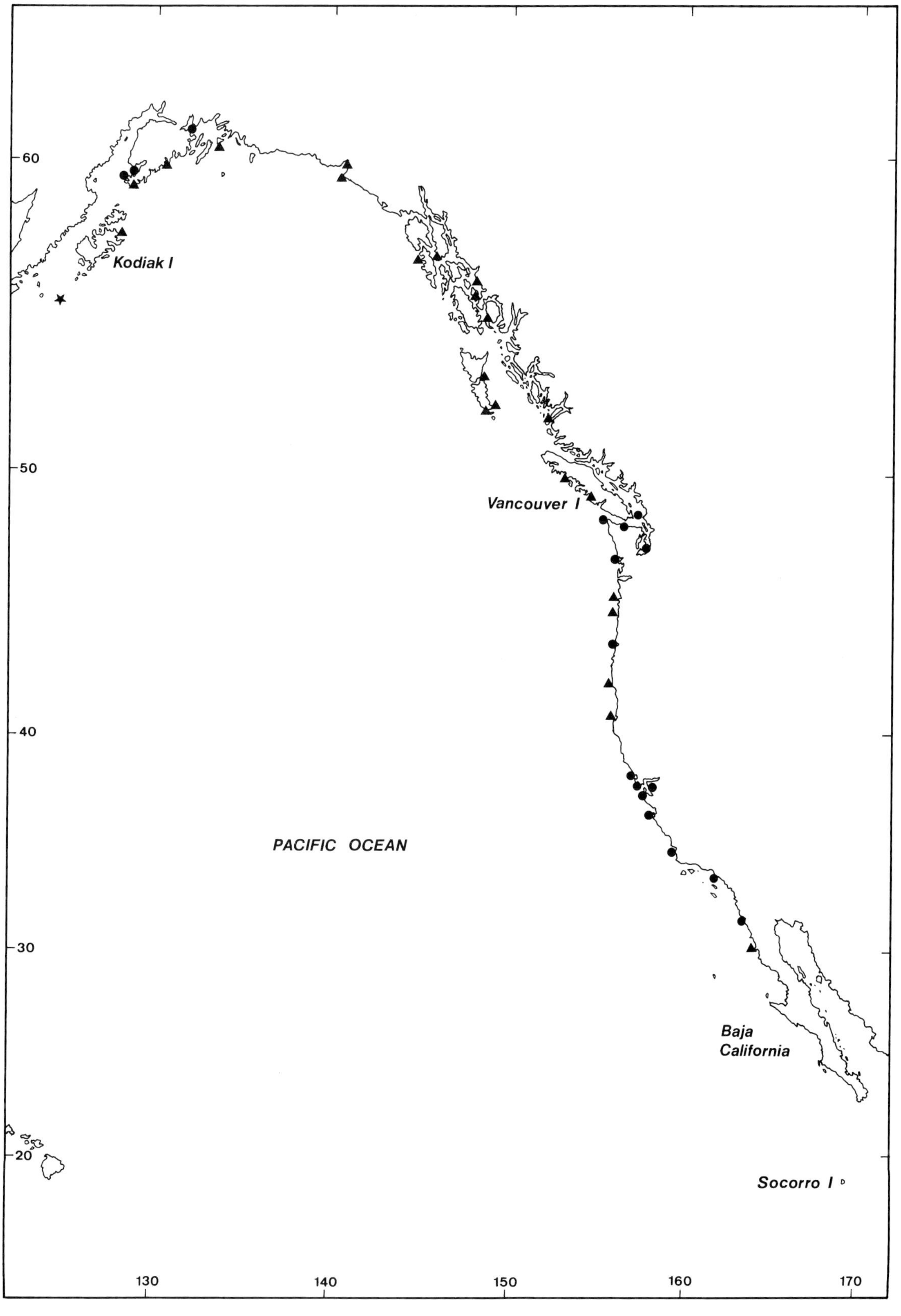

Figure 25. *Littorina (Littorina) plena*: distribution map. Key: circles, anatomical material examined; triangles, probable determination based on examination of shells only; asterisk, record from literature (Zimmerman, Hanson, Fujioka, Calvin, Gharrett & MacKinnon, 1977).

Throughout most of its range *L. plena* is sympatric with *L. scutulata*. Only a few ecological studies have distinguished these two species (see remarks on *L. scutulata*, p. 67), and no differences in zonation or microhabitat have yet been recorded. On most shores they occur together, but their relative abundance appears to vary with wave exposure: *L. scutulata* usually predominates on the more exposed shores and *L. plena* on the more sheltered ones. In a salt marsh at Westport, Washington (pers. obs.), in the sounds around Wrangell in Alaska (USNM, LACM), and in some sheltered localities in San Francisco Bay (BMNH), *L. plena* appears to occur alone. Working in Vancouver Island, Sacchi & Voltolina (1987) found *L. plena* to be infrequent, but more common in the estuary of the Sooke River.

No detailed studies of the diet of *L. plena* have yet been published. According to Chow (1987b, 1989), it resembles *L. scutulata*, grazing primarily on microalgal films of red, green and blue-green algae, diatoms and macroalgal sporelings, but also on macroalgae in the lower parts of its vertical range.

Range (Fig. 25). In his rediscovery of *L. plena*, Murray (1979) recorded it from Newport, Oregon, south to San Diego, and this range was extended north to Vancouver by Mastro *et al.* (1982). In the present study the range of material confirmed by dissection is from Prince William Sound in Alaska (61°N; BMNH) to Ensenada, Baja California (32°N; BMNH). Based on shells alone, the most southerly reliable record is from San Martin Island, Baja California (30°N; LACM 73–2). Two dry lots labelled from Socorro Island (18°N; USNM 207285 and 207286; on which Dall's (1921) record of *L. scutulata s.l.* was probably based) are considered unreliable, for several reasons. This locality is much further south than others on the Mexican coast; the species has not been recorded on offshore islands such as the Channel Islands or Guadalupe, so occurrence on the much more distant Socorro Island is unlikely; collections from Socorro Island in LACM and CAS do not include *L. plena*; the shells in lot USNM 207285 are exceptionally large (to 18.7 mm), and in this respect, as in shape and coloration, resemble shells known from only two localities, San Quintin Bay, Baja California, and the vicinity of San Francisco (compare Fig. 24N and Fig. 24O, P). At the northwestern extreme of the range, dry shells have been seen from Kodiak Island (USNM 47067; LACM 84–65). A record of *L. scutulata s.l.* from Chirikof Island (Zimmerman, Hanson, Fujioka, Calvin, Gharrett & MacKinnon, 1977) is probably the most southwesterly collection of *L. plena* in the Gulf of Alaska. Despite statements to the contrary (Murray, 1979; Strathmann, 1987), this species does occur in the San Juan Archipelago, Washington (pers. obs.; BMNH). Its absence from offshore islands, and presence at sheltered, turbid and estuarine sites, indicates a distribution of the continental type.

Fossil record. As noted under *L. scutulata* (p. 71), shells from the Pliocene San Diego formation resemble these two similar species. According to Carpenter (1864), *Littorina pedroana* Conrad, 1855, a Pleistocene fossil from San Pedro, California, is *L. plena*. However, the original figure shows a markedly low-spired shell, and both Gabb (1869) and Grant & Gale (1931) have synonymized *L. pedroana* with a *Lacuna* species; the type is missing from USNM.

Discussion

Intraspecific variation. This species shows greater variation in shell characters, on both geographical and local scales, than does *L. scutulata*. As in that species, a study of allozyme variation has found a surprisingly high level of interpopulation differentiation for a planktotrophic species, in which dispersal and gene flow might have been expected to be high (Mastro *et al.*, 1982, recalculated by Ward, 1990). Either strong selection or restricted dispersal might explain this anomalous result.

Phylogenetic relationships and similar species. The close relationship of this species with *L. scutulata*, and the means of distinguishing them, have been discussed in the remarks on *L. scutulata* (p. 71; Table 2).

Littorina (Littorina) squalida Broderip & Sowerby, 1829

(Figs 26–30)

Synonymy

Littorina squalida Broderip & Sowerby, 1829: 370 (in Oceano Boreali [northern ocean]; type locality here restricted to Icy Cape, Alaska; neotype (here designated) BMNH 1858.5.12.216, 29.3 mm, seen, Fig. 28A; see remarks under Taxonomic History, p. 81). Gray, 1839: 139; pl. 34, fig. 12. Deshayes & Milne Edwards, 1843:

203–204. Hirase, 1934: 47; pl. 79, fig. 7. Habe, 1951: 87; pl. 14, fig. 6. Ushakov, 1953. Hirase & Taki, 1954: pl. 79, fig. 7. Galkin & Skarlato, 1955: 171; pl. 44, fig. 19. Kojima, 1958b: 233–237; fig. 1 (spawn). Kojima, 1958d: 39–41; fig. 1 (spawn). Golikov, 1959: 1335–1343; fig. 2. Altukhov & Kalabushkin, 1974: 1477–1480. Kalabushkin, 1976: 369–377; fig. 2. Foster, 1981: 101–102. Gladenkov, Gladikova, Kafanov, Konova, Krishtofovich, Sinelnikova & Popov, 1984: 238; pl. 60, fig. 5; pl. 62, fig. 3 (Miocene). Okutani, 1986: 70–71; unnumbered plate; text-fig. (spawn). Baxter, 1987: 49. Golikov, Gulbin & Sirenko, 1987: 32; pl. 2, fig. 20. Repenning, Brouwers, Carter, Marinkovich & Ager, 1987: 15–17 (Pliocene). Egorov, 1992: 169; fig. 1B.

Litorina squalida—Philippi, 1846, **2**: 101–102; *Litorina* pl. 1, fig. 6. Küster, 1856: 5–6; pl. 1, figs 5, 6 (1853). Weinkauff, 1878: 27. Weinkauff, 1883: 214.

Littorina (Littorina) squalida—Tryon, 1887: 240; pl. 41, fig. 81. Dall, 1923: 3. Rosewater, 1970: 423. Reid, 1989a: 101; fig. 11s (spawn). Reid, 1990b: 1–19; fig. 1i; fig. 3d (spawn); fig. 4e (radula). Taylor & Reid, 1990: 199–215 (mineralogy).

Littorina (Algaroda) squalida—Dall, 1921: 153. Oldroyd, 1927: 59. Golikov & Kusakin, 1962: 269–271. Golikov & Skarlato, 1967: 26–27. Golikov & Gulbin, 1978: 190–191. Golikov & Kusakin, 1978: 84–85; fig. 50a; fig. 50b (spawn). Volova, Golikov & Kusakin, 1979: 67–68; fig. 44. Golikov & Skarlato, 1985: 384.

Littorina (Algaroda) aff. *squalida*—MacNeil, Mertie & Pilsbry, 1943: 84 (Pleistocene).

Ezolittorina squalida—Habe, 1958a: 9; pl. 1, fig. 22; pl. 4, fig. 9 (radula). Yamamoto & Habe, 1962: 15; pl. 3, figs 20, 21 (spawn). Amio, 1963: 303 (spawn). Habe, 1964: 28; pl. 9, fig. 29. Higo, 1973: 46.

Littorina (Ezolittorina) squalida—Oyama & Takemura, 1963: *Neritrema, Littorina* pl., fig. 8. Habe & Ito, 1965: 25; pl. 7, fig. 8. Habe & Kosuge, 1967: 29; pl. 12, fig. 8; text-fig. (spawn). Higo & Goto, 1993: 73.

Littorina littorea—Wood, 1842: 532 (as *littoreus*; in part, includes *L. littorea*). Wood, 1848: 118–119 (in part, includes *L. littorea*). Jeffreys, 1869: 206 (as *litorea*; in part, includes *L. littorea*). Mørch, 1871b: 396 (not *Turbo littoreus* Linnaeus, 1758; Pliocene). Dunker, 1874: 105 (not Linnaeus, 1758). Aurivillius, 1885: 325 (not Linnaeus, 1758). Dautzenberg & Fischer, 1912: 181–187 (in part, includes *L. littorea*). Dodge, 1959: 216–218 (in part, includes *L. littorea*). Gladenkov, Norton & Spaink, 1980: 65–66; pl. 13, figs 8, 8a, 9 (in part, other figs are *L. islandica*; not Linnaeus, 1758; Pliocene).

Littorina grandis Middendorff, 1848: columns 241–242 (Sinus Tugur, maris Ochot.; ins. Schantar [Tugur Bay and Shantar Islands, Okhotsk Sea]; 7 syntypes ZISP, seen, Middendorff, 1851: pl. 11, figs 4–10; 5 syntypes BMNH 1992161, seen, Fig. 28J; 5 syntypes BMNH 1992162, seen, Reeve, 1857: pl. 3, fig. 16a, b, and Fig. 28I herein; syntypes MCZ 119380, not seen; 4 syntypes MCZ 51998, not seen. Middendorff, 1849: 385. Middendorff, 1851: 198–201; pl. 11, figs 4–10. Reeve, 1857: sp. 16; pl. 3, fig. 16a, b. Pilsbry, 1895: 61.

Litorina grandis—Schrenck, 1867: 319–321. Weinkauff, 1882: 51–52; pl. 6, figs 10, 11. Dunker, 1882: 110. Weinkauff, 1883: 214.

Littorina (Littorina) grandis—Tryon, 1887: 240; pl. 41, fig. 80.

Littorina palliata—Schlesch, 1924: 28–29; pl. 10, fig. 6 (not *Turbo palliatus* Say, 1822=*L. obtusata*; Pliocene). Ilina, 1939: 47–48; pl. 14, figs 6, 6a (not Say, 1822; according to Gladenkov *et al.*, 1984).

Littorina sp. Bárdarson, 1925: 44 (Pliocene).

Natica janthostoma—Ilina, 1939: pl. 6, figs 6, 6a (not Deshayes, 1841; according to Gladenkov *et al.*, 1984).

Littorina rudis—Norton, 1975: 109 (not *Turbo rudis* Maton, 1797=*L. saxatilis*).

Littorina saxatilis—Norton, 1975: 109 (not *Turbo saxatilis* Olivi, 1792).

Littorina etolonensis Sinelnikova, in Gladenkov & Sinelnikova, 1990: 99; pl. 29, fig. 8a, b (Cape Bolshoi (label accompanying holotype gives additional details: W coast of Pendginsk Bay, between mouth of Pustaya River and Cape Bolshoi), Rekinniksk Gulf, western Kamchatka; Etolonian stage, Middle Miocene; holotype PMM N495/41, seen, Fig. 30A; 2 paratypes BMNH GG22523, seen).

Littorina praesqualida Sinelnikova, in Gladenkov & Sinelnikova, 1990: 100; pl. 31, fig. 9 (between mouths of Kheisheveem and Keicheveem Rivers, Kavrano-Ukholoksk Gulf, western Kamchatka; Etolonian stage, Middle Miocene; holotype PMM N182/84, seen, Fig. 30G; paratype BMNH GG22522, seen, Fig. 30E).

Taxonomic history

The original description of *L. squalida* by Broderip & Sowerby (1829) was brief and inadequate; it was based on material from Beechey's voyage, in the museum of the Zoological Society of London, and the locality was given only as 'northern ocean'. The identity of the species was confirmed by Gray's (1839) more complete description and figure, in the *Zoology of Captain Beechey's Voyage*. This figured specimen is lost, but two specimens in the BMNH (reg. no. 1858.5.12.216), without locality information, were originally purchased from the Zoological Society. The neotype is selected from these possible syntypes. Gray (1839) cited the locality as Icy Cape (Alaska), and the type locality is here restricted accordingly.

In 1848 Middendorff described material from the Okhotsk Sea as *L. grandis*; he recognized the similarity to Gray's (1839) figure of *L. squalida*, but justified his decision by differences in colour, size and sculpture, and in particular because *L. squalida* was supposed to be from New Zealand. This confusion arose because Philippi (1846) had figured *L. squalida* from the Cuming Collection (BMNH) with the erroneous locality New Zealand; independently, Deshayes & Milne Edwards (1843) had misquoted the type locality as 'southern ocean'. The two names, *L. squalida* supposedly from the Southern Ocean and *L. grandis* from the northern Pacific, were thus perpetuated in the literature (Küster, 1856; Weinkauff, 1878, 1883; Tryon, 1887) until synonymized by Dall (1921).

The resemblance to *L. littorea* was noted even by Broderip & Sowerby (1829), and several workers have synonymized *L. squalida* as a northern Pacific form of it (e.g. Wood, 1842, 1848; Jeffreys, 1869; Aurivillius, 1885; Dautzenberg & Fischer, 1912). This close relationship has often been emphasized by placing both species in the subgenus *Littorina*, or by Dall (1921) and many Russian authors in the subgenus *Algaroda* (an objective synonym of *Littorina*). Habe (1958) proposed the new genus *Ezolittorina* for *L. squalida*, on the basis of its spawn type.

As discussed in the section on Fossil Record (p. 89), the earliest Atlantic fossil *Littorina*, from Iceland, previously assigned to *L. littorea* (Mørch, 1871b; Gladenkov, Norton & Spaink, 1980), are here identified as *L. squalida*, because of their characteristically rounded whorls. Confusingly, Gladenkov *et al.* (1980) included in their concept of *L. littorea* from Iceland carinate shells from higher levels in the sequence, here described as a new species (p. 246). *Littorina squalida* has been recorded several times, under a variety of names, from the Miocene of Kamchatka (Gladenkov, Gladikova, Kafanov, Konova, Krishtofovich, Sinelnikova & Popov, 1984; Ilina, 1939). Recently, two new species, *L. etolonensis* and *L. praesqualida*, have been described from the Middle Miocene of this region (Gladenkov & Sinelnikova, 1990) and are here synonymized with *L. squalida*.

Diagnosis

Mature shell height 8.8–43.7 mm. Shell high-turbinate with rounded whorls; last whorl sometimes swollen. Mineralogy of external irregular-prismatic calcite and internal crossed-lamellar aragonite. Sculpture of numerous narrow ribs and fine spiral microstriae over whole surface, often worn away on largest specimens. Aperture round, sometimes lirate at margin. Colour white to grey, usually with 4–7 brown lines coinciding with slightly enlarged ribs; aperture and columella white, outer edge of aperture with brown lines, brown within, with indistinct pale basal band. Penis with 10–100 mamilliform glands; broad base, minute filament. Pallial oviduct of two consecutive loops of albumen gland followed by large spiral loop of capsule gland; straight section not enlarged or septate; bursa small, near anterior end of straight section. Spawn of large, pelagic, biconvex capsules with peripheral rim, containing 14–15 eggs; development planktotrophic. Distribution: northwestern Pacific from northern Sea of Japan to Bering Strait.

Anatomical description

Head-foot (Fig. 26N). Head and tentacles covered with dense transverse black lines; sides of foot with dense black lines or reticulation. Opercular ratio 0.48–0.62.

Penis (Fig. 26A–G). Base wrinkled, partly or entirely pigmented, distal flange on outer side; 10–100 mamilliform glands, in single row distally and dense mass 3–4 glands wide near base. Filament minute. Sperm groove open, ending terminally. Reduced or absent outside breeding season.

Sperm. Euspermatozoa 51–57 µm. Paraspermatozoa 9–14 µm diameter, round or slightly oval; vesicles large, distinct; nucleus not visible.

Oviduct (Fig. 26J–M). Two consecutive loops of albumen gland followed by large spiral loop of capsule gland with distal reddish translucent portion; straight section without enlarged or septate jelly gland. Bursa small, variable in position, but usually originating nearer anterior end of straight section. Seminal receptacle simple.

Spawn and development (Fig. 26H, I). Pelagic, symmetrically biconvex capsules with peripheral rim; diameter 870–880 µm, containing 14–15 eggs 95 µm in diameter (Kojima, 1958b, d). Development planktotrophic. Protoconch not well preserved in available specimens, but approximately 340 µm.

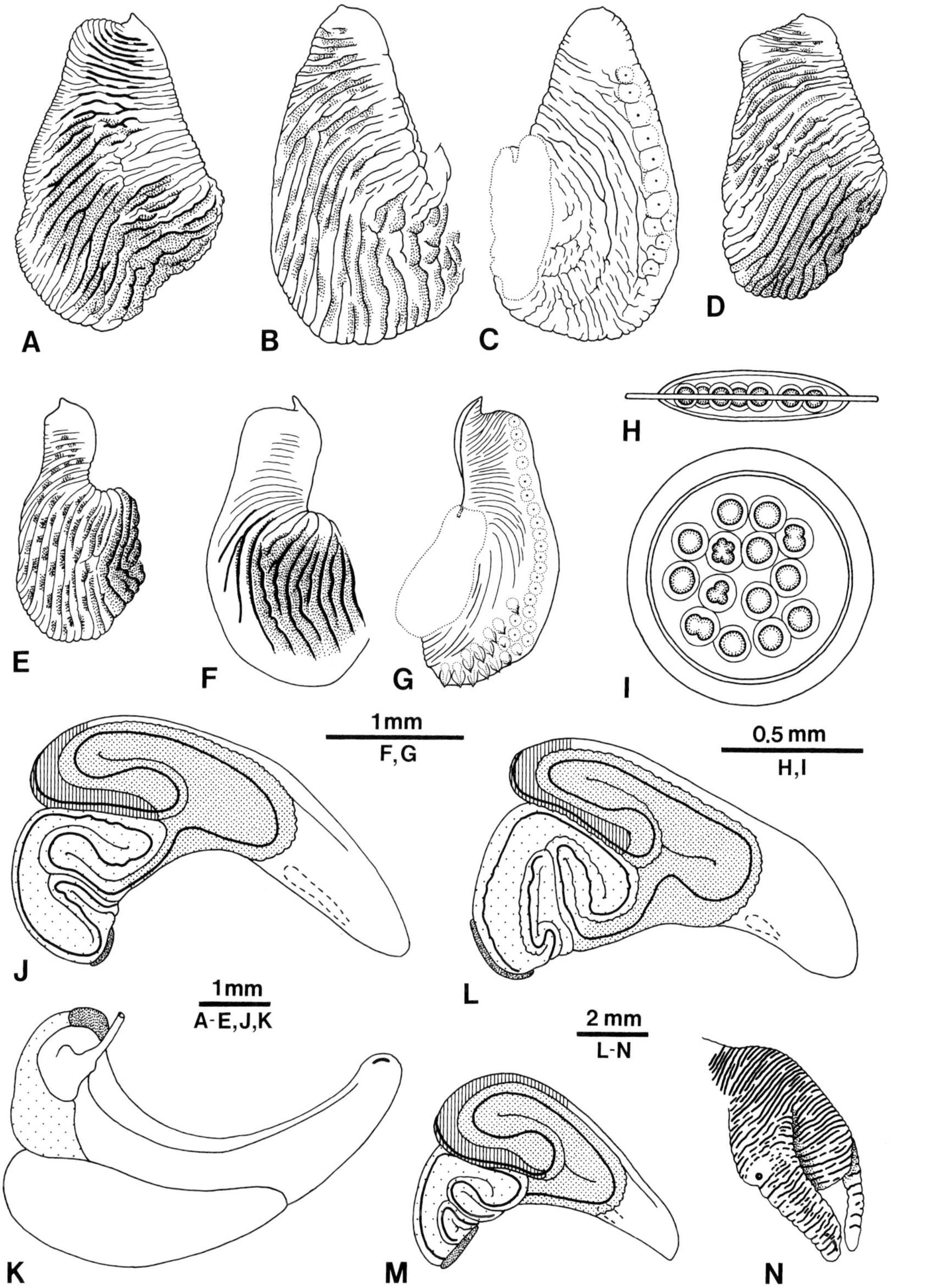

Figure 26. *Littorina (Littorina) squalida*: **A–G**, penes; **B**, **C** and **F**, **G**, two views of same penis; **H**, **I**, pelagic egg capsule (after Kojima, 1958b); **J–M**, pallial oviducts; **J**, **K**, lateral and medial views of same pallial oviduct; **N**, head pigmentation. **A**, Peter the Great Bay, Russia; **B–G**, **J**, **K**, **M**, **N**, Abashiri, Hokkaido; **H**, **I**, Akkeshi, Hokkaido; **L**, Ulbanskiy Bay, Okhotsk Sea, Russia. Shell heights: **A**, 17 mm; **B**, **C**, 16 mm; **D**, 15 mm; **E**, 11.8 mm; **F**, **G**, 9.8 mm; **J**, **K**, 16 mm; **L**, 32.6 mm; **M**, 19.7 mm; **N**, 16 mm. For shading conventions of pallial oviducts see p. 6.

Breeding season: not recorded; Kojima (1958b, d) obtained spawn in July at Akkeshi, Hokkaido; museum collections from Hokkaido, Peter the Great Bay and Sakhalin made in May, June and September contain mature individuals; those from the Okhotsk sea, Kamchatka and Sakhalin made in July and August are immature (pers. obs.).

Radula and foregut (Fig. 27C–H). Relative radular length 1.01–1.95. Rachidian: length/width 0.73–1.23; central cusp rounded, blunt or somewhat pointed; one pointed cusp and one outer denticle on each side. Lateral: four cusps, third from inside largest and bluntly rounded, others pointed. Inner marginal: four cusps, third from inside largest and bluntly rounded, others pointed. Outer marginal: 4–6 blunt or rounded cusps; narrow neck, projecting flange on inside of base only. Salivary glands enlarged.

Shell variation (Figs 27A, B; 28)

Geographical variation. Among *Littorina* species, this attains a size second only to *L. littorea*. The largest specimens (over 30 mm in height) that have been seen are all from northern Hokkaido, the Okhotsk Sea and Kamchatka, which may be a reflection of the high primary productivity of this region (Vermeij, 1989).

Ecotypic variation. In the relatively unfavourable conditions of lagoons, subject to freshwater influence, temperature fluctuation and periodic emersion, the shells are much smaller, because they grow more slowly (Golikov, 1959), at least after reaching maturity (Golikov & Kusakin, 1962). In Busse Lagoon, in southern Sakhalin, adults are from 8 to 13 mm in height, and the spire may be a little taller than is usual (Fig. 28H), reminiscent of the dwarf brackish-water form of *L. littorea* (p. 106).

Sculptural variation. Shells range from rather smooth with incised spiral lines (Fig. 28H) to strongly sculptured with numerous narrow but prominent ribs of unequal size (Fig. 28J); in a few cases the larger ribs may become raised, giving a somewhat carinate effect (Fig. 28D, E). This variability can also be seen in juvenile shells (Fig. 27A, B). Frequently, the largest shells are almost smooth, as a result of erosion (Fig. 28A, I). Some large shells bear conspicuous axial growth interruptions; if, as is likely, these are formed annually, the 40 mm shell in Figure 28I was 14 years old.

Sexual dimorphism. This is often pronounced in *L. squalida*, males (Fig. 28D, E) being smaller, with a more inflated final whorl, relatively larger aperture and lower spire than females (Fig. 28C, F).

Colour variation. Many populations are not obviously polymorphic. However, in a lagoon in southern Sakhalin, three discrete colour classes have been recognized: unbanded, pale banded and dark banded, and their relative proportions suggest that colour is determined by a two-allele system with incomplete dominance. The frequency of the pale allele among adults is lower in the middle and upper littoral than lower on the shore, but is uniform in juveniles, because the phenotypes differ in their tolerance of environmental factors (Altukhov & Kalabushkin, 1974; Kalabushkin & Altukhov, 1974; Kalabushkin, 1976; Sergievsky, 1989).

Distribution

Habitat. Typically, *L. squalida* is a low-shore and sublittoral species, which can occur on a range of substrates, usually rock, but also including pebbles, gravel, sand, macroalgae and seagrasses. On rocky shores in the warmer, more southerly parts of the range, such as southern Primor'e and the southern Kurile Islands, it occurs in the mid and low eulittoral (in the *Chthamalus* zone and among *Fucus*, *Enteromorpha*, *Pelvetia* and *Gigartina*) and in the sublittoral to a depth of 3–4 m (among *Scytosiphon*, *Punctaria*, *Laminaria* and *Phyllospadix*) (Mokyevsky, 1960; Golikov & Kusakin, 1962, 1978; Golikov & Gulbin, 1978; Volova, Golikov & Kusakin, 1979). On these shores it is often one of the dominant species in the eulittoral, reaching densities of tens per m^2 (Golikov & Kusakin, 1978) and up to 1140 per m^2 (Golikov & Skarlato, 1967), while in the sublittoral up to 15 per m^2 have been recorded (Golikov & Kusakin, 1978). In Hokkaido, Habe (1958a) noted it to be common at low tide at Akkeshi, and at Abashiri it was common in low-tide pools on large algae and coralline turf (pers. obs.). Golikov

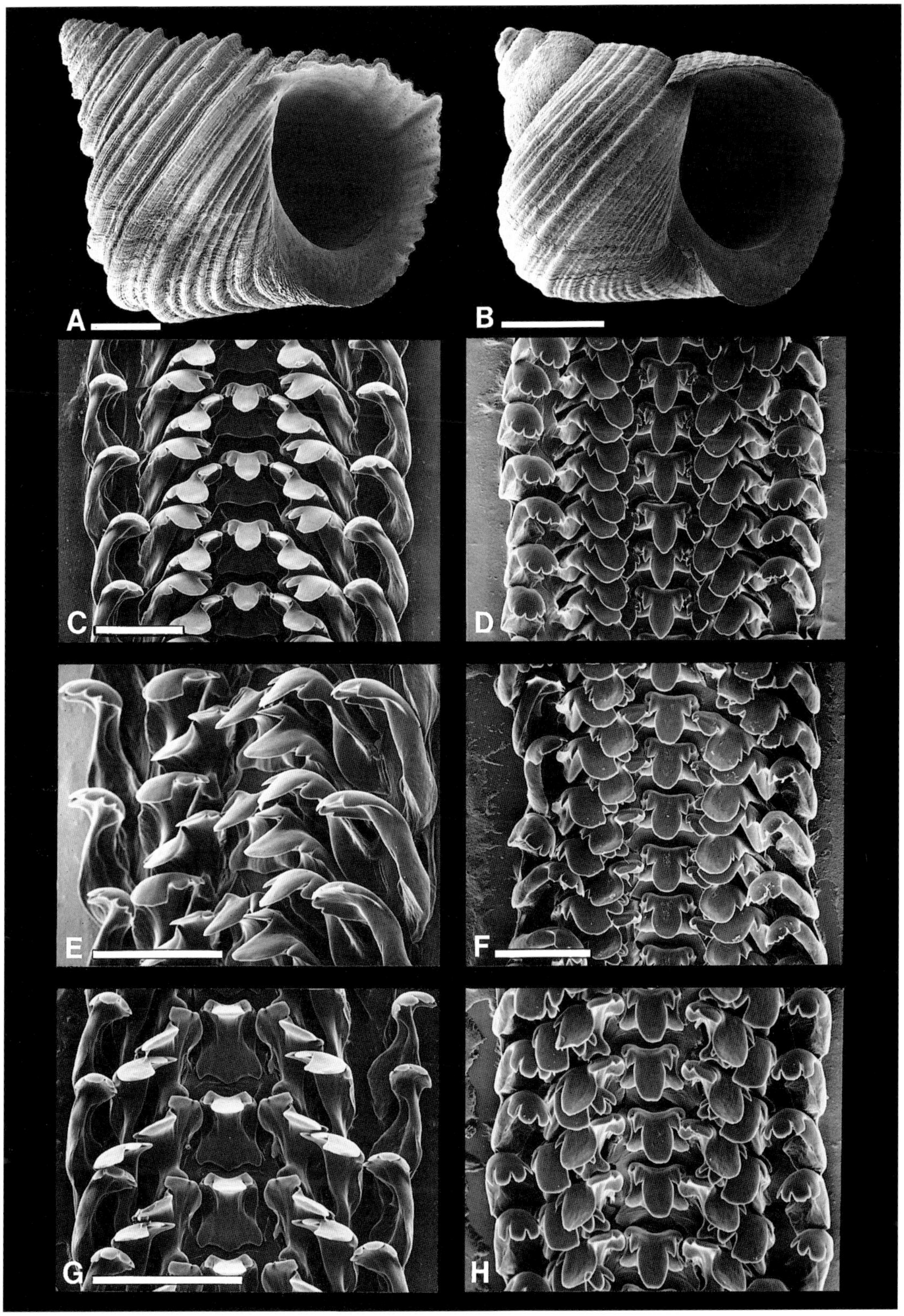

Figure 27. *Littorina (Littorina) squalida*: **A, B,** juvenile shells; **C–E,** three views of radula (shell H=11.8 mm), flat (**C**), at 45° (**D**), side view (**E**); **F,** radula (shell H=11.7 mm), flat; **G, H,** two views of radula of juvenile shell (shell H=7.9 mm), flat (**G**), at 45° (**H**). **A,** Pos'et Bay, Russia; **B,** Karaginskiy Bay, NE Kamchatka, Russia; **C–E,** Abashiri, Hokkaido; **F,** Kronotskiy Bay, NE Kamchatka, Russia; **G,** Ulbanskiy Bay, Okhotsk Sea, Russia. Scale bars: **A**=1 mm; **B**=1 mm; **C, D**=100 µm; **E**=100 µm; **F**=100 µm; **G, H**=100 µm.

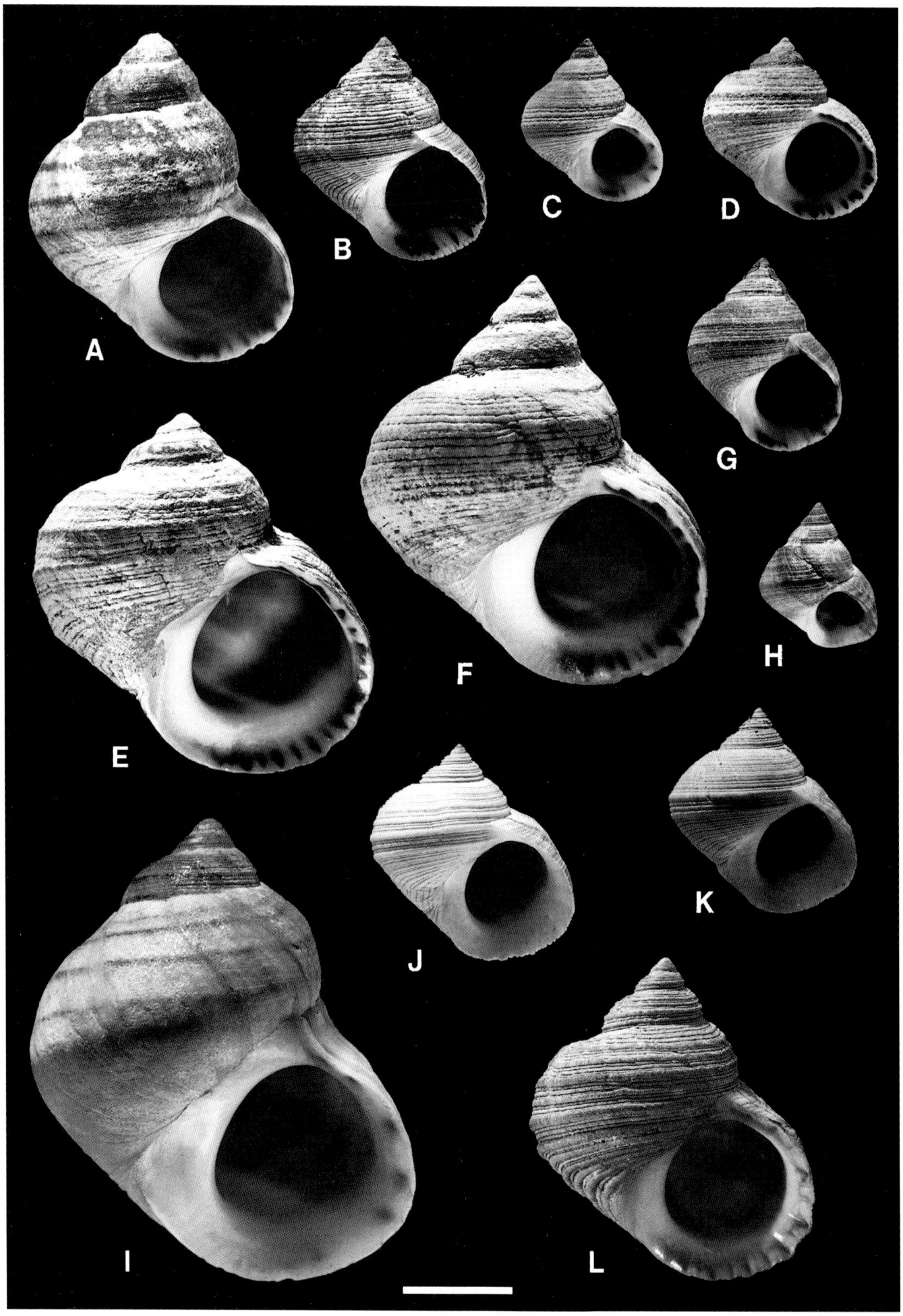

Figure 28. *Littorina (Littorina) squalida.* **A,** *Littorina squalida* Broderip & Sowerby, 1829; neotype, BMNH 1858.5.12.216; unlocalized. **B,** Ohmu, Hokkaido. **C, D,** Abashiri, Hokkaido. **E, F,** Simanova, Sakhalin Island, Russia. **G,** Peter the Great Bay, Russia. **H,** Busse Lagoon, Aniva Gulf, Sakhalin Island, Russia. **I,** *Littorina grandis* Middendorff, 1848; syntype, BMNH 1992162; Shantar Islands, Okhotsk Sea, Russia. **J,** *Littorina grandis* Middendorff, 1848; syntype, BMNH 1992161; Okhotsk Sea. **K,** Ulbanskiy Bay, Okhotsk Sea, Russia. **L,** Japan. Sexual dimorphism: **C, F,** females; **D, E,** males. Scale bar=10 mm.

& Kusakin (1962) reported that *L. squalida* remains on the shore all year round in the southern Kurile Islands, but only in pools and damp crevices, and does not form clusters in winter. Kojima (1960) mentioned a migration from low to near high-tide level during the breeding season in Hokkaido. The species can be found under a range of wave action, from surf-washed shores to protected bays (Golikov & Kusakin, 1962, 1978). It has often been recorded in sheltered, shallow lagoons, for example on the leaves of the seagrass *Zostera* at a density of 522 per m^2 at 1 m depth in Busse Lagoon in southern Sakhalin (Golikov & Skarlato, 1985; see also Golikov & Kusakin, 1962; Rowland, 1973, quoted in Foster, 1981); shells in this habitat are smaller, as described earlier (p. 84). These lagoons may be somewhat brackish, and the species is relatively tolerant of lowered salinity (Golikov, 1959; V. Ya. Berger, 1973, 1978b; Berger & Kovaleva, 1974). There is less information for the more northerly parts of the range but, where the species occurs on the shore at all, it is in the low eulittoral, among *Laminaria* and *Balanus* (Spassky, 1961). Here it is probably mainly sublittoral, maximum depths of 10–15 m having been recorded in southern Kamchatka (Golikov & Gulbin, 1978) and 15–20 m in other unspecified areas (Golikov & Kusakin, 1978).

Littorina squalida is sympatric with several congeners throughout most of its range, but is always the species occupying the lowest shore level. The relative zonation of *L. sitkana, L. brevicula, L. mandshurica* and *L. squalida* in the northwestern Sea of Japan and Tatar Strait has been described by Mokyevsky (1960). The diet of *L. squalida* has not been recorded.

Range (Fig. 29). *Littorina squalida* has a wide distribution across the northern Pacific, from the northern Sea of Japan to Hokkaido, the Okhotsk Sea, Kamchatka, the northern Bering Sea and Alaska. It is abundant in Pos'et Bay in southern Primor'e (Golikov & Skarlato, 1967), but no specimens or records have been traced from further west. Literature records are available for several localities on the Russian coast of the northern Sea of Japan (Mokyevsky, 1960), although it is reported to be only occasional in the Tatar Strait (Golikov & Skarlato, 1985). In Japan there are no records for the west coast (also Kuroda & Habe, 1952), but it occurs from Wakkanai around the northeastern and southern coasts of Hokkaido, and the southernmost record is from Iwate Prefecture in northeastern Honshu (NSMT). It is absent from most of the Kurile chain, occurring only on the Okhotsk Sea coast of the southern island of Iturup, and on Kunashir, Shikotan and Yuriy nearer Hokkaido (Golikov & Kusakin, 1962; Golikov & Gulbin, 1978). Specimens have been seen from the western and northern Okhotsk Sea, and the eastern side of Kamchatka as far north as Olyutorskiy Gulf. Much of the remaining Russian coast of the Bering Sea has been poorly collected, but the apparent absence of the species from the Chukotskiy Peninsula (Kusakin & Ivanova, 1978) is noteworthy. There are several specimens from Bering Island in the Komandor group (USNM, ZISP) but, although the range has been said to include the Aleutian and Pribiloff Islands (Dall, 1921; Golikov & Kusakin, 1978; Volova *et al.*, 1979; Reid, 1990c), only two beachworn specimens have been seen from this well-collected region, with the (therefore possibly unreliable) locality Nazan Bay, Atka Island (UWP). There are no modern records from the southern Bering Sea, and a recent expedition to the Aleutian Islands failed to find it (Vermeij, Palmer & Lindberg, 1990). In Alaska there are records from Norton Sound and the Seward Peninsula, as far north as Cape Prince of Wales on the Bering Strait (USNM). The localities Sitka and Kodiak Island (ZMB) are considered to be erroneous. There is some uncertainty about its occurrence north of the Bering Strait. A range limit of 72°N given by Kuroda & Habe (1952) is perhaps a typographical error. However, Gray (1939), describing material from Beechey's voyage, recorded the locality Icy Cape, which at 70°20'N is far north of the Bering Strait, on the northern Alaskan coast. No labelled specimen exists to corroborate the record, but the *Blossom* did indeed visit Icy Cape (Beechey, 1831), and this has here been selected as the type locality. It has been said that because it is an 'intertidal to shallow subtidal gastropod that cannot tolerate thick accumulations of sea ice, which can occur virtually year round along the northern Alaskan coastline', *L. squalida* does not now occur north of the Bering Strait (Repenning, Brouwers, Carter, Marincovich & Ager, 1987). Nevertheless, in view of its recorded occurrences to depths of 15–20 m in Russian waters (Golikov & Kusakin, 1978), it does seem possible that populations could survive in the presence of ice cover, so that Gray's (1939) record may in fact be reliable. All records from the Southern Ocean are in error (see Taxonomic History, p. 81).

With the exceptions of the records from Bering Island and the southernmost Kurile Islands, all the occurrences of *L. squalida* are on continental shores, so that this species displays a typically continental type of distribution.

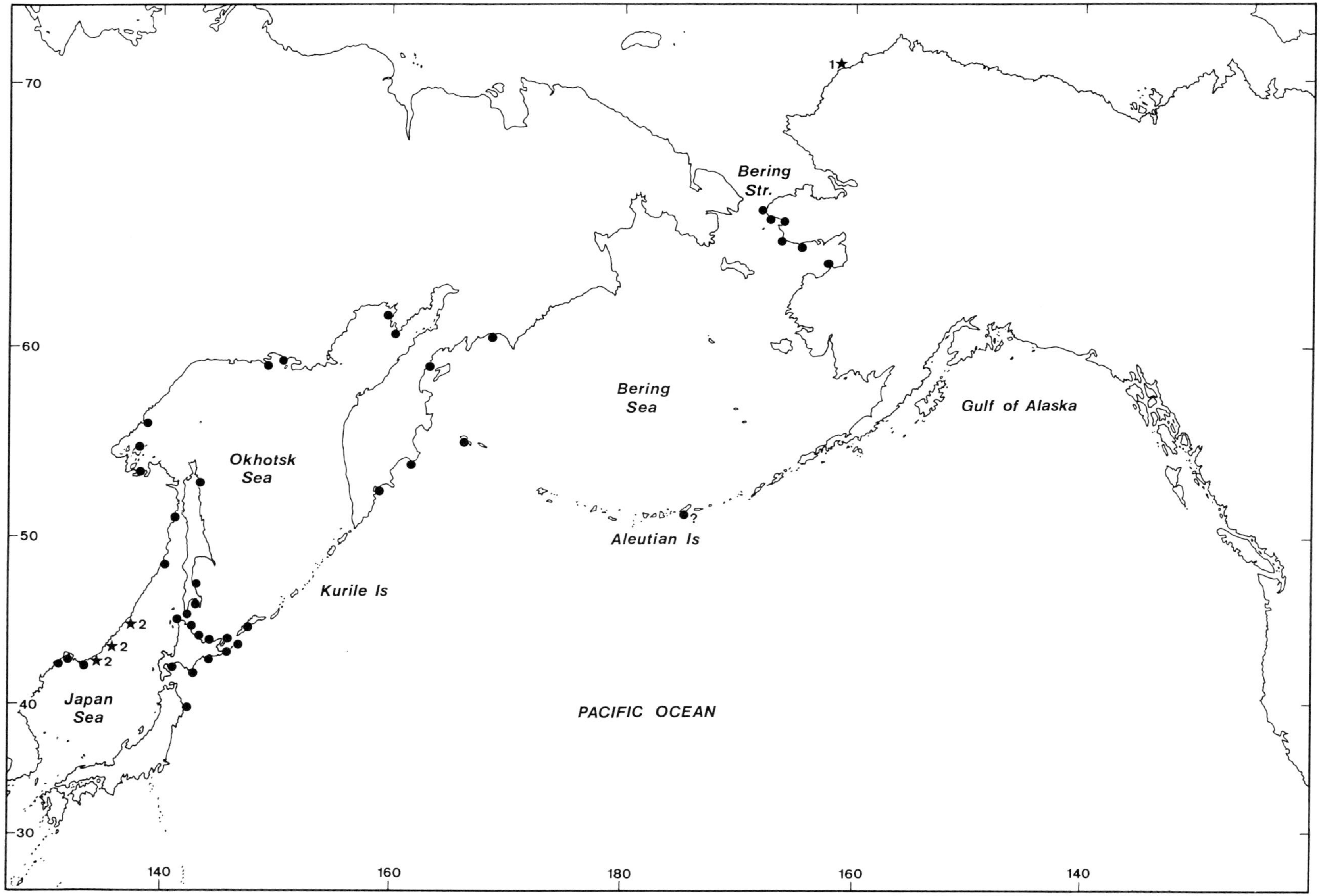

Figure 29. *Littorina (Littorina) squalida*: distribution map. Key: circles, specimens examined (record from Atka Island is the only one from the Aleutian Islands, and therefore of doubtful provenance); asterisks, records from literature (1, Gray, 1839; 2, Mokyevsky, 1960).

Fossil record (Fig. 30). *Littorina squalida* has been recorded from the Etolonian and Kakertian stages of the Middle Miocene of Kamchatka (Ilina, 1939; Gladenkov *et al.*, 1984; Gladenkov & Sinelnikova, 1990; see Taxonomic History, p. 82), dated at 10–12 and 12–15 Ma respectively (Gladenkov & Shanster, 1990). Two new species have recently been based upon this material (Gladenkov & Sinelnikova, 1990). The holotype of *L. etolonensis* (Fig. 30A) is well preserved, with a marked shoulder and a peripheral keel on the last whorl; this shape is unusual in modern *L. squalida*, but is sometimes encountered (e.g. Fig. 28H). Its sculpture of fine spiral riblets is typical of *L. squalida*, and the reticulate effect of the intersection of growth lines and riblets, mentioned in the original description (Gladenkov & Sinelnikova, 1990), is apparent in neither the holotype nor in the two paratypes that have been seen. *Littorina etolonensis* therefore falls within the morphological range of modern *L. squalida*. Neither the holotype (Fig. 30G) nor paratype (Fig. 30E) of *L. praesqualida* show either the broad spiral ribs or irregular tubercles mentioned by Sinelnikova (Gladenkov & Sinelnikova, 1990), but instead display fine, close riblets. In the holotype the aperture is elongate, unlike any Recent *L. squalida*, but this is perhaps an abnormality caused by a succession of breakages just behind the lip, because the penultimate whorl is more rounded, as is the last whorl of the paratype. This species is therefore also included in the synonymy of *L. squalida*.

Littorina squalida has been recognized as one of the Pacific species which took part in the trans-Arctic migration which occurred following the opening of the Bering Strait about 3.5 to 4 million years ago (Golikov & Tzvetkova, 1972; Reid, 1990c; Vermeij, 1991). *Littorina littorea* is generally believed to be the descendant of *L. squalida* in the Atlantic. As discussed earlier (p. 87), the modern occurrence of *L. squalida* north of the Bering Strait is uncertain, but there is fossil evidence for its northward extension in the past. On the northern Alaskan coastal plain *L. squalida* is abundant in deposits from the Bigbendian and Fishcreekian transgressions, which have been dated as late Pliocene, 2.4 to 3 Ma (Carter, Brigham-Grette, Marincovich, Pease & Hillhouse, 1986; Repenning *et al.*, 1987). These shells (Fig. 30F) closely resemble modern examples, and their occurrence so far north has been taken as evidence of a milder climate, without the presence of the almost perennial sea ice that now occurs in the region.

Littorina squalida has not hitherto been recorded from deposits in the Atlantic, fossils from this region generally having been identified as *L. littorea* (e.g. Gladenkov *et al.*, 1980). However, modern shells of these two species are quite distinctive, and on the basis of shell resemblance, rather than biogeographic convenience, the earliest fossils in the Atlantic must be classified as *L. squalida*. These fossils are from Bárdarson's (1925) horizon 11 near the top of the *Mactra* zone of the Tjörnes sequence in northern Iceland (Fig. 30B–D). (Other records from higher in the sequence are here referred to a new species of *Littorina*, p. 246). There is disagreement about the precise palaeontological correlation of the Tjörnes sequence (Gladenkov *et al.*, 1980), but both palaeomagnetic (Einarsson, Hopkins & Doell, 1967) and radiometric techniques (review by Gladenkov, 1981) suggest an age of 3 to 3.5 Ma for the top of the *Mactra* zone. Specimens from Iceland (Fig. 30B–D) are entirely comparable in shape both with fossils from northern Alaska (Fig. 30F) and with modern shells from the Pacific (Fig. 28), and in some the fine sculpture is well preserved (Fig. 30D).

Identification of shells in the slightly later Red Crag Formation of Britain is less clear-cut, perhaps demonstrating the anagenetic origin of *L. littorea* from *L. squalida*. The majority of these shells appear to be typical *L. littorea* (Fig. 37C), but a few do somewhat resemble *L. squalida* (Fig. 37D). These are, however, all included with *L. littorea* (p. 118).

A more recent fossil record in the Pacific is that of MacNeil, Mertie & Pilsbry (1943) from deposits of the Pelukian transgression at Nome, Alaska, dated at 100 000 BP (Hopkins, 1967a).

Discussion

Intraspecific variation. In common with other *Littorina* species with pelagic eggs and planktotrophic larvae, *L. squalida* shows a relatively low degree of variation in shell characters. Such differences as there are may be explained at least in part by ecophenotypic effects; large size in parts of the range may be connected with environmental productivity, and dwarf, tall-spired shells occur in lagoonal situations. Genetic variation on a large scale has not yet been studied in this species, but some differentiation within a lagoon was found by Kalabushkin, Zhivotovskii & Afanas'ev (1985). Interestingly, there is a report of a relatively low level of genetic variation in this species (mean expected heterozygosity per locus=0.068, Zaslavskaya, 1989; Tatarenkov, 1992); a similar finding in

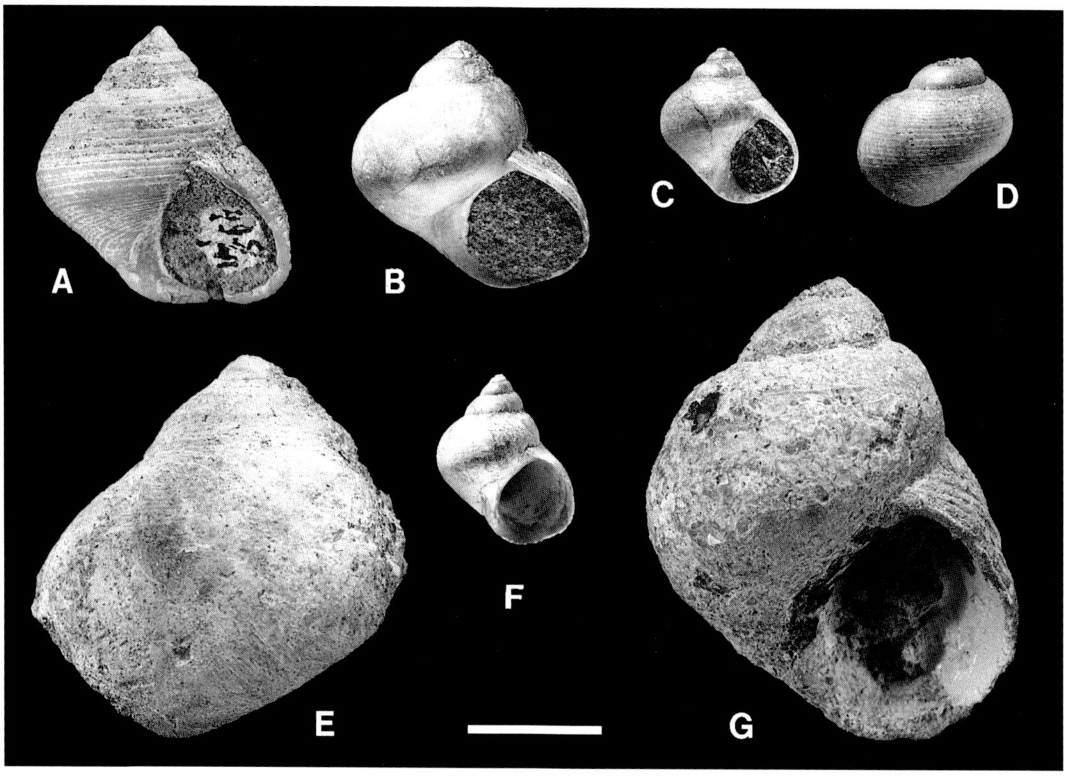

Figure 30. *Littorina (Littorina) squalida*: fossils. **A,** *Littorina etolonensis* Sinelnikova, in Gladenkov & Sinelnikova, 1990; holotype, PMM N495/41; W coast Pendginsk Bay, between mouth of Pustaya River and Cape Bolshoi, W Kamchatka, Russia; Etolonian stage, Middle Miocene. **B,** Husavik, Iceland; Pliocene; BMNH G1700. **C, D,** Tjörnes, Iceland; Pliocene, Bárdarson (1925) horizon 11; IMNH 3976. **E,** *Littorina praesqualida* Sinelnikova, in Gladenkov & Sinelnikova, 1990; paratype, BMNH GG22522; Cape Bolshoi, W Kamchatka, Russia; Etolonian stage, Middle Miocene. **F,** Fish Creek, N Alaska; Gubik Formation, Pliocene. **G,** *Littorina praesqualida* Sinelnikova, in Gladenkov & Sinelnikova, 1990; holotype, PMM N182/84; between mouths of Kheisheveem and Keicheveem Rivers, Kavrano-Ukholoksk Gulf, W Kamchatka; Etolonian stage, Middle Miocene. Scale bar=10 mm.

L. littorea has been explained as the result of a severe genetic bottleneck in the past (Johannesson, 1992).

Phylogenetic relationships. Among Recent species, there is no doubt that the closest relative of *L. squalida* is *L. littorea*, as first suggested by the obvious resemblance between their shells (Rosewater, 1963; Golikov & Tzvetkova, 1972), with characteristically fine spiral ribs not seen in other living species. There are anatomical similarities too, but these are mostly plesiomorphic or subjective (e.g. similarity of head pigmentation), so that cladistic analysis provides no unequivocal support for this sister-species relationship (Reid, 1990b; p. 351). There are in fact some significant anatomical differences, as listed in Table 3. Nevertheless, phylogenetic analyses of allozyme frequencies (Zaslavskaya, Sergievsky & Tatarenkov, 1992a, b; p. 356) and of mitochondrial DNA sequences (Reid, Rumbak & Thomas, 1996; p. 360) do provide strong evidence that these two are indeed sister taxa.

As discussed earlier (p. 89), shells that are here identified as *L. squalida* occurred in Iceland 3 to 3.5 million years ago, shortly after the opening of the Bering Strait at 3.5–4 Ma. *Littorina squalida* therefore reached Iceland rather soon after the opening of the Strait, and indeed the main influx of Pacific species is not recorded until the *Serripes* zone, overlying the *Mactra* zone in the Tjörnes beds (Einarsson *et al.*, 1967; note that their suggestion of an earlier, late Miocene, migration from the Pacific has been discredited, Vermeij, 1991). At this time the species may have had a continuous distribution across the Arctic coast of North America, connecting the Atlantic and Pacific populations, since the Arctic climate was then warmer than at present (Herman & Hopkins, 1980; Carter *et al.*, 1986). This is supported by the occurrence of *L. squalida* on the north coast of Alaska about 2.4 to 3 Ma (Repenning *et al.*, 1987). One possible interpretation of the phylogenetic and fossil evidence is that vicariant speciation occurred (p. 378), the Atlantic isolate giving rise to *L. littorea*. Vicariance may have occurred as a result of climatic cooling, at some time between the onset of widespread

Table 3. Comparison of *Littorina squalida* and *L. littorea*.

Character	*L. squalida*	*L. littorea*	% accuracy	References
Penis	Fig. 26A–G	Fig. 33A–F	100	This study
Filament	Minute	30–50% total length, tapering or rounded at tip		
Mamilliform glands	10–100	10–42		
Pallial oviduct	Fig. 26J–M	Fig. 33G–O	100	This study
Copulatory bursa	Small	Large		
Spawn	Fig. 26H, I	Fig. 33Q–S	100	Caullery & Pelseneer, 1910; Tattersall, 1920; Linke, 1933; Lebour, 1935,1937; Kojima, 1958b, d
Shape	Symmetrically biconvex	Asymmetrically biconvex		
Number of ova	14–15	1–9		
Radula	Fig. 27C–H	Fig. 34		This study
Cusps of 5 central teeth	Rounded to blunt	Pointed		
Shell	Figs 27A, B; 28, 30	Figs 35–37	99	This study
Shape	Slightly broader (H/B=1.02–1.31); sutures impressed; last whorl swollen	Slightly narrower (H/B=1.11–1.48); sutures less deep; last whorl less swollen		
Sexual dimorphism	Pronounced	Not usually evident		
Sculpture	Numerous fine, usually raised, ribs of irregular size	Numerous fine, usually flat, ribs of regular size		
Recent distribution	Fig. 30. Northern Pacific	Fig. 38. Northern Atlantic	100	This study

glaciation at about 2.4 Ma (Shackleton *et al.*, 1984; Loubere, 1988) and the appearance of perennial sea ice in the Arctic Ocean at about 0.7 Ma (Herman & Hopkins, 1980). Periodic emergence of the Bering Land Bridge (Hopkins, 1967a) may also have contributed.

This scenario implies a relatively recent separation of *L. littorea* from *L. squalida*. If this is the case, the extent of the morphological differences between them (Table 3) is surprising. Even the penial shape and the size of the copulatory bursa, which may be species recognition characters or isolating mechanisms (pp. 14, 18), are conspicuously different, which is unexpected if they have differentiated in complete isolation. The evidence from the analysis of mitochondrial DNA sequences also casts some doubt on the above interpretation of the speciation event, for the differences between these two species are often greater than expected for a separation of this age; estimates of the age of their divergence (assuming a constant rate of sequence divergence across the genus) range between 3.28 and 10.07 Ma (mean 6.42 Ma) (Reid *et al.*, 1996; p. 365). The molecular evidence implies either a faster rate of divergence in the lineage of *L. squalida* and *L. littorea*, or a separation between them that preceded the opening of the Bering Strait and migration into the Atlantic. If the latter is the case, the apparent occurrence of *L. squalida* in Iceland suggests either that the two species co-occurred in the Atlantic during the Upper Pliocene, or that these shells were of *L. littorea*, still at that stage morphologically similar to *L. squalida*. Further fossil material from both oceans will be required to investigate these possibilities. From their genetic distance, calculated from allozyme frequencies, Zaslavskaya *et al.* (1992a) estimated the age of separation of *L. squalida* and *L. littorea* as 3.45 Ma, which is more in keeping with allopatric speciation following trans-Arctic migration.

One extinct species also has a bearing on this question. *Littorina petricola* is known only from the Pliocene of California and Oregon. Its fine spiral ribs and apertural form support a close relationship with *L. squalida*, although the differences in shape of the whorls, spire height, size and likely environmental temperature are here regarded as sufficient to warrant its specific status (p. 95). Subjective assessment of the resemblances between the three species, as well as their geographical distribution, suggests that *L. petricola* was derived from *L. squalida* and was not involved in either the trans-Arctic migration or the origin of *L. littorea*. However, fossil *Littorina* species are scarce, and this hypothesis may require re-evaluation if further material becomes available. The identity of *L. kozaiensis*, described from a single specimen from the Middle Miocene of Japan, is uncertain, but it may also be related to the *L. squalida* species-group (p. 344).

Similar species. *Littorina squalida* is distinguished from all sympatric *Littorina* species by its numerous, fine, spiral ribs separated by narrow incised lines, quite different from the coarse ribs of *L. brevicula*, the two to three low ribs or angulations of *L. mandshurica*, and the smooth or carinate shell of *L. sitkana*. Small, worn examples could perhaps be confused with *L. mandshurica*, but in that species the spire is lower and the maximum size only 21 mm.

The only really similar Recent species is *L. littorea* from the Atlantic. Differences are summarized in Table 3; the shape of the penis, size of the copulatory bursa, and probably the spawn capsule (only once described for *L. squalida*) are all diagnostic, and there are also differences in the radular teeth. Shells can almost always be readily separated; in *L. squalida* the whorls are more swollen, the sutures more distinct, the spire straight-sided to slightly concave, lacking the onion-shaped profile of *L. littorea*, and the final whorl is a little enlarged, with a round aperture. The spiral ribs of *L. squalida* are usually more prominently raised and less uniform in size than in *L. littorea*. A very few shells of *L. squalida* from lagoonal localities (e.g. Fig. 28H) could be confused with unusual forms of *L. littorea*. Interestingly, sexual dimorphism is obvious only in *L. squalida*.

The fossil *L. petricola* is similar but smaller, the whorls less rounded, the last whorl not swollen, and the spire often taller; it is restricted to the Pliocene of California and Oregon (see below).

Littorina (Littorina) petricola **Arnold, 1908**

(Fig. 31)

Synonymy

Litorina planaxis—Arnold, 1906: 30 (see Addicott, 1969; not Philippi, 1847=*Littorina keenae*).
Littorina sp. Arnold, 1908: 355.
Littorina petricola 'Dall' Arnold, 1908: pl. 37, fig. 7 and caption (Merced formation, upper Pliocene, locality no. 21, between Felt Lake and Los Trancos Creek, $2\frac{1}{2}$ miles south-southwest of Stanford University, California; lectotype, here designated, CAS Geology 61995.01, seen, Fig. 31J; not *Turbo petricola* Leach, 1852=*Melarhaphe neritoides*). Branner, Newsom & Arnold, 1909: fig. 78. Keen & Bentson, 1944: 167.
Littorina petricola Dall, 1909: 80; pl. 4, fig. 9 (Pleistocene of Fossil Rock, Coos Bay, Oregon [Coos conglomerate, Pliocene; Howe, 1922]; lectotype, here designated, USNM 153991, seen, Fig. 31D; 6 paralectotypes, USNM 153923, seen, Fig. 31K). Howe, 1922: 93. Grant & Gale, 1931: 781. Weaver, 1942: 329; pl. 67, fig. 20. Addicott, 1969a: 59, 62, 64, 65; pl. 1, figs 2–5. Addicott, 1976: 111; pl. 5, fig. 20. Vermeij, 1989: 355.
Littorina cf. *L. petricola*—Woodring & Bramlette, 1951: 70; pl. 19, fig. 12.
Littorina mariana Arnold, 1910: 86–87; pl. 29, fig. 1 (United States Geological Survey loc. 4718, main ridge of Kettleman Hills, central part, on summit of 1,020-foot hill, 8 miles northwest of Light's place, east side of sec. 11, T 23 S, R 18 E, Coalinga district, Kings Co., California, Upper *Mya* horizon at top of Etchegoin formation, top of Upper Miocene [San Joaquin formation, Upper Pliocene; Woodring, Stewart & Richards, 1940]; holotype USNM 165481, seen, Fig. 31A). Arnold & Anderson, 1910: 137; pl. 51, fig. 1. Nomlund, 1917: 221. Grant & Gale, 1931: 781. Woodring, Stewart & Richards, 1940: 84–85; pl. 4, figs 4–6; pl. 8, figs 1–6; pl. 29, figs 8, 9. Keen & Bentson, 1944: 167. Adegoke, 1969: 160.
Littorina mariana var. *alta* Arnold, 1910: 87; pl. 29, fig. 2 (United States Geological Survey loc. 4730, in central part of Kettleman Hills, $9\frac{3}{4}$ miles northwest of Light's place, centre of sec. 3, T 23 S, R 18 E, Coalinga district, Kings Co., California, Upper *Mya* horizon at top of Etchegoin formation, top of Upper Miocene [San Joaquin formation, Upper Pliocene; Woodring *et al.*, 1940]; holotype USNM 165487, seen, Fig. 31B). Arnold & Anderson, 1910: 137; pl. 51, fig. 2. Grant & Gale, 1931: 781. Keen & Bentson, 1944: 167.
Littorina mariana alta—Nomlund, 1917: 221.
Littorina remondi—Arnold & Hannibal, 1913: 594 (not *L. remondii* Gabb, 1866).

Taxonomic history

The formal description of *L. petricola* was first published by Dall (1909), but unfortunately the name was used for a figure published by Arnold (1908) in the previous year. Although the name was there credited to Dall, it was Arnold who first satisfied the criteria of publication of the new name, and he must therefore be cited as its author. The original specimen of Arnold's figure is here designated as the lectotype of *L. petricola* Arnold, 1908, while *L. petricola* Dall, 1909, becomes a homonym and subjective synonym, with its own type specimens.

In 1910 Arnold described *L. mariana* and its var. *alta*, recognizing its similarity to *L. petricola*, but noting that the new species was smaller, thinner in texture, more slender and with less prominent sculpture. As discussed below, this is considered to be within the likely range of intraspecific variation of *L. petricola*.

There has been some confusion with another Californian fossil, *L. remondii* Gabb, 1866, from the Upper Miocene. Arnold & Hannibal (1913) used that name for specimens from the type locality of Dall's *L. petricola* in Oregon, while Woodring *et al.* (1940) suggested that *L. remondii* might prove to be an earlier name for *L. mariana*.

Diagnosis

Shell height up to 21 mm. Shell turbinate to high-turbinate; sutures only moderately impressed, spire whorls rather flat-sided; last whorl rounded, with slight peripheral keel in juveniles. Mineralogy of external calcite and internal aragonite layers. Sculpture of 19–36 fine spiral ribs on last whorl, with incised lines between them. Aperture round, sometimes lirate at margin; columellar pillar uniformly curving around aperture. Distribution: Pliocene of California and Oregon.

Shell variation (Fig. 31)

Twenty-six specimens have been seen from Dall's type locality at Coos Bay, Oregon. These are solid, turbinate shells (SH=1.12 to estimated 1.4), often of large size for the species, and if well-preserved show conspicuous but narrow spiral ribs (Fig. 31D, K). Similar shells occur in Santa Barbara County (Woodring & Bramlette, 1951) and in Santa Clara County (Addicott, 1969), both in central California. These shells are extremely similar to the Recent *L. squalida* (see Discussion, p. 95), and shell variation in *L. petricola* is interpreted in the light of variation in that species (p. 84). The specimens from Oregon are analogous to those of *L. squalida* from fully marine situations. Large examples of *L. petricola* are sometimes smooth, and have been considered a smooth form (Woodring & Bramlette, 1951; Addicott, 1969), but as in *L. squalida* are believed to be the result of erosion during life.

In comparison, the 66 specimens examined from localities in Kings County, California, are often less thick-walled, usually with taller spires (SH=1.12–1.66), attain a smaller maximum size (estimated 17.4 mm) and when well-preserved show less prominent, but equally numerous, ribs (Fig. 31A–C, E–I). Consequently, while authors have compared these shells with *L. petricola*, they have generally recognized them as a distinct species, *L. mariana* (Arnold, 1910; Woodring *et al.*, 1940; Woodring & Bramlette, 1951; Addicott, 1969). The most tall-spired shells were named *L. mariana* var. *alta* by Arnold (1910), but as illustrated by Woodring *et al.* (1940), large collections show intergradation between this extreme form and typical *L. mariana*. However, there is some overlap in the shapes of shells from Oregon and Kings County; for example, Figures 31E-I illustrate shells from a single locality in King's County, and at least one (Fig. 31H) is virtually indistinguishable from shells from Oregon. It is therefore considered that shells from these regions represent extremes of variation of a single species. This degree of intraspecific variation is only a little greater than that seen in the Recent analogue *L. squalida*, and comparable with that in another closely related Recent species, *L. littorea* (p. 103). In both these living species shells that are unusually small, thin-walled and tall-spired are found in brackish and lagoonal habitats. Both the Etchegoin and San Joaquin formations of the San Joaquin Basin, in which *L. mariana* is found, were laid down in a shallow, and probably brackish, inland sea (Adegoke, 1969), so the same pattern of ecotypic variation seen in the modern species may account for intraspecific variation in *L. petricola*.

Distribution

Habitat. The relative abundance of *L. petricola* in fossil assemblages characteristic of sheltered, depositional environments (Addicott, 1969a; Adegoke, 1969), suggests that it was able to inhabit sandy, muddy and lagoonal habitats. This is unusual among *Littorina* species, but is also the case in the closely related Recent species *L. squalida* and *L. littorea*, which are both likewise fairly frequent as fossils.

Fossil record. In Oregon *L. petricola* occurs in the Coos conglomerate (Pliocene), but not in the underlying Empire formation (Howe, 1922), and is said to be abundant (Dall, 1909). In Santa Clara

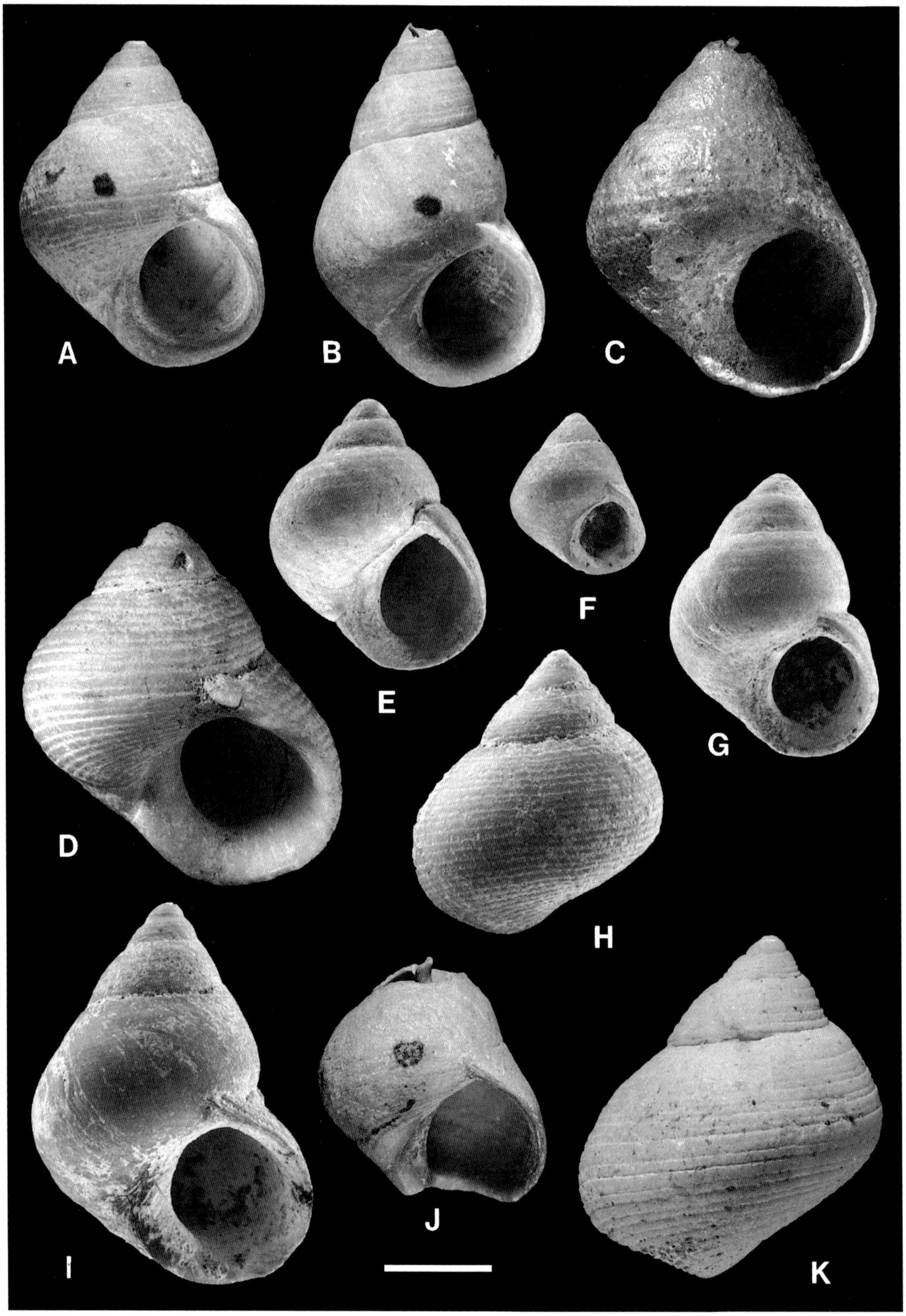

Figure 31. *Littorina (Littorina) petricola*. **A,** *Littorina mariana* Arnold, 1910; holotype, USNM 165481; Kettleman Hills, Kings Co., California; San Joaquin formation, Upper Pliocene. **B,** *Littorina mariana* var. *alta* Arnold, 1910; holotype, USNM 165487; Kettleman Hills, Kings Co., California; San Joaquin formation, Upper Pliocene. **C,** Bed F, North Dome, Kings Co., California (UCLA loc. 697); Pliocene. **D, K,** *Littorina petricola* Dall, 1909; lectotype (**D**), USNM 153991; paralectotype (**K**), USNM 153923; Fossil Rock, Coos Bay, Oregon; Coos conglomerate, Pliocene. **E–I,** La Cima Quad., Kings Co., California (UCMP loc. D-6369); *Littorina* zone, Etchegoin formation, Middle Pliocene. **J,** *Littorina petricola* Arnold, 1908; lectotype, CAS Geology 61995.01; 2.5 miles SSW of Stanford University, between Felt Lake and Los Trancos Creek, Santa Clara Co., California; Merced formation, Upper Pliocene. Scale bar = 5 mm.

County, California, it is found in the Merced formation (Upper Pliocene) (Arnold, 1908; Addicott, 1969a). In the San Joaquin Basin the form known as *L. mariana* is frequent in the Kettleman Hills of Kings County, where it is recorded from the *Littorina* zone and upper *Pseudocardium* zones of the Etchegoin formation (Middle Pliocene), upper *Mya* zone and *Pecten* zones of the San Joaquin formation (Upper Pliocene) and in the Tulare formation (Upper Pliocene and perhaps Pleistocene) (Woodring *et al.*, 1940). Further south, there is a record from the Graciosa member of the Careaga sandstone (Upper Pliocene) of Santa Barabara County (Woodring & Bramlette, 1951), and in Santa Clara County it is present in the Merced formation (Upper Pliocene) (Addicott, 1969a). Some small and damaged specimens from the Pliocene San Diego formation (LACM 323) also appear to be referrable to this species. Addicott (1976) reported this species in addition from the Wishkahan stage of the Upper Miocene in the northwestern United States, but gave no further details.

Discussion

Several authors have remarked upon the close relationship between this species and the Recent *L. squalida* (Dall, 1909; Arnold, 1910; Arnold & Anderson, 1910), and Vermeij (1989) has even suggested that the two might be conspecific. *Littorina squalida* has a fossil record extending back to the Middle Miocene of Kamchatka, and a Recent distribution across the northwestern Pacific from the northern Sea of Japan to the Bering Strait. Some examples of *L. petricola* are very similar to *L. squalida* (e.g. Fig. 31D, H, K) in outline, fine sculpture, and the sometimes lirate apertural margin. However, in the species as a whole there seem to be sufficient differences to support its separate status. It is not known to achieve the large size of *L. squalida* (up to 43.7 mm); the whorls are usually a little less rounded and the last whorl not unduly swollen; the tendency to produce tall-spired forms (suggested to have inhabited brackish or lagoonal environments) is more marked than in *L. squalida*; the sculptural ribs are of uniform size and prominence, without the tendency for some to become enlarged, which sometimes gives the whorl an angled profile in *L. squalida*. Finally, comparative faunistic studies suggest that *L. petricola* lived in climates similar to, or a little warmer than, those prevailing in the same geographical regions at present (Addicott, 1969a, 1970; Adegoke, 1969), and therefore considerably warmer than regions now inhabited by *L. squalida*. Since *L. squalida* has the older fossil record, available evidence suggests that *L. petricola* was derived from it as a relatively warm-adapted, northeastern Pacific offshoot. Whether this occurred by dispersal across the northern Pacific, or by vicariance of an extensive ancestral range, is not known. Vermeij (1989) has explained the extinction of *L. petricola* in the northeastern Pacific, and survival of the larger *L. squalida* in the northwestern Pacific (where seasonal climatic fluctuation is more extreme), by the higher productivity in the latter region.

From the fossil evidence at present available, it seems unlikely that *L. petricola* was involved in the origin of *L. littorea*; Pliocene fossils from northern Alaska and from Iceland resemble *L. squalida* rather than this species, so that there is no evidence that it was involved in the trans-Arctic migration. However, shells in this group are sufficiently variable that they may be an unreliable guide to phylogenetic affinities. The evidence from mitochondrial DNA sequences (p. 365) suggests that *L. squalida* and *L. littorea* may both have occurred in the Pacific before the opening of the Bering Strait, in which case a relationship between *L. petricola* and *L. littorea* cannot be dismissed on geographical grounds alone. Further fossil material is highly desirable to investigate this question.

Littorina (Littorina) littorea (Linnaeus, 1758)

(Figs 3D, 5, 6B, 32–38)

Synonymy

Turbo littoreus Linnaeus, 1758: 761 (O. Europaeo, frequens ad littora Norvegiae; here restricted to Marstrand, Sweden; types lost (Dance, 1967); lectotype (here designated) Linnaeus, 1747: pl. 5, fig. 4, Fig. 32 herein; see remarks under Taxonomic History, p. 99; official list of specific names in Zoology, no. 2714). Baster, 1761: 110–111; pl. 14, fig. 1. Linnaeus, 1761: 524–525. Linnaeus, 1767: 1232. Pennant, 1777: 128; pl. 81, fig. 109. da Costa, 1778: 98–100; pl. 6, fig. 1 (in part, includes *L. saxatilis s.l.*). Chemnitz, 1781: 230–233; pl. 185, fig. 1852, nos 1–8. Schröter, 1784: 5–7. Gmelin, 1791: 3588 (in part, includes refs to *L. saxatilis*). Röding, 1798: 88. Pulteney, 1799: 45. Donovan, 1800: pl. 33, fig. 1 (in part, other figs are *L. saxatilis*).

Montagu, 1803: 301–303 (in part, includes refs to *L. obtusata* and *L. saxatilis*). Link, 1807: 134. Maton & Rackett, 1807: 158–159; pl. 4, fig. 9 (in part, other figs are *L. saxatilis*). Pennant, 1812: 293–294; pl. 84, fig. 1. Pulteney, 1813: 49; pl. 17, fig. 1; pl. 19, fig. 3 (in part, other fig. is *L. saxatilis*). Sowerby, 1814: 163; pl. 71, fig. 1 (Pliocene). Dillwyn, 1817: 817 (in part, includes *L. saxatilis s.l.*). Lamarck, 1822: 47–48. Brown, 1827: pl. 46, figs 1–3 (in part, other figs probably *L. saxatilis*). Blainville, 1828: 106. Blainville, 1830: 298–299; pl. 12, figs 3, 3a, 4, 5 (in part, includes *L. saxatilis*). Brown, 1837a: 15; pl. 10, figs 1–3 (in part, other figs *L. saxatilis*). Thorpe, 1844: 165. Leach, 1852: 186–187 (in part, includes refs to *L. obtusata* and *L. saxatilis*). Hanley, 1855: 326–327.

Nerita littorea—Müller, 1776: 244.

[*Paludina (Littorina) littorea*]—Férussac, 1822: ix, xi, xxxiv (see Kadolsky, 1970; Melville, 1980; discussion p. 38).

Littorina littorea—Lyell, 1835: 34, 37 (fossil). Johnston, 1842: 267. Wood, 1842: 532 (in part, includes ref. to *L. squalida*; as *littoreus*; Pliocene). Deshayes & Milne Edwards, 1843: 199–208, 214. MacGillivray, 1843: 136–137. Brown, 1844a: 128; pl. 10, figs 1–3 (in part, other figs are *L. saxatilis*). Alder, 1848: 149. Wood, 1848: 118–119; pl. 10, figs 14, 14a–k (in part, includes refs to *L. squalida*; Pliocene). Middendorff, 1849: 382–383. Clark, 1850a: 361–362. Deshayes, 1850: 42; pl. 69, figs 3–5. Forbes & Hanley, 1850: 29–32, 52–53; pl. 83, figs 8, 9 (not 7, 8, as in text); pl. GG, fig. 3 (head). Souleyet, 1852: 551–558; pl. 33 (anatomy, including penis, oviduct, radula; as Littorine littorale). H. & A. Adams, 1854, **1**: 312; **3**: pl. 33, figs 1, 1a–c. Clark, 1855: 340–342. Gray, 1857: 87–89; fig. 49 (radula). Reeve, 1857: sp. 18; pl. 4, fig. 18a–d. Sowerby, 1859: pl. 12, figs 14, 15. Mørch, 1868: 78. Petit de la Saussaye, 1869: 121. Gould, 1870: 308–309; fig. 577. Mørch, 1871a: 185. Meyer & Möbius, 1872: 10–13; figs facing p. 14: 1 (animal); 2; 3 (operculum); 4 (radula). Sars, 1878: 164, 447; anatomical pl. 6, fig. 3 (radula). Collin, 1884: 56–57. Herzenstein, 1885: 678. Nobre, 1886: 20. Locard, 1886: 285, 580. Sparre-Schneider, 1886: 5. Sowerby, 1887: pl. 12, figs 14, 15. Kobelt, 1888: 167. Locard, 1891: 189; fig. 165. Provancher, 1891: 91–92. Bumpus, 1898: 247–259. Marshall, 1898: 123. Nobre, 1898: 166. Teilman-Friis, 1898: 211–213; fig. 1b (radula). Dautzenberg & Durouchoux, 1900: 45. Brøgger, 1901: pl. 9, fig. 12; pl. 12, figs 8, 10, 11, 15 (post-Pleistocene). Johansen, 1901: 295–298; figs 1–3 (radula). Marshall, 1903: 54. Kobelt, 1907: 55–57; pl. 110, figs 1–5. Caullery & Pelseneer, 1910: 357–360; pl. 9, figs 1, 2 (spawn); pl. 9, figs 5, 6, 8–11 (embryology, larva). Pelseneer, 1911: 9–11, 15–17; pl. 3, figs 1, 2 (spawn); pl. 3, figs 3, 4 (larva). Dautzenberg & Fischer, 1912: 181–187 (in part, includes refs to *L. squalida*). Dautzenberg & Durouchoux, 1913: 21. Delsman, 1913: 53–66; fig. 1a; fig. 4 (spawn); figs 5, 6 (embryo, larva). Marshall, 1915: 329. Hidalgo, 1917: 390–392. Harmer, 1920: 645–652; pl. 52, figs 1–8 (typical), 9–26 (varieties) (in part, includes *L. squalida*; Pliocene). Tattersall, 1920: 1–11; figs 1–9 (spawn, larva). Winckworth, 1922: 96. Dautzenberg & Fischer, 1925: 51–52. Baden-Powell, 1930: 71, 74 (Pleistocene). Bøggild, 1930: 309; pl. 10, fig. 6 (mineralogy). Nobre, 1931: 144–145; pl. 33, figs 4–6. Winckworth, 1932: 221. van Benthem Jutting, 1933: 57–62; figs 23 (radula); 24; 25; 26 (spawn). Linke, 1933a: 1–60 (anatomy); pl. 2, fig. 22 (penis); pl. 3, fig. 36, pl. 6, fig. 60 (oviduct); pl. 6, fig. 64, pl. 7, figs 69, 75 (spawn). Spärck & Thorson, 1933: 14. Lebour, 1935: 373–378; figs 7, 8 (spawn), 9 (larva). Ankel, 1936a: 174–182; figs 5–7 (radula). Ankel, 1936b: 47; figs 18, 19 (nervous system); fig. 95; figs 98, 112B (anatomy); fig. 169 (spawn); figs 194, 200, 204, 206B. Nobre, 1936: 101; pl. 82, fig. 6. Lebour, 1937: 127; fig. 2d (spawn). Ankel, 1938: 317–333; figs 4, 6, 7, 8 (radula). Johansson, 1939: 344–396 (anatomy). Thorson, 1941: 33–34. Kaas & ten Broek, 1942: 143–144; pl. 12, fig. 21. Smidt, 1944: 266–267; fig. 12 (larva). Thorson, 1946: 174–176; fig. 96 (larva). Ankel, 1950: 19–27 (radula). Lecointre, 1950: 241; figs 18, 19 (Pleistocene). Lecointre, 1952: 96–97; pl. 27, fig. 1a, b (in part, includes *L. squalida*; Pleistocene). La Roque, 1953: 166. Abbott, 1954a: 132; pl. 19, fig. b. van Regteren Altena, Bloklander & Pouderoyen, 1954: 61; pl. 3, fig. 29a (Pleistocene). Feyling-Hanssen, 1955: 160; pl. 26, figs 4–8 (post-Pleistocene). Dodge, 1959: 216–218 (in part, includes refs to *L. squalida*). White, 1959: 106; pl. 39, fig. 14 (Pliocene). Bousfield, 1960: 16; pl. 2, fig. 16. Kuznetsov, 1960: 192–197; pl. 55, figs 2a, 2b (spawn). Mileikovsky, 1961: 178–179; fig. 4 (larva). Fretter & Graham, 1962: 14–48; figs 1–5, 7–12, 14–23, 43B, 124, 148, 201C, 201F, 235 (anatomy, spawn, larva). Richter, 1962: 391–406; fig. 24b (radula). Lecointre, 1963: 58–59; pl. 1, figs 10–11 (Pleistocene). Sacchi & Rastelli, 1966: fig. 2F (penis). Ziegelmeier, 1966: 41; pl. 10, fig. 4a–c. James, 1968b: 633–634; figs 8, 17; fig. 21 (head); figs 29–31. Fretter & Pilkington, 1970: 14; fig. 14a, b (larva). Davis, 1971: 1–5; fig. 1, fig. 4:1. von Moor, 1973: 546–573; fig. 1 (spawn). Óskarsson, 1973: 154–155. Rasmussen, 1973: 238–241. Abbott, 1974: 67; text-fig. 549. Bandel, 1974: 105; figs 18–21, 23 (radula). Matveeva, 1974: 121–128; fig. 19 (spawn). Sacchi, 1974: 88. Bandel, 1975: 4–5; figs 1–5 (larva). Hubendick & Warén, 1975: 40; fig. 266. Daguzan, 1977: 17–40. Fish & Fish, 1977: 495–504; figs 1b, 2b; pl. 1, c–e (larva). Wagner, 1977: 1305–1321; pl. 1, fig. 1 (Pleistocene). Merklin, Zarkhidze & Ilina, 1979: 61; pl. 11, figs 5–7 (Pleistocene). Sneli & van Marion, 1979: 4–8; fig. 1d. Dudley, 1980: 108–112. Fretter, 1980: 148–153; fig. 2 (oviduct). Fretter & Graham, 1980: 255–258; fig. 204. Jüch & Boekschoten, 1980: figs 5–7 (radula). Steneck & Watling, 1982: 305; fig. 6 (radula). Vermeij, 1982: 561–580. Bandel, 1984: 15, 19, 20; fig. 8 (radula). Jardine, 1984: 5; fig. 2a. Kemp & Bertness, 1984: 811–813. Rolán, 1984: 106–107; fig. Fish & Sharp, 1985: 148; fig. 10.2 F (larval shell). Reid, 1986a: fig. 11 (oviduct). Jagnow & Gosselck, 1987: 211; pl. 2, fig. 13. Janson, 1987a: 245–256. Janssen & Triebskorn, 1987: 73–82; figs 3–6 (radula). Graham, 1988: 166–167; figs 59A, 60. Warmoes, Backeljau & de Bruyn, 1988:

51–70; fig. 4A (penis). Hawkins, Watson, Hill, Harding, Kyriakides, Hutchinson & Norton, 1989: 151–165; fig. 2B, F (radula). Hayward, Wigham & Yonow, 1990: 659; fig. 12.9. de Bruyne, 1991: 99; fig. 62. Poppe & Goto, 1991: 92; pl. 11, figs 1–3. Rózycki, 1991: 386–387; fig. 151. Crothers, 1992: 91–97. Egorov, 1992: 169; fig. 1A. Finet, Wüest & Mareda, 1992: 38; pl. 11a, b; c–e (radula). Bauer, Fioroni, Ide, Liebe, Oehlmann, Stroben & Watermann, 1995: 15–27; figs 1–4 (oviduct).

Paludina littorea—Jeffreys, 1839: 33.

Littorina (Littorina) littorea—Récluz, 1869: 44. Fischer, 1885: 707–708; text fig. 476 (radula); pl. 9, fig. 10. Tryon, 1887: 240; pl. 41, figs 82–85. Bequaert, 1943: 3–5; pl. 1, figs 1–11. Rosewater, 1970: 423. Smith, 1982: 535–539. Reid, 1989a: 101; pl. 2, fig. 2o; fig. 5k (head); fig. 7o (penis); fig. 10n (oviduct); fig. 11q (spawn); fig. 14q (radula). Reid, 1990b: 1–19; fig. 1j; fig. 2f (oviduct); fig. 3f (spawn). Taylor & Reid, 1990: 199–215; figs 11–16, 26 (mineralogy). Smith & Heppell, 1991: 17.

Litorina littorea—Johnson, 1915: 120–121.

Littorina (Algaroda) littorea—Beets, 1946a: 32–33 (Pliocene, Pleistocene). Winckworth, 1951: 131. Glibert, 1962: 21 (Pliocene, Pleistocene). Nordsieck, 1968: 40; pl. 6, fig. 23.10. Golikov & Kusakin, 1978: 82–84; fig. 49. Nordsieck, 1982: 61; pl. 26, fig. 23.10. Golikov, 1987: 78–79; fig. 30.

Turbo littoralis—Schröter, 1779: 344; pl. 8, fig. 5; pl. 11, fig. 5 (not *Nerita littoralis* Linnaeus, 1758; error for *Turbo littoreus* Linnaeus, 1758). Blainville, 1824: 226 (not *Nerita littoralis* Linnaeus, 1758; error for *Turbo littoreus* Linnaeus, 1758, as correctly cited by Blainville, 1828).

Littorina littoralis—Rang, 1829: 185 (possibly an error for *Littorina littorea*; Herrmannsen, 1846; Kadolsky, 1970). Petit de la Saussaye, 1852: 182 (not *Nerita littoralis* Linnaeus, 1758; error for *Littorina littorea*).

Turbo rudis—Sowerby, 1814: 164; pl. 71, fig. 2 (not Maton, 1797=*L. saxatilis*; Pliocene). Blainville, 1830: 297–298; pl. 12, fig. 2 (not Maton, 1797; in part, includes *L. saxatilis*).

Littorina rudis—Brown, 1843b: 74; pl. 38, figs 57, 58 (not *Turbo rudis* Maton, 1797=*L. saxatilis*). Wood, 1872: 79; pl. 5, figs 9, 10a, b (not Maton, 1797; Pliocene). Harmer, 1921: 653–661; pl. 53, figs 1, 4, 7–9, 11, 13, 15, 18 (in part, includes *L. saxatilis*; not Maton, 1797; Pliocene).

Turbo litoreus Turton, 1819: 196 (unjustified emendation of *Turbo littoreus* Linnaeus, 1758, because on p. 126 *litoralis* is spelled in same way, ICZN, 1985: art. 33b; however, the two spellings are deemed identical and are homonyms, ICZN, 1985: art. 58; in part, includes *L. saxatilis s.l.*).

Litorina litorea—Menke, 1828: 25. Menke, 1830: 45. Anton, 1838: 52. Menke, 1845: 49–53. Philippi, 1846a, **2**: 102–103, 105; *Litorina* pl. 1, figs 7–10. Lovén, 1846: 154. Philippi, 1848, **3**: 64–65; *Litorina* pl. 7, fig. 12. Küster, 1853: pl. 1, figs 1–4; 1856: 4–5. Woodward, 1856: 456 (radula). Weinkauff, 1878: 26. Weinkauff, 1883: 214.

Littorina litorea—Jeffreys, 1865, **3**: 368–376; 1869, **5**: 206; pl. 65, figs 4, 4a (in part, includes *L. squalida*). Tryon, 1873–1874: 80; fig. 147. Jeffreys, 1883: 113 (in part, includes *L. squalida* and *L. mandshurica*).

Litorina (Litorina) litorea—Kobelt, 1878: 124; pl. 46, fig. 5.

Littorina (Algaroda) litorea—Dall, 1918: 137.

Turbo ustulatus Lamarck, 1822: 48 (no loc.; 2 syntypes MHNG 1096/68, photograph seen). Blainville, 1828: 107. Blainville, 1830: 301 (in part, fig. is *L. obtusata*).

Litorina ustulata—Anton, 1838: 52.

Littorina vulgaris var. *ustulata*—Maitland, 1854: 82.

Littorina vulgaris J. Sowerby, 1832: pl. 211, fig. 1 (no loc.; type not traced). Sowerby, 1839: 58; fig. 363. Reeve, 1842: 170; pl. 222, fig. 1. Maitland, 1854: 82 (in part, includes refs to *L. saxatilis*). Hanley, 1855: 326–327.

Littorina littorea var. *vulgaris*—Wood, 1848: pl. 10, fig. 14. Dautzenberg & Durouchoux, 1913: 21. Harmer, 1920: 647–648; pl. 52, figs 9–11 (Pliocene). Dautzenberg & Fischer, 1925: 52.

Delphinula ? carinata Woodward, 1833: 44; pl. 3, fig. 9 (Thorpe and Bramerton [Norfolk, England]; Pliocene; type not traced).

Turbo carinatus Woodward, 1833: 44; pl. 3, fig. 11 (Bramerton and Thorpe; Pliocene; type not traced; not Born, 1778).

Littorina littorea var. *carinata*—Wood, 1848: pl. 10, fig. 14i (Pliocene). Harmer, 1920: 651–652; pl. 52, figs 21, 22 (Pliocene).

Turbo ventricosus Woodward, 1833: 44; pl. 3, fig. 12 (Bramerton and Thorpe; Pliocene; type not traced).

Turbo bicarinatus Woodward, 1833: 44; pl. 3, fig. 13 (Bramerton and Thorpe; Pliocene; type not traced; not Sowerby, 1825).

Littorina littorea var. *bicarinata*—Wood, 1848: pl. 10, fig. 14e (Pliocene). Harmer, 1920: 652; pl. 52, figs 23, 24 (Pliocene).

Turbo sulcatus Woodward, 1833: 44; pl. 3, figs 14, 15 (Bramerton and Thorpe; Pliocene; type not traced; not Gmelin, 1791).

Littorina littorea var. *sulcata*—Wood, 1848: pl. 10, fig. 14b (Pliocene). Harmer, 1920: 649; pl. 52, figs 14, 15 (Pliocene).

Turbo elongatus Woodward, 1833: 44; pl. 3, figs 16–18 (Bramerton and Thorpe; Pliocene; type not traced; not Defrance, 1827).

Littorina elongata—Wood, 1842: 532 (Pliocene).

Littorina littorea var. *elongata*—Wood, 1848: pl. 10, fig. 14k (Pliocene). Harmer, 1920: 650–651 (Pliocene).

Littorina bartonensis Brown, 1843b: 74; pl. 38, figs 54, 55 (crag at Bramerton Hill, near Norwich [Norfolk, England]; Pliocene; type not traced).

Littorina communis Brown, 1843b: 74; pl. 38, fig. 56 (crag at Bramerton Hill, near Norwich; Pliocene; type not traced).

Littorina littorea var. *antiqua* Wood, 1848: pl. 10, fig. 14a (mammaliferous [Norwich] crag, Bramerton [Norfolk, England]; holotype BMNH G. 1979, seen, Fig. 37K). Marshall, 1916: 44. Harmer, 1920: 648–649; pl. 52, fig. 13 (Pliocene).

Littorina littorea var. *picta* Wood, 1848: pl. 10, fig. 14c (mammaliferous crag, Bramerton; Pliocene; 2 syntypes BMNH G. 1984, seen; not *Litorina obtusata* var. *picta* Menke, 1845).

Littorina littorea var. *elegans* Wood, 1848: pl. 10, fig. 14d (mammaliferous crag, Bramerton (here designated); Pliocene; type in Fitch Collection, not traced; not *Littorina elegans* Deshayes, 1842). Marshall, 1903: 54. Marshall, 1916: 44. Harmer, 1920: 650; pl. 52, fig. 18 (Pliocene).

Littorina littorea var. *pyramidata* Wood, 1848: pl. 10, fig. 14f (mammaliferous crag, Bramerton; Pliocene; 5 syntypes BMNH G. 1983, seen). Harmer, 1920: 651 (Pliocene).

Littorina littorea var. *delphinula* Wood, 1848: pl. 10, fig. 14g (mammaliferous crag, Bramerton; Pliocene; 6 syntypes BMNH G. 1980, seen). Harmer, 1920: 650 (Pliocene).

Littorina littorea var. *deformis* Wood, 1848: pl. 10, fig. 14h (mammaliferous crag, Thorpe [Norfolk, England]; Pliocene; holotype BMNH G. 1978, seen).

Littorina littorea [var. ex col.] *fuscofasciatus* Middendorff, 1849: 383 ([Russian Lappland]; as *fusco fasciatus*; infrasubspecific).

Littorina littorea [var. ex col.] *pallidefasciatus* Middendorff, 1849: 383 ([Russian Lappland]; as *pallide fasciatus*; infrasubspecific).

Littorina litorea var. *paupercula* Jeffreys, 1865, **3**: 369–370 ([Britain]; types not traced (Warén, 1980)).

Littorina litorea var. *brevicula* Jeffreys, 1865, **3**: 369–370 (Llanrhidian salt-marsh near Swansea [Wales]; Southend [England]; Christiania fiord [Norway]; types not traced (Warén, 1980); not *Turbo breviculus* Philippi, 1844= *L. brevicula*). Dautzenberg & Durouchoux, 1900: 45.

Littorina litorea var. *turrita* Jeffreys, 1865, **3**: 369–370 (Loch Carron [Scotland]; types not traced (Warén, 1980)).

Littorina littorea monstr. *turrita*—Marshall, 1898: 123.

Littorina litorea var. *sinistrorsa* Jeffreys, 1865, **3**: 370; 1869, **5**: pl. 65, fig. 4a (obtained from Billingsgate Market, London; 2 syntypes USNM 185553, 185524, seen; 1 syntype Scarborough Museum, ex Leckenby, seen).

Littorina littorea monstr. *sinistrorsum*—Marshall, 1898: 123. Marshall, 1916: 44.

Littorina rudis var. *jugosa*—Wood, 1872: pl. 5, fig. 9 (not *Turbo jugosus* Montagu, 1803=*L. saxatilis*; Pliocene).

Littorina armoricana Locard, 1886: 285, 580 (Langrune, France (here restricted); lectotype (here selected) MNHNP, 1 spec. from Langrune, 26 mm, seen). Locard, 1891: 189–190.

Littorina sphaeroidalis Locard, 1886: 285–286, 580 (Granville, France (here restricted); lectotype (here selected) MNHNP, 1 spec. from Granville, 29 mm, seen). Locard, 1891: 190.

Littorina littorea var. *unicarinata* Raeymaekers, 1889: xxxvi–xxxviii; fig. (Zierickzée (bouches de l'Escaut) [Netherlands]; monstrosity, unavailable).

Littorina parva 'Mørch' Teilman-Friis, 1898: 211–213; fig. 1b (radula) (Refsnæs, Holbæk, Svendborg, Esbjerg and Bornholm [Denmark]; type not traced. Attribution to Mørch is based on a misreading of Mørch (1871a: 185), in which 'parva' is used as an adjective, and is not intended as the name of a new variety).

Littorina littorea var. ex colore *pallida* Dautzenberg & Durouchoux, 1900: 45 (Saint-Servan (Fours-à-Chaux) [France]; infrasubspecific). Dautzenberg & Fischer, 1912: 186. Dautzenberg & Fischer, 1925: 52.

Littorina littorea var. ex colore *miniata* Dautzenberg & Durouchoux, 1900: 45 (Fours-à-Chaux (Saint-Servan) [France]; original spec. IRSNB, 23.1 mm, seen; infrasubspecific). Dautzenberg & Fischer, 1912: 186.

Littorina littorea form. *intermedia* Brøgger, 1901: pl. 12, fig. 6a–c (Myabanke, sandstuen, Ryenvarde, Kristiania [Oslo, Norway]; Pleistocene; type not traced; not *L. intermedia* Philippi, 1846=*Littoraria intermedia*).

Littorina littorea monstr. *elongata* Martel, 1901: 221; pl. 1, fig. 3 (Port-Briac [France]; monstrosity, unavailable; not *Turbo elongatus* Woodward, 1833=*L. littorea*).

Littorina littorea var. *sanguinea* Dautzenberg & Durouchoux, 1906: 44–45 (La Guimorais [St-Malo, France]; original spec. INSNB, 18.1 mm, seen; infrasubspecific; not *L. rudis* var. ex col. *sanguinea* Dautzenberg & Durouchoux, 1900=*L. saxatilis*). Dautzenberg & Fischer, 1912: 186. Dautzenberg & Fischer, 1925: 52.

Littorina littorea var. *balteata* Dautzenberg & Fischer, 1912: 186 (no locality; original spec. IRSNB, coll. Récluz, 26.0 mm, seen; infrasubspecific; not *Littorina balteata* Reeve, 1857=*L. brevicula*).

Littorina littorea var. ex forma *major* Dautzenberg & Durouchoux, 1913: 21 (Saint-Malo [France]; infrasubspecific; not *L. saxatilis rudis* var. *major* Dautzenberg & Fischer, 1912=*L. saxatilis*). Dautzenberg & Fischer, 1925: 52.

Littorina littorea var. *parva* Harmer, 1920: 648; pl. 52, fig. 12 (Icenian [Norwich] Crag: Bramerton [Norfolk, England]; Pliocene; holotype SMC C7434, seen; not *L. parva* Teilman-Friis, 1898).

Littorina littorea var. *distorta* Harmer, 1920: 649; pl. 52, fig. 16 (Icenian Crag: Bramerton [Norfolk, England]; Pliocene; holotype NCM, seen).
Littorina littorea var. *conica* Harmer, 1920: 649–650; pl. 52, fig. 17 (Icenian Crag: Bramerton [Norfolk, England]; Pliocene; holotype NCM, seen, Fig. 37I; not *L. conica* Philippi, 1846=*Littoraria conica*).
Littorina littorea var. *complanata* Harmer, 1920: 651; pl. 52, figs 19, 20 (Icenian Crag: Thorpe [Norfolk, England] (here restricted); Pliocene; 1 syntype (Harmer, 1920: pl. 52, fig. 19) SMC C7636, seen; original of fig. 20 not traced in NCM).
Littorina littorea var. *angulata* Harmer, 1920: 652; pl. 52, fig. 25 (Icenian Crag: Bramerton [Norfolk, England]; Pliocene; holotype NCM, seen).
Littorina littorea var. *truncata* Harmer, 1920: 652–653; pl. 52, fig. 26 (Icenian Crag: Bramerton [Norfolk, England]; Pliocene; holotype NCM, seen, Fig. 37J).
Littorina rudis var. *laevis*—Harmer, 1921: 655–656; pl. 53, fig. 4 (in part, other fig. is *L. saxatilis*; not Jeffreys, 1865=*L. saxatilis*; Pliocene).
Littorina rudis var. *globosa*—Harmer, 1921: 656–657; pl. 53, figs 7, 8 (in part, other fig. is *L. saxatilis*; not Jeffreys, 1865=*L. saxatilis*; Pliocene).
Littorina rudis var. *nigrolineata*—Harmer, 1921: 657; pl. 53, fig. 9 (in part, includes refs to *L. saxatilis*, *L. compressa*; not *L. nigrolineata* Gray, 1839=*L. saxatilis*; Pliocene).
Littorina rudis var. *rudissima*—Harmer, 1921: 657–658; pl. 53, fig. 11 (in part, other fig. is *L. saxatilis*; not *L. rudissima* Johnston, 1842=*L. saxatilis*; Pliocene). Harmer, 1925: 874; pl. 65, fig. 29 (in part, other fig. is *L. saxatilis*; not Johnston, 1842; Pliocene).
Littorina rudis var. *groenlandica*—Harmer, 1921: 658; pl. 53, fig. 13 (in part, other fig. is *L. saxatilis*; not *Litorina groenlandica* Menke, 1830=*L. saxatilis*; Pliocene).
Littorina rudis var. *reevei* Harmer, 1921: 659; pl. 53, fig. 15 (Icenian Crag: Thorpe, near Norwich [Norfolk, England] (here restricted); Pliocene; type not traced in NCM).
Littorina rudis var. *tenebrosa*—Harmer, 1921: 660–661; pl. 53, fig. 18 (in part, other fig. is *L. saxatilis*; not *Turbo tenebrosus* Montagu, 1803=*L. saxatilis*; Pliocene).
Littorina obtusata—Harmer, 1921: 661–662; pl. 52, fig. 29 (in part, other figs are *L. obtusata*; not *Turbo obtusatus* Linnaeus, 1758=*L. obtusata*; Pliocene).
Littorina palliata—Harmer, 1921: 663–664; pl. 53, figs 22, 23 (figs probably *L. littorea*; in part, includes *L. obtusata*; not *Turbo palliatus* Say, 1822=*L. obtusata*; Pliocene).
Littorina littorea var. ex col. *aurantia* Dautzenberg & Fischer, 1925: 52 (Grève de Roscoff [France]; infrasubspecific; not *L. rudis* var. ex col. *aurantia* Dautzenberg, 1887=*L. saxatilis s.l.*).

Taxonomic history

The original description by Linnaeus (1758) of his *Turbo littoreus* was brief and inadequate. Although specimens of the species now identified with this name are present in the Linnean shell collection (Hanley, 1855; Dodge, 1959), they have not been authenticated as Linnaeus' original material, and the types must be considered lost (Dance, 1967). Nevertheless, there is no doubt about the identity of the species, because Linnaeus (1758) referred to three published figures. The first of these figures was incorrectly cited as '*It. wgoth.* 169, 199. *t.* 5. *f.* 3' (i.e. Linnaeus, 1747: 169, 199, pl. 5, fig. 3; see Boss, 1988), but this illustrates a *Mya*, and it is clear from the page references that plate 5 figure 4 was intended. Although printed in reverse, this figure is clearly recognizable (Fig. 32 herein) and accompanied by a reasonable description and a locality, Marstrand in Sweden (Linnaeus, 1747: 169). The two other figures cited (Lister, 1678: pl. 3, fig. 9; Gualtieri, 1742: pl. 45, fig. G), although inferior, are also recognizable as the same species. In the twelfth edition of the *Systema Naturae* Linnaeus (1767) added three additional references, all of which are to this same species. The figure of Linnaeus (1747) is here selected as a representation of the lectotype, and Marstrand as the type locality.

Since the time of Linnaeus this species has generally been correctly identified, and most authors have used his specific name. There have been two minor nomenclatural complications. At least three authors have used the specific name *littoralis* (Schröter, 1779; Blainville, 1824; Petit de la Saussaye, 1852; possibly Rang, 1829), leading to the possibility of confusion with *Nerita littoralis* Linnaeus, 1758 (a name that has sometimes been applied to *Littorina obtusata*, see p. 202), but in each case it is clear from the context that these were merely errors for *littoreus* or *littorea* and not formal emendations. Turton's (1819) spelling of the specific name as *litoreus* must be treated as an unjustified emendation, because he also spelled *litoralis* (derived from the same Latin word, *littus* or *litus*, the sea shore) in this way in the same work (ICZN, 1985: art. 33b); however, these two spellings are deemed to be identical variants for nomenclatural purposes (ICZN, 1985: art. 58). It should be noted that, in

Figure 32. *Littorina (Littorina) littorea.* Lectotype of *Turbo littoreus* Linnaeus, 1758 (Linnaeus, 1747: pl. 5, fig. 4; actual length of shell=27 mm).

contrast, a one-letter difference is sufficient to prevent homonymy of generic names (ICZN, 1985: art. 56b), so that Menke's (1828) use of the combination *Litorina litorea* and repeated spelling of *Litorina*, constituting an unjustified emendation of the generic name, did create an available name. These emendations were followed by a number of authors, especially by the German school in the nineteenth century, and inexplicably some apparently accepted only the emendation of the specific name (e.g. Jeffreys, 1865; Dall, 1918). There have, in addition, been a number of incorrect subsequent spellings which are not enumerated here, since they do not enter into homonymy (ICZN, 1985: art. 33c).

The synonymy includes only five additional species names, eight shape varieties, and seven colour varieties that have been proposed for Recent specimens, which is an indication of the relatively low degree of variation in *L. littorea* (in comparison with *L. (Neritrema)* species), and none of these has been widely used. In contrast, Pliocene fossils from eastern England include numerous extraordinary monstrosities, upon which no less than eight species and 13 varieties have been based (Woodward, 1833; Brown, 1843b; Wood, 1848; Harmer, 1920, 1921); these distorted specimens are discussed later (p. 108). Harmer (1920, 1921) identified some of these Pliocene fossils as *L. rudis* (=*L. saxatilis*) and *L. obtusata*, but examination of his collection in SMC has shown that all are small, eroded or distorted examples of *L. littorea*.

Throughout its history, most authors have had a concept of this species similar to that presented here. In the eighteenth and early nineteenth centuries *L. saxatilis s.l.* was often included in the concept of *Turbo littoreus* (e.g. da Costa, 1778; Donovan, 1800; Maton & Rackett, 1807; Dillwyn, 1817; Turton, 1819; Blainville, 1830; Brown, 1837). Although there is no evidence, it seems possible that Linnaeus himself shared this broad concept of the species, since it is surprising that he failed to describe so common a species as *L. saxatilis* during his Scandinavian travels. Most later authors correctly excluded *L. saxatilis* and other European species, but there were some who believed that *L. littorea* had a circum-boreal distribution, and therefore included *L. squalida* in the synonymy (Wood, 1842, 1848; Jeffreys, 1865; Dautzenberg & Fischer, 1912; Harmer, 1920; Dodge, 1959). Jeffreys (1883) even included *L. mandshurica* as well. Only a few workers divided Recent *L. littorea* into more than one species. Lamarck (1822; followed by Blainville, 1828, 1830, and Anton, 1838) had two, *littorea* and *ustulata*. Locard (1886, 1891) had three, *littorea*, *armoricana* and *sphaeroidalis*. Teilman-Friis (1898) attempted to distinguish his *L. parva* from *L. littorea* by adult shell size, relative radular length and small differences in the rachidian tooth, but this small form is here regarded as an ecotype of *L. littorea*, characteristic of brackish environments (see Ecotypic Variation, p. 106).

Turbo littoreus is the type species of *Littorina* Férussac, 1822, by subsequent designation of Anton (1838), as established in Opinion 1159 of the ICZN (1980). It may be noted that Anton actually used the emendation *Litorina litorea*. Dall (1918) erected the new section *Algaroda* of *Littorina*, with this species as type, although a correction of *L. (Algaroda) squalida* to *L. (Littorina) squalida* in 1923 suggests that he recognized the synonymy.

Diagnosis

Mature shell height 10.6–52.8 mm. Shell high-turbinate; sutures not deep, giving rather flat-sided spire whorls; spire outline onion-shaped. Mineralogy of external irregular-prismatic calcite and internal crossed-lamellar aragonite. Sculpture of numerous low narrow ribs and fine spiral microstriae over whole surface. Colour varies from uniform cream to brown, black, orange or red, but usually pale brown with 8–25 narrow black to dark brown lines; columella white, outer edge of aperture brown or with brown lines, brown within, occasionally with indistinct pale basal band. Penis with 10–42 mamilliform glands; tapering or rounded filament 30–50% total length. Pallial oviduct of two consecutive loops of albumen gland followed by large spiral loop of capsule gland; straight section not enlarged or septate; bursa large, at posterior end of straight section. Spawn of large, pelagic, biconvex capsules with peripheral rim, containing 1–9 eggs; development planktotrophic. Distribution: eastern Atlantic from White Sea to southern Portugal; western Atlantic from Newfoundland to Virginia.

Anatomical description

Head-foot (Fig. 33P). Head and tentacles covered with dense transverse black lines, or sometimes all black; sides of foot with dense black lines or reticulation. Opercular ratio 0.41–0.53.

Penis (Fig. 33A–F). Base wrinkled, pigmented in folds only or entirely black; 10–42 mamilliform glands in irregular single or double row. Filament tapering, or rounded at tip, usually unpigmented, but sometimes black; 30–50% total length. Sperm groove open, ending terminally. It has long been known that the penis appears reduced in size outside the breeding season (Tattersall, 1920; Pelseneer, 1926; Linke, 1933; Jenner, 1956), and this has been ascribed to a process of regression. However, some authors have observed, in the laboratory, that necrosis of the penial insertion is followed by shedding of the penis and its subsequent regrowth, and have suggested that this is the normal process in the field (Grahame, 1969; Le Breton, 1970, 1971; Daguzan, 1976b). It may be that the penis is either resorbed or shed in different populations (Grahame, 1969), or that shedding is a response to stress that is rare in natural, unpolluted habitats (Deutsch & Fioroni, 1992). In some gastropods females may show abnormal development of a penis (pseudohermaphroditism) under certain conditions (Fioroni, Oehlmann & Stroben, 1990), including presence of the anti-fouling pollutant tributyltin (TBT). The sensitivity of *L. littorea* to TBT is relatively low (Deutsch & Brick, 1993), but at high concentrations females may occasionally develop a penis (Fioroni, Deutsch, Stroben & Oehlmann, 1992; Bauer, Fioroni, Ide, Liebe, Oehlmann, Stroben & Watermann, 1995).

Sperm (Fig. 3D). Euspermatozoa 50–56 µm. Paraspermatozoa 11–14 µm diameter, round or slightly oval; vesicles large, distinct; nucleus not visible.

Oviduct (Fig. 33G–O). Two consecutive loops of albumen gland followed by large spiral loop of capsule gland with distal reddish translucent portion; straight section without enlarged or septate jelly gland. Bursa large, sometimes originating towards posterior end of straight section (but origin difficult to characterize if straight section is short), not completely separated from lumen of pallial oviduct until rather far back. Seminal receptacle simple. The oviduct has previously been described by Linke (1933), Fretter & Graham (1962) and Fretter (1980), but none of these correctly described the spiral path of the egg groove, and the terminology of the glands differed from that used here (p. 17). In populations exposed to TBT pollution (for example in proximity to harbours) the pallial oviduct may be reduced in size, and at very high levels of TBT the oviduct may be malformed, exhibiting a ventral split and eventually complete transformation into a prostate gland; a penis and sperm groove may also develop (Bauer *et al.*, 1995).

Spawn and development (Fig. 33Q–S). Although this is such a common, well-studied species, its spawn was not described until early this century (Tattersall, 1908a, b; Caullery & Pelseneer, 1910); in earlier accounts the egg masses of *L. obtusata* and *Lacuna pallidula* were incorrectly ascribed to *L. littorea* (as discussed by Caullery & Pelseneer, 1910; Hertling & Ankel, 1927; Goodwin, 1979). *Littorina littorea* produces pelagic, asymmetrically biconvex capsules with a flat peripheral rim; total diameter of the capsule varies from 310 to 1100 µm according to the number of contained eggs, but is usually in the range 600 to 1000 µm (Caullery & Pelseneer, 1910; Pelseneer, 1911; Delsman, 1913; Tattersall, 1920; Hayes, 1927, 1929; Linke, 1933; Lebour, 1935, 1937; Ankel, 1936b; Wright, 1936; Matveeva, 1974). The usual number of eggs per capsule is one to three (Caullery & Pelseneer, 1910; Tattersall, 1920; Hayes, 1927; Linke, 1933; Ankel, 1936b; Thorson, 1946; Fish, 1979; Alifierakis & Berry, 1980), but more have been recorded (Hayes, 1927; Wright, 1936; Lebour, 1937; Rasmussen, 1973; Bandel, 1974; Matveeva, 1974), up to a maximum of nine (Linke, 1933). Measurements of egg cell diameter are 120–160 µm (Delsman, 1913; Tattersall, 1920; Hayes, 1927, 1929; Linke, 1933;

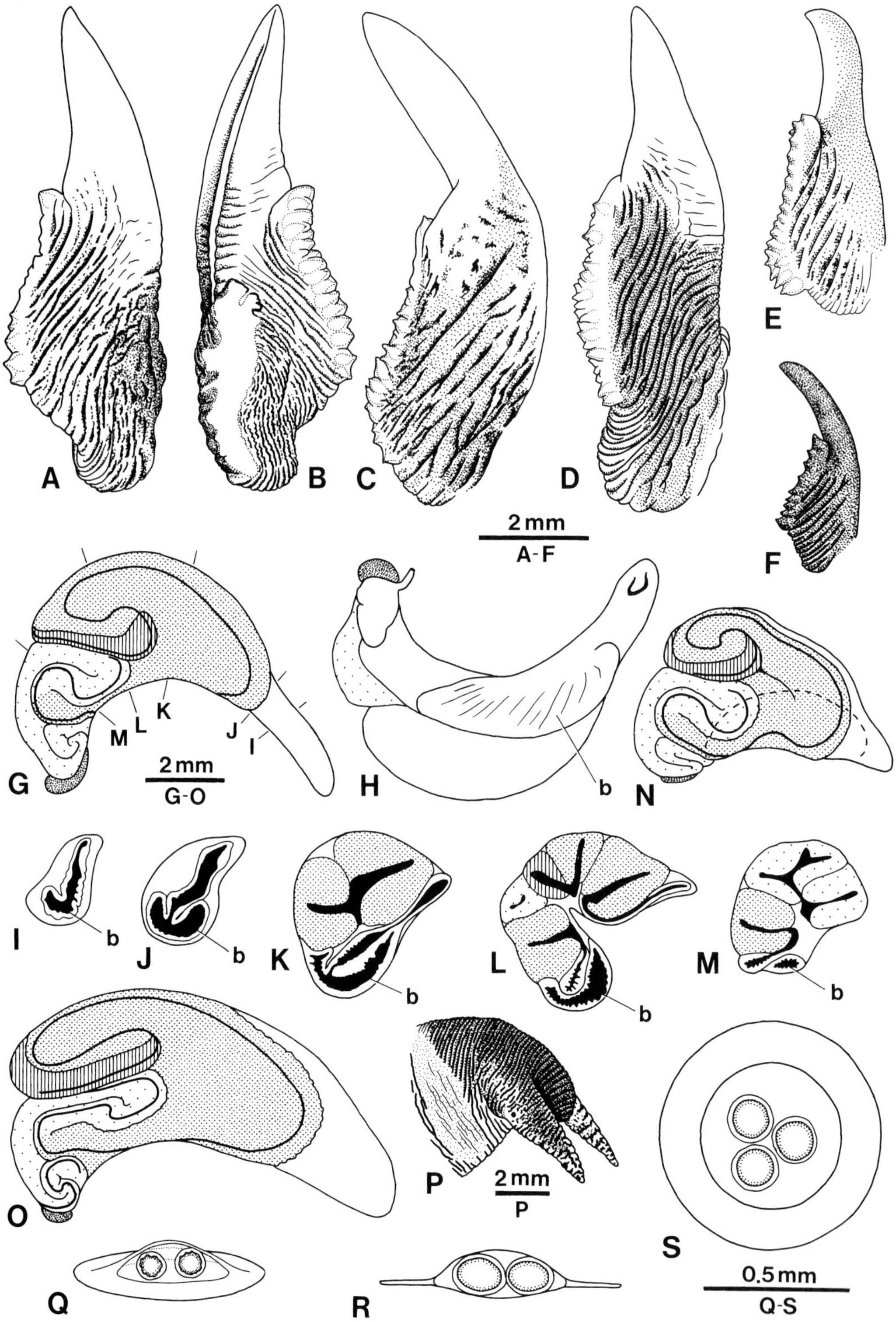

Figure 33. *Littorina (Littorina) littorea*: **A–F**, penes; **A**, **B**, two views of same penis; **G–O**, pallial oviducts; **G–M**, lateral view, medial view and transverse serial sections (taken at levels indicated in **G**, viewed from anterior end, right lateral side of animal towards top of page) of same oviduct; **P**, head pigmentation; **Q–S**, pelagic egg capsules (**Q**, after Ankel, 1936b; **R**, schematic section after Linke, 1933; **S**, after Tattersall, 1920). **A**, **B**, **G–M**, 1 km NW Trearddur, Anglesey, Wales; **C**, Trescott, Cobscook Bay, Maine; **D**, Porto, Portugal; **E**, Solovetskiye Island, White Sea, Russia; **F**, Skovshovedhavn, Øresund, Denmark (ZMK) (sheltered/brackish ecotype); **N**, Appledore Island, Isles of Shoals, Maine; **O**, Smelror, 2 km N Vardø, Norway; **P**, Old Head of Kinsale, Co. Cork, Ireland; **Q**, German North Sea; **R**, Helgoland, Germany; **S**, Galway, Ireland. Shell heights: **A**, **B**, 22.4 mm; **C**, 20.5 mm; **D**, 22.9 mm; **E**, 19.0 mm; **F**, 11.1 mm; **G–M**, 23.5 mm; **N**, 20 mm; **O**, 25.1 mm; **P**, 23 mm. For shading conventions of pallial oviducts see p. 6. Abbreviation: b, copulatory bursa.

Wright, 1936; von Moor, 1973), although Lebour (1937) gave 205 μm. Development is planktotrophic; at metamorphosis the larval shell measures 275 μm high by 375–390 μm in diameter, consists of about 3 whorls (Thorson, 1946) and has been illustrated as smooth, with scattered minute tubercles (1–2 μm in diameter) that become arranged in spiral bands, and an apertural beak (J. D. & S. Fish, 1977; these authors note the possibility of confusion with the veliger of *Hydrobia ulvae*, which lacks an apertural beak). In some relatively mild localities spawning has been recorded all year round (Plymouth, Devon: Lebour, 1947; Whitstable, Kent: Smith & Newell, 1955). More commonly, there is a defined breeding season, with peak spawning in late winter and spring and most adults being spent by late summer and autumn. In some cases the onset of breeding occurs later in the colder climates of northerly latitudes. Examples of spawning seasons are: October to April at Morbihan, southern Brittany (Daguzan, 1976b); November or December to July in Wales (Williams, 1964; Fish, 1972); December to June in the Firth of Forth, Scotland (Berry & Smith, 1987); January to June at Helgoland (Linke, 1933) and Galway (Tattersall, 1920); April to October at Oosterschelde, Netherlands (Bandel, 1974); February to June in Isefjord, Denmark (Rasmussen, 1973); April to November in New Brunswick, southern Canada (Chase & Thomas, 1995a, b); June to August in the Barents and White Seas (Kuznetsov, 1960; Mileikovsky, 1960; Matveeva, 1974).

Radula and foregut (Fig. 34). Relative radular length 1.49–2.15 (to 2.63, Teilman-Friis, 1897). Rachidian: length/width 0.96–1.12; central cusp leaf-shaped and pointed; one pointed cusp and one outer denticle on each side. Lateral: four cusps, second from inside rounded, third from inside largest and asymmetrically pointed, others pointed. Inner marginal: four cusps, second from inside rounded to pointed, third from inside largest and asymmetrically pointed, others pointed. Outer marginal: 4–5 blunt or rounded cusps (4–7 according to Johansen, 1901); narrow neck, projecting flange on inside of base only. All cusps more elongate in a juvenile examined (Fig. 34G, H). Salivary glands enlarged.

Shell variation (Figs 35–37)

Geographical variation. There are no clearly marked geographical varieties in *L. littorea*, but some general trends in size, shape and thickness have been observed. This is the largest species in the genus, and the shells of greatest size (exceeding 40 mm) occur in the north of the range, in northern Scotland and the Shetland Islands (Jeffreys, 1865), in northern Norway and in Newfoundland (Dautzenberg & Fischer, 1912). The maximum recorded size is 52.8 mm in a specimen from Scotland (BMNH).

Bumpus (1898) compared large samples from north America and Britain; no consistent differences in shape were found, but he claimed that the American shells were more variable because, as an introduced species, they were freed from some of the constraints of selection acting in their native habitats. This has not been confirmed, and his results have been criticised for not taking account of the effects of shell erosion (Bigelow & Rathbun, 1903). Dautzenberg & Fischer (1912) remarked that French shells were both smaller and more globular in shape than those from more northern localities, and these were described as a new species, *L. sphaeroidalis*, by Locard (1886). Such a trend has not yet been verified, but could perhaps be explained by the relationship between growth rate and shell shape (Kemp & Bertness, 1984; p. 106).

Patterns in shell thickness on a geographical scale have been found by Vermeij (1982). The thickest specimens were from the English Channel. In America shells were thinner to the south of Cape Cod, again perhaps because of the relationship between temperature, growth rate and shell form. In an earlier, more limited, study Dudley (1980) claimed an opposite result, with samples from south of Cape Cod being thicker than those from the north, and this was explained by an inhibition of calcification at low temperatures. However, it may be that Dudley's results were affected by the inclusion of shells from a marshland habitat in the north, since shells from brackish habitats are known to be thinner (see Ecotypic Variation, p. 106).

These geographical patterns do not appear to reflect gradients in predation. The crab *Carcinus maenas* is a predator of *L. littorea* and since the beginning of this century has extended its western Atlantic range north of Cape Cod. Vermeij (1982) tested the hypothesis that *L. littorea* could have responded to the arrival of this crushing predator by an increase in shell thickness. Although there was some evidence of an increase in unsuccessful shell breakage, he was unable to detect the expected

Figure 34. *Littorina (Littorina) littorea*: **A–C,** three views of radula (shell H=23.5 mm), flat (**A**), at 45° (**B**) and side view (**C**); **D,** radula (shell H=25.1 mm); **E, F,** two views of radula (shell H=20.5 mm), flat (**E**) and at 45° (**F**); **G, H,** two views of radula of juvenile (shell H=3.4 mm), flat (**G**) and at 45° (**H**). **A–C,** Trearddur, Anglesey, Wales; **D,** Smelror, Vardø, Norway; **E, F,** Cobscook Bay, Maine; **G, H,** Poulsallagh, Co. Clare, Ireland. Scale bars: **A, B**=100 μm; **C**=100 μm; **D**=100 μm; **E, F**=100 μm; **G, H**= 20 μm.

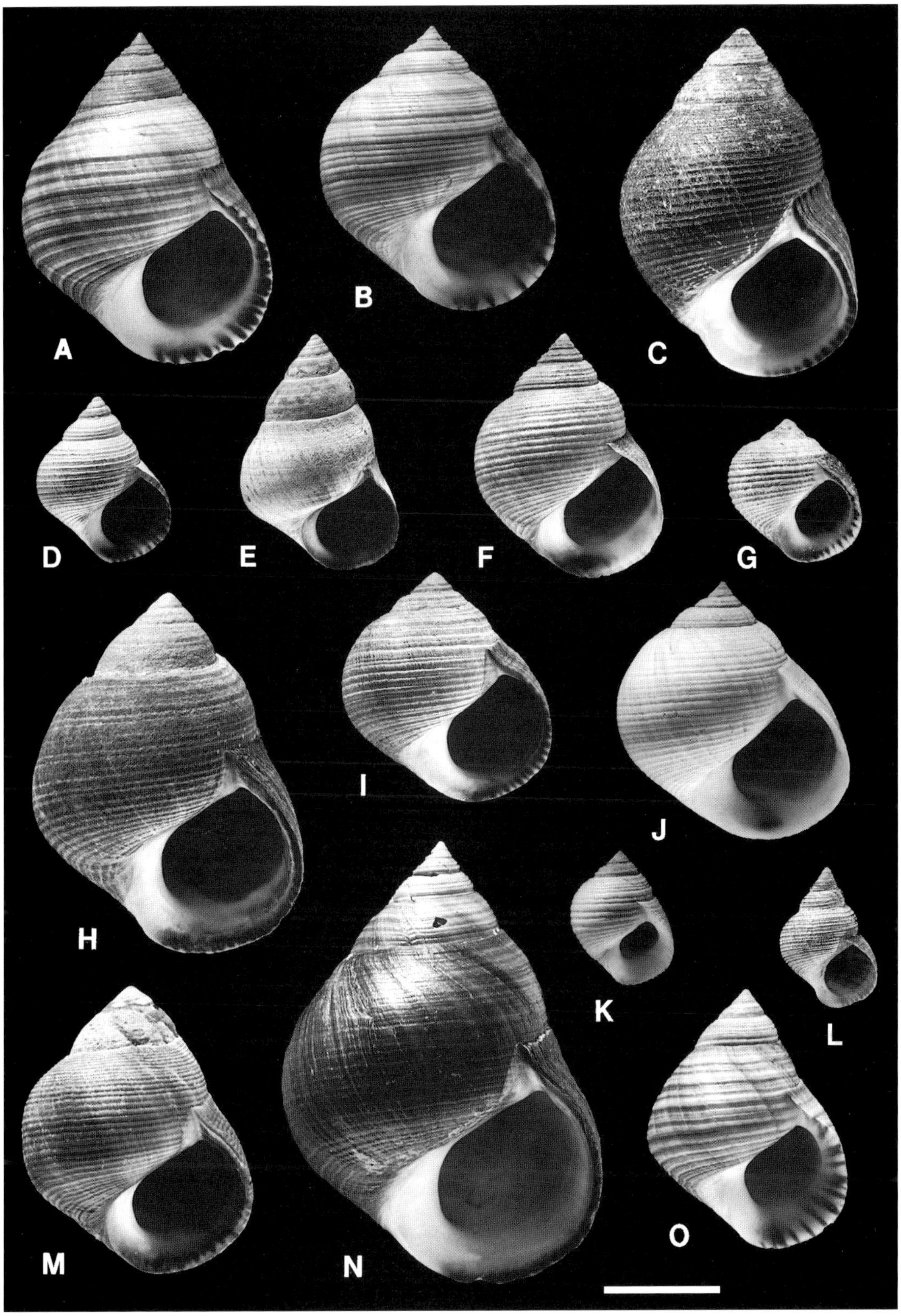

Figure 35. *Littorina (Littorina) littorea*. **A,** Lyme Regis, Dorset, England. **B,** St Margaret's Bay, Kent, England. **C,** Burtonport, Co. Donegal, Ireland. **D, E,** Solovetskiye Island, White Sea, Russia. **F,** Andernos, Arcachon Lagoon, France. **G,** Porto, Portugal. **H,** Appledore Island, Isles of Shoals, Maine. **I,** Penobscot, near Bucksport, Maine. **J,** Birchington, Kent, England. **K,** Seaford, Sussex, England. **L,** Kjose, Sjæland, Denmark (sheltered/brackish ecotype); SMNH. **M,** Porto, Portugal. **N,** Balta Sound, Shetland Islands, Scotland. **O,** Wigtownshire, Scotland. Scale bar=10 mm.

differences in thickness between samples collected before and after the arrival of the crab. He suggested that if *L. littorea* is itself a recent immigrant from Europe, it may already have been adapted to high levels of predation and therefore showed no phenotypic response to the selection pressure. Alternatively, extensive gene flow in this planktotrophic species may have prevented local adaptation to different levels of predation (Vermeij, 1982).

Ecotypic variation. Shell shape and thickness are predictably correlated with the exposure of the habitat to wave action in some species of *Littorina (Neritrema)*, but in *L. (Littorina) littorea* such correlations are weak or absent. Janson (1987a) found almost identical shapes on an exposed rocky shore and a boulder shore nearby, but Crothers (1992) claimed slightly larger apertures and lower spires on more exposed shores. Shell thickness and strength do not usually show predictable differences between shores of contrasting exposure (Currey & Hughes, 1982; Ash, 1989). However, Hylleberg & Christensen (1977) noted relatively thicker shells on the more exposed parts of a breakwater. These authors also found that shells were larger where exposed to waves, and suggested that this was due not to more favourable conditions for growth, but to differential migration of larger snails which were more tolerant of wave impact (Hylleberg & Christensen, 1978; Rasmussen, 1973). In one study of variation with tidal level on a single shore, samples from upper and lower zones did not differ in shape, but those from the lower level were more thick-walled (Tresierra-Aguilar, 1985).

In sheltered, brackish inlets a dwarf form occurs, that was described as var. *paupercula* by Jeffreys (1865) and as a distinct species *L. parva* by Teilman-Friis (1897). This form is frequent in lagoons in Denmark (Fig. 35L) and has a tall-spired shell only 10 to 15 mm in height. Teilman-Friis (1897) distinguished his species by slight differences in the relative length of the radula and in the form of the rachidian tooth, but the differences are not consistent (Johansen, 1901). Anatomically, it does not appear to differ from larger, typical examples (Fig. 33F). Shells from brackish habitats are not only smaller, but are also thinner in texture (Ankel, 1936b; Jaeckel, 1951; Vermeij, 1982). This form could be regarded as an ecotype, of similar appearance to the sheltered/brackish ecotype of *Neritrema* species, although probably not produced by genotypic adaptation.

When compared with the nonplanktotrophic species of the subgenus *Neritrema*, *L. littorea* shows relatively little differentiation between habitats over small distances. Its pelagic spawn and planktotrophic development lead to widespread dispersal and genetic uniformity (Janson, 1987a; Johannesson, 1992), so that any ecotypic variation has generally been explained by ecophenotypic effects rather than by natural selection leading to genetic differentiation. A striking experimental demonstration of this morphological plasticity was by Kemp & Bertness (1984), who showed that under conditions of rapid growth (achieved by reducing population density) a thinner shell, with more rounded whorls and lower spire, was produced than when growing slowly. Assuming that there was a fixed maximum rate of calcium carbonate secretion, they proposed that the thin, globose shell morphology with a larger internal volume permitted (or resulted from) a higher rate of body growth. This is an important result that could account for much of the local phenotypic differentiation in this species, in the absence of detectable genotypic differences. It explains not only the morphological differences between shells from populations of low and high density (Kemp & Bertness, 1984), but could also account for the thinner (Vermeij, 1982) and more globose (Dautzenberg & Fischer, 1912) shells at southerly latitudes (where higher temperatures may produce more rapid growth). However, a thinner shell is not always more globose; under brackish conditions shells are smaller, thinner and more tall-spired, which cannot be explained simply as the result of a connection between growth rate, shell deposition and shell shape. Morphological plasticity in *L. littorea* has also been demonstrated by measurements of growth rates in snails transplanted between geographically distant sites (Behrens Yamada, 1987). Natural selection by predators has been shown in this species; snails with relatively smaller apertures were more likely to survive attacks by oystercatchers in an experimental population (Robertson, 1992), but in the face of extensive gene flow the influence of local effects of this kind must be small.

Sculptural variation. The most striking sculptural variation in *L. littorea* is ontogenetic; in small shells spiral ribs are much more pronounced (Fig. 36A, B) and there are minute bristle-like projections on the periostracum (Fig. 36D). Raffaelli (1982) suggested a possible anti-predatory function for the coarse juvenile sculpture. Some fossil shells are more variable in sculpture (see Norwich Crag Monstrosities, p. 108; Fossil Record, p. 118).

Figure 36. *Littorina (Littorina) littorea*: **A, B,** juvenile shells; **C,** detail of adult shell sculpture; **D,** detail of juvenile shell sculpture. **A,** Poulsallagh, Co. Clare, Ireland; **B,** Loud's Island, Round Pond, Maine; **C,** Ballyvaughan, Co. Clare, Ireland; **D,** Robin Hood's Bay, Yorkshire, England. Scale bars: **A, B**=1.0 mm; **C**=1.0 mm; **D**=0.2 mm.

Sexual dimorphism. Jeffreys (1865) believed that male shells were narrower, with less rounded whorls and a more contracted aperture than females, and Daguzan (1977) also found the length of the aperture to be slightly smaller relative to shell height in females. Females have sometimes been found to be larger, either because they grow faster (Moore, 1937) or live longer (Daguzan, 1977), but neither Pelseneer (1926) nor Saur (1990) found significant dimorphism in shell height. If it exists at all, dimorphism in size and shape is slight, only detectable by morphometric analysis. Sexual selection for female size has been reported in Sweden, where males mate preferentially and for longer with larger (and therefore more fecund) females (Erlandsson & Johannesson, 1994).

Recent monstrosities. Since this species is so common, and has been collected commercially in great quantities for food, there have been many reports of occasional monstrous or malformed shells. The most commonly described abnormality is a single sharp keel on the shoulder (Philippi, 1848; Raeymaekers, 1889; Walrecht, 1950; Le Faucheux, 1961; Reed, 1982) (Fig. 37M, O). Elongate, turreted forms with rounded whorls are also well known (Jeffreys, 1865; Sykes, 1892; Marshall, 1898; Davis, 1972; Reed, 1982) (Fig. 37A). The rarest and most celebrated monstrosities show a reversal of coiling (Fig. 37B), and at least six such specimens are known to exist (Jeffreys, 1865; Marshall, 1916; one in BMNH; one in SM; three in USNM). In the majority of all these cases the cause of the abnormality is a severe breakage of the shell aperture, which interrupts the previously normal growth pattern (Fig. 37O), although if the damage occurred at small size it may no longer be visible high on the spire of an eroded adult. This is even the case in two of the five shells with reversed coiling that have been examined (Fig. 37B), the apical whorls being normal; these two shells are therefore not truly 'sinistral', but dextral and hyperstrophic. Other peculiarities following damage include the formation of a second aperture (Jeffreys, 1865). Cooke (1895) illustrated a shell with two apertures, which was claimed to be the result of fusion of two individuals, but this seems improbable. In the case of the turreted form, 'colonies' have been mentioned in Belfast Lough, Ireland and in Loch

Carron, Scotland (Jeffreys, 1865; Sykes, 1892; Marshall, 1898), suggesting that there might be a genetic basis for this particular abnormality.

Another obvious extrinsic cause of malformation is settlement of barnacles on the parietal area, columella or at the end of the suture, any of which may distort the growing aperture, although this is unusual (Beets, 1946b; Hermans, 1977). Despite living in the eulittoral zone and below, fouling of the shell is uncommon, but encrustation by barnacles (Roskell, 1962), bivalves (Seed, 1977) and coralline algae (van Marion, 1978b) have been recorded. In some localities fouling is heavier where snail populations are sparse, for at high snail densities the settlement of barnacles may be inhibited by grazing, bulldozing or foot mucus (Wahl & Sönnichsen, 1992).

There are several records of pearls up to 2.5 mm in diameter in *L. littorea* (Jeffreys, 1865; Coates, 1916; Elliott, 1921), but perhaps the most remarkable abnormality of all is the complete absence of a shell. Read (1965, 1966) found five shell-less, but otherwise intact, animals in seagrass beds in New England, and speculated that they may have been mutants that survived in the absence of predators, or that they may have been diseased.

Norwich Crag monstrosities. In the Pliocene Norwich Crag formation of Norfolk, England, there is a most interesting and peculiar case of abundant shell monstrosities that occur together with normal shells that are identical in appearance to Recent specimens. Some of the abnormalities are clearly the result of barnacle attachment (e.g. Harmer, 1920: pl. 52, fig. 16; see Spaink, 1974; probably also Wood, 1872: pl. 5, fig. 10, as *L. rudis*) or mechanical damage. These have parallels in Recent specimens, and are not exceptional. However, in the majority of the fossil monstrosities the abnormality cannot be ascribed to breakage or epizoic growth. These shells are depressed, elongate, variously carinate, angled at the periphery, with a concave whorl profile between shoulder and periphery, or may show combinations of these traits, and so far as can be seen the form of the whorls is constant right to the apex (Fig. 37E–K). This range of variation far exceeds that shown by Recent examples of *L. littorea*, and none of these forms has an exact Recent equivalent. Some of these curious shells were initially described as different species (six by Woodward, 1833; one by Brown, 1843), but soon correctly recognized as varieties of *L. littorea*, with which they are connected by intermediates (Wood, 1842, 1848). Many varietal names were used to describe the variation, ten by Wood (1848) and 14 by Harmer (1920). These two authors also confused some of the variants with *L. saxatilis* and even *L. obtusata* (Wood, 1872; Harmer, 1920, 1921, 1925; see Synonymy, p. 95), but personal examination of their own and other material has shown that only the single species *L. littorea* occurs in the Norwich Crag formation.

The stratigraphic and geographical range of the Norwich Crag monstrosities appears to be very limited. In the earlier Red Crag formation *L. littorea* is a rare fossil, but shells are mainly of normal appearance and no monstrosities have been reported. The Norwich Crag formation covers a wide area of East Anglia (Zalasiewicz & Gibbard, 1988), and yet the monstrosities have been found only at Bramerton, Thorpe, Postwick Grove and Whitlingham (Harmer, 1920–1921; Cambridge, 1975), over a distance of about 5 km. Elsewhere in Britain Mansel-Pleydell (1898) quotes Sykes that some specimens from the Portland Raised Beach in Dorset show 'a curious flattening of the whorls and a tendency to become keeled', although at the same locality Baden-Powell (1930) found *L. littorea* less variable than two other *Littorina* species. In The Netherlands van Regteren Altena, Bloklander & Pouderoyen (1954) illustrated a Pleistocene fossil with three carinae, but it does not exactly resemble the Norwich Crag monstrosities, and Beets (1946a) stated that comparable monstrosities were not represented in collections from The Netherlands. Strongly ribbed specimens have been seen from the Middle Tiglian stage (equivalent to the lower Norwich Crag) in The Netherlands (see Fossil Record, p. 118), but these are not carinate and are otherwise of normal shape. Norton (1976) mentioned carinate *Littorina* from the Pliocene of Iceland, but these belong to a different species (p. 246). The varietal names of Norwich Crag forms have occasionally been applied to Recent shells (Marshall, 1903, 1916, without illustration), but in almost 700 Recent lots of *L. littorea* examined during the present study no shells exactly equivalent to these monstrosities have been found. The closest are some elongate shells from brackish habitats in Denmark and the White Sea (Fig. 35E, L), but in these examples the whorl outline is normal, without keel or carination, and the degree of elongation is less extreme. In modern turreted shells the whorls are rounded and the sutures deep (Fig. 37A), unlike most of the elongate fossils (Fig. 37G–I, K). Modern carinate shells produced as a result of mechanical damage (Fig. 37M, O) are not similar to the carinate fossils (Fig. 37E, F, J, K). There is therefore no

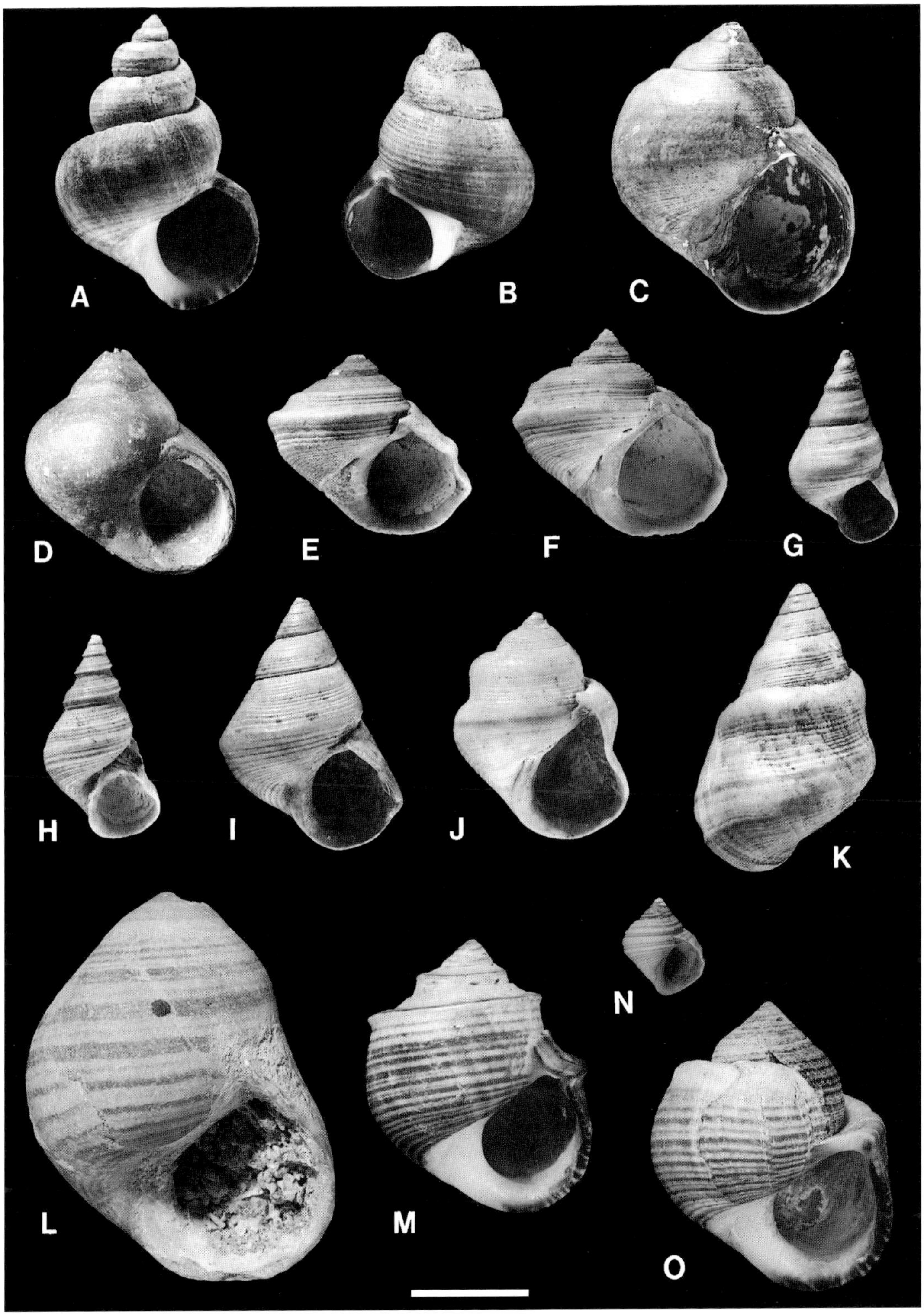

Figure 37. *Littorina (Littorina) littorea*: monstrosities and fossils. **A,** Belfast, Northern Ireland. **B,** Britain; BMNH 1937.7.12.1. **C,** Sutton, Suffolk, England; Red Crag, Pliocene; BMNH GG6832. **D,** Stratton Hall Pit, Levington, Suffolk, England; Red Crag, Pliocene; BMNH GG14583. **E,** Bramerton, Norfolk, England; Norwich Crag, Pliocene; BMNH G1981. **F, H,** Bramerton, Norfolk, England; Norwich Crag, Pliocene; NCM. **G,** Thorpe, Norfolk, England; Norwich Crag, Pliocene; BMNH G25580. **I,** *Littorina littorea* var. *conica* Harmer, 1920; holotype, NCM; Bramerton, Norfolk, England; Norwich Crag, Pliocene. **J,** *Littorina littorea* var. *truncata* Harmer, 1920; holotype, NCM; Bramerton, Norfolk, England; Norwich Crag, Pliocene. **K,** *Littorina littorea* var. *antiqua* Wood, 1848; holotype, BMNH G1979; Bramerton, Norfolk, England; Norwich Crag, Pliocene. **L,** bed G_1, Sidi Abd er Rahmane, Casablanca, Morocco; Post-Maarifien, Pleistocene; MNHNP. **M,** locality unknown; BMNH, Fowler Collection. **N,** Roozendaal, The Netherlands; Maassluis formation, Middle Tiglian, Pliocene; RGD 49E/0005. **O,** purchased at Billingsgate market, London. Scale bar=10 mm.

firm evidence for the occurrence of the Norwich Crag monstrosities outside the confines of this formation in a small area of Norfolk.

Detailed descriptions of two sections at Bramerton (Blake's Pit and Bramerton Common), most famous of the localities of the Norwich Crag monstrosities, have been given by Funnell (1961), Cambridge (1975, 1979) and Funnell, Norton & West (1979). There is, nevertheless, some doubt about the precise occurrence of the monstrosities. At each site, there are a lower and an upper shell bed, and Funnell *et al.* (1979) remarked that distorted specimens were abundant in both beds in both pits. However, at Blake's Pit Cambridge (1975) believed that monstrosities were confined to the upper shell bed. At the Common Pit Norton (1967) recorded '*L. littorea* var. *carinata*' only in the upper shell bed, whereas Cambridge (1979) did not find monstrosities here at all. Correlations between the two pits are not entirely clear (Funnell *et al.*, 1979), but foraminiferan assemblages suggest that the upper shell beds may correspond. Cambridge (1975) correlated the upper shell beds at the two sites partly on the basis of the occurrence of monstrosities of *L. littorea* in each. This confusion may be partly explained by the fact that authors have not distinguished between specimens distorted by barnacles or breakage and the distinctive Norwich Crag monstrosities. In summary, it appears that the latter are probably confined to the upper shell bed at Bramerton and are only common at Blake's Pit. At this site, Cambridge (1975) recorded that in a sample of 725 shells (some also identified as *L. rudis*, but only *L. littorea* is present in his own collection, pers. obs.) no less than 11% showed 'minor variations', and 12% were 'gross monstrosities'. Furthermore, in the same shell bed are found monstrosities and extreme varieties of *Nucella*, *Neptunea* and *Spisula* (Cambridge, 1975; Funnell *et al.*, 1979). The fossil assemblage of the upper shell bed suggests littoral sand or mudflats in a sheltered estuarine environment, with stony and shelly banks on which *L. littorea* lived, in a boreal climate (Cambridge, 1975; Funnell *et al.*, 1979).

Many authors have speculated upon the possible cause of the Norwich Crag monstrosities. Early suggestions were that lowered salinity in an estuary, occasional cold or icebergs could have been responsible (Wood, 1842, 1848; Forbes & Hanley, 1850), although Harmer (1920–1921) considered that this 'could hardly explain' the great variation in the fossils. Funnell *et al.* (1979) discussed the evidence supporting climatic deterioration and increasingly estuarine conditions in the upper parts of the Bramerton sections, and suggested that enfeeblement of the gastropods by low salinity or ice could have allowed barnacles to settle on their shells, causing distortions. Carination was said to be a response to rough water, but with no evidence whatever. Interestingly, at Bramerton Common Funnell (1961) reported distorted foraminifera at the level of the upper shell bed and below, indicating stagnant water at the head of a bay, possibly combined with freshwater influx.

Any explanation must take into account the extremely localized distribution of the monstrosities, their occurrence in a single shell bed, the lack of an obvious cause in the form of breakage or fouling in most specimens, and the absence of any similar modern examples of *L. littorea*. Low salinity does appear to have an effect on the shape of *Littorina* shells, being correlated with increased spire height (p. 29), but the Norwich Crag monstrosities are not simply elongated like some modern examples from brackish water (Fig. 35E, L). If any slow or sudden salinity change had been responsible, the effect would probably have been duplicated in some modern habitat, and yet modern parallels are unknown throughout the wide occurrence of this species in the estuarine habitats of the Baltic, The Netherlands and elsewhere. Analysis of contemporaneous pollen and foraminifera (Funnell *et al.*, 1979) shows that the Norwich Crag monstrosities lived in a period of climatic deterioration culminating in a cold or early glacial substage. Nevertheless, there is no evidence that temperature shock can produce malformations of shell growth in littorinids; if it could do so it would again be expected to be seen in some living populations, perhaps near the northern extreme of the modern range, but *L. littorea* from northern Norway and the White Sea are normal. However, the very limited geographical and stratigraphical range of the Norwich Crag monstrosities could imply that the environmental conditions under which they were formed were highly localized and perhaps of rare occurrence.

In fact a modern parallel does exist in the extraordinary Arctic monstrosities of *L. obtusata* (p. 216, Fig. 79C, D), which in their combinations of elongated, carinate and concave whorls do closely resemble the Norwich Crag shells. They are not connected at all with estuarine habitats, but occur at frequencies of less than 10% in otherwise normal populations in fully marine habitats, and are found only in parts of Iceland and northern Norway in the far north of the range of the species. No detailed investigation of these forms of *L. obtusata* has been made, but the animals are not obviously parasitized or otherwise abnormal, and a direct influence of low temperature upon shell growth is unlikely, because of their absence from other Arctic regions. It is suggested that they might be extreme genotypes of restricted geographical occurrence, which survive only where selection on shell form is relaxed, as in the absence of crab predators. A similar explanation could account for the Norwich Crag monstrosities of *L. littorea*. As pointed out by Funnell *et al.* (1979), the upper shell bed of Bramerton formed in a restricted marine basin, in which the population may have been genetically impoverished and isolated, and the shells could be those of a few or even a single generation. Genetic events in a small and isolated population could have resulted in the appearance of novel or extreme genotypes, which may have survived for a short while under peculiar environmental conditions.

Colour variation (Fig. 5). Five colour varieties have been named by Dautzenberg & Fischer (1912, 1925, see Synonymy, p. 95), which are respectively white, red-brown, red, orange, and with a single white band. The red phenotype is widely distributed from Brittany, to the British Isles, Sweden and north Norway (pers. obs.), but is always rare, never more than 5% of a population (Sacchi, 1974), and has not been recorded from North America. This highly distinctive form presumably has a genetic basis, and is comparable with the rare red morphs of some other species (e.g. *L. horikawai*, *L. sitkana*, *L. saxatilis*). The unbanded whitish phenotype is unusually frequent at some localities in southeastern England between Norfolk and the Isle of Wight, but is rare in Brittany (Dautzenberg & Durouchoux, 1900), Belgium (Warmoes, Backeljau & de Bruyn, 1988), the Barents Sea and Maine (pers. obs.). Bumpus (1898) believed American shells to have a less clearly banded pattern than British ones.

Distribution

Habitat. Most ecological studies of *L. littorea* have been made on rocky shores, where it is a characteristic inhabitant of levels between the high-water mark of neap tides and the extreme low-water mark of spring tides, being found in the zones of *Fucus vesiculosus* and *Ascophyllum nodosum*, of *Fucus serratus*, and extending into the *Laminaria* zone (Gowanloch & Hayes, 1926; Colman, 1933; Moore, 1937, 1940; Evans, 1947; T. A. & A. Stephenson, 1954a, b; Barsanova, 1961; Moyse & Nelson-Smith, 1963; Lewis, 1964; James, 1968b; Crapp, 1973; Underwood, 1973; Finet, 1978; Golikov & Kusakin, 1978; Bertness, Yund & Brown, 1983; Janke, 1986, 1990; Golikov, 1987). At these levels it occurs on various hard substrates, usually rock (although it does not favour chalk or limestone; Evans, 1947; Smith & Newell, 1955), but also wood and occasionally on larger algae such as *Laminaria* (pers. obs.). It is most abundant on sheltered shores and generally absent from the most exposed ones (Evans, 1947; T. A. & A. Stephenson, 1954a, b; Kuznetsov, 1960; Lewis, 1964; James, 1968b; Crapp, 1973; Finet, 1978; Fretter & Graham, 1980; Little & Williams, 1989). The upper limit of this species can be raised by wave splash (Moore, 1940; Evans, 1947; Daguzan, 1976a); one extreme report was of a colony among yellow lichens at 120 m above sea level on a cliff constantly wetted by spray on the exposed northeastern Atlantic island of St Kilda (Waterston & Taylor, 1906; Hewitt, 1907), although it could be that the animals were misidentified *L. saxatilis* (Cooke, 1914; Dacie, 1917) and a subsequent visit failed to locate them (Bagenal, 1957). On shores with protective clefts, stones and cover of *Pelvetia* and *Fucus spiralis* at high levels, *L. littorea* can extend just into the *Verrucaria* zone of the littoral fringe (Lewis, 1964), and it can be common among *Fucus* in pools in the littoral fringe (Barker & Chapman, 1990). Nevertheless, caging experiments (Behrens Yamada & Mansour, 1987) and observations of peculiar field conditions (Fischer-Piette & Seoane-Camba, 1962) have suggested that the upper limit may be set by food availability rather than tolerance of emersion. The lower limit is less clearly defined, since most studies of zonation have not extended into the sublittoral. The records of occurrence below the immediate sublittoral have often been from the colder and more northerly parts of the range (T. A. & A. Stephenson, 1954b; Lambert & Farley, 1968; Mathieson, Penniman & Harris, 1991), with maximum depths of 36 m in the Gulf of St Lawrence (Huntsman, 1918), 15–20 m in Russia (Golikov & Kusakin, 1978) and 60 m in northern Britain (Fretter & Graham, 1962, 1980) and the Skagerrak (SMNH). Predation may sometimes impose the normal lower depth limit (Huntsman, 1918; T. A. & A. Stephenson, 1954b).

Settlement of juveniles on rocky shores has, at some localities, been observed over a vertical range as wide as or wider than that occupied by the adults (Williams, 1964; James, 1968b; Crapp, 1973). Elsewhere, settlement appears to occur at lower levels or subtidally, and migration establishes the adult zonation pattern (Kuznetsov, 1960; Smith & Newell, 1965; Lambert & Farley, 1968; Matveeva, 1974; Daguzan, 1976a). However, Chase & Thomas (1995b) found recruitment on an artificial substrate only on the upper shore at a locality in southern Canada. The micro-topography of the shore may affect settlement patterns, for juveniles are often particularly abundant among barnacles (Williams, 1964; Crapp, 1973; Warmoes, Backeljau & de Bruyn, 1988; Little & Williams, 1989) and *Lichina* (Colman, 1940). The size distribution of the larger individuals can be complex and variable (Moore, 1940; Rasmussen, 1973; Gendron, 1977; Hylleberg & Christensen, 1978). Although *L. littorea* can tolerate freezing on the shore (Kanwisher, 1966; Murphy & Johnson, 1980; but see Matveeva, 1974), in harsh climates some or most individuals migrate to low-tide level or below during winter (Batchelder, 1915; Gowanloch & Hayes, 1926; Lambert & Farley, 1968; Rasmussen, 1973; Matveeva, 1974; Gendron, 1977; Bertness *et al.*, 1983; Vadas, 1992). In Britain there was no seasonal migration at a southern site in Kent (Smith & Newell, 1955), but some downshore movement was observed in winter at a northern site in Yorkshire (Williams & Ellis, 1975). At Plymouth a population dispersed both down and across the shore between January and July (Underwood, 1973). On the most favourable shores this relatively large species can be present in remarkable densities. In Europe densities do not often exceed 200 per m^2 (Moore, 1940; Lein, 1980; Norton, Hawkins, Manley, Williams & Watson, 1990), but under exceptional conditions, as in Scottish firths, may reach mean densities of 1200 to 3398 per m^2, and a maximum of 17500 per m^2 (Watson & Norton, 1985; Berry & Smith, 1987). In New England densities appear to be higher, even on average shores, often exceeding 800 per m^2 (Dexter, 1945, 1947; Lubchenco & Menge, 1978; Behrens Yamada & Mansour, 1987; Barker & Chapman, 1990; Norton *et al.*, 1990), although the highest figure found in the literature is 2436 per m^2 (Dippolito, Sienko & Smith, 1975). In the far north of the range densities are lower, reaching a maximum of only 72 per m^2, and usually much less, in the White Sea (Kuznetsov, 1960; Golikov & Kusakin, 1978; Golikov, 1987).

Overlap with other *Littorina* species is small. *Littorina saxatilis* is abundant on rocky shores throughout the geographical range of *L. littorea*, but the optimum level of the former species is generally further up the shore (T. A. & A. Stephenson, 1954b; James, 1968b); overlap may occur in atidal regions (e.g. western Sweden, K. Johannesson, pers. comm.) and in lagoonal environments where populations of *L. saxatilis* sometimes occur subtidally (Gilkinson & Methven, 1991). *Littorina compressa* has a restricted range in the northeastern Atlantic and overlaps with *L. littorea* in the upper eulittoral. Two further species, *L. obtusata* and *L. fabalis*, also occupy the mid and lower eulittoral of Atlantic shores, but are both typically found in association with larger algae rather than, like *L. littorea*, on the rock surface.

An unusual feature of the ecology of *L. littorea* is that, unlike most other species of *Littorina*, it is not confined to hard substrates, but is able to crawl over sand (Newell, 1958) and even soft mud (Häntzschel, 1938, 1941; Graf, 1956). Nevertheless, it is not usually found on large areas of unconsolidated sediment, but only on sheltered shores where scattered stones, wood, mussel beds, weed or worm tubes provide refuges of firm attachment (Thamdrup, 1935; Wright, 1936; Moore, 1937, 1940; Spooner & Moore, 1940; Dexter, 1945; Jaekel, 1952; Smith & Newell, 1955; Bandel, 1974; Finet, 1978; Brenchley & Carlton, 1983; Janke, 1986; Frid, 1988; Wilhelmsen & Reise, 1994). From such islands of hard substrate the animals make feeding excursions over the sediment (Newell, 1958), and without them are rolled around by the tide, although apparently not killed (Brenchley & Carlton, 1983). In such habitats *L. littorea* may reach densities of 350 to 480 per m^2 (Spooner & Moore, 1940; Wilhelmsen & Reise, 1994). At the lower levels of sand and mudflats, beds of the seagrass *Zostera* may also harbour large populations (Wright, 1936; Wohlenberg, 1937; Moore, 1937; Jaeckel, 1952; T. A. & A. Stephenson, 1954b; Kuznetsov, 1960; Fischer-Piette & Seoane-Camba, 1962; Rasmussen, 1973; Finet, 1978; Brenchley & Carlton, 1983; Golikov, 1987). Further up the shore *L. littorea* occurs among *Spartina* salt-marsh vegetation (Dexter, 1945; Brenchley & Carlton, 1983), where densities exceeding 1000 per m^2 have been reported (Dexter, 1947). As on rocky shores, the populations on sedimentary substrates may extend into the sublittoral, to 15 m near the mouth of the Baltic Sea, for example (Rasmussen, 1973; Hylleberg & Christensen, 1978; Jagnow & Gosselck, 1987). In winter some populations on tidal flats may migrate into pools or deeper water (Wright, 1936; but see Thamdrup, 1935), and animals have even been reported to burrow in soft sediments where they remain inactive for long periods (Rasmussen, 1973). Being tolerant of sedimentary conditions, *L. littorea* frequently occurs at river mouths, and may penetrate far into estuaries and fjords, surviving at salinities of 9.5‰ (Fischer-Piette, 1931; Spooner & Moore, 1940; Jaeckel, 1951; Brattegard, 1966; Hardwick-Witman & Mathieson, 1983; Jagnow & Gosselck, 1987).

Littorina littorea is an omnivorous grazer, including a wide range of algae in its diet, but displaying a marked preference for the small, ephemeral species of early successional stages, such as the green

Cladophora, Enteromorpha and *Ulva*, the brown *Ectocarpus* and *Scytosiphon*, and the red *Porphyra* (Hayes, 1929; Lubchenco, 1978, 1983; Lubchenco & Menge, 1978; Lein, 1980; Watson & Norton, 1985a, b; Imrie, Hawkins & McCrohan, 1989; Norton *et al.*, 1990; Vadas, 1992; Wilhelmsen & Reise, 1994). There is a much lower preference for robust algae such as *Fucus*, although reproductive tissues may be selected, and as germlings the plants are highly susceptible (Lubchenco, 1980, 1983; Watson & Norton, 1985a; Barker & Chapman, 1990). As an adult plant, the fucoid *Ascophyllum* appears repellant (Watson & Norton, 1985a, b), but even this is grazed when a sporeling (Miller & Vadas, 1984). Encrusting algae such as *Hildenbrandia* and *Ralfsia* can be eaten, but cause substantial wear to the radular teeth (Bertness *et al.*, 1983). Perennial reds such as *Mastocarpus, Chondrus* and *Corallina* are hardly touched, protected by structural defences, but epiphytes may be grazed from their surface (Lubchenco & Menge, 1978; Watson & Norton, 1985a; Norton *et al.*, 1990). On the mid and low shore the snails graze mainly on the rock surface and on the understorey of ephemerals, but at higher levels and on more exposed shores drift seaweeds may be an important food source (Watson & Norton, 1985a). In addition, aufwuchs and diatom films are consumed (Hunter & Russell-Hunter, 1983; Hawkins, Watson, Hill, Harding, Kyriakides, Hutchinson & Norton, 1989). On sediments diatoms are grazed from the surface of particles, so *L. littorea* is not primarily a deposit feeder in such environments (Brenchley, 1987; Frid & James, 1988), although in salt marshes it eats leaves and rhizomes of *Spartina* (Bertness, 1984), and also detritus (Pourreau, 1979). While grazing on rocks, small animals such as *Mytilus* larvae, barnacle cyprids and ostracods may inadvertently be ingested, but whether they can be digested is unclear (Blevgad, 1915; Hayes, 1929; Dexter, 1947; Hylleberg & Christensen, 1978; Brenchley & Carlton, 1983). Snails may gather on dead fish (Hayes, 1929) and have been reported to feed on dead crabs (Berry, 1961). On mudflats the eggs of the gastropod *Ilyanassa* are ingested, apparently inadvertently, since the capsule walls cannot be digested (Brenchley, 1982). Mucous trails induce a feeding response, and may be enriched by the adhesion of phytoplankton and organic matter (Imrie, 1992; Davies, Hawkins & Jones, 1992).

Range (Fig. 38). *Littorina littorea* is distributed from the White Sea (and perhaps Spitzbergen) to southern Portugal in the eastern Atlantic, and in the western Atlantic from Labrador to Virginia. Since it is a large and familiar species, its distributional limits have been rather precisely documented, to the extent that changes in its range during historic time can be followed. In particular, there is an extensive literature on its possible introduction and subsequent spread in North America, which is discussed in the following section (p. 116). However, on a smaller scale the distributional pattern has sometimes been confused by the commercial collection and transport of this species for food in Europe, and there are also several undoubted cases of introduction by man.

In the eastern Atlantic the most northerly record is southern Spitzbergen, where it was recently recorded for the first time and said to be rare (Rózycki, 1991). This record is surprising and has not been verified, but the accompanying figure is evidently of this species. On the Kola Peninsula it is rare or absent on open coasts, but more frequent in enclosed bays (Kuznetsov, 1960; Matveeva, 1974). The species is common in inlets throughout the White Sea, but absent from the Zimniy Coast and Mezenskiy Gulf in the northeast (Kuznetzov, 1960; Golikov & Kusakin, 1978). It is also frequent in northern Norway. Although *L. littorea* apparently occurred in Iceland during the post-glacial warm period (Thoroddsen, 1892), there are no reliable modern records of it living there. If not incorrectly localized (Mørch, 1869), shells from the coasts of Iceland may be subfossil specimens (Thorson, 1941), or may have been transported on ships from Denmark (Óskarsson, 1973). From the Faeroes there are three nineteenth century records, but despite collecting efforts it has not been found since (Mørch, 1868; Spärck & Thorson, 1933). The first record from Greenland (Fabricius, 1780) was a misidentification of *L. saxatilis* (Møller, 1842), and another (Philippi, 1846a) was probably based on incorrectly localized material, since not even post-glacial subfossils are known from Greenland (Spjeldnaes & Henningsmoen, 1963a, b). *Littorina littorea* penetrates the Baltic as far as the west coast of Rügen and Bornholm (Schlesch, 1937; Fretter & Graham, 1980; Jagnow & Gosselck, 1987; Wallentinus, 1991).

In the British Isles this species occurs all around Britain, Ireland, the Outer Hebrides, Orkneys and Shetlands. Interestingly, it is very rare on certain small offshore islands, including the Isles of Scilly (Smart & Cooke, 1885; Marshall, 1898; Smith & Newell, 1955; Crisp & Southward, 1958), Lundy (Picton, 1978; Hawkins & Hiscock, 1983; Cole, 1991) and St Kilda (Bagenal, 1957). In the Channel Islands it is frequent only on Alderney, closest to the French mainland, and is otherwise extremely

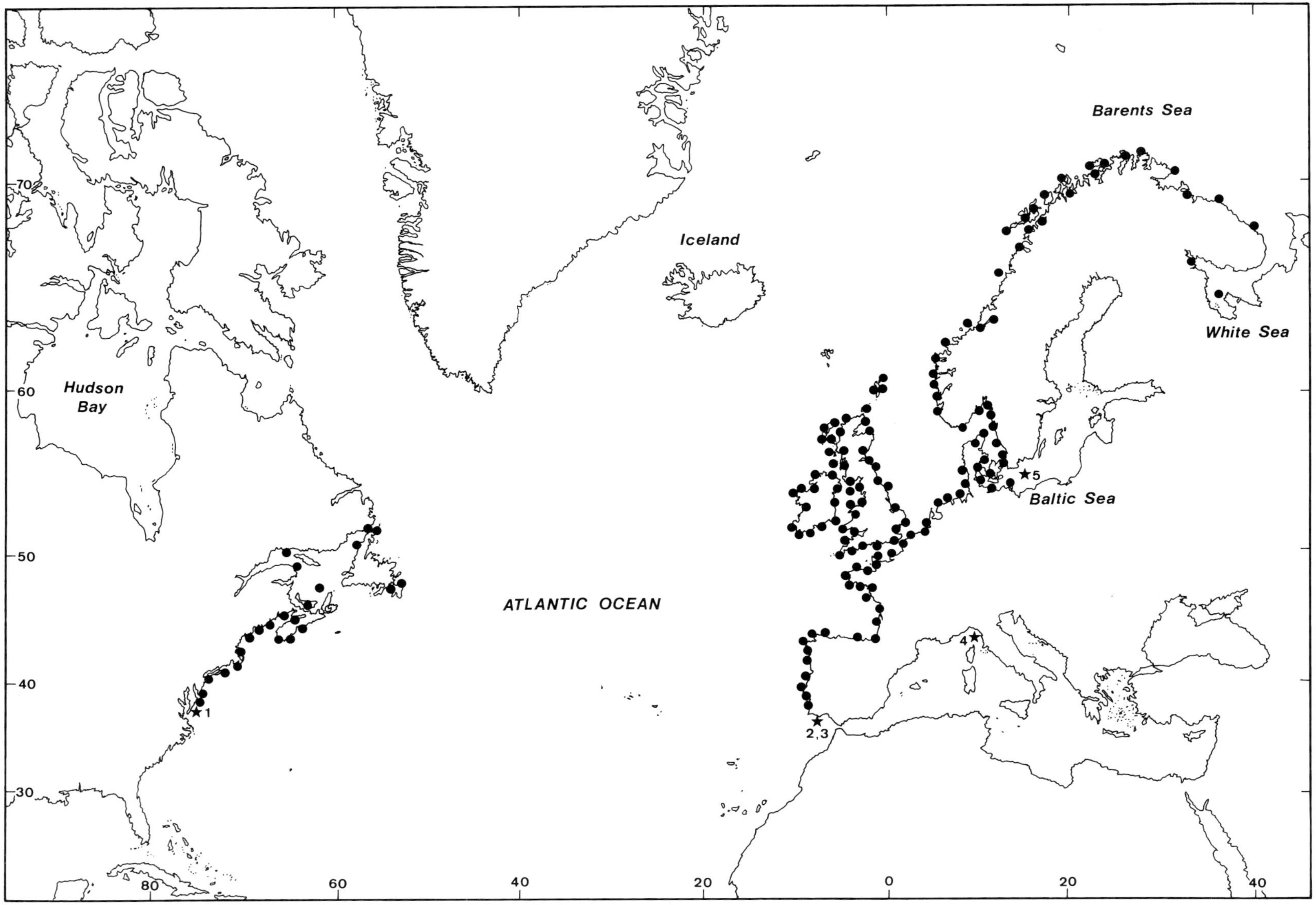

Figure 38. *Littorina (Littorina) littorea*: distribution map. Key: circles, specimens examined; asterisks, records from literature (1, Vermeij, 1982; 2, Nobre, 1931; 3, Carvalho, 1945; 4, Barsotti & Campani, 1982, an introduction, now extinct; 5, Jagnow & Gosselck, 1987). There is in addition a literature record from southern Spitzbergen (Rózycki, 1991).

rare (Marshall, 1915; Crisp & Southward, 1958); dead shells found on Jersey have probably been imported for food (Marshall, 1898). Habitats apparently suitable for the species are present at all these island locations. This scarcity on small offshore islands has been explained by a combination of the probable low rates of colonization by plantotrophic larvae from mainland populations, and the difficulty of maintaining populations on islands if endogenous larvae are swept away by currents (Johannesson, 1988). Estimates of the length of the pelagic life of *L. littorea* range from a month (Thorson, 1946) to six or seven weeks (Williams, 1964); it may be extended by up to six weeks under unfavourable conditions (Thorson, 1946) and metamorphosis may take place before settlement (Hayes, 1927; Smith & Newell, 1955). Despite this length of planktonic existence, eggs and larvae of *L. littorea* are only common in the inshore plankton (Caullery & Pelseneer, 1910; Hayes, 1927); for example, Fretter & Shale (1973) recorded the veligers in plankton tows taken at one mile, but not at three miles, from the shore near Plymouth. It is therefore probable that few planktonic larvae reach offshore islands. In the absence of strong currents, eggs and larvae may develop on the bottom among detritus (Fish, 1972), but this is unlikely on small islands. The occurrence of *L. littorea* on islands is probably a function of distance from the mainland, strength and direction of prevailing currents, and of hydrographic conditions around the islands.

Further south the species is common in Brittany, and as far as the Île d'Oléron. Thereafter, it is not found on the open sandy coasts between the mouth of the Gironde and the Spanish border, but does occur in the Arcachon Lagoon (Crisp & Fischer-Piette, 1959; DGR, 1987). Over the past century its distribution in the Bay of Biscay appears to have fluctuated. Nineteenth century authors recorded it from the mouth of the Gironde and from three localities on open rocky coasts in the vicinity of Biarritz at the head of the Bay of Biscay, but by 1955 it was absent from these sites, and near Biarritz was collected only in an estuary at St-Jean-de-Luz (Crisp & Fischer-Piette, 1959). A similar pattern has been repeated along the north coast of Spain. Hidalgo (1917) recorded it from the whole north coast, but in 1949 Fischer-Piette (1953, 1955) failed to find it between the French border and the Ría de Betanzos in Galicia. By 1956, however, the situation had changed again, and *L. littorea* was common at Plencia near Bilbao (Fischer-Piette, 1956). More recently, it has been found at Ribadeo, Laredo and Castro Urdiales in northern Spain (DGR, 1987). Further west the species is abundant in Galicia (Fischer-Piette, 1953; DGR, 1987). The present southern limit is uncertain. Working with collections made at or before the beginning of the twentieth century (Fischer-Piette, 1957), Nobre (1886, 1898, 1931) and Hidalgo (1917) reported many localities around the coast of Portugal, and there seems no doubt that it was then living as far south as the Algarve (also undated records by Carvalho, 1945). Hidalgo (1917) gave Gibraltar as the southern limit, but there is no confirmation of its occurrence there. Working in 1956 and 1957, Fischer-Piette (1957a, 1958) found it no further south than Vila Nova de Milfontes. It is possible that this northward retreat has continued since then, for *L. littorea* was not found at this locality, at several localities near Setúbal, nor at Sesimbra, Estoril, Peniche or Nazaré in 1987 (pers. obs.), where it was formerly common (Nobre, 1886, 1898, 1931; Hidalgo, 1917). In the BMNH there is a collection made at Porto in 1984. The explanation for these changing patterns is not clear, but presumably some climatic effect is involved. The scarcity of the species at the head of the Bay of Biscay can be related to the Mediterranean climate of the area, whereas in Galicia the climate is cooler, although the effect cannot be simple, since it survives in the enclosed, warm waters of the Arcachon Lagoon (Crisp & Fischer-Piette, 1959; Wells, 1965). The adults are tolerant of high temperatures (up to 46°C, Evans, 1948), so the action of temperature may be indirect, perhaps on predators or on recruitment of larvae (Crisp & Fischer-Piette, 1959; Wells, 1965). At least in North America small populations near the southern limit are apparently dependent on sporadic recruitment that occurs only in colder than average years (Wells, 1965; Kraeuter, 1974). The movement of the southern limit of *L. littorea* is striking, and this species might repay investigation as a sensitive indicator of climatic change.

There are a number of early records from the Mediterranean, including Corsica (Requien, 1848), Sicily (Philippi, 1836, 1844a, 1846a), the Adriatic (Jeffreys, 1869) and Algiers (Weinkauff, 1868), but at least some of these were of dead shells transported from the Atlantic, perhaps for food (Aradas & Benoit, 1870) and others may have been incorrectly localized. However, in 1981 a breeding population of about 400 animals was discovered in a large rock pool at Castiglioncello in northwestern Italy (Barsotti & Campani, 1982). Their origin is unknown, but presumably was a deliberate or accidental introduction, and over a few years the population subsequently disappeared (Campani, quoted in Johannesson, 1988). Another unexplained record is of 34 shells collected at

Lanzarote in the Canary Islands, although it is not known if these were from a living population (Moolenbeek, 1989), and transport as a food item is again possible.

Spread in North America. There has been much debate as to whether *L. littorea* is native or introduced in North America, but in either event its rapid spread along the northwestern Atlantic coast following its discovery early in the nineteenth century provides one of the most well-documented examples of the dispersal of a marine species. The known facts are as follows. There is only one Pleistocene record, based on a single shell from southwestern Nova Scotia, collected apparently in situ in a deposit of mid-Wisconsinan age (33000–44000 BP) (Wagner, 1977). At the site of a Norse settlement in northern Newfoundland, two specimens have been found in undisturbed material several feet beneath the surface of the beach terrace on which the Norse houses were built, and therefore presumably predate the settlement (Bird, 1968). A number of shells have also been found in middens of two Indian sites in Nova Scotia, and two more in New Brunswick, which, from radiocarbon ages of the shells themselves, have been dated to between 1000 and 1500 AD (Clarke & Erskine, 1961; Clarke, 1963a, 1971). The earliest records following European settlement are attributed to J. W. Dawson, who, 30 years after the event, recollected finding it at Pictou and Prince Edward Island in the southern Gulf of St Lawrence in about 1840 (Dawson & Harrington, 1871; Verrill, 1874; Ganong, 1887). During the first half of the nineteenth century the fauna of Nova Scotia was poorly known, so there are no other published records or specimens to substantiate these occurrences. That of New England, however, was relatively well studied, and there is no doubt that *L. littorea* was then absent (e.g. Gould, 1841; Stimpson, 1851; see Ganong, 1886). The first detailed account of its occurrence and spread in Nova Scotia was by Willis (1863), who found it first at Halifax in 1857. Thereafter, its southward spread along the coast was documented by many authors. It reached the Bay of Fundy in 1861 or 1862; Portland, Maine, in 1870; the north side of Cape Cod between 1870 and 1872; Woods Hole south of Cape Cod in 1875; Long Island Sound in 1879; Atlantic City, New Jersey, in 1888; and Cape May in 1890 (Morse, 1880; Verrill, 1880; Ganong, 1886; Bequaert, 1943; Alexander, 1947; Clarke & Erskine, 1961; Dexter, 1961; Clarke, 1971; Carlton, 1982; and references therein). In about 40 years it had spread from the Gulf of St Lawrence to New York, a distance of some 1500 km, and had become the dominant intertidal herbivore on many rocky shores in the region.

There are several possible explanations for these facts. The single Pleistocene occurrence is interesting, but not directly relevant to the 'native or introduced' debate, since all the rocky shore fauna of the northwestern Atlantic probably became extinct during the last glaciation. Ice then covered rocky habitats as far south as Long Island, and further south the coastline is of inhospitable sandy beaches. Repopulation from Europe must then have occurred during the 13000 years since the glacial maximum (Ingólfsson, 1992). If correctly dated, the record does show that in the past *L. littorea* inhabited North America in the absence of human activity. It has been doubted that the shells from Indian middens imply the existence of an established population predating European settlement, and suggested that they represent chance arrival of drifted adults (Johannesson, 1988). The rarity of the finds and absence of any reports from raised beaches are indeed peculiar, when contrasted with the frequency of *L. littorea* in Pleistocene and post-glacial deposits in Europe (see Fossil Record, p. 118). Nevertheless, the numbers of shells (at least 12 from one site, Clarke & Erskine, 1961), their presence at four sites and apparent utilization as food, suggest that local populations were established before European settlement.

Whether or not these Indian shells were from established populations, it is interesting to speculate how the species first arrived in North America. Drifting of larvae is one obvious means but, as mentioned earlier (p. 115), eggs and larvae of *L. littorea* are only common close to the shore. Even with a possible pelagic lifespan of ten weeks (Thorson, 1946), the direct Atlantic crossing of 3000 km from the British Isles to Newfoundland is too far for transport of larvae, with currents averaging about 10 km per day (Kraeuter, 1974). However, stepping-stone migration via Iceland and Greenland is not impossible, and could have occurred in the past (Kraeuter, 1974). The species does not now live on either, but occurred on Iceland during the post-glacial climatic optimum, about 7000–5000 BP (Thoroddsen, 1892). Nevertheless, there are as yet no post-glacial records from Greenland, which has been taken to indicate that no population was established either there or in North America (Spjeldnaes & Henningsmoen, 1963a, b; but see Clarke, 1963b). As an alternative, some authors favour the rafting of adults, possibly on driftwood, as is known to have occurred with iron nails (Bird, 1968; Clarke, 1971; Johannesson, 1988). Here again, the great distance makes this unlikely, and the difficulty of establishing a population of a planktotrophic species from a small number of colonists must be considered (Johannesson, 1988). One genetic comparison between populations on either side of the Atlantic showed surprising differences and has been interpreted as evidence for a native North American population (Berger, 1977; Vermeij, 1982). This result has since been questioned, for another study found the populations to be remarkably similar (Morris, 1979, quoted in Johannesson, 1992).

The alternative to these natural means of crossing the Atlantic is introduction by man. Vikings could have carried the snails deliberately as food or accidentally with ballast, either directly from Norway or from the population existing on Iceland during the secondary climatic optimum (1000–1200 AD) (Spjeldnaes & Henningsmoen, 1963a, b). Against this possibility is the discovery of two shells apparently pre-dating a Viking settlement in Newfoundland (Bird, 1968). The problem with any hypothesis that requires a native population, or one predating European settlement, is to explain why an established population in the Gulf of St Lawrence failed to spread southwards until about 1850, coinciding with increased transatlantic shipping. Several suggestions have been made. Clarke (1971) suggested that the native population was a warm-stenothermal race, which inhabited enclosed bays in Nova Scotia and the Northumberland Strait. It is known that the southwestern Gulf of St Lawrence is a refugium for some warmer-water species (Bousfield & Thomas, 1975). Around 1870 there was a peak in summer temperatures, which could have allowed invasion of the outer coast of Nova Scotia and New England (Clarke, 1971). Current patterns may also have been unfavourable for dispersal from the Northumberland Strait to the outer coast of Nova Scotia (Clarke & Erskine, 1961). Once outside the Gulf of St Lawrence the prevailing currents are southerly and along the coast (Mathieson *et al.*, 1991), and would carry larvae towards New England. Another idea is that the native population became extinct as a result of climatic deterioration at the end of the thirteenth century (Carlton, 1982, 1992; Brenchley & Carlton, 1983). If it is accepted that there never was an established native population, or that it became extinct, the remaining possibility is introduction by European settlers, which has been favoured by a number of authors (Ganong, 1886, 1887; Carlton, 1982, 1992; Johannesson, 1988). Introduction might have been inadvertent, for example with ballast, or deliberate, since *L. littorea* was at that time a common item of food in northwestern Europe (e.g. Jeffreys, 1865; see also Bequaert, 1943). Finally, it may be that these various ideas are not mutually exclusive; for example, a native race might have been dispersed from the Gulf of St Lawrence by European shipping, or genetically altered by introductions from Europe (Clarke & Erskine, 1961; Clarke, 1971). In summary, there is insufficient evidence to choose between the competing explanations, since all depend on the interpretation of a handful of American shells that predate European arrival. The problem may be resolved by further information about the presence or absence of the species in natural postglacial deposits in the northwestern Atlantic, and about its genetic structure.

The present day northern limit of *L. littorea* in North America is not well established; collections have been seen from as far north as the Strait of Belle Isle in southern Labrador (as also recorded by Bequaert, 1943; Vermeij, 1982), where either temperature or the southerly Labrador current (Mathieson *et al.*, 1991) might be limiting. The southernmost occurrences have been reported in some detail. In New England the most southerly large area of suitable habitat (rocky shore and estuary) is on the Connecticut coast, and this is the probable source of larvae for settlements further south. Southerly currents in this area are in some years sufficient to transport larvae to mid-New Jersey and the eastern side of the Delmarva Peninsula, where on an otherwise unsuitable sandy coast settlement occasionally occurs on pilings and rock jetties (Richards, 1929; Alexander, 1947; Wells, 1965; Kraeuter, 1974). The years in which settlement has been recorded on the Delmarva Peninsula correlate partly with cooler than average air temperatures and strong river outflows, and can be explained by known circulation patterns (Kraeuter, 1974). The southernmost record in the literature appears to be Wachapreague, Virginia (37°36'N; Vermeij, 1982). Wells (1965) suggested that settlement is in some way inhibited above about 21°C, the summer water temperature at the southern limit on both sides of the Atlantic.

There are also four records from the west coast of North America, one from Puget Sound (Hanna, 1966), two from San Francisco Bay and one from Trinidad Bay in California (Carlton, 1969, 1992). These probably represent chance introductions with *Crassostrea* or *Mercenaria* from the Atlantic coast, and have not produced established populations (Carlton, 1969, 1992).

Fossil record (Fig. 37C–L). In contrast to the majority of *Littorina* species, *L. littorea* is common as a fossil, presumably owing to its ability to live on sheltered, sedimentary shores. The oldest fossils to have been identified as *L. littorea* are from the Tjörnes sequence in northern Iceland, dated at about 3 to 3.5 Ma (Gladenkov, Norton & Spaink, 1980; Gladenkov, 1981). However, these are here assigned to the sister species *L. squalida*, at present restricted to the Pacific (p. 89).

Fossil shells identical to modern *L. littorea* first appear in the Red Crag formation of East Anglia. The stratigraphy of this coarse, sandy deposit is complex, with much reworking, and there is disagreement about both dating and assignment in relation to the Pliocene/Pleistocene boundary. Recent estimates date the Red Crag formation to between 2.4 and 3.2 Ma, and some authors place the Pliocene/Pleistocene boundary between its lower Walton Crag and upper Butley Crag members, at the point where there is palynological evidence for climatic cooling (reviews by Zalasiewicz & Gibbard, 1988; Jones & Keen, 1993). If, however, the start of the Pleistocene is fixed by reference to its type locality in Calabria, Italy, at about 1.6 Ma, then not only the entire Red Crag formation, but also the overlying Norwich Crag formation, should be assigned to the Pliocene; this view is followed here. *Littorina littorea* is a rare fossil in the Butley Crag of the late Red Crag formation (Fig. 37C), and several specimens are also known from the Newbourn Crag (Fig. 37D), which may be slightly earlier (Harmer, 1920–1921; pers. obs. of material in SMC, NCM, IM, collections of P. E. Long and P. G. Cambridge). The shell illustrated in Figure 37D bears some resemblance to *L. squalida* in its swollen final whorl. Deposits of comparable age to the Red Crag also occur in continental Europe. Fragments of *L. littorea* have been seen from the Ouwerkerk borehole in Zeeland (83–94.5 m depth; NNML), of an age corresponding to the Belgian Merksemian stage, although the possibility of contamination from overlying deposits cannot be ruled out (A. W. Janssen, pers. comm.).

The Norwich Crag formation overlies the Red Crag in East Anglia and is dated at about 1.6 to 2.0 Ma (Zalasiewicz & Gibbard, 1988). Here *L. littorea* is abundant in spasmodic beds of shelly sand and silt, deposited on intertidal or shallow subtidal mud or sand flats, in both coastal and estuarine conditions (Harmer, 1920–1921; Norton, 1967; Cambridge, 1975, 1979; Funnell, 1979; Funnell *et al.*, 1979). The local occurrence of monstrous shells (Fig. 37E–K) in the Bramertonian stage of the Norwich Crag has been described earlier (p. 108). This species is also a common fossil in boreholes of a similar age in the southern Netherlands, in the vicinity of Rotterdam, Breda and Tilburg. Of particular interest is a sample of about 200 shells from the Maassluis formation, of the Middle Tiglian stage, at Roozendaal (72–82.5 m below surface, RGD 49E/0005). Many of these are entirely typical, but in about one third the sculpture is unusually well developed; alternate ribs above the periphery are enlarged, so that four ribs are prominent and rounded. This type of sculpture is particularly distinctive in juveniles (Fig. 37N), and adult shells of this form do not exceed 15 mm in height. In isolation, such shells could be mistaken for *L. saxatilis s.l.* However, the identification is based on the narrow grooves between the ribs, the concave spire outline, the colour pattern of brown lines on the ribs and within the aperture, and the presence of the complete range of intermediates connecting these shells with the typical form. In modern shells, slight enlargement of alternate ribs is occasionally seen, but never to this degree.

Elsewhere in Britain, *L. littorea* is common in some raised beach deposits from the Late Pleistocene (e.g. at Portland, Baden-Powell, 1930; Gower Peninsula, George, 1932, and Chatwin, 1932). *Littorina littorea* is also a common Pleistocene fossil in The Netherlands (Beets, 1946; van Regteren Altena *et al.*, 1954; RGD), in the vicinity of the White and Barents Seas (Knipowitsch, 1900; Merklin, Zarkhidze & Ilina, 1979), and in the latest glacial and post-glacial deposits of southern Scandinavia (Brøgger, 1901; Antevs, 1928). Since the last glaciation the salinity of the Baltic Sea has fluctuated; in the deposits formed during a period of salinity higher than that prevailing at the present, dated at 4500–7000 BP, *L. littorea* penetrated far into the Baltic and is so common that this stage in the evolution of the sea has been named the 'Litorina Sea' (e.g. Munthe, 1931; Kessel & Raukas, 1967; Ignatius, Axberg, Niemistö & Winterhalter, 1981).

There is only a single Pleistocene record from North America, based on a shell from Nova Scotia of mid-Wisconsinan age (33000–40000 BP) (Wagner, 1977). In addition, specimens dated at 500–1000 BP are known from Indian middens in the Canadian Maritimes, prompting the debate as to whether *L. littorea* is or is not native to North America, as reviewed previously (p. 116).

Fossils are also known from outside the present-day latitudinal limits of this species, demonstrating shifts in geographical range with climatic change. In Morocco *L. littorea* occurred as far south as Cape Rhir (30°40'N) during cold periods of the Anfatien and Maarifien stages (Neuville & Ruhlmann, 1941;

Lecointre, 1950, 1952, 1963; Brebion, 1979; Fig. 37L), about 0.5–1 Ma (Biberson, 1977). In contrast, during warm intervals the range was greatly extended northwards. A specimen in ZISP shows that *L. littorea* reached the Taymyr Peninsula in northernmost Siberia (76°N) about 100000 BP during the Kazansevskaya Transgression. In Spitzbergen, occurrence of *L. littorea* characterizes deposits from the post-glacial warm period (2500–6500 BP; Feyling-Hanssen, 1955). Fossils have also been recorded from Iceland during post-glacial warm periods (Thoroddsen, 1892; Spjeldnaes & Henningsmoen, 1963a; see also Thorson, 1941).

Discussion

Intraspecific variation. In comparison with some other *Littorina* species, *L. littorea* shows a low degree of morphological variation. There have been several studies of genetic variation in this species, and again almost all show a remarkably low level of differentiation between populations (E. M. Berger, 1973; Wilkins, O'Regan & Moynihan, 1978; Wilkins & O'Regan, 1980; Janson, 1987a). In the most detailed study Johannesson (1992) found values of Nei's (1972) genetic identity between 0.986 and 1.000 in 13 populations from Norway to Spain, with no clear geographical pattern evident. Assuming that allozyme variation is selectively neutral, the genetic uniformity may be explained by high gene flow between populations; this suggests that *L. littorea* does disperse relatively widely, as predicted from its planktotrophic development. This dispersal probably takes place mainly in inshore waters, as indicated by the scarcity of larvae in offshore plankton samples and of adults on offshore islands (p. 115).

Nevertheless, two results conflict with this picture of genetic uniformity. Berger (1977) compared a population from Cape Cod with one from Roscoff, Brittany, and found considerable allelic differences. Berger's results give a Nei's (1972) genetic distance of 1.02, which is above the average for pairs of sibling species (Janson, 1987a). However, not all populations on either side of the Atlantic are so different; a comparison between samples from Wales and Maine gave a genetic distance of only 0.012 (Morris, 1979, quoted in Janson, 1987a). A further anomalous result was reported by Johannesson (1992), who found ten 'mysterious' individuals at Oostende, Belgium, with unusual combinations of otherwise rare alleles. Johannesson made a possible connection between these two anomalous findings, with the suggestion that a sibling species could be involved, explaining the peculiarities of both the Roscoff population and the Oostende individuals. There is as yet no further genetic information on this possibility, nor has the present morphological study suggested that more than one species is involved, but further work is desirable.

Another feature of the genetic results is the unusually low level of heterozygosity in *L. littorea* as a whole (Berger, 1977; Beardmore & Morris, 1978; Wilkins *et al.*, 1978; Wilkins & O'Regan, 1980; Fevolden & Garner, 1987; Janson, 1987a; Ward, 1990; Foltz, Shumway & Crisp, 1993; mean expected heterozygosity per locus from 4 studies=0.048, Tatarenkov, 1992). This low level of variation suggests that not only the American populations (Berger, 1977), but most of the European ones too, have at some time in the past suffered a severe genetic bottleneck (Johannesson, 1992). One possibility is that during the Pleistocene glaciations the species was reduced to a relatively small population. It is notable that Vuilleumier & Matteo (1972) found some evidence of greater genetic variation in American than French populations, supporting Bumpus' (1898) idea of increased variability in a species undergoing rapid range expansion.

Spread in North America. As discussed in detail earlier (p. 116), the possible introduction and observed spread of *L. littorea* in North America is one of the most well-documented cases of invasion by a marine invertebrate. This has provided the opportunity for a number of studies on the impact of the invader on the native species and environment. *Littorina littorea* is now one of the dominant intertidal herbivores throughout its range in the northwestern Atlantic, and here it exerts a more profound effect on the composition of the algal flora than in Europe (reviews by Hawkins, Hartnoll, Kain & Norton, 1992; Vadas & Elner, 1992). Growth studies have provided some evidence of competitive interactions between *L. littorea* and the native *L. saxatilis* (Behrens Yamada & Mansour, 1987), and the gastropod *Ilyanassa obsoleta* has been partly displaced from mudflats by *L. littorea* (Brenchley & Carlton, 1983). The invader may even cause a direct modification of the physical habitat by removing sediment and clearing salt marshes (Bertness, 1984).

Phylogenetic relationships. Shell characters, some anatomical details, allozyme frequencies and mitochondrial DNA sequences all suggest that among Recent species the sister-taxon of *L. littorea* is *L. squalida*, now restricted to the Pacific (see p. 90 for detailed discussion). Although the Pacific ancestry of *L. littorea* is not in doubt, it is not clear whether the two species differentiated as a result of vicariant speciation following trans-Arctic migration and subsequent climatic cooling, or whether their separation preceded the opening of the Bering Strait (3.5–4 Ma). The fossil record is yet insufficiently complete to provide details of the speciation process itself. The shells in the Butley Crag of the Red Crag formation are indistinguishable from modern *L. littorea* (Fig. 37C), but at least one (Fig. 37D) is known, among the very few from the possibly earlier Newbourn Crag of the same formation, which in its swollen last whorl recalls the shape of *L. squalida* (cf. Fig. 28). Taken together with the presence of shells that appear to be *L. squalida* in the Upper Pliocene of Iceland, this supports the anagenetic transformation of an Atlantic isolate of *L. squalida* into *L. littorea*. However, analysis of mitochondrial DNA sequences has cast doubt on this scenario of vicariant allopatric speciation; the molecular estimates of the age of separation of these two species range from 3.28 to 10.07 Ma (mean 6.42 Ma), suggesting that the origin of *L. littorea* might predate the opening of the Bering Strait (Reid, Rumbak & Thomas, 1996; p. 365).

Littorina littorea differs from *L. squalida* in several anatomical characters (Table 3). These include the shape of the penis and size of the copulatory bursa; the latter, at least, may possibly be a species-isolating mechanism, since differences between sympatric pairs of sister-species are usually more pronounced than between allopatric ones (pp. 14, 18). If this is so, the marked differences between the bursae of *L. littorea* and *L. squalida* could imply that at some time the two species were at least partially sympatric. If regarded as species-recognition characters, however, the origin of these anatomical differences in complete isolation (following vicariance of the ancestral range) is more easily explained.

The fossil *L. petricola* (p. 91), from the Pliocene of California and Oregon, also appears to be related to *L. littorea* and *L. squalida*.

Similar species. Adult specimens of *L. littorea* are usually readily distinguished from sympatric *L. saxatilis* on shell charcters alone; larger size, fine and regular sculpture, shallow sutures, onion-shaped profile and the common colour pattern of many narrow brown bands are all characteristic of *L. littorea*. Small juveniles, however, look very different from adults (Fig. 36A, B); the shells are whitish with sculpture relatively stronger than in adults; at a similar size *L. saxatilis* is coloured as in adults, is more globose, with less regular sculpture and a larger, smooth protoconch of the nonplanktotrophic type. *Littorina compressa* is sympatric with *L. littorea* in parts of northwestern Europe; it can usually be told apart by its smaller size, yellow or white shell, often with black lines in the grooves between the relatively larger and more strongly raised ribs, and by its slightly shouldered final whorl. However, a few small, pale examples of *L. littorea* (Fig. 35K) could be mistaken for *L. compressa*. Characters of the oviduct and penis separate *L. littorea* from all sympatric species.

More similar is the sister-species *L. squalida* from the Pacific, and differences are summarized in Table 3. Anatomically, the large copulatory bursa, long penial filament, and asymmetrically convex egg capsules containing fewer eggs are diagnostic of *L. littorea*. Differences in the shells are subtle, but identification is nevertheless unequivocal in almost all cases. The shell of *L. littorea* is sometimes more elongate; the sutures are not so much impressed, so that the spire whorls appear more flat-sided; the shell often has an onion-shaped profile, with a concavity near the apex and a convexity at the final whorl, which is not swollen; the aperture is oval and may even appear slightly constricted; and the sculpture of fine, low ribs is more regular, with no ribs enlarged or carinate. Sexual dimorphism is pronounced only in *L. squalida*.

Littorina (Littorina) kasatka Reid, Zaslavskaya & Sergievsky, 1991

(Figs 39–42)

Synonymy

? *Littorina kurila* m. *tesselata* Ryabinina, in Spassky, 1961: 298 (Cape Zavodskoy and Kopyto Peninsula, Kamchatka; infrasubspecific; not *Litorina tessellata* Philippi, 1847=*Littoraria tessellata*).

Littorina (Littorina) kurila—Golikov & Kusakin, 1978: 87–89 (in part, includes *L. sitkana*, *L. subrotundata*, *L. aleutica*, *L. natica*; not Middendorff, 1848=*L. sitkana*).

Littorina (Littorina) kasatka Reid, Zaslavskaya & Sergievsky, 1991: 1–6; figs 1–4; figs 8–11 (radulae); figs 12, 13 (oviducts); figs 14–18 (penes); figs 19, 20 (head pigmentation) (Kasatka Point, Iturup Island, Kurile Islands; holotype BMNH 1990050, seen, Reid *et al.*, 1991: fig. 1, Fig. 41A herein; 9 paratypes BMNH 1990051, seen, Reid *et al.*, 1991: fig. 2, Fig. 41B herein; 38 paratypes BMNH 1990052 (alcohol), seen; 10 paratypes ZISP (alcohol), seen. Reid & Golikov, 1991: 11, 13; fig. 36 (oviduct).
Littorina kasatka—Egorov, 1992: 170; fig. 1H.

Taxonomic history

This recently described species was discovered during a genetic study of *L. sitkana* by Zaslavskaya & Sergievsky, and independently during anatomical studies of material from ZISP by Reid. The few specimens in ZISP had previously been misidentified as *L. kurila*, and some literature references to *L. kurila* (e.g. Golikov & Kusakin, 1978, and others listed in full in Synonymy of *L. sitkana*, p. 146) may have included this species.

The identity of *L. kurila* morph *tesselata* Ryabinina, in Spassky, has not been verified by voucher specimens. It was described as a small shell, presumably smooth (since the name *kurila* has generally been applied only to smooth shells), with a tessellated pattern of pale spots on a dark background, found in the supralittoral zone. Among the three sympatric *Littorina (Neritrema)* species which can resemble *L. kasatka* in shell characters (see Similar Species, p. 127), only *L. subrotundata* is known to produce a tessellated pattern, and then only rarely. It seems likely that Ryabinina's name applies to *L. kasatka*, but since it was proposed as an infrasubspecific 'morph', it is unavailable (ICZN, 1985: arts 1b, 16).

Diagnosis

Mature shell height 2.4–11.0 mm. Shell turbinate to globular. Mineralogy of external irregular-prismatic calcite and internal crossed-lamellar aragonite. Surface smooth, lacking spiral grooves and strong microstriae. Colour mid-brown, often with 5–16 rows of smudged whitish dashes, more prominent at periphery and on base, occasionally forming tessellated pattern or spiral bands; aperture purplish brown, columella white to pinkish brown. Penis with small, triangular filament, 10–20% total length; single or no mamilliform gland. Pallial oviduct of two consecutive loops of albumen gland followed by loop of capsule gland; straight section short; bursa small, near anterior end of straight section. Spawn and development not known, but probably pelagic capsules and planktotrophic development. Distribution: northeastern Okhotsk Sea, Kamchatka, Kurile Islands and Hokkaido.

Anatomical description

Head-foot (Fig. 39G–I). Head and tentacles black, pale tips to tentacles and snout, unpigmented stripe across eye and partly or entirely across tentacle base; sides of foot pale to darkly mottled or unpigmented. Operculum yellow-brown, becoming dark brown only in largest specimens; opercular ratio 0.46–0.62.

Penis (Fig. 39A–H). Base wrinkled, unpigmented; single large, distal, mamilliform gland in populations from Akkeshi and Nemuro in Hokkaido, but absent in samples from Kurile Islands and Gulf of Shelikov. Filament short, 10–20% of total length, triangular. Sperm groove open, ending terminally. Vestigial penes were present in some females in a sample from Koen-misaki (3 km NE of Nemuro, Hokkaido; DGR, 1993); whether this might be an effect of TBT pollution, as in some other littorinids (Fioroni, Oehlmann & Stroben, 1990; Oehlmann, Liebe, Watermann, Stroben, Fioroni & Deutsch, 1994), is unknown.

Sperm. Euspermatozoa 32–36 µm. Paraspermatozoa 9–11 µm diameter, oval; vesicles large, distinct; nucleus sometimes visible (measurements on material fixed in formalin and stored in alcohol).

Oviduct (Fig. 39J–M). Large renal oviduct adjacent to albumen gland. Large initial loop and smaller second loop of albumen gland, followed by loop of capsule gland; straight section short, jelly gland not enlarged or septate. The extent of the capsule gland is not clear; the densely stippled area in Figure

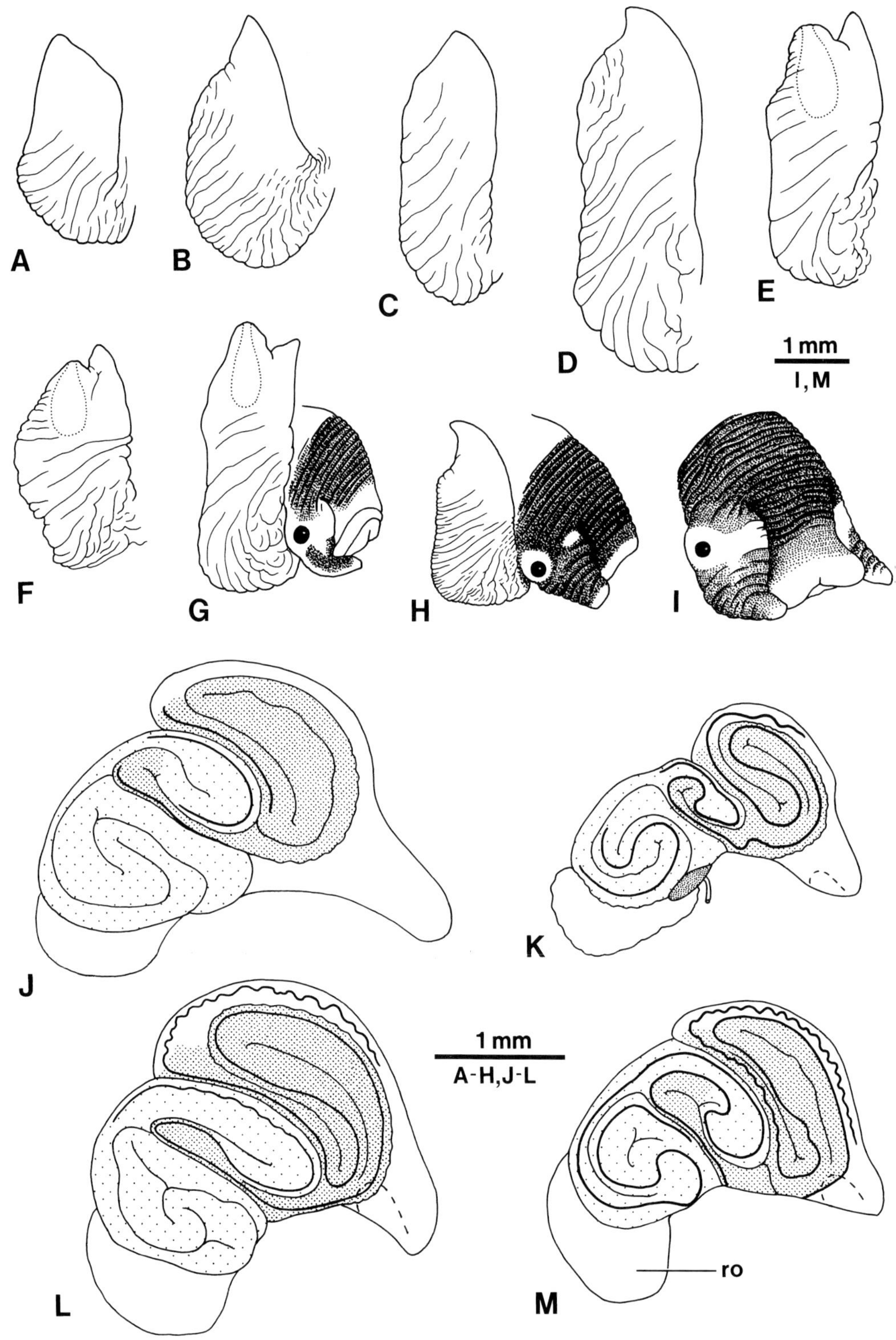

Figure 39. *Littorina (Littorina) kasatka*: **A–H,** penes; **G–I,** head pigmentation; **J–M,** pallial oviducts. **A, M,** Nataliya Bay, Urup Island, Kurile Islands, Russia; **B, C, I, L,** Kasatka Point, Iturup Island, Kurile Islands, Russia; **D, J,** Kasatka Bay, Iturup Island, Russia; **E–G,** Aikappu, Akkeshi, Hokkaido; **H,** Paramushir Island, Kurile Islands, Russia; **K,** Koen-misaki, 3 km NE of Nemuro, Hokkaido. Shell heights: **A,** 6.8 mm; **B,** 7 mm; **C,** 7.7 mm; **D,** 7.8 mm; **E,** 3.4 mm; **F,** 3.1 mm; **G,** 3.3 mm; **H,** 5 mm; **I,** 9.5 mm; **J,** 7.5 mm; **K,** 5 mm; **L,** 8.1 mm; **M,** 6.6 mm. For shading conventions of pallial oviducts see p. 6. Abbreviation: ro, renal oviduct. Bursa visible by dissection only in **K**; positions indicated in **L** and **M** reconstructed from serial histological sections.

39J–M is opaque pale fawn in living and preserved specimens, whereas the preceding small second loop of the albumen gland is opaque cream or white, and might also be part of the capsule gland. In some specimens the most distal part of the capsule gland is slightly brownish, and may correspond to a translucent portion of the gland; since this is indistinct, it has not been shown in the figures. Histological preparations did not stain sufficiently clearly to allow identification of the glandular components of the oviduct. Bursa small (visible only in histological sections or in life when filled with sperm), originating near anterior end of straight section. Seminal receptacle simple. The renal oviduct was not included in the figures by Reid, Zaslavskaya & Sergievsky (1990).

Spawn and development. Not known, but large capsule gland and short straight section of pallial oviduct without enlarged jelly gland indicate pelagic egg capsules. Duration of breeding season unknown, but available samples were collected between May and August, and all were mature. Protoconch not seen, but development is likely to be planktotrophic, as in all other *Littorina* species with pelagic eggs.

Radula and foregut (Fig. 40). Relative radular length 1.42–3.50. The radula of this species is unusually variable in the form of the rachidian tooth and shape of the major cusps of the teeth. Rachidian: length/width 0.82–1.50; central cusp pointed or leaf-shaped; one smaller pointed cusp and one outer denticle on each side. Lateral: four pointed cusps, third from inside largest. Inner marginal: four pointed cusps, third from inside largest. The largest cusp on each of the five central teeth is sometimes long and a little rounded (Fig. 40B). Outer marginal: 6–8 pointed cusps; narrow neck, projecting flange on inside of base only. Eleven radulae have been examined and appear to show regional differences. In four specimens from the type locality, Iturup in the southern Kurile Islands, the rachidian is rather square (length/width 0.82–0.93), its central cusp about twice the size of the flanking pair, and the cusps of all teeth are pointed (Fig. 40E–G). In three examples from Paramushir and the Gulf of Shelikov, in the northern part of the range, the rachidian is more rectangular (length/width 1.33–1.50) and all cusps less pointed (Fig. 40H). Three radulae from Akkeshi, Hokkaido, show a rachidian of intermediate shape, but all main cusps are elongate and rounded (Fig. 40A, B), quite unlike the examples from Iturup. Nevertheless, a single radula from Nemuro, Hokkaido, (Fig. 40C, D) connects these extremes of variation. The somewhat recurved cusps of one radula illustrated by Reid *et al.* (1990: figs 10, 11) are an artefact produced by heating during preparation. Salivary glands enlarged.

Shell variation (Fig. 41)

So far, this species is known from only 12 collections from eight localities, and the shells are rather uniform in shape, SH varying between 1.38 and 1.57. Although the whitish dashes on the shell are often inconspicuous, and may be absent, they are a useful recognition character.

Distribution

Habitat. Data with collections in ZISP indicate that *L. kasatka* occurs on rocks and boulders in the upper and middle littoral zone on moderately sheltered shores (Reid *et al.*, 1991). However, at both Akkeshi and two localities near Nemuro, Hokkaido, the species was found only on bare rocks in the littoral fringe, on exposed headlands; it did not extend down into the eulittoral among sparse barnacles and tufts of fucoid algae, and did not occur on adjacent shores more protected from wave action (pers. obs.). In all the museum collections of *L. kasatka* seen so far, the species was mixed with *L. sitkana*; at the sites in Hokkaido the two were collected on the same shores, but with little overlap between them, *L. sitkana* being present mainly in the upper eulittoral zone. The diet has not been recorded.

Range (Fig. 42). Reid *et al.* (1991) recorded *L. kasatka* from an untraceable locality in northwestern Kamchatka on the Gulf of Shelikov (between Nepropusk and River Gank), and from Paramushir, Ryponkich, Urup and Iturup Islands in the Kurile chain. This range is extended with new records from Aikappu near Akkeshi, and from two localities just east of Nemuro in northwestern Hokkaido (BMNH). These Japanese localities lie in an area influenced by a prevailing cold current from the north.

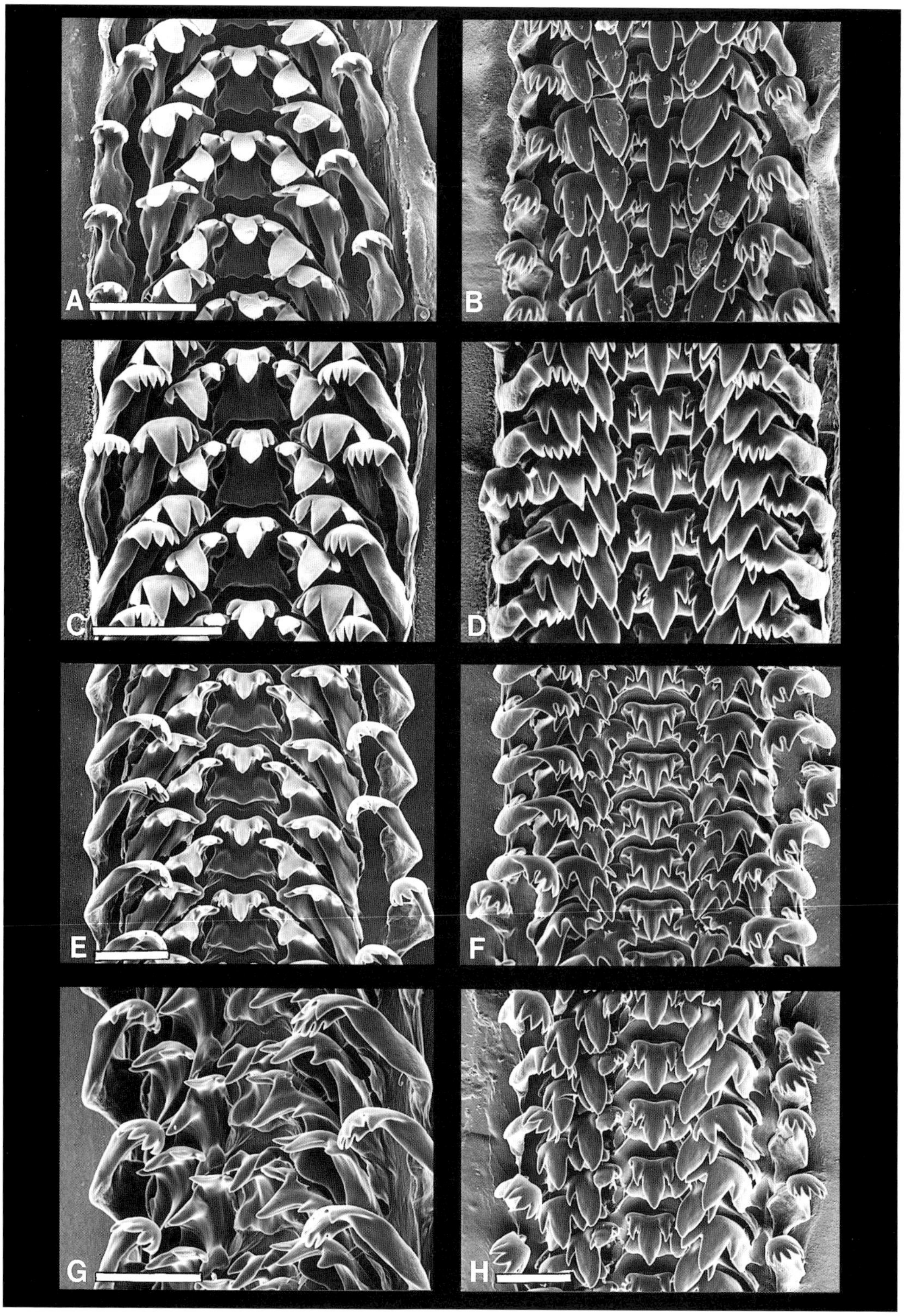

Figure 40. *Littorina (Littorina) kasatka*: **A, B**, two views of radula (shell H=4.4 mm), flat (**A**) and at 45° (**B**); **C, D**, two views of radula (shell H=4.5 mm), flat (**C**) and at 45° (**D**); **E–G**, three views of radula (shell H=8.1 mm), flat (**E**), at 45° (**F**) and side view (**G**); **H**, radula (shell H=7.0 mm), at 45°. **A, B**, Aikappu, Akkeshi, Hokkaido; **C, D**, 5 km NE of Nemuro, Hokkaido; **E–G**, Kasatka Point, Iturup Island, Kurile Islands, Russia; **H**, Paramushir Island, Kurile Islands, Russia. Scale bars: **A, B**=50 μm; **C, D**=50 μm; **E, F**=50 μm; **G**=50 μm; **H**=50 μm.

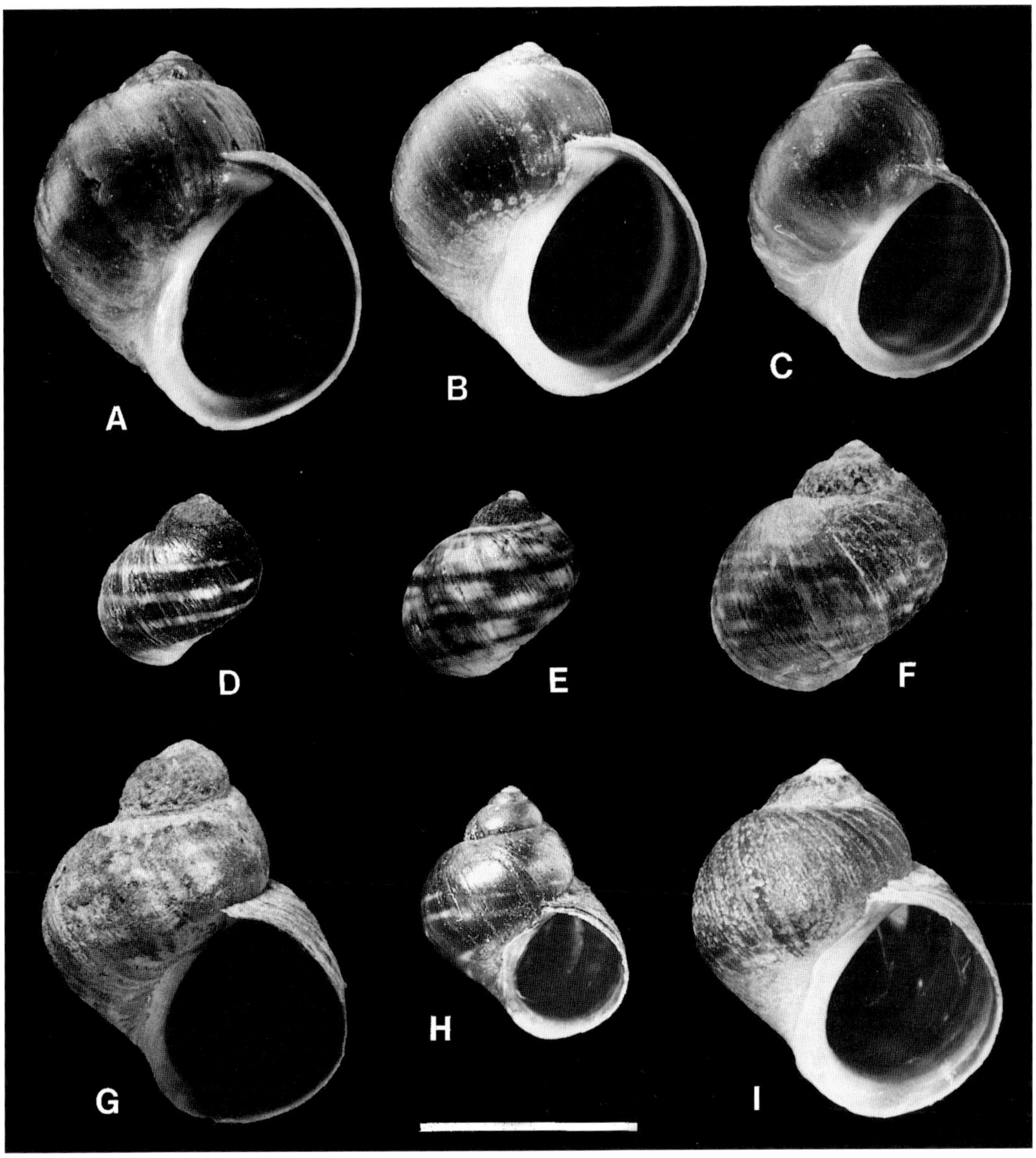

Figure 41. *Littorina (Littorina) kasatka.* **A, B,** *Littorina (Littorina) kasatka* Reid, Zaslavskaya & Sergievsky, 1991; holotype (**A**), BMNH 1990050; paratype (**B**), BMNH 1990051; Kasatka Point, Iturup Island, Kurile Islands, Russia. **C,** Nataliya Bay, Urup Island, Kurile Islands, Russia. **D, E, G,** Aikappu, Akkeshi, Hokkaido. **F,** Paramushir Island, Kurile Islands, Russia. **H,** Koen-misaki, 3 km NE of Nemuro, Hokkaido. **I,** Kasatka Bay, Iturup Island, Kurile Islands, Russia. Scale bar=5 mm.

Discussion

Intraspecific variation. Although relatively uniform in shell features and in the diagnostic form of the pallial oviduct, this species shows unusual variability in the shape of the radular teeth. In addition, a single mamilliform penial gland is present in the samples from Hokkaido, but this is absent from the other collections examined. Future collections may clarify the patterns of variation in these two characters, but at this stage there seems no strong reason to divide the species. Although penial form is often considered diagnostic in littorinids, such variation is not unprecedented. In all *Littorina* species in which mamilliform glands occur, their number is variable; when the number is small, this variability is more noticeable, 1 to 3 in *L. keenae* and 0 to 7 in *L. brevicula*, for example. In these cases the overall shape of the penial filament and base are relatively constant. This is also true in *L. kasatka*, and furthermore in those specimens lacking a gland the area in which it would have been present is puckered, suggesting that not all trace of the structure has been lost. Seasonal variation in penial

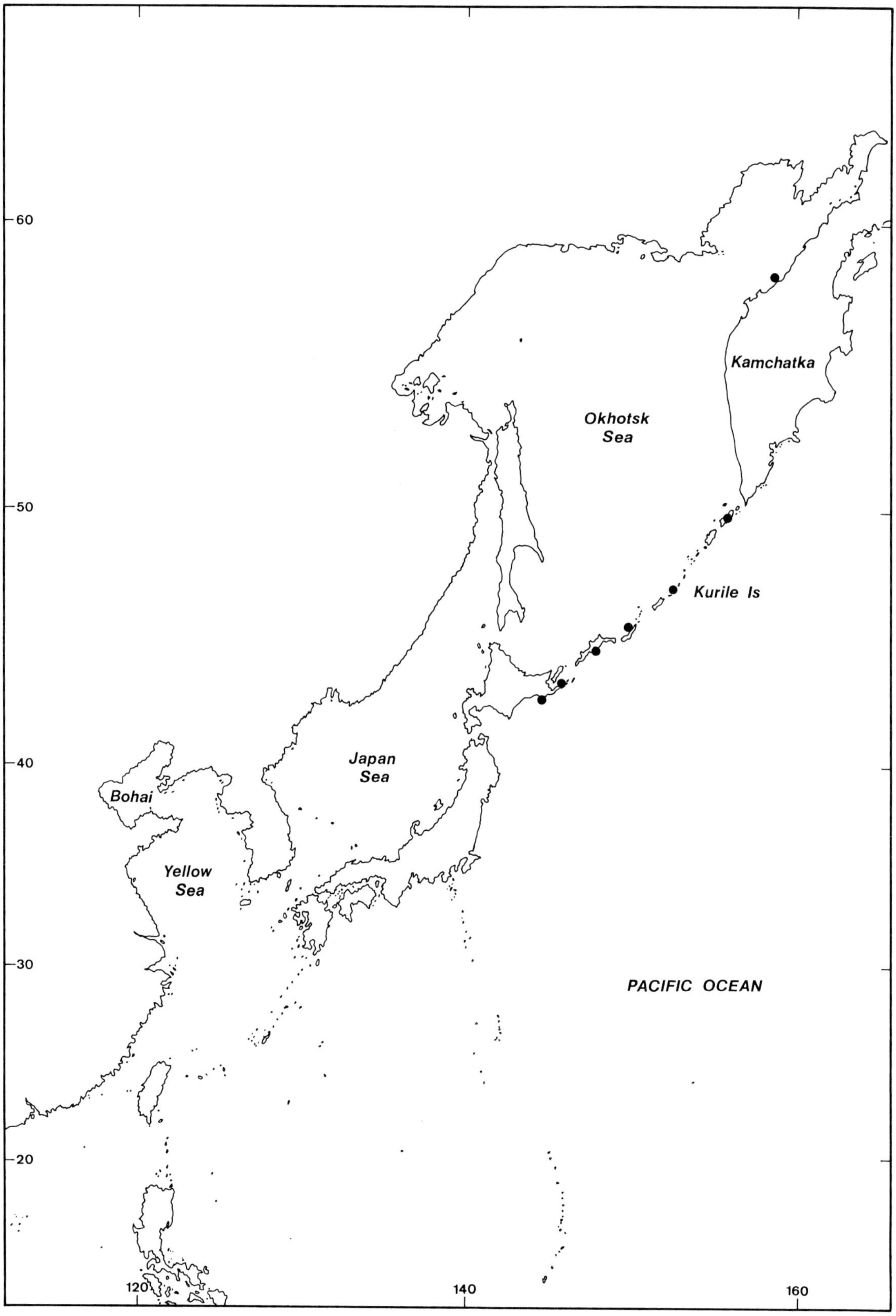

Figure 42. *Littorina (Littorina) kasatka*: distribution map. Key: circles, specimens examined (note that record from northwestern Kamchatka is not precisely localized).

glands has not been reported in other species and is considered unlikely here. It is possible that the variation is on a geographical scale; a comparable example is the presence and absence of the mamilliform gland in *Peasiella roepstorffiana* (Nevill) (Reid, 1989b). The variation of the radula is striking, but intermediates link the two extremes; radular tooth shape is known to be variable in some other *Littorina* species and may show allometry, both teeth and cusps being more elongated in smaller specimens (p. 21).

Phylogenetic relationships. Superficially, the shell of *L. kasatka* resembles smooth examples of *L. sitkana* and *L. subrotundata*, but allozyme studies have shown that there is nevertheless a large genetic distance between them (Zaslavskaya, 1995; Fig. 116C). This is not unexpected, since details of the pallial oviduct show that *L. kasatka* probably does not produce a benthic spawn and is therefore excluded from the subgenus *Neritrema*. Cladistic analysis of morphological characters places *L. kasatka* in a polychotomy among *L. squalida*, *L. littorea*, *L. brevicula*, *L. mandshurica* and the subgenus *Neritrema* (p. 351). The single or absent penial gland could suggest a basal position within this group, since there has been a trend towards an increasing number of glands in the genus as a whole, but this character is too variable for much reliance to be placed upon it. Phylogenetic analysis of mitochondrial DNA sequences has supported the position of this species in the polychotomy, without demonstrating any close relationship with other living species (Reid, Rumbak & Thomas, 1996; p. 360).

Similar species. This is one of a group of five *Littorina* species occurring in the northern Pacific, which often have smooth, brown shells, the others being *L. sitkana*, *L. subrotundata*, *L. natica* and *L. aleutica*. Their distinguishing characters are summarized in Table 5 (pp. 160–161). All except *L. aleutica* are sympatric with *L. kasatka* over at least part of its range. Of these five, *L. kasatka* is by far the least frequent in collections. Its anatomical characters are also the most distinctive, the penis and pallial oviduct both being diagnostic. In the others the mamilliform glands are numerous and the jelly gland of the pallial oviduct is swollen and septate. It is more difficult to identify *L. kasatka* on shell characters alone. *Littorina natica*, *L. aleutica* and *L. subrotundata* are usually more patulous, often with a pattern of dark lines on a pale ground colour, and in the last two species spiral sculpture may be present. Smooth forms of *L. sitkana* are the shells most similar to those of *L. kasatka*; the two are often found on the same shores and their shells are sometimes indistinguishable. *Littorina sitkana* is larger (up to 23 mm), the last whorl slightly more inflated, the columella relatively wider, and never shows a flecked or tessellated shell pattern. In addition, the pigmentation of the head-foot is darker in *L. sitkana*, which (once identification has been confirmed anatomically) can provide a rapid means of separating mixed samples. In mixed samples from Japan the operculum of *L. sitkana* is orange-brown and that of *L. kasatka* a paler yellow-brown; however, in larger specimens of *L. kasatka* from the Kurile Islands the operculum is as dark as in *L. sitkana*.

Littorina (Littorina) brevicula (Philippi, 1844)

(Figs 3E, 5, 43–48)

Synonymy

Turbo (Litorina) breviculus Philippi, 1844b: 166 (China, ad ostium fluminis Yang-tse-kiang [mouth of the Yangtze Jiang]; lectotype (Rosewater, 1970) Philippi, 1847, **2**: *Litorina* pl. 3, fig. 10, Fig. 43 herein).

Litorina brevicula—Philippi, 1847, **2**: 161–162; *Litorina* pl. 3, fig. 10. Dunker, 1861: 12. Schrenck, 1867: 330–333. Lischke, 1869: 77. Dunker, 1882: 111. Weinkauff, 1882: 100–101; pl. 14, figs 10, 11. Weinkauff, 1883: 215.

Tectarius breviculus—H. & A. Adams, 1854, **1**: 315.

Littorina brevicula—Reeve, 1857: sp. 51; pl. 10, fig. 51a, b. Crosse & Debeaux, 1863: 259–261. Pilsbry, 1895: 61. Yen, 1936: 192–193; pl. 16, figs 18, 18a, b. Yen, 1942: 195. Hirase & Taki, 1954: pl. 79, fig. 8. Hayasaka, 1955: 35–37. Habe, 1958a: 7–8; pl. 2, fig. 24 (in part, includes ref. to *L. sitkana*). Kira, 1962: 23; pl. 12, fig. 27. Yamamoto & Habe, 1962: 14; pl. 3, fig. 16; pl. 3, figs 32, 33 (spawn). Amio, 1963: fig. 21 (spawn, larva). Oyama & Takemura, 1963: *Neritrema, Littorina* pl., fig. 7. Habe & Ito, 1965: 25; pl. 7, fig. 7. Habe & Kosuge, 1967: 29; pl. 12, figs 3–6; text-fig. (spawn). Yamaguchi, 1967: 73–76; fig. 2 (spawn). Kuroda, Habe & Oyama, 1971: Japanese p. 88; English p. 57; pl. 16, figs 29–31 (in part, includes ref. to *L. mandshurica*). Higo, 1973: 46. Oyama, 1973: 18–19; pl. 1, fig. 16 (Pliocene). Kartavtsev, 1976: 93–98. Volova, Golikov & Kusakin, 1979: 66; fig. 42. Ohtsuka & Yoshioka, 1985: 231–232; fig. 4A (spawn). Kizaki, 1986: 71–75; fig. 1 (penis).

Okutani, 1986: 70–71; unnumbered plate; text-fig. (spawn). Golikov, Gulbin & Sirenko, 1987: 31; pl. 2, fig. 18. Masuda, Handa, Kanno, Miyasaka, Nemoto, Sasaki, Takaizumi & Wako, 1990: 4; pl. 4, fig. 11A, B (Pliocene). Cai, 1991: 50–51; fig. 42. Choe, 1992: 287–288, 584–589; pl. 54, fig. 80a, b, text-fig. 53 (in part, includes *L. mandshurica*). Egorov, 1992: 170; fig. 1G. Fukuda, Mashino & Sugimura, 1992: 57; pl. 9, fig. 123a, b.

Littorina (Littorina) sitchana var. *brevicula*—Tryon, 1887: 241; pl. 41, figs 89, 90; pl. 45, fig. 12.

Littorina (Tectaria) brevicula—Grabau & King, 1928: 229; pl. 10, fig. 103.

Littorivaga brevicula—Kuroda, 1931: 71; pl. 9, figs 64, 65 (Miocene). Habe, 1951: 88; pl. 14, figs 15, 17, 18 (in part, includes ref. to *L. sitkana*). Kojima, 1957: 59–62; fig. 1 (spawn). Kojima, 1958a: 224–229; fig. 1 (spawn).

Littorina (Littorivaga) brevicula—Hirase, 1934: 47; pl. 79, fig. 8. Nomura, 1938: 272; pl. 33, fig. 1a, b (Pliocene). Oyama, 1953: 26–33. Taki & Oyama, 1954: 7; pl. 6, fig. 15 (Pliocene). Golikov & Kusakin, 1962: 278. Sawada, 1962: 44; pl. 5, figs 15, 16 (Pliocene). Golikov & Skarlato, 1967: 25–26; fig. 17. Golikov & Gulbin, 1978: 189.

Littorina (Littorina) brevicula—Fischer, 1967: 8–11. Rosewater, 1970: 433–434; pl. 329; pl. 330A (radula); pl. 330B (penis) (in part, includes *L. mandshurica*). Golikov & Kusakin, 1978: 91–92; fig. 55a; fig. 55b (spawn). Reid, 1989a: 101; fig. 11t (spawn). Reid, 1990b: 1–19; fig. 1g; fig. 3b (spawn); fig. 4d (radula). Taylor & Reid, 1990: 199–215 (mineralogy). Reid, 1992: 204; pl. 4, figs h, j-l; fig. 1l (penis); fig. 2k (oviduct). Higo & Goto, 1993: 73.

Brevilittorina brevicula—Golikov, Dolgolenko, Maximovich & Skarlato, 1990: 295.

Littorina balteata Reeve, 1857: sp. 71; pl. 14, fig. 71a, b (China; lectotype (Rosewater, 1970: pl. 329, figs 7, 8)+2 paralectotypes BMNH 1968316, seen).

Littorina (Tectaria) balteata—Grabau & King, 1928: 229; pl. 10, fig. 102.

Littorina brevicula balteata—Yen, 1942: 195; pl 14, fig. 38.

Littorina souverbiana Crosse, 1862: 53–54; pl. 1, figs 6, 7 (in sinu Talienwhanensi [Dalian Bay, China]; lectotype (Yen, 1942: pl. 14, fig. 39)+2 paralectotypes BMNH 1968372, seen. Crosse described this species from the Cuming Collection, now in BMNH. Some doubt surrounds Yen's (1942) designation of a lectotype, however, since although the three supposed syntypes are from the Cuming Collection, a (non-original) label bears Crosse's specific name and the locality Peiho [near Tianjin]. The lectotype is probably not the figured specimen, but resembles it closely. Two lots in MNHNP, of 20 and 3 shells, from which Rosewater (1970) selected and figured a 'holotype', and which Fischer-Piette (1950) listed in his catalogue of types in the collection of the *Journal de Conchyliologie*, are not syntypes, but are probably the additional material mentioned by Crosse & Debeaux (1863)). Yen, 1942: 195; pl. 14, fig. 39.

Litorina brevicula forma *elatior* Schrenck, 1867: 331 (infrasubspecific; not *L. rudis* f. *elatior* Middendorff, 1849= *L. saxatilis*).

Litorina brevicula forma *depressior* Schrenck, 1867: 331 (infrasubspecific; not *Litorina tenebrosa* f. *depressior* Schrenck, 1867=*L. sitkana*).

Litorina brevicula var. *costulata* Schrenck, 1867: 332 (infrasubspecific; not *L. tenebrosa* var. *costulata* Middendorff, 1849=*L. saxatilis*).

Litorina brevicula var. *laevigata* Schrenck, 1867: 332 (infrasubspecific; not *Litorina tenebrosa* var. *laevigata* Schrenck, 1867=*L. sitkana*).

Littorina mandshurica—Pilsbry, 1895: 61 (as *L. mandschurica*; in part, includes *L. mandshurica*). Oyama & Takemura, 1963: *Neritrema, Littorina* pl., fig. 5 (as *L. mantchuriana*; in part, other fig. is *L. mandshurica*).

Littorina sitchana—Bavay, 1920: 254 (in part, includes *L. sitkana, L. mandshurica, L. aleutica*). Nomura, 1926a: 53–55 (not *L. sitkana* Philippi, 1846).

Leptothyra cf. *paucicostata*—Yokoyama, 1920: 86–87; pl. 5, fig. 15a, b (not Dall, 1871; Pliocene).

Leptothyra sakaensis Yokoyama, 1925: 8; pl. 1, fig. 8 (Sakae [near Nagano, Honshu]; Lower Pliocene; type in Geological Institute, Tokyo University, not seen).

Littorina (Littorivaga) sakaensis—Oyama, 1953: 32.

Littorina heterospiralis Grabau & King, 1928: 230; pl. 10, fig. 104 (Peitaiho, China; types not traced; this is probably *L. brevicula*, but description and figure are poor).

Littorina itelmenica Sinelnikova, in Gladenkov & Sinelnikova, 1990: 99; pl. 30, fig. 2 (Cape Bolshoi [label accompanying holotype gives additional details: coast between mouth of Etolona River and Cape Bolshoi, Pendginskaya Bay], Rekinniksk Gulf, western Kamchatka; Etolonian stage, Middle Miocene; holotype PMM N670/00, seen, Fig. 47C; 2 paratypes BMNH GG22520, GG22521, seen).

Taxonomic history

The variability of this shell explains why it received the additional names *L. balteata* Reeve, 1857, and *L. souverbiana* Crosse, 1862, soon after the original description and figure given by Philippi (1844b, 1847). However, as more specimens became available these three were soon synonymized by Crosse & Debeaux (1863), and thereafter almost all authors have used only the name *brevicula*. Grabau & King

Fig. 10.

Figure 43. *Littorina (Littorina) brevicula.* Lectotype of *Turbo (Litorina) breviculus* Philippi, 1844 (Philippi, 1847, **2**: *Litorina* pl. 3, fig. 10; actual length of shell=11 mm).

(1928) were exceptions, retaining both *L. brevicula* and *L. balteata*, and describing a further species, *L. heterospiralis*, the identity of which is somewhat doubtful, since the description and figure were poor and the type has not been located.

Some confusion has arisen with the highly variable northern Pacific species *L. sitkana*. Tryon (1887) reduced *brevicula* to a variety of *L. sitkana*, and Bavay (1920) also had a broad concept of that species. *Littorina brevicula* has also sometimes been misidentified as *L. mandshurica*.

The shell of *L. brevicula* is never truly nodulose, but the colour pattern and intersection of growth lines can give this impression. For this reason *L. brevicula* has occasionally been placed in the genus or subgenus *Tectarius* (H. & A. Adams, 1854; Grabau & King, 1928). The genus or subgenus *Littorivaga* was mistakenly used, first by Kuroda (1931), then by other Japanese and Russian authors, until description of the pelagic spawn made it clear that the species should be retained in *Littorina* (Habe, 1958a; Golikov & Kusakin, 1978). Golikov, Dolgolenko, Maximovich & Skarlato (1990) used the new generic name *Brevilittorina* for both *L. brevicula* and *L. mandshurica*, but since the name was not accompanied by a definition or fixation of a type species, it must be treated as unavailable (ICZN, 1985: art. 13).

In Japan fossil specimens were mistakenly described as species of the turbinid genus *Leptothyra* (Yokoyama, 1920, 1925). Recently, the new species *L. itelmenica* has been described from the Middle Miocene of Kamchatka (Sinelnikova & Gladenkov, 1990), and is here synonymized with *L. brevicula* (see Fossil Record, p. 135).

Diagnosis

Mature shell height 6.3–20.1 mm (fossils to 28.7 mm). Shell turbinate, spire outline often concave apically. Mineralogy of external irregular-prismatic calcite and internal crossed-lamellar aragonite. Sculpture of 3–5 (usually 4, rarely up to 9) carinate ribs on last whorl, 4–6 smaller ribs on base; fine spiral microstriae in wide spaces between ribs. Colour black to brown; ribs usually with prominent yellow-brown to white dashes, or rarely ribs are all white giving banded pattern; aperture brown with whitish basal band and spotted margin. Head black, tentacles with unpigmented stripe partly or entirely across base. Penis with 0–7 mamilliform glands; base blackish; tapering filament 40–60% total length. Pallial oviduct of two consecutive loops of albumen gland followed by large spiral loop of capsule gland; straight section short, not enlarged or septate; bursa at posterior end of straight section, with anterior and posterior arms. Spawn of small pelagic biconvex capsules, usually containing single eggs; development planktotrophic. Distribution: Japan, Korea, China as far south as Hong Kong.

Anatomical description

Head-foot (Fig. 44H–K). Head black; tentacles black with pale tips, usually with unpigmented stripe across eye and partly or entirely across tentacle base; sides of foot black or mottled. Opercular ratio 0.50–0.72.

Penis (Fig. 44A–I). Base wrinkled, usually pigmented; mamilliform glands sometimes absent, but usually 1–7 in short row. Filament tapering, 40–60% total length. Sperm groove open, ending terminally. Penis shed in May to June, following breeding (Kizaki, 1986).

Sperm (Fig. 3E). Euspermatozoa 49–54 μm. Paraspermatozoa 8–14 μm diameter, spherical to oval; vesicles large, distinct; nucleus sometimes visible.

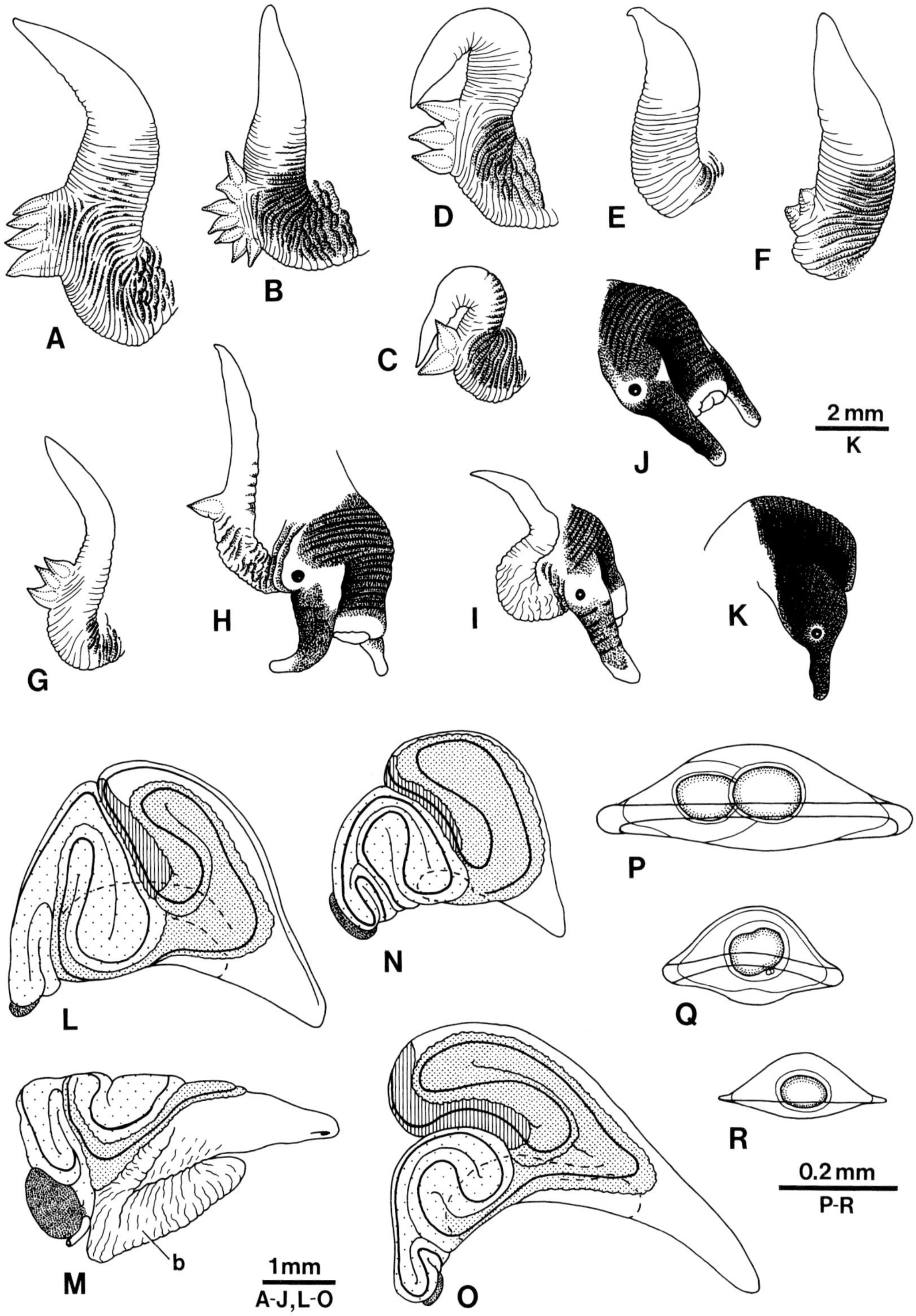

Figure 44. *Littorina (Littorina) brevicula*: **A–I,** penes; **H–K,** head pigmentation; **L–O,** pallial oviducts; **L, M,** lateral and ventral views of same pallial oviduct; **P–R,** pelagic egg capsules (**P**, after Yamaguchi, 1967; **Q,** after Kojima, 1957; **R,** after Amio, 1963). **A–D, O,** Abashiri, Hokkaido; **E–F,** Muroran, Hokkaido; **G–I, N,** Akkeshi, Hokkaido; **J,** Oshoro, Otaru, Hokkaido; **K, Q,** Asamushi, Honshu; **L, M,** Cape Notori, near Abashiri, Hokkaido; **P,** Misaki, Honshu; **R,** Shimonoseki, Honshu. Shell heights: **A,** 16.6 mm; **B,** 13.0 mm; **C,** 12.0 mm; **D,** 14.0 mm; **E,** 11.0 mm; **F,** 15.5 mm; **G,** 6.9 mm; **H,** 7.1 mm; **I,** 6.7 mm; **J,** 8.0 mm; **K,** 14.3 mm; **L, M,** 15.6 mm; **N,** 9.3 mm; **O,** 16.0 mm. For shading conventions of pallial oviducts see p. 6. Abbreviation: b, copulatory bursa.

Oviduct (Fig. 44L–O). Two consecutive loops of albumen gland followed by large spiral loop of capsule gland with distal reddish translucent portion; straight section relatively short, jelly gland not enlarged or septate. Bursa well-developed, originating at posterior end of straight section, with anterior and posterior arms. Seminal receptacle simple.

Spawn and development (Fig. 44P–R). Pelagic, asymmetrically biconvex capsules with peripheral rim; diameter usually 310–450 μm, containing single egg 84–114 μm diameter; capsules sometimes contain 2 eggs, and are then larger, up to 540 μm; rarely up to 5 eggs (Kojima, 1957, 1958a; Amio, 1963; Yamaguchi, 1967; Ohtsuka & Yoshioka, 1985). Veligers hatch at shell size of 130 μm; development planktotrophic (Amio, 1963). Protoconch 350–360 μm diameter, 2.2 whorls. Spawning season similar throughout Japan: February to April at Asamushi (Kojima, 1957, 1958a; Hirai, 1963); January to April at Misaki (Yamaguchi, 1967); February and March at Shirahama (Ohgaki, 1981); January to March at Mukaishima (Ohtsuka & Yoshioka, 1985).

Radula and foregut (Fig. 45C–H). Relative radular length 1.15–6.64; varies seasonally, at a maximum in September-October (Watanabe, 1985; Kizaki, 1987). Rachidian: length/width 1.00–1.20; central cusp pointed; one pointed cusp and one outer denticle on each side. Lateral: four cusps, third from inside largest; all pointed, but second from inside sometimes more rounded. Inner marginal: four cusps, third from inside largest; all pointed. Outer marginal: 5–7 pointed cusps; narrow neck, projecting flange on inside of base only. Salivary glands enlarged.

Shell variation (Figs 5, 45A, B; 46, 47)

Geographical variation. Shell shape varies most noticeably on a geographical scale, and two principal varieties can be distinguished:

> Geog. var. *typica* (Figs 46A–D, F; 47A, B, D): H usually less than 15 mm, but up to 17.8 mm; H/B=1.10–1.31; SH=1.48–1.56. Shell high-turbinate; sculpture of 3–9 narrow but prominent ribs above and including peripheral rib, sometimes with secondary ribs between; colour often grey, pale brown or white with indistinct markings, but may have dark and light dashes on ribs or white ribs on dark background. Range: Hong Kong to southern Bohai Sea, Kyushu and Ryukyu Islands.
>
> Geog. var. *souverbiana* (Figs 5, 46E, G–N): H up to 20.1 mm; H/B=0.95–1.10; SH=1.23–1.42. Shell low-turbinate; sculpture of 3–5 broad, carinate ribs on last whorl; anterior part of aperture frequently produced into a lip; colour brown to black, usually with conspicuous white to yellow-brown dashes on ribs. Range: Qingdao to northern Bohai Sea, Korea, Honshu and Hokkaido.

The variety *souverbiana* is a larger, heavier, form with fewer, more strongly carinate ribs. The profile is more low-spired, but this is partly a consequence of erosion; in young specimens and those from more sheltered habitats the concave apical part of the spire is more projecting. Intermediates between the two varieties occur in the Shandong region and in Kyushu. Adult size is reduced in the southern part of the range of *L. brevicula*, reaching only 10 mm in Hong Kong. In addition to being small, some shells from Hong Kong are also unusually tall-spired (Fig. 46C). These two traits might be adaptive in relation to convective heat loss and reduction of desiccation in the warm climate at the southern extreme of the range, in agreement with the observations of Vermeij (1973) on other littorinids.

Ecotypic variation. This has not been examined in detail, but does not seem to be marked. Nomura (1926) claimed that shells were more elongate at the more exposed of two localities near Asamushi, northern Honshu.

Colour variation (Fig. 5). In the species as a whole, variation in colour is considerable, from completely white, through banded forms, the common dashed pattern, and to black. However, most samples from Japan are dark with a dashed pattern. The most variable samples are from central and northern China. The strikingly banded form (Fig. 46B) is generally rare, but most examples come from central and northern China. Kartavtsev (1976) distinguished three distinct pattern types in Vostok Bay, black, banded and spotted, and interpreted this as a two-allele system with incomplete dominance.

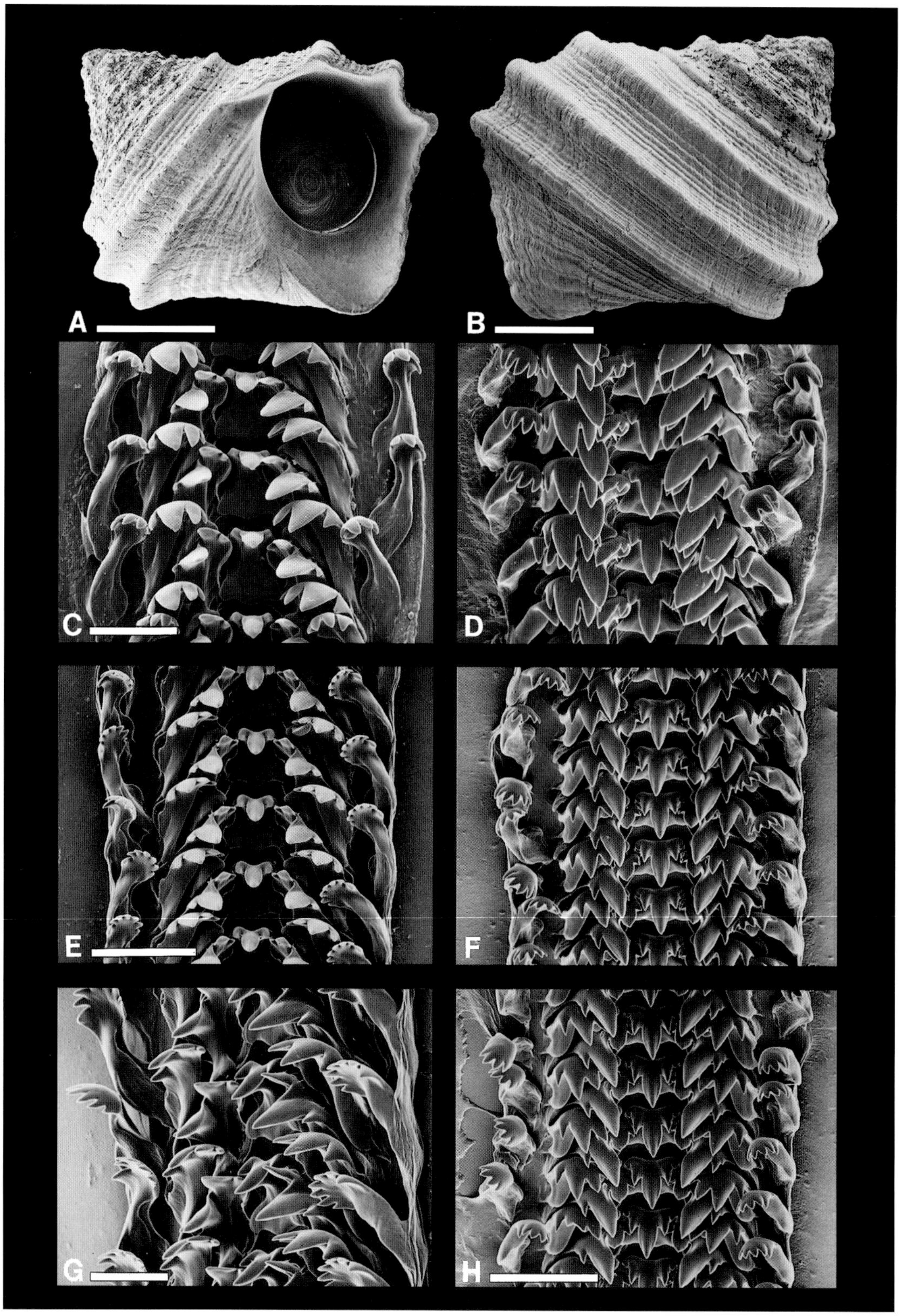

Figure 45. *Littorina (Littorina) brevicula*: **A, B,** juvenile shells; **C, D,** two views of radula (shell H=6.3 mm), flat (**C**), at 45° (**D**); **E–G,** three views of radula (shell H=12.8 mm), flat (**E**), at 45° (**F**), side view (**G**); **H,** radula (shell H=11.7 mm). **A, B,** Asamushi, Honshu; **C, D,** Akkeshi, Hokkaido; **E–G,** east of Cape Chifu, Yantai, China; **H,** Pos'et Bay, Russia. Scale bars: **A**=1 mm; **B**=1 mm; **C, D**=50 μm; **E, F**=100 μm; **G**=50 μm; **H**=100 μm.

Distribution

Habitat. *Littorina brevicula* is typically found from the littoral fringe down to the mid-eulittoral zone, on rocky, boulder and cobble shores. Smaller snails occur near the mean high-tide level, with larger individuals extending over a wider tidal range (Kojima, 1959). The vertical range varies seasonally, some or all of the larger, adult snails migrating downshore to low-tide level during the winter breeding season from January to April (Abe, 1935; Kojima, 1959; Luckens, 1970; Golikov & Skarlato, 1971; Takada & Kikuchi, 1991; Takada, 1992, 1995). At Amakusa, Japan, the differing migratory behaviour of individuals is consistent and possibly genetically determined (Takada, 1995). On more sheltered shores the range extends only into the barnacle zone, lower than in wave-exposed localities (Morton, 1990b). The snails congregate in sheltered microhabitats, between boulders, in crevices and among barnacles (Golikov & Skarlato, 1967; Tsuchiya, 1979, 1984; Morton & Morton, 1983; Mori, Nishihama & Tanaka, 1985; Britton, 1990), and very dense aggregations can occur (Abe, 1935). Golikov & Skarlato (1967) recorded a density of 32200 per m^2 in Pos'et Bay. At the top of the shore, individuals can be found among supralittoral grasses (Morton & Morton, 1983), and at lower levels sometimes on algae (Abe, 1935; Golikov & Kusakin, 1978). Occasional animals occur in the sublittoral, down to 1 to 3 m depth, but this is probably accidental (Golikov & Skarlato, 1967; Golikov & Kusakin, 1978; Volova, Golikov & Kusakin, 1979). The species tolerates a wide range of exposures to wave action, but is most abundant on moderately exposed to sheltered shores (Ohgaki, 1983; Morton, 1990b). At the southern extreme of its range, in Hong Kong, it occurs only in sheltered bays (Morton & Morton, 1983; Ohgaki, 1985). It does not, however, extend far into the brackish water of estuaries (Berger & Kovaleva, 1974; Golikov & Kusakin, 1978; Yi & Li, 1990). Despite its temperate to warm-water distribution, *L. brevicula* survives snow cover (Luckens, 1970) and freezing in ice (Golikov & Skarlato, 1971).

Over much of its range, from Hong Kong to northern Honshu, *L. brevicula* is sympatric with several *Nodilittorina* species, of which *N. radiata* (Eydoux & Souleyet) (synonym *N. exigua* (Dunker)) is the most frequent. The *Nodilittorina* species overlap with *L. brevicula* in the littoral fringe, but extend to higher levels (Ohgushi, 1956; Tsuchiya, 1979; Morton & Morton, 1985; Mori, Tanaka & Nishihama, 1985; Ohgaki, 1985; Dudgeon & Yipp, 1990; Yi & Li, 1990; You, 1990). In the far northern part of its range, in Hokkaido and the vicinity of Peter the Great Bay, *L. brevicula* occurs on the same shores as *L. sitkana* and *L. mandshurica*. Its zonation level is similar to that of *L. sitkana*, but the latter is more abundant on exposed shores; *L. mandshurica* occurs lower in the eulittoral zone (Golikov & Kusakin, 1978; pers. obs.).

No detailed studies of the diet of *L. brevicula* have been located, but it has been observed feeding on seaweeds (Abe, 1935), and specifically on *Ulva*, *Enteromorpha*, *Monostroma* and *Porphyra* by Kizaki (1987). When these algae become scarce and disappear in the summer months the snails do not feed, and the radula, being unused, increases in length (Watanabe, 1985; Kizaki, 1987).

Range (Fig. 48). In the present study specimens have been seen from Hong Kong, northwards around the coast of China and Korea, as far as Peter the Great Bay (BMNH), and from Okinawa (NSMT 39518), throughout the four main Japanese Islands, to Kunashir Island, the southernmost of the Kurile Chain (ZISP). This is in broad agreement with distributions given by Kuroda & Habe (1952), Rosewater (1970) and Reid (1990a); Golikov & Kusakin (1978) add northern Taiwan to this range. Hidalgo (1904–1905) recorded the species from Mindanao in the Philippines, but this is far to the south of known occurrences and has not been confirmed since, so is regarded as unreliable. In Hong Kong *L. brevicula* is found only in sheltered parts of Tolo Channel; Chambers (1980) found it commonly, but when the area was resurveyed by Ohgaki (1985) it was scarce and juveniles were absent, leading him to suggest that annual recruitment was unstable here at the southern limit of the range. In the western Sea of Japan the species is abundant at Pos'et Bay, but rare on Putyatin Island and at Vrangel Bay, and is absent further north on this coast (Mokyevsky, 1960; Golikov & Skarlato, 1967; Golikov & Kusakin, 1978). It is common throughout Hokkaido, except on the exposed southeastern side under the influence of a cold current, where it is restricted to sheltered bays (Ohgaki, 1983). It just reaches the southern part of Kunashir Island, two specimens having been recorded from Izmeny Bay (Golikov & Kusakin, 1962; Golikov & Gulbin, 1978). Golikov, Gulbin & Sirenko (1987) noted it from Moneron Island, between Sakhalin and Hokkaido; both they and Golikov & Kusakin (1978) mentioned its occurrence in southwestern Sakhalin, but no specimens are present

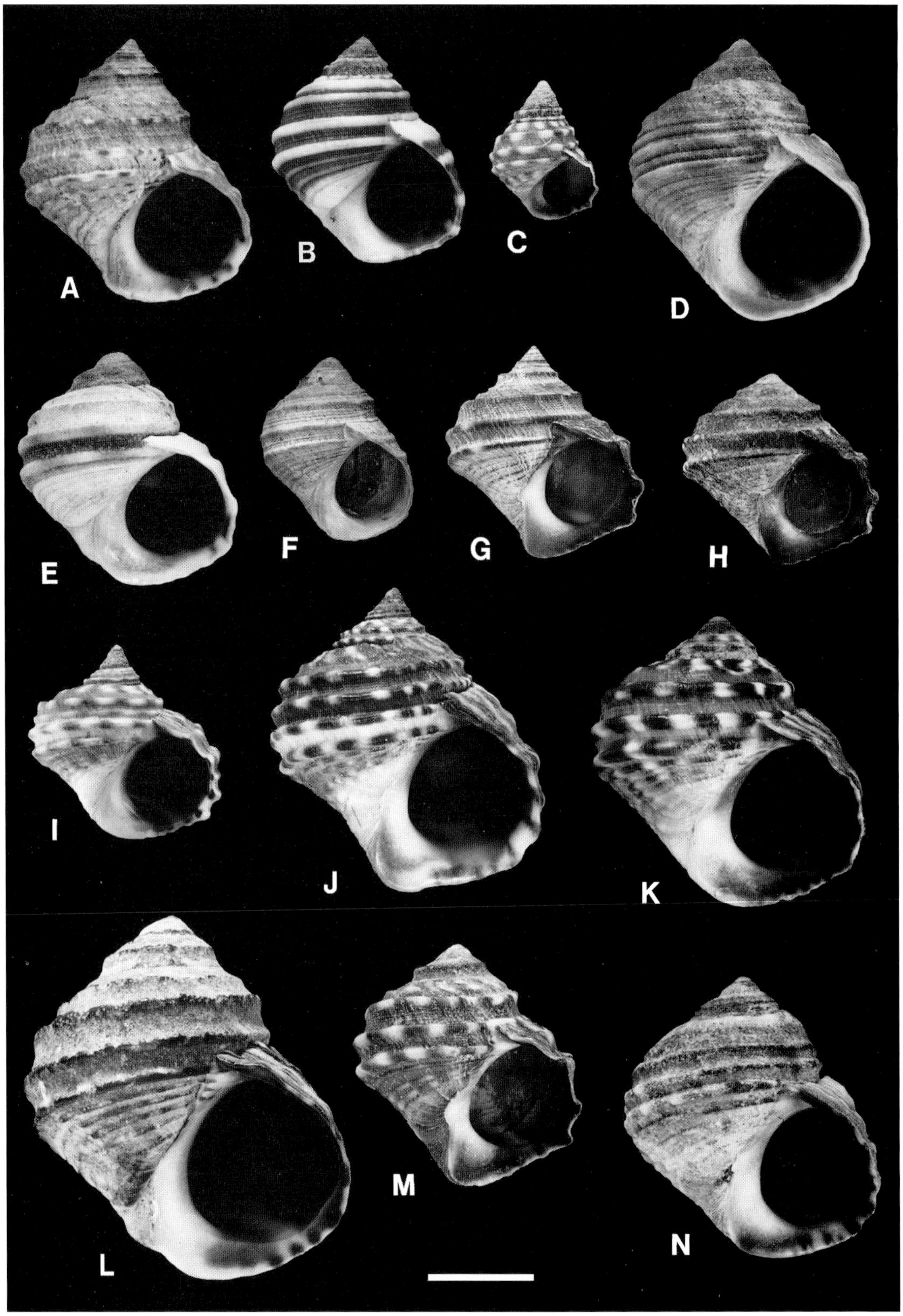

Figure 46. *Littorina (Littorina) brevicula*: **A–D**, **F**, geog. var. *typica*; **E**, **G–N**, geog. var. *souverbiana*. **A**, Xiamen, China. **B**, **D**, east of Cape Chifu, Yantai, China. **C**, Hoi Sing Wan, Hong Kong. **E**, Qingdao, China. **F**, Fuzhou, China. **G**, **M**, Shakotan-misaki, Hokkaido. **H**, Asamushi, Honshu. **I**, **J**, Yokohama, Honshu. **K**, **N**, Muroran, Hokkaido. **L**, Cape Notori, Hokkaido. Scale bar=5 mm.

Figure 47. *Littorina (Littorina) brevicula.* **A, B, D,** geog. var. *typica*; **C,** fossil. **A, D,** Matsuura, Kyushu. **B,** Kafuri, Maebaru, Kyushu. **C,** *Littorina itelmenica* Sinelnikova, in Gladenkov & Sinelnikova, 1990; holotype, PMM N670/00; coast between mouth of Etolona River and Cape Bolshoi, Pendginskaya Bay, Kamchatka, Russia; Etolonian stage, Middle Miocene. Scale bar=5 mm.

in ZISP from this locality. *Littorina brevicula* has been recorded from Willapa Bay, Washington, introduced with a shipment of Japanese oysters, but has not become established there (Hanna, 1966).

Fossil record (Fig. 47C). There are a number of fossil records of this species, from Kyushu, Honshu and Hokkaido. Most of them are of Pliocene age (Yokoyama, 1920, as *Leptothyra* cf. *paucicostata*; Kuroda, 1931; Nomura, 1938; Taki & Oyama, 1954; Sawada, 1962; Oyama, 1973; Masuda, Handa, Kanno, Miyasaka, Nemoto, Sasaki, Takaizumi & Wako, 1990), and at least the record by Yokoyama (1925, as *Leptothyra sakaensis*) may be from the early Pliocene (Hatai & Nisiyama, 1952: 14). Kuroda (1931) described and figured a shell from the Upper Miocene Ogawa formation (Hatai & Nisiyama, 1952). If *L. itelmenica* from the Etolonian stage of Kamchatka (Gladenkov & Sinelnikova, 1990) is correctly synonymized with *L. brevicula*, the fossil record may be extended back to the Middle Miocene (10–12 Ma, Gladenkov & Shanster, 1990). Although not explicitly compared with *L. brevicula* by Sinelnikova, the three major ribs at and above the periphery of the last whorl and the somewhat produced and angular base of the columella of *L. itelmenica* (Fig. 47C) suggest that they are closely allied. The main points of difference are the larger size of the fossil form (height of holotype 23.9 mm,

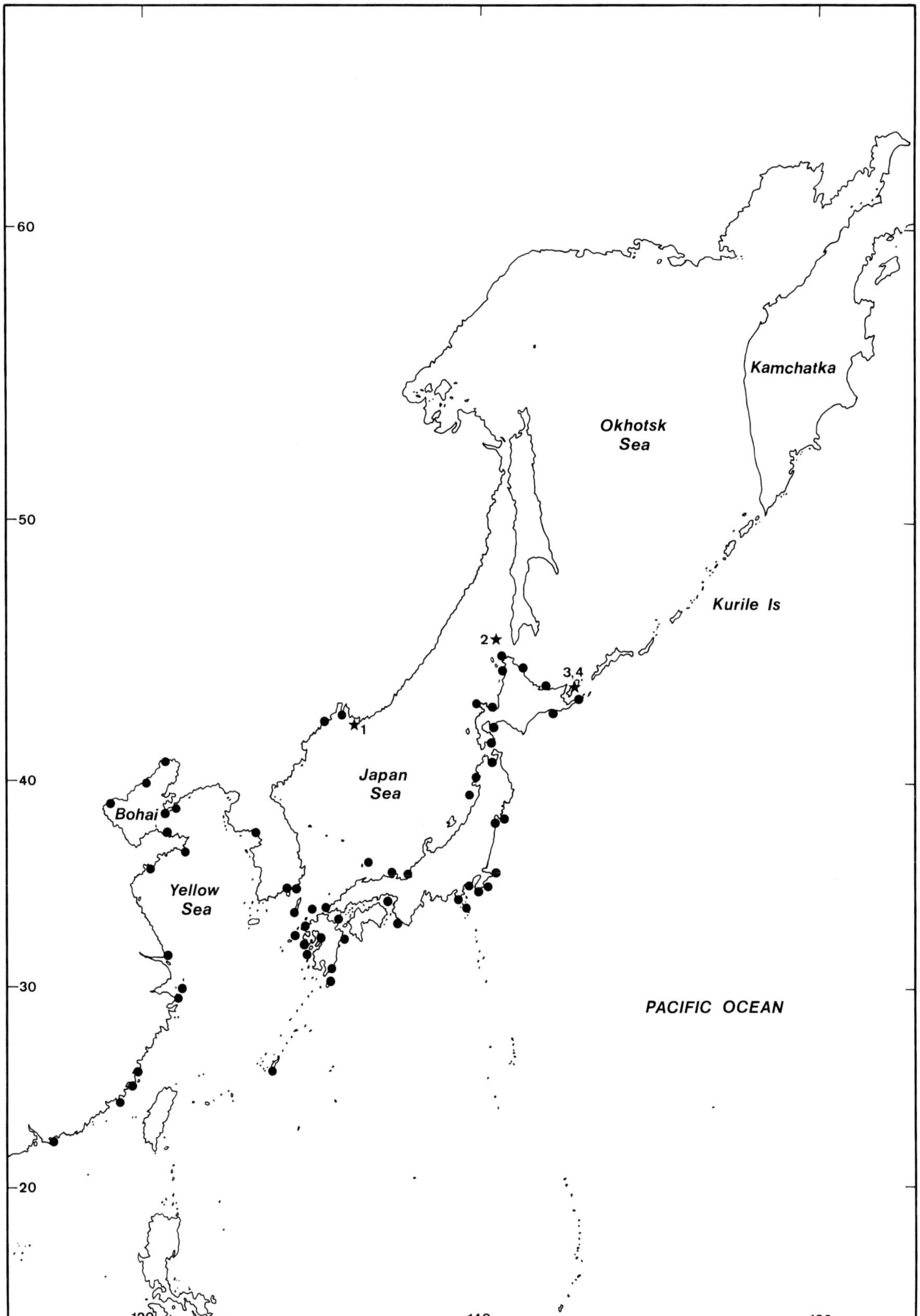

Figure 48. *Littorina (Littorina) brevicula*: distribution map. Key: circles, specimens examined; asterisks, records from literature (1, Mokyevsky, 1960; 2, Golikov, Gulbin & Sirenko, 1987; 3, Golikov & Kusakin, 1962; 4, Golikov & Gulbin, 1978).

Table 4. Comparison of *Littorina brevicula* and *L. mandshurica*.

Character	*L. brevicula*	*L. mandshurica*	% accuracy	References
Cephalic tentacles	Fig. 44H–K. Black, usually with unpigmented stripe across eye and tentacle base	Entirely black	96	This study
Penis	Fig. 44A–I	Fig. 49A–I	93	This study
Filament	40–60% total length, elongate, tapering	20–30% total length, triangular		
Pigmentation	Black on base only, or unpigmented	Base and filament black		
Pallial oviduct	Fig. 44L–O	Fig. 49J–L		This study
Copulatory bursa	Relatively large, with anterior and posterior arms	Very small	100	
Spiral glands	Two loops of albumen gland followed by one large loop of capsule gland	One loop of albumen gland followed by one small and one large loop of capsule gland	100	
Spawn	Fig. 44P–R	Fig. 49M, N	100	Kojima, 1957, 1958a, e; Amio, 1963; Yamaguchi, 1967; Ohtsuka & Yoshioka, 1985
Capsule diameter	310–450 μm	1000 μm		
Shape	Asymmetrically biconvex	Helmet-shaped		
Number of ova	1, sometimes 2–5	9–12		
Radula	Fig. 45C–H	Fig. 50C–F		This study
Relative length	1.15–6.64	1.23–2.43		
Rachidian length/width	1.00–1.20	0.86–1.07		
Shell	Figs 45A, B; 46, 47	Figs 50A, B; 51		This study
Sculpture of last whorl	3–9 prominent ribs	2–3 low ribs, sometimes obsolete	95	
Columella	Pillar relatively narrow, often with anterior lip	Pillar relatively wider, rounded anteriorly		
Colour	Black to brown, with pale spots on ribs; rarely white ribs	Purple–black; sometimes grey with dark ribs	95	
Distribution				This study
Latitude	Fig. 48. 22–45°N	Fig. 53. 41–54°N		

and of larger paratype in BMNH 28.7 mm; maximum size of Recent shells 20.1 mm) and the sculpture of the base with six narrow, closely spaced riblets, which is at the extreme of the range of Recent examples, in which basal sculpture is usually coarser. The apical whorls are missing in the three available shells of *L. itelmenica* and therefore cannot be compared with the distinctive concave profile of the upper part of the spire in Recent shells. *Littorina itelmenica* appears insufficiently distinct morphologically to justify separation from *L. brevicula*.

Discussion

Intraspecific variation. As in other species of *Littorina* with planktotrophic development, local ecotypic variation in shell morphology is insignificant, and shell variation is apparent only on a large geographical scale. This can be explained by the homogenizing effect of presumably substantial gene flow in a widely dispersing species. There have been several studies of allozyme variation within individual populations (Kartavtsev & Efremov, 1981; Zaslavskaya, 1989; Zaslavskaya, Sergievsky & Tatarenkov, 1992a, b; Tatarenkov, 1992). Comparison between populations has supported the expectation of relatively low genetic differentiation between them, although some loci have been found to show significant differences in allele frequencies, even between samples from the same shores, probably as a result of selection (Tatarenkov, 1995b). Genetic heterogeneity within shores might also be promoted by the breeding migration of this species, which in one studied case appears to result in partial reproductive isolation between migratory and sedentary subpopulations (Takada, 1995).

Phylogenetic relationships. Several authors have recognized a close relationship between *L. brevicula* and *L. mandshurica*, Golikov *et al.* (1990) even going so far as to use a new (but unavailable) generic name for these two species, while Ushakov (1953) and Mokyevsky (1960) treated them as subspecies. This relationship is suggested by the morphological similarity of their shells, and also of their egg capsules, oviducts, penes and radulae, although these resemblances are either plesiomorphic or subjective, so that cladistic analysis has not revealed any synapomorphies of the two (Reid, 1990b; p. 351). Estimates of genetic distance from allozyme frequencies (Kartavtsev & Efremov, 1981; Zaslavskaya, 1989; Zaslavskaya *et al.*, 1992a, b) also support a close relationship. Analysis of mitochondrial DNA sequencies has shown that these are sister-species, with an estimated age of divergence of about 5.84 Ma (but a wide range, 1.99–15.62 Ma; Reid, Rumbak & Thomas, 1996; p. 365). Another estimate of their age of divergence, from their genetic distance measured by allozyme frequencies, is 2.85 Ma (Zaslavskaya *et al.*, 1992a). The modern distributions of these two species overlap only in Hokkaido and Peter the Great Bay, or over about 30% of the range of the more narrowly distributed *L. mandshurica*, and vicariance induced by climatic change may have been involved in their speciation (p. 378, 387).

Similar species. Among the sympatric species, *L. mandshurica* is most similar; differences are summarized in Table 4. Shell characters are usually sufficient to separate the two, *L. brevicula* having more numerous ribs and usually a spotted pattern. Identification can be confirmed using pigmentation of the penis and cephalic tentacles, and size of the copulatory bursa. Confusion with *L. sitkana* is less likely; in the areas of sympatry *L. sitkana* usually has a smooth shell, often with a heavily ridged base, and never shows a spotted pattern; oviduct and penis are quite different. In the southern part of its range some small pale forms of the typical variety of *L. brevicula* (Fig. 46F) could possible be confused with *Nodilittorina radiata*. The latter is 5–11 mm in height, with 17–23 narrow ribs on the last whorl, the larger ribs are usually nodulose, the colour is whitish, the penis has a bifurcate base, with a single mamilliform gland and adjacent glandular disc, and the pallial oviduct consists of a single, simple spiral of albumen and capsule glands (Reid, 1992: figs 1i, 2h, plate 3d-h).

Littorina (Littorina) mandshurica (Schrenck, 1861)

(Figs 49–53)

Synonymy

Litorina mandshurica Schrenck, 1861: 409 (Sinus de Castries [De Kastri, Tatar Strait, Russia]; lectotype (here designated) 8.7x8.2 mm+39 paralectotypes, ZISP, seen; 12 paralectotypes BMNH 1992163, seen, Fig. 51A; paralectotypes MCZ 52007, not seen). Dunker, 1882: 111.

Littorina mandshurica—Pilsbry, 1895: 61 (as *mandschurica*; in part, includes refs to *L. brevicula*). Habe, 1951: 89; pl. 14, fig. 10; text-fig. 2 (as *mandschurica*). Habe, 1958a: 8; pl. 2, fig. 6 (as *mandschurica*). Yamamoto & Habe, 1962: 14–15; pl. 3, fig. 14; pl. 3, figs 25, 26 (spawn) (as *mandschurica*). Oyama & Takemura, 1963: *Neritrema, Littorina* pl., fig. 6 (as *mantchuriana*; in part, other fig. is *L. brevicula*). Habe, 1964: 28; pl. 9, fig. 24 (as *mandchurica*). Habe & Ito, 1965: 25; pl. 7, fig. 6 (as *mandschurica*). Habe & Kosuge, 1967: 29; pl. 12, fig. 7; text-fig. (spawn) (as *mandchnrica*). Higo, 1973: 46 (as *mandschurica*). Volova, Golikov & Kusakin, 1979: 66–67; fig. 43. Okutani, 1986: 70–71; unnumbered plate; text-fig. (spawn) (as *mandschurica*). Egorov, 1992: 170; fig. 1I.

Littorina brevicula mandshurica—Ushakov, 1953. Mokyevsky, 1960: 257.

Littorivaga mandshurica—Kojima, 1958e: 81–86; fig. 1 (spawn) (as *mandschurica*).

Littorina (Littorivaga) mandshurica—Golikov & Kusakin, 1962: 276–278. Golikov & Skarlato, 1967: 24–25; fig. 16. Golikov & Gulbin, 1978: 190.

Littorina (Littorina) mandshurica—Golikov & Kusakin, 1978: 92–93; fig. 56a; fig. 56b (spawn). Golikov & Skarlato, 1985: 383–384. Reid, 1989a: 101. Reid, 1990b: 1–19; fig. 1h; fig. 3e (spawn). Taylor & Reid, 1990: 199–215 (mineralogy). Higo & Goto, 1993: 73 (as *mandchurica*).

Brevilittorina mandshurica—Golikov, Dolgolenko, Maximovich & Skarlato, 1990: 295.

Litorina mandshurica forma *elatior* Schrenck, 1867: 334–335 (infrasubspecific; not *L. rudis* f. *elatior* Middendorff, 1849=*L. saxatilis*).

Litorina mandshurica forma *depressior* Schrenck, 1867: 335 (infrasubspecific; not *Litorina tenebrosa* f. *depressior* Schrenck, 1867=*L. sitkana*).

Litorina mandshurica var. *carinata* Schrenck, 1867: 338 (infrasubspecific; not *Delphinula carinata* Woodward, 1833=*L. littorea*).

Litorina mandshurica var. *laevigata* Schrenck, 1867: 338 (infrasubspecific; not *Litorina tenebrosa* var. *laevigata* Schrenck, 1867=*L. sitkana*).
Littorina litorea—Jeffreys, 1883: 113 (in part, includes *L. littorea* and *L. squalida*).
Littorina (Littorina) sitchana—Tryon, 1887: 240–241; pl. 41, fig. 86 (in part, includes *L. sitkana, L. brevicula, L. aleutica*).
Littorina sitchana—Bavay, 1920: 254 (in part, includes *L. sitkana, L. brevicula, L. aleutica*).
? *Littorina palliata*—Dall, 1920: 29–30; pl. 5, fig. 12 (Pleistocene). Dall, 1921: 152 (Pleistocene). MacNeil, Mertie & Pilsbry, 1943: 84; pl. 11, fig. 13 (Pleistocene).
Littorina (Littorina) brevicula—Rosewater, 1970: 433–434 (in part, includes *L. brevicula*).
Littorina brevicula—Kuroda, Habe & Oyama, 1971: English p. 57 (in part, includes *L. brevicula*).

Taxonomic history

The history of this species is straightforward, with no synonyms. It has, however, been treated as a subspecies (Ushakov, 1953) or as a synonym (Rosewater, 1970; Kuroda, Habe & Oyama, 1971) of the closely related *L. brevicula*. Both Tryon (1887) and Bavay (1920) had a broad concept of the variable species *L. sitkana*, in which they included *L. mandshurica*. Jeffreys (1883) synonymized *L. mandshurica* with *L. littorea*. Its inclusion in the subgenus *Littorivaga* by several authors was based on shell resemblance; the spawn of *L. mandshurica* is pelagic and unlike the benthic egg mass of *L. sitkana*, type species of *Littorivaga*. The generic name *Brevilittorina* is unavailable, as discussed in the remarks on *L. brevicula* (p. 129).

Diagnosis

Mature shell height 7.7–21.1 mm. Shell turbinate. Mineralogy of external irregular-prismatic calcite and internal crossed-lamellar aragonite. Sculpture of 2–3 ribs on last whorl, up to 6 small ribs on base; fine spiral microstriae over whole surface; all sculpture becoming obsolete on last whorl of large and worn shells. Colour usually purple-black, sometimes with pale basal band, occasionally cream to grey with darker ribs giving banded pattern; aperture purple-brown with whitish basal band. Head and tentacles black. Penis with 1–5 mamilliform glands; short, triangular filament 20–30% total length; base and filament black. Pallial oviduct of one loop of albumen gland followed by one small and one large spiral loop of capsule gland; straight section short, not enlarged or septate; bursa very small, at posterior end of straight section. Spawn of large, pelagic, helmet-shaped capsules containing 9–12 eggs; development planktotrophic. Distribution: Hokkaido, Sakhalin, northern Sea of Japan.

Anatomical description

Head-foot. Head black, sometimes with pale tips to tentacles; sides of foot black. Opercular ratio 0.60–0.68.

Penis (Fig. 49A–I). Base and filament darkly pigmented. Base wrinkled, 1–5 mamilliform glands in short row. Filament triangular, 20–30% total length. Sperm groove open, ending terminally. Reduced or shed outside breeding season.

Sperm. Euspermatozoa 56–58 μm. Paraspermatozoa 11–18 μm diameter, oval or rounded; vesicles large, distinct; nucleus not visible.

Oviduct (Fig. 49J–L). One loop of albumen gland followed by one small and one large spiral loop of capsule gland with distal reddish translucent portion; straight section relatively short, jelly gland not enlarged or septate. Bursa very small, originating at posterior end of straight section. Seminal receptacle simple.

Spawn and development (Fig. 49M, N). Pelagic, helmet-shaped (i.e. flat base, domed upper side and peripheral brim) capsules; diameter 1000 μm, containing 9–12 eggs 120 μm diameter (Kojima, 1958e). Development planktotrophic. Protoconch not seen. Breeding season: February to April at Asamushi, Honshu (Kojima, 1958e, 1960; Hirai, 1963); males mature but females spent in June at Akkeshi, Hokkaido (pers. obs.).

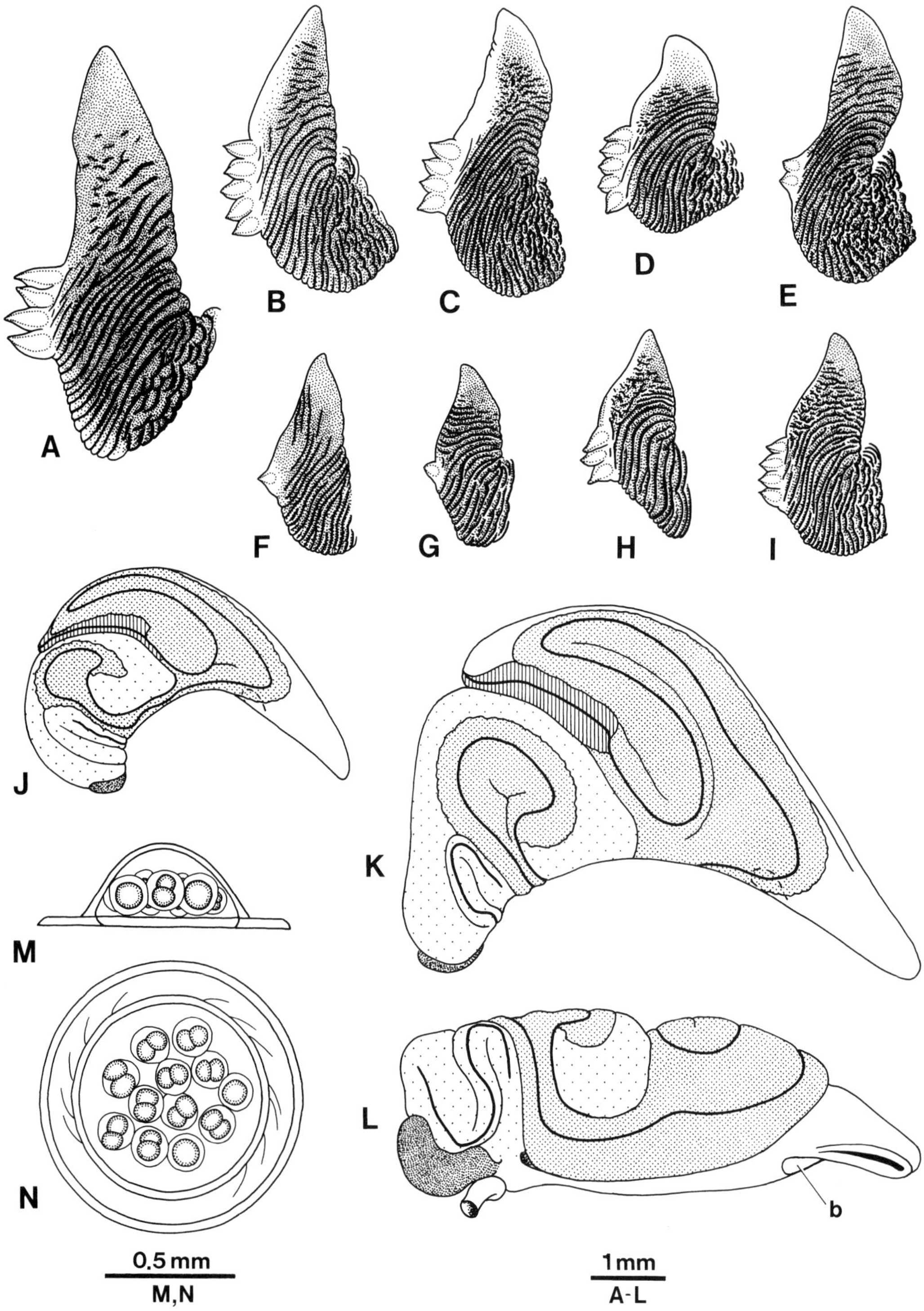

Figure 49. *Littorina (Littorina) mandshurica*: **A–I,** penes; **J–L,** pallial oviducts; **K, L,** lateral and ventral views of same pallial oviduct; **M, N,** pelagic egg capsule (after Kojima, 1958e). **A–E, J–L,** Akkeshi, Hokkaido; **F,** Ohmu, Hokkaido; **G–I,** Abashiri, Hokkaido; **M, N,** Tsuchiya, near Asamushi, Honshu. Shell heights: **A,** 18 mm; **B,** 15 mm; **C,** 15 mm; **D,** 13 mm; **E,** 14 mm; **F,** 12.4 mm; **J,** 18.6 mm; **K, L,** 15.7 mm. For shading conventions of pallial oviducts see p. 6. Abbreviation: b, copulatory bursa.

Radula and foregut (Fig. 50C–F). Relative radular length 1.23–2.43. Rachidian: length/width 0.86–1.07; central cusp pointed; one pointed cusp and one outer denticle on each side. Lateral: four cusps, third from inside largest, all pointed. Inner marginal: four cusps, third from inside largest, all pointed. Outer marginal: 6–7 pointed cusps; narrow neck, projecting flange on inside of base only. Salivary glands enlarged.

Shell variation (Figs 50A, B; 51)

This species is rather uniform in appearance. Sculpture is most prominent in young shells (Fig. 50A), older ones becoming more smooth and frequently eroded. The colour does not vary greatly; it is usually dark purple-black, but some shells are paler with darker ribs, producing a banded pattern.

Distribution

Habitat. Individuals can be found from the littoral fringe down to depths of 0.5 to 1.0 m in the sublittoral, but the majority occur in the mid to low eulittoral. In northern Honshu Kojima (1960) reported that *L. mandshurica* undertakes the same vertical migration as *L. brevicula*, moving from high to low-tide level in the breeding season, from January to April, but gave no further details. Golikov & Skarlato (1971) observed aggregations below datum level during the winter at Pos'et Bay in southern Primor'e, before upward migration in April. However, other Russian workers have not noted seasonal changes in distribution, but have found that *L. mandshurica* remains in the littoral zone throughout the year (Golikov & Kusakin, 1962, 1978). It tolerates a range of exposure to waves, from relatively exposed to sheltered shores (Golikov & Skarlato, 1967). On rocky, boulder and cobble shores it is found mainly in the fucoid zone, remaining in crevices and beneath boulders during low tide, but also lives among barnacles and, lower on the shore, with coralline algae (Mokyevsky, 1960; Golikov & Kusakin, 1962; Golikov & Skarlato, 1985). In sheltered bays it can occur on muddy sand, on shell debris and among *Zostera* (Golikov & Kusakin, 1962; Golikov & Gulbin, 1978). Golikov & Kusakin (1962) and Golikov & Skarlato (1967) reported that *L. mandshurica* is tolerant of freshwater influence, but in physiological experiments its tolerance is less than that of other Pacific *Littorina* species (V. Ya. Berger, 1973; Berger & Kovaleva, 1974).

In Hokkaido and southern Primor'e *L. mandshurica* usually occurs with *L. brevicula*, but occupies a lower level on the shore. Throughout its range it is sympatric with *L. sitkana*, a high-level species, and with *L. squalida*, which is found in the sublittoral. It is generally less common than these congeners. Its diet has not been studied.

Range (Fig. 52). This species occurs in the northern Sea of Japan, Hokkaido, in southeastern Sakhalin and the Tatar Straight. In this study the southernmost records are Pos'et Bay in southern Primor'e (BMNH, ZISP) and Mutsu Bay in northern Honshu (USNM, NSMT). Although few collections have been seen from the area, the species does occur commonly in the northwestern Sea of Japan (Mokyevsky, 1960). It has not previously been recorded from north of De Kastri Bay in the Tatar Straight (Schrenck, 1861; Kuroda & Habe, 1952; Mokyevsky, 1960), but in ZISP there is a collection from Litke in the Sakhalinsky Gulf (54°N). On the east coast of Sakhalin it does not appear to occur north of Terpeniya Gulf (ZISP). It is common all around Hokkaido, and just reaches the southern Kurile Islands, being recorded from Kunashir, Shikotan and Iturup (Golikov & Kusakin, 1962).

Fossil record. A single shell from near Nome, Alaska, from deposits of the Pelukian transgression that have been dated at 100 000 BP (Hopkins, 1967a), was identified by Dall (1920, 1921) and by MacNeil, Mertie & Pilsbry (1943) as *L. palliata* (=*L. obtusata*). Hopkins (1967a) compared it with both *L. palliata* and *L. mandshurica*, and Reid (1990c) tentatively identified it as *L. mandshurica*. Unequivocal identification of this worn and slightly damaged shell (Fig. 53) is not possible. The relative size of the aperture is too small for *L. sitkana*, and the lack of any trace of fine spiral sculpture suggest that it is not *L. squalida*. There is indeed some resemblance to geog. var. *palliata* of *L. obtusata*, but occurrence of this Atlantic species in the northern Pacific, even during a warm climatic interval, is considered unlikely. The shell is most similar to *L. mandshurica* in its thickness, columellar shape and presence of three slight angulations on the last whorl. If correct, this would represent a considerable extension to the known range of *L. mandshurica*. The species has been recorded from a Holocene deposit in Primor'e (Jull, Kuz'min, Lutaenko, Orlova, Popov, Rakov & Sulerzhitskii, 1994).

Figure 50. *Littorina (Littorina) mandshurica*: **A,** juvenile shell; **B,** detail of shell microsculpture; **C,** radula (shell H=10.7 mm), at 45°; **D–F,** three views of radula (shell H=15.7 mm), at 45° (**D**), flat (**E**), side view (**F**). **A,** Sergeevka, Sakhalin Island, Russia; **B,** Abashiri, Hokkaido; **C,** Peter the Great Bay, Russia; **D–F,** Akkeshi, Hokkaido. Scale bars: **A**=1 mm; **B**=1 mm; **C**=100 μm; **D, E**=100 μm; **F**=100 μm.

Discussion

Intraspecific variation. The shell is morphologically rather uniform in *L. mandshurica*, as expected in a planktotrophic species with presumably widespread dispersal and gene flow. There have been several studies of allozyme frequencies in individual populations, which have revealed an unusually low level of genetic variation (mean expected heterozygosity per locus 0.008 to 0.069; Kartavtsev & Efremov, 1981; Zaslavskaya, 1989; Ward, 1990; Tatarenkov, 1992); this might be due to a severe genetic bottleneck in the past, as has been suggested for *L. littorea* (Janson, 1987a).

Phylogenetic relationships. This is the sister-species of *L. brevicula* (see discussion p. 137).

Similar species. Confusion is most likely with *L. brevicula*, with which *L. mandshurica* is often found; differences are summarized in Table 4. Anatomically, the sympatric *L. sitkana* is quite different, but the

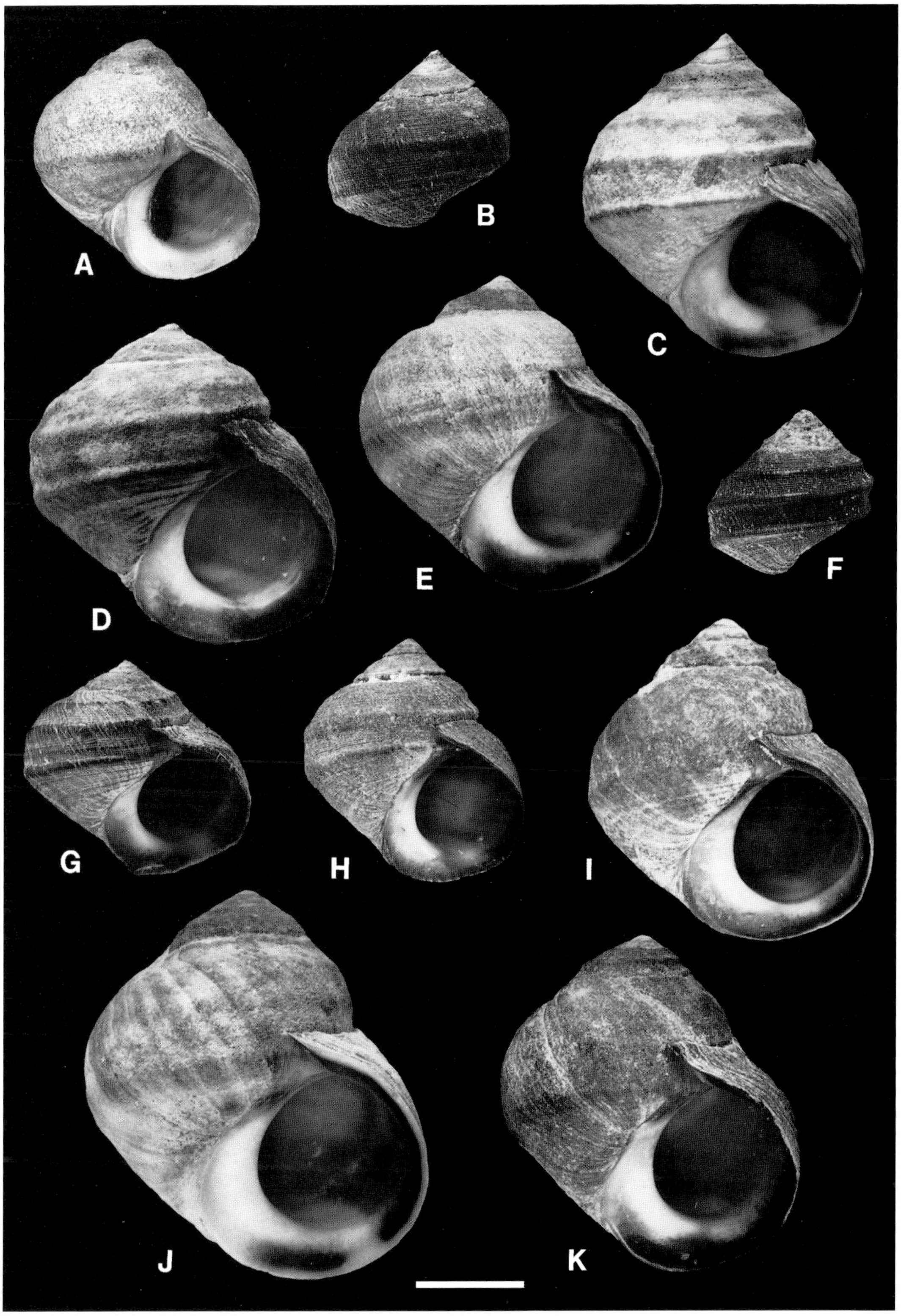

Figure 51. *Littorina (Littorina) mandshurica.* **A,** *Litorina mandshurica* Schrenck, 1861; paralectotype, BMNH 1992163; De Kastri, Tatar Strait, Russia. **B, F, I,** Abashiri, Hokkaido. **C,** Ohmu, Hokkaido. **D,** Muroran, Hokkaido. **E, G, J, K,** Akkeshi, Hokkaido. **H,** Bakkai, Wakkanai, Hokkaido. Scale bar=5 mm.

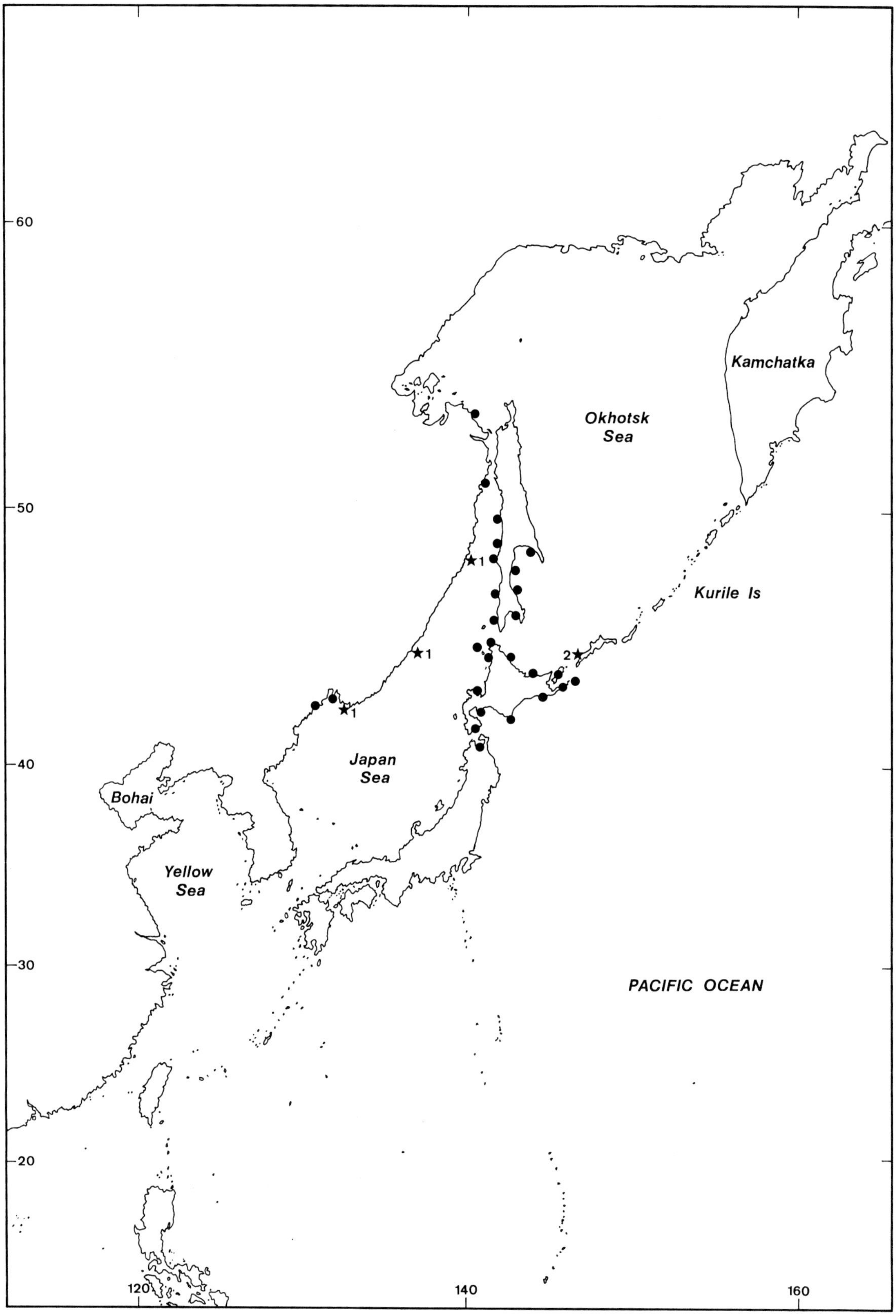

Figure 52. *Littorina (Littorina) mandshurica*: distribution map. Key: circles, specimens examined; asterisks, records from literature (1, Mokyevsky, 1960; 2, Golikov & Kusakin, 1962).

Figure 53. Possible specimen of *Littorina (Littorina) mandshurica*, Second Beach, near Nome, Alaska; Pleistocene; USNM 324319. Scale bar=5 mm.

shells can be similar. *Littorina sitkana* has a more thin-walled shell and relatively larger aperture. In the areas of sympatry the shell of *L. sitkana* is usually completely smooth or has ribs only on the base.

Subgenus *Neritrema* Récluz, 1869

Synonymy

Neritoides Brown, 1827: index p. iv; pl. 43, figs 14, 15, 21, 22 (type species *Neritoides littoralis* (i.e. *Nerita littoralis* Linnaeus, 1758); this is a misidentification of the figured species, see p. 202; figs 14, 21, 22 are *L. obtusata*, fig. 15 is *L. fabalis*; in the interests of stability, *Turbo obtusatus* Linnaeus, 1758, should be regarded as the type species, but designation requires use of ICZN plenary powers, ICZN, 1985: art. 70b; not *Neritoides* Meuschen, 1779, see Kadolsky, 1970, 1971, Heppell, 1971).

? *Littorelaea* Leach, in Gray, 1847a: 271 (in combination *Littorelaea pultneyii* Leach, in Gray, 1847, *nomen nudum*, but a possible synonym of *L. obtusata*).

Neritrema Récluz, 1869: 43, 46 (type species *Littorina obtusata* Linnaeus (i.e. *Turbo obtusatus* Linnaeus, 1758) by subsequent designation (Dall, 1909); proposed as subgenus of *Littorina*; misspelled *Neritotrema* by Wenz, 1938).

Littorivaga Dall, 1918: 137 (type species *Littorina sitchana* Philippi (=*Littorina sitkana* Philippi, 1846) by original designation; proposed as section of *Littorina*).

Algamorda Dall, 1918: 137 (type species *Littorina newcombiana* Hemphill (i.e. *Paludinella newcombiana* Hemphill, 1877=*L. subrotundata*) by original designation; proposed as subgenus of *Littorina*).

Taxonomic history

The smooth, globose shells of *L. obtusata* and *L. fabalis* are very distinctive among *Littorina* species, and the genus *Neritoides* was first proposed to accommodate them by Brown as early as 1827. This name was retained at subgeneric rank for *L. obtusata s.l.* (and sometimes including a few low-spired species now classified as *Littoraria* or *Nodilittorina*) by several subsequent authors (H. & A. Adams, 1854; Tryon, 1883, 1887; Dautzenberg & Durouchoux, 1900; Kobelt, 1907; Dall, 1923; Dautzenberg & Fischer, 1925; Winckworth, 1922, 1932). This same group was named *Neritrema* by Récluz (1869) and, since *Neritoides* Brown was preoccupied, the later name was adopted by Bequaert (1943). However, those authors who took *L. obtusata* to be the type species of *Littorina* (p. 39) used the nominal subgenus for this same group of neritiform shells (Dall, 1909, 1918, 1921; Thiele, 1929; Wenz, 1938; Nordsieck, 1968, 1982; Palmer, 1989, at generic rank).

This grouping, initially based upon shell characters, was thought to be supported by spawn characteristics; Dall (1909) mistakenly stated the group to be viviparous, but Winckworth (1922) correctly noted their benthic gelatinous egg masses and nonplanktotrophic development. Dall (1918) named a new section, *Littorivaga*, for another group that he believed to be viviparous (Dall, 1909), with *L. sitkana* as its type species. In fact this species has benthic gelatinous egg masses, and accordingly Habe (1958a) synonymized *Littorivaga* with *Neritrema*. Nevertheless, those who wished to distinguish the genuinely or supposedly ovoviviparous species at subgeneric level employed *Littorivaga* (Winckworth, 1922, 1932; Thiele, 1929; Wenz, 1938; Bequaert, 1943; Nordsieck, 1968, 1982). According to the classification proposed here, the ovoviviparous forms are regarded as a single species, *L. saxatilis*. Smith (1982) used the subgenus *Littorivaga* in a slightly different sense, for the *L. saxatilis* species-group, and Palmer (1989) in this same sense, but at generic rank.

Dall (1918) also described a new subgenus, *Algamorda*, with *L. newcombiana* as its type, and in 1921 assigned two additional species to this group; all three are here regarded as synonyms of *L. subrotundata*. This subgenus was presumably established in recognition of the small, tall-spired, thin-walled shells of the salt-marsh ecotypes that were available to Dall. Since the species was poorly known, *Algamorda* remained in the literature as a subgenus (Oldroyd, 1927; Thiele, 1929; Palmer, 1958) or genus (Wenz, 1938; Keen, 1970; Abbott, 1974; Kozloff, 1987) until synonymized with *Neritrema* by Reid (1989a).

The cladistic analyses of morphological characters by Reid (1989a, 1990b) showed *Neritrema* as a monophyletic group, although all the synapomorphies were connected with benthic spawn, nonplanktotrophic development and associated characters of the reproductive system, as in the revised analysis presented here (p. 351). Phylogenetic analyses of mitochondrial DNA sequences have produced equivocal results, with information from different genes either not resolving, weakly supporting or disagreeing with the monophyly of *Neritrema* (Rumbak, Reid & Thomas, 1994; Reid, Rumbak & Thomas, 1996; p. 360). However, two separate phylogenetic analyses of allozyme frequencies in a restricted number of Pacific and Atlantic *Littorina* species have clustered *L. sitkana* with Atlantic species, thus adding some support for the monophyly of *Neritrema* (Zaslavskaya, Sergievsky & Tatarenkov, 1992a, b; p. 356).

The anatomical differences between the oviparous species and the ovoviviparous *L. saxatilis* are small (transformation of jelly gland into brood pouch, reduction of capsule gland, loss of egg capsule). Both allozyme (e.g. Ward, 1990; Backeljau & Warmoes, 1992) and DNA studies (Rumbak *et al.*, 1994; Reid *et al.*, 1996) support the very close phylogenetic relationship between *L. saxatilis* and the oviparous members of the *L. saxatilis* group, and also with the *L. obtusata* group. Therefore, recognition of a separate subgenus for *L. saxatilis* on the basis of its ovoviviparity is not warranted.

Diagnosis

Shell: turbinate or globose; with spiral grooves or smooth, occasionally nodulose; protoconch of the nonplanktotrophic type; aragonitic with thick outer layer of irregular-prismatic calcite. Male: 0–68 mamilliform penial glands. Female: egg groove coiled in two or three successive spiral loops through albumen and capsule glands of pallial oviduct; jelly gland enlarged, sometimes forming brood pouch. Spawn and development: benthic gelatinous mass with ovoid capsules containing single ova, or ovoviviparous with brooding in pallial oviduct; nonplanktotrophic with intracapsular metamorphosis. Radula: outer marginal teeth with reduced outer basal projection and short cusps; cusps of five central teeth short, either pointed or blunt. Alimentary system: salivary glands enlarged.

Littorina (Neritrema) sitkana **Philippi, 1846**

(Frontispiece, Figs 5, 54–58)

Synonymy

Littorina sitkana Philippi, 1846b: 140 (Sitka, Nova Albion [Alaska]; lectotype (here designated, 12.2 mm) BMNH 1968219, seen, Philippi, 1847, **3**: *Litorina* pl. 6, fig. 13, Fig. 56A herein; 1 paralectotype BMNH 1968219, seen). Dall, 1925: 19, 38; pl. 25, fig. 7. Keep & Baily, 1935: 198–199. Abbott, 1954a: 134; pl. 20, fig. b. Urban, 1962: 320–323. Buckland-Nicks, Chia & Behrens, 1973: 359–365; figs 2, 4–15 (spawn, embryo). Abbott, 1974: 67; text-fig. 552. Rosewater, 1980: 5–8; figs 1, 2 (radula). Foster, 1981: 100–102 (*s.l.*). Baxter, 1987: 49 (*s.l.*; in part, includes *L. aleutica*). Sacchi & Voltolina, 1987: 209–234; fig. 2a, b; fig. 3d (penis); fig. 3e (oviduct); fig. 5 (radula) (*s.l.*). Buckland-Nicks & Chia, 1990: 159–176; fig. 2 (oviduct); fig. 3 (spawn).

Egorov, 1992: 170; fig. 1D. Boulding, Buckland-Nicks & van Alstyne, 1993: 43–68; fig. 1b, e; fig. 3b (penis); fig. 4a, d (spawn); fig. 5b (radula).

Litorina sitkana—Lischke, 1869: 77–79. Dall, 1886b: 211.

Littorivaga sitkana—Kuroda & Koba, 1933: 153–154. Habe, 1951: 87–88; text-fig. 4.

Littorina (Littorivaga) sitkana—Dall, 1918: 137. Dall, 1923: 3. Burch, 1945: 11, 36. La Roque, 1953: 168. Golikov & Gulbin, 1978: 190.

Neritrema sitkana—Habe, 1958a: 8–9; pl. 2, figs 21, 25. Yamamoto & Habe, 1962: 15; pl. 2, fig. 25; pl. 3, figs 2, 29. Oyama & Takemura, 1963: *Neritrema, Littorina* pl., fig. 4. Habe, 1964: 28; pl. 9, fig. 25 (as *sitkanum*). Habe & Ito, 1965: 25; pl. 7, figs 4, 5. Higo, 1973: 46. Higo & Goto, 1993: 73.

Littorina (Littorina) sitkana—Rosewater, 1970: 423.

Littorina (Neritrema) sitkana—Reid, 1989a: 102. Reid, 1990b: 1–19; fig. 1l; fig. 2h (oviduct); fig. 3h (spawn); fig. 4f (radula). Taylor & Reid, 1990: 199–215 (mineralogy). Reid, Zaslavskaya & Sergievsky, 1991: 1–6; figs 5–7; fig. 21 (oviduct); fig. 22 (penis). Reid & Golikov, 1991: 7–15; fig. 39 (oviduct).

Litorina sitchana Philippi, 1847, **3**: 13; *Litorina* pl. 6, figs 13, 18 (unjustified emendation of *Littorina sitkana* Philippi, 1846). Carpenter, 1864: 655. Weinkauff, 1882: 77–78; pl. 10, figs 10, 11. Dunker, 1882: 111. Weinkauff, 1883: 216.

Littorina sitchana—Middendorff, 1849: 392–393; pl. 8, figs 13–15. Reeve, 1857: sp. 49; pl. 10, fig. 49a, b. Pilsbry, 1895: 62. Bavay, 1920: 254 (in part, includes *L. aleutica, L. brevicula, L. mandshurica*). Golikov & Skarlato, 1955: 171; pl. 44, fig. 20.

Littorina (Littorina) sitchana—Tryon, 1887: 240–241; pl. 41, figs 87, 88 (in part, other figs are *L. mandshurica, L. brevicula, L. aleutica*). Golikov & Kusakin, 1978: 90; fig. 54.

Littorina (Littorivaga) sitchana—Dall, 1918: 137. Dall, 1921: 153. Oldroyd, 1924: 149. Oldroyd, 1927: 60. Golikov & Kusakin, 1962: 275–276.

Neritrema sitchana—Okutani, 1986: 70–71; unnumbered plate.

Littorina cincta Gould, 1847: 252 (Puget Sound [Washington]; types lost; not Quoy & Gaimard, 1833).

Littorina kurila Middendorff, 1848: column 242 (Mare Ochot., ins. Kurilae [Ochotsk Sea, Kurile Islands]; lectotype (here designated, 8.9x7.9 mm)+approx. 100 paralectotypes ZISP *L. kurila* no. 77 (alcohol; Schantar Islands), seen; 62 paralectotypes ZISP *L. kurila* nos 71 to 76 and unnumbered, seen; many paralectotypes ZISP *L. kurila* nos 162 and 163 (alcohol), seen; 22 paralectotypes BMNH 1993002 (alcohol; from same lot as lectotype), seen, Fig. 56E; 25 paralectotypes BMNH 1993003 (alcohol), seen; paralectotypes MCZ 52009, not seen). Middendorff, 1849: 392. Middendorff, 1851: 201–202; pl. 11, figs 13, 14. Golikov, Gulbin & Sirenko, 1987: 31–32; pl. 2, fig. 19. Egorov, 1992: 170; fig. 1E. Selin, 1994: 126–136 (*s.l.*). Selin, 1995: 90–91 (*s.l.*).

Littorina (Littorina) kurila—Golikov & Kusakin, 1962: 272–275 (*s.l.*). Golikov & Skarlato, 1967: 23–24; fig. 15. Golikov & Gulbin, 1978: 189 (*s.l.*). Golikov & Kusakin, 1978: 87–89; fig. 53 (*s.l.*). Volova, Golikov & Kusakin, 1979: 65–66; fig. 41. Golikov & Skarlato, 1985: 383.

Neritrema sitkana form *kurila*—Habe & Ito, 1965: 25; pl. 7, fig. 5 (not fig. 4 as cited).

Neritrema sitkana kurila—Habe & Kosuge, 1967: 29; pl. 12, figs 1, 2. Higo, 1973: 46. Higo & Goto, 1993: 73.

Littorina subtenebrosa Middendorff, 1848: columns 242–243 (Ins. Urup, Kurilarum [Urup Island, Kurile Islands]; neotype (here designated, 8.7x8.2 mm, from lot of 10 specimens, Lopatka, Kamchatka) ZISP *L. kurila* no. 70, seen; 3 specimens from same lot as neotype BMNH 1993004. Although collected in 1847 by Wosnessenski, who supplied Middendorff with specimens, and accompanied by Middendorff's original identification label, this lot was not specifically mentioned by the author and therefore cannot be considered as a syntypic series). Middendorff, 1849: 392. Middendorff, 1851: 202–203; pl. 11, figs 11, 12. Pilsbry, 1895: 62.

Litorina sitchana var. *subtenebrosa*—Carpenter, 1864: 655.

Litorina subtenebrosa—Dunker, 1882: 111.

Litorina sitkana var. *subtenebrosa*—Dall, 1886b: 218.

Littorina sitchana subtenebrosa—Ushakov, 1953. Mokyevsky, 1960: 257.

Littorina (Littorina) kurila var. *subtenebrosa*—Golikov & Kusakin, 1962: 272–275; fig. 5.

Littorina sitchana [var. ex col.] *castaneus* Middendorff, 1849: 393 (infrasubspecific; not *L. castanea* Deshayes, 1843=*L. saxatilis*).

Littorina sitchana [var. ex col.] *zonatus* Middendorff, 1849: 393 (infrasubspecific; not *Litorina obtusata* var. *zonata* Menke, 1845=*L. obtusata*).

Littorina castanea A. Adams & Reeve, 1850: 49; pl. 11, fig. 8 (Eastern Seas; 4 syntypes BMNH 1968322, seen; not Deshayes, 1843=*L. saxatilis*). Reeve, 1857: sp. 26; pl. 6, fig. 26a, b.

Litorina castanea—Weinkauff, 1882: 47–48; pl. 6, figs 1, 4.

Littorina kurila [var. ex sculpt.] *costulata* Middendorff, 1851: 201 (infrasubspecific; not *L. tenebrosa* var. *costulata* Middendorff, 1849=*L. saxatilis*).

Littorina kurila [var. ex col.] *fuscus* Middendorff, 1851: 201 (infrasubspecific; not *Littorina fusca* Pfeiffer, 1840=*Nodilittorina mespillum* (Mühlfeldt, 1824)).

Littorina kurila [var. ex col.] *griseolacteus* Middendorff, 1851: 201 (as *griseo-lacteus*; infrasubspecific; not *L. tenebrosa* var. ex col. *griseolacteus* Middendorff, 1849=*L. saxatilis*).

Littorina kurila [var. ex col.] *zonatus* Middendorff, 1851: 202 (infrasubspecific; not *Litorina obtusata* var. *zonata* Menke, 1845=*L. obtusata*).

Littorina kurila [var. ex col.] *fasciatus* Middendorff, 1851: 202 (infrasubspecific; not *L. fasciata* Gray, 1839= *Littoraria fasciata*).

Littorina arctica—Reeve, 1857: sp. 69; pl. 13, fig. 69 (in part, includes *L. saxatilis* and ref. to *L. obtusata*; not Møller, 1842=*L. obtusata*). Hanna & Hertlein, 1959: 82–83; pl. 10 (in part, includes *L. obtusata*; not Møller, 1842).

Littorina rudis—Cooper, 1860: 374 (in part, includes *L. saxatilis*; not *Turbo rudis* Maton, 1797=*L. saxatilis*). Jeffreys, 1865, **3**: 367; 1869, **5**: 206 (in part, includes *L. saxatilis*, *L. compressa*, *L. arcana*; not Maton, 1797). Jeffreys, 1874: 106 (not Maton, 1797). Keep, 1888: 68–69 (not Maton, 1797). Keep, 1904: 216–217 (not Maton, 1797). Keep & Baily, 1935: 198 (not Maton, 1797).

Littorina (Littorivaga) rudis—Dall, 1921: 153 (in part, includes *L. saxatilis*; not Maton, 1797). Oldroyd, 1924: 149 (in part, includes *L. saxatilis*; not Maton, 1797). Oldroyd, 1927: 61 (in part, includes *L. saxatilis*; not Maton, 1797). Burch, 1945: 11–12, 37 (in part, includes *L. saxatilis*; not Maton, 1797).

Litorina sulcata 'Gould' Carpenter, 1864: 598, 655 (error for *L. cincta* Gould, 1847).

Litorina sitchana var. *modesta*—Carpenter, 1864: 655 (not *Littorina modesta* Philippi, 1846=*Nodilittorina modesta*).

Litorina tenebrosa—Schrenck, 1867: 321–330 (in part, includes refs to *L. saxatilis*; not *Turbo tenebrosus* Montagu, 1803=*L. saxatilis*).

Litorina tenebrosa var. *costulata*—Schrenck, 1867: 322 (not Middendorff, 1849=*L. saxatilis*).

Litorina tenebrosa var. *laevigata* Schrenck, 1867: 322–323 (infrasubspecific).

Litorina tenebrosa forma *depressior* Schrenck, 1867: 323–326 (infrasubspecific).

Litorina tenebrosa forma *elatior*—Schrenck, 1867: 323–326 (not *L. tenebrosa* var. *elatior* Middendorff, 1849= *L. saxatilis*).

Litorina tenebrosa var. *fusca* Schrenck, 1867: 327 (infrasubspecific; not *Littorina fusca* Pfeiffer, 1840=*Nodilittorina mespillum* (Mühlfeldt, 1824)).

Litorina tenebrosa var. *zonata*—Schrenck, 1867: 327–328 (not *L. tenebrosa* var. *zonatus* Middendorff, 1849= *L. saxatilis*).

Litorina tenebrosa var. *fasciata* Schrenck, 1867: 328 (infrasubspecific; not *L. fasciata* Gray, 1839=*Littoraria fasciata*).

Litorina tenebrosa var. *albida* Schrenck, 1867: 328 (infrasubspecific; not *Litorina albida* Philippi, 1848=*Nodilittorina modesta* (Philippi, 1846)).

Litorina gouldiana Weinkauff, 1882: 105 (new name for *Littorina cincta* Gould, 1847, not Quoy & Gaimard, 1833).

Littorina tenebrosa var. *obtusataea*—Aurivillius, 1885: 325; pl. 12, figs 4, 5 (in part, includes *L. saxatilis*; not Middendorff, 1849=*L. obtusata*).

Littorina tenebrosa var. *costulata*—Aurivillius, 1885: 325–326; pl. 12, fig. 6 (in part, includes ref. to *L. saxatilis*; not Middendorff, 1849=*L. saxatilis*).

Littorina sitchensis Nevill, 1885: 136 (unjustified emendation of *L. sitkana* Philippi, 1846).

? *Littorina sitchensis* subvar. *sublaevis* Nevill, 1885: 136 (*nomen nudum*).

? *Littorina sitchensis* subvar. *fasciata* Nevill, 1885: 136 (*nomen nudum*).

Litorina sitkana var. *atkana* Dall, 1886b: 211–212 (Bering Island; Atka Island [Aleutian Islands]; Kyska [Aleutian Islands]; lectotype (here designated) Kyska Harbor, USNM 108986, seen, Fig. 56G; 11 paralectotypes BMNH 1951.11.1.173–183, seen; paralectotypes MCZ 187512, ANSP 190934, ANSP 190950, DMNH 043607, none seen).

Litorina atkana—Dall, 1902: 551; pl. 39, fig. 11.

Littorina atkana—Keep, 1904: 217; fig. 235. Keep & Baily, 1935: 199; fig. 182. Hirase & Taki, 1954: pl. 79, fig. 9.

Littorina (Littorivaga) atkana—Dall, 1921: 153. Oldroyd, 1927: 60–61. Hirase, 1934: 47; pl. 79, fig. 9. Burch, 1945: 11, 36. La Roque, 1953: 167.

Littorivaga atkana—Kuroda & Koba, 1933: 153; pl. 1. Habe, 1951: 88; pl. 14, figs 2, 9. Kojima, 1958a: 224–229; fig. 2 (spawn). Kojima, 1958c: 35–37; fig. 1 (spawn).

Neritrema sitkana atkana—Oyama & Takemura, 1963: *Neritrema*, *Littorina* pl., figs 1–3. Higo, 1973: 46. Baxter, 1987: 49 (*s.l.*).

Neritrema sitkana form *atkana*—Habe & Ito, 1965: 25; pl. 7, fig. 4 (not fig. 5 as cited). Higo & Goto, 1993: 73.

Littorina (Littorina) atkana—Rosewater, 1970: 423.

Littorina (Littorivaga) grönlandica—Dall, 1921: 153 (in part, includes *L. saxatilis*; not *Litorina groenlandica* Menke, 1830=*L. saxatilis*). Oldroyd, 1927: 61 (in part, includes *L. saxatilis*; not Menke, 1830). Burch, 1945: 11 (in part, includes *L. saxatilis*; not Menke, 1830). La Roque, 1953: 167 (in part, includes *L. saxatilis*; not Menke, 1830).

Littorina grönlandica—Dall, 1925: 19, 38; pl. 25, fig. 2 (in part, includes ref. to *L. saxatilis*; not Menke, 1830).

Littorina (Littorivaga) saxatile saxatile La Roque, 1953: 168 (in part, includes *L. saxatilis*; unjustified emendation of *Turbo saxatilis* Olivi, 1792).

Littorina sitkana morphs *albida*, *lutea*, *aurantia*, *olivacea*, *fulva*, *fusca*, *nigra*, *zonata*, *sanguinea* Sacchi & Voltolina, 1987: 218–220 (infrasubspecific).

Note: in the synonymy '*s.l.*' (*sensu lato*) indicates that the author's concept of the species probably included some or all of the other smooth-shelled species (*L. subrotundata, L. aleutica, L. natica, L. kasatka*), but without specific reference to, or figures of, these species.

Taxonomic history

The history of this species is complicated, as a result of its extreme variability in shell form, resemblance to Atlantic species and because different nomenclature has prevailed in several parts of its wide range in North America, Russia and Japan. It received six specific names between 1846 and 1850. First described from Sitka in Alaska in 1846 by Philippi, the author himself emended the name to *L. sitchana* in the following year. Unaware of Philippi's work, Gould (1847) gave it the name *L. cincta*, but this was not used subsequently. Middendorff made large collections on the Russian shores of the Okhotsk Sea and was interested in shell variation, developing a complex system for the designation of form, sculpture and colour varieties (Middendorff, 1848, 1851). He used the name *L. sitkana* for the larger, carinate shells from the Aleutians and Alaska (Middendorff, 1849), and described the smaller, variably coloured, smooth or ridged form from the Okhotsk Sea as *L. kurila*, and the larger, dark brown, smooth form from the Kurile Islands as *L. subtenebrosa* (Middendorff, 1848, 1849, 1851).

These three forms, *L. sitkana, L. kurila* and *L. subtenebrosa*, were united as intraspecific variants by Schrenck (1867), but using the name *L. tenebrosa* (=*L. saxatilis*, an Atlantic species). This was corrected by Lischke (1869), who gave an extensive synonymy under the name *L. sitchana* and recorded its extreme variability and wide distribution from Japan to northwestern America. With the exception of Weinkauff (1882), few other authors of the period held to this correct view, however. Tryon (1887; followed by Bavay, 1920) also had a broad concept of the species, but included *L. brevicula, L. mandshurica* and *L. aleutica*.

In the Japanese literature, carinate and smooth shells were commonly separated as two species, *L. sitchana* and *L. subtenebrosa* (Dunker, 1882; Pilsbry, 1895), and later as *L. sitkana* and *L. atkana* (Kuroda & Koba, 1933; Habe, 1951). Only following Habe's (1958a) revision were they synonymized as *L. sitkana*, or treated as subspecies or varieties.

In the Russian literature, carinate and smooth shells were also separated, as *L. sitkana* and *L. kurila* (e.g. Golikov & Kusakin, 1962, 1978; Egorov, 1992). Only recently has the synonymy of these two been appreciated (Reid, 1990b; Reid & Golikov, 1991). The shell of the geographical form of *L. sitkana* that was named *L. kurila* by Middendorff is frequently indistinguishable from that of the moderate ecotype of another species, *L. subrotundata* (p. 170). However, some of Middendorff's original specimens, from the Shantar Islands in the western Okhotsk Sea, survive in alcohol in ZISP and dissection has shown that their oviducts are of the type found in *L. sitkana* (Reid, 1990b).

The North American tradition was different again. Dall (1886b) initially described a large, smooth variety as *L. sitkana* var. *atkana*, but later considered it to be a separate species (Dall, 1902). In his widely followed northwestern American list of 1921, Dall gave four species now recognized as *L. sitkana*: *L. sitchana, L. atkana, L. grönlandica* and *L. rudis*, of which the last three names were applied to smooth-shelled forms. Both *L. groenlandica* and *L. rudis* were originally described from the Atlantic (being synonyms of *L. saxatilis*) and Dall evidently believed them to be common to the Pacific and Atlantic Oceans. This use of synonyms of *L. saxatilis* for *L. sitkana* had had a long history. Cooper (1860) and Keep (1888, 1904) used *L. rudis*, and Jeffreys (1865, 1874) accepted that *L. rudis* occurred also in the Pacific. In the northwestern Pacific both Schrenck (1867) and Aurivillius (1885) had used the name *L. tenebrosa*, another synonym of *L. saxatilis*. In North America, Dall's separation of four species was repeated by Oldroyd (1927), Burch (1945) and La Roque (1953), although Burch did quote doubts about the status of *L. rudis* by A. M. Keen. The synonymy of the northeastern Pacific *L. rudis* with *L. sitkana* was finally established by Ricketts, Calvin & Hedgpeth (1952) and Urban (1962). Smooth-shelled *L. sitkana* were erroneously identified as another Atlantic species, *L. arctica* (=*L. obtusata*), by Reeve (1857) and Hanna & Hertlein (1959). *Littorina atkana* was still considered a distinct species by Rosewater (1970).

A further complication arises because five separate species have recently been recognized among the smooth-shelled *Littorina* species of the northwestern Pacific (Reid & Golikov, 1991). These are *L. sitkana, L. subrotundata, L. aleutica, L. natica* and *L. kasatka*, and their shells are often so similar that they are frequently intermingled in museum collections. Their respective geographical distributions enable many literature references to smooth-shelled forms to be reassigned, but inevitably some authors must have included more than one in their concepts of smooth-shelled species such as *L. kurila*, *L. subtenebrosa*, *L. atkana*, *L. tenebrosa*, *L. rudis* and *L. groenlandica*, and instances where this is known (from the authors' material) or highly likely, are indicated by '*s.l.*' in the preceding Synonymy.

Dall (1918) introduced the subgenus *Littorivaga*, with *L. sitkana* as its type species, but this was correctly recognized as a synonym of *Neritrema* by Habe (1958a) (p. 145).

Diagnosis

Mature shell height 4.7–25 mm. Shell low-turbinate to high-turbinate. Mineralogy of external irregular-prismatic calcite and internal crossed-lamellar aragonite. Sculpture variable: smooth; or with 4–6 ribs on base only; or with 8–11 flattened ribs on last whorl, the interspaces 1–2 times rib width, often with a narrow riblet and always with spiral microstriae. Colour variable: usually blackish or purple-brown, but may be entirely white, or white or orange variously banded or lined with blackish brown; columella white to purple-brown; interior purple-brown. Head black. Penis with short, triangular filament, 10–30% total length; 3–22 mamilliform penial glands; base often with some black pigment. Pallial oviduct of single loop of albumen gland, large loop of capsule gland, and swollen, septate jelly gland. Spawn a benthic mass of singly encapsulated eggs in gelatinous matrix; development nonplanktotrophic with intracapsular metamorphosis. Distribution: northern Pacific from Oregon to Bering Sea, Okhotsk Sea and northern Sea of Japan.

Anatomical description

Head-foot. Head and tentacles usually all black, but for pale tentacle tips; sometimes an unpigmented patch just behind eye and occasionally at inside of base of tentacle; sides of foot black. Opercular ratio 0.50–0.61.

Penis (Fig. 54A–J). Base wrinkled, usually with some grey to black pigmentation on outside surface and occasionally in sperm groove, but sometimes unpigmented; 3–19 (up to 22, Boulding, Buckland-Nicks & van Alstyne, 1993) mamilliform penial glands usually in single row, but occasionally several glands irregularly arranged or in a shorter second row. Filament short, triangular, 10–30% total length; subepithelial glandular tissue sometimes visible in preserved specimens. Sperm groove open, ending terminally. Penis reduced or absent outside breeding season (pers. obs.); no regression when breeding season is continuous (Sacchi & Voltolina, 1987).

Sperm. Euspermatozoa 48–51 μm. Paraspermatozoa 11–14 μm diameter, round to oval; vesicles large, distinct; nucleus not visible.

Oviduct (Fig. 54L–R). Single loop of albumen gland, followed by relatively large loop of capsule gland (folded back and alongside albumen gland) in which a small, distal, reddish, translucent part is sometimes visible, and finally a swollen, septate, jelly gland. Bursa of moderate size, extending back just to anterior part of capsule gland, opening near anterior end of jelly gland. Seminal receptacle simple. The ultrastructure of the glandular elements of the pallial oviduct, and their contributions to the egg capsule, have been described by Buckland-Nicks & Chia (1990) (see p. 17 for comparison of terminology).

Spawn and development (Fig. 54K). Individual females lay spawn masses 5–15 mm long (Buckland-Nicks, Chia & Behrens, 1973) or up to 37 by 25 mm, containing about 120 singly-encapsulated eggs (Kojima, 1958c). Spawn is deposited under stones or on seaweed in the eulittoral, and several females frequently spawn together, producing communal egg masses up to 380 by 120 by 40 mm, containing more than 2000 eggs (Kojima, 1958a, c; Buckland-Nicks *et al.*, 1973; Boulding *et al.*, 1993). There is some variation in the size and shape of the capsules within the gelatinous spawn matrix. In Hokkaido the capsules are a compressed spherical shape, with maximum diameter 550 μm (Kojima, 1958a, c) or 570–600 μm (pers. obs.; Fig. 54K). In Washington the capsules are larger, 910–1000 μm in diameter (Buckland-Nicks *et al.*, 1973; Boulding *et al.*, 1993); two drawings of these capsules (Buckland-Nicks *et al.*, 1973: fig. 2; Buckland-Nicks & Chia, 1990: fig. 3b) show a biconvex lens-shape with a narrow peripheral flange, but accompanying photographs (like those of Boulding *et al.*, 1993)

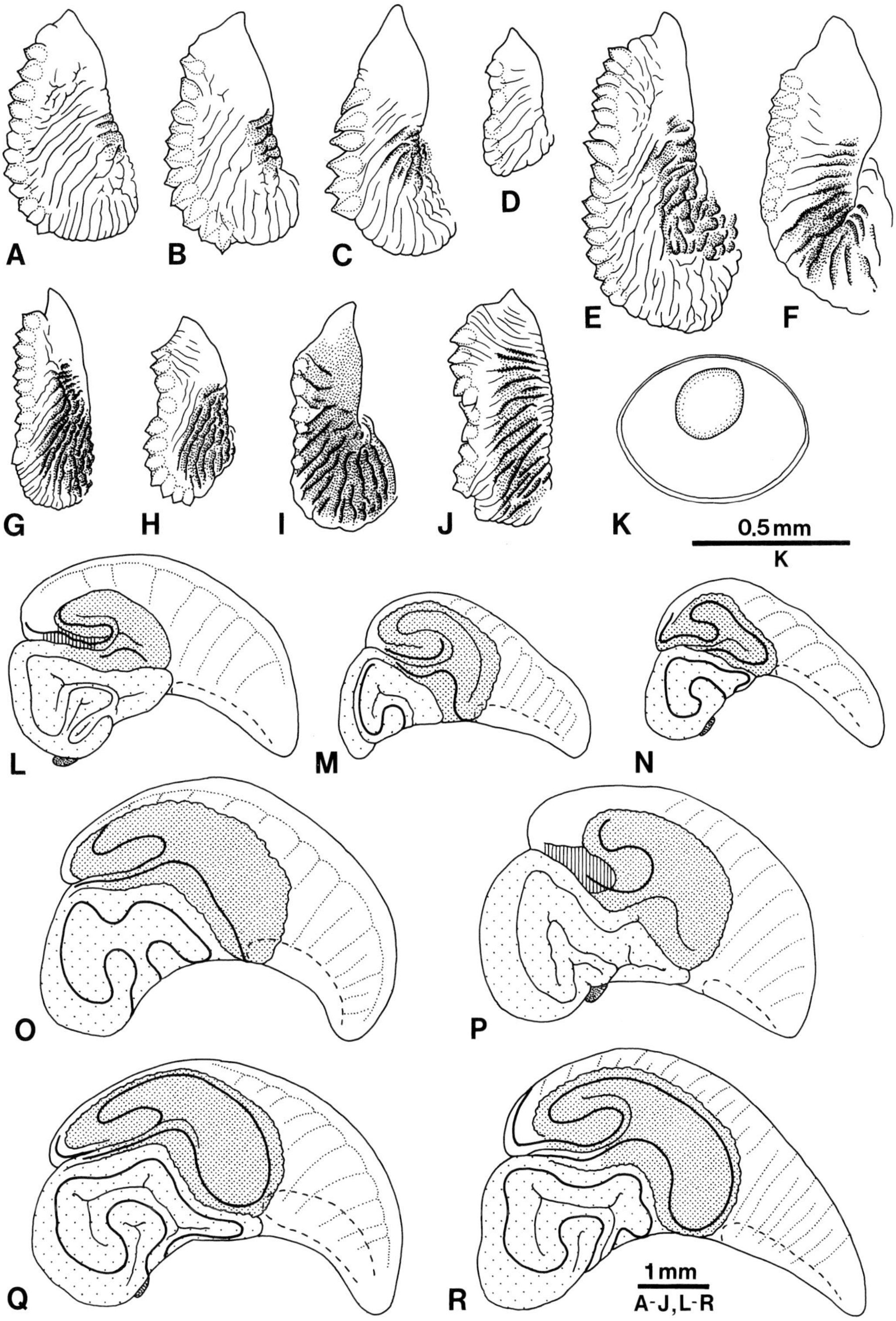

Figure 54. *Littorina (Neritrema) sitkana*: **A–J,** penes; **K,** egg capsule from benthic spawn; **L–R,** pallial oviducts. **A–C, Q,** Seldovia, Alaska; **D,** Abashiri, Hokkaido; **E, R,** Neah Bay, Washington; **F,** Andrew (Shagak) Bay, Adak Island, Aleutian Islands; **G,** Furugelm Island, Peter the Great Bay, Russia; **H,** Krabovaya Bay, Shikotan Island, Kurile Islands, Russia; **I,** Oshoro, Otaru City, Hokkaido; **J,** Vancouver Island, British Columbia, Canada; **K, N,** Akkeshi, Hokkaido; **L,** Ryponkich Island, Kurile Islands, Russia; **M,** Muroran, Hokkaido; **O,** 20 km W of Port Angeles, Washington; **P,** Shakotan Misaki, Hokkaido. All are moderate ecotypes. Shell heights: **A,** 10.1 mm; **B,** 10 mm; **C,** 9.8 mm; **D,** 6 mm; **E,** 11.5 mm; **F,** 16.2 mm; **G,** 10.1 mm; **H,** 7.5 mm; **I,** 13.9 mm; **J,** 12.8 mm; **L,** 10.6 mm; **M,** 11.0 mm; **N,** 7.6 mm; **O,** 10.4 mm; **P,** 13.0 mm; **Q,** 10.9 mm; **R,** 14.5 mm. For shading conventions of pallial oviducts see p. 6.

appear to resemble Fig. 54K; either the drawings were exaggerated or there may be variation in capsule form. In spawn from both Hokkaido and Washington the two halves of the capsule are of different thickness and composition (Buckland-Nicks & Chia, 1990; Boulding *et al.*, 1993; pers. obs.). The diameter of the egg is 175–200 µm (Kojima, 1958a, c; Buckland-Nicks *et al.*, 1973); only rarely are two eggs found within a capsule (Strathmann, 1987). Development nonplanktotrophic with intracapsular metamorphosis (Buckland-Nicks *et al.*, 1973; Buckland-Nicks & Chia, 1990). Juvenile shells measure 450 by 575 µm at hatching (Buckland-Nicks *et al.*, 1973; Strathmann, 1987). The breeding season is variable over the geographical range. In the vicinity of Washington spawning occurs year-round, but with peaks in autumn and spring (San Juan and Vancouver Islands) or in summer (Oregon) (Buckland-Nicks *et al.*, 1973; McCormack, 1982; Sacchi & Voltolina, 1987; Behrens Yamada, 1989). Kojima (1958a, c) recorded spawning from January to March at Hakodate, Hokkaido, whereas spawning was observed in mid May at Akkeshi, Hokkaido (pers. obs.). In Russian waters spawning occurs later in the year in colder water, beginning in April in Pos'et Bay, May in the southern Kurile Islands, and only in July to August in the colder Kurile Islands such as Simushir, and lasting for 2 to 3 months (Golikov & Kusakin, 1962, 1978, as *L. kurila*).

Radula and foregut (Fig. 55). Relative radular length 1.00–1.82. Rachidian: length/width 0.88–1.04; central cusp pointed, one pointed cusp and one outer denticle on each side. Lateral: four pointed cusps, third from inside largest. Inner marginal: four pointed cusps, third from inside largest. Outer marginal: 5–8 blunt to rounded or pointed cusps (sometimes showing irregularities or malformations); narrow neck, projecting flange on inside of base only. Salivary glands enlarged.

Shell variation (Figs 5, 56, 57)

Geographical variation. Large shell samples can usually be reliably assigned to one of three geographical varieties:

> Geog. var. *typica* (Figs 5, 56A–D, F, G): H up to 25 mm. Whorls rounded, sutures distinct. Sculpture variable, usually with 8–11 flattened ribs on last whorl, but these may be poorly developed or shell may be smooth. Colour variable. Distribution: Oregon to Alaska, eastern Bering Sea, Aleutian and Komandor Islands.
>
> Geog. var. *japonica* (Fig. 56H, J, M–Q): H up to 18.7 mm. Whorls rounded or slightly flattened, sutures moderately distinct. Sculpture variable: smooth, or often with 4–6 strong ribs on base only, or sometimes with 8–11 ribs covering whole surface. Colour variable. Distribution: Peter the Great Bay to southern Sakhalin, northern Honshu, Hokkaido, Kurile Islands and Kamchatka.
>
> Geog. var. *okhotensis* (Fig. 56E, I, K, L): H up to 13.2 mm. Sutures indistinct. Sculpture variable: usually smooth, or sometimes with 8–11 ribs covering whole surface. Colour: pale brown with 4–9 brown to blackish spiral lines, often a peripheral pale or white band. Distribution: northern Sakhalin, Okhotsk Sea from Shantar Islands to Gulf of Shelikov.

Strongly ribbed shells similar to Philippi's type of *L. sitkana* predominate throughout most of the range of the typical variety (Fig. 56A, B, D). Some smooth shells do occur in the south-east of the range (Fig. 56C), together with partially ribbed forms (Fig. 56F), but smooth shells become more frequent in Alaska, the Aleutian and Komandor Islands; shells also attain largest size in these areas. Dall's *L. atkana* (Fig. 56G) was a large, smooth, banded example. Growth rate has a marked influence on development of shell sculpture in this species, which may partly explain geographical trends in sculpture (see Sculptural Variation, p. 155).

In Peter the Great Bay, southern Sakhalin, northern Honshu, and western and northern Hokkaido most examples of geog. var. *japonica* show a very distinctive form with strong ridges on the base only, and rather flattened whorls (Fig. 56O, P), unlike shells from any other part of the range. In southeastern Hokkaido there is a locally high frequency of completely ribbed shells (Fig. 56Q; Ohgaki, 1983). On the colder east coast of Hokkaido and northwards through the Kurile Islands and Kamchatka the most common shell form is completely smooth (Fig. 56H, M, N); Middendorff's *L. subtenebrosa* was an example of this type. There is no clear distinction between these shells and smooth examples of geog. var. *typica* from the Komandor and Aleutian Islands.

Figure 55. *Littorina (Neritrema) sitkana*: **A–C,** three views of radula (shell H=15.0 mm), flat (**A**), at 45° (**B**), side view (**C**); **D,** radula (shell H=9.8 mm), at 45°; **E, F,** two views of radula (shell H=12.4 mm), flat (**E**), at 45° (**F**); **G,** radula (shell H= 7.9 mm), flat; **H,** radula (shell H=14.4 mm), at 45°. **A–C,** Nizki Island, Aleutian Islands; **D,** Seldovia, Alaska; **E, F,** Akkeshi, Hokkaido; **G,** Magadan, Okhotsk Sea, Russia; **H,** Neah Bay, Washington. Scale bars: **A, B**=100 μm; **C, D**=100 μm; **E, F**= 100 μm; **G**=100 μm; **H**=100 μm.

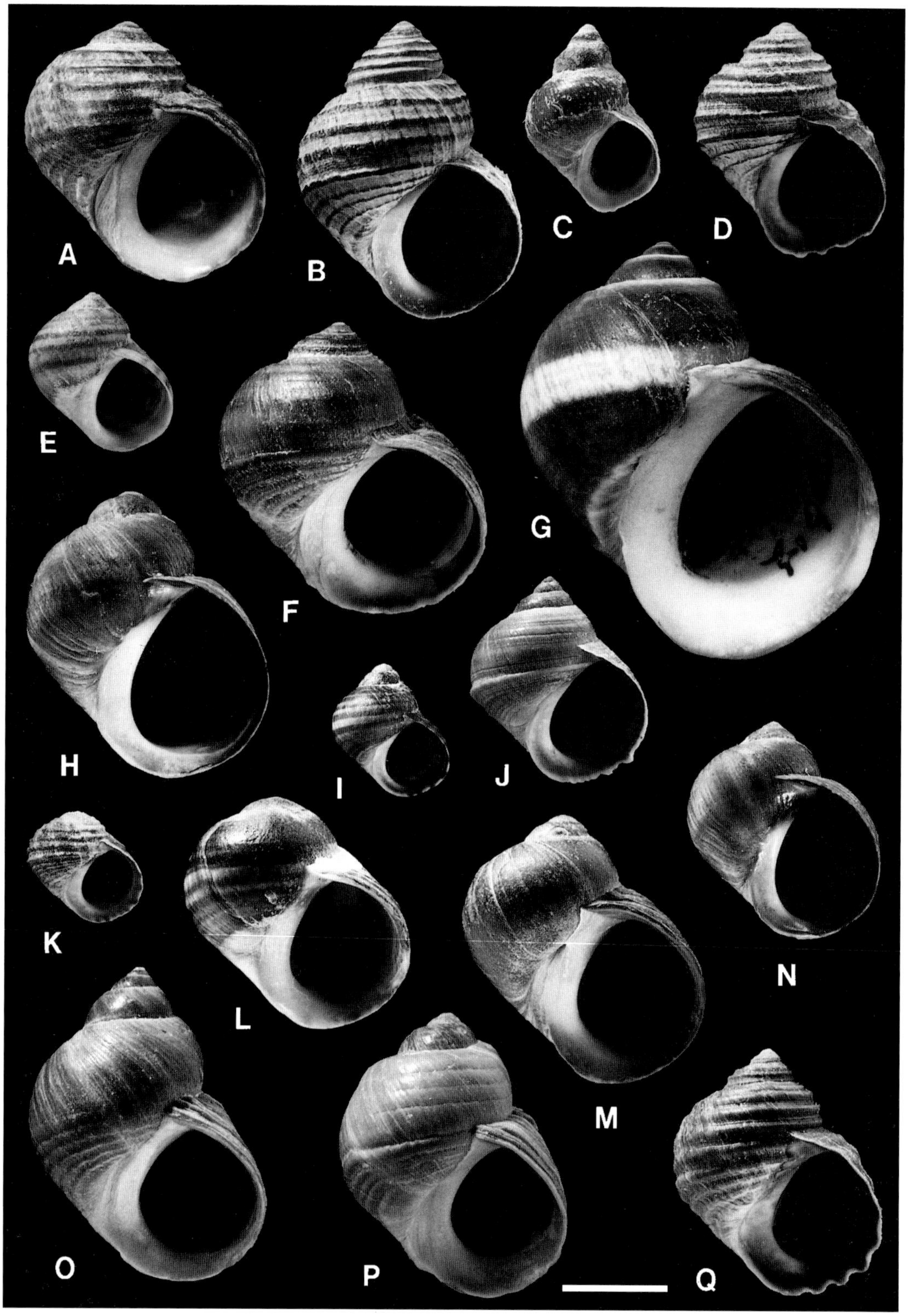

Figure 56. *Littorina (Neritrema) sitkana*: **A–D, F, G,** geog. var. *typica*; **E, I, K, L,** geog. var. *okhotensis*; **H, J, M–Q,** geog. var. *japonica*; **A, B, D–J, M–Q,** moderate ecotype; **C,** sheltered/brackish ecotype; **K, L,** wave-exposed ecotype. **A,** *Littorina sitkana* Philippi, 1846; lectotype, BMNH 1968219; Sitka, Alaska. **B,** Valdez, Alaska. **C,** mouth of Sooke River, Vancouver Island, Canada. **D,** Seldovia, Alaska. **E,** *Littorina kurila* Middendorff, 1848; paralectotype, BMNH 1993002; Schantar Islands, Okhotsk Sea, Russia. **F,** Sooke, Vancouver Island, Canada. **G,** *Litorina sitkana* var. *atkana* Dall, 1886; lectotype, USNM 108986; Little Kyska Island, Kyska Harbor, Aleutian Islands. **H, N,** Kasatka Point, Iturup Island, Kurile Islands, Russia. **I,** Nagayeva Bay, Magadan, Okhotsk Sea, Russia. **J,** Tserlovnaya Bay, Shikotan Island, Kurile Islands, Russia. **K,** Ayan, Okhotsk Sea, Russia. **L,** Cape Yelizavety, Sakhalin Island, Russia. **M,** Akkeshi, Hokkaido. **O, P,** Furugelm Island, Primor'e, Russia. **Q,** Muroran, Hokkaido. Scale bar = 5 mm.

The most distinctive geographical form is var. *okhotensis*, usually a smaller, smooth shell with shallower sutures, characterized by a lined colour pattern (Fig. 56E, I, L). This form was described as *L. kurila* by Middendorff (1848, 1849, 1851); the name was unfortunate, since the variety for which the name is here used does not appear to occur in the Kurile Islands, but only in the northern and western Okhotsk Sea. Ribbed examples do occasionally occur (Fig. 56K).

Some other aspects of geographical variation may be mentioned. Frank (1979) stated that the largest shells occurred in the centre of the latitudinal range in the eastern Pacific. Golikov & Kusakin (1978) noted larger shells from the Kurile Islands than from the southern extreme of the range at Pos'et Bay. Over the entire geographical range, however, the largest sizes are attained in the Aleutian, Pribilof and Komandor Islands; of the 245 lots seen in the present study, the maximum size in a lot exceeded 16.0 mm in 34 cases, of which 24 were from the Aleutian, Pribilof and Komandor Islands, 8 from the northeastern Pacific and the remaining 2 from the Kurile Islands. The maximum size recorded is 25 mm (Boulding *et al.*, 1993, locality unspecified).

Vermeij (1992b) found thicker shells in Hokkaido than in the Aleutian Islands which, because of a higher incidence of shell breakage in the former, he attributed to selection by the crushing crabs and fish that are more common in the southern part of the range. There do not, however, appear to be any striking geographical trends in shell shape consistent with latitudinal gradients in predation.

Ecotypic variation. This is less marked than in some other species of *Neritrema*, because *L. sitkana* does not occupy such a wide habitat range. Ecotypic variation has not yet been studied in detail in *L. sitkana*, but there is some predictable variation in shell shape between habitats. Most shells are of the moderate ecotype, but in extreme habitats two others can be distinguished:

Moderate ecotype (Fig. 56A, B, D–J, M–Q): the great majority of examples; even on wave-exposed shores many shells tend to be of this type, since they occupy sheltered microhabitats.

Wave-exposed ecotype (Fig. 56K, L): relative spire height tends to be lower on more exposed shores, but examples with SH as low as 1.31–1.43 are uncommon in the species as a whole.

Sheltered/brackish ecotype (Fig. 56C): this thin, tall-spired ecotype is occasionally found in brackish salt-marsh habitats. In the illustrated example, from the mouth of the Sooke River, Vancouver Island, SH=1.95. (In fully marine inlets and salt marshes, the species is usually of the moderate ecotype).

Selin (1994, 1995) recorded shells with relatively lower spires on more exposed shores in the Kurile Islands, but more than one species may have been included. There may be an ecophenotypic component to shell shape, since spire height was found to be lower when cultured at faster growth rates in the laboratory (Boulding *et al.*, 1993). Humphrey & Macy (1931) measured correlations between shell size and environmental variables in tide pools.

Sculptural variation. One of the most striking aspects of variation in *L. sitkana* is shell sculpture (Fig. 57). At first sight it is hard to believe that the strongly ribbed shells that are familiar in northwestern America (Fig. 56B, D) are the same species as the smooth shells commonly found in the Kurile Islands and Kamchatka (Fig. 56H, N). Indeed, as discussed earlier (p. 149), many authors have regarded these two forms as separate species. However, examination of large series shows that in many populations sculpture is continuously variable between these extremes, even if one or other form tends to predominate. On a geographical scale shells with ribs only on the base predominate in the Sea of Japan, smooth shells in the Kurile Islands and Kamchatka, and strongly ribbed shells in the Aleutian Islands and northeastern Pacific. In the Kurile Islands (p. 152) the sporadic occurrence of ribbed shells (as *L. sitkana*) and abundance of smooth ones (as *L. kurila*) was noted by Golikov & Kusakin (1962). On a smaller scale Ohgaki (1983) recorded the distribution of these three sculptural types around Hokkaido and found some correlations with environmental variables. Smooth shells predominate along the east coast where a cold current impinges, while in Akkeshi Bay smooth shells occur at the mouth, but those with ribs only on the base (Fig. 56O) become increasingly common towards the head of the bay. On Vancouver Island Sacchi & Voltolina (1987) recorded the proportions of smooth and ridged shells, but found no clear correlations with wave-exposure. However, Golikov & Kusakin (1962, 1978) noted that on Russian coasts smooth shells (as *L. kurila*) predominate on exposed coasts and ridged

Figure 57. *Littorina (Neritrema) sitkana*: **A–E,** juvenile shells; **F, G,** shell sculpture. **A,** Avacha Bay, Kamchatka, Russia; **B,** Shirokaya Pad, Sakhalin Island, Russia; **C,** 51°5' N, 156°42' E, S Kamchatka, Russia; **D,** Ozerskiy, Sakhalin Island, Russia; **E, F,** Neah Bay, Washington; **G,** 20 km W of Port Angeles, Washington. Scale bars: **A–E**=1 mm; **F, G**=1 mm.

shells (as *L. sitkana*) in protected bays (but note that their smooth category probably included smaller numbers of several similar species, see Taxonomic History, p. 149).

By rearing siblings at different densities, Boulding *et al.* (1993) demonstrated that at low growth rates shells were more likely to be ribbed, whereas at high growth rates they were smooth, even though the parents were ribbed. Although a genetic component cannot yet be entirely ruled out, this environmental effect on shell phenotype may partly explain both geographical and local variation in sculpture. There is some evidence that primary productivity is higher in the northwestern than in the northeastern Pacific (Vermeij, 1989) and the smooth shells commonly found in the former region might therefore be a result of higher growth rates there. In Hokkaido the greatest densities of this species are recorded on the cold east coast, suggesting favourable conditions for growth, and it is here that Ohgaki (1983) found the highest proportions of smooth shells. (However, Ohgaki himself explained the distribution in terms of Vermeij's (1973) hypothesis of heat dissipation by sculptured shells). Differences in rate of growth may also account for more local effects; Golikov & Kusakin (1978), for example, noted higher growth rates and larger size on exposed shores, and smaller shells with slower growth in inlets.

Sexual dimorphism. Sacchi & Voltolina (1987) found that females were significantly larger in some of the populations studied on Vancouver Island.

Colour variation (Fig. 5). In most populations of geographical varieties *typica* and *japonica* the great majority of shells are uniformly blackish brown. However, a wide range of differently coloured forms can be found, including shells externally entirely white, yellow, olive, orange, red or variously banded with dark brown (Fig. 56; Harger, 1972). Sacchi & Voltolina (1987) used a system of named morphs, like that applied to *L. saxatilis*, and found the most polychromatic populations in estuarine environments on Vancouver Island. Some colour forms are geographically restricted; for example, Ohgaki (1983) found his 'white-line' morph only in southern Hokkaido. Another example is the prevalence of the lineated pattern in geog. var. *okhotensis* in the Okhotsk Sea. Shorebirds do prey upon *L. sitkana*, for shells have been found at bird anvils (Spight, 1976), but whether visual predation has any influence on colour variation has not been investigated.

Distribution

Habitat (Frontispiece). *Littorina sitkana* is a common species on rocky and boulder shores, in crevices, pools and on fronds of macroalgae. It can be found from mean low water up to the littoral fringe, but on sheltered shores is most frequent in the upper eulittoral, often among *Fucus*, more rarely on red algae and *Laminaria*. On shores more exposed to waves it occurs in the barnacle zone and extends into the high splash zone, and may then be absent from most of the eulittoral (Rigg & Miller, 1949; Mokyevsky, 1960; Spassky, 1961; T. A. & A. Stephenson, 1961a, b; Urban, 1962; Nybakken, 1967; McCormack, 1982; Ohgaki, 1983; Behrens Yamada, 1992). In Russian waters it extends from 2.5 m depth to the littoral fringe; in the Kurile Islands a downshore migration occurs in November and December, and in winter the animals congregate under boulders in the mid-eulittoral, before most return to the upper eulittoral in April and May (Golikov & Kusakin, 1962, 1978; Kusakin, 1958; as *L. kurila*). There are several records from deeper water, 20 m at Paramushir Island, 40 m at Urup (Golikov & Gulbin, 1978) and 25 m at Moneron Island (Golikov, Gulbin & Sirenko, 1987; Gulbin, Golikov & Sirenko, 1993; as *L. kurila*), but these probably do not represent the normal habitat. Larger animals occur higher on the shore, because predation and wave action reduce survival at lower levels (McCormack, 1982; Behrens Yamada, 1992).

In the region of Washington and British Columbia several authors have observed that this species is most frequent in protected and humid microhabitats on relatively sheltered shores. It is rare or absent on wave-exposed headlands without suitable crevices, because adults are dislodged by waves, and also on shores without shady microhabitats to protect egg masses from desiccation (Harger, 1972; Behrens, 1972; Sacchi & Voltolina, 1987; Voltolina & Sacchi, 1990; Behrens Yamada, 1992; Boulding & van Alstyne, 1993). This contrasts with reports (together with corresponding collections in ZISP) from the Kurile Islands and Pos'et Bay, where the smooth-shelled form is one of the dominant species on wave-swept shores (Golikov & Kusakin, 1962, 1978; Golikov & Skarlato, 1967). On the most sheltered shores *L. sitkana* has been reported from mudflats (Behrens, 1972) and from the leaves of the seagrasses *Zostera* and *Phyllospadix* (Golikov & Kusakin, 1962; Golikov & Skarlato, 1985), but is uncommon in salt marshes and other estuarine situations (Ohgaki, 1983; pers. obs.). This species is surprisingly tolerant of environmental acidity in areas of volcanic activity (Selin, 1995).

In the Kurile Islands, Okhotsk Sea, northwestern Sea of Japan and eastern Hokkaido this species is often extraordinarily abundant, occurring in close-packed aggregations several shells thick (Kuroda & Koba, 1933; Mokyevsky, 1960; Ohgaki, 1983; Kusakin, 1985; pers. obs. at Akkeshi, Hokkaido). The animals cluster on the upper sides of rocks in dull weather, or between and beneath boulders during sunny weather at low tide, and in winter. The highest density recorded is 100 000 per m^2 (a biomass of 15.5 kg per m^2) in the upper eulittoral of Iturup in the Kurile Islands (Golikov & Kusakin, 1962, 1978, as *L. kurila*). Such great densities have not been found in the Aleutian Islands or Alaska (Nybakken, 1967; O'Clair, 1977; Zimmerman, Hanson, Fujioka, Calvin, Gharrett & MacKinnon, 1977).

Over its wide geographical range *L. sitkana* is sympatric with a number of other *Littorina* species. In the northeastern Pacific it is usually found with *L. scutulata* and *L. plena*; their relative zonation is variable, but *L. scutulata* is more tolerant of wave-exposure and survives on bare, sunny rocks in the littoral fringe (Urban, 1962; Harger, 1972; Behrens, 1972; Sacchi & Voltolina, 1987; Voltolina & Sacchi, 1990; Behrens Yamada, 1992). On the most wave-exposed headlands in Washington and British Columbia *L. sitkana* is replaced by *L. subrotundata* (Boulding, 1990; Boulding & van Alstyne, 1993; *L. subrotundata* as *L.* sp.), and the latter species also predominates in salt marshes. In Alaska, the Aleutian Islands, Kamchatka and the Kurile Islands *L. sitkana* commonly occurs with *L. subrotundata* on shores of moderate wave-exposure, in the Aleutian Islands also with *L. aleutica*, and in Kamchatka and the Kurile Islands also with *L. natica* and *L. kasatka*. Museum collections suggest that *L. sitkana* is much more common than these other species, but there may be a bias of collections towards more accessible, relatively sheltered shores. Very few ecological studies have yet distinguished between these conchologically similar species. In Hokkaido, Sakhalin and the northern Sea of Japan *L. sitkana* is found with *L. mandshurica*, which is zoned lower on the shore, and in Hokkaido, Honshu and Peter the Great Bay together with *L. brevicula*, which occurs at the same level as *L. sitkana*, but is less tolerant of wave action.

Littorina sitkana is an opportunistic omnivore, its diet including plant litter, black lichens, microalgae, macroalgae and epiphytes (Sacchi & Voltolina, 1987; van Alstyne, 1988; Voltolina & Sacchi, 1990; Behrens Yamada, 1992). In the Kurile Islands the huge aggregations of this species feed on macroalgae washed up on the shore (Golikov & Kusakin, 1962). It seems unable to utilize the sparse epilithic and endolithic microalgal growth on rocks of the littoral fringe (Voltolina & Sacchi, 1990).

Range (Fig. 58). *Littorina sitkana* occurs throughout the northern Pacific, from Oregon to the Bering Sea, Okhotsk Sea and northern Sea of Japan. In the eastern Pacific its southern limit is at Cape Arago, Oregon, and further extension of the range may be prevented by a combination of unsuitable exposed sandy coasts, desiccation of egg masses and predation by crabs (Behrens Yamada, 1977a, b, 1992). Abbott (1974) mentioned rare occurrences in northern California, but these are unsubstantiated, and might relate to *L. subrotundata*. In the Bering Sea it is common throughout the Aleutian chain, in the Pribilof and Komandor Islands and in Kamchatka; there are two records from northern Bristol Bay (Foster, 1981) and the most northerly records are from Nunivak Island (USNM 220156), St Lawrence Island (SMNH 891, oviduct examined) and the Olyutorskiy Gulf (ZISP, BMNH, oviducts examined). Kusakin & Ivanova (1978) recorded the species from localities on the Chukotsk Peninsula, but no specimens of *L. sitkana* from this area are present in the collection of ZISP, and the records probably refer to *L. aleutica* and *L. natica*. It occurs throughout the Okhotsk Sea and the Kurile Islands, in Sakhalin and Hokkaido. In Japan the southernmost records are Kinkasan Island in Miyagi Prefecture on the Pacific coast of Honshu (BMNH) and at 39°N on the Japan Sea coast (Kuroda & Habe, 1952). It is found on the northwestern shore of the Japan Sea (Mokyevsky, 1960), but has not been mentioned south of Pos'et Bay, and even here is absent from enclosed bays that are warm in summer (Golikov & Skarlato, 1967; Golikov & Kusakin, 1978; as *L. kurila*).

Fossil record. No fossil records have been traced, although this species has been found in Holocene deposits in Primor'e (Jull, Kuz'min, Lutaenko, Orlova, Popov & Sulerzhitskii, 1994) and Hokkaido (Matsushima & Kawakami, 1995).

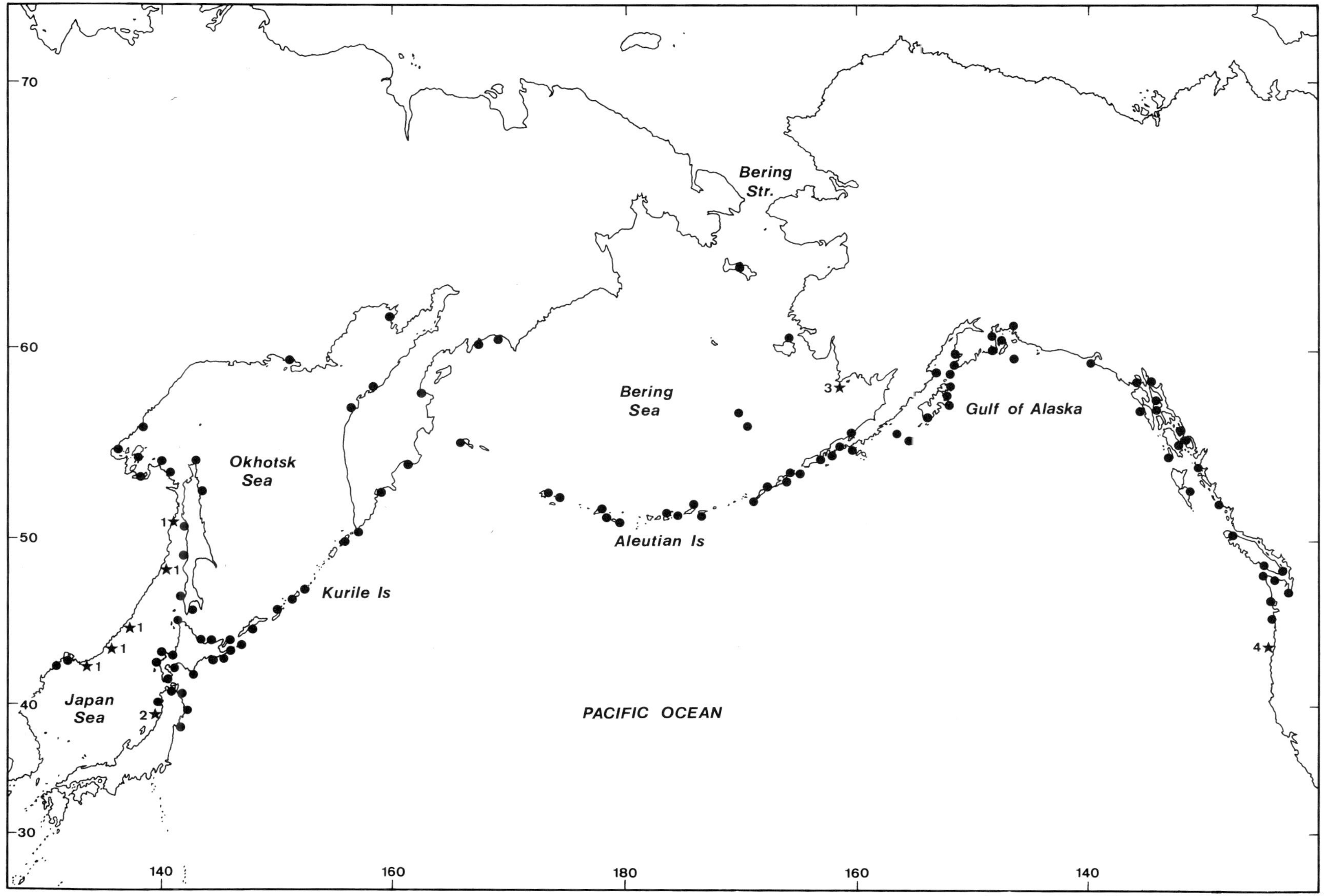

Figure 58. *Littorina (Neritrema) sitkana*: distribution map. Key: circle, specimens examined; asterisks, records from literature (1, Mokyevsky, 1960; 2, Kuroda & Habe, 1952; 3, Foster, 1981; 4, Behrens Yamada, 1977a, b, 1992).

Table 5. Comparison of *Littorina sitkana*, *L. subrotundata*, *L. aleutica*, *L. natica* and *L. kasatka*.

Character	*L. sitkana*	*L. subrotundata*
Head-foot pigmentation	Dark; tentacles usually all black	Fig. 63M–P. Paler; tentacles usually with unpigmented patch across base
Penis	Fig. 54A–J	Fig. 63A–L
Pigmentation	Usually grey to black base	Usually unpigmented, sometimes grey base
Length of filament	10–30% total length	20–40% total length
Mamilliform glands	3–22	5–17
Pallial oviduct	Fig. 54L–R	Fig. 63Q–X
Albumen gland	Simple loop, concealing renal oviduct	Simple loop, concealing renal oviduct
Capsule gland	Single loop folded back beside albumen gland; about equal in area to albumen gland	Single loop between albumen and jelly gland; 0.5–1 times area of albumen gland
Jelly gland	Large, swollen, septate	Large, swollen, septate
Copulatory bursa	Large; extends back to anterior part of capsule gland	Large; extends back beneath capsule gland
Spawn and development	Fig. 54K. Benthic gelatinous mass with embedded singly-encapsulated ova. Nonplanktotrophic	Benthic gelatinous mass with embedded singly-encapsulated ova. Nonplanktotrophic
Radula	Fig. 55. Cusps of 5 central teeth pointed	Fig. 64. Cusps of 5 central teeth pointed or blunt
Shell	Figs 56, 57	Figs 65, 66
Adult size	5–25 mm	4–18 mm
Shape	Turbinate, rarely patulous	Turbinate to high-turbinate, rarely patulous
Sculpture	Smooth, or strongly ribbed, or ribs on base only	Smooth, or with incised lines, or rounded ribs
Microsculpture	Absent in smooth shells; otherwise present between ribs	Absent
Colour	Uniformly dark, banded or lined; never tessellated	Uniformly dark, banded or lined; rarely tessellated
Distribution	Fig. 58. N Sea of Japan, Okhotsk Sea, Bering Sea, Aleutian Is to Oregon	Fig. 67. Kurile Is, Kamchatka, Aleutian Is to N California

Discussion

Intraspecific variation. The degree of shell variation in this species is considerable, but is nevertheless comparable with that of other *Littorina (Neritrema)* species which have similarly wide ranges of geographical distribution and habitat. Characters of the reproductive anatomy and of the radula are relatively uniform, although limited observations suggest that there is geographical variation in size (and perhaps in shape) of the egg capsules (p. 150).

As expected in a species without a pelagic dispersal phase, and therefore with low gene flow, genetic differentiation between populations is marked. Boulding *et al.* (1993) reported an unbiased genetic identity (Nei, 1978) of 0.868 between populations in the Aleutian Islands and Oregon, but found that hybridization was possible, producing fertile eggs. In the northwestern Pacific Zaslavskaya (1995) used 16 allozyme loci to measure a genetic distance (Nei, 1978) of 0.161 between populations in Peter the Great Bay and the Komandor Islands, and believed them to be conspecific. However, at Kasatka Bay, an exposed shore on Iturup Island in the southern Kurile chain, she found two morphologically distinct, microsympatric populations with a genetic distance of 0.786. One of these was genetically very similar to *L. sitkana*, and she suggested that the other may be a separate species, with an average genetic distance from other populations of *L. sitkana* of 0.622. Specimens of this genetically distinct type have been examined; the shells were smooth, dark brown, up to 6.7 mm in height, and relatively low-spired (resembling Fig. 56L in shape); they were unfortunately sexually immature and no penes

Table 5. (*Continued*)

L. aleutica	*L. natica*	*L. kasatka*
Dark; tentacles usually all black	Fig. 72R–T. Dark; tentacles all black or with unpigmented patch across base	Fig. 39G–I. Paler; tentacles usually with unpigmented patch across base
Fig. 68A–K	Fig. 72A–K	Fig. 39A–H
Usually unpigmented, sometimes grey base	Usually unpigmented, sometimes grey base	Unpigmented
30–50% total length	20–30% total length	10–20% total length
7–12	4–17	0 or 1
Fig. 68L–P	Fig. 72L–Q	Fig. 39J–M
Additional first loop posteriorly, concealing renal oviduct	Additional first loop posteriorly, concealing renal oviduct	Normal first loop, small second loop, enlarged renal oviduct visible posteriorly
Two loops; rectangular or trapezoid shape; about equal in area to albumen gland	Two loops; large, usually circular shape; about twice area of albumen gland	Small first and large second loop; about equal in area to albumen gland
Large, swollen, septate	Large, swollen, septate	Very small
Extends half length of jelly gland	Extends half length of jelly gland	Small, seldom visible by dissection
Benthic gelatinous mass with embedded singly-encapsulated ova. Nonplanktotrophic	Benthic gelatinous mass with embedded singly-encapsulated ova. Nonplanktotrophic	Not known, but planktonic capsules and planktotrophic development are predicted
Fig. 69C–H. Cusps of 5 central teeth blunt	Fig. 73. Cusps of 5 central teeth blunt	Fig. 40. Cusps of central teeth pointed to elongate
Figs 69A, B; 70	Fig. 74	Fig. 41
5–14 mm	6–16 mm	2–11 mm
Turbinate to low-spired and patulous	Turbinate to low-spired and patulous	Turbinate
Smooth or with 3–4 nodulose ridges	Smooth	Smooth
Microstriae over whole surface	Absent	Absent
Uniformly dark, banded or lined; never tessellated	Usually lined; never tessellated	Uniformly dark, or tessellated with white dashes or bands
Fig. 71. Aleutian Is, islands of Bering Sea, Provideniya	Fig. 75. Kurile Is, Kamchatka, NW Bering Sea	Fig. 42. NE Hokkaido, Kurile Is, Kamchatka

were present, but small oviducts were of the same shape as in typical *L. sitkana* (pers. obs.). Further investigation will be necessary to determine whether these do indeed represent an additional species related to *L. sitkana*. Clearly this degree of local genetic variation is much greater than that recorded in another nonplanktotrophic *Neritrema* species, *L. saxatilis* (e.g. K. Johannesson, B. Johannesson & Rolán-Alvarez, 1993; K. Johannesson, B. Johannesson & Lundgren, 1995), in which genetic distances of only 0.01–0.05 have been found between microhabitats on single shores.

Phylogenetic relationships. In the cladistic analysis of morphological characters, *L. sitkana* forms a clade together with *L. horikawai* and *L. subrotundata* (p. 351); however, this is weakly supported, with no unequivocal synapomorphies, although in some trees the form of the pallial oviduct does constitute a synapomorphy. These three species are very similar in other morphological attributes, for example in their range of shell sculpture and shape of the penis. Two allozyme analyses have included both *L. sitkana* and *L. subrotundata*; that by Boulding *et al.* (1993) failed to show them as sister-species, but was based on too few loci to be reliable; that of Zaslavskaya (1995) found a rather large genetic distance (Nei, 1978) between them, of 0.981. Analyses of mitochondrial DNA gene sequences have identified *L. horikawai* as the sister-species of *L. sitkana*, with an estimated age of divergence of 3.99 Ma, but either provide no support for, or contradict, the clustering of *L. subrotundata* with these two (Reid, Rumbak & Thomas, 1996; p. 360). The modern distributions of *L. sitkana* and *L. horikawai* suggest that the latter was formed by peripheral isolates (peripatric) speciation, perhaps as a result of vicariance induced by climatic change (p. 379).

Similar species. Littorina sitkana is sympatric with other *Littorina* species throughout its extensive range, and can be confused with a number of them. Strongly ribbed forms are the most easily identified; only *L. subrotundata* can be similar, but lacks microstriae in the grooves between the ribs. *Littorina brevicula* is also ribbed, but is distinguished by its spotted pattern, more concave spire outline and produced anterior end of the aperture; it is sympatric only in Hokkaido and the Sea of Japan. Most confusion will arise with smooth-shelled forms, which can strongly resemble four other Pacific species, *L. subrotundata, L. aleutica, L. natica* and *L. kasatka*. Shells alone are sometimes indistinguishable; for example, compare a lined, exposed-shore ecotype (Fig. 56L) with *L. natica* (Fig. 74), or the small, lined geog. var. *okhotensis* (Fig. 56E, I) with *L. subrotundata* (Fig. 65L, Q), or a small geog. var. *japonica* (Fig. 56N) with *L. kasatka* (Fig. 41A–C). Characters of the shell, anatomy and geographical distributions of these five problematical species are summarized in Table 5. Some shell characters are useful, for example only *L. subrotundata* and *L. kasatka* ever develop a tessellated pattern, and only *L. aleutica* is ever nodulose. However, in many cases anatomical examination is required for unequivocal identification. The most important feature is the pallial oviduct, which in this group of species is diagnostic for *L. sitkana, L. subrotundata* and *L. kasatka*, and distinguishes most examples of *L. aleutica* and *L. natica* from each other. The penis is diagnostic in *L. kasatka*, with one or no mamilliform gland, and that of *L. aleutica*, with a long filament, is characteristic. However, among the other three species the overall shape and number of glands of the penis are similar and other characters must be used. Geographical distributions are known with some precision, and may assist identification in some cases. In Hokkaido and the Sea of Japan *L. sitkana* could also be confused with *L. mandshurica*, which has a more solid shell with lower spire, shallower sutures and 2–3 indistinct ribs; both penis and oviduct also separate the two. The sister-species of *L. sitkana* is *L. horikawai*, but although anatomically similar, shells cannot be confused and the two are not sympatric.

Littorina (Neritrema) horikawai Matsubayashi & Habe, in Habe, 1979

(Figs 59–62)

Synonymy

Littorina mononema Pilsbry (according to Habe, 1979; unavailable manuscript name).

Littorina horikawai Matsubayashi & Habe, in Habe, 1979: 2–3 (Nejiko, Hirado Island, Nagasaki Prefecture, Japan; holotype and paratypes NSMT 58042, seen; 8 paratypes USNM 777928, seen, Fig. 61A, B).

Taxonomic history

The description of this species appeared in Habe's (1979) list of 'Marine molluscan species described from Hirado, Nagasaki Prefecture, as type locality', under the heading '*Littorina mononema* Pilsbry MS'. Immediately below, the 'present scientific name' was given as '*Littorina horikawai* Matsubayashi & Habe, 1980', and there followed a detailed description, with designation of a holotype and type locality. Finally, it was stated that a full description would appear in the *Venus* in 1980, but it did not do so, nor has it been redescribed since. Pilsbry's manuscript name is not available, but it could be argued that Habe's use of the name validates it and further that *L. horikawai*, being published effectively in synonymy, is not available. Nevertheless, the clear intention was to rename the species (Pilsbry's name being thought inappropriate), and *L. horikawai* is therefore considered a valid name following its description in 1979.

Diagnosis

Mature shell height 3.6–7.2 mm. Shell high-turbinate to conical. Mineralogy of external irregular-prismatic calcite and internal crossed-lamellar aragonite. Sculpture variable: smooth with indistinct spiral microstriae or fine incised lines; or with 5–8 spiral ribs on last whorl, with interspaces up to twice rib width and containing microstriae. Colour variable: dark olive-brown indistinctly marbled with whitish flecks, often with three paler orange or cream bands at suture, periphery and on base; orange-cream with black lines in grooves; olive-brown with 6–11 grey or whitish spiral lines; bright vermilion-orange with 6–11 paler spiral lines; columella brown, purple or orange. Penis with short, pointed filament, 15–25% total length; 4–6 mamilliform penial glands. Pallial oviduct of single loop of albumen gland, large loop of capsule gland, and long, swollen, septate, jelly gland. Spawn probably a benthic mass of encapsulated eggs in a gelatinous matrix; development probably nonplanktotrophic with intracapsular metamorphosis. Distribution: islands off Kyushu, Japan.

Anatomical description

Head-foot (Fig. 59C, D, F, G). Head black; pale tip to snout; tentacles pale, with two short black longitudinal stripes, sometimes fused near base or joining head pigmentation; sides of foot speckled with black or unpigmented. Opercular ratio 0.47–0.53.

Penis (Fig. 59A–F). Base wrinkled, sometimes with a little grey pigmentation; 4–6 mamilliform penial glands in single row. Filament pointed, 15–25% total length; subepithelial glandular tissue sometimes visible near sperm groove. Sperm groove open, ending terminally. Penis absent outside breeding season.

Sperm. Paraspermatozoa 13–14 μm diameter, round; vesicles large, distinct; nucleus sometimes visible.

Oviduct (Fig. 59H). Single loop of albumen gland, followed by relatively large loop of capsule gland (folded back and alongside albumen gland); no translucent portion visible in capsule gland; finally a long, swollen, septate, jelly gland. Bursa extending back nearly to anterior part of capsule gland, opening near anterior end of jelly gland. Seminal receptacle simple.

Spawn and development (Fig. 60B). Not known. Since the capsule gland is large and the jelly gland swollen and septate, the eggs are predicted to be encapsulated and embedded in a gelatinous matrix. By analogy with *L. sitkana* and other oviparous members of the subgenus *Neritrema*, the spawn mass is probably benthic and development nonplanktotrophic with intracapsular metamorphosis. This type of development is also indicated by the large (390–440 μm diameter), smooth protoconch of 1.7 whorls. Breeding season not known, but probably winter; a sample collected at Hirado Island in early November was mature, but all animals collected in May lacked penes or well-formed oviducts.

Radula and foregut (Fig. 60C–H). Relative radular length 2.00–2.67. Rachidian: length/width 1.11–1.21; central cusp large, bluntly rounded to pointed, one pointed cusp and one outer denticle on each side. Lateral: four cusps, third from inside largest and rectangular or pointed, others pointed. Inner marginal: four cusps, third from inside largest and rectangular or pointed, others pointed. Outer marginal: 7–8 narrowly pointed cusps; narrow neck, projecting flange on inside of base only. Salivary glands enlarged.

Shell variation (Figs 60A, 61)

Sculptural variation. Shell sculpture varies from smooth (without ribs, but with fine striations, Fig. 61F, G) to strongly ribbed (Fig. 61B–D, K–M). From the limited number of collections available there appears to be a geographical trend to this variation. Populations in the north of the range (Tsushima Island; Hado-misaki, Karatsu; Uku Island) are composed entirely of smooth shells, whereas those from the south (Shimo-jima; Osumi-shoto islands) are all ribbed. However, samples from the type locality (Nejiko, Hirado Island, off northwestern Kyushu) show the complete range of sculpture; here there is a correlation with habitat. On sheltered rocks in a small cove the shells were mostly smooth, but some showed a peripheral keel or 3 to 5 ribs (Fig. 61I, N). Among barnacles on a more exposed shore just 40 m away, the shells were entirely of the ribbed form and also differed in colour (Fig. 61H, K; see below; pers. obs.).

Colour variation. Differences between the lined (Fig. 61F, G, J, K–M) and marbled (Fig. 61A, E, H, N) forms are striking, but intermediates showing elongation and fusion of the pale flecks of marbled shells

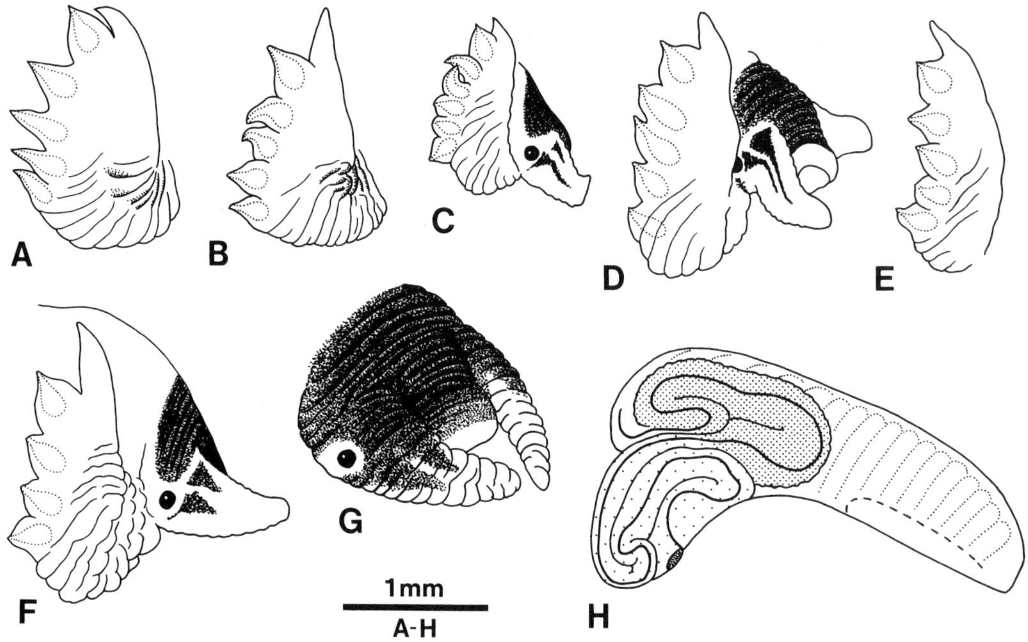

Figure 59. *Littorina (Neritrema) horikawai*: **A–F**, penes; **C**, **D**, **F**, **G**, head pigmentation; **H**, pallial oviduct. **A**, **B**, **G**, **H**, Nejishi, Hirado Island, off Kyushu; **C–F**, Shijiki, Hirado Island, off Kyushu (USNM 777926). Shell heights: **A**, 6.1 mm; **B**, 6.1 mm; **C**, 4.0 mm; **D**, 4.5 mm; **E**, 4.6 mm; **F**, 4.5 mm; **G**, 6.3 mm; **H**, 6.3 mm. For shading conventions of pallial oviduct see p. 6.

are transitional between the two. The pale spiral lines sometimes anastomose to give a zigzag pattern (Fig. 61G).

At Nejiko, Hirado Island, the shells from dark basaltic rocks with few barnacles, in a sheltered cove, were almost all of a predominantly dark olive-brown colour, indistinctly marbled with grey and ochre (Fig. 61I). At the more exposed site 40 m away, with abundant barnacles, the shells were mostly strongly patterned, with black lines and spots on a cream to orange ground colour (Fig. 61H, K; DGR); this form is analogous to the barnacle ecotype of other *Neritrema* species. Both samples appeared visually cryptic against their background and this could suggest visual selection by predators.

Samples from near Tsutsu, at the southern tip of Tsushima Island, were mostly of dark olive-brown or orange-brown shells, with regular pale grey lines (Fig. 61F, G, J). However, over a small part (about 50 m) of coastline opposite a lighthouse at the tip of the peninsula some brilliant vermilion-orange shells (with the same paler pattern of regular lines) were found, making up about 1% of the otherwise brown population. There were no barnacles with the snails and, while the brown shells were cryptic against the dark slate rock, the red forms were highly conspicuous and did not resemble any element of the background (DGR). As in *L. horikawai*, the red colour morph of *L. saxatilis* has a restricted and unpredictable distribution (Fischer-Piette & Gaillard, 1968) which suggests the operation of genetic drift or founder effects rather than selection (p. 323). In contrast, in polymorphic *Littoraria* species which occur on the visually-varied background of mangrove foliage conspicuous red or orange shells are persistently present, but at low frequency, which has been ascribed to apostatic selection (Reid, 1987).

Distribution

Habitat. Details are available from only two localities (pers. obs.). At the type locality on Hirado Island this species was found abundantly on bare basaltic rocks in the upper eulittoral zone in a moderately sheltered cove, and also in barnacle beds on a more exposed shore nearby. At Tsutsu, Tsushima Island, it was absent from sheltered rocks at the head of the bay, and from the most exposed western side of the peninsula; at intermediate exposures it was abundant in crevices of bare slate rock in the upper eulittoral, in a narrow zone above tufts of the red alga *Gloiopeltis furcata* lower on the shore, but below *Nodilittorina radiata* in the littoral fringe. *Littorina horikawai* is clearly an oceanic species,

Figure 60. *Littorina (Neritrema) horikawai*: **A**, shell sculpture; **B**, protoconch; **C–E**, three views of radula (shell H=4.3 mm), flat (**C**), at 45° (**D**) and side view (**E**); **F**, radula (shell H=6.0 mm), at 45°; **G**, radula (shell H=5.0 mm), at 45°; **H**, radula (shell H=4.9 mm), at 45°. **A**, **B**, **G**, Tsutsu, Tsushima Island, N of Kyushu; **C–E**, Nejiko, Hirado Island, off Kyushu; **F**, Nejishi, Hirado Island, off Kyushu; **H**, Shijiki, Hirado Island, off Kyushu. Scale bars: **A**=1.0 mm; **B**=200 µm; **C**, **D**=50 µm; **E**=50 µm; **F**=50 µm; **G**=50 µm; **H**=50 µm.

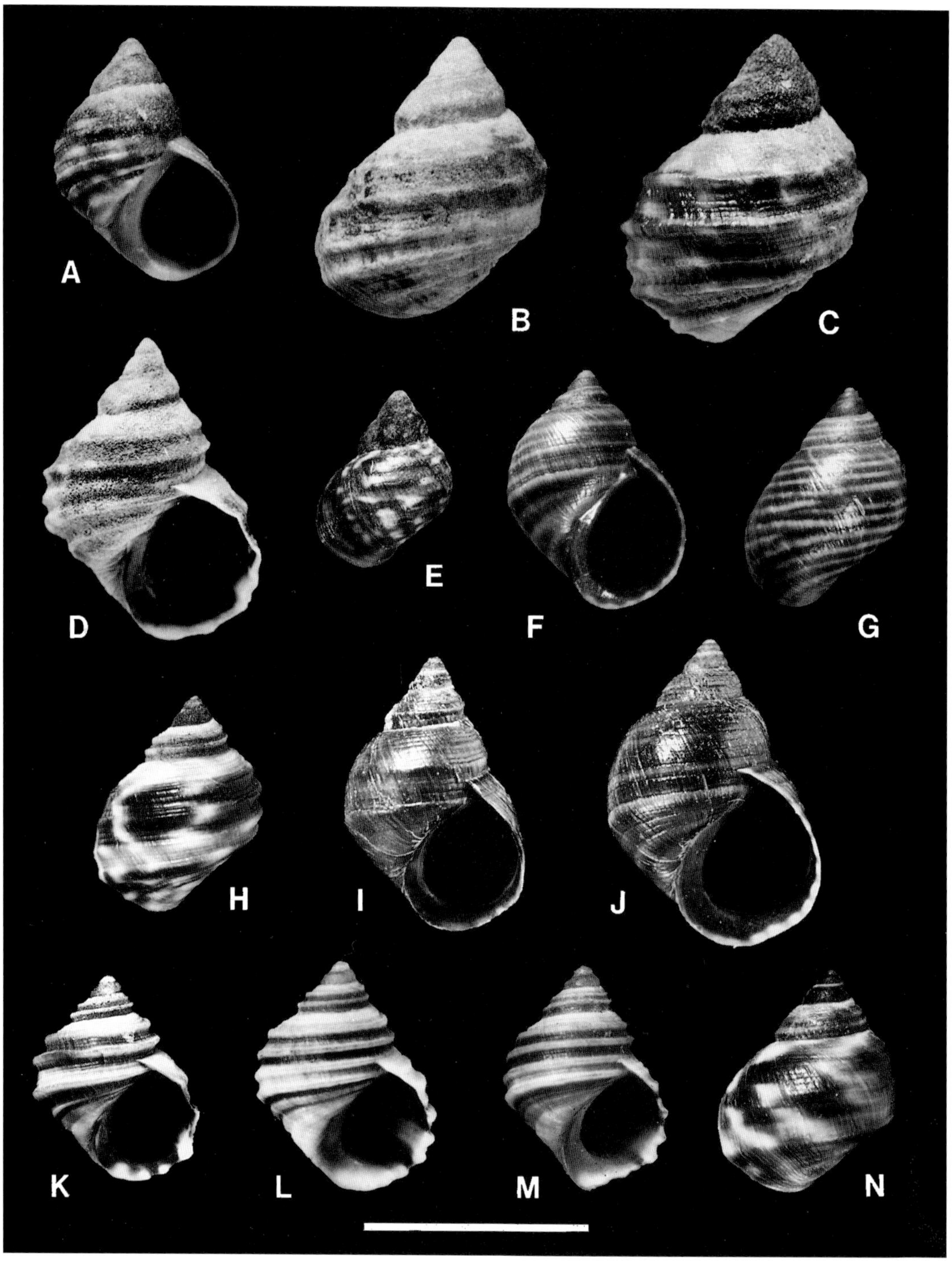

Figure 61. *Littorina (Neritrema) horikawai*. **A, B,** *Littorina horikawai* Matsubayashi & Habe, in Habe, 1979; paratypes, USNM 777928; Nejiko, Hirado Island, Nagasaki Prefecture, Japan. **C, D,** Nejishi, Hirado Island, off Kyushu. **E,** Shijiki, Hirado Island, off Kyushu; USNM 777926. **F, G,** Tsutsu, Tsushima Island, N of Kyushu; USNM 777927. **H, I, K, N,** Nejiko, Hirado Island, off Kyushu. **J,** Tsutsu, Tsushima Island, N of Kyushu. **L, M,** Kuro-shima Island, Kagoshima Prefecture, Japan. Scale bar=5 mm.

being found only on islands, and at the tips of promontories on mainland coasts, at sites with clear oceanic water.

Range (Fig. 62). The recorded range of this species is the most restricted of any of the known species of the genus *Littorina*. So far, it is known only from the offshore islands of Iki (Habe, 1979), Uku and Tsushima (BMNH), all off the northwestern coast of Kyushu, and from the offshore islands of Tanega-shima (Y. Yukita, pers. comm.), Kuro-shima and Yaku-shima (BMNH), to the south of Kyushu. The malacofauna of Kyushu is relatively well known and well collected, and yet this species is virtually unknown from mainland sites; the sole record is from Hado-misaki, a peninsula near Karatsu. The remaining localities are at Nejiko and Shishiki on Hirado-shima and Ushibuka on Shimo-shima; these are large islands close to the mainland of Kyushu, but in each case the localities are on the southwestern side facing the open sea. Further collecting may extend this range. The known localities fall within the warm-temperate and subtropical marine zoogeographical divisions of the Japanese region (Ogasawara, 1994).

Fossil record. Unknown.

Discussion

Geographical distribution. The spawn and development of *L. horikawai* have not yet been described, but direct development in benthic egg masses is predicted from the anatomy of the pallial oviduct and the form of the protoconch. If this is correct, it is noteworthy that no other similarly oviparous *Neritrema* extends to such a southerly latitude, and that desiccation of egg masses and juveniles in the intertidal zone has been considered a likely limiting factor to the southerly spread of *L. (Neritrema) sitkana* (Behrens Yamada, 1977a, b, 1992). Climatic factors may therefore be important in controlling the local and geographical distribution of *L. horikawai*.

Intraspecific variation. The range of sculpture from smooth to strongly ribbed is reminiscent of that in *L. sitkana*, in which ribs develop in animals reared at slow growth rates (Boulding, Buckland-Nicks & van Alstyne, 1993). The variation in colour also finds parallels in other species of *Littorina*. The cream and black patterned form found among barnacles is analogous in coloration to the barnacle ecotype of some other *Neritrema* species, although in these this ecotype is usually smooth-shelled. Discrete red or orange morphs are known in some other *Littorina* species, in which they are likewise generally uncommon (e.g. *L. littorea*, *L. sitkana*, *L. obtusata*, *L. saxatilis*).

Phylogenetic relationships. Analysis of mitochondrial DNA sequences has shown that *L. horikawai* is the sister-species of *L. sitkana* (pp. 161, 360). The narrow geographical range of *L. horikawai*, 1200 km south of the closest occurrence of *L. sitkana* in northern Honshu, suggests that it may have been formed by peripheral isolates (peripatric) speciation (p. 379). Comparing the shells of these two species, that of *L. horikawai* is smaller and taller-spired; these traits may be adaptive in relation to convective heat loss and desiccation resistance in a warmer climate, as has been suggested in other littorinids (Vermeij, 1973).

Similar species. Anatomical similarities to *L. sitkana* include the form of the capsule gland, the size and position of the copulatory bursa, and the appearance of the penis. However, there are differences in head pigmentation, in the spiral pattern of the albumen gland, and in *L. horikawai* the jelly gland is relatively larger and the penial filament more slender. There are also differences in the radulae; in *L. horikawai* the rachidian is more elongate, the largest cusp on each of the five central teeth is large and often rectangular rather than pointed, and the cusps on the outer marginals are longer and more pointed. Despite the anatomical resemblances between the two, their shells are not similar, and no confusion should arise; *L. sitkana* has a larger, broader shell with more rounded whorls, and never shows a marbled pattern. *Littorina sitkana* does not occur south of northern Honshu.

From shells alone, confusion is possible between the strongly ribbed form of *L. horikawai* and the small, narrow form of *L. brevicula* that occurs in Hong Kong (Fig. 46C), although in the former the ribs are usually narrower and the base of the columella not flared or angled. Anatomically, the two species are quite different.

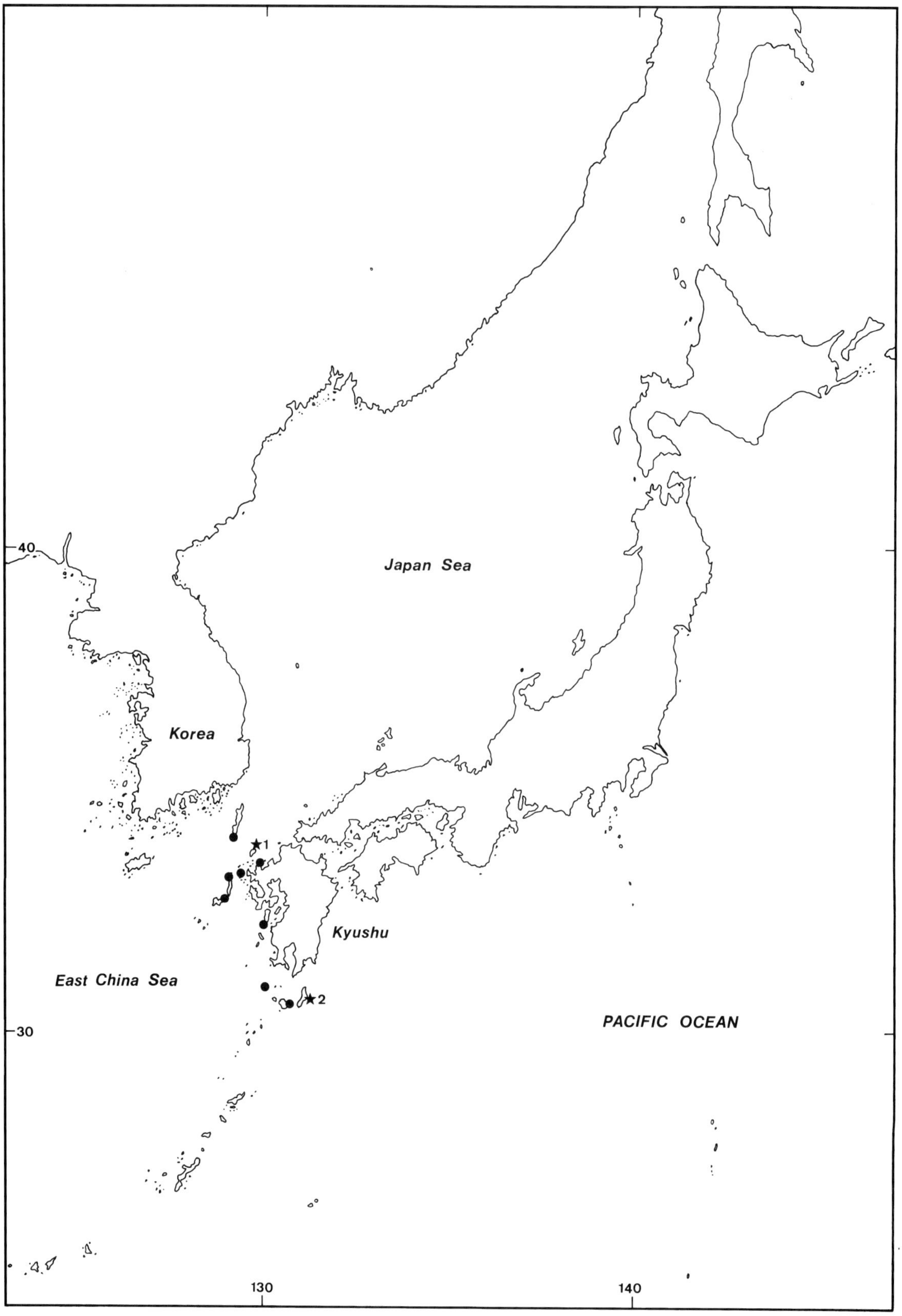

Figure 62. *Littorina (Neritrema) horikawai*: distribution map. Key: circles, specimens examined; asterisks, other records (1, Habe, 1979; 2, Y. Yukita, pers. comm.).

Littorina (Neritrema) subrotundata (Carpenter, 1864)

(Figs 3F, 4, 7A, 63–67)

Synonymy

Assiminea [?] *subrotundata* Carpenter, 1864: 628, 656 (Neeah Bay [Neah Bay, Washington]; holotype USNM 15586, seen, Palmer, 1958: pl. 18, figs 1, 2, Fig. 65A herein).

Assiminea subrotundata—Newcombe, 1893: 63. Taylor, 1895: 81 (as *Assiminia*).

Littorina (?Algamorda) subrotundata—Dall, 1921: 153.

Littorina (Algamorda) subrotundata—Oldroyd, 1927: 63. Burch, 1945: 13. La Roque, 1953: 167. Palmer, 1958: 159; pl. 18, figs 1, 2. Taylor, 1981: 151–152.

Algamorda subrotundata—Kozloff, 1987: 209; fig. 12:26.

Littorina (Neritrema) subrotundata—Reid, 1989a: 102. Reid, 1990b: 1–19; fig. 1a. Taylor & Reid, 1990: 199–215 (mineralogy). Vermeij, Palmer & Lindberg, 1990: 348. Reid & Golikov, 1991: 7–15; fig. 14; fig. 40 (oviduct). Vermeij, 1992b: 136.

Littorina subrotundata—Boulding, Buckland-Nicks & van Alstyne, 1993: 43–68.

Paludinella [?] *castanea* Carpenter, 1865: 28 (Neeah Bay [Neah Bay, Washington]; holotype USNM 16290, seen, Palmer, 1958: pl. 18, figs 7, 8; not *Littorina castanea* Deshayes, 1843=*L. saxatilis*).

Paludinella castanea—Taylor, 1895: 81.

Littorina (?Algamorda) castanea—Dall, 1921: 153. Palmer, 1958: 158; pl. 18, figs 7, 8.

Littorina (Algamorda) castanea—Oldroyd, 1927: 63. Burch, 1945: 13. La Roque, 1953: 167.

Paludinella newcombiana Hemphill, 1877: 49 (Humboldt Bay, California; 8 syntypes USNM 32725, seen, Fig. 65B; 10 syntypes USNM 127340, seen; syntypes CAS 058872, University of Colorado Museum (UCM 20878, 23545), San Diego Natural History Museum (SDNHM 780, 780A), Santa Barbara Museum of Natural History (SBMNH 33995), none seen, listed in Coan & Roth, 1987). Keep, 1888: 66. Keep, 1904: 223.

Littorina (Algamorda) newcombiana—Dall, 1918: 137. Dall, 1921: 153; pl. 5, fig. 5. Oldroyd, 1927: 62. Keep & Baily, 1935: 200; fig. 184. Burch, 1945: 13. Rosewater, 1970: 424. Coan & Roth, 1987: 332.

Algamorda newcombiana—Wenz, 1938: 521; fig. 1374. Baily, 1939: 3. Keen, 1970: 52. Abbott, 1974: 69; text-fig. 569.

Littorina (Littorina) kurila—Golikov & Kusakin, 1978: 87–89 (in part, includes *L. sitkana*, *L. natica*, *L. aleutica*, *L. kasatka*; not Middendorff, 1848=*L. sitkana*).

Littorina (Neritrema) kurila—Reid, 1989a: 102 (not Middendorff, 1848). Reid, 1990b: 1–19; fig. 1m; fig. 2i (oviduct) (not Middendorff, 1848).

Littorina kurila subsp. nov. Boulding, 1990: 42–43.

Littorina (Littorina) kurila—Taylor & Reid, 1990: 199–215 (mineralogy; error for *L. (Neritrema) kurila*; not Middendorff, 1848).

Littorina 'kurila'—Boulding, Buckland-Nicks & van Alstyne, 1993: 43–68; fig. 1c, f; fig. 3a (penis); fig. 4c (spawn); fig. 5a (radula) (not Middendorff, 1848).

Littorina sp. Boulding, Buckland-Nicks & van Alstyne, 1993: 43–68; fig. 1a, d, g-i; fig. 3c (penis); fig. 4b, e (spawn); fig. 5c (radula).

Note: for references in which *L. subrotundata* and *L. sitkana* were not differentiated, and which may have included both species, see Synonymy of *L. sitkana* (p. 146).

Taxonomic history

This species has had a confused history, as a consequence of the considerable ecotypic variation shown by the shell and also because of its close resemblance to the sympatric species *L. sitkana*. Carpenter described *Assiminea subrotundata* in 1864 from Neah Bay in northwestern Washington. The holotype is a worn and broken, but nevertheless recognizable, specimen of the form of this species found in salt marshes and other sheltered habitats. In the following year he named a better specimen of this same ecotypic form, from the same locality, as *Paludinella castanea*. For many years these species remained largely unknown, mentioned only in a few faunal lists (Newcombe, 1893; Taylor, 1895), but not figured. Meanwhile, in 1877 Hemphill had described *Paludinella newcombiana* from the sheltered locality of Humboldt Bay in northern California. This species achieved wider notice (Keep, 1888, 1904), and was transferred to *Littorina* by Dall (1918) and figured by him for the first time in 1921. In Dall's influential catalogue of northwestern American molluscs (1921) all three species were listed under *Littorina*, and this was followed by Oldroyd (1927), Burch (1945) and La Roque (1953). Not until Palmer's (1958) catalogue were the types of Carpenter's two species figured. Both Rosewater (1970) and Abbott (1974) listed only *L. newcombiana*, but the name was replaced with the earlier

L. subrotundata by Taylor (1981). Thus, gradually, was established the concept of a small, inconspicuous *Littorina* of very local distribution in saltmarshes between northern California and Washington (Burch, 1946; Talmadge, 1962; Duggan, 1963; MacDonald, 1969a, b; Keen, 1970; Kozloff, 1987; Reid, 1989a, 1990b, c).

More recently, the concept of this species has been broadened, as the anatomy of northern Pacific *Littorina* species has been examined for the first time. Reid (1990b) recognized two species, *L. subrotundata* and *L. kurila* (*sensu* Reid, not Middendorff), distinguishing them from *L. sitkana* on the basis of shell characters and the form of the pallial oviduct, but found no anatomical differences between the two. *Littorina kurila* (*sensu* Reid) was believed to be a widespread northern Pacific species (Reid, 1990c) with a more low-spired shell, found on open and exposed rocky coasts, in contrast to *L. subrotundata* from salt marshes in the northwestern United States. Confusingly, *L. kurila* (*sensu* Middendorff) was then synonymized with *L. sitkana* after anatomical examination of its type specimens (Reid, 1990b, note added in proof). Shortly afterwards, the difficult group of northern Pacific *Littorina* species conchologically resembling *L. sitkana* was revised, using much new material from Russia, Alaska and Washington, and *L. kurila* (*sensu* Reid, 1989a, 1990b, c) was recognized as consisting of ecotypes of *L. subrotundata* from rocky shores (Reid & Golikov, 1991).

The ecotypes from rocky shores had escaped notice by most earlier workers, because throughout the wide geographical range of the species it is broadly sympatric, and often microsympatric on rocky shores, with *L. sitkana*, and shells of these two species can be extremely similar. Thus, for example, almost all material in ZISP from the Kurile Islands and Kamchatka had been identified as *L. kurila*, and specimens in USNM from the Aleutian Islands and Alaska were variously labelled as *L. sitkana*, *L. atkana* or *L. kurila*, and in both museums were mixed with the more numerous smooth form of *L. sitkana*. Literature references to *L. sitkana* and its synonyms that may perhaps have included *L. subrotundata* are indicated in the Synonymy of *L. sitkana* (p. 146). In the United States *L. subrotundata* is rare on rocky shores, occurring at only a few localities in Washington and Oregon. Here it has recently been brought to attention by the detailed studies of Boulding and others (Boulding, 1990; Boulding, Buckland-Nicks & van Alstyne, 1993; Boulding & Hay, 1993; Boulding & van Alstyne, 1993), who demonstrated its differences from *L. sitkana* in ecology, shell morphology, body pigmentation and allozyme frequencies, and carried out unsuccessful hybridization experiments between the two species. At first, this form was considered as a new, but unnamed, subspecies of *L. kurila* (*sensu* Reid; Boulding, 1990). Later, samples from the Strait of Juan de Fuca were referred to only as *Littorina* sp., and allozyme frequencies were compared with a sample designated as *L. 'kurila'* from the Aleutian Islands and with *L. subrotundata* from salt marshes in Washington, showing a close relationship among them (Boulding *et al.*, 1993). Nevertheless, it was suggested that the three might not be conspecific; this view is considered in more detail later (see Shell Variation, p. 174).

Various generic combinations have been used for *L. subrotundata*. The shells of the salt-marsh ecotype are small, delicate and tall-spired, leading Carpenter (1864) first to assign it, with a query, to *Assiminea*, then (Carpenter, 1865) to suggest that it might prove to be a *Hydrobia*. His *castanea* was placed, again with a query, in *Paludinella* (Carpenter, 1865), and Hemphill (1877) also described his *newcombiana* as a *Paludinella*. Dall (1918) transferred *newcombiana* to *Littorina*, as the type species of the new subgenus *Algamorda*, and in 1921 listed all three as species of *Littorina (Algamorda)*, but in both cases without explanation. A note on the radula by Baily (1939) confirmed its littorinid affinity, but like Wenz (1938) he treated *Algamorda* as a full genus, as did Keen (1970) and Abbott (1974). The species was included in the subgenus *Neritrema* once its anatomy had been examined (Reid, 1989a, 1990b).

Diagnosis

Mature shell height 3.7–17.7 mm. Shell turbinate to conical, whorls rounded. Mineralogy of external irregular-prismatic calcite and internal crossed-lamellar aragonite. Sculpture variable: smooth; or with 7–15 irregularly spaced incised lines; or with 5–8 regular rounded ribs; spiral microstriae absent. Colour variable: commonly brown or yellow-brown with 5–12 spiral dark brown lines; may be entirely white, blackish brown, orange or variously marked with broad bands; occasionally tessellated with whitish dashes on dark background; columella white to purple-brown; interior brown. Head grey to black, usually with unpigmented band across tentacle base. Penis with short, triangular filament, 20–40% total length; 5–17 mamilliform penial glands. Pallial oviduct of single large loop of albumen gland, small loop of capsule gland, and swollen, septate jelly gland. Spawn a benthic mass of singly-encapsulated eggs in gelatinous matrix; development nonplanktotrophic with intracapsular metamorphosis. Distribution: northern Pacific from northern California through Aleutian Islands to Kurile Islands.

Anatomical description

Head-foot (Fig. 63M–P). Head and tentacles grey to black; pale tip to snout, sometimes with pale transverse band behind; tentacles with pale tips, usually an unpigmented band across eye and tentacle base, sometimes continued as a pale stripe along length of tentacle; sides of foot black, grey mottled, or unpigmented. Opercular ratio 0.44–0.59.

Penis (Fig. 63A–L). Base wrinkled, occasionally with slight black pigmentation on lower part and rarely in sperm groove, but usually unpigmented; 5–17 mamilliform penial glands in single row. Filament short, triangular, 20–40% total length; subepithelial glandular tissue sometimes visible in preserved specimens. Sperm groove open, ending terminally.

Sperm (Fig. 3F). Euspermatozoa 39–49 µm. Paraspermatozoa 11–15 µm diameter, round to oval; vesicles large, distinct; nucleus usually visible.

Oviduct (Fig. 63Q–X). Relatively large spiral loop of albumen gland, followed by loop of capsule gland of equal size or smaller, and finally a swollen, septate jelly gland. The pattern of the loops of the capsule gland is variable, even within populations, ranging from a simple narrow loop (Fig. 63S) to a spiral loop (Fig. 63Q), in either case often showing a small additional flexure adjacent to the albumen gland (Fig. 63U, V). The path of the egg groove through the albumen gland also shows some variation. No distal translucent part visible in capsule gland. Bursa moderately large, opening near anterior end of jelly gland. Seminal receptacle simple.

Spawn and development. Two spawn masses from Humboldt Bay, California, preserved in alcohol (USNM 794251) both measured 2.0 by 2.7 mm and contained about 60 eggs, enclosed singly in slightly compressed spherical capsules, embedded in jelly. Matthews (1978) recorded 15–40 eggs in spawn masses attached to *Fucus*, *Salicornia*, *Distichlis* and the mud surface, in salt marshes near Coos Bay, Oregon. Boulding *et al.* (1993), working with populations from rocky shores, illustrated small embryos about 360 µm in greatest diameter, within capsules 890–960 µm in diameter, and described the capsule walls as uniformly thin; egg masses were laid singly (not in communal clusters), attached to algae or barnacles (Boulding & van Alstyne, 1993). Development nonplanktotrophic with intracapsular metamorphosis (Matthews, 1978; Boulding *et al.*, 1993). Hatching occurs at a shell length of 525 µm (Boulding & van Alstyne, 1993). Breeding season: year-round spawning in Washington, but egg masses common at Tatoosh Island only from May to September; spawning only in late June and early July in captive population from Adak Island, Aleutian Islands (Boulding *et al.*, 1993; Boulding & van Alstyne, 1993).

Radula and foregut (Fig. 64). Relative radular length 0.89–2.33. Rachidian: length/width 0.61–1.20; central cusp pointed, leaf-shaped or bluntly rounded, one pointed cusp and one outer denticle on each side. Lateral: four pointed cusps, third from inside largest; two central cusps sometimes blunt or rounded. Inner marginal: four pointed cusps, third from inside largest and sometimes blunt or rounded. Outer marginal: 5–9 pointed (or occasionally bluntly rounded) cusps; narrow neck, projecting flange on inside of base only. Salivary glands enlarged.

There is considerable variation in the radula of this species, particularly in the proportions of the rachidian tooth (compare Fig. 64A and E) and the shapes of the major cusps on the five central teeth (compare Fig. 64B and D). A total of 20 radulae have been examined, but no clear patterns of variation have emerged. Specimens from the western part of the range (Bering and Simushir Islands) showed short and broad rachidian teeth (length/width=0.61–0.84, N=4; Fig. 64A), and yet at the type locality, Neah Bay in Washington, the length/width ratio ranged from 0.75 to 1.15 (N=3; Fig. 64E). There was some suggestion of a correlation between cusp shape and habitat; specimens from sheltered localities such as Valdez (Alaska) and Neah Bay showed bluntly rounded cusps on the

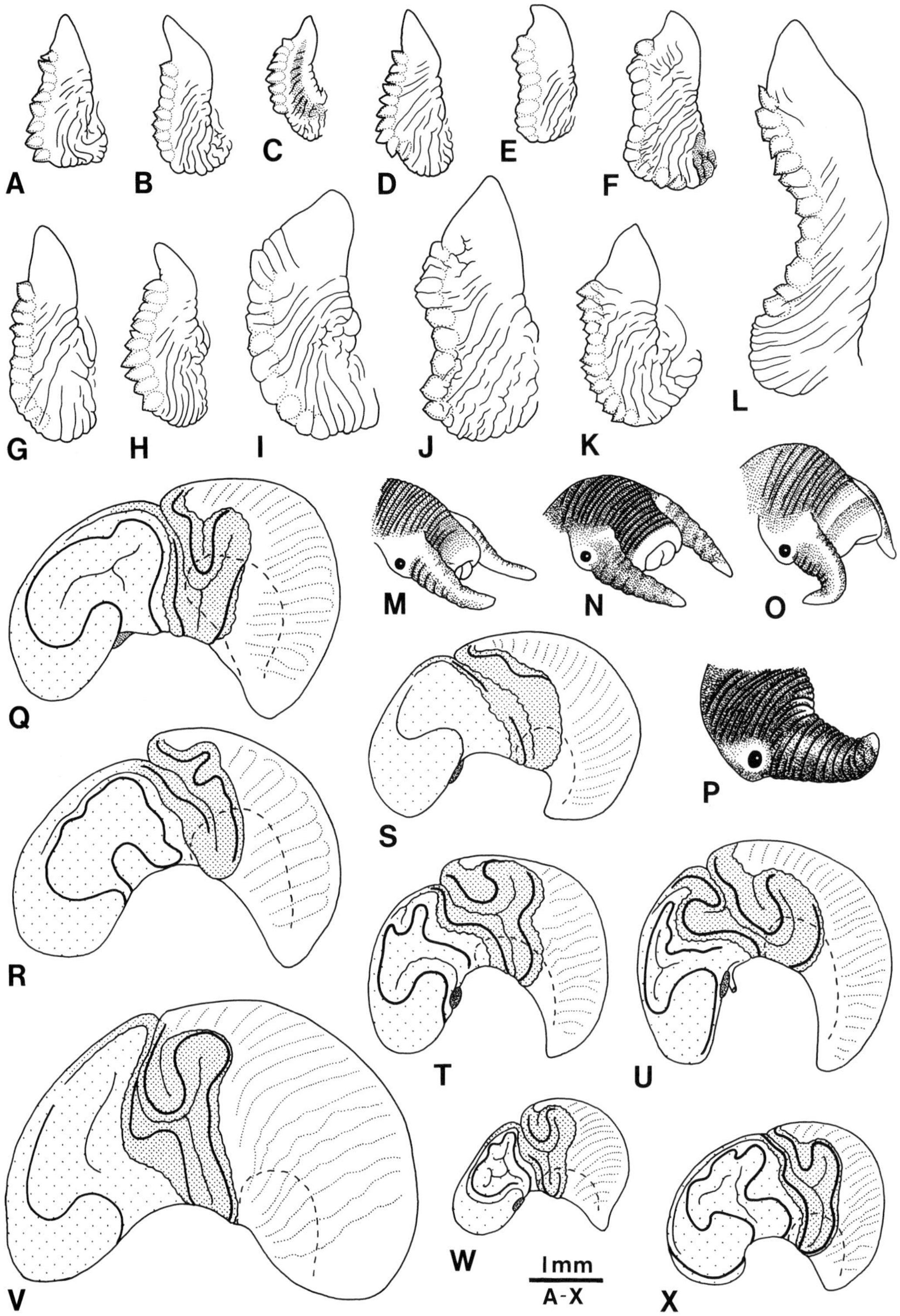

Figure 63. *Littorina (Neritrema) subrotundata*: **A–L**, penes; **M–P**, head pigmentation; **Q–X**, pallial oviducts. **A, B, M–O, U**, mouth of Waatch River, near Neah Bay, Washington (sheltered/brackish ecotype); **C, T**, Westport, Gray's Harbor, Washington (sheltered/brackish ecotype); **D**, Tongue Point, 20 km W of Port Angeles, Washington (barnacle ecotype); **E, W**, Tatoosh Island, Washington (barnacle ecotype); **F, P**, Andrew (Shagak) Bay, Adak Island, Aleutian Islands (USNM 796480; wave-exposed ecotype); **G, H, Q, S**, Seldovia, Alaska (barnacle ecotype); **I**, Spaseniya Bay, Simushir Island, Kurile Islands, Russia (wave-exposed ecotype); **J, V**, Finger Cove, Adak Island, Aleutian Islands (USNM 796498; moderate ecotype); **K, R**, Valdez, Alaska (moderate ecotype); **L**, Casco Cove, Attu Island, Aleutian Islands (moderate ecotype); **X**, 1 mile N of Samoa, Humboldt Bay, California (USNM 794251; sheltered/brackish ecotype). Shell heights: **A**, 4.9 mm; **B**, 6.2 mm; **C**, 4.4 mm; **D**, 4.5 mm; **E**, 4.8 mm; **F**, 6.8 mm; **G**, 5.5 mm; **H**, 5 mm; **I**, 11 mm; **J**, 9.1 mm; **K**, 9.5 mm; **L**, 9.2 mm; **M**, 5.7 mm; **N**, 6.3 mm; **O**, 7.6 mm; **P**, 11.5 mm; **Q**, 6.8 mm; **R**, 9.9 mm; **S**, 6.0 mm; **T**, 6.2 mm; **U**, 7.6 mm; **V**, 10.2 mm; **W**, 5.6 mm; **X**, 6 mm. For shading conventions of pallial oviducts see p. 6.

Figure 64. *Littorina (Neritrema) subrotundata*: **A–C,** three views of radula (shell H=12 mm), flat (**A**), at 45° (**B**) and side view (**C**); **D,** radula (shell H=10.5 mm); **E, F,** two views of radula (shell H=6.6 mm), flat (**E**) and at 45° (**F**); **G, H,** two views of radula (shell H=6.0 mm), flat (**G**) and at 45° (**H**). **A–C,** Milne Bay, Simushir Island, Kurile Islands, Russia (wave-exposed ecotype); **D,** Valdez, Alaska (moderate ecotype); **E, F,** mouth of Waatch River, near Neah Bay, Washington (sheltered/brackish ecotype); **G, H,** Bamfield, British Columbia, Canada (barnacle ecotype). Scale bars: **A, B**=100 µm; **C, D**=100 µm; **E, F**=50 µm; **G, H**=100 µm.

five central teeth (Fig. 64D, F), while at more exposed localities such as Tatoosh Island (Washington) and Bamfield (British Columbia) the cusps were pointed (Fig. 64H). However, there were exceptions (e.g. pointed cusps in one specimen from the sheltered locality of Samoa, California) and for many museum collections the exposure of the locality was not recorded, so it is not clear that this is a general trend.

Shell variation (Figs 4, 65, 66)

Geographical variation. The sheltered/brackish ecotype has so far only been recorded between Vancouver Island and Humboldt Bay (northern California), but it is not known whether its absence elsewhere is real, or whether it is a reflection of collecting effort or of the absence of suitable salt-marsh habitats further north. The species reaches greatest size in the northern and western parts of its range, from Alaska to the Kurile Islands.

Boulding *et al.* (1993) have made a detailed morphological and allozymic comparison among two taxa, referred to as *L.* sp., from the Strait of Juan de Fuca, and *L. 'kurila'*, from Adak Island in the Aleutians (these two are here regarded as examples of respectively the barnacle and moderate ecotypes of *L. subrotundata*, p. 176), and *L. sitkana* from several northeastern Pacific localities. Their results clearly showed that the two were morphologically and genetically distinct from *L. sitkana*. However, they were not able unequivocally to resolve the status of *L. 'kurila'* and *L.* sp., and suggested that they might be either separate taxa or extremes of a geographical cline. There were differences in shell size and shape (greater than 9 mm in *L. 'kurila'*, less than 5 mm in *L.* sp.); operculum colour (dark in *L. 'kurila'*, light tan in *L.* sp.), number of penial glands (15–17 in *L. 'kurila'*, 7–11 in *L.* sp.) and numbers of cusps on the outer marginal radular teeth (6–7 in *L. 'kurila'*, 9–10 in *L.* sp.). The darker pigmentation on the penis and body of *L. 'kurila'* was found to be heritable, and four hybridization attempts between the two were unsuccessful. Nevertheless, the genetic data showed closely similar allozyme frequencies. The authors pointed out the impossibility of making a decision about the status of two such geographically distant populations without examining material from intervening areas. For some comparisons their sample sizes were small, 14 penes and 7 radulae, for example. In the present study much more material has been seen (Table 1, p. 6), including samples from the populations used by Boulding *et al.* Shell shape in this species is highly variable, with both genetic and ecophenotypic components (Boulding & Hay, 1993). Shell size shows geographical variation, increasing towards the north, and penial gland number (p. 14) may be a correlate of size. The colour of the operculum simply reflects its thickness and is also correlated with size. Pigmentation of the penis is rare in the species as a whole, having been seen only in a few specimens from Adak Island, Valdez (Alaska) and Westport (Washington) (Fig. 63C, F), but is known to be variable in most *Neritrema* species. Numbers of cusps on the outer marginal teeth are variable in this species; counts of between 5 and 9 have been made in the 20 radulae examined in the present study and show no evidence of clinal change. The failure of the hybridization experiments is inconclusive in view of the different breeding seasons (in which geographical variation occurs in other *Littorina* species) of *L. 'kurila'* and *L.* sp., as noted by Boulding *et al.* It appears, therefore, that at present there is no sound reason for supposing that the *L. 'kurila'* and *L.* sp. of Boulding *et al.* (1993) are specifically distinct from each other, although further investigation is desirable.

Ecotypic variation. Littorina subrotundata occupies a broad range of habitats and, as expected in a nonplanktotrophic species, shows a corresponding range of shell form. Four ecotypes can be distinguished:

> Moderate ecotype (Fig. 65G, H, J, M): occurs on rocks on sheltered and moderately exposed coasts; shell height up to 17.7 mm; spire height moderate to tall (H/B=1.14–1.34; SH=1.48–1.74); shell of moderate thickness; sculpture varies from smooth to strongly ridged; colour variable.
>
> Wave-exposed ecotype (Fig. 65L, N–P): occurs on rocks on wave-exposed coasts; shell height up to 13.4 mm; spire height relatively low and aperture large (H/B=1.08–1.22; SH=1.24–1.44); shell of moderate thickness; sculpture smooth in available specimens; colour variable.
>
> Barnacle ecotype (Fig. 65I, K, Q): occurs among barnacles, in crevices and in beds of small foliose algae on strongly wave-exposed rocks; shell small (H up to 6.7 mm); spire height

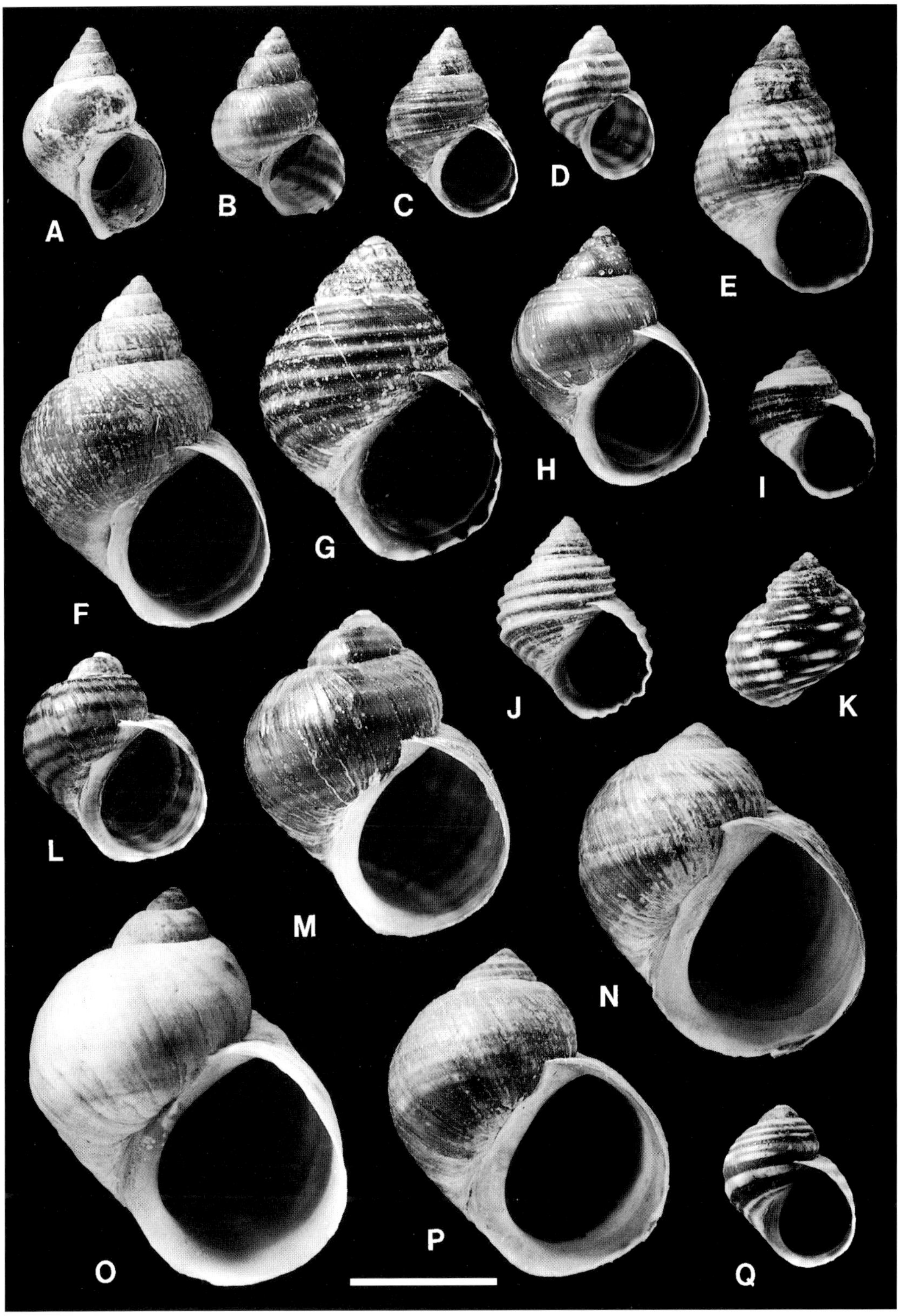

Figure 65. *Littorina (Neritrema) subrotundata*: **A–F**, sheltered/brackish ecotype; **G**, **H**, **J**, **M**, moderate ecotype; **I**, **K**, **Q**, barnacle ecotype; **L**, **N–P**, wave-exposed ecotype. **A**, *Assiminea subrotundata* Carpenter, 1864; holotype, USNM 15586; Neah Bay, Washington. **B**, *Paludinella newcombiana* Hemphill, 1877; syntype, USNM 32725; Humboldt Bay, California. **C**, **E**, **F**, mouth of Waatch River, near Neah Bay, Washington. **D**, Westport, Gray's Harbor, Washington. **G**, **H**, Valdez, Alaska. **I**, Tongue Point, 20 km W of Port Angeles, Washington. **J**, Victoria, Vancouver Island, British Columbia, Canada. **K**, Homer, Alaska. **L**, Andrew (Shagak) Bay, Adak Island, Aleutian Islands. **M**, Finger Cove, Adak Island, Aleutian Islands. **N**, Spaseniya Bay, Simushir Island, Kurile Islands, Russia. **O**, **P**, Bering Island, Russia. **Q**, Seldovia, Alaska. Scale bar=5 mm.

relatively low (H/B=1.09–1.18; SH=1.38–1.43); shell of moderate thickness; sculpture smooth or lightly ridged; colour variable, but often black, or banded or tessellated with white.

Sheltered/brackish ecotype (Fig. 65A–F): occurs in salt marshes, on driftwood and occasionally on extremely sheltered rocks; shell height up to 11.5 mm; spire tall (H/B=1.34–1.43; SH=1.67–1.89); shell thin-walled; sculpture smooth or sometimes lightly ribbed; colour cream to pale brown, usually with opaque cream dashes giving indistinctly tessellated pattern, or with 5–12 dark brown spiral lines.

This range of ecotypes is similar to that found in the most variable of *Neritrema* species, *L. saxatilis*, in which ecotypic variation and its causes have been most thoroughly investigated (Fig. 4; p. 305), and similar adaptive explanations probably apply. The barnacle ecotype of *L. subrotundata* has been studied on the most exposed rocky shores of Washington, where it shelters in rock crevices, among barnacles and in beds of the alga *Iridaea cornucopiae* (Boulding & van Alstyne, 1993; pers. obs.). In this microhabitat there is likely to be selection for maturity at small size and possibly also for cryptic or disruptive coloration. Similar forms are known from the Seward Peninsula in Alaska (DGR), but not yet from other parts of the range, possibly because of the difficulty of collecting on very exposed coasts.

The sheltered/brackish ecotype is also well known in Washington and further south, and shells are almost identical to those of the corresponding ecotype of *L. saxatilis* (p. 316). The shells of this ecotype are larger, more thin-walled and tall-spired than in other ecotypes of *L. subrotundata*. While there is a strong genetic component of shell form, there are also direct environmental effects on a plastic phenotype; shells of animals reared at low density and high growth rate are larger, more thin-walled and more tall-spired than those grown slowly at high density (Boulding & Hay, 1993). More favourable conditions for rapid growth in salt marshes might therefore partially explain the characteristic shell form. The three specific names that this species has received were all based on the sheltered/brackish ecotype. Typical examples of this ecotype have only been seen from localities between California and British Columbia, although taller-spired forms of the moderate ecotype, which are very similar, are found in other parts of the range on sheltered shores (Fig. 65G, H).

Interestingly, *L. subrotundata* is rare or absent on most sheltered or moderately exposed rocky shores in Washington and further south, probably as a result of its susceptibility to crab predators (see Habitat, p. 178). This species never seems to achieve the thick, crab-resistant shell of *L. sitkana*, which replaces it in moderate habitats in this region. Elsewhere in its range the moderate ecotype is the most common form. On more exposed coasts relative spire height decreases and the aperture is enlarged, producing the wave-exposed ecotype, perhaps the result of selection for a relatively larger foot for adherence. These relatively large, patulous shells are known only from the Aleutian and Kurile Islands and Kamchatka, and in only a few of the available museum collections (USNM, ZISP, BMNH) is an exposed habitat explicitly recorded.

So striking are the habitat segregation and conchological differences between the sheltered/brackish and barnacle ecotypes in the northwestern United States, that there have been suggestions that they are different species (Behrens Yamada, 1992; Boulding *et al.*, 1993). However, in the present study no consistent anatomical differences have been detected between the two varieties, in penial or oviduct morphology, radular dentition or body pigmentation. Although typical examples of these two shell types are distinctive (compare Fig. 65B, C with Fig. 65I, Q), intermediate forms were found to be common on shores of moderate exposure in Alaska (Fig. 65G, H) and are also now known from a few collections in Washington and British Columbia (Fig. 65J). The shells of both ecotypes are often smooth, but sculpture is present to some degree in many examples of each; this is of irregularly-spaced incised lines (Fig. 66A), grading into flattened (Fig. 66B) or rounded ribs without microstriae between them (Fig. 66C), and such sculpture is unique within the group of closely related species resembling *L. sitkana*, of which *L. subrotundata* is a member. Boulding *et al.* (1993) did not record any anatomical or conchological observations on the sheltered/brackish ecotype, but based their suggestion that it may be distinct (from their *L.* sp. and *L. 'kurila'*, see Geographical Variation, p. 174) only on comparison of allozyme frequencies. At nine loci the frequencies were almost identical in two populations of the barnacle ecotype (as *L.* sp.) from Tatoosh Island and Bamfield, Vancouver Island, and two populations of the sheltered/brackish ecotype (as *L. subrotundata*) from Gray's Harbor and Waatch River, Washington. At the tenth locus, *Pep-3*, a fixed allelic difference was said to separate the two forms,

Figure 66. *Littorina (Neritrema) subrotundata*: shell sculpture. **A, B,** mouth of Waatch River, near Neah Bay, Washington (sheltered/brackish ecotype); **C,** Tongue Point, 20 km W of Port Angeles, Washington (barnacle ecotype). Scale bar=1 mm.

implying a lack of genetic exchange between them. However, this result should be treated with caution. The two alleles concerned are so similar (a maximum of 2 mm apart) that in a previous interpretation of the gels (quoting Boulding, 1990, PhD Thesis) they were not distinguished, and even in the reinterpretation were not always clear; data from gels in which the alleles were not distinguishable were discarded. One allele typical of the barnacle ecotype was in fact included for an individual of the sheltered/brackish ecotype, but this was said to be an error. In any genetic data of this kind it is always difficult to interpret significant allozyme differences as evidence of reproductive isolation when the populations are not sympatric, since clines can occur. This is more likely in species with limited dispersal, such as nonplanktotrophic *Neritrema* species. Indeed, in *L. saxatilis* significant differences in allozyme frequencies have been detected over distances of a few metres (e.g. K. & B. Johannesson, 1990, 1993). The two closest populations of the two varieties used by Boulding *et al.* were more than 10 km apart (Waatch River and Tatoosh Island), but the species does occur between these two points (DGR, 4 km southeast of Cape Flattery). Clearly, further investigation of the genetic structure of *L. subrotundata* is required, in particular of the *Pep-3* locus in the two forms here regarded as ecotypes, and of their potential for interbreeding.

Sculptural variation. In most ecotypes shell sculpture is variable. In the sheltered/brackish ecotype most specimens are smooth (Fig. 65A, B, D, E), or show only irregularly-spaced incised lines (Figs 65F, 66A), but rounded ribs are sometimes developed (Figs 65C, 66B). Smooth shells also predominate in all available samples of the wave-exposed ecotype and in most populations of the barnacle and moderate ecotypes (Fig. 65L–Q), but lined (Fig. 65H) or ribbed (Figs 65G, I, J, K, 66C) shells are not uncommon. Shells differing widely in sculpture can occur together in the same microhabitat (e.g. Fig. 65G, H). As in *L. sitkana*, Boulding *et al.* (1993) have demonstrated an ecophenotypic component, finding that shell sculpture is more likely to develop in slow-growing individuals reared in the laboratory.

Colour variation. A wide range of colours and patterns can be found, but populations are not usually strikingly polymorphic. The shells of the sheltered/brackish ecotype are cream to pale brown, usually with a coarsely but indistinctly tessellated pattern of interrupted cream bands or dashes, or with 5–12 continuous and conspicuous brown lines (Fig. 65B, D). These colour forms are similar to those of salt-marsh and brackish-water forms of some other *Littorina* species (p. 322).

Many populations of the moderate and wave-exposed ecotypes are dull brown with darker lines (Fig. 65L, P) (corresponding to the shell ribs, if present), but with occasional white (Fig. 65O) or orange shells, or shells variously patterned with a few broad white bands. Boulding *et al.* (1993) have illustrated some common patterns in a population of the barnacle ecotype from Tatoosh Island, Washington. The more conspicuously white or banded forms (Fig. 65I, Q) are often common in the barnacle ecotype, perhaps suggesting visual selection for crypsis against a background of pale barnacle tests. Tessellated shells are rare in examples of this species from rocky shores, but do occasionally occur in the barnacle ecotype (Fig. 65K).

Distribution

Habitat. The sheltered/brackish ecotype occurs in salt marshes, on the lower stems of *Salicornia* and other marsh plants such as *Glaux* and *Distichlis*, and on the firm mud beneath the vegetation (Talmadge, 1962; Keen, 1970; Matthews, 1978; pers. obs.). It is also found on tufts of *Fucus* in runnels in marshes and on hard substrates such as driftwood, rock (pers. obs.) and oyster shells (Duggan, 1963) in muddy inlets. The snails occur at and just above the mean tide level (Talmadge, 1962; Keen, 1970), often together with *Assiminea californica* and slightly below the level occupied by *Ovatella myosotis* (Matthews, 1978; Berman & Carlton, 1991). The only other littorinid commonly encountered in the same habitat is *L. plena*, while *L. sitkana* is occasional. *Littorina subrotundata* can be abundant in salt marshes, densities of up to 501 per m^2 having been recorded at Gray's Harbor, Washington (MacDonald, 1969a). In Humboldt Bay it is most common in the less brackish areas (MacDonald, 1969b). In a salt marsh at Coos Bay, Oregon, the diet included bacteria, diatoms, *Enteromorpha* and the epithelial cells of halophytes (Berman & Carlton, 1991).

Less is known about the habitat of this species on rocky shores. On moderately exposed coasts at Seldovia, Alaska, and near Port Angeles, Washington, it was common in crevices at the top of the barnacle zone and just above, and on a more sheltered shore at Valdez, Alaska, among tufts of weed and on rocks at the top of the *Fucus* zone (pers. obs.). Labels accompanying material from Russian localities (ZISP, BMNH) also suggest occurrence in the upper eulittoral zone. On shores of moderate to extreme wave exposure at Tatoosh Island, Washington, the barnacle ecotype was found in beds of the red alga *Iridaea cornucopiae*, among barnacles and in rock crevices, reaching densities of almost 3000 per m^2 in the summer months, but declining in winter (Boulding, 1990; Boulding & van Alstyne, 1993). It may also occur lower on the shore, for example among *Pollicipes* and *Mytilus* on an exposed headland near Cape Flattery, Washington (pers. obs.) and at depths of 1 to 2 m on rocks with *Alaria* at Yankich Island in the Kurile chain (BMNH). Throughout most of its range it is sympatric with *L. sitkana*, and on shores of low to moderate wave exposure the two may be found together, but with *L. subrotundata* predominating at higher levels (pers. obs., Seward Peninsula, Alaska). *Littorina subrotundata* is more tolerant of strong wave action than *L. sitkana* and occurs alone on the most wave-exposed shores of Washington (Boulding, 1990; Boulding *et al.*, 1993; Boulding & van Alstyne, 1993). In the laboratory the snails were found to feed on diatom films (Boulding *et al.*, 1993) and in the field eat tissue from the tips of the foliose alga *Iridaea cornucopiae* (Boulding & van Alstyne, 1993).

In a recent study of this species, Boulding *et al.* (1993) considered it unlikely that the sheltered/brackish ecotype (as *L. subrotundata*) and the barnacle ecotype (as *L.* sp.) were conspecific. Although this was based largely on electrophoretic evidence (see Ecotypic Variation, p. 176), they were probably influenced by the apparent restriction of the former to salt marshes and of the latter to shores of strong wave-exposure. This habitat segregation in Washington and southwards has also been noted by Reid & Golikov (1991). It is, however, not so complete as has been thought; material has recently come to light from intermediate habitats, rocky shores of lower exposure, in Washington (4 km south-east of Cape Flattery; Tongue Point, 20 km west of Port Angeles; BMNH), but in general *L. subrotundata* appears to be rare in these habitats. Further north, however, this is not the case; in the northern Gulf of Alaska *L. subrotundata* is frequent on rocky shores in wave-exposed through to sheltered conditions (pers. obs.), and in the Aleutian Islands too it occurs on shores of moderate exposure (Boulding *et al.*, 1993, as *L. 'kurila'*). The explanation may lie in the distribution of crab predators, which can easily crush the shell of *L. subrotundata* (Boulding, 1990; Boulding & van Alstyne, 1993). Both exposed coasts and salt marshes might be refuges from crab predation where the species can survive in significant numbers (Reid & Golikov, 1991). At Tatoosh Island the crab *Hemigrapsus nudus* was not found on the most exposed shores, but was abundant at protected sites where heavy predation apparently excluded *L. subrotundata* (Boulding & van Alstyne, 1993). Further north, in Alaska and the Aleutian Islands, molluscivorous crabs become scarce (Vermeij, 1992b), which may permit *L. subrotundata* to persist on any suitable shore. In Britain, it has been found that predation by crabs on *L. saxatilis* is more severe on shores of intermediate exposure than on exposed shores or in salt marshes (review by Raffaelli, 1982).

Range (Fig. 67). During the time when this species was known by the name *L. newcombiana*, it was for long recorded only from the type locality, Humboldt Bay in northern California. Burch (1946) 'rediscovered' this 'long-lost' species at Samoa in Humboldt Bay and it was believed to be endemic to the locality (Talmadge, 1962; Keen, 1970). New records were made by Duggan (1963) from Willapa Bay, Washington, and by MacDonald (1969a) from Coos Bay, Oregon, and Gray's Harbor, Washington. Matthews (1978) added new records from Oregon, and it appears that the species is locally common in suitable salt marshes from Humboldt Bay northwards. Despite subsequent searching (Burch, 1945) it was not collected at Neah Bay, Washington (type locality of *L. subrotundata*) until found at Waatch River in 1988 (DGR; same material used by Boulding *et al.*, 1993).

Using the name *L. kurila* for the forms from rocky shores, Reid (1990c) extended the range to Alaska, the Aleutian Islands, Bering Sea and Kurile Islands. Following the broader concept of *L. subrotundata* introduced by Reid & Golikov (1991), Vermeij (1992b) listed records from the Aleutian Islands. Boulding *et al.* (1993) recorded their *L. 'kurila'* from Adak Island, and their *L.* sp. from Cape Arago, Oregon, to Sitka, Alaska (both taxa here considered as ecotypes of *L. subrotundata*). They also identified shells from St Lawrence Island as *L.* sp., but the specimens concerned (USNM 424625) are probably *L. aleutica*.

The present study gives a more detailed picture of the distribution of this species: from Humboldt Bay northwards to the Gulf of Alaska; throughout the Aleutian chain, the Pribilof and Komandor Islands; the east coast of Kamchatka; the Kurile Islands south to Iturup (ZISP). In addition there is an isolated record from Ulbanskiy Bay in the western Okhotsk Sea (ZISP, duplicate material in BMNH; identification confirmed by dissection).

Coan & Roth (1987) quoted J. T. Carlton's suggestion that the salt-marsh form of *L. subrotundata* could be an introduced Atlantic species, possibly *L. saxatilis*. The present anatomical study clearly shows that *L. subrotundata* is a distinct species, and its wide northern Pacific distribution gives no reason to suspect that recent introductions have been involved.

Fossil record. No fossils have been recorded. Wang (1982) described *Algamorda parva*, a Pliocene fossil from Hainan Island, China, but the shells illustrated are umbilicate and this is therefore probably not a *Littorina* species.

Discussion

Intraspecific variation. This species, as here defined, shows a wide range of variation in both shell and anatomical characters, and the patterns and causes of this variation are still poorly understood. In particular, there has been controversy about the interpretation of the status of widely separated geographical forms and of the distinctive sheltered/brackish ecotype (Boulding *et al.*, 1993). A parallel degree of intraspecific variation in shell form is known in several other *Neritrema* species that span a habitat range from exposed rocky coasts to protected shores and salt marshes, of which the best known is *L. saxatilis* (Fig. 4; p. 305). In such cases examples from the very sheltered, brackish habitats are generally taller-spired, thinner in texture and often tessellated (p. 29), as in the corresponding ecotype of *L. subrotundata*. Interestingly, like *L. subrotundata*, *L. saxatilis* also shows a polarization of habitats at the southern extreme of its range, where it occurs only in salt marshes or lagoons and on open coasts (Gofas, 1975). However, the explanation may be different, since *L. saxatilis* is capable of producing a thicker shell than *L. subrotundata* and coexists with crab predators over much of its range.

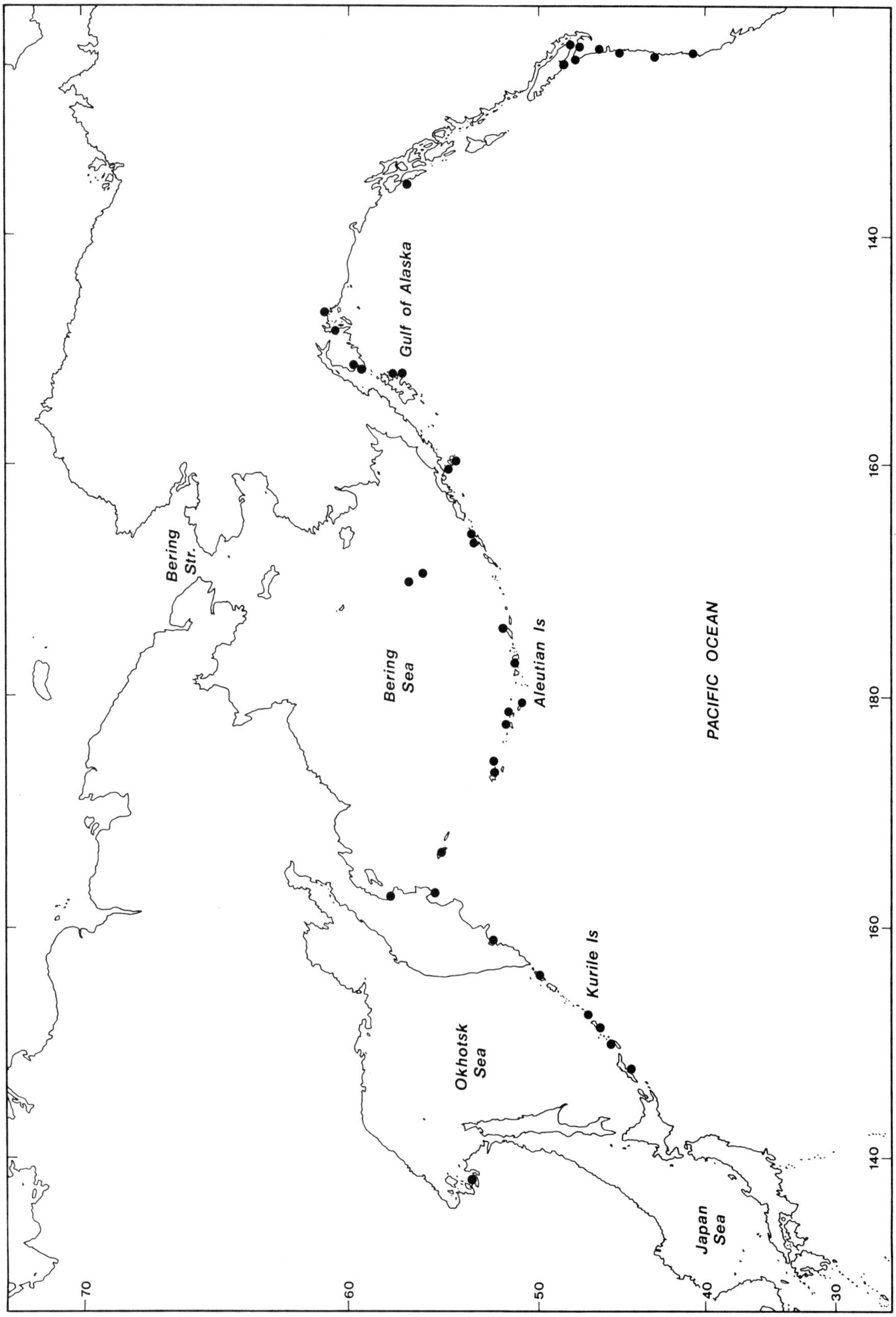

Figure 67. *Littorina (Neritrema) subrotundata*: distribution map. Key: circles, specimens examined.

Boulding & Hay (1993) have made an important quantitative study of the genetic and environmental determination of shell form in *L. subrotundata*, using the barnacle ecotype from an exposed site in Washington. By rearing broods at different densities (and therefore growth rates) they demonstrated phenotypic plasticity in shell shape and sculpture. Despite probable strong selection in its natural habitat, they also found significant additive genetic variance for components of shell shape, and discussed both how this might be maintained and how it would permit a rapid genetic response to changing selection pressures.

Phylogenetic relationships. Although there are morphological similarities in shell shape, penis, spawn and radula between this species and *L. sitkana*, there are no unequivocal morphological synapomorphies (p. 351), nor do the analyses of mitochondrial DNA sequences (p. 360) or of allozyme frequencies (p. 356) support a very close relationship between these two (p. 161). The DNA data suggest an age of separation between them of about 6–10 Ma (Reid, Rumbak & Thomas, 1996).

Boulding (1990) suggested that different selection pressures on the wave-exposed shores occupied by the barnacle ecotype of *L. subrotundata* (as *L. kurila* new subspecies) and on sheltered shores frequented by *L. sitkana* may have been responsible for genetic differences leading to their speciation. This was modified by Boulding *et al.* (1993), with their recognition that *L. 'kurila'* co-occurred with *L. sitkana* on a less exposed shore at Adak Island in the absence of crab predators; they suggested that further south, in the presence of crabs, existing divergence may simply have increased. If, as proposed here, *L. subrotundata* occupies a wide range of habitats from wave-exposed coasts to salt marshes, and over most of its range co-occurs with *L. sitkana*, this model of sympatric (or micro-parapatric) speciation is not appropriate in this particular case.

Endangered status. When known only as the ecotype from salt marshes, this species achieved notoriety as the only endangered marine mollusc from the west coast of North America mentioned in a review by Keen (1970, as *Algamorda newcombiana*). It was then believed to be endemic to Humboldt Bay, where its salt-marsh habitat was being threatened by accumulation of sawdust from nearby sawmills. Abbott (1974) noted that it was possibly extinct; Taylor (1981) knew of four other populations outside California, but still considered it threatened. In fact, the sheltered/brackish ecotype is widespread and locally abundant in the northwestern United States, and the species as a whole has a great range across the northern Pacific, so it is not endangered.

Similar species. As stressed throughout the remarks on this species, it is closely similar to *L. sitkana* in many respects; penes, radulae and shells are each sometimes indistinguishable. Reid & Golikov (1991) separated them by shell microsculpture and the pallial oviduct. In a more detailed comparison, Boulding *et al.* (1993) found shell sculpture and body pigmentation to be useful for field identification, and also noted differences in egg capsules, and fixed allozymic differences. Morphological characters of these species (and of the other similar northern Pacific taxa *L. aleutica*, *L. natica* and *L. kasatka*) are listed in Table 5 (pp. 160–161). Most useful is the form of the pallial oviduct, in which the capsule gland is relatively smaller in *L. subrotundata*, a character that distinguishes all females of the two species. At any one locality this character can be used to separate individuals, which can then be examined for other consistent features such as shell form and coloration, and body pigmentation, which will permit identification of males. At many, but not all, localities *L. subrotundata* has paler body pigmentation than *L. sitkana* and lacks the pigmentation on the penis often found in *L. sitkana*. Shell shape is highly variable in both species, as is coloration, but a pattern of 9–12 brown lines is common only in *L. subrotundata* and it alone can show a tessellated pattern. Either may be completely lacking in spiral sculpture, but when it is present the 7–15 incised lines, or rounded ribs without microstriae in the interspaces, of *L. subrotundata* are distinctive (Fig. 66), and can be contrasted with the generally more prominent ribs with microstriae between that are common in *L. sitkana* (Fig. 57).

Shells of *L. natica* are generally more patulous than those of *L. subrotundata*, although sometimes almost indistinguishable; females can be distinguished with ease from their pallial oviducts, but there are no diagnostic characters for males.

Confusion could also occur with *L. plena*, which can be found together with *L. subrotundata* in salt marshes. The former has a more solid shell, flatter whorls, finer and more marked tessellation, and reproductive anatomy is quite different. On open, rocky shores both *L. scutulata* and *L. plena* have taller spires and shells more conspicuously tessellated than are seen in forms of *L. subrotundata* from this habitat.

Littorina (Neritrema) aleutica Dall, 1872

(Figs 7B, 68–71)

Synonymy

Littorina aleutica Dall, 1872: 1 (Gull Rocks, Akutan Pass, Aleutian Islands; lectotype (here designated) USNM 214721, seen, Fig. 70A; 12 paralectotypes USNM 635480, seen). Dall, 1873: 271; pl. 1, figs 3, 3a. Keep, 1904: 217; fig. 234. Keep & Baily, 1935: 199; figs 180, 181. Higo, 1973: 46 (*s.l.*). Foster, 1981: 99, 102. Egorov, 1992: 169–170; fig. 1J.

Litorina aleutica—Weinkauff, 1882: 106. Weinkauff, 1883: 216. Dall, 1902: 551; pl. 39, figs 4, 6.

Littorina (Littorina) sitchana var. *aleutica*—Tryon, 1887: 241; pl. 41, fig. 91.

Littorina (Littorivaga) aleutica—Dall, 1921: 153. Oldroyd, 1927: 60. Burch, 1945: 11, 36. La Roque, 1953: 167.

Littorivaga aleutica—Habe, 1951: 89; text fig. 3 (*s.l.*).

Littorina (Littorina) aleutica—Fischer, 1967: 11. Rosewater, 1970: 423. Golikov & Kusakin, 1978: 85–86; fig. 51.

Littorina (Neritrema) aleutica—Reid, 1989a: 102. Reid, 1990b: 1–19; fig. 1k; fig. 2g (oviduct). Taylor & Reid, 1990: 199–215 (mineralogy). Reid & Golikov, 1991: 7–15; fig. 13; figs 24–26 (penis); figs 27, 38 (oviduct).

Littorina sitchana—Bavay, 1920: 254 (in part, includes *L. sitkana, L. brevicula, L. mandshurica*; not Philippi, 1847=*L. sitkana*). Kusakin & Ivanova, 1978: 27–36 (not Philippi, 1847).

Littorina sitkana—Baxter, 1987: 48 (in part, includes *L. sitkana*; not Philippi, 1846).

Littorina (Littorina) kurila—Golikov & Kusakin, 1978: 87–89 (in part, includes *L. sitkana, L. natica, L. subrotundata, L. kasatka*; not Middendorff, 1848=*L. sitkana*).

Littorina kurila—Kusakin & Ivanova, 1978: 27–36 (not Middendorff, 1848).

Neritrema aleutica—Higo & Goto, 1993: 73.

Note: in the synonymy '*s.l.*' (*sensu lato*) indicates that the author's concept of the species probably included *L. natica*, but without specific reference to, or figure of, that species.

Taxonomic history

As a consequence of its remote distribution in the Bering Sea, this species has been collected relatively infrequently and only described under the one name. Several authors have considered it a variety (Tryon, 1887; also suggested as a possibility by Weinkauff, 1883) or a synonym (Bavay, 1920; Baxter, 1987) of *L. sitkana*, a species known for its variable shell form. It is now known (Reid & Golikov, 1991) that *L. aleutica* is itself variable in shell sculpture. In the Russian literature, and in the collections of ZISP, smooth-shelled forms have previously been included (together with four other similar species) under the name *L. kurila* (e.g. Golikov & Kusakin, 1978; see p. 149). More rarely, these forms have been misidentified as *L. sitkana* (Kusakin & Ivanova, 1978; ZISP collections). Reid (1989a, 1990b) transferred *L. aleutica* to the subgenus *Neritrema*, since its pallial oviduct indicated a benthic spawn, here described for the first time.

Diagnosis

Mature shell height 5.0–14.3 mm. Shell turbinate to low-spired and patulous; columella sometimes separates from coiling axis towards end of final whorl to give pseudo-umbilicus. Mineralogy of external irregular-prismatic calcite and internal crossed-lamellar aragonite. Sculpture variable: smooth; or with 3 evenly-spaced indistinct ridges or angulations (one at periphery, two above); or with 3–4 narrow, strongly raised, nodulose ridges at and above periphery and 0–3 cords on base; in all cases whole surface with fine spiral microstriae. Colour variable: white, yellow, orange or brown, sometimes with 1–3 brown or white bands; columella white to purplish; interior darker. Penis with elongate, often rounded, filament, 30–50% total length; 7–12 mamilliform penial glands. Pallial oviduct of unusual double loop of albumen gland, double loop of capsule gland, and large, swollen, septate, jelly gland. Spawn a benthic mass of singly-encapsulated eggs in a gelatinous matrix; development nonplanktotrophic with intracapsular metamorphosis. Distribution: Aleutian and other islands of the Bering Sea, Provideniya.

Anatomical description

Head-foot. Head and tentacles black, sometimes with pale tips to tentacles and snout and small unpigmented patch just behind eye and occasionally at inside of base of tentacle also; sides of foot black or mottled. Opercular ratio 0.43–0.60.

Penis (Fig. 68A–K). Base wrinkled, usually unpigmented, sometimes some grey to black pigmentation on outside surface; 7–12 mamilliform penial glands in single row. Filament fairly long, usually rounded at tip, 30–50% total length. Sperm groove open, ending terminally.

Sperm. Euspermatozoa 43–46 μm. Paraspermatozoa 11–16 μm diameter, round; vesicles large, distinct, nucleus sometimes visible.

Oviduct (Fig. 68L–P). Albumen gland appears superficially to consist of the single loop seen in most other *Littorina* species, but close examination reveals an additional first loop on the medial side (adjacent to the renal oviduct and rectum) that is usually partly visible at the posterior edge of the pallial oviduct. The shape of this extra loop, the route of the egg groove through it and its relation to the renal oviduct are the same as in *L. natica* (Fig. 72O), but in most specimens the egg groove is difficult to follow. Capsule gland of two consecutive loops (the first often slightly the larger), compressed antero-posteriorly into a rectangular or trapezoid shape, with a small, distal, reddish or brownish, translucent portion; overall area of capsule gland approximately equal to that of albumen gland. Jelly gland large, swollen, septate; approximately equal in area to combined albumen and capsule glands. Bursa relatively small, not extending back as far as anterior part of capsule gland, opening near anterior end of jelly gland. Seminal receptacle simple.

Spawn and development. Spawn known only from one collection in ZISP (*L. kurila* no. 332, in alcohol, duplicate material in BMNH) from Provideniya in the northern Bering Sea. Egg masses up to 8 by 7 mm, consisting of single eggs within spherical capsules 900 μm in diameter, embedded in a gelatinous matrix. Breeding season not recorded, but eggs were collected in July; available collections were all made between June and September and adults were mostly mature.

Radula and foregut (Fig. 69C–H). Relative radular length 1.17–1.81. Rachidian: length/width 1.00–1.33; central cusp blunt to bluntly rounded, one pointed cusp and one outer denticle on each side. Lateral: four cusps, third from inside largest, largest only or two largest cusps rectangular or bluntly rounded, others pointed. Inner marginal: four cusps, third from inside largest, largest only or two largest cusps rectangular or bluntly rounded, others pointed. Outer marginal: 6–8 triangular to rounded cusps; narrow neck, projecting flange on inside of base only. Salivary glands enlarged.

Shell variation (Figs 69A, B; 70)

Ecotypic variation. Shell shape ranges from almost flat-spired (H/B=0.86, SH=1.10) to turbinate (H/B=1.24, SH=1.42). These extremes can be found in both nodulose and smooth forms, although taller-spired shells are more frequent in the smooth-shelled populations in the northern Bering Sea. It is not known whether taller-spired shells are found in more sheltered habitats, but by analogy with other *Neritrema* species this is likely. In samples from Provideniya there is a distinction between large (to 14 mm), well-preserved shells (Fig. 70H, I, K–M) and smaller (less than 7 mm), eroded shells with more inflated whorls and globular shape (Fig. 70J). Anatomically, the small forms do not differ from the larger shells (Fig. 68K, N). There is some evidence that the small forms occur in the harsher environment at the upper limits of the intertidal range of the species.

Sculptural variation. The most well-known form of this species bears 3–4 strongly nodulose ridges which are unique in the genus (Figs 69B, 70A–F). This is the shell type that is most frequent throughout the Aleutian and Komandor Islands, but even here sculpture is variable and in some shells the nodulose ridges are lacking or reduced to angulations of the whorls. In a sample from St Lawrence Island (SMNH 891) the juveniles are slightly nodulose, but the adults are virtually smooth. Nine samples have been seen from Provideniya, consisting of several hundred shells, and in these the nodules are invariable absent, the ridges only slightly developed or represented only by dark bands on the shell (Figs 69A, 70H–M).

Colour variation. More northerly populations are uniformly dull brown, with indistinct darker lines corresponding to the position of the poorly-developed spiral ridges. Some samples from the Aleutian Islands are strikingly polymorphic, with brown, orange, yellow or white shells variously banded with brown or white.

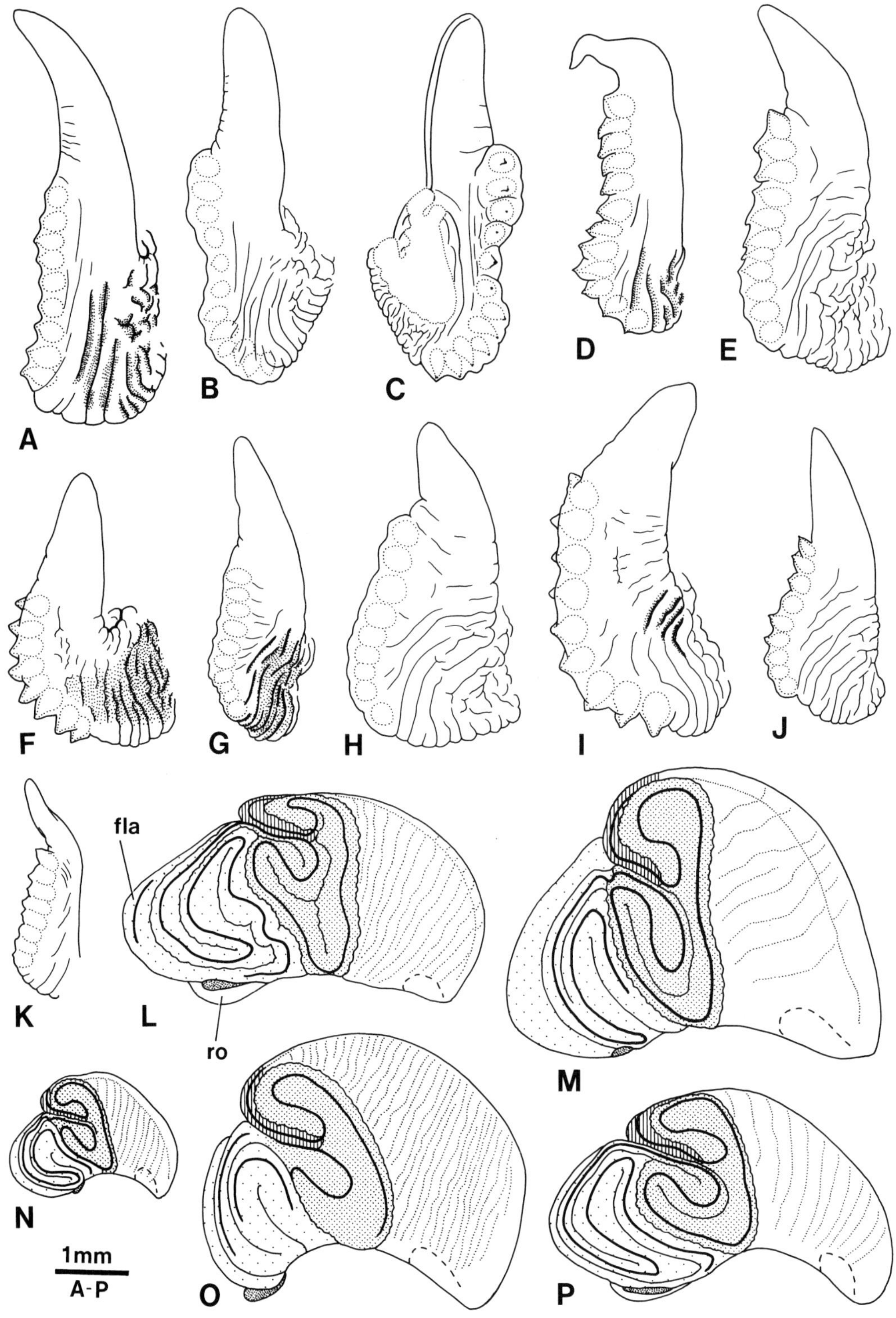

Figure 68. *Littorina (Neritrema) aleutica*: **A–K,** penes; **B, C,** two views of same penis; **L–P,** pallial oviducts. **A,** St Lawrence Island, Bering Sea, Alaska (SMNH 891); **B, C,** Adak Island, Aleutian Islands (USNM 796481); **D, O,** Nizki Island, Aleutian Islands; **E–J, L, M, P,** Provideniya, Russia; **K, N,** Lihacheva, Provideniya, Russia. Shell heights: **A,** 10.2 mm; **B, C,** 7.3 mm; **D,** 7.2 mm; **E,** 8.9 mm; **F,** 8 mm; **G,** 9.9 mm; **H,** 8.5 mm; **I,** 8.5 mm; **J,** 9.0 mm; **K,** 5.5 mm; **L,** 9.2 mm; **M,** 13 mm; **N,** 5.7 mm; **O,** 9.8 mm; **P,** 8.3 mm. Abbreviations: fla, first loop of albumen gland; ro, renal oviduct. For shading conventions of pallial oviducts see p. 6.

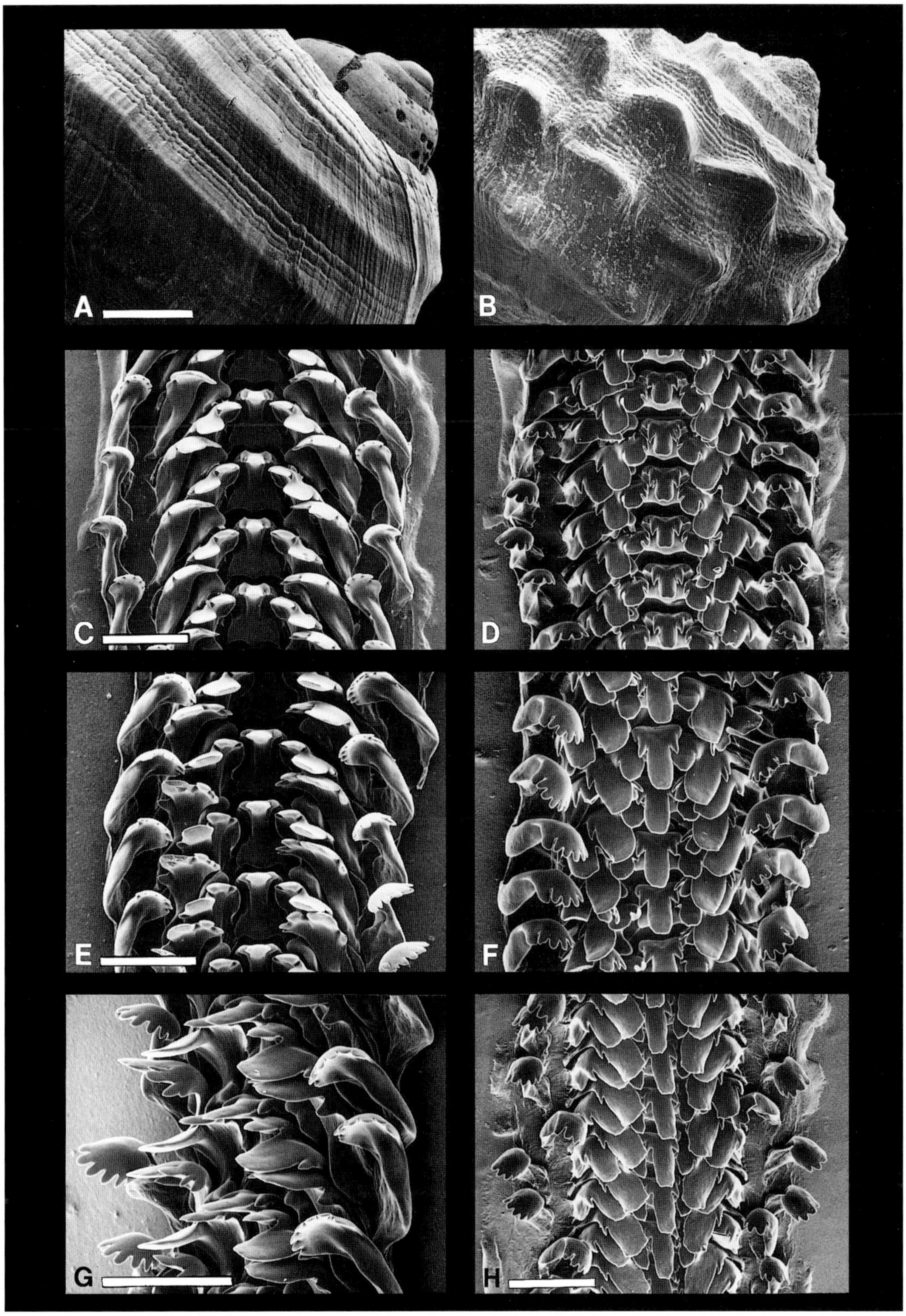

Figure 69. *Littorina (Neritrema) aleutica*: **A, B,** shell sculpture; **C, D,** two views of radula (shell H=9.8 mm), flat (**A**) and at 45° (**B**); **E–G,** three views of radula (shell H=11.6 mm), flat (**E**), at 45° (**F**) and side view (**G**); **H,** radula (shell H=9.9 mm), at 45°. **A,** Lihacheva, Provideniya, Russia; **B–D,** Nizki Island, Aleutian Islands; **E–G,** St Lawrence Island, Bering Sea, Alaska; **H,** Provideniya, Russia. Scale bars: **A, B**=1.0 mm; **C, D**=100 μm; **E, F**=100 μm; **G**=100 μm; **H**=100 μm.

Figure 70. *Littorina (Neritrema) aleutica*. **A,** *Littorina aleutica* Dall, 1872; lectotype, USNM 214721; Gull Rocks, Akutan Pass, Aleutian Islands. **B,** West Adak Island, Aleutian Islands, USNM 220152. **C,** Atka Island, Aleutian Islands, USNM 130623. **D–F,** Nizki Island, Aleutian Islands. **G,** St George Island, Pribilof Islands, Russia, USNM 271739. **H–J, L, M,** Provideniya, Bering Sea, Russia. **K,** Lihacheva, Provideniya, Russia. Scale bar = 5 mm.

Distribution

Habitat. On Amchitka in the Aleutian Islands *L. aleutica* is found from the mid to upper eulittoral and in the littoral fringe, and is one of the most characteristic invertebrates of the *Halosaccion* and *Fucus* community, where it reaches densities of 129 per 0.25 m^2 and occurs together with *L. sitkana* (O'Clair & Chew, 1971; O'Clair, 1977). From the material in ZISP it appears that *L. aleutica* is the sole *Littorina* species at Provideniya. Here it has been recorded among cobbles in the upper eulittoral (A. J. Sutcliffe, pers. comm.) and to a depth of at least 0.3 m below zero datum (ZISP). Kusakin & Ivanova's (1978) records of both *L. kurila* and *L. sitkana* in their survey of shores around Provideniya probably refer only to forms of *L. aleutica*, and include occurrences in association with *Halosaccion, Alaria, Chordaria, Acrosiphonia, Pylaiella* and *Fucus*, up to a maximum density of 1300 per m^2. *Littorina aleutica* has been collected on both exposed (Zimmerman, Hanson, Fujioka, Calvin, Gharrett & MacKinnon, 1977) and sheltered shores (A. J. Sutcliffe, pers. comm.). Except at Provideniya, this species has not been recorded on mainland coasts, but only on islands.

Range (Fig. 71). *Littorina aleutica* occurs at Bering Island in the Komandor Islands (ZISP) and throughout most of the Aleutian chain. Vermeij, Palmer & Lindberg (1990) found it commonly on most of the Aleutian Islands in their survey, but noted an apparent absence on Atka. No records are available from the Alaskan Peninsula, but it is recorded from Afognak Island (Eyerdam, 1960) and from Chirikov Island and Sud Island (south and north of Kodiak Island respectively, the latter the most easterly record; Zimmerman *et al.*, 1977; Foster, 1981). Elsewhere in the Bering Sea it occurs in the Pribilof Islands and at St Lawrence Island, and the most northerly records are the collections from Provideniya on the Chukotsk Peninsula. The distribution is of a typically oceanic type, the species occurring only on islands and a mainland peninsula. Kuroda & Habe (1952) gave the range of *L. aleutica* as between 46 and 56°N in the northwestern Pacific; it is unclear to what species they were referring, but perhaps to *L. natica*.

Fossil record. Unknown.

Discussion

Phylogenetic relationships. Littorina aleutica resembles *L. natica* in many respects, including the patulous shell, the form of the pallial oviduct, shape of the penis and blunt radula tooth cusps. Despite some uncertainty, these two do appear to be distinct species (p. 193). The cladistic analysis of morphological characters (p. 351) identifies the form of the egg groove and the additional first loop of the albumen gland as synapomorphies uniting *L. aleutica* and *L. natica* as sister-species, while the blunt tooth cusps unite this clade in a polychotomy with all the Atlantic *Neritrema* species. In contrast, analysis of base sequences from a single mitochondrial gene (16S rRNA) contradict this topology, instead placing *L. natica* as the sister-species of the five Atlantic *Neritrema* species, and *L. aleutica* as the sister-species to all six (Reid, Rumbak & Thomas, 1996). However, the molecular evidence consists of only a small number of base-pair differences (1–7 differences among these seven species) and the alternative (morphological) topology is but one step longer. Further molecular data are desirable to resolve this conflict, but for the present the morphological topology is accepted (p. 362).

In terms of speciation events, the morphological topology implies that divergence of *L. aleutica* and *L. natica* took place after trans-Arctic migration through the Bering Strait and subsequent vicariance had given rise to the Atlantic clade (assuming that migration took place once only, in one direction, i.e. that the Atlantic species are monophyletic). The molecular topology, however, implies that the separation of the two Pacific species took place first, and that only the common ancestor of *L. natica* and the Atlantic species was involved in the trans-Arctic migration. The likelihood of these alternatives is related to the timing of geological events in the Bering Sea and their possible influence on speciation processes. The opening of the Bering Strait has been dated to between 3.5 and 4 Ma (Hopkins, 1967b) and the modern volcanic islands of the Aleutian Chain may have begun to appear at about 3 Ma (Scholl, Buffington & Marlow, 1975). In their modern distributions in the Bering Sea *L. natica* is restricted to the Siberian mainland, whereas *L. aleutica* shows an oceanic distribution on islands and the Chukotsk Peninsula. If the Aleutian Islands did indeed appear after the opening of the Bering Strait (and assuming that their formation was important in the origin of the oceanic species), this is

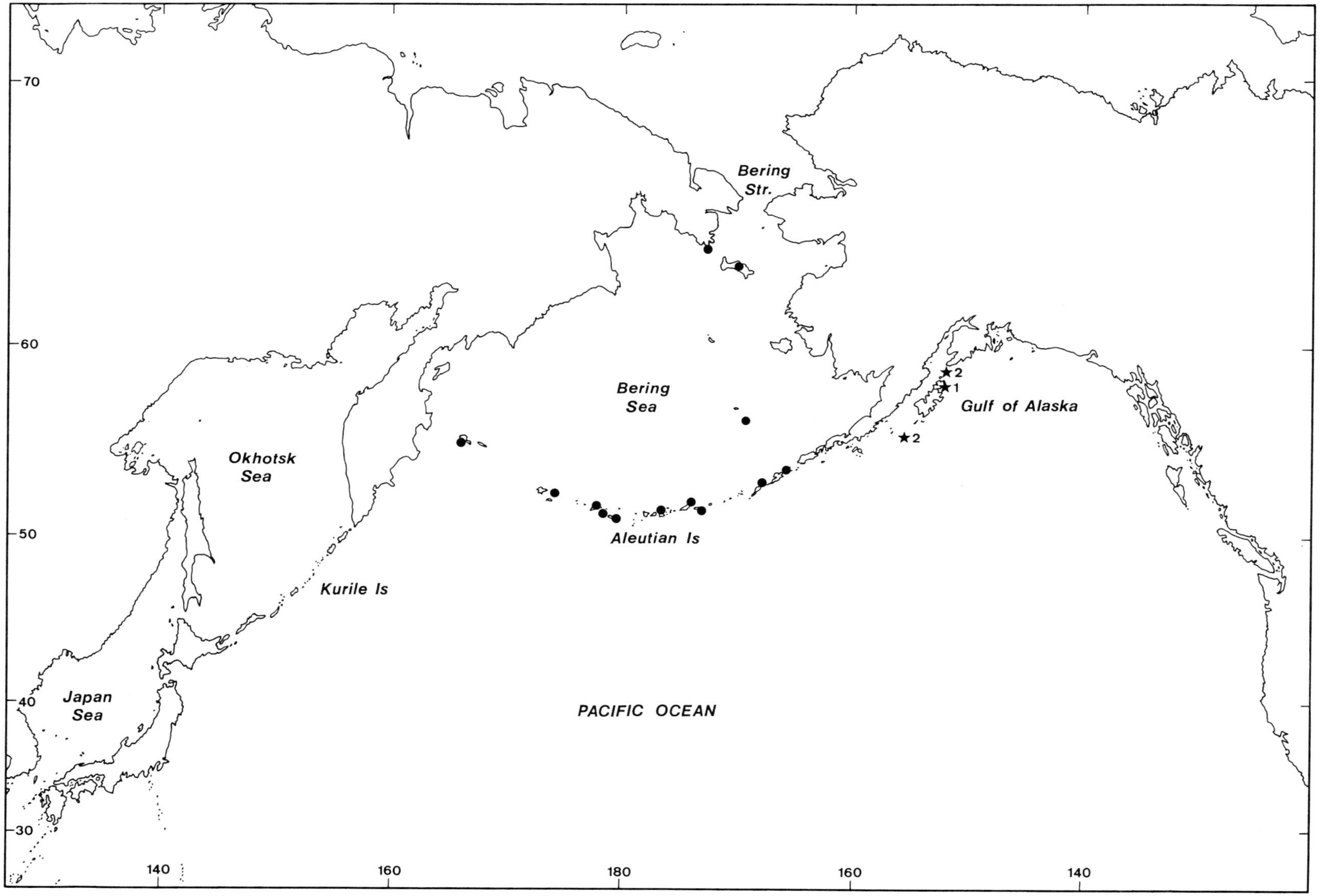

Figure 71. *Littorina (Neritrema) aleutica*: distribution map. Key: circles, specimens examined; asterisks, records from literature (1, Eyerdam, 1960; 2, Zimmerman, Hanson, Fujioka, Calvin, Gharrett & MacKinnon, 1977).

consistent with the morphological topology. Any species undertaking the trans-Arctic migration route must have been able to exist on continental shores; whether this species was the common ancestor of *L. aleutica* and *L. natica*, or of *L. natica* and the Atlantic clade, a parsimony argument (p. 389) supports this assertion. If the two Pacific species diverged before the opening of the Strait, and had then their modern habitat range, it would not be expected that the oceanic species, *L. aleutica*, would have been capable of making the trans-Arctic migration, and indeed neither tree topology suggests this.

The nature of the speciation event that gave rise to the oceanic *L. aleutica* from a continental ancestor (whether its sister-group is *L. natica* alone or the clade *L. natica* plus the Atlantic species) is unknown. The habitat difference taken together with the almost contiguous modern distributions of the two (separated by at most 250–350 km) might suggest a parapatric process (p. 380). Alternatively, waif-dispersal from the Siberian mainland to the newly formed Aleutian Islands could have occurred (peripheral isolates speciation mode, p. 379).

Similar species. When the three or four nodulose ribs are well-developed *L. aleutica* cannot be confused with any other species. Elsewhere in the genus only *L. striata* ever shows nodulose sculpture, but there the nodules are finer and the spiral ribs more numerous. Difficulty may be experienced in identifying shells without nodules, since four other northern Pacific species are potentially very similar in shell characters. These are *L. natica*, *L. sitkana*, *L. subrotundata* and *L. kasatka*, each of which can develop a relatively smooth, brown and sometimes patulous shell (see Table 5 for a summary of characters, pp. 160–161).

The most similar is the sister-species, *L. natica*. If shells are not eroded, the presence of strong spiral microstriae in *L. aleutica* is useful; these are absent or only faint in *L. natica*. Penial and oviduct characters are usually reliable, and their geographical distributions are not known to overlap. The distinction between these two species is discussed in detail in the Discussion of *L. natica* (p. 193).

Sculptured shells of *L. sitkana* are quite different, and most smooth forms are more tall-spired, with deeper sutures than in *L. aleutica*. The longer penial filament and two loops of the capsule gland serve to separate *L. aleutica* from this frequently sympatric species.

Although strongly sculptured and tall-spired examples of *L. subrotundata* cannot be confused with *L. aleutica*, it can also show smooth, patulous shells (Fig. 65N, P) of which the identity must be confirmed by dissection of penis or oviduct. Both of these anatomical characters are diagnostic, the penial filament of *L. subrotundata* being shorter, and its capsule gland smaller and consisting of a single loop. The ranges of these two species overlap throughout the Aleutian Islands.

Littorina kasatka is a small, smooth species which might be confused with some of the smaller, globular examples of *L. aleutica*, although the geographical distributions of the two species do not overlap. The absence of a swollen, septate jelly gland, and the single or absent mamilliform penial gland of *L. kasatka*, distinguish these two species.

Littorina (Neritrema) natica new name

(Figs 72–75)

Synonymy

Littorina (Littorina) kurila—Golikov & Kusakin, 1978: 87–89 (in part, includes *L. sitkana*, *L. aleutica*, *L. subrotundata*, *L. kasatka*; not Middendorff, 1848=*L. sitkana*).

Littorina kurila—Kusakin & Ivanova, 1978: 15–27 (not Middendorff, 1848).

Littorina sitchana—Kusakin & Ivanova, 1978: 15–27 (not Philippi, 1847=*L. sitkana*).

Littorina (Neritrema) naticoides Reid & Golikov, 1991: 7–15; figs 1–12; figs 15–20 (penis); figs 21–23, 37 (oviduct); figs 28–30 (head pigmentation); (Milne Bay, Simushir Island, Kurile Islands; holotype BMNH 1990053, seen, Reid & Golikov, 1991: fig. 1, Fig. 74A herein; 2 paratypes BMNH 1990054, seen; 6 paratypes BMNH 1990055 (alcohol), seen, Fig. 74I herein; 2 paratypes ZISP, seen; not d'Orbigny, 1841=*Nodilittorina mespillum* (Mühlfeldt, 1824); here renamed).

Littorina naticoides—Egorov, 1992: 170–171; fig. 1K.

Taxonomic history

Before its recent description this species had been included under the name *L. kurila*, together with smooth-shelled forms of four other species (e.g. Golikov & Kusakin, 1978; see Synonymy of *L. sitkana*,

p. 146). All material examined in ZISP (including that from Kresta Bay, the locality surveyed by Kusakin & Ivanova, 1978) had previously been identified either as *L. kurila*, or occasionally as *L. sitkana*. This species was first described as *L. naticoides* Reid & Golikov, 1991, but since that name is a junior primary homonym of *L. naticoides* d'Orbigny, 1841, it is here renamed.

Diagnosis

Mature shell height 6.2–15.9 mm. Shell rounded-turbinate to low-spired and patulous. Mineralogy of external irregular-prismatic calcite and internal crossed-lamellar aragonite. Sculpture smooth, sometimes with faint spiral striations, but no strong microstriae. Colour orange brown to dark purple brown, usually with 5–12 dark brown or black spiral lines, sometimes a pale spiral band on base; columella white to purplish; interior brown. Penis with short, bluntly triangular filament, 20–30 (rarely 40)% total length; 4–17 mamilliform penial glands. Pallial oviduct of unusual double loop of albumen gland, double loop of capsule gland, and large, swollen, septate jelly gland. Spawn a benthic mass of singly-encapsulated eggs in gelatinous matrix; development nonplanktotrophic with intracapsular metamorphosis. Distribution: Kurile Islands, Kamchatka, northwestern Bering Sea.

Anatomical description

Head-foot (Fig. 72R–T). Head and tentacles black, with pale tips to tentacles and snout; unpigmented patch around eye and usually at inside of tentacle base also, sometimes extending across tentacle base; sides of foot may be unpigmented, faintly mottled, grey or occasionally blackish. Opercular ratio 0.45–0.50.

Penis (Fig. 72A–K). Base wrinkled, usually unpigmented, sometimes some grey to black pigmentation on outside surface; 4–17 mamilliform penial glands in single row (1 abnormal specimen had only 1). Filament short, bluntly triangular, 20–30% total ength (rarely 40%, in examples with few penial glands). Sperm groove open, ending terminally.

Sperm. Not known.

Oviduct (Fig. 72L–Q). As in *L. aleutica*, the albumen gland consists of two loops, the first on the medial side (adjacent to the renal oviduct and rectum), usually partly visible at the posterior edge of the normal second loop of the albumen gland, although sometimes concealed by it. In the available preserved specimens the albumen gland cannot be distinguished from the renal oviduct by its appearance, and often the route of the egg groove through it is difficult to trace. However, if (as in Fig. 72O) the groove contains black pigment its path can be clearly seen; the renal oviduct is then distinguished as the glandular region before the junction with the seminal receptacle (the presumed point of fertilization) and the albumen gland as the glandular region following the junction. Capsule gland of two consecutive loops (the second often slightly the larger), forming a large mass of almost circular outline, about twice the area of the albumen gland; a small, distal, reddish or brownish, translucent portion can usually be distinguished. Jelly gland large, swollen, septate; usually less than half the area of the combined albumen and capsule glands. Bursa of small to moderate size, sometimes extending back as far as anterior part of capsule gland, opening near anterior end of jelly gland. Seminal receptacle simple.

Spawn and development. Spawn known from two collections from Anadyrskiy Gulf (ZISP *L. kurila* no. 210–224, and *L. kurila* no. 215, in alcohol; duplicate material in BMNH) and one from Simushir (ZISP *L. kurila* no. 479, in alcohol; duplicate material in BMNH). Egg masses 5–7 mm in diameter, consisting of single eggs within spherical capsules 700–1000 μm in diameter, embedded in a gelatinous matrix. Breeding season not recorded; eggs were collected in August and adults were all mature in the available collections, made between July and September.

Radula and foregut (Fig. 73). Relative radular length 1.09–2.39. Rachidian: length/width 0.77–1.41, central cusp blunt to rounded, one pointed cusp and one outer denticle on each side. Lateral: four cusps, third from inside largest, largest only or two largest cusps rectangular or bluntly rounded,

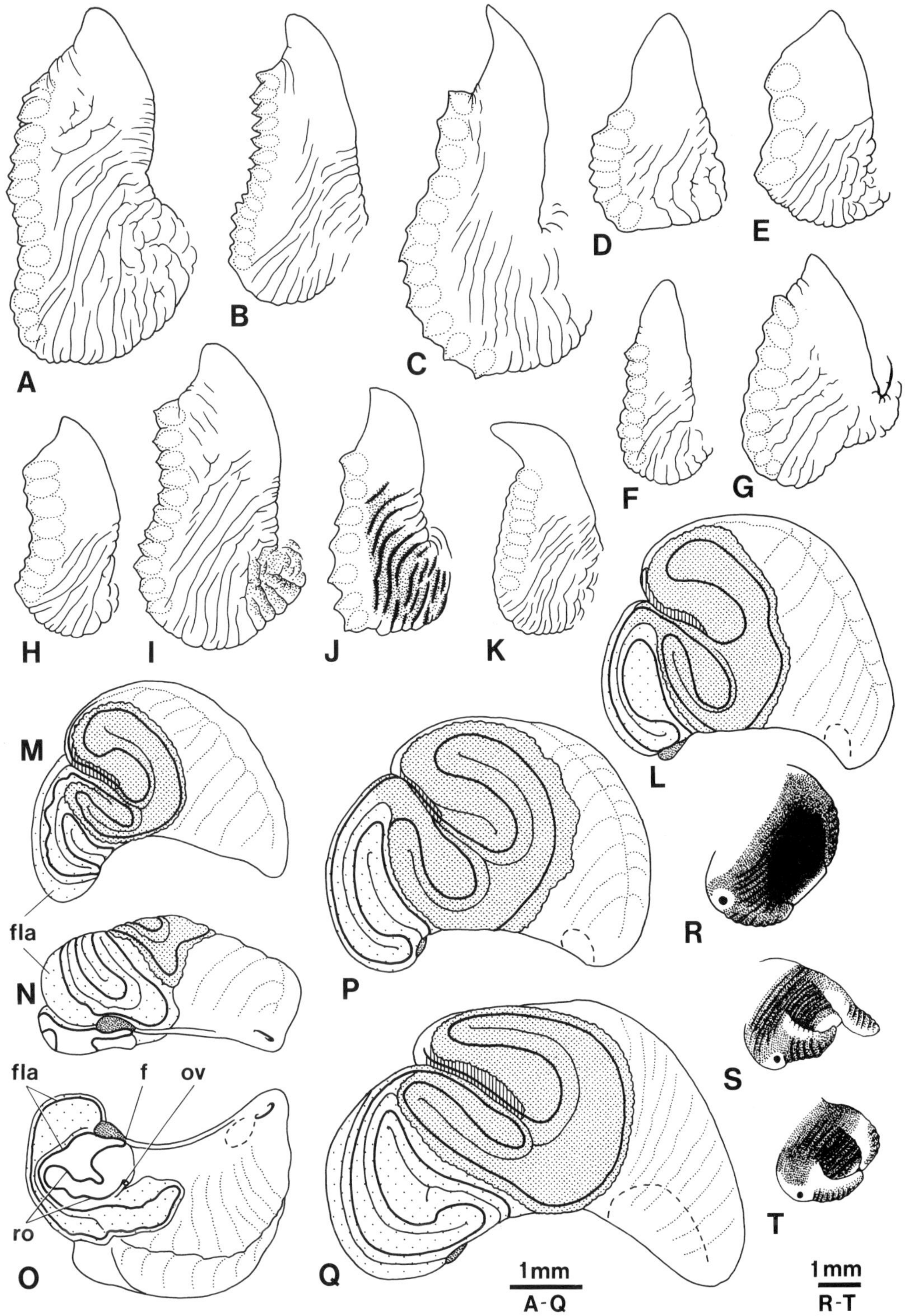

Figure 72. *Littorina (Neritrema) natica*: **A–K,** penes; **L–Q,** pallial oviducts; **M, N, O,** lateral, ventral and medial views of same pallial oviduct; **R–T,** head pigmentation. **A,** holotype, BMNH 1990053, Milne Bay, Simushir Island, Kurile Islands, Russia; **B, Q,** Nataliya Bay, Urup Island, Kurile Islands, Russia; **C, J, R,** Ozernovskiy, Kamchatka, Russia; **D–G, M–O, S,** Kresta Bay, Anadyrskiy Gulf, Russia; **H,** Egvekinot Inlet, Anadyrskiy Gulf, Russia; **I,** Milne Bay, Simushir Island, Kurile Islands, Russia (paratype, BMNH 1990055); **K, P, T,** Kronotskiy Bay, Kamchatka, Russia; **L,** 10 km E of Cape Kamchatskiy, Kamchatka, Russia. Shell heights: **A,** 13.3 mm; **B,** 7.9 mm; **C,** 10.9 mm; **D,** 8 mm; **E,** 8 mm; **F,** 7.9 mm; **G,** 9.6 mm; **H,** 9 mm; **I,** 9.6 mm; **J,** 11.7 mm; **K,** 9 mm; **L,** 7.6 mm; **M–O,** 7.9 mm; **P,** 9 mm; **Q,** 9.2 mm; **R,** 13.4 mm; **S,** 9 mm; **T,** 9.4 mm. Abbreviations: f, presumed point of fertilization, at junction of renal oviduct, first loop of albumen gland and seminal receptacle; fla, first loop of albumen gland with pigmented egg groove; ov, cut end of ovarian oviduct from ovary; ro, pigmented egg passage running through renal oviduct. For shading conventions of pallial oviducts see p. 6.

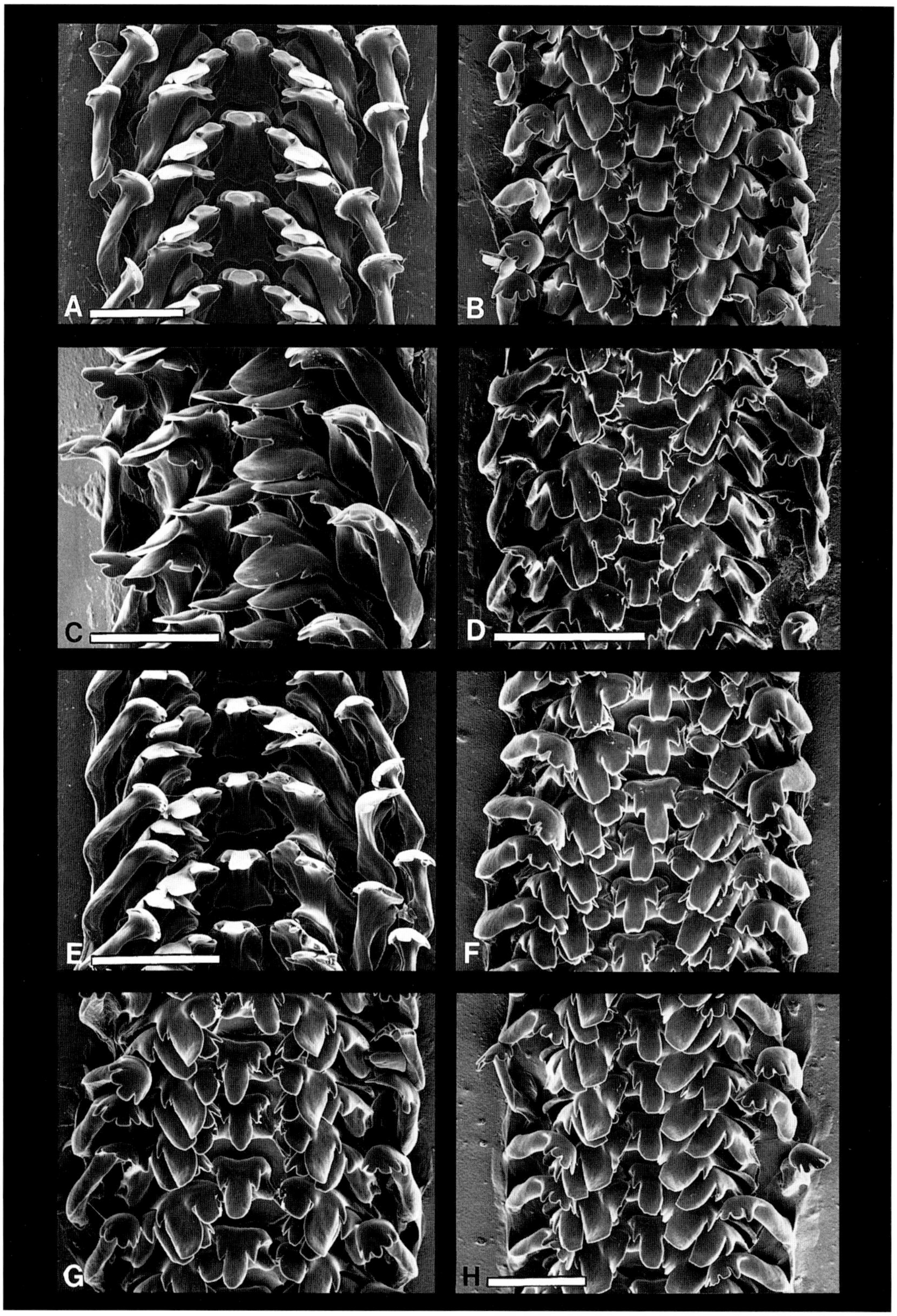

Figure 73. *Littorina (Neritrema) natica*: **A–C,** three views of radula (shell H=9.2 mm), flat (**A**), at 45° (**B**) and side view (**C**); **D,** radula (shell H=7.5 mm), at 45°; **E, F,** two views of radula (shell H=9.6 mm), flat (**E**) and at 45° (**F**); **G,** radula (shell H= 7.6 mm), at 45°; **H,** radula (shell H=9.2 mm), at 45°. **A–C,** Ozerniy River estuary, Kamchatka, Russia; **D,** Kronotskiy Bay, Kamchatka, Russia; **E, F,** Krista Bay, Anadyrskiy Bay, Russia; **G,** 10 km E of Cape Kamchatskiy, Kamchatka, Russia; **H,** Nataliya Bay, Urup Island, Kurile Islands, Russia. Scale bars: **A, B**=100 µm; **C**=100 µm; **D**=100 µm; **E–G**=100 µm; **H**=100 µm.

others pointed. Inner marginal: four cusps, third from inside largest, largest only or two largest cusps rectangular or bluntly rounded, others pointed. Outer marginal: 5–7 triangular to rounded cusps; narrow neck, projecting flange on inside of base only. Radulae illustrated by Reid & Golikov (1991) showed cusps slightly recurved, a distortion produced by heating during the cleaning process. Salivary glands enlarged.

Shell variation (Fig. 74)

Ecotypic variation. Shell shape varies from low-spired and rather patulous (H/B=1.01, SH=1.13) to turbinate (H/B=1.11, SH=1.45). Few habitat details are noted for the available museum specimens, but the most low-spired, wide-mouthed forms (Fig. 74B, H) are from a locality recorded as exposed, and the most tall-spired forms (Fig. 74E, J, K) from a sheltered inlet. This suggests that wave-exposed and moderate ecotypes might be recognized.

Colour variation. The shell colour is rather constant in this species, shades of brown usually with 5–12 darker spiral lines, and sometimes a pale band on the base. No polymorphic samples have been seen.

Distribution

Habitat. Available museum records (ZISP, BMNH) indicate a range of intertidal habitats: an estuary (Ozerniy River, western Kamchatka), on dead shells in a sheltered inlet (Egvekinot Inlet, Anadyrskiy Gulf), with the brown algae *Alaria* and *Fucus* near low tide (Kresta Bay, Anadyrskiy Gulf; Urup Island), on the red alga *Halosaccion* in the mid-eulittoral (eastern Kamchatka), in the *Mytilus* zone on exposed shores (eastern Kamchatka). Kusakin & Ivanova (1978) recorded the abundance of this species (as *L. kurila* and *L. sitchana*) in a variety of intertidal algal associations at Kresta Bay; these included *Pylaiella, Acrosiphonia, Stictyosiphon, Alaria, Rhodomela, Fucus* and *Laminaria.* The highest biomass (1780 individuals per m^2, 240 g per m^2) occurred in an association of *Rhodomela, Alaria* and *Fucus*, low in the eulittoral.

Range (Fig. 75). The distribution of this species was given by Reid & Golikov (1991), based on material in ZISP and BMNH. It occurs through the Kurile Islands, from Urup in the south to Paramushir in the north. In Kamchatka it is recorded from Ozernovskiy in the south west (another western locality, the Ozerniy River estuary, could not be traced) and from several eastern localities as far north as Cape Kamchatskiy. There is then a considerable gap along the poorly-collected coast of the northwestern Bering Sea until the several collections from Kresta Bay on the Anadyrskiy Gulf. At least in Kamchatka and the Bering Sea this species shows a typically continental distribution pattern.

Fossil record. Unknown.

Discussion

Specific status. Littorina natica is clearly closely related to *L. aleutica*, showing similarities in penis, radula and shell shape, and (according to the cladistic analysis, p. 351) possessing two unequivocal synapomorphies in the form of the loops of the capsule gland, and the double loop of the albumen gland. In the absence of sympatric occurrences, these morphological differences must be carefully evaluated; they are summarized in Table 5 (pp. 160–161). To estimate the accuracy of discrimination using anatomical characters, the two taxa were initially separated on the basis of their apparently allopatric distributions (see Table 1, p. 6, for numbers of specimens examined). The two most useful characters were found to be the shape of the penis (filament 20–30% total length in *L. natica*; 30–50% in *L. aleutica*) and the form of the capsule gland (circular shape, about twice area of albumen gland in *L. natica*; rectangular or trapezoid shape, about equal to area of albumen gland in *L. aleutica*). These two characters gave correct identification in 84% and 93% of cases respectively. The presence of strong shell microsculpture in *L. aleutica* appears to be a reliable character in well-preserved shells, but since the surface is sometimes eroded, it cannot always be used. Other less reliable differences include a generally paler body pigmentation and a slightly smaller jelly gland in *L. natica*. Neither these shell, penial nor oviduct differences alone could be considered good evidence for the separation of two species, but the strong correlation between these three characters suggests that it may be justified.

Figure 74. *Littorina (Neritrema) natica.* **A**, *Littorina (Neritrema) naticoides* Reid & Golikov, 1991; holotype, BMNH 1990053; Milne Bay, Simushir Island, Kurile Islands, Russia. **B**, **H**, 10 km E of Cape Kamchatskiy, Kamchatka, Russia. **C**, **F**, Ozerniy River estuary, Kamchatka, Russia. **D**, Kronotskiy Gulf, Kamchatka, Russia. **E**, **J**, **K**, Egvekinot Inlet, Anadyrskiy Gulf, Russia. **I**, **L**, Ozernovskiy, Kamchatka, Russia. Scale bar = 5 mm.

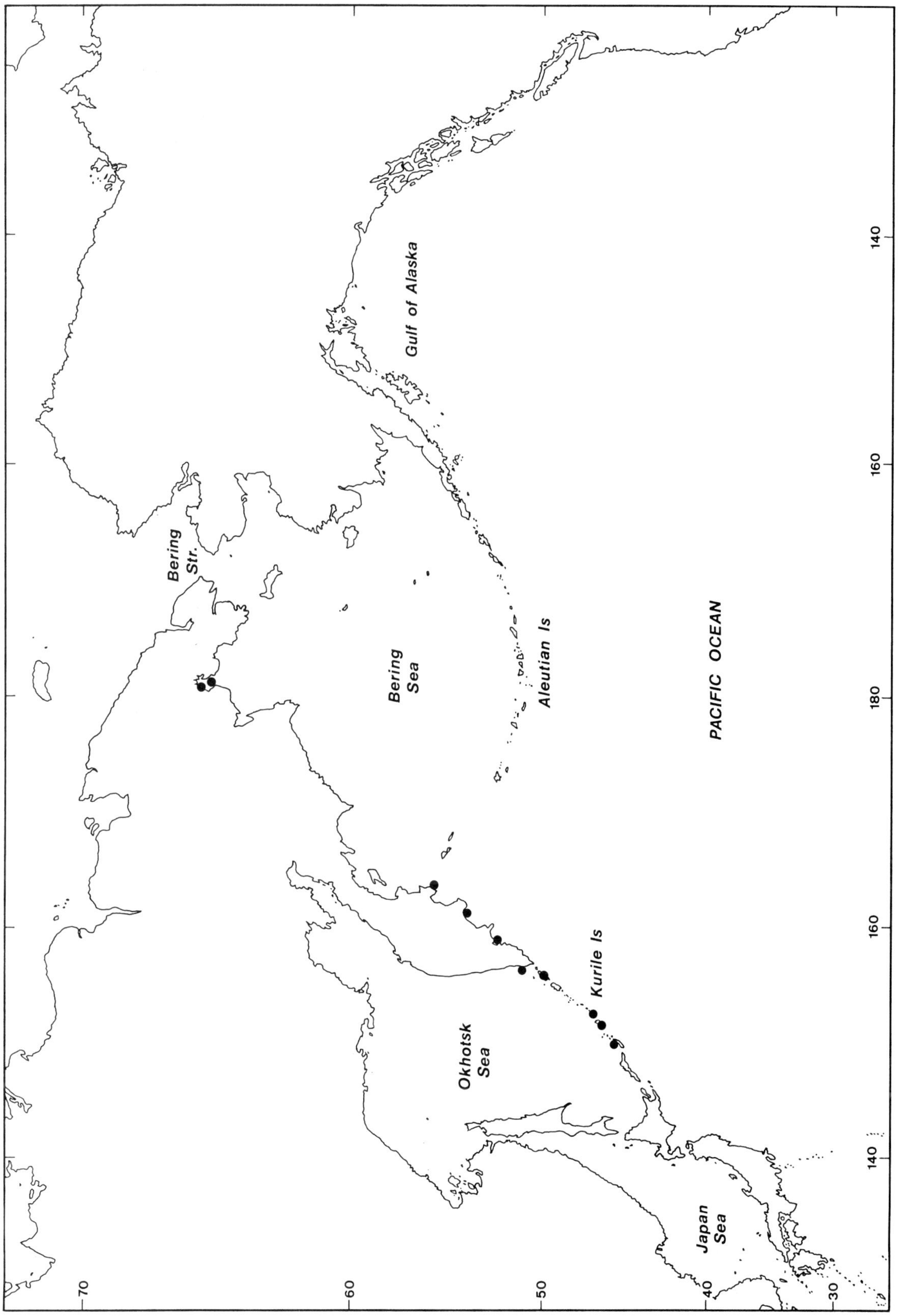

Figure 75. *Littorina (Neritrema) natica*: distribution map. Key: circles, specimens examined.

In addition there appears to be an ecological distinction, between *L. natica* from the continental shores of the Bering Sea (although also from the Kurile Islands) and *L. aleutica* from islands in the Bering Sea (only recorded from the mainland at Provideniya, at the tip of the Chukotsk Peninsula). There are several similar examples among tropical littorinids of pairs of closely related congeners with allopatric 'continental' and 'oceanic' distributions (*Littoraria*, Reid, 1986a; *Peasiella*, Reid, 1989b), and among *Littorina* species there is another example in the pair *L. sitkana* and *L. horikawai*. The separate distribution of the two species in the Kurile Islands, Kamchatka and the Aleutian Islands is probably real, for there are extensive collections from all these areas. *Littorina aleutica* occurs on Bering Island, only 250 km from the nearest recorded occurrence of *L. natica* in eastern Kamchatka. However, the coast of the northwestern Bering Sea is poorly known, and there is a gap of about 1700 km between the records of *L. natica* in eastern Kamchatka and in the northern Anadyrskiy Gulf. A gap of about 350 km separates the latter from the closest record of *L. aleutica*, at Provideniya, and collecting in this region is desirable, to determine whether or not the two species remain distinct in an area of possible sympatry.

There is also some molecular evidence to support the recognition of these two species. DNA sequences for both are available only for the 16S rRNA mitochondrial gene, and from a single individual of each. These differ at five positions out of 475 (1.05%), which is comparable to the degree of difference among members of the Atlantic radiation of *Neritrema* (Reid, Rumbak & Thomas, 1996).

Phylogenetic relationships. Cladistic analysis of morphological characters suggests that *L. aleutica* and *L. natica* are sister-species, but the molecular evidence suggests that *L. natica* is the sister-taxon to all the Atlantic *Neritrema* species. These alternative topologies and possible speciation mechanisms are discussed in the remarks on *L. aleutica* (p. 187).

Similar species. *Littorina natica* is most similar to *L. aleutica* (p. 193). Over its range in Kamchatka and the Kurile Islands it is sympatric with three other species that can appear very similar in shell characters (see Table 5 for a summary, pp. 160–161). *Littorina kasatka* is usually smaller, often with a white-flecked or tessellated pattern, and identification is readily confirmed anatomically by the oviduct with small jelly gland and by the penis with only one or no mamilliform glands. Smooth-shelled forms of *L. sitkana* and *L. subrotundata* are usually taller-spired than *L. natica*, but can sometimes be impossible to distinguish. In such cases females of the three can be separated by the form of the capsule gland, but since penial shape is similar there are no diagnostic characters for males.

Littorina (Neritrema) obtusata (Linnaeus, 1758)

(Figs 3G, 4, 5, 76–81)

Synonymy

Turbo obtusatus Linnaeus, 1758: 761 (O. Septentrionali [Northern Ocean]; restricted to coast of Lappland, Norway (Bequaert, 1943); lectotype (here designated) LSL 13.3x12.7 mm, Hanley, 1855: pl. 3, fig. 6, Fig. 79A herein+10 paralectotypes, seen). Linnaeus, 1767: 1232. Müller, 1776: 242 (*s.l.*). Chemnitz, 1781: 234; pl. 185, fig. 1854 a-d (non-binominal, *Turbo obtusatus* in synonymy; in part, other figs probably *L. fabalis*). Schröter, 1784: 3–4 (*s.l.*). Gmelin, 1791: 3588 (*s.l.*). Pulteney, 1799: 45 (in part, includes *L. saxatilis*). Dillwyn, 1817: 815 (*s.l.*). Wood, 1818: 142 (*s.l.*). Lamarck, 1822: 49. Blainville, 1828: 109. Wood, 1828: 142; pl. 30, *Turbo* fig. 1 (*s.l.*). Blainville, 1830: 302. Hanley, 1855: 325–326; pl. 3, fig. 6.

[*Paludina (Littorina) obtusata*]—Férussac, 1822: ix, xi, xxxiv (see Kadolsky, 1970; Melville, 1980; discussion p. 38).

Litorina obtusata—Menke, 1828: 25. Menke, 1830: 45. Anton, 1838: 53. Menke, 1845: 55–60 (in part, includes *L. fabalis*). Lovén, 1846: 154 (*s.l.*). Philippi, 1846a: **2**: 104–105; *Litorina* pl. 1, figs 20–22 (*s.l.*; fig. 22 is possibly *L. fabalis*). Küster, 1853: pl. 2, figs 1–7; 1856: 11–13. Weinkauff, 1878: 28, 29 (in part, includes refs to *L. fabalis*). Weinkauff, 1883: 216–217 (in part, includes refs to *L. fabalis*). Bárdarson, 1919: 63 (*s.l.*). Ankel, 1930: 599–600; figs 60, 61 (sperm).

Littorina obtusata—Middendorff, 1849: 385–388; pl. 8, figs 24–28 (*s.l.*). Sars, 1851: 181 (*s.l.*). Hanley, 1855: 325–326; pl. 3, fig. 6. Mørch, 1857: 80 (*s.l.*). Jeffreys, 1865: **3**: 356–361; 1869: **5**: 205–206; pl. 65, figs 1, 1a; pl. 101, fig. 8 (in part, includes *L. fabalis*). Mørch, 1868: 79 (*s.l.*). Weinkauff, 1868: 270–272. Mørch, 1869: 206–207. Brusina, 1870: 235–236 (*s.l.*). Mørch, 1871a: 186 (in part, includes refs to *L. fabalis*). Meyer & Möbius, 1872: 15–16 (in part, includes *L. fabalis*). Jeffreys, 1877: 238 (*s.l.*). Sars, 1878: 167–168, 447; anatomical pl. 6, fig. 6 (radula, perhaps *L. fabalis*) (*s.l.*). Daniel, 1883: 341 (*s.l.*). Jeffreys, 1883: 112 (in part, includes refs to *L. fabalis*). Collin, 1884: 61–62 (*s.l.*). Herzenstein, 1885: 680 (*s.l.*). Locard, 1886: 281–282, 577–578 (as *obtusa*). Sparre-Schneider, 1886: 10–12 (*s.l.*). Sowerby, 1887: pl. 12, figs 20, 21 (in part, other fig. is *L. fabalis*). Kobelt, 1888: 168 (in part, includes ref. to *L. fabalis*). Hidalgo, 1890: pl. 89, figs 16–19 (in part, other fig. is *L. fabalis*). Locard, 1891: 187; fig. 163 (*s.l.*; as *obtusa*). Dautzenberg, 1893: 20 (*s.l.*). Marshall, 1898: 120–121. Posselt & Jensen, 1898: 229–230 (*s.l.*). Nobre, 1900: 79–80 (*s.l.*). Brøgger, 1901: pl. 9, fig. 15 (Pleistocene). Johansen, 1901: 303–305; fig. 7; fig. 8 (radula). Norman, 1902: 355 (*s.l.*). Caullery & Pelseneer, 1910: pl. 9, fig. 4 (spawn) (*s.l.*). Pelseneer, 1911: 9–11; pl. 2, fig. 1 (spawn); figs 2, 3 (embryo) (*s.l.*). Delsman, 1914: 170–340 (ontogeny) (*s.l.*). Dautzenberg & Fischer, 1915; 87–128; text-figs 1–3; pl. 2, figs 1–10, 13–15; pl. 3, figs 1–4, 8, 10–15 (fig. 8 is possibly *L. fabalis*); pl. 4 (colour chart) (in part, includes *L. fabalis*). Marshall, 1915: 328–329 (*s.l.*). Hidalgo, 1917: 391, 395–396 (in part, includes refs to *L. fabalis*). Bárdarson, 1920: 89–90 (*s.l.*). Harmer, 1921: 661–662; pl. 52, figs 27, 28 (in part, other fig. is *L. littorea*). Hertling & Ankel, 1927: 3–6; figs 1–5 (spawn) (*s.l.*). Hertling, 1928: 44–47; figs 23, 26 (spawn) (*s.l.*). Nobre, 1931: 146; pl. 33, fig. 7 (*s.l.*). Colman, 1932: 226–241; figs 3, 6, 7, 10. van Benthem Jutting, 1933: 63. Linke, 1933: 1–60 (anatomy); pl. 1, fig. 2 (penis); pl. 5, fig. 56, pl. 6, fig. 59 (oviduct) (in part, includes *L. fabalis*). Spärck & Thorson, 1933: 15 (*s.l.*). Winckworth, 1933: 335–336. Ankel, 1936b: 47; fig. 132 (penis); fig. 140 (sperm); fig. 146 (spawn); fig. 206A (*s.l.*). Nobre, 1936: 102; pl. 82, fig. 10 (*s.l.*). Johansson, 1939: 344–396 (anatomy) (*s.l.*). Thorson, 1941: 30–32 (*s.l.*). Thorson, 1946: 173–174; fig. 95 (spawn, embryo) (*s.l.*). Knudsen, 1949: 247–255. Thorson, 1951: 28 (*s.l.*). La Roque, 1953: 167. Abbott, 1954a: 133; pl. 19, fig. f. Dodge, 1959: 214–216. Bousfield, 1960: 16; pl. 2, fig. 17. Sacchi, 1961: 271–290 (*s.l.*). Sacchi, 1962: 517–528 (*s.l.*). Sacchi, 1963: 299–313 (*s.l.*). Sacchi, 1964: 93–156 (*s.l.*). Sacchi, 1966a: 2370–2372; fig. 1 (penis) (in part, includes *L. fabalis*). Sacchi, 1966b: 281–294 (*s.l.*). Sacchi, 1966c: 521–541 (*s.l.*). Sacchi & Rastelli, 1966: 351–369; figs 1, 2A–C, E, pl. 11, figs 1, 3, 4 (penis). Ziegelmeier, 1966: 41; pl. 10, fig. 1a, b (*s.l.*). Sacchi, 1967a: 1504–1506. Sacchi, 1967b: 339–355. Lucas, 1968: (3): 11–14; (4): 3–8 (in part, includes *L. fabalis*). Gratton, 1969: fig. 1 (penis). Sacchi, 1969a: fig. 6a–d (penis); fig. 7a–d. Sacchi, 1969b: 33–47. Davis, 1971: 2–5; fig. 4: 11–13. Macpherson, 1971: 31–32. Sacchi, 1972: 105–137. Rasmussen, 1973: 235–238 (*s.l.*). Abbott, 1974: 67; text-fig. 550; pl. 3, fig. 550. Bandel, 1974: 106; figs 18–21, 28, 29 (radula) (*s.l.*; fig. 28 perhaps *L. fabalis*). Matveeva, 1974: 117–121; fig. 18 (spawn) (*s.l.*). Sacchi, 1974: 64–75. Bandel, 1975: 5–6; figs 6–8 (protoconch) (*s.l.*). Hubendick & Warén, 1975: 10; fig. 270 (*s.l.*). Smith, 1976: 114–135 (*s.l.*). Goodwin & Fish, 1977: 241–254; fig. 5 (radula); fig. 7, nos 1, 3, 4. Goodwin, 1979: 1–11; fig. 11 (spawn). Sneli & van Marion, 1979: fig. 1f; fig. 1j (penis). Fretter, 1980: 148–153; fig. 1B (oviduct). Fretter & Graham, 1980: 258–260; fig. 205A–C. Moyse, Thorpe & Al-Hamadani, 1982: 7–15; fig. 2 (penis). Reimchen, 1982: 687–695. Backeljau, de Meyer, Janssens & Proesmans, 1984: 3; fig. 3. Jardine, 1984: 5; fig. 2c. Rolán, 1984: 110; figs. Sacchi, 1984: 209–232. Sergievsky, 1984: 96–102. Sergievsky, 1985c: 44–53. Janke, 1986: 10–11. Seeley, 1986: 6897–6901; fig. 1. Warmoes, 1986. Jagnow & Gosselck, 1987: 211 (in part, fig. is *L. fabalis*). Rolán & Templado, 1987: 71–85; fig. 2A (penis). Watson & Norton, 1987: fig. 2A (radula). Graham, 1988: 170–171; fig. 62A–D. Sacchi & Sconfietti, 1988: 285–304. Warmoes, Backeljau & de Bruyn, 1988: 51–70; fig. 4N, O (penis); fig. 5E, G. Hawkins, Watson, Hill, Harding, Kyriakides, Hutchinson & Norton, 1989: figs 2G, 3C. Hayward, Wigham & Yonow, 1990: 659–660; fig. 12.9. Mikhailova & Naumov, 1990: 131–134. Reid, 1990a: 113–114; fig. 1A (oviduct). de Bruyne, 1991: 99; fig. 63 (right) (in part, other fig. is *L. fabalis*). Finet, Wüest & Mareda, 1991: 38; pl. 12a, b; pl. 12c, d (radula) (*s.l.*; pl. 12e is probably *L. fabalis* radula). Poppe & Goto, 1991: 92–93; pl. 11, figs 11–13. Egorov, 1992: 171; fig. 1C. Sergievsky, 1992: 235–245. Trussell, Johnson, Rudolph & Gilfillan, 1993: 135–144. Lewis & Williams, 1995: 161–166.

Littorina (Neritoides) obtusata—H. & A. Adams, 1854: **1**: 314. Kobelt, 1878: 124; pl. 46, fig. 9. Fischer, 1885: 708. Nobre, 1886: 21 (*s.l.*). Tryon, 1887: 252–253; pl. 46, figs 20, 21, 23 (in part, other fig. is *L. fabalis*). Dautzenberg & Durouchoux, 1900: 46 (*s.l.*). Kobelt, 1907: 64–66; pl. 110, figs 14–18 (in part, other fig. is *L. fabalis*). Dautzenberg & Fischer, 1925: 55–58 (*s.l.*).

Littorina (Neritrema) obtusata—Récluz, 1869: 46 (*s.l.*). Bequaert, 1943: 18–19; pl. 6, figs 1–6. Smith, 1982: 535–539. Reid, 1989a: 102; pl. 2, fig. 2p; fig. 4g, h (operculum); fig. 5l (head); fig. 7p (penis); fig. 8d (paraspermatozoa); fig. 10o (oviduct); fig. 14r (radula). Reid, 1990b: 1–19; fig. 1o; fig. 2j (oviduct); fig. 4g (radula). Taylor & Reid, 1990: 199–215 (mineralogy). Smith & Heppell, 1991: 17, 81.

Neritoides obtusata—Mansel-Pleydell, 1898: 41 (*s.l.*). Winckworth, 1922: 97. Chatwin, 1932: 317–318.

Littorina obtusata [subsp. *obtusata*] forma typica—Dautzenberg & Fischer, 1915: 105–106; pl. 2, figs 1, 2.

Littorina (Littorina) obtusata—Nordsieck, 1968: 40; pl. 6, fig. 23.00 (*s.l.*). Rosewater, 1970: 423 (*s.l.*). Golikov & Kusakin, 1978: 93–94; fig. 57a (spawn); fig. 57b (*s.l.*). Nordsieck, 1982: 61; pl. 26, fig. 23.00b, d (in part, other figs are *L. fabalis*). Golikov, 1987: 77–78; fig. 29 (*s.l.*). Sabelli, Giannuzzi-Savelli & Bedulli, 1990: 19, 146.

Nerita littoralis—Linnaeus, 1758: 777 (in part, includes *Nerita littoralis* Linnaeus, 1758=*Theodoxus fluviatilis* (Linnaeus, 1758), see Taxonomic History, p. 202). Linnaeus, 1761: 533 (in part, includes *T. fluviatilis*; not Linnaeus, 1758). Linnaeus, 1767: 1253 (in part, includes *T. fluviatilis*; not Linnaeus, 1758). Müller, 1776: 244 (*s.l.*; not Linnaeus, 1758). Pennant, 1777: 141; pl. 87, fig. 143 (*s.l.*; not Linnaeus, 1758). da Costa, 1778: 50–51; pl. 3, figs 13–16 (not 7 as in text); pl. 4, figs 2, 3 (*s.l.*; not Linnaeus, 1758). Fabricius, 1780: 402–403 (*s.l.*; not Linnaeus, 1758). Schröter, 1784: 287–288 (*s.l.*; not Linnaeus, 1758). Gmelin, 1791: 3677 (*s.l.*; not Linnaeus, 1758). Donovan, 1800: pl. 20, fig. 2 (*s.l.*; not Linnaeus, 1758). Montagu, 1803: 467–468 (*s.l.*; not Linnaeus, 1758). Maton & Rackett, 1807: 226–227; pl. 5, fig. 15 (*s.l.*; not Linnaeus, 1758). Pennant, 1812: 346; pl. 90, fig. 3 (*s.l.*; not Linnaeus, 1758). Pulteney, 1813: 57; pl. 16, figs 13–16; pl. 20, figs 2, 3 (*s.l.*; not Linnaeus, 1758). Dillwyn, 1817: 989 (*s.l.*; not Linnaeus, 1758). Wood, 1818: 171 (*s.l.*; not Linnaeus, 1758). Wood, 1828: 171; pl. 35, *Nerita* fig. 27 (*s.l.*; not Linnaeus, 1758). Leach, 1852: 180–181 (*s.l.*; not Linnaeus, 1758).

Neritoides littoralis—Brown, 1827: pl. 43, figs 14, 21, 22 (in part, other fig. is *L. fabalis*; not *Nerita littoralis* Linnaeus, 1758). Brown, 1843a: 25–26; pl. 13, figs 14, 21, 22 (in part, other fig. is *L. fabalis*; not Linnaeus, 1758). Winckworth, 1922: 97 (*s.l.*; not Linnaeus, 1758).

Littorina littoralis—Rang, 1829: 185 (*s.l.*; not *Nerita littoralis* Linnaeus, 1758; possibly *L. littorea* was intended, Kadolsky, 1970). Clark, 1850b: 29–31 (*s.l.*; not Linnaeus, 1758). Forbes & Hanley, 1850: 45–48, 53; pl. 84 (not 74 as cited), figs 3–7 (*s.l.*; not Linnaeus, 1758). Clark, 1855: 347–348 (*s.l.*; not Linnaeus, 1758). Reeve, 1857: sp. 34; pl. 7, fig. 34a, b (not Linnaeus, 1758). Jeffreys, 1859: 195 (in part, includes *L. fabalis*; not Linnaeus, 1758). Sowerby, 1859: pl. 12, figs 20, 21 (not Linnaeus, 1758). Smith, 1860: 9–10 (not Linnaeus, 1758). Jeffreys, 1865: **3**: 360–361 (not Linnaeus, 1758). Petit de la Saussaye, 1869: 121 (*s.l.*; not Linnaeus, 1758). Smith & Prime, 1870: 393 (not Linnaeus, 1758). Houssay, 1884: 180–186; pl. 7, figs 1–7 (operculum and foot; not Linnaeus, 1758). Winckworth, 1933: 335–336 (*s.l.*; not Linnaeus, 1758). Fretter & Graham, 1962: figs 202, 220 (spawn, embryo) (*s.l.*; not Linnaeus, 1758). James, 1968b: 633; fig. 9; fig. 22 (head pigmentation); fig. 41 (*s.l.*; not Linnaeus, 1758). Daguzan, 1977: 17–40 (*s.l.*; not Linnaeus, 1758).

Littorina (Littorina) littoralis—Tryon, 1887: 242; pl. 41, figs 5–8 (not *Nerita littoralis* Linnaeus, 1758).

Littorina obtusata littoralis—Dautzenberg & Durouchoux, 1913: 24 (not *Nerita littoralis* Linnaeus, 1758). Dautzenberg & Fischer, 1915: 111–112; pl. 2, fig. 10 (not Linnaeus, 1758). van Benthem Jutting, 1933: 64–66; fig. 28; fig. 29 (spawn) (*s.l.*; not Linnaeus, 1758). Kaas & ten Broek, 1942: 147–148; pl. 11, fig. 13a, b (*s.l.*; not Linnaeus, 1758).

Littorina (Neritoides) obtusata littoralis—Dautzenberg & Fischer, 1925: 56–58 (*s.l.*; not *Nerita littoralis* Linnaeus, 1758).

Littorina (Neritoides) littoralis—Winckworth, 1932: 221 (*s.l.*; not *Nerita littoralis* Linnaeus, 1758). Lebour, 1937: 128 (*s.l.*; not Linnaeus, 1758).

Turbo neritoides—Chemnitz, 1781: 234–235; pl. 185, figs 1854, 1–8, 11 (non-binominal, *Turbo neritoides* in synonymy; in part, other figs probably *L. fabalis*; not Linnaeus, 1758=*Melarhaphe neritoides*). Schröter, 1784: 4–5 (*s.l.*; not Linnaeus, 1758). Gmelin, 1791: 3588 (*s.l.*; not Linnaeus, 1758). Pulteney, 1799: 44–45 (*s.l.*; not Linnaeus, 1758). Link, 1807: 134 (*s.l.*; not Linnaeus, 1758; as *neritoideus*). Lamarck, 1822: 48 (*s.l.*; not Linnaeus, 1758). Blainville, 1828: 109 (*s.l.*; not Linnaeus, 1758). Blainville, 1830: 300–301; pl. 12, fig. 8 (*s.l.*; not Linnaeus, 1758). Thorpe, 1844: 171–172 (*s.l.*; not Linnaeus, 1758).

Litorina neritoides—Menke, 1828: 25 (presumably not *Turbo neritoides* Linnaeus, 1758). Menke, 1830: 45 (not Linnaeus, 1758). Anton, 1838: 53 (*s.l.*; not Linnaeus, 1758).

Littorina neritoides—Johnston, 1842: 269 (*s.l.*; not Linnaeus, 1758). Deshayes & Milne Edwards, 1843: 214–216 (*s.l.*; not Linnaeus, 1758). de Kay, 1843: 105–106; pl. 6, figs 109a, b, 110a, b, 111a, b (not Linnaeus, 1758). MacGillivray, 1843: 140 (*s.l.*; not Linnaeus, 1758). Brown, 1844a: 128; pl. 13, figs 14, 21, 22 (in part, other figs are *L. fabalis*; not Linnaeus, 1758). Deshayes, 1850: 42; pl. 69, figs 6, 7 (*s.l.*; not Linnaeus, 1758). Petit de la Saussaye, 1852: 183 (*s.l.*; not Linnaeus, 1758).

Turbo littoreus—Montagu, 1803: 301–303 (*s.l.*; in part, includes *L. littorea*; not Linnaeus, 1758).

Nerita litoralis Turton, 1819: 126 (unjustified emendation of *Nerita littoralis* Linnaeus, 1758, because on p. 196 *litorea* is spelled in the same way, ICZN, 1985: art. 33b; however, the two spellings are deemed identical and are homonyms, ICZN, 1985: art. 58; *s.l.*; not *Nerita littoralis* Linnaeus, 1758).

Litorina litoralis—Weinkauff, 1878: 29–30 (in part, includes ref. to *L. fabalis*; not *Nerita littoralis* Linnaeus, 1758). Weinkauff, 1883: 216 (not Linnaeus, 1758).

Turbo retusus Lamarck, 1822: 48–49 (les côtes de la Manche, près de Calais [France]; 4 syntypes MHNG 1096/71, photograph of syntype 2 seen; not Brocchi, 1814). Blainville, 1828: 109. Blainville, 1830: 299–300; pl. 12, fig. 6.

Litorina retusa—Menke, 1828: 25. Menke, 1830: 45.

Paludina retusa—Jeffreys, 1839: 33 (*s.l.*).

Littorina retusa—Deshayes & Milne Edwards, 1843: 215. Alder, 1848: 151 (*s.l.*). Petit de la Saussaye, 1852: 182 (*s.l.*).

Littorina obtusata littoralis var. ex forma *retusa*—Dautzenberg & Durouchoux, 1913: 24. Dautzenberg & Fischer, 1915: 113–114; pl. 2, figs 13, 14. Kaas & ten Broek, 1942: 147; pl. 11, figs 14a, b, 15, 16.

Turbo palliatus Say, 1822: 240–241 (shores of the New England states; lectotype (here designated) 9.0 mm, cream, Fig. 79M,+2 paralectotypes, ANSP 18420, seen; one of paralectotypes is a tessellated *L. saxatilis*).

Littorina palliata—Gould, 1841: 260–262; fig. 177* (not 176 as cited). de Kay, 1843: 106; pl. 6, fig. 105a–c. Forbes & Hanley, 1850: 51–52; pl. 84, figs 8–10. Forbes & Hanley, 1852: 262. Thompson, 1852: 75–76. Reeve, 1857: sp. 31; pl. 7, fig. 31a, b. Sowerby, 1859: pl. 12, fig. 24. Gould, 1870: 309–310; fig. 578. Tryon,

1873–74: 80; fig. 148. Sars, 1878: 165–167, 447; pl. 9, fig. 9a, b; anatomical pl. 6, fig. 5 (radula). Herzenstein, 1885: 680. Locard, 1886: 282–283. Kobelt, 1888: 168 (in part, includes ref. to *L. fabalis*). Locard, 1891: 187–188. Provancher, 1891: 92–93. Posselt & Jensen, 1898: 230–231. Kobelt, 1907: 62–63; pl. 110, figs 21, 22; pl. 112, figs 7–11, 21 (in part, other fig. is *L. fabalis*). Schlesch, 1916: 279–280 (*s.l.*; in part, includes *L. saxatilis*). Bárdarson, 1920: 89–90 (*s.l.*). Harmer, 1921: 663–664 (in part, figs are probably *L. littorea*). Thorson, 1941: 32–33. Fischer, 1942: 187. Hubendick & Warén, 1975: 10; fig. 273. Backeljau, de Meyer, Janssens & Proesmans, 1984: 3; fig. 4A.

Litorina palliata—Philippi, 1848: **3**: 68; *Litorina* pl. 7, figs 27–29. Bárdarson, 1919: 63 (*s.l.*).

Littorina (Neritrema) palliata—Récluz, 1869: 46 (*s.l.*).

Littorina obtusata var. *palliata*—Daniel, 1883: 341 (*s.l.*). Sparre-Schneider, 1886: 10–12. Norman, 1902: 355.

Littorina obtusata [subsp. *obtusata*] var. ex forma *palliata*—Dautzenberg & Fischer, 1915: 107; pl. 2, fig. 4.

Litorina obtusata palliata—Johnson, 1915: 121.

Littorina (Neritoides) palliata—Dall, 1923: 3.

Littorina (Littorina) obtusata var. *palliata*—Nordsieck, 1968: 40.

Littorina (Littorina) obtusata form *palliata*—Nordsieck, 1982: 61; pl. 26, fig. 23.00c.

Littorina (Neritrema) obtusata var. *palliata*—Smith & Heppell, 1991: 17, 81.

Turbo neritiformis Brown, 1827: pl. 46, fig. 24 (near Downpatrick, Lough Strangford, Ireland; types not traced; probable identification, see Taxonomic History, p. 205). Brown, 1837a: 17; pl. 10, fig. 24.

Littorina neritiforma—Brown, 1844a: 128; pl. 10, fig. 24.

Littorina obtusata var. *neritiformis*—Jeffreys, 1865: **3**: 357.

Littorina (Neritoides) obtusata var. ex forma *neritiformis*—Dautzenberg & Durouchoux, 1900: 46.

Turbo ustulatus—Blainville, 1830: 301; pl. 12, fig. 7 (in part, includes *L. littorea*; not Lamarck, 1822=*L. littorea*).

Littorina ustulata—Locard, 1886: 282, 578 (in part, includes ref. to *L. littorea*; not *Turbo ustulatus* Lamarck, 1822). Locard, 1891: 187 (not Lamarck, 1822).

Littorina (Littorina) obtusata var. *ustulata*—Nordsieck, 1968: 40 (not *Turbo ustulatus* Lamarck, 1822).

Paludina retusa (var. *zonata*) Jeffreys, 1839: 33 (*nomen nudum*; *s.l.*).

Litorina arctica Møller, 1842: 82 (Greenland; lectotype (here designated) 10.9 mm, Schiøtte & Warén, 1992: fig. 24,+4 paralectotypes, ZMK GAS-2, seen, all are *L. obtusata*; 5 paralectotypes BMNH 1843.6.30.90–94, seen, all are *L. obtusata*; 2 possible paralectotypes SMNH 3844, seen, 1 is *L. obtusata*, 1 is *L. fabalis*). Philippi, 1848: **3**: 68; *Litorina* pl. 7, figs 24, 25 (in part, other fig. is *L. saxatilis*). Küster, 1853: pl. 2, figs 14, 15; 1856: 15.

Littorina arctica—Sars, 1851: 180–181 (*s.l.*). Reeve, 1857: sp. 69 (in part, fig. is *L. sitkana*, includes *L. saxatilis*). Kobelt, 1888: 168–169. Kobelt, 1907: 63–64; pl. 111, fig. 6.

Littorina (Neritoides) arctica—H. & A. Adams, 1854: **1**: 314.

Littorina (Neritrema) arctica—Récluz, 1869: 46–47.

Littorina (Neritrema) obtusata var. *arctica*—Smith & Heppell, 1991: 17, 81.

Litorina obtusata [var.] *unicolor* Menke, 1845: 55–58 (Helgoland [Germany]; almost all figures cited are of *L. obtusata*; lectotype (here designated) Chemnitz, 1781: pl. 185, fig. 1853).

Litorina obtusata [var.] *fasciata* Menke, 1845: 56–58 (Helgoland [Germany]; figures cited are mostly of certain *L. obtusata*; lectotype (here designated) Chemnitz, 1781: pl. 185, fig. 1854c, d; not *Littorina fasciata* Gray, 1839=*Littoraria fasciata*).

Litorina obtusata [var.] *zonata* Menke, 1845: 56–58 (Helgoland [Germany]; figures cited are mostly of certain *L. obtusata*; lectotype (here designated) Klein, 1753: pl. 1, fig. 25).

Littorina obtusata littoralis var. ex col. *zonata*—Dautzenberg & Fischer, 1915: 120–121; pl. 3, fig. 1; pl. 4, fig. 6 (colour chart).

Litorina ioides Menke, 1845: 57 (Helgoland [Germany]; in synonymy of *Litorina obtusata*; therefore unavailable).

Littorelaea pultneyii Leach, in Gray, 1847a: 271 (possibly intended as a new name for *Turbo neritoides* Pulteney, 1813, not Linnaeus, 1758; *nomen nudum*).

Littorina obtusata forma *globosa* Middendorff, 1849: 386–387 (refers to de Kay, 1843: pl. 6, figs 110, 111; infrasubspecific; not *Litorina globosa* Dunker, 1845=*L. striata*).

Littorina obtusata forma *elatiorangulata* Middendorff, 1849: 386–387 ; pl. 8, figs 26–28 (as *elatior angulata*; infrasubspecific).

Littorina tenebrosa var. hybr. *obtusataea* Middendorff, 1849: 391; pl. 8, figs 1–5 (infrasubspecific).

Littorina tenebrosa forma *elatiorcrassa* Middendorff, 1849: 598; pl. 8, figs 22, 23 (as *elatior crassa*; infrasubspecific).

Littorina rudis—Clark, 1850a: 362–364 (in part, includes *L. saxatilis* and *L. fabalis*; not *Turbo rudis* Maton, 1797=*L. saxatilis*).

Littorina arctica var. *elatior* M. Sars, 1851: 181 (Tromsö and Havösund [Norway]; types not traced; not *L. rudis* forma *elatior* Middendorff, 1849 (infrasubspecific)=*L. saxatilis*).

Littorina palliata forma *elatior*—Sars, 1878: 166; pl. 21, fig. 19. Backeljau, de Meyer, Janssens & Proesmans, 1984: 3; fig. 4B, C.

Littorina obtusata var. *palliata* forma *elatior*—Norman, 1902: 355.

Littorina obtusata [subsp. *obtusata*] var. ex forma *elatior*—Dautzenberg & Fischer, 1915: 106; pl. 2, fig. 3.

Littorina palliata var. *elatior*—Schlesch, 1916: 279.
Littorina (Littorina) obtusata form *elatior*—Nordsieck, 1982: 61; pl. 26, fig. 23.00b.
Littorina arctica var. *angulata* M. Sars, 1851: 181 (Tromsö [Norway]; types not traced).
Littorina obtusata monst. *angulatum*—Schlesch, 1930: 217 (probably not intended as introduction of a new taxon, since no type mentioned).
Littorina groenlandica var. *laevior* Mørch, 1857: 80 (Greenland; refers to *Nerita littoralis* of Müller, 1776, and Fabricius, 1780; types not traced; *s.l.*).
Littorina peconica S. Smith, 1860: 9–10 (Peconic Bay, Long Island, New York; types not present in AMNH or Staten Island Institute, perhaps destroyed by fire in collection of Lyceum of Natural History of New York in 1866, W. K. Emerson and W. E. Sage, pers. comm.).
Littorina littoralis var. *peconica*—Smith & Prime, 1870: 393.
Littorina obtusata var. *ornata* Jeffreys, 1865: **3**: 357–358; 1869: **5**: pl. 65, fig. 1a (Isle of Wight and Southampton [England]; 2 syntypes USNM 185826, seen; *s.l.*). Daniel, 1883: 341 (*s.l.*). Sowerby, 1887: pl. 12, fig. 24.
Littorina obtusata var. ex col. *ornata*—Dautzenberg, 1893: 20 (*s.l.*). Dautzenberg & Durouchoux, 1900: 46.
Littorina obtusata [subsp. *obtusata*] var. ex col. *ornata*—Dautzenberg & Fischer, 1915: 109–110; pl. 2, fig. 7; pl. 4, fig. 6 (colour chart).
Littorina obtusata var. *aestuarii* Jeffreys, 1869: **5**: 205–206; pl. 101, fig. 8 (River Deben at Shottisham Creek near Sutton [Suffolk, England]; Manningtree, Suffolk; lectotype (here designated) USNM 185836, 9.6 mm, Jeffreys, 1869: pl. 101, fig. 8, Fig. 78J herein,+1 paralectotype, seen, no locality; about 100 paralectotypes, USNM 185837, seen, 1 is *L. saxatilis*; 6 paralectotypes, USNM 185838, seen; 32 paralectotypes, USNM 185839, seen; 10 paralectotypes, USNM 185850, seen, 2 are *L. saxatilis*; 2 paralectotypes, USNM 206059, seen; 4 paralectotypes, MCZ 89668, not seen; 8 paralectotypes, BMNH 1868.5.16.1, seen). Tryon, 1887: 253; pl. 46, fig. 23. Marshall, 1898: 121.
Littorina aestuarii—Jeffreys, 1869: **5**: caption, pl. 101, fig. 8. Doughty, 1934: 284. Fretter & Graham, 1980: 262.
Neritoides aestuarii—Winckworth, 1922: 97.
Littorina (Neritoides) aestuarii—Winckworth, 1932: 221.
Littorina (Neritrema) obtusata var. *aestuarii*—Smith & Heppell, 1991: 17, 81.
Littorina palliata monstr. *coarctata* G. O. Sars, 1878: 166; pl. 21, fig. 20 ([Arctic Norway]; monstrosity, therefore unavailable). Schlesch, 1916: 280. Óskarsson, 1973: 155–157; figs 2, 3.
Littorina obtusata var. *palliata* forma *coarctata*—Norman, 1902: 355.
Littorina obtusata [subsp. *obtusata*] monstr. *coarctatum*—Dautzenberg & Fischer, 1915: 126; pl. 3, figs 10, 11; text-fig. 2.
Littorina palliata form *coarctata*—Backeljau, de Meyer, Janssens & Proesmans, 1984: 3; fig. 5C, D.
Littorina obtusata var. *orana* 'Jeffreys' Norman, 1888: 14 (error for *ornata*; unavailable).
Littorina obtusata var. ex colore *citrina* Dautzenberg, 1893a: 20 (Granville, St-Pair [France]; infrasubspecific; *s.l.*).
Littorina obtusata littoralis coloration typique [var. ex col. *citrina*]—Dautzenberg & Fischer, 1915: 116–117; pl. 4, fig. 1 (colour chart).
Littorina obtusata var. ex colore *aurantia* Dautzenberg, 1893a: 20 (Granville, St-Pair [France]; infrasubspecific; not *L. rudis* var. ex col. *aurantia* Dautzenberg, 1887=*L. saxatilis*; *s.l.*).
Littorina obtusata littoralis var. ex col. *aurantia*—Dautzenberg & Fischer, 1915: 117–118; pl. 4, fig. 2 (colour chart) (not *L. rudis* var. ex col. *aurantia* Dautzenberg, 1887; *s.l.*).
Littorina obtusata monstr. *planispira* Martel, 1901: 111, 221; pl. 1, fig. 7 (Cancale [France]; 2 original specimens IRSNB, seen; monstrosity, therefore unavailable).
Littorina obtusata littoralis monstr. *planispira*—Dautzenberg & Fischer, 1915: 127; text-fig. 3.
Littorina obtusata [subsp. *obtusata*] var. ex col. *lutea* Dautzenberg & Fischer, 1915: 108; pl. 4, fig. 2 (colour chart) (Casco-Bay, Maine; Saint-Pierre-Miquelon; infrasubspecific; not *Litorina scabra* var. *lutea* Philippi, 1847=*Littoraria lutea*; not *L. rudis* var. ex col. *lutea* Dautzenberg & Durouchoux, 1900=*L. saxatilis*).
Littorina obtusata [subsp. *obtusata*] var. ex col. *castanea* Dautzenberg & Fischer, 1915: 108–109; pl. 4, fig. 3 (colour chart) (Tromsö; Vardö; Greenland; Saint-Pierre-Miquelon; infrasubspecific; not *L. castanea* Deshayes, 1843=*L. saxatilis*).
Littorina obtusata [subsp. *obtusata*] var. ex col. *balteata* Dautzenberg & Fischer, 1915: 109; pl. 2, fig. 5; pl. 4, fig. 4 (colour chart) (Vardö [Norway]; 1 spec., original of fig. 5, in MNHNP, seen; infrasubspecific; not *L. balteata* Reeve, 1857=*L. brevicula*).
Littorina obtusata [subsp. *obtusata*] var. ex col. *ambigua* Dautzenberg & Fischer, 1915: 109; pl. 2, fig. 6; pl. 4, fig. 5 (colour chart) (Hamerfest [Hammerfest, Norway]; Tromsö [Norway]; 1 spec., original of fig. 6, in MNHNP, seen; infrasubspecific; not *Litorina ambigua* 'Nuttall' Philippi, 1848=*Littoraria intermedia* (Philippi, 1846)).
Littorina obtusata [subsp. *obtusata*] var. ex col. *lineolata* Dautzenberg & Fischer, 1915: 110; pl. 2, fig. 8; pl. 4, fig. 7 (colour chart) (Hamerfest; Tromsö; Casco-Bay [Maine]; 1 spec., original of fig. 8, in MNHNP, seen; infrasubspecific; not *L. lineolata* d'Orbigny, 1840=*Nodilittorina lineolata*).
Littorina obtusata [subsp. *obtusata*] var. ex col. *tessellata* Dautzenberg & Fischer, 1915: 110–111; pl. 2, fig. 9; pl. 4,

fig. 8 (colour chart) (Vardö; Tromsö; Saint-Pierre-Miquelon; 1 spec., original of fig. 9, in MNHNP, seen; infrasubspecific; not *Litorina tessellata* Philippi, 1847=*Littoraria tessellata*; not *L. tenebrosa* var. ex col. *tessellatus* Middendorff, 1849=*L. saxatilis*).

Littorina obtusata littoralis var ex forma *compacta*—Dautzenberg & Fischer, 1915: 114–115; pl. 2, fig. 15 (not *L. obtusata* var. *compacta* Jeffreys, 1865, which is probably *L. fabalis*).

Littorina obtusata littoralis var. ex col. *rubens* 'Menke' Dautzenberg & Fischer, 1915: 118; pl. 4, fig. 3 (colour chart) (Baie de Saint-Malo [France]; Jersey; l'Île Dumet; Hellebock, Denmark; infrasubspecific; *s.l.*).

Littorina obtusata littoralis var. ex col. *fusca* 'Menke' Dautzenberg & Fischer, 1915: 118–119; pl. 2, fig. 15; pl. 4, fig. 4 (colour chart) (Croisic; La-Trinité-sur-Mer; Saint-Lunaire; Saint-Malo; Cherbourg; Saint-Vaast; Ouistreham; Dieppe [all France]; Jersey; Svolvaer, Lofoten Isles [Norway]; 1 spec. from Porsanger Fjord [Norway], original of fig. 15, MNHNP, seen; in part, other fig. is probably *L. fabalis*; infrasubspecific; not *L. fusca* Pfeiffer, 1840=*Nodilittorina mespillum* (Mühlfeldt, 1824)).

Littorina obtusata littoralis var. ex col. *olivacea* Dautzenberg & Fischer, 1915: 119–120; pl. 2, figs 13, 14; pl. 4, fig. 5 (colour chart) (Val-André; Saint-Malo; la Guimorais [all France]; Lofoten Isles; 2 specs from Le Pouliguen [France], originals of figs 13, 14, MNHNP, seen; infrasubspecific).

Littorina obtusata littoralis var. ex col. *alternata* Dautzenberg & Fischer, 1915: 121; pl. 3, fig. 2; pl. 4, fig. 7 (colour chart) (La Trinité-sur-Mer; Croisic; Locmariaker; Saint-Lunaire; l'Île Harbour; l'embouchure de la Rance; la Pointe de Rochebonne [all France]; Lofoten Isles; 1 spec., original of fig. 2, MNHNP, seen; infrasubspecific).

Littorina obtusata littoralis var. ex col. *inversicolor* Dautzenberg & Fischer, 1915: 122–123; pl. 3, fig. 3; pl. 4, fig. 8 (colour chart) (La Trinité-sur-Mer [France]; 1 spec., original of fig. 3, MNHNP, seen; in part, other figs are *L. fabalis*; infrasubspecific).

Littorina obtusata littoralis var. ex col. *rhabdota* Dautzenberg & Fischer, 1915: 123; pl. 3, fig. 4; pl. 4, fig. 9 (colour chart) (Lofoten Isles; 1 spec., original of fig. 4, MNHNP, seen; infrasubspecific).

Littorina obtusata littoralis monstr. *contabulatum* Dautzenberg & Fischer, 1915: 127; pl. 3, fig. 12 (Quiberon [France]; 1 spec., original of fig. 12, MNHNP, seen; monstrosity, therefore unavailable; not *L. rudis* var. *globosa* monstr. *contabulata* Martel, 1901=*L. saxatilis*).

Littorina palliata form *contabulata*—Backeljau, de Meyer, Janssens & Proesmans, 1984: 3; fig. 5A, B.

Littorina obtusata littoralis monstr. *carinatum* Dautzenberg & Fischer, 1915: 127–128; pl. 3, fig. 13 (La Trinité-sur-Mer [France]; 1 spec., original of fig. 13, MNHNP, seen; monstrosity, therefore unavailable; not *Delphinula carinata* Woodward, 1833, nor *Turbo carinata* Woodward, 1833, both=*L. littorea*).

Littorina obtusata littoralis monstr. *exsertum* Dautzenberg & Fischer, 1915: 128; pl. 3, fig. 14 (no loc.; 1 spec., original of fig. 14, MNHNP, seen; monstrosity, therefore unavailable).

Littorina obtusata littoralis monstr. *umbilicatum* Dautzenberg & Fischer, 1915: 128; pl. 3, fig. 15 (Locmariaker [France]; 1 spec., original of fig. 15, MNHNP, seen; monstrosity, therefore unavailable; not *L. umbilicata* d'Orbigny, 1840).

Littorina palliata var. *carinata* Schlesch, 1916: 279 (Isafjördur, Iceland; 1 syntype NHMR, seen; 1 syntype TM, seen; further syntypes may be in Hull Museum; not *Delphinula carinata* Woodward, 1833, nor *Turbo carinata* Woodward, 1833, both=*L. littorea*).

Littorina palliata var. *auricularia* Schlesch, 1916: 279 (Isafjördur, Iceland; types may be in Hull Museum).

Littorina (Neritoides) obtusata littoralis var. ex-col. *albescens* Dautzenberg & Fischer, 1925: 56 (Ile Verte, Roscoff; Île de Sieck [both France]; infrasubspecific; *s.l.*).

Littorina obtusata var. ex col. *virescens* 'Dautzenberg' Schlesch, 1930: 216–217 (Húnaflói, Iceland; infrasubspecific).

Littorina obtusata var. ex col. *albida* 'Dautzenberg' Schlesch, 1930: 217 (Húnaflói, Iceland; infrasubspecific; not *Litorina albida* Philippi, 1848=*Nodilittorina modesta* (Philippi, 1846); not *Litorina tenebrosa* var. *albida* Schrenck, 1867=*L. sitkana*; *s.l.*).

Littorina obtusata var. ex col. *pulchra* 'Dautzenberg' Schlesch, 1930: 217 (Húnaflói, Iceland; infrasubspecific; not *Littorina pulchra* Sowerby, 1832=*Littoraria zebra* (Donovan, 1825); *s.l.*).

Littorina obtusata form *magnei* Sacchi, 1961: 273–274; fig. 1, top left (not as in caption) (Roscoff, France; infrasubspecific; *s.l.*).

Littorina obtusata form *torquata* Sacchi, 1963: 310–312; fig. 2 (Roscoff, France; infrasubspecific).

Littorina mariae—Poppe & Goto, 1991: 92; pl. 11, figs 4, 8 (in part, other figs are *L. fabalis*; not Sacchi & Rastelli, 1966=*L. fabalis*).

Note: In the synonymy '*s.l.*' (*sensu lato*) following a reference indicates that the author's concept of *L. obtusata* probably included *L. fabalis*, but without specific reference to, or a figure of, the latter. In the text, '*L. obtusata s.l.*' is used where both species may have been included, in contrast to '*L. obtusata*' or, for emphasis, '*L. obtusata s.s.*' (*sensu stricto*) where identification is not in doubt.

Taxonomic history

The synonymy of *L. obtusata* is a complicated one, for several reasons. There has been persistent confusion of the Linnaean taxa *Turbo obtusatus*, *T. neritoides* and *Nerita littoralis*. The species is highly variable in shape and colour, so that the synonymy contains six species-level taxa, and 19 subspecies, forms, varieties and monstrosities based on variations in shape, and 25 subspecific and infrasubspecific taxa based on colour varieties. In particular, the relatively tall-spired northern, American and estuarine forms have often been treated as taxonomically distinct. To add to the confusion, in 1966 it was recognized that another conchologically-similar species, *L. fabalis* (named *L. mariae* by Sacchi & Rastelli, 1966), was present in Europe, which had been overlooked or included with *L. obtusata* by the majority of previous authors.

Linnaeus listed his *Turbo obtusatus* as the first of the four species in the section 'Neritoidei' at the beginning of the genus *Turbo*, the other three being (with their modern genera) *Melarhaphe neritoides*, *Littorina littorea* and *Cenchritis muricatus* (all members of the Littorinidae). The description of the species was brief: 'shell roundish, smooth, very obtuse and more ventricose above; margin of the pillar flat' (translation of the Latin by Dillwyn, 1817) and the locality simply noted as 'Northern Ocean'. No references to other published works were supplied, and nothing further was added in the twelfth edition of the *Systema Naturae* (Linnaeus, 1767). Nevertheless, these details were apparently sufficient for its recognition, since even the earliest authors (e.g. Müller, 1776; Chemnitz, 1781; Schröter, 1784) applied the name correctly. This concept of the species was established beyond doubt by Hanley (1855), who figured the 'marked specimen' from the Linnaean shell collection. The LSL collection actually contains 11 shells, and the one marked '526' (the number of the species in edition 10 of the *Systema Naturae*; whether or not actually inscribed by Linnaeus himself, Dance, 1967) is here designated the lectotype (Fig. 79A). The type series consists of shells of the taller-spired northern form, in agreement with the locality given by Linnaeus and with the subsequent restriction of the type locality to the coast of Lappland (Bequaert, 1943).

Turbo neritoides was also given an inadequate diagnosis, but the adjective 'minute' in the subdescription, the locality in the Mediterranean, and the reference to Gualtieri's acceptable figure (1742: pl. 45, fig. F; see Boss, 1988, for explanation of references cited by Linnaeus) are entirely sufficient to identify the species now classified as *Melarhaphe neritoides*. There are three possible syntypes in the Linnaean shell collection.

The determination of *Nerita littoralis* presents greater difficulty. It was listed in the section of *Nerita* 'imperforatae labio edentule', together with species now assigned to the genera *Clithon*, *Neritopsis*, *Neritodryas*, *Theodoxus* and *Thais*. Significantly, it was placed between the species *fluviatilis* and *lacustris*, both now recognized as forms of the fresh-water and brackish-water neritid *Theodoxus fluviatilis*. Throughout his arrangement of the molluscs, Linnaeus tended to group similar species together within his genera. Again the description was brief: 'shell smooth, apex carious, lips toothless'; in the subdescription the habitat was given as rocky shores of European seas, with the additional note 'variegated in innumerable ways'. The description of a carious (or decayed) apex is noteworthy (although dismissed by both Schröter, 1784, and Gmelin, 1791); the apex of *Theodoxus fluviatilis* is frequently eroded away, whereas that of *L. obtusata* is usually intact. *Theodoxus fluviatilis* also has a variegated pattern, whereas that of *L. obtusata*, while variable, is not commonly variegated. In the subdescription of the following species, *Nerita lacustris*, the shell was stated to be similar to that of *N. littoralis*. The specifically marine habitat given by Linnaeus for *N. littoralis* has misled subsequent authors. However, it should be recognized that Linnaeus' local shores on the east coast of Sweden faced the Baltic Sea, here too brackish for the survival of *Littorina* species, but abundantly populated by *T. fluviatilis*. Perhaps Linnaeus supposed that other European shores were similarly inhabited. Among the five references supplied by Linnaeus (1758) for *N. littoralis*, he quoted two of his own works (see Boss, 1988). The reference to the *Fauna Suecica* (Linnaeus, 1746) indicated two species, one from lakes and one marine, which in Sweden could apply to *T. fluviatilis*. Linnaeus there described the animal as having 'two bristles, like as many feet, projecting on each side'; this odd description misled several later authors to suppose that the foot bore cirrous projections (Fabricius, 1780; Jeffreys, 1865; Winckworth, 1933), but it could well apply to the long, slender, cephalic tentacles of *T. fluviatilis*. The reference to the *Gothländska Resa* (Linnaeus, 1745) is unambiguous, for only *T. fluviatilis*, and not *L. obtusata*, is found on the shores of Gotland (Mørch, 1864). The citation of '*List. angl.* 164. *t.* 2. *f.* 3' by Linnaeus (1758) refers to Lister (1678), on page 164 of which is described '*Nerita ex fusco viridescens...* ', but this species is actually figured on pl. 3, fig. 11, which is a recognizable (but reversed) engraving of *L. obtusata*. Lister's pl. 2, fig. 3, is a *Cepaea nemoralis*, and this was obviously misquoted by Linnaeus (1758). This error was noted by both Hanley (1855) and Dodge (1959) in their studies of the Linnaean species, but neither recognized the correct citation of pl. 3, fig. 11. The second reference to Lister given by Linnaeus was '*List. conch. t.* 4. *s.* 8. *f.* 39' (i.e. Lister, 1688). Again the figure was misquoted, for that indicated is a nodulose trochid or littorinid from Barbados, and again both Hanley (1855) and Dodge (1959) noted the error, but failed to correct it. The reference to Lister (1688) should clearly have been to section 6 (devoted to nerites), fig. 39, which is a *Littorina*, almost certainly *L. obtusata*. The final reference of Linnaeus (1758) under *N. littoralis* was to Petiver (1695–1703) and indicates '*Nerita anglicus maritimus*...', with the *correct* references to the two figures of Lister (1678, 1688), which it is here argued were the ones intended by Linnaeus himself. In the twelfth edition of the *Systema Naturae*, Linnaeus (1767) added two new references under *N. littoralis*. The first was to the new binominal edition of the *Fauna Suecica* (Linnaeus, 1761) and again indicates both a marine and a similar lacustrine variety of *N. littoralis*, suggesting that Linnaeus recognized the similarity of the freshwater and marine (Baltic) forms of *T. fluviatilis*. The second new reference was to Gualtieri (1742: pl. 4, fig. LL), in which both figures and text clearly indicate a freshwater *Theodoxus* species. Unfortunately, there is no specimen of *Nerita littoralis* among the authenticated material in the Linnaean shell collection (Dodge, 1959; Dance, 1967; although Hanley, 1855, misidentified '*Littorina (Turbo) neritoides* of Lamarck' (=*L. obtusata*) as this species). In summary, therefore, it appears that Linnaeus based his *Nerita littoralis* upon his own experience of the species now widely known under the name *Theodoxus fluviatilis*, familiar to him on the Baltic shores of Sweden. In searching for figures of this shell, he apparently made the incorrect assumption that it would also be found on other European shores, accounting for his citation of Lister's figures of the superficially similar *Littorina obtusata*. That Linnaeus did not recognize these figures as his *Turbo obtusatus* may be explained by the fact that his specimens of *T. obtusatus* were of the relatively tall-spired northern form (p. 204), the more nerite-like southern form illustrated by Lister being unknown to him. Strictly, therefore, Linnaeus indicated two species, currently recognized as *Theodoxus fluviatilis* and *Littorina obtusata*, in his description of *Nerita littoralis*, although it is here argued that he intended only the former. To stabilize this confusing state of affairs, a neotype of *Nerita littoralis* is here designated (7.9 mm, from Slesvik, Gotland, deposited in LSL). Surprisingly, the synonymy of *N. littoralis* with *N. fluviatilis* has been recognized by rather few previous authors (Mørch, 1864; Récluz, 1869; Westerlund, 1871, 1873; Johansen, 1901; Rasmussen, 1973; Fretter & Graham, 1978).

The separate identity of the Linnaean species *Turbo obtusatus*, *T. neritoides* and *Nerita littoralis* can thus be established from Linnaeus' own descriptions, material and references. Nevertheless, in the subsequent literature there was much confusion between these three. Both Müller (1776) and Fabricius (1780) applied the name *N. littoralis* to a species from Greenland (i.e. *L. obtusata*), presumably misled by Linnaeus' (1758) description of the species as marine and by his references to Lister (1668, 1678). This was followed by Pennant (1777), beginning a long tradition of using the specific name *littoralis* rather than (or in addition to) *obtusata* (e.g. da Costa, 1778; Donovan, 1800; Montagu, 1803; Maton & Rackett, 1807; Pennant, 1812; Pulteney, 1813; Turton, 1819; Brown, 1827, 1843a; Forbes & Hanley, 1850; Reeve, 1857; Sowerby, 1859; Smith, 1860; Petit de la Saussaye, 1869; Tryon, 1887). This was a largely British and to some extent French tradition, which continued well into the twentieth century (e.g. Dautzenberg & Fischer, 1915; Winckworth, 1933; Fretter & Graham, 1962; Daguzan, 1977).

The work of Chemnitz (1781) was non-binominal, but gave Linnaean names in the synonymies, and was influential and widely quoted because of its coloured and reasonably accurate figures. Chemnitz was the first to misidentify the present species as *Turbo neritoides*, which he distinguished from *Turbo obtusatus* by its less obtuse spire (although this difference is scarcely apparent from his figures, which appear to show a mixture of *L. obtusata* and *L. fabalis* under both the preceding names). In this, Chemnitz was followed by Schröter (1784) and Gmelin (1791), both of whom further suggested that Linnaeus had described the same species twice, under the names *Turbo neritoides* and *Nerita littoralis*. This misuse of the name *neritoides* for the species now known as *L. obtusata* was continued for some time, mainly by some British (e.g. Pulteney, 1799; Johnston, 1842; MacGillivray, 1843; Brown, 1844a; Thorpe, 1844) and French authors (e.g. Lamarck, 1822; Blainville, 1828, 1838; Deshayes & Milne Edwards, 1843; Petit de la Saussaye, 1852). Correct usage of *Turbo neritoides* was established by Dillwyn (1817) and followed by Wood (1818, 1828), although not until the works of Menke (1845) and Philippi (1847) did it gain general acceptance.

Meanwhile, some of these early nineteenth century authors continued to employ, in addition to *neritoides* or *littoralis*, the name *Turbo obtusatus* (or *L. obtusata*) (e.g. Müller, 1776; Chemnitz, 1781; Schröter, 1784; Gmelin, 1791; Pulteney, 1799; Dillwyn, 1817; Lamarck, 1822; Wood, 1818, 1828; Blainville, 1828, 1830; Menke, 1828, 1830; Anton, 1838). Following Chemnitz (1781), the name *T. obtusatus* was generally used for shells with flatter spires, although Gmelin (1791) and Lamarck (1822) limited it to the pale colour form with two brown bands. It is possible that in distinguishing the taller from the flatter shells (as *littoralis* or *neritoides*, and *obtusatus*, respectively) some of these authors may in fact have been separating the species now known as *L. obtusata* and *L. fabalis*, but in the absence of unequivocal descriptions or figures this cannot be ascertained. A broader use of the name *Littorina* (or *Litorina*) *obtusata* for the form of this species common on western European shores was established by Menke (1845, giving *littoralis* as a synonym) and Philippi (1846), and thereafter passed into general use (although, as discussed previously, *L. littoralis* was still favoured by some British and French workers).

In addition to these nomenclatural difficulties, the variation in shape of *L. obtusata* gave rise to taxonomic disagreements. A large but slightly immature form was named *Turbo retusus* by Lamarck (1822), but its separate recognition was not widely followed (e.g. Blainville, 1830; Deshayes & Milne Edwards, 1843). Perhaps the most striking aspect of shell variation in *L. obtusata* is the occurrence of a relatively taller-spired form in northern parts of the range (North America north of New England, Greenland, Iceland, Norway north of the Lofoten Isles, and the White Sea; see detailed discussion under Shell Variation, p. 208). This form was described from New England as *Turbo palliatus* by Say (1822) and from Greenland as *Litorina arctica* by Møller (1842). (*Litorina limata* of Lovén (1846) also included this form, but designation of a lectotype restricts the name to *L. fabalis*, as discussed under that species, p. 230). Many subsequent authors have considered that this northern form represents one or more distinct species. Philippi (1846–48), for example, recognized *L. obtusata*, *L. palliata* and *L. arctica* as three species from Europe, North America and the Arctic respectively, as did Reeve (1857), Récluz (1869) and Kobelt (1888, 1907). American authors also used the name *L. palliata* during the mid-nineteenth century (e.g. Gould, 1841, 1870; Tryon, 1873–74), while in Europe the northern form was at this time called *L. arctica* (Sars, 1851; Kuster, 1853, 1856). Despite similarity of the shells of the northern form to some examples of *L. saxatilis*, there was surprisingly little confusion between the two, although Middendorff (1849) did describe typical northern forms as hybrid variety *obtusataea* of *L. tenebrosa* (=*L. saxatilis*) and Mørch (1857) as variety *laevior* of *L. groenlandica* (=*L. saxatilis*). Jeffreys (1865) appears to have been the first to suggest that the northern American and Arctic forms were a single species (for which he used the name *L. littoralis*), distinct from the European form. This nomenclature was followed by Weinkauff (1878) and Tryon (1887), although most other authors of the time used the name *L. palliata* for the single northern species (Sars, 1878; Herzenstein, 1885; Posselt & Jensen, 1898). Some authors recorded typical *L. obtusata* together with this northern species in Arctic regions (e.g. Middendorff, 1849; Mørch, 1857; Sars, 1878; Herzenstein, 1885; Sparre-Schneider, 1886; Posselt & Jensen, 1898; Schlesch, 1916; Bárdarson, 1919, 1920; Thorson, 1941), and it is possible that some of these workers may have been discriminating between the species now recognized as *L. obtusata* and *L. fabalis*, but descriptions and figures are not adequate to show this. Eventually, it was realized that the northern taller-spired shells were conspecific with the common European form (Jeffreys, 1877; Sparre-Schneider, 1886). In a detailed description of shell variation in *L. obtusata s.l.*, Dautzenberg & Fischer (1915) recognized two subspecies, *L. obtusata obtusata* (the northern form) and *L. obtusata littoralis* (the southern form). (Their identification of the northern form as the typical subspecies rested upon Hanley's (1855) figure of one of Linnaeus' syntypes, which correctly showed that it was of the northern form, a fact overlooked by intervening authors, who mostly employed *L. obtusata* for the southern form). This classification was not, however, widely followed (although Winckworth, 1922, 1933, used the same names for northern and southern forms, considering them full species). In fact, since 1915, the majority of authors have used the name *L. obtusata* for this species throughout its range (see Synonymy), rarely mentioning the northern form, and then usually as *L. palliata* (Schlesch, 1916; Bárdarson, 1919, 1920; Dall, 1923; Thorson, 1941; Fischer, 1942; Óskarsson, 1973; Hubendick & Warén, 1975; Backeljau, de Meyer, Janssens & Proesmans, 1984) or a subspecies, variety or form of the same name (Johnson, 1915; Nordsieck, 1968, 1982; Smith & Heppell, 1991). The conclusion of conspecificity of northern and southern forms has been reinforced by radular studies (Johansen, 1901), shell morphometrics (Colman, 1932; Knudsen, 1949; Warmoes, 1986), and most recently by allozyme analysis (Seeley, 1986; Lewis & Williams, 1995) and a demonstration of natural selection acting on shell shape (Seeley, 1986).

In addition to the taller-spired northern form, *L. obtusata* can also exhibit taller-spired shells in estuarine habitats throughout its geographical range. This estuarine form is probably that first described as *Turbo neritiformis* by Brown (1827) from the inner parts of Strangford Lough, in northern Ireland. (The original figure is apparently poorly drawn, and the types are lost, so this identification is tentative, but other authors have also concluded that it is a form of *L. obtusata*, e.g. Forbes & Hanley, 1850; Jeffreys, 1865; the name has been little used). The relatively high spire of the estuarine form gives it a resemblance to the northern form, and indeed the name *L. palliata* has sometimes been applied to it in Europe (Forbes & Hanley, 1850; Thompson, 1852; Sowerby, 1859; Locard, 1886, 1891; Kobelt, 1888). Used in this sense, *L. palliata* was synonymized with *L. rudis* (=*L. saxatilis*) by Clarke (1855), whereas Thompson (1852) had suggested that it was a hybrid between *L. saxatilis* and *L. obtusata*. In 1865 Jeffreys named a banded shell from southern England with a produced spire *L. obtusata* var. *ornata* and this name has also been used for the estuarine form (Sowerby, 1887). Jeffreys (1869) also described the estuarine form as *L. obtusata* var. *aestuarii*, under which name it has been most commonly known (Tryon, 1887; Marshall, 1898; Smith & Heppell, 1991). Some authors have considered *aestuarii* to be a distinct species (Winckworth, 1922, 1932; Doughty, 1934; Fretter & Graham, 1980), but the allozyme analysis of Moyse, Thorpe & Al-Hamadani (1982) did not distinguish it from *L. obtusata*.

Even more extreme variations in shell shape occur commonly in some Arctic populations of *L. obtusata*, and some have been named as varieties or forms (Middendorff, 1849; M. Sars, 1851; G. O. Sars, 1878; Schlesch, 1916); these monstrosities are discussed later (p. 216). Occasional monstrous forms have also been named in southern populations (Martel, 1901; Dautzenberg & Fischer, 1915). In addition, the extreme polymorphism of shell colour and pattern in this species has led to a proliferation of varietal names, as listed in the Synonymy (p. 196). The pigmentation of the animal is also variable, and Smith (1860) based his *L. peconica* on a form with dark pigmentation.

The taxonomic history of *L. obtusata* has been further complicated by the recognition of the similar and largely sympatric species, *L. fabalis*. Although the latter was first described by Turton as long ago as 1825, and has received a number of other specific and varietal names, it was not until its description as *L. mariae* by Sacchi & Rastelli (1966) that it became widely recognized. The rediscovery of *L. fabalis* by Sacchi & Rastelli was prompted mainly by the observation of its distinctive penial shape, and at first it was considered difficult or impossible to distinguish from *L. obtusata* by shell characters. However, the range of shell variation in *L. fabalis* is now better known, and with experience the great majority of adult shells can be distinguished on sight. The differences in the shells are subtle and none alone is diagnostic (see p. 225), but they are sufficient to allow determination of many published figures and thus to construct synonymies of the two species. Before 1966 few workers used anatomical characters in their descriptions of these forms, and none appears to have separated them into two groupings corresponding entirely with the two species as currently recognized. In many cases it is clear from descriptions, figures or references that an author's concept of a particular taxon embraced both species, and these cases are here listed in the synonymies of both. Where determination is not clear (either because figures are ambiguous or descriptions too generalized) and in those monographic and faunistic works where it is considered probable that both species were included, the references are listed only in the synonymy of *L. obtusata*, with the note '*s.l.*'. The history of the discrimination of these two species is discussed more fully under *L. fabalis* (p. 229). Here it may simply be noted that before 1966 almost all authors working in the northern and northeastern Atlantic (where the two species are sympatric) confused *L. obtusata* and *L. fabalis* to some extent. From the mid-nineteenth to the early twentieth century, when fine subdivision of shells into species and varieties was the fashion, some authors recognized taxa comprising only *L. obtusata* or *L. fabalis* (but not in their entirety) (e.g. Brown, 1837a; Thorpe, 1844; Menke, 1845; Philippi, 1846–48; Forbes & Hanley, 1850; Küster, 1853, 1856; Reeve, 1857; Jeffreys, 1865; Tryon, 1887; Dautzenberg & Fischer, 1915). After 1915 the two species were not discriminated in any way and only *L. obtusata* (and its synonyms) appeared in the literature until 1966. Even following the description of *L. mariae* (=*fabalis*) (Sacchi & Rastelli, 1966) the recognition of both species gained acceptance slowly among systematists, and became widespread only after the publications of Goodwin & Fish (1977) and Fretter & Graham (1980).

Littorina obtusata and *L. fabalis* together form a distinctive group, and *L. obtusata* has seldom been confused with any other species. Pulteney (1799) apparently included *L. saxatilis* under the name

Turbo obtusatus, since he described it as having a sulcate shell, and Montagu (1803) included some forms of *L. obtusata* under *Turbo littoreus*. Harmer's (1921) Pleistocene records of *L. obtusata* and *L. palliata* are of worn specimens of *L. littorea*. Understandably, the taller-spired northern and estuarine forms of *L. obtusata* have occasionally been classified as varieties or synonyms of *L. saxatilis* (Middendorff, 1849; Clarke, 1855; Mørch, 1857) and vice versa (Schlesch, 1916), or have proved difficult to identify from shells alone (Herschberg, 1992). Fossils of the Pacific species *L. mandshurica* and *L. squalida* have occasionally been misidentified as *L. palliata*, and *L. sitkana* as *L. arctica*.

On account of its obtuse, nerite-like shell, *L. obtusata* has twice been made the type of a new genus, first of *Neritoides* (Brown, 1827, not Meuschen, 1779, see Kadolsky, 1971) and then of *Neritrema* (Récluz, 1869) (p. 145).

Diagnosis

Mature shell height 4.6–19.8 mm. Shell low-spired, globular or turbinate, shoulder often angled to give square profile to last whorl; relative columellar width (C/LA) less than 0.29. Mineralogy of external irregular-prismatic calcite and internal crossed-lamellar aragonite. Sculpture smooth or with microstriae over whole surface. Colour highly variable: olive-brown, yellow, orange, brown or blackish, sometimes with a fine reticulate or zigzag pattern, numerous fine spiral lines or occasionally one or two broad bands of contrasting colour; columella white, often tinged with dull purple-brown. Penis with short, triangular filament, 10–25% total length, 10–54 mamilliform penial glands. Pallial oviduct of single loop of albumen gland, single loop of capsule gland and swollen, septate jelly gland; bursa long, extending back to capsule gland. Spawn a benthic mass of singly-encapsulated eggs in gelatinous matrix; development nonplanktotrophic with intracapsular metamorphosis. Distribution: White Sea to southern Portugal, Iceland, southwestern Greenland, Labrador to Long Island.

Anatomical description

Head-foot. Pigmentation of head, tentacles and sides of foot varies from unpigmented to grey to entirely black. In lightly pigmented animals head and tentacles are usually darker than foot. Tentacles usually unpatterned, but can show diffuse transverse banding or a single dark longitudinal stripe. The degree of pigmentation shows some correlation with shell colour; dark reticulated, brown and blackish shells contain black animals, olive-brown shells usually have dark but sometimes unpigmented animals, and yellow shells usually unpigmented but rarely grey or black animals (Sacchi, 1961, 1969a, 1974; Smith, 1976; Rolán & Templado, 1987). Opercular ratio 0.44–0.62.

Penis (Fig. 76A–L). Base wrinkled, often with grey to black pigmentation on outer surface, sometimes unpigmented; 10–54 mamilliform penial glands in irregular double row, or close-packed. Filament short, triangular, 10–25% total length, unpigmented. Sperm groove open, ending terminally. Subepithelial glandular tissue around sperm groove is clearly visible only in living specimens. There is no evidence for seasonal regression or shedding of the penis in *L. obtusata* (Sacchi, 1969a; Daguzan, 1976b; Goodwin, 1978).

Sperm (Fig. 3G). Euspermatozoa 39–47 μm. Paraspermatozoa 11–16 μm diameter, round; vesicles large (to 6μm diameter), distinct; nucleus not visible.

Oviduct (Fig. 76M–Q). Single loop of albumen gland; followed by single loop of capsule gland of about half to the same area as albumen gland, translucent part of capsule gland small, not usually visible except in histological sections; finally a swollen, septate jelly gland, unpigmented or containing grey to blackish pigment. Bursa large, opening at anterior end of jelly gland and extending back at least to capsule gland (Reid, 1990a). Seminal receptacle often lobulate. The oviduct has previously been described by Linke (1933) and by Fretter (1980), but neither correctly described the spiral path of the egg groove, and the terminology of the glands differed from that used here.

Spawn and development. In Europe, where both *L. obtusata* and the similar *L. fabalis* occur, only in recent studies has care been taken to distinguish the spawn of each. These accounts show that females lay individual spawn masses that are oval (occasionally round and rarely kidney-shaped),

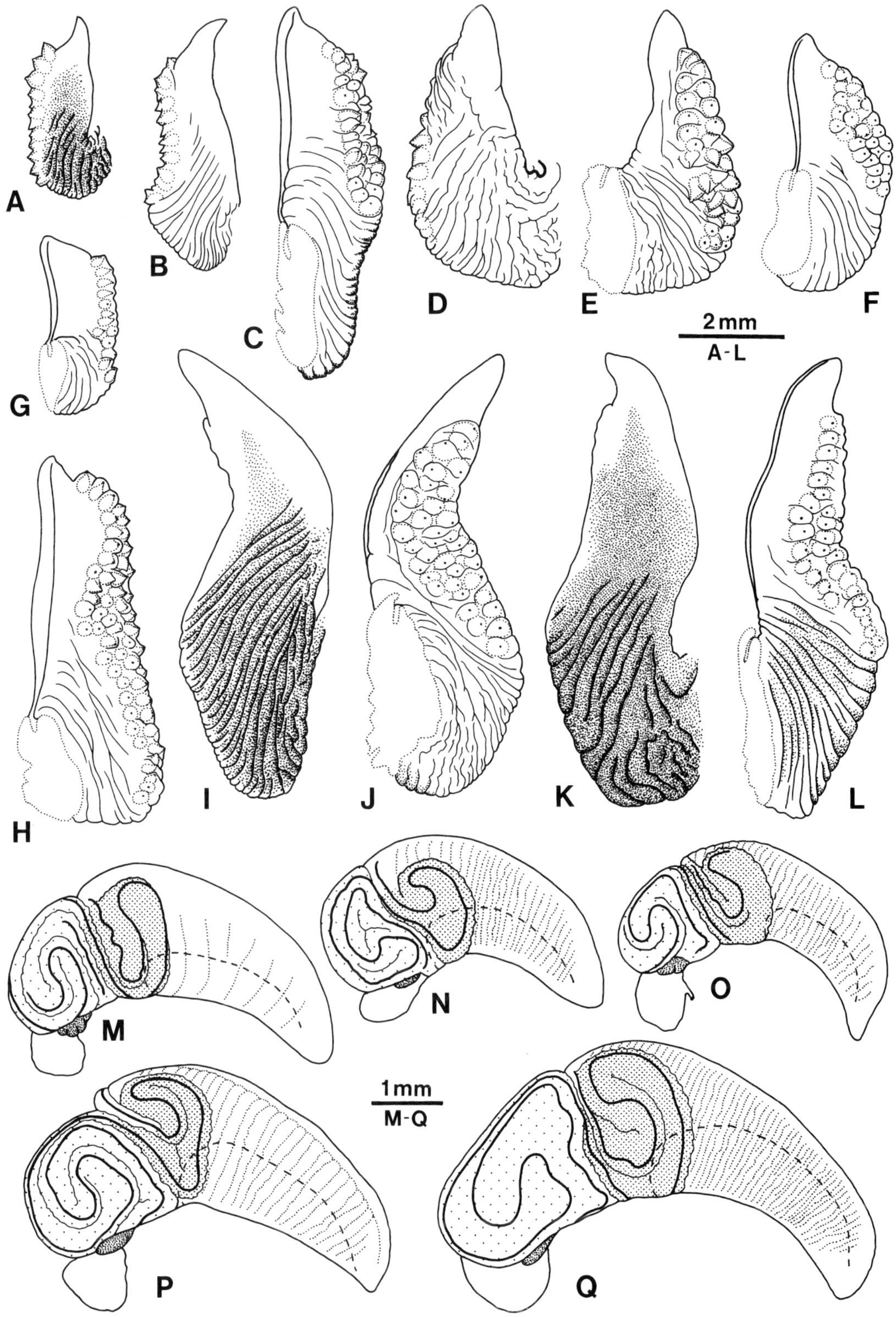

Figure 76. *Littorina (Neritrema) obtusata*: **A–L,** penes; **D, E** and **I, J** and **K, L** are each lateral and medial views of same penis; **M–Q,** pallial oviducts. **A, N,** Ivanov Navolok, Chupa Bay, White Sea (moderate ecotype); **B, Q,** Cabo Estai, Canido, Ría de Vigo, Spain (wave-exposed ecotype); **C,** south side of Black Head, Co. Clare, Ireland (moderate ecotype); **D, E,** Appledore Island, Isles of Shoals, Maine (moderate ecotype); **F, M,** Loud's Island, off Round Pond, Maine (moderate ecotype); **G,** Machias Bay, Maine (moderate ecotype); **H,** Porth Cressa, St Mary's, Scilly Isles, England (moderate ecotype); **I, J,** Flatey, Breidafjördur, Iceland (moderate ecotype); **K, L,** Yesnaby, Orkney, Scotland (moderate ecotype); **O,** Locheport, North Uist, Outer Hebrides, Scotland (sheltered/brackish ecotype); **P,** Grindavík, Iceland (wave-exposed ecotype). Shell heights: **A,** 8.5 mm; **B,** 8.6 mm; **C,** 13.1 mm; **D, E,** 10 mm; **F,** 10.1 mm; **G,** 11.7 mm; **H,** 11.9 mm; **I, J,** 11.6 mm; **K, L,** 12.4 mm; **M,** 10.8 mm; **N,** 9.6 mm; **O,** 7.9 mm; **P,** 8.9 mm; **Q,** 9.9 mm. For shading conventions of pallial oviducts see p. 6.

6–17 mm long, but on average 10 by 5.5 mm (Goodwin & Fish, 1977; Goodwin, 1979; Rolán & Templado, 1987). The spawn masses are lightly attached to the lower parts of the thallus, or to the stipe or holdfast, of the algae *Fucus vesiculosus*, *F. spiralis* or *Ascophyllum nodosum*, or may occasionally be found on the rock beneath (Sacchi, 1969a; Goodwin & Fish, 1977; Goodwin, 1979). The firm gelatinous matrix contains 51–425 eggs, with a mean of about 186 (Goodwin, 1979; Gallardo & Götting, 1985; Rolán & Templado, 1987). The mean egg diameter is 210–255 μm and each is singly enclosed in a thin, almost spherical, capsule 628–682 μm in diameter (Goodwin, 1979, capsule described as an 'egg covering membrane'). Development nonplanktotrophic with intracapsular metamorphosis. On hatching the juvenile shell measures about 500 μm (Goodwin, 1977, in Wales), or 422 μm by 514 μm in length (Lima & Lutz, 1990, in New England). The protoconch, of 1.4–1.5 whorls, is smooth but for faint spiral lines and has been described in detail by Lima & Lutz (1990). At least in the southern parts of the range, animals are mature throughout the year and spawn can be found in most months, although most abundantly in spring and summer (Sumner, Osburn & Cole, 1913, Massachusetts; Sacchi, 1969a, Brittany and Galicia; Goodwin & Fish, 1977, and Goodwin, 1978, Wales; Gallardo & Götting, 1985, Helgoland and Brittany).

Some eighteenth and nineteenth century authors ascribed the spawn of this species to *L. littorea*, and this error was only corrected by Caullery & Pelseneer in 1910. There has also been some confusion with the similar spawn of *Lacuna pallidula* (Hertling & Ankel, 1927; Hertling, 1928) in which larger eggs (270–300 μm) are enclosed in polygonal covering membranes, and the whole gelatinous mass is covered by a firm rind (Goodwin, 1979). In the older European studies it cannot be certain that *L. fabalis* was not also included, and these descriptions of the spawn of *L. obtusata s.l.* are listed in the Synonymy (p. 196). The spawning season of *L. obtusata s.l.* is shorter and later in the year at more northerly latitudes: February to September in Scotland (Elmhirst, 1923), March to October in Helgoland (Linke, 1933), March to September in Denmark (Rasmussen, 1973; Thorson, 1946), February to July in the Barents Sea, June to July in the White Sea (Kuznetsov, 1960; Matveeva, 1974). Embryonic development was described by Delsman (1914). In Ireland Tattersall (1908a, b, 1920) claimed that the eggs hatched into fully-formed free-swimming veligers, but this has not been observed by other workers.

Radula and foregut (Fig. 77). Relative radular length 1.09–1.61. Rachidian: length/width 0.93–1.00; central cusp bluntly rounded; one pointed cusp and one outer denticle on each side. Lateral: usually five (range 4–7) cusps; from the inside these are: small pointed to rounded cusp; two short, blunt cusps (sometimes represented by a single short, broad cusp (Fig. 77G), or rarely by up to four short, blunt cusps; number can be inconsistent even within the same column of teeth); largest, square cusp; outermost small pointed cusp. Inner marginal: five (occasionally 4) cusps; from the inside these are: small, pointed cusp; three blunt cusps, of which inner two sometimes represented by a single cusp, third always largest; outermost pointed cusp. Outer marginal: 5–7 short, blunt cusps, more prominent in smaller shells (Fig. 77F), but sometimes only just indenting tooth margin (Fig. 77B, H); narrow neck, projecting flange on inside of base only. Intraspecific variation in radular tooth cusps has also been illustrated by Goodwin & Fish (1977). Salivary glands greatly enlarged, enveloping mid-oesophagus and radular sac. Oesophageal pouches small, less than one third length of buccal mass.

Shell variation (Figs 4, 5, 78–80)

Geographical variation. As reviewed in the Taxonomic History (p. 204), it has long been known that shells from the more northerly parts of the range are strikingly different in shell shape, with relatively taller spires than southern shells. Superimposed on this geographical variation is a considerable degree of ecotypic and individual variation, and the following diagnoses exclude the most extreme variants (monstrosities, gerontic forms and shells from the more exposed or brackish habitats):

> Geog. var. *retusa* (Figs 78A–O, 79P): H/B=0.79–1.01; SH=0.95–1.22. Globular shape, usually with a distinctly angled shoulder; spire outline strongly domed and suture descending towards base close to end of last whorl; relatively low-spired to almost flat-spired; relatively thick-walled. Range: Europe south of Lofoten Islands, southwest Iceland, Maine to New Jersey.
>
> Geog. var. *palliata* (Fig. 79A–O): H/B=0.95–1.25; SH=1.18–1.50. Shape turbinate, whorls rounded, sometimes slightly concave below suture; doming less pronounced; spire projecting;

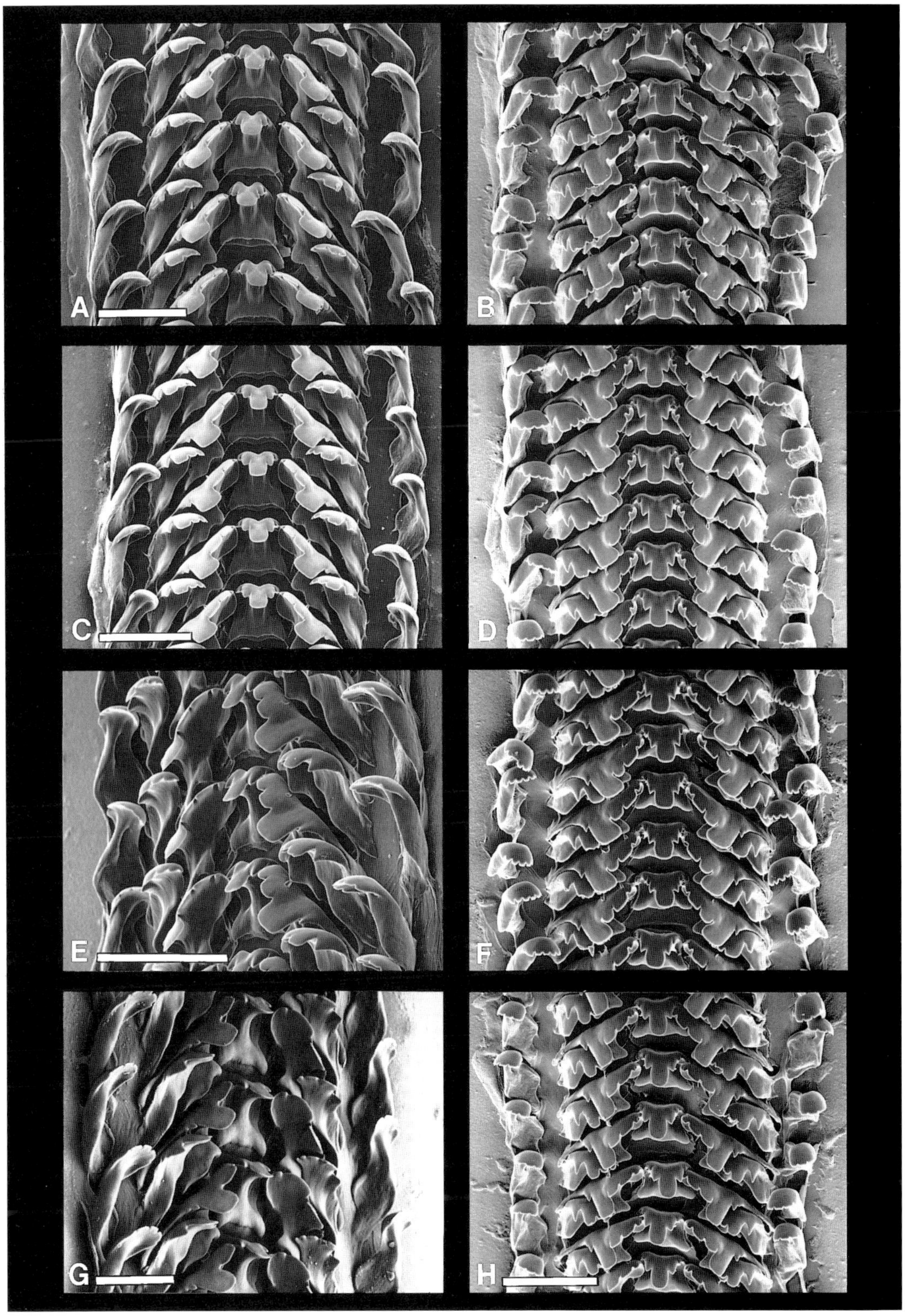

Figure 77. *Littorina (Neritrema) obtusata*: **A, B,** two views of radula (shell H=15.3 mm), flat (**A**) and at 45° (**B**); **C–E,** three views of radula (shell H=13.0 mm), flat (**C**), at 45° (**D**) and side view (**E**); **F,** radula (shell H=8.6 mm), at 45°; **G,** radula (shell H=15.8 mm), side view; **H,** radula (shell H=12.1 mm), at 45°. **A, B,** Flatey, Breidafjördur, Iceland (moderate ecotype); **C–E,** Golam Head, Co. Galway, Ireland (moderate ecotype); **F,** Canido, Vigo, Spain (wave-exposed ecotype); **G,** Yesnaby, Orkney Islands, Scotland (moderate ecotype); **H,** Machias Bay, Maine (moderate ecotype). Scale bars: **A, B**=100μm; **C, D**=100 μm; **E, F**= 100 μm; **G**=100 μm; **H**=100 μm.

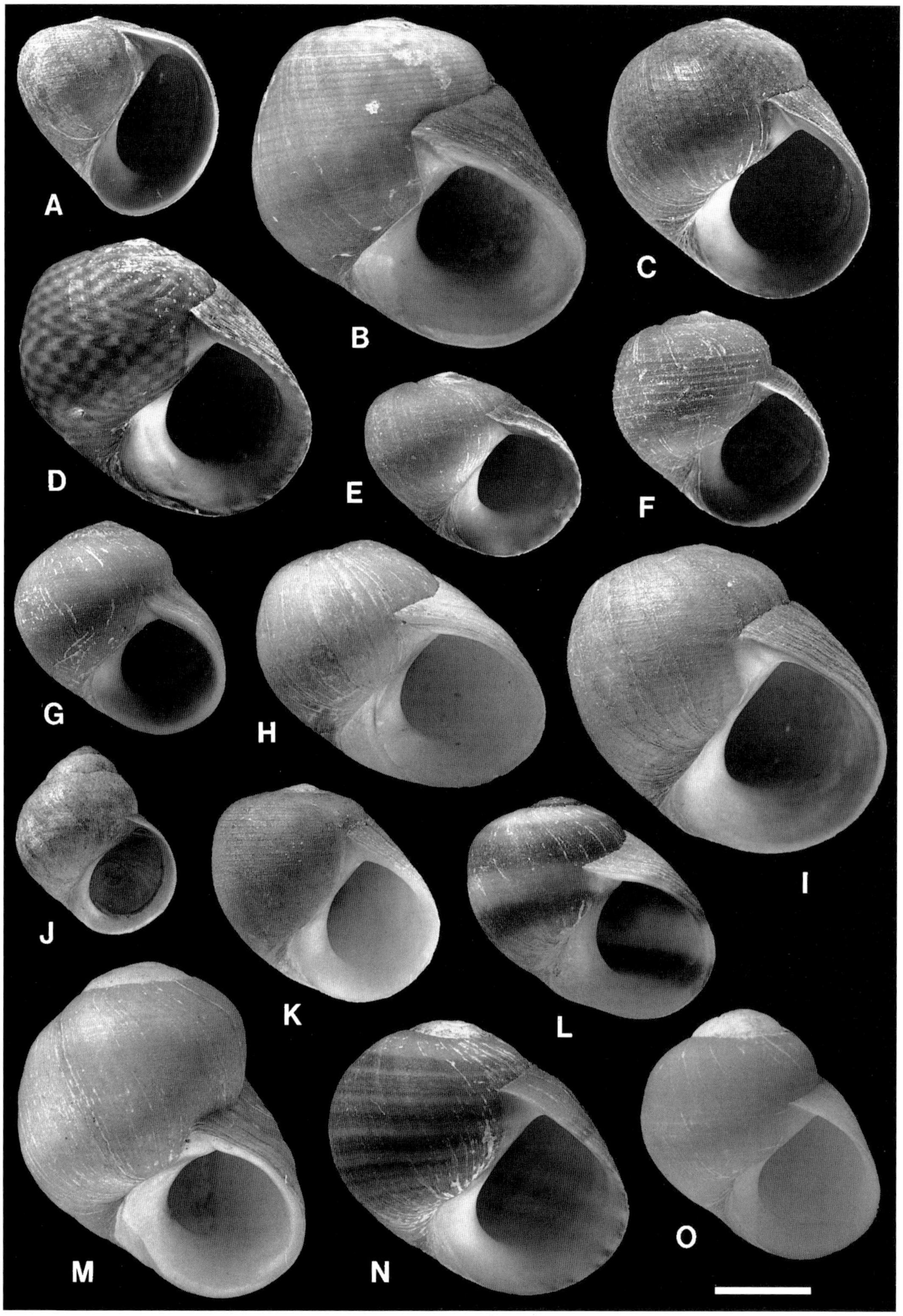

Figure 78. *Littorina (Neritrema) obtusata*: geog. var. *retusa*: **A–D, I, K, M–O,** moderate ecotype (**A,** juvenile); **E, H, L,** wave-exposed ecotype; **F, G, J,** sheltered/brackish ecotype. **A,** Robin Hood's Bay, Yorkshire, England. **B,** Balta Sound, Shetland Islands, Scotland. **C,** Camusnagaul, Little Loch Broom, Scotland. **D,** Golam Head, Co. Galway, Ireland. **E,** Cabo Estai, Ría de Vigo, Spain. **F,** Laredo, Spain. **G,** North Locheynort, South Uist, Outer Hebrides, Scotland. **H, L,** Rubh' Aird-mhicheil, South Uist, Outer Hebrides, Scotland. **I,** Ramsgate, Kent, England. **J,** *Littorina obtusata* var. *aestuarii* Jeffreys, 1869; lectotype, USNM 185836; River Deben at Shottisham Creek near Sutton, and Manningtree, Suffolk, England. **K,** Ribadeo, Spain. **M,** Firth of Clyde, Scotland. **N,** Trondheim, Norway. **O,** Barna, Co. Galway, Ireland. Scale bar = 5 mm.

Figure 79. *Littorina (Neritrema) obtusata*: **A–O,** geog. var. *palliata*; **P,** geog. var. *retusa*; **A, E, G–N, P,** moderate ecotype; **B,** wave-exposed ecotype; **C, D,** monstrosities of geog. var. *palliata*; **F, O,** sheltered/brackish ecotype. **A,** *Turbo obtusatus* Linnaeus, 1758; lectotype, LSL. **B, D, G, I,** shore by Tromsø Museum, Tromsø, Norway. **C,** Dal'niye Zelentsy Bay, Barents Sea, Russia. **E,** Ingøy, Norway. **F,** Ivan Navolok, Chupa Bay, White Sea, Russia. **H,** Flatey, Breidafjördur, Iceland. **J,** Kangerdluarssuk Fjord, Narssaq, Greenland. **K,** 1.5 km ESE of Valadalstorfa lighthouse, Tjörnes, Iceland. **L,** Quoddy Head, S of Lubec, Maine. **M,** *Turbo palliatus* Say, 1822; lectotype ANSP 18420; New England. **N,** Machias Bay, Maine. **O,** Carrying Place Cove, Eastport, Maine. **P,** Loud's Island, off Round Pond, Maine. Scale bar=5 mm.

relatively thin-walled. Range: Lofoten Islands to White Sea, Iceland, Greenland, Maine to Quebec and Labrador.

Although the southern variety is usually considered the 'typical' form, simply because it is more familiar to most workers, the name *'typica'* cannot be applied to it, because Linnaeus' type of *Turbo obtusatus* (Fig. 79A) was in fact of the northern variety. The name *retusa* is chosen instead, since it is descriptive of the shell shape. The well-known name *palliata* is retained for the northern variety.

The boundaries between the two forms are relatively abrupt, in each case with a zone of about 150 to 200 km within which both varieties as well as intermediates have been recorded. In Europe geog. var. *retusa* is known as far north as Ure (Lofoten Islands; ZMO, undated), Skarberget (BMNH, 1988) and Narvik (DGR, 1989); *palliata* is recorded as far south as Tromsø (DGR, 1989), Bjerkvik and Grindjord (both near Narvik; MNHNP, 1950s), Svolvær (Lofoten Islands; ZMK, 1916; but Norman, 1902, recorded *retusa* there) and Flakstad (Lofoten Islands; ZMO, undated); with intermediate forms from Bodø (TM, 1925), Inhavet (MNHNP, 1950s) and Bogen (ZMO, 1965). In Iceland *retusa* is known only from the Vestmannaeyjar (DGR, 1989), Ståkkseyri (NHMR, 1911), Grindavík (DGR, 1989) and Gardskagi (NHMR, undated); a range of shell types including intermediates have been found at Thorlákshöfn (DGR, 1989), Útskálar (NHMR, 1907) and Seltjarnarnes (DGR, 1989); *palliata* alone is known from Kollafjördur (NHMR, 1911), Hvalfjördur (SMNH, 1987), Borgarnes (DGR, 1989), and all around the coast further north and as far as Höfn (DGR, 1989) in the south east (the species is absent from the rest of the exposed, sandy, southern coast). In Maine *retusa* has been found as far north as Loud's Island (DGR, 1985), Ellsworth (DGR, 1985) and Perry (1984, see Seeley, 1986), and at Yarmouth County in Nova Scotia (NMC, 1977); the most southerly records for *palliata* are Queen's County, Nova Scotia (NMC, 1977) and Penobscot, Maine (DGR, 1985) (see Seeley, 1985, for further details of the distribution in New England). More detailed investigation of these zones is needed to show whether there is a mosaic distribution, or a cline, between the two varieties or, as shown by Seeley (1986), a change in the position of the boundary over time (for this reason only twentieth century samples are listed here, with dates of collection).

Not surprisingly, many authors have interpreted this pattern of geographical variation as indicating two distinct species (see Taxonomic History, p. 204). The most persuasive was Thorson (1941), who argued that since intermediates were 'very rare' in Iceland, and both forms lived commonly together in Faxaflói, they could not be considered as geographical races. Some of the material available to Thorson has now been reexamined, showing that his concept of *'L. obtusata'* included not only geog. var. *retusa*, but also *L. fabalis*. The co-occurrence of *L. fabalis* and *L. obtusata* geog. var. *palliata* in some of the samples undoubtedly contributed to his conviction that the low-spired and high-spired forms were sympatric species. This probable confusion of *L. obtusata* and *L. fabalis* renders most of the early studies of geographical variation in the former invalid. These include the statistical comparisons of shell shapes by Colman (1932) and Knudsen (1949), which found broad overlap between the shell ratios of northern and southern forms and therefore concluded that they were conspecific. This work has, however, been repeated with pure samples of *L. obtusata*, demonstrating intermediates in the zones of intergradation between *retusa* and *palliata* (Seeley, 1986; Warmoes, 1986; Lewis & Williams, 1995; the latter two using multivariate morphometric techniques).

Anatomical criteria have also been used to justify the specific separation of *retusa* and *palliata*. Both Sars (1878) and Teilman-Friis (1899) described small differences in the cusps of the lateral teeth of the radulae, but again these supposedly consistent differences could be explained if the low-spired shells were *L. fabalis*. There is in fact variation in this character in *L. obtusata* (Goodwin & Fish, 1977; p. 208), and Johansen (1901) found intermediate types in *palliata*, therefore concluding that the two varieties were conspecific. Winckworth (1933) recommended anatomical investigation of the varieties, citing the (incorrect) report of cirri around the foot in *palliata* (see discussion of *Nerita littoralis*, p. 202). Recent studies of radular variation and of reproductive anatomy have failed to find any significant differences between *palliata* and *retusa* (Goodwin & Fish, 1977; this study). Allozyme studies have found negligible differences between high-spired and low-spired varieties at one locality in Maine (Seeley, 1986) and between Icelandic *palliata* and British *retusa* (Lewis & Williams, 1995; Nei's, 1972, genetic identity greater than 0.983). One final, if indirect, piece of evidence that the two varieties are not two species is the new observation of parallel geographical variation in spire height in the closely related species *L. fabalis* (p. 235). Some common explanation should be sought for these stepped clines.

Knudsen (1949) pointed out the similar correlation between increasing spire height in more northerly samples and sea temperature in samples from both Iceland and New England, but did not claim a causative link. A general explanation for the stepped clines has, however, been suggested by the work of Seeley (1986) in New England. She noticed that *L. obtusata* collected from northern New England before 1900 were all of the high-spired, thin-walled, *palliata* type, but that this was now restricted to parts of northern Maine, having been replaced in the south by *retusa*. This she explained as the result of intense directional selection on local populations, caused by the predatory crab *Carcinus maenas* which did not occur north of Cape Cod until 1900, but then rapidly expanded its range into northern New England. This hypothesis was supported by field and laboratory experiments showing that the *palliata* form was more vulnerable to crushing by crabs than the low-spired, thick-walled *retusa*. In northern Maine there are some localities with few or no crabs, where *palliata* shells can still be found, while at nearby localities with abundant crabs *retusa* predominates. Seeley found that the morphological transition from one form to the other could be surprisingly rapid, only 17 years separating the collection of *palliata* and of intermediate forms at Nahant, Massachusetts. Seeley (pers. comm.) has pointed out that the same correlation of the presence of *Carcinus* with *retusa* and of the absence of the crab with *palliata* is found in Iceland (see Stephensen, 1939, for occurrence of *Carcinus* only in southwestern Iceland). The correlation also holds in northern Norway, where *Carcinus* is common south of the Lofoten Islands, but of rare occurrence further north, in Finnmark (Vader, 1979). It could be objected that the mechanism proposed by Seeley (1986), of intense natural selection on local populations, has not been fully demonstrated. Assuming genetic determination of shell shape, migration rather than selection of the *retusa* genotype could have accomplished the morphological transition (although no significant genetic differences between the two varieties were detected at allozyme loci). Alternatively, the effect might be an ecophenotypic one, for effluents from crabs can influence shell morphology of gastropods (e.g. in *Nucella*, Palmer, 1990). A recent study of additive genetic variance in the shell shape of *L. subrotundata* has, however, supported Seeley's genetic interpretation (Boulding & Hay, 1993). Nevertheless, it seems clear that the low-spired, thick-walled, shell morphology of var. *retusa* is more resistant to crushing predation by crabs, and is therefore functional and probably adaptive. Since the habitat of *L. fabalis* partly overlaps that of *L. obtusata*, it is quite likely to respond to the same selective or environmental influences as its sister-species, so that crab predation may explain the clines in both.

Interestingly, there are late Pleistocene fossil records of *palliata* in Scotland (Jeffreys, 1865) and Bohuslän, Sweden (Sars, 1878), where it does not now occur. Possibly the boundary between *retusa* and *palliata* has also shifted in the eastern Atlantic, in response to the changing distribution of crabs brought about by cycles of glaciation. More recently it has evidently remained fairly stable, several authors having reported the boundary between the two forms at the Lofoten Islands (Sars, 1851; Norman, 1902; Dautzenberg & Fischer, 1915).

Another aspect of geographical variation is the tendency for southern shells to attain smaller sizes. This was noted in comparisons between Brittany and Galicia (Sacchi, 1967b, 1969a; Sacchi & Rastelli, 1966), but can be extended further northwards, for the very largest shells of var. *retusa* are from northwestern Scotland.

Ecotypic variation. The same relative shell variations occur between habitats in both geographical varieties, and three principal ecotypes can be distinguished:

Moderate ecotype (Figs 78A–D, I, K, M–O; 79A, E, G–N, P): these are shells from habitats that are neither very wave-exposed nor extremely sheltered or brackish; they reach large size (maximum H=18.2 mm in *palliata*, 19.8 mm in *retusa*); spire heights differ in the two geographical varieties.

Wave-exposed ecotype (Figs 78E, H, L; 79B): *L. obtusata* seldom occurs on truly exposed shores, but on shores near its exposure limit shells are reduced in size (H as little as 4.6 mm), slightly lower-spired and with the aperture relatively a little larger than on more sheltered shores in the vicinity.

Sheltered/brackish ecotype (Figs 78F, G, J; 79F, O): shells are smaller in size (H as little as 5.6 mm), generally more thin-walled, sometimes with relatively smaller apertures, and have taller spires (SH up to 1.67 in River Deben, Suffolk) than typical forms.

The wave-exposed ecotype is uncommon, because *L. obtusata* is usually absent from exposed shores. Several authors have recorded that shell size decreases on more exposed shores (Sacchi, 1962, 1964, 1969a; Goodwin & Fish, 1977; Reimchen, 1982; Payne, 1983; Sacchi & Sconfietti, 1988; Trussell, Johnson, Rudolph & Gilfillam, 1993; Williams, 1994), although the small difference in shell shape have been noted only by Newkirk & Doyle (1975) and measured by Trussell *et al.* (1993). Assuming that the variation between shores is genotypic, it may be the result of selection. On wave-exposed shores the shells are smaller and lower (when flat on a surface), which may be adaptive for sheltering in crevices, while their more patulous shape may accommodate a relatively larger foot for strong attachment (Trussell *et al.*, 1993). On sheltered shores crab predators are more abundant and may select for larger shells that are more resistant to attack (Reimchen, 1982). However, when corrected for this difference in size, shells from sheltered shores are in fact weaker than those from exposed shores. Fletcher (1995) suggested two possible explanations for this anomaly; these shells may attain a size at which they are invulnerable to the common shore crab (*Carcinus maenas*), or alternatively the thinner shell may be an ecophenotypic effect of faster growth. Desiccation resistance may perhaps be another consideration on sheltered shores, with corresponding selection for reduced foot and aperture size in order to limit water loss when exposed at low tide (Trussell *et al.*, 1993). There is also variation with tidal level on the same shore; at lower levels shells are both larger (Reimchen, 1982) and relatively stronger (Fletcher, 1995). This trend is paralleled by the increasing importance of crushing predation at lower levels, but whether it is the result of selection, migration, differential survival or ecophenotypic effects is yet unknown.

The sheltered/brackish ecotype is also relatively uncommon in collections, because it is found only near the extreme brackish-water limits of the species. As reviewed under Taxonomic History (p. 205), several authors have considered this a distinct species, sometimes confusing it with the tall-spired geog. var. *palliata*. The most well-known name for this ecotype is *aestuarii*, described (as a variety of *L. obtusata*) by Jeffreys (1869) from the River Deben in Suffolk, England, and spire height in shells from this locality is more extreme than in any others of geog. var. *retusa* (Fig. 78J). Examples are now very rare at Jeffreys' type locality, but Moyse *et al.* (1982) collected a few living animals and found no significant differences in penial shape or in allele frequencies between these and the typical ecotypes found at the mouth of the river. This ecotype in fact has a wide geographical distribution, for example occurring in lagoons in Spain (Fig. 78F; Sacchi, 1968b), in brackish creeks in the Outer Hebrides (Fig. 78G) and Maine (Fig. 79O), and indeed most of the small shells from the White Sea (where salinity is low, Zenkevitch, 1963; Fig. 79F) could be included. The smaller and thinner-walled shells in sheltered, muddy localities have also been noted by Sacchi (1969a) and Sacchi & Sconfietti (1988).

Sculptural variation (Fig. 80). *Littorina obtusata* and *L. fabalis* are unique among members of the subgenus *Neritrema* in never developing true spiral ribs or carinae (but see discussion of Monstrosities, p. 216). Nevertheless, *L. obtusata* shows variation in the degree of microsculpture of the surface. The periostracum is unusually thick for *Littorina* species; microstriae of the shell surface are generally periostracal, and are more strongly developed in juveniles, becoming obsolete in adults (Fig. 80A–D). Although both geographical varieties can be almost smooth or show microstriae, the strongest microstriae are shown in some examples of var. *palliata* (Fig. 79E, G, I) and in these cases the striae are incised into the shell itself (Fig. 80E).

Ontogenetic variation. Juvenile shells differ from adults not only in the thin growing edge of the aperture, but also in the more produced anterior lip (Figs 78A, 80C–F). Since the spire outline is markedly domed (i.e. apical angle decreases with size), juveniles are also relatively lower-spired. In geog. var. *retusa* this process accelerates close to the end of the final whorl, whose point of attachment to the preceding whorl descends towards the base. In the most extreme gerontic forms the result is a peculiar, somewhat cylindrical (Fig. 78B) or even turreted (Fig. 78M) outline, which is not uncommon in some populations. Doming is much less pronounced in geog. var. *palliata*, in which the final whorl descends hardly at all. The combination of strong doming and considerable thickening of the aperture in adults means that shell growth is apparently more determinate in this species than in other members of the genus. The pattern of allometric growth in shape and thickness, and its relation to shell strength and resistance to crushing predation, has been investigated by Lowell, Fletcher, Grahame & Mill (1994) and by Fletcher (1995). In some examples, particularly of the wave-exposed ecotype, during the final quarter of the last whorl the columella separates from the axis of coiling to give a pseudo-umbilicus (Figs 78H, 79B).

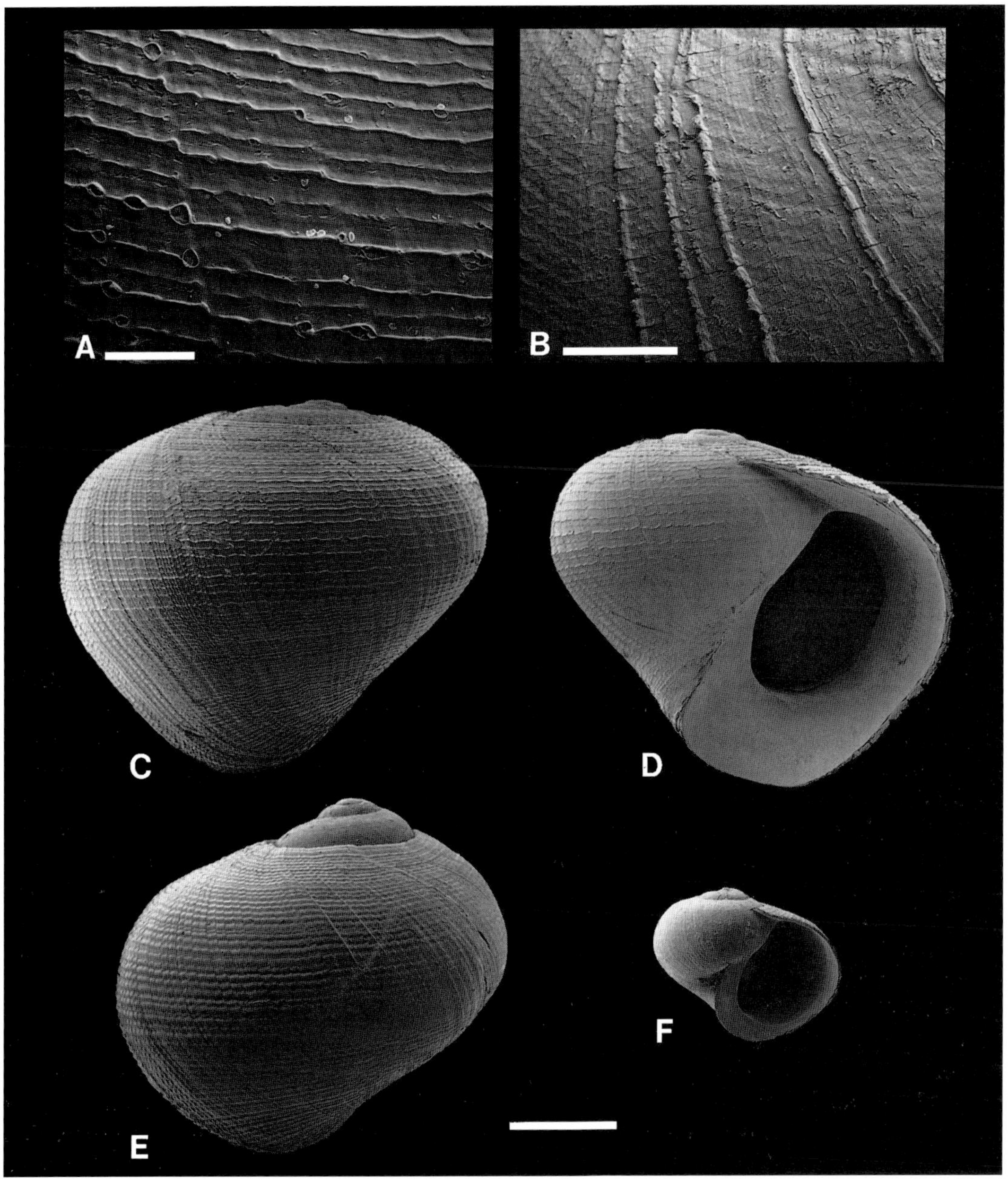

Figure 80. *Littorina (Neritrema) obtusata*: **A**, high-power detail of surface of relatively strongly sculptured juvenile shell as in **C**; **B**, low-power view of surface of relatively smooth adult shell, showing periostracum; **C–F**, juvenile shells; **A–D**, **F**, geog. var. *retusa*; **E**, geog. var. *palliata*. **A–C**, **F**, Yesnaby, Orkney Islands, Scotland; **D**, North Locheynort, South Uist, Outer Hebrides, Scotland; **E**, Flatey, Breidafjördur, Iceland. Scale bars: **A**=0.2 mm; **B**=2.0 mm; **C–F**=1.0 mm.

Specimens of geog. var. *palliata* from the most northerly parts of its range often show pronounced growth lines, corresponding to winter interruptions to growth (Fig. 79J, O). In the White Sea these lines have been used in demographic studies, showing that animals live for 8–12 years (Sergievsky, Granovitch & Mikhailova, 1991; Sergievsky, 1992).

Sexual dimorphism. Since dimorphism is more pronounced in *L. fabalis*, reports of dimorphism in *L. obtusata s.l.* are unreliable (Jeffreys, 1865; Pelseneer, 1926; Daguzan, 1977). Nevertheless, in most (but not all) populations of *L. obtusata s.s.* females are slightly, but significantly, larger in shell height (Sacchi, 1967b, 1968b, 1969b; Goodwin & Fish, 1977; Sacchi & Sconfietti, 1988; Williams, 1994). Dimorphism in shape is even less pronounced, but in some populations females have more globose shells (Sacchi, 1967b, 1969b; Sacchi & Sconfietti, 1988). Fletcher (1995) found no significant sexual dimorphism in shell strength, which is determined largely by relative mass. Pelseneer (1926) suggested that females of *L. obtusata s.l.* are larger because they live longer than males (judged by epiphytic growths), but in view of the deterministic nature of shell growth in this species, this seems unlikely.

Monstrosities. In geog. var. *retusa* several malformations have been described, with single carinae on the last whorl, or abnormal coiling resulting in a cylindrical spire or a small umbilicus (Martel, 1901; Dautzenberg & Fischer, 1915). Umbilication may occur in a few of the most extreme gerontic forms (Dautzenberg & Fischer, 1915: pl. 3, fig. 15). Severe damage to the growing edge may result in abnormal coiling of the regrown shell, as in an elongate shell illustrated by Dautzenberg & Fischer (1915: pl. 3, fig. 14). The very rare carinae (Dautzenberg & Fischer, 1915: pl. 3, figs 12, 13) are not analogous to the regular ribs of sculptured *Littorina* species, but are perhaps also a result of early damage to the growing edge or mantle. Epibionts such as barnacles and spirorbid worms are relatively unusual in *L. obtusata* (Sacchi, 1970), but even in occasional heavily fouled populations have not been recorded as causing shell malformations (Sacchi, 1985; Warmoes, Backeljau & de Bruyn, 1988). Parasitism by trematodes has not been found to cause abnormalities (Williams & Brailsford, 1990). In addition to the commonly observed higher spire of the estuarine ecotype, Rasmussen (1973) recorded that 7% of shells of *L. obtusata s.l.* from the interior of Isefjord, Denmark, were 'planorbid' (i.e. with planispiral coiling). Unfortunately, these were not illustrated and no similar specimens or reports have been discovered.

Perhaps the most extraordinary of all the shell variants of *L. obtusata* are found in some populations of geog. var. *palliata* from northern Norway and Iceland. These have been given varietal names by M. Sars (1851), G. O. Sars (1878) and Schlesch (1916), and have been described and illustrated by a number of other workers (Dautzenberg & Fischer, 1915; Schlesch, 1930; Óskarsson, 1973; Backeljau *et al.*, 1984). There are two main types: rare shells with spires within the normal range of height for *palliata*, but with a single large, rounded carina on the shoulder of the last whorl (Fig. 79C); and more common types with greatly elongated spires (SH up to 2.04) and often an exagerated concavity near the suture on the last whorl (Fig. 79D) which, were it not for the occurrence of intermediates, would hardly be thought conspecific with typical examples of *palliata*. These monstrosities are known only from populations between Cape Svetoi Nos on the Kola Peninsula and Tromsø in northern Norway, and in Iceland from the north and west coasts (Tjörnes, Ísafjördur, Breidafjördur and Seltjarnarnes) and are absent from large collections available from the White Sea, Greenland and North America. They are never very numerous; at Flatey in Breidafjördur monstrosities (with SH>1.50 and carinate forms) made up only 8% of a representative collection of 107 shells, at Ísafjördur 6% (of 202) and at Tromsø 9% (of 244) (all DGR).

The cause of these monstrosities is unclear. There are no obvious breakages, interruptions to growth, or epizoic encrustations on the spire; there are no peculiarities of the reproductive organs; the incidence of trematode parasitism seems no higher than in normal *palliata* forms; and they occur together with normal forms in fully marine habitats. The fact that comparable monstrosities are unknown in geog. var. *retusa* suggests some connection with an Arctic climate, but low temperatures seem unlikely to be a direct cause, since the monstrosities are not found in Greenland or Canada, with equally severe climates. The geographical restriction of these forms to northern Norway and Iceland (known to share a close biogeographical relationship, Ingólfsson, 1992) might imply a regional genotypic effect. It may be that these monstrosities, although so striking, should not be regarded as monstrous forms at all, but simply extreme examples of the normal distribution of shell shapes in geog. var. *palliata* at these localities, for they are connected with 'normal' forms by an entire range of intermediates. They may well have a genetic basis, but presumably the genotypes only persist in populations in which selection for shell form is relaxed, as in these northern localities where crab predators are rare or absent. The similarity of these odd shells to the Norwich Crag monstrosities of *L. littorea* should be mentioned (p. 108).

Colour variation (Fig. 5). Although some other *Littorina* species may be able to match the wide range of shell colour and pattern shown by *L. obtusata* and *L. fabalis*, none of the others displays such a consistently high diversity of colour polymorphism within local populations. This visual polymorphism is one of the most characteristic features of this pair of sister-species.

The shell colour morphs have been described and classified in various ways. Dautzenberg & Fischer (1915) adopted a purely descriptive approach, employing existing Latin names and creating new ones, to designate eight colour varieties ('var. ex colore') for what is here termed geog. var. *palliata*, and twelve others for geog. var. *retusa*; some of these names were based wholly or in part on what is now recognized as *L. fabalis*, but the range of colour forms is comparable in both (see Synonymy, p. 196, and that of *L. fabalis*, p. 227). Six further names were later added to this system (Dautzenberg & Fischer, 1925; Schlesch, 1930; Sacchi, 1961, 1963). This classification has been used by most subsequent workers, although since few studies have been made of geog. var. *palliata*, and because similar colour forms occur in both geographical varieties, only the varietal names for *retusa* have gained acceptance (Barkman, 1955; Bakker, 1959; Sacchi, 1961, 1962, 1963, 1964, 1966b, c, 1976a, b, 1969a, 1972a, 1974, 1984, 1985; Smith, 1976; Goodwin & Fish, 1977; Warmoes *et al.*, 1988). Smith (1976) simplified the system, recognizing only those varieties which he found to be discrete and non-intergrading in a study of large collections from around Britain. The three most common were: olivacea (olive-green to olive-brown, purplish interior; 55% of all samples combined; Fig. 5, second from left), reticulata (yellow to brown background with darker reticulation or zigzag markings; Fig. 5, fifth from left; Figs 78C, D; 79B, D, H, O; 33%) and citrina (yellow; white interior; 10%; Fig. 5, first from left). Rarer morphs (each less than 1%) were: fusca (dark brown to black; Fig. 5, third from left), aurantia (orange), rubens (red), inversicolor (two broad dark bands; Fig. 5, fourth from left; Figs 78L, 79K), zonata (single pale peripheral band) and alternata (two pale bands).

This classification has been found useful in the British Isles and southern Europe, and is also applicable in New England. Some workers have, even here, found difficulty in classifying morphs (Bakker, 1959; Gill, Ramsay & Smith, 1976), but after exclusion of *L. fabalis* and boiling in alkali to remove algal stains, these groups do seem discrete (Smith, 1976). However, this system may not be of universal application. Throughout much of the range of geog. var. *palliata* there is a preponderance of dull brownish and faintly reticulated shells, olive colours are rare or absent, and differentiation of discrete morphs is more difficult (Dautzenberg & Fischer, 1915; pers. obs.). Working with *palliata* in the White Sea, where some populations are highly polymorphic, Sergievsky (1983a, 1984, 1992) devised a different classification scheme, based on the genetic model of Pettitt (1973a) for *L. saxatilis*. For each shell, three characteristics (which could be taken as genetic loci, although there is no evidence for this) were scored: background colour ('locus' C, 'alleles' purple p, brown b, orange o, yellow y, purple-orange po, white w), presence of spots or reticulation (present S^+, absent S^-) and presence of broad bands (one band B^1, two bands B^2, bands absent B^0). A citrina shell could thus be designated $C^yS^-B^0$; all but one of the 36 possible combinations were found.

The structural basis of shell pigmentation was first considered by Fischer (1941) who, after treating yellow and olive shells with various acids, alkalis and solvents, claimed that a green pigment was present in the periostracum (and that this was not chlorophyll from algae on the surface), while a yellow pigment was localized in the shell surface, the superposition of the two being responsible for the overall colour of the shell. This cannot be reconciled with the method of Smith (1976), who scored the colour of his samples after removing surface algae (and presumably also the periostracum) with boiling alkali. Gill, Ramsay & Smith (1976) did not detect pigments in the shells, but claimed to find similar pigments (including chlorophyll) in the algal substrate, the snail tissues and the periostracum, and implied that pigments from the food influenced coloration; here, epizoic algae were not excluded. In a more rigorous approach Mikhailova & Naumov (1990) sectioned shells and showed that in the internal (i.e. crossed-lamellar aragonite) layer a purple pigment could be present or absent, while in the outer (i.e. irregular-prismatic calcite) layer any of three pigments, purple, yellow or orange, could be present or absent, the possible combinations giving rise to the full range of shell coloration. If this work could be combined with studies of heritability, there would seem to be a real possibility of revising Sergievsky's description of colour morphs to produce a genetically meaningful classification.

So far, only Reimchen (1974; quoted in Smith, 1976) has undertaken the necessary breeding experiments to show that shell colour is indeed under genetic control. This assumption is, of course, essential before any explanation of the origin and maintenance of the polymorphism in terms of natural selection is possible. Various indirect lines of evidence have been used to argue that this is so, for example the often clear-cut differences between morphs (Smith, 1976), the apparent disequilibria between colour and banding patterns (Sergievsky, 1992), and possible 'founder effects' in isolated populations with high frequencies of normally rare morphs (Sacchi, 1974). Further indirect evidence comes from the restricted geographical distribution of some morphs, for example the uncommon, but distinctive, morph named lineolata by Dautzenberg & Fischer (1915) (narrow dark spiral lines on a grey-green background; Fig. 79I; which does not intergrade with reticulata and should therefore be added to Smith's (1976) total of nine discrete morphs), which is probably restricted to Arctic Norway, the Barents Sea, Iceland and New England.

At least in Britain and southern Europe almost all populations of *L. obtusata* are polymorphic to some degree (Sacchi, 1974; Smith, 1976), and the origin, maintenance and adaptive significance of this colour polymorphism must be considered. The possible factors involved in the maintenance of molluscan shell colour polymorphisms have been set out by Cook & Garbett (1992). Firstly, it is clear that the polymorphism is not neutral, but is in some way selected, for systematic patterns can be demonstrated in the distribution of morphs. It was first shown by Sacchi (1961, 1962, 1963, 1964, 1966b, c, all with *L. obtusata s.l.*, then 1967b, 1969a, 1972a, 1974, 1984, with *L. obtusata s.s.*) that in Brittany, Galicia and southern England the morph olivacea predominates in sheltered habitats, and morph reticulata in wave-exposed ones, while the proportion of citrina remains little affected by turbulence. This widespread pattern has been confirmed throughout Britain (Smith, 1976; Goodwin & Fish, 1977) and in northern Norway (Payne, 1983). Secondly, since there is no evidence for assortative mating based on shell colour (Sergievsky, 1983b), some kind of balancing selection must operate. It is the nature of such selection that is debated, with emphasis on either non-visual effects (e.g. Sacchi, 1966c, 1974; Sergievsky, 1989, 1992) or on the visible differences between morphs (Smith, 1976). The possibility of heterosis should also be mentioned, although in the absence of genetic information there can be no evidence for this.

Sacchi (1961) considered that shell colour itself was of doubtful adaptive significance and therefore sought physiological differences between the morphs, which by pleiotropy or linkage could influence the visual polymorphism. He found slight differences in desiccation resistance, darker colours often being more resistant (Sacchi, 1969a), and in thermal tolerance, olivacea being the most eurythermic and reticulata stenothermic (Sacchi, 1972a). There were also small differences in the sex ratios in the three principal morphs (Sacchi, 1967a, b, 1984) and in infestation by epibionts (Sacchi, 1985). He also noted that citrina and other pale colours were more frequent on the exposed surfaces of algae (Sacchi, 1969a, 1974), because they were more active under bright illumination (Sacchi, 1961, 1963, 1964, 1966b, in *L. obtusata s.l.*). These pale colours are conspicuous to the human eye and were presumed to be so to natural predators, and indeed this was one reason why Sacchi (1974) did not consider visual predation to be significant in determining the distribution of the morphs. Nevertheless, none of these physiological and behavioural observations satisfactorily accounted for the predictably different morph frequencies in exposed and sheltered habitats, and it could only be suggested that perhaps reticulata was in some way more tolerant of wave action (Sacchi, 1974).

Sergievsky (1989, 1992) also stressed the non-visual aspects of the polymorphism. Working in the very different, and extreme, environment of the White Sea, with lowered salinity and ice cover in winter, he was impressed by correlations between proportions of certain phenotypes and environmental variables such as salinity and temperature (Sergievsky, 1982a, 1983c, 1984, 1985c, 1987). Physiological experiments showed, for example, that the morph fusca, which increases in frequency from the usual level of 2–10% to 45–50% in brackish estuaries, is more resistant than others to dilution (Sergievsky & Berger, 1984; Sergievsky, 1985c, 1992; Berger & Sergievsky, 1986). On a smaller scale, yellow phenotypes predominated near low water and brown ones at higher levels, corresponding to the greater resistance of the latter to overheating (Sergievsky, 1987, 1989; but note that Smith, 1976, did not find a comparable morph distribution with tidal level in Britain). Trematode parasites were shown to be another likely selective force, for in populations with a high (more than 20%) incidence of infection, colour phenotypes differed consistently in their level of infection (particularly by parthenitae of the *Microphallus pygmaeus* group) (Sergievsky, 1982b, 1985a, b, 1986, 1989, 1992). The basis of this resistance (e.g. likelihood of infection, or susceptibility) was not identified, but that selection was occurring was suggested by significant increases in resistant phenotypes in older age-classes. Some behavioural differences among morphs in algal-choice experiments were also found (Sergievsky, 1989); this had earlier been investigated by Barkman (1955), who almost certainly included *L. fabalis* in his experiments. As emphasized by Sergievsky (1992), shell colour is evidently connected with a number of adaptive systems, which in different environments may vary in their contribution to the overall maintenance of the visible polymorphism. While not denying the possible contribution of visual predation, the only suggestion of this was the fact that the proportion of apparently conspicuous banded morphs tended to fall in older age-classes (Sergievsky, 1989).

Despite this absence of direct evidence for visual predation, it should not be dismissed throughout the geographical range of *L. obtusata*. The White Sea is an unusual habitat, in which climatic and hydrological conditions are extreme and large crab predators absent. It seems of some significance that *L. obtusata* and *L. fabalis* together show the highest diversity of intrapopulation polymorphism of all *Littorina* species, and that they alone are almost exclusively epiphytic in habit, living in association with fucoid and other algae, on a more visually diverse background than that of rocky substrates. Elsewhere in the Littorinidae a connection has been noted between a visually diverse habitat (in this case the foliage of mangrove trees) and maintenance of shell colour polymorphism in *Littoraria* species by visual predation (Cook, 1983; Cook & Garbett, 1992; Reid, 1987). Unfortunately, direct field evidence for the selective action of visual predation is hard to obtain in marine environments, and often only weak evidence is available in the form of associations between morph frequencies and visual characteristics of the habitat. Smith (1976) analysed populations from all around Britain and attempted to segregate the confounding variables of algal substrate, tidal level and wave exposure. He found that citrina was most common on *Fucus spiralis*, olivacea on *F. vesiculosus* and usually on *Ascophyllum nodosum*, and reticulata on *F. serratus* and sometimes on *Ascophyllum*. Phenotypic diversity was highest where algal diversity (and therefore also the visual diversity of the habitat) was greatest, and lowest in brackish and polluted sites (where possibly the physiological effects studied by Sergievsky may be important; see Warmoes *et al.*, 1988, for examples in Belgium). As an explanation Smith proposed disruptive selection by visual predators. He suggested that the three common morphs were all cryptic, citrina resembling yellow receptacles of *F. spiralis*, olivacea the greenish bladders and receptacles of *Ascophyllum* and *F. vesiculosus*, and reticulata the brown lamina and black stipe of *F. serratus*. Morph ratios could be balanced by selection acting disruptively over a spatially and temporally variable habitat. Interestingly, Smith found that juveniles contributed most to the differential distribution of morphs on algal species. Although he suggested that adults might spawn where their offspring would be most fit, it is perhaps more likely that visual selection simply operates most strongly on juveniles. Some geographical clines could also be accounted for by Smith's hypothesis, for example the scarcity of citrina in northeastern England, where *F. spiralis* is less common.

The close resemblance of the cryptic phenotypes to the bladders and receptacles of the fucoids is important; it has even been claimed that there is a precise matching for size and shape of bladders in different localities (Gill *et al.*, 1976). This type of crypsis has been termed 'masquerade' by Endler (1981) and could lead to the polymorphism being maintained by a type of frequency-dependent selection, acting to preserve the morphs in the same proportions as the elements of the background that they imitate (as suggested by Cook, 1983, for mangrove-dwelling *Littoraria* species). Another possible role for frequency-dependent effects is in the selection of the rarer banded, orange and red morphs, which appear more conspicuous, and could be maintained by apostatic advantage (Smith, 1976).

The predators that might be involved in visual selection of *L. obtusata* have yet to be demonstrated, but *Littorina* species are known to be consumed by a range of fish, crabs and birds, all of which could possess the necessary colour vision (Pettitt, 1975; Reimchen, 1982). Even in northern regions, where the important predatory crab *Carcinus maenas* is absent, birds may be significant predators; it has been estimated that in Iceland a single knot (*Calidris canutus*) consumes 354 *Littorina* (mostly *L. obtusata*) per day while foraging in the seaweed belt, and that over the course of a year predation by these shorebirds accounts for 1–4% of the snail population (Alerstam, Gudmundsson & Johannesson, 1992). The very best evidence for visual predation on littorinids comes from laboratory experiments with the blenny *Lipophrys pholis* feeding on *L. fabalis*, which may explain the different colour morph frequencies in sheltered and wave-exposed habitats (Reimchen, 1979; p. 241). It seems possible that similar visual selection could take place in *L. obtusata*, which also shows an increased frequency of reticulata on exposed shores. On the other hand, *L. obtusata* occupies a higher intertidal zone than *L. fabalis*, and here predation by fish and crabs may be relatively less significant and that by birds more so.

There remains the possibility of selection related to visible differences between morphs, but unconnected with predation. Colours may affect thermal properties of shells (Sergievsky, 1992), but this has not been investigated in *L. obtusata* and is likely to be unimportant in a mid-shore to low-shore species that shelters among algae during low tide.

In summary, there are various explanations for the maintenance and adaptive significance of the shell colour polymorphism of *L. obtusata*. Under extreme or marginal environmental conditions, as in Arctic regions, in estuaries or polluted sites, or where there is a high incidence of parasitism, the physiological differences between the colour morphs (produced by pleiotropy or linkage) may be important. In the southern parts of the range visible differences between morphs are more pronounced and there are systematic associations of morph frequencies with wave-exposure, and also with the visual properties of the algal substrate, which could suggest a role of visual predation, although there is as yet no strong evidence for this.

Distribution

Habitat. Since *L. obtusata* and *L. fabalis* are largely sympatric in the eastern Atlantic, many of the earlier European references to the ecology of *L. obtusata*, dating from before the discrimination of these two sister-species (Sacchi & Rastelli, 1966), must be interpreted with caution (e.g. Colman, 1933; Moore, 1940; Evans, 1947; Barkman, 1955; Bakker, 1959; Moyse & Nelson-Smith, 1963; Lewis, 1964; Golikov & Kusakin, 1978). However, in North America, where only *L. obtusata s.s.* occurs, this problem does not arise.

Characteristically, *L. obtusata* is a species of the mid-eulittoral zone, found on sheltered shores and almost always in close association with macrophytes, principally fucoids. On a typical shore of low wave-exposure, with well-developed algal zonation, it is sparse or absent among *Pelvetia* near the top of the eulittoral, common on *Fucus spiralis* lower down, reaches maximum abundance on *F. vesiculosus* and *Ascophyllum nodosum*, and is sparse among *F. serratus* and at the upper limit of *Laminaria* low on the shore. This pattern is repeated with only minor variations throughout the range, in the British Isles (Smith, 1976; Watson & Norton, 1987; Little & Williams, 1989; Williams, 1990b, 1994, 1995; Fletcher, 1992), Brittany (Sacchi, 1969a), Galicia (Rolán & Templado, 1987), Scandinavia (Nielsen, 1980; Ekendahl, 1994), the White Sea (Kuznetsov, 1960; Matveeva, 1974; Golikov & Kusakin, 1978; Golikov, 1987; all *s.l.*, but *L. fabalis* is rare in this area), New England and the Canadian Maritimes (Dexter, 1947; T. A. & A. Stephenson, 1954a, b; Seeley, 1982a, b; Chapman, 1989; Barker & Chapman, 1990). The lower tidal limit is usually considered to be at about low water of spring tides or lower if *Fucus* extends sublittorally (T. A. & A. Stephenson, 1954a, b), and may often be set by crab predation (Williams, 1992a). Greater depths have been recorded, especially in the northern parts of the range, to 12 m in the Kieler Bucht (Jaeckel, 1952, *s.l.*), 47 m or more in Iceland (Posselt & Jensen, 1898; Bárdarson, 1919; Thorson, 1941; all *s.l.*), 150 m in Greenland (Thorson, 1941, *s.l.*), 90–140 m in Finnmark (SMNH) and 300 m in the Barents Sea (ZISP). Whether the species regularly lives at such depths, or the living animals have been transported offshore on floating weed, is unknown. Individuals of *L. obtusata s.l.* are commonly found on clumps of weed drifting off the coast of Iceland (Ingólfsson, 1995). In the range of geog. var. *retusa* maximum abundances on eulittoral algae range from 200 to 800 per m^2 (Dexter, 1947; Lubchenco, 1983; Williams, 1992a), but in northern regions *palliata* reaches 800 to 1200 per m^2 (Kuznetsov, 1960; Golikov & Kusakin, 1978; McCook & Chapman, 1991; Alerstam *et al.*, 1992), or exceptionally 14000 per m^2 (Matveeva, 1974), perhaps because the predatory crab *Carcinus maenas* is absent (Alerstam *et al.*, 1992). Although typically found on fucoids, *L. obtusata* does also sometimes occur on other algae, including *Chondrus*, *Mastocarpus* and *Ulva* (Seeley, 1982a, b; Rolán & Templado, 1987). It is much less common on the rock surface beneath its host plants, and there predation by crabs and *Nucella* is more severe than in the algal canopy (Seeley, 1982b, 1983). Where algae are removed, for example by ice scour (Mathieson, Penniman & Harris, 1991) or pollution (Sacchi, 1968a), *L. obtusata* is rare or absent.

On shores of increasing exposure to wave action algal cover is reduced and *L. obtusata* becomes scarce (Sacchi, 1969a; Williams, 1990b; Fletcher, 1992). At steeply sloping sites it is said to be less common or absent (Fletcher, 1992). It is absent from most exposed coasts, but occasional populations are adapted to strong wave action. One case has been described on an exposed point in Maine, where snails (of the wave-exposed ecotype, p. 214) live on bare rock, in crevices and among barnacles and mussels (Trussell *et al.*, 1993). A similar population has been found at Grindavík in southwestern Iceland, where snails cling tightly to the rock among sparse *Fucus* tufts on a wave-beaten shore (pers. obs.). At the opposite extreme of the exposure scale, *L. obtusata* can be found among *Fucus* in the runnels of salt marshes and on the bases of *Spartina* (Bequaert, 1943; Dexter, 1945, 1947; Moyse *et al.*, 1982). It penetrates quite far into estuaries (Fretter & Graham, 1980; Little & Smith, 1980; Hardwick-Witman & Mathieson, 1983). It is not found directly on soft sediments, but can occur on weeds in lagoons and embayments. Occasionally, *L. obtusata s.l.* has been recorded on *Zostera*, particularly near the mouth of the Baltic (Ziegelmeier, 1966; Rasmussen, 1973), and some of these records apply to *L. obtusata s.s.* (Nielsen, 1980). However, Sacchi (1984) did not find it in *Zostera* meadows in Galicia.

Accounts differ as to whether *L. obtusata* shows any migratory behaviour. In southern Britain there is no evidence for seasonal or breeding migrations (Smith, 1976; Goodwin, 1978; Fretter & Graham, 1980; Fletcher, 1992). However, Watson & Norton (1987) observed that in Scotland snails left the canopy and moved to the rock surface in winter, and in the colder climate of New England there is a downshore migration to avoid shore ice (Batchelder, 1915; Pohley, 1976). In the White Sea it is said to hide under stones or even to bury in sand in the winter (Matveeva, 1974). Over the tidal cycle there is evidence that the snails crawl in the weed canopy when submerged and hide lower in the plants during low tide (Wieser, 1952, *s.l.*; Hazlett & Seed, 1976), but this behaviour is affected by shell colour (p. 218) and parasitism by trematodes (Mikhailova, Granovitch & Sergievsky, 1988). A breeding migration down the shore has sometimes been claimed (Sacchi, 1966c, 1969a), while others have found juveniles throughout the same levels as adults (Smith, 1976; Fretter & Graham, 1980; Williams, 1995). Juveniles may be more common in small, sheltered microhabitats, within damaged fucoid bladders, in rock crevices, among barnacles, coralline algae and *Lichina* (Colman, 1940; Lewis, 1964; Underwood, 1973; Watson & Norton, 1987; all *s.l.*).

The only other *Littorina* to inhabit fucoid algae is *L. fabalis*, which is sympatric with *L. obtusata* throughout the range of the latter, except in North America. On tidal shores the zones of the two species show some overlap, but *L. fabalis* is generally most abundant lower on the shore, in the zone of *Fucus serratus*. Under atidal conditions the overlap may be more extensive.

In Britain the principal food of adult *L. obtusata* is the thallus, and particularly the soft tissue below the epidermis, of the fucoids on which it lives (Hawkins, Watson, Hill, Harding, Kyriakides, Hutchinson & Norton, 1989; Norton, Hawkins, Manley, Williams & Watson, 1990; Williams, 1990b). Feeding trials indicate that *Fucus* species are usually preferred over *Ascophyllum* (Watson & Norton, 1987) and that growth is slower on a diet of the latter (Seeley, 1983). Nevertheless, *L. obtusata* is clearly more tolerant of the polyphenols in *Ascophyllum* than other grazers such as *L. littorea*, which find it unpalatable (Norton *et al.*, 1990). The order of preference for the different *Fucus* species is variable, and germlings are eaten as well as adult plants (Watson & Norton, 1987; Barker & Chapman, 1990). In North America the host fucoids are seasonally heavily infected with epiphytes, which may then be a food source (Lubchenco, 1982), but in Britain epiphytes are scarce at the tidal levels occupied by *L. obtusata*, and epiphytic algae are considered a significant food source only for juvenile snails (Watson & Norton, 1987; Williams, 1990a, b).

Range (Fig. 81). *Littorina obtusata* is widely distributed in the northern Atlantic, from the White Sea to Portugal, in Iceland, Greenland, New England and Canada. At its northeastern limit in the White Sea it occurs in Onezhsky and Kandalakshsky Bays, and on the shore of the Kola Peninsula (BMNH; Kuznetsov, 1960). Records from Novaya Zemlya and Spitzbergen (Lech, 1878; Thorson, 1941, 1951) were probably based on misidentifications of *L. saxatilis*, which is the only *Littorina* that has been seen from these localities. *Littorina obtusata* is found all around the coast of Norway and in the Kattegat. Although there are literature records of *L. obtusata s.l.* from Kieler Bucht and Lübecker Bucht in the extreme southwest of the Baltic (Schlesch, 1937; Barkman, 1955; Fretter & Graham, 1980; Jagnow & Gosselck, 1987), no museum collections have been seen from beyond Helsingør and Odense Fjord (ZMK) and these published records probably apply to *L. fabalis* (which is known to penetrate further into brackish water in Denmark, Nielsen, 1980). It occurs in the Faroes, and all around Iceland with the exception of the southeastern coast between Höfn and the Vestmannaeyjar, which is too exposed and sandy to be suitable. Further north Becher (1886) recorded the species from Jan Mayen, and since the specimens have been examined by Sneli & Steinnes (1975) the identification is probably reliable. In Greenland it is found from Julianehaab to Upernavik (ZMK) on the west coast, but is absent from the east coast.

Further south in Europe *L. obtusata* is common in The Netherlands, Belgium and around the north coast of France to Brittany (Warmoes, Dumoulin & Reid, 1992) and also throughout the British Isles (Seaward, 1990). On the French coast of the Bay of Biscay it occurs as far south as the Île d'Oléron, and there is an isolated locality in the Arcachon Lagoon (Crisp & Fischer-Piette, 1959; MNHNP). Occurrence further south was considered in detail by Fischer-Piette, but unfortunately the two sister-species were not distinguished at that time. At the end of the nineteenth century *L. obtusata s.l.* occurred at the head of the Bay of Biscay, at Guéthary and Biarritz, as well as at localities all around

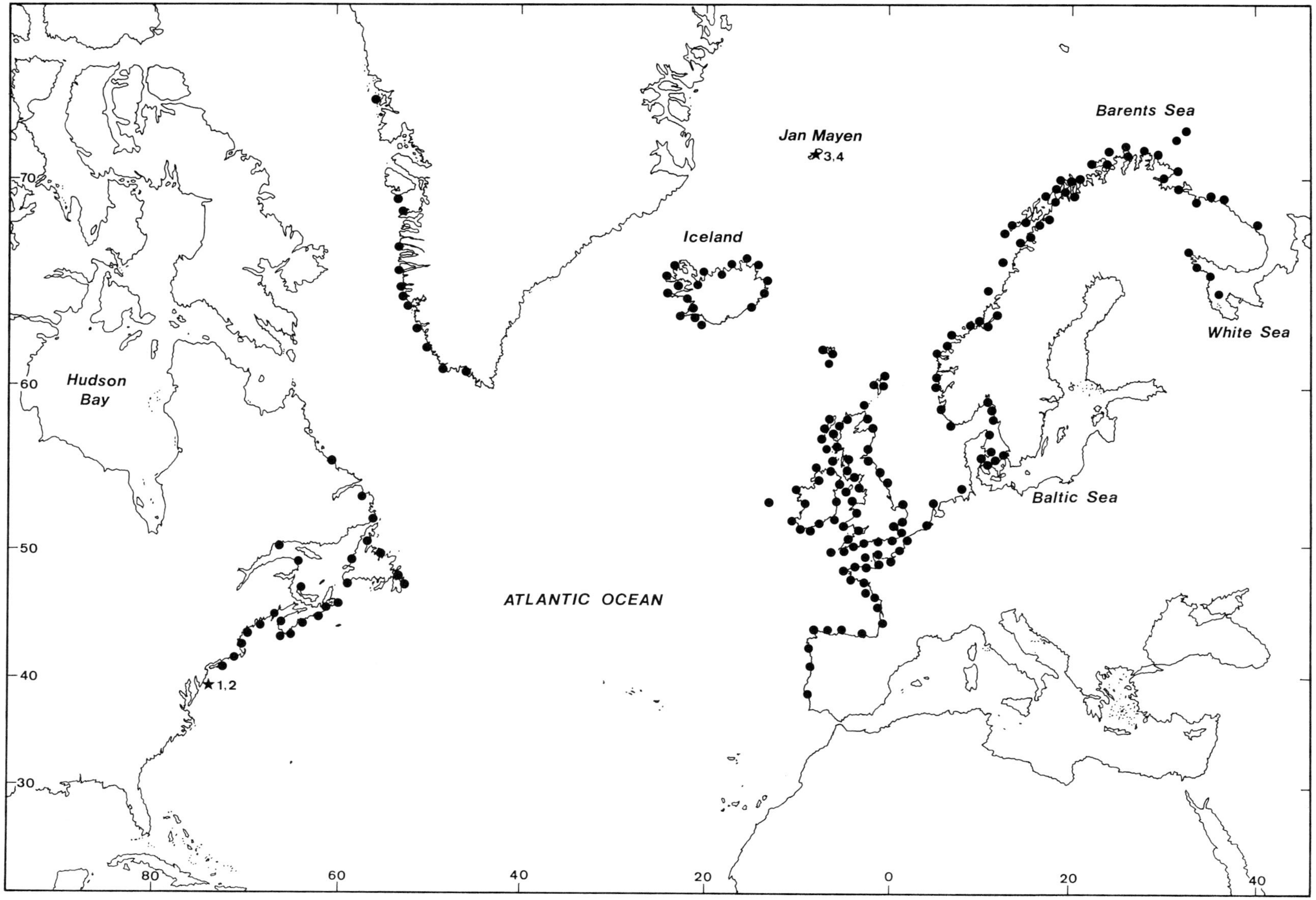

Figure 81. *Littorina (Neritrema) obtusata*: distribution map. Key: circles, specimens examined; asterisks, records from literature (1, Richards, 1929; 2, Bequaert, 1943; 3, Becher, 1886; 4, Sneli & Steinnes, 1975). There are in addition unconfirmed records from Hudson Strait (Halkett, 1898) and Hudson Bay (Pelletier, Wagner & Grant, 1968).

the coasts of Spain and Portugal as far as Gibraltar (Hidalgo, 1917; Fischer-Piette, 1953). By 1949 this range had contracted, so that it was not found east of Santander, and in 1956 the southern limit in Portugal was Leixoes (770 km north of Gibraltar) (Fischer-Piette, 1955, 1957a). By 1958, however, it had advanced 350 km southwards to Setúbal (Fischer-Piette, 1958). These fluctuations are presumably connected with climatic change, affecting the snails directly or their host algae. The scarcity of *L. obtusata s.l.* on the Basque coast is matched by the rarity of fucoids in the Mediterranean climate prevailing there. More recently *L. obtusata s.s.* has been collected in the lagoon at Laredo (DGR, 1987), at scattered localities further west along the northern Spanish coast, and is abundant in Galicia. *Littorina obtusata s.s.* is rare in Portugal, with one recent record from Aveiro (Coll. L. Burnay, pers. obs.) and another from Setúbal (USNM 837394, 1976), which thus seems to be the present southern limit. There are several eighteenth and nineteenth century records of *L. obtusata* from the Mediterranean and the Azores (Olivi, 1792; Philippi, 1836, 1844a; Requien, 1848; Hidalgo, 1867, 1917; Weinkauff, 1868; Jeffreys, 1883), but all of these must be considered as cases of human transport or incorrect localities, since there are no reliable records of living specimens (Hidalgo, 1867; Aradas & Benoit, 1870; Brusina, 1870; Coen, 1933; Dautzenberg & Fischer, 1915; Sabelli, Gianuzzi-Savelli & Bedulli, 1992). The absence of *L. obtusata s.l.* from the Mediterranean may be explained by the almost total lack of *Fucus* and absence of *Ascophyllum* there, as well as by high sea temperatures; the fact that it does not now occur on the Moroccan Atlantic coast is less easily explained (Barkman, 1955).

Littorina obtusata is native in North America, because it occurs in pre-colonial Indian middens (Bequaert, 1943; see Fossil Record, below), although the species must have recolonized the region from Europe following the last glaciation 13000 years ago (Ingólfsson, 1992). Rafting on floating algae provides one likely means of dispersal (Ingólfsson, 1995). The southernmost record is Cape May, New Jersey (Richards, 1929; Bequaert, 1943), but the northern range is unclear. In the present study specimens have been seen from throughout New England, Nova Scotia, the Gulf of St Lawrence, Newfoundland, and in Labrador as far north as Hopedale (BMNH, USNM). Packard (1867) recorded it as abundant in Labrador, but gave no further details. In a review of the marine molluscs of Arctic Canada, Macpherson (1971) listed only two records from further north and west. One was by Halkett (1898), who dredged the species in Hudson Strait between Ungava Bay and King George Sound; the other was by Pelletier, Wagner & Grant (1968) who recorded it living in southwestern Hudson Bay and in James Bay (though it was not found in the muddy, brackish waters of James Bay by Richards, 1936). No specimens could be traced in NMC from either Hudson Strait or Hudson Bay, but since the authors of both these records listed *L. saxatilis* as well, they are presumably reliable, and not misidentifications. *Littorina obtusata* does not occur in the Pacific, records of *L. arctica* by Reeve (1857) and Hanna & Hertlein (1959) being based on misidentifications of *L. sitkana*.

Fossil record. *Littorina obtusata* is common in latest Pleistocene and post-glacial deposits, throughout much of its present range, in Britain (Jeffreys, 1865, *s.l.*; Harmer, 1921; Berry & Smith, 1987, *s.l.*), Scandinavia (Jeffreys, 1865, *s.l.*; Brøgger, 1901; Antevs, 1928, *s.l.*), Iceland (Thoroddsen, 1892, *s.l.*; Bárdarson, 1923, *s.l.*), Greenland (Laursen, 1944, 1950, *s.l.*), the White Sea (Knipowitsch, 1900, *s.l.*), New England (Knight, 1934) and Hudson Bay (Pelletier *et al.*, 1968) (note that many of these records apply to *L. obtusata s.l.*). Fossils from raised beaches at Portland in Dorset (Baden-Powell, 1930, *s.l.*; *L. obtusata s.s.* D. Kadolsky, pers. comm.) and on the Gower Peninsula in Wales (Chatwin, 1932, *s.l.*; George, 1932, *s.l.*) date from 100 000 to 200 000 BP (Jones & Keen, 1993). At the latter site specimens were so common that the deposit was named the Neritoides Beach; samples from this deposit at Minchin Hole (120 000 BP) have been examined (BMNH) and, while most specimens were immature, a few were adults and could be confirmed as *L. obtusata*. In view of this frequent occurrence in the later Pleistocene, it is interesting that no earlier records have been traced. The species does not, for instance, occur in the Pliocene deposits at Tjörnes in Iceland (the record of *L. palliata* by Schlesch, 1924, was probably a misidentification of *L. squalida*), nor in the Plio-Pleistocene Crag formations of East Anglia (records by Harmer, 1921, were misidentifications of *L. littorea*). Records from the early Pleistocene of Morocco (Neuville & Ruhlmann, 1941; Lecointre, 1950; Brebion, 1979) and from archaeological sites in the Mediterranean (Cataliotti-Valdina, 1984) are referred to *L. fabalis* (p. 244). Fossil records of *L. palliata* from the Pacific are misidentifications of *L. squalida* (Ilina, 1939) and perhaps of *L. mandshurica* (Dall, 1920; MacNeil, Mertie & Pilsbry, 1943).

Discussion

Intraspecific variation. With its combination of colour polymorphism and morphometric variability, *L. obtusata* is among the most variable of all *Littorina* species. So far, the genetic basis of shell colour is poorly understood and that of shell shape has not been investigated. The few genetic studies of allozyme frequencies (e.g. E. M. Berger, 1973; Wilkins, O'Regan & Moynihan, 1978; Newkirk & Doyle, 1979; Rolán-Alvarez, 1993) support the idea that in this species with nonplanktotrophic development, and therefore limited dispersal, there is more genetic divergence among populations than in planktotrophic *Littorina* species. Nevertheless, Berger (1977) found a surprisingly high degree of genetic similarity between populations on either side of the Atlantic, at Cape Cod and Roscoff. It should be recalled that North America was probably only repopulated from Europe after the last glacial maximum 13 000 years ago (Ingólfsson, 1992). Since this species usually lives and spawns on fucoids, it is possible that dispersal may take place on floating weed. A clinal change at an esterase locus along an estuary has been interpreted as evidence of selection (Newkirk & Doyle, 1979).

In a wider context, *L. obtusata* is a suitable subject for studies of natural selection and other microevolutionary processes. Seeley (1986) argued that the replacement of the high-spired *palliata* form by the low-spired *retusa* form in historic times in New England was a result of intense directional selection by a predatory crab, so that in the fossil record rapid morphological transitions may not always correspond to speciation events. This particular case of intraspecific variation is a most interesting one, but further investigation of its genetic basis is required to confirm the interpretation. The colour polymorphism of *L. obtusata* also offers the potential for further experimental work, to investigate the relationship between the physiological correlates of colour and the visual appearance of the shell to predators, and the likelihood that these considerations vary over the wide latitudinal and habitat range. Comparisons with *L. fabalis* may be useful, for this sister-species occupies a lower intertidal zone where different conditions of microclimate and predation prevail. It has been suggested that shell shape and size in *L. obtusata* and *L. fabalis* represent two contrasting responses to the selection imposed by crab predation, the former achieving a degree of immunity by rapid growth to large size and the latter by attaining sexual maturity (and with it a thickened shell) at small size (Reimchen, 1982; Lowell *et al.*, 1994; Fletcher, 1995). Despite the difference in adult size, the strength of the shell is similar in the two species (Fletcher, 1995).

Epiphytic habit. In a genus of largely rock-dwelling snails it is remarkable that this pair of sister-species is so closely associated with macrophytes. Vermeij (1992a) has reviewed the origins of the dozen known specialized relationships between northern marine plants and herbivorous molluscs. He argued that fucoid algae probably originated in the northern Atlantic and, since the subgenus *Neritrema* was of Pacific origin, the association of *L. obtusata* and *L. fabalis* with fucoids must have arisen relatively recently, after the trans-Arctic interchange between the two oceans in the Pliocene. Snails may suffer less predation if they live on weeds rather than on the rock surface, which may explain the initial selective advantage of the association (Seeley, 1982b, 1983; Vermeij, 1992a). For both *Littorina* species, specializations for the association include behavioural attraction to certain fucoids (Barkman, 1955; van Dongen, 1956; Bakker, 1959), spawning on the algae, hiding in broken bladders by juveniles, and cryptic coloration. It is not known whether the apparent resemblance in size and shape of the littorinids to the bladders and receptacles of the fucoids is itself the result of selection (Gill *et al.*, 1976; Smith, 1976) or is a fortuitous result of selection for a low-spired shell form resistant to crushing by crabs. *Littorina fabalis* shows the additional specialization of synchrony between its annual life cycle and development of the lamina of *Fucus serratus*, at some localities (Williams, 1990a, 1992a). Whereas *L. fabalis* primarily grazes microalgae from the surface of its host plant, *L. obtusata* consumes the fucoid tissues as well (Watson & Norton, 1987; Williams, 1990a, b, 1992a, b). Since the Pacific *Neritrema* species are opportunistic, unspecialized grazers, it is unclear whether microalgal grazing is really the ancestral type, and *L. obtusata* the more specialized, as suggested by Vermeij (1992a). *Littorina obtusata* certainly appears to be physiologically adapted to its diet, for it is tolerant of the polyphenols that render *Fucus* and especially *Ascophyllum* unpalatable to *L. littorea* (Geiselman & McConnell, 1981; Seeley, 1983; Norton *et al.*, 1990). Some authors have gone further and suggested 'coevolution' between the littorinids and their host plants, citing the removal of deleterious epiphytes by *L. fabalis* (and perhaps by *L. obtusata*, at least in New England) as being of potential benefit to the hosts (Norton *et al.*, 1990; Williams, 1990a, b, 1992b). Nevertheless, there is no evidence for mutual selection in these cases and any benefit to the host may be coincidental. Furthermore, in neither case is the association with fucoids an obligate one, for both littorinids can survive on a variety of red algae and on seagrasses.

Phylogenetic relationships. Almost all the phylogenetic analyses of allozyme frequencies of *Littorina* species support the sister-species relationship of *L. obtusata* and *L. fabalis* (Knight & Ward, 1986; Ward, 1990; Zaslavskaya, Sergievsky & Tatarenkov, 1992a, b; Backeljau & Warmoes, 1992; p. 356). However, estimates of Nei's genetic distance vary greatly, from 0.087 (Morris, 1979, quoted in Ward, 1990) to 0.501 (Moyse *et al.*, 1982; review by Tatarenkov, 1995a). The cladistic analysis of morphological characters (p. 351) identifies the low-spired, globular shell shape as a synapomorphy supporting this relationship, and to this may be added other shell characters (absence of ribs, determinate growth, and an apparently homologous colour polymorphism) and the presence of five cusps on the inner marginal tooth. The close relationship of these two species is also supported by phylogenetic analyses of mitochondrial DNA sequences (Reid, Rumbak & Thomas, 1996; p. 360).

Since the modern geographical distributions of the two species are almost totally sympatric, nothing can be said about possible allopatric speciation. They show partial differentiation of habitat, diet and life history (summarized below), but it is not clear how this might have contributed to a sympatric speciation process. Crab predation seems to be an important influence on the shell morphology of these species; on sheltered shores more intense crab predation may contribute to the size divergence between them (Reimchen, 1982; see below) but, because of the considerable overlap in habitat between the two, it is difficult to envisage that disruptive selection of this kind can have been involved in the speciation process. The very limited fossil record of these species suggests that *L. fabalis* may perhaps be the older (p. 244) and that *L. obtusata* has been derived from it, in agreement with the arguably more specialized diet of the latter. From the divergence of their mitochondrial DNA sequences, Reid *et al.* (1996) estimated the age of separation between *L. obtusata* and *L. fabalis* as 1.32 Ma (range 0–3.49 Ma). Allozyme data suggest an older divergence, about 1.25 Ma (Tatarenkov, 1995a).

Similar species. Throughout most of its geographical range, *L. obtusata* is sympatric with its sister-species, *L. fabalis*, and there has been much confusion between the two in the past (see Taxonomic Histories, pp. 202, 229). While there are numerous morphological and ecological differences between these taxa (summarized in Table 6), only the characters of the adult reproductive system are unequivocal. In males the penis of *L. obtusata* has a short, triangular filament (10–25% of the total length of the penis) and 10–54 mamilliform penial glands in more than one row, whereas in *L. fabalis* the penial filament is long and vermiform (30–60% total length) and there are 3–17 glands in a single row. This character was for long considered to be the only reliable, morphological, means of separation of the two species (Sacchi, 1966a; Sacchi & Rastelli, 1966; Goodwin & Fish, 1977; Sneli & van Marion, 1979; Nielsen, 1980; Moyse *et al.*, 1982). A similarly unequivocal character has now been found for the separation of females, for in *L. obtusata* the copulatory bursa is large, extending back at least to the capsule gland, whereas in *L. fabalis* it is only half that length or smaller (Reid, 1990a). Although the penis is easy to observe, examination of the bursa requires dissection and is less straightforward.

In practice, however, with a little experience the two species can almost always be readily separated even in the field, for at most localities there are clear differences in the shells. Nevertheless, these differences are not constant; for field identification it is recommended that shells from the upper *Fucus vesiculosus* and *Ascophyllum nodosum* zone (where *L. obtusata* predominates) should be compared with those from lower on the same shore in the *F. serratus* zone (where *L. fabalis* predominates), in order to determine which shell characters are appropriate for discrimination, and that identifications using them should be confirmed by dissection before routine application in the field. The most obvious shell character is size; *L. fabalis* was first noticed as a 'dwarf' form of *L. obtusata s.l.* (Sacchi, 1961, 1962, 1963, 1964, 1966b, c), and at many localities it is indeed smaller, with little overlap in adult size (adults being recognized by the determinate shell growth; see Ontogenetic Variation, p. 214) (Sacchi & Rastelli, 1966; Sacchi, 1967b; Goodwin & Fish, 1977; Fletcher, 1995). This contrast in size is less marked on moderately exposed than on sheltered shores (Goodwin & Fish, 1977; Reimchen, 1982), although there are local exceptions (Sacchi, 1969a; Sacchi & Sconfietti, 1992) and indeed on a few shores it is even reversed, *L. fabalis* then being the larger (Janke, 1986, in Helgoland; some shores on the Channel coast of southeastern England, pers. obs.). Shell shape is a much more reliable guide;

Table 6. Comparison of *Littorina obtusata* and *L. fabalis*.

Character	*L. obtusata*	*L. fabalis*	*% accuracy*	*References*
Head-foot pigmentation	Darker	Paler	Variable	Sacchi, 1969a; Goodwin & Fish, 1977
Ovipositor	Usually pigmented	Usually unpigmented		
Penis	Fig. 76A–L	Fig. 82A–M	100	Sacchi & Rastelli, 1966; Goodwin & Fish, 1977; this study
Filament	Short, triangular, 10–25% total length	Long, vermiform, 30–60% total length		
Mamilliform glands	10–54, in irregular double row	3–17, in single row		
Paraspermatozoa	Fig. 3G. 11–16 μm diameter, vesicles to 6 μm diameter	Fig. 3H. 14–21 μm diameter, vesicles to 9 μm diameter		This study
Pallial oviduct	Fig. 76M–Q	Fig. 82N–R	100	Reid, 1990a
Copulatory bursa	Long, extending back to capsule gland	Short, less than half length of jelly gland		
Spawn				Goodwin & Fish, 1977
Shape of egg mass	Oval, or rarely kidney-shaped	Oval, or often kidney-shaped		
Mean ovum diameter	210–255 μm	195–200 μm		
Radula	Fig. 77	Fig. 83	87	Goodwin & Fish, 1977; Watson & Norton, 1987; this study
Lateral tooth	Usually 5 cusps	4 cusps		
Shell	Figs 78–80	Figs 84–86	Variable, but for adults >95% at most localities	Sacchi & Rastelli, 1966; Sacchi, 1967b, 1968b, 1969a, 1974; Goodwin & Fish, 1977; this study
Adult size	Usually larger than *L. fabalis* where sympatric	Usually smaller than *L. obtusata* where sympatric		
Relative columellar width	Thinner, C/LA usually <0.29	Thicker, C/LA usually >0.29		
Sexual dimorphism	Slight	Pronounced		
Shell colour	Olivacea common	Olivacea rare		
Ecology				Sacchi, 1969a; Watson & Norton, 1987; Williams, 1990b
Zone of max. abundance	*Fucus vesiculosus* and *Ascophyllum*	*Fucus serratus*		
Wave-exposure	Sheltered	Sheltered to moderately exposed		
Distribution	Fig. 81. Includes North America	Fig. 87. Excludes North America		This study

L. obtusata often has an angulation at the shoulder, giving the final whorl a square outline, and the aperture appears slightly constricted, whereas *L. fabalis* has a more rounded outline and is often slightly patulous (i.e. with flaring final whorl). These differences vary both with wave-exposure and with geographical region. A more useful characteristic is the relative width of the columella; if this is measured as columella width divided by aperture length (C/LA, see Fig. 2B), then for the 59 adult shells illustrated in Figs 78, 79, 84 and 85, the 95% confidence interval for the mean of *L. obtusata* is 0.221–0.256 and for *L. fabalis* 0.289–0.323 ($t=5.61$, $P<0.001$). For this sample of 59 shells (which necessarily includes many extreme geographical and ecotypic variants), if C/LA greater than or equal to 0.29 is taken as diagnostic of *L. fabalis*, then 75% of the shells are correctly determined; for less extreme samples, this accuracy is improved. Not only the columella, but also the entire shell of *L. fabalis* is often thicker, so that the internal dimensions of the aperture are relatively reduced (Goodwin & Fish, 1977). Another aspect of shell shape is sexual dimorphism, which is much more pronounced in *L. fabalis* (Sacchi, 1968b; Goodwin & Fish, 1977; Sacchi & Sconfietti, 1992).

In the southern geographical varieties of the two species (i.e. *retusa* and *typica*) shell colour can also be helpful, for on sheltered shores the dominant colour morph of *L. obtusata* is olivacea (which is extremely rare in most populations of *L. fabalis*) and of *L. fabalis* is citrina (Sacchi, 1967b, 1969a, 1974; Sacchi & Rastelli, 1966; Goodwin & Fish, 1977). However, on moderately exposed shores both often show high frequencies of reticulata. Shell sculpture has also been mentioned as a useful character; in Wales, Williams (1990b) found adult shells to be smooth in *L. obtusata* and sculptured with microstriae in *L. fabalis*. As shown earlier (p. 214), shell microsculpture is in fact rather variable. In the northern parts of the range *L. obtusata* is usually more strongly sculptured than *L. fabalis*, in both juveniles and adults (compare Figs 80E and 86C, D). In many cases, however, juveniles of the two species are indistinguishable by their shape and sculpture (compare Figs 80C, D and 86A, B). In these cases shell colour can sometimes be a useful guide, if adults show different colour frequencies at the locality. It can sometimes be difficult to separate juvenile citrina and olivacea shells using exterior colour, but in citrina it will be found that the interior is always white, in olivacea it is tinged purple or brownish. Reimchen (1989) described the white-spiral juvenile pattern only in *L. fabalis* (p. 241); it is indeed much more common in this species, but has occasionally been seen in *L. obtusata*. In summary, shells alone are usually sufficient for identification, especially in large samples in which both species can be compared from a single locality. There remain some geographical regions where discrimination of shells is difficult, for example the coast of Denmark where brackish ecotypes are common (Nielsen, 1980) and the south coast of England between Dorset and Kent.

In the species as a whole *L. obtusata* more frequently has a darkly pigmented animal (Sacchi, 1969a), and in some regions coloration of the head-foot can be useful for discrimination. In Wales Goodwin & Fish (1977) found the ovipositor to be colourless in *L. fabalis*, but to varying degrees blackish in *L. obtusata*, giving 99% accuracy of identification. This character has been used successfully in Galicia (Rolán & Templado, 1987) and Ireland (Matthews & Montgomery, 1987), but is not reliable elsewhere (Nielsen, 1980; Jardine, 1984; Warmoes *et al.*, 1988; Reid, 1990a). Other morphological differences have been described, but are likely to be of limited use for identification. The spawn mass of *L. fabalis* is more often kidney-shaped, and the eggs are larger (Goodwin & Fish, 1977). The paraspermatozoa of *L. fabalis* are larger (this study). There are subtle differences in the radula; in *L. fabalis* there are four cusps on the lateral tooth, in *L. obtusata* usually five; supposed differences in the prominence of the cusps of the outer marginal teeth (Goodwin & Fish, 1977; Watson & Norton, 1987) seem mainly correlated with shell size (this study).

Ecological differences between the two species are now well known, but there is much overlap between them. Typically, in areas with a significant tidal range, *L. fabalis* occupies a lower level on the shore, reaching maximum abundance in the *Fucus serratus* zone, whereas *L. obtusata* is most frequent among *Fucus vesiculosus* and *Ascophyllum* (Sacchi, 1969a; Janke, 1986; Watson & Norton, 1987; Williams, 1990b, 1994, 1995). On the atidal shores of Denmark (Nielsen, 1980) and Sweden (Ekendahl, 1994), however, differential zonation is less clear or even absent, and the degree of wave-exposure is then the important factor determining the occurrence of the species. *Littorina fabalis* is usually more tolerant of wave-exposure (Sacchi & Rastelli, 1966; Williams, 1990b; Ekendahl, 1994), but at the opposite extreme sometimes extends further into brackish habitats than *L. obtusata* (Nielsen, 1980; but see Smith & Little, 1980, and Williams, 1994, for an opposite result in the Severn Estuary). At least on some shores *L. fabalis* has a shorter life cycle (Sacchi & Rastelli, 1966; Williams, 1992a). Differences have also been noted in behaviour (Sacchi, 1969a), environmental tolerance (Sacchi & Rastelli, 1966; Sacchi, 1967b, 1969a, 1972a, b) and sex ratio (Sacchi, 1969a, 1984). Several studies have found fixation of different alleles at certain genetic loci (Moyse *et al.*, 1982; Zaslavskaya *et al.*, 1992a, b; Rolán-Alvarez, Zapata & Alvarez, 1995a; Tatarenkov, 1995a).

In the southern parts of the range, where both species are usually low-spired, confusion with other *Littorina* species is unlikely. The sheltered/brackish ecotypes with taller spires could possibly be confused with smooth shells of *L. saxatilis*, but there the spire is never domed to such a degree. Both penial shape and oviduct characters also clearly separate *L. saxatilis*. In northern regions confusion is more likely, for spires of *L. obtusata* and *L. fabalis* are taller and doming less marked. Even here discrimination is usually straightforward, for *L. saxatilis* is mostly ribbed and lacks the strong, regular microstriae of *L. obtusata*.

Littorina (Neritrema) fabalis (W. Turton, 1825)

(Figs 3H, 82–88)

Synonymy

Turbo obtusatus—Chemnitz, 1781: 234; pl. 185, fig. 1854 e, f (probably *L. fabalis*; non-binominal, *Turbo obtusatus* in synonymy; in part, other figs *L. obtusata*; not Linnaeus, 1758=*L. obtusata*).

Littorina obtusata—Troschel, 1858: 134; pl. 11, fig. 1 (radula, probably *L. fabalis*) (not *Turbo obtusatus* Linnaeus, 1758). Meyer & Möbius, 1872: 15–16; figs (facing p. 14) 6, 7 (animal); figs 8, 9; fig. 10 (operculum); fig. 11 (radula) (in part, includes *L. obtusata*; not Linnaeus, 1758). Sowerby, 1887: pl. 12, fig. 22 (in part, other figs are *L. obtusata*; not Linnaeus, 1758). Hidalgo, 1890: pl. 89, fig. 15 (in part, other figs are *L. obtusata*; not Linnaeus, 1758). Linke, 1933: 1–60 (anatomy); pl. 2, figs 20, 23 (penis) (in part, includes *L. obtusata*; not Linnaeus, 1758). Neuville & Ruhlmann, 1941: 129; pl. 8, fig. 5 (probably *L. fabalis*; Pleistocene; not Linnaeus, 1758). Sacchi, 1966a: 2370–2372; fig. 2 (penis) (in part, includes *L. obtusata*; not Linnaeus, 1758). Lucas, 1968: (3): 11–14; (4): 3–8 (in part, includes *L. obtusata*; not Linnaeus, 1758). Cataliotti-Valdina, 1984: 213–215; fig. 1: 9, 10 (probably *L. fabalis*) (includes *L. obtusata*; not Linnaeus, 1758; Pleistocene). Jagnow & Gosselck, 1987: 211; pl. 2, fig. 10 (in part, includes *L. obtusata*; not Linnaeus, 1758). de Bruyne, 1991: 99; fig. 63 (left) (in part, other fig. is *L. obtusata*; not Linnaeus, 1758).

Littorina (Neritoides) obtusata—Kobelt, 1907: 64–66; pl. 111, fig. 9 (in part, other figs are *L. obtusata*; not *Turbo obtusatus* Linnaeus, 1758).

Littorina (Littorina) obtusata—Nordsieck, 1982: 61; pl. 26, fig. 23.00a, d (in part, other figs are *L. obtusata*; not *Turbo obtusatus* Linnaeus, 1758).

Turbo neritoides—Chemnitz, 1781: 234–235; pl. 185, fig. 1854: 9, 10 (probably *L. fabalis*; non-binominal, *Turbo neritoides* in synonymy; in part, other figs *L. obtusata*; not Linnaeus, 1758=*Melarhaphe neritoides*).

Littorina neritoides—Brown, 1844a: 128, 141; pl. 13, figs 15, 23 (in part, other figs are *L. obtusata*; not Linnaeus, 1758).

Turbo fabalis Turton, 1825: 366; pl. 13, fig. 10 (Scarborough [Yorkshire, England]; lectotype (here designated, 3.6x4.3 mm, Fig. 84B)+3 paralectotypes, BMNH 1968356, seen; 7 possible paralectotypes, USNM 185758, seen). Brown, 1827: pl. 45, figs 38, 39. Brown, 1837a: 17; pl. 10, figs 38, 39. Thorpe, 1844: 167.

Littorina fabalis—Brown, 1844a: 128; pl. 10, figs 38, 39. Forbes & Hanley, 1850: 49–50; pl. 86, figs 2, 3; 1852: 262. Reeve, 1857: sp. 90; pl. 16, fig. 90a, b. Sowerby, 1859: pl. 12, fig. 22.

Litorina fabalis—Philippi, 1848: **3**: 66; *Litorina* pl. 7, fig. 18.

Littorina littoralis var. *fabalis*—Jeffreys, 1859: 195.

Littorina obtusata var. *fabalis*—Jeffreys, 1865: **3**: 357–358. Mørch, 1869: 207. Marshall, 1898: 121.

Littorina (Neritrema) fabalis—Récluz, 1869: 47. Smith, 1982: 535–539.

Littorina (Neritoides) obtusata var. *fabalis*—Tryon, 1887: 253; pl. 46, fig. 22.

Littorina obtusata var. ex col. *fabalis*—Dautzenberg, 1893: 20. Dautzenberg & Durouchoux, 1900: 46.

Littorina obtusata littoralis var. ex forma et colore *fabalis*—Dautzenberg & Durouchoux, 1913: 24.

Neritoides littoralis—Brown, 1827: pl. 43, fig. 15 (in part, other figs are *L. obtusata*; not *Nerita littoralis* Linnaeus, 1758=*Theodoxus fluviatilis* (Linnaeus, 1758)). Brown, 1843a: 25–26; pl. 13, fig. 15 (in part, other figs are *L. obtusata*; not Linnaeus, 1758).

Littorina obtusata littoralis—Lecointre, 1950: 241; figs 15, 16 (Pleistocene; probably *L. fabalis*; not *Nerita littoralis* Linnaeus, 1758). Lecointre, 1952: 97; pl. 27, figs 2, 3 (Pleistocene; probably *L. fabalis*; not Linnaeus, 1758).

Littorina litoralis—Warén, 1983: 166 (not *Nerita littoralis* Linnaeus, 1758).

Turbo expansus Brown, in J. Smith, 1839: 101–102; pl. 1, figs 12, 13 (Dalmuir [Clyde, Scotland]; types not traced; Pleistocene). Brown, 1849: 254; pl. 33*. figs 54, 55 (Pleistocene; as *expansa*).

Litorina arctica var. *retusa* Møller, 1842: 82 (Greenland; 5 original specimens ZMK GAS-86, seen, 4 including largest are *L. fabalis*, 1 is *L. obtusata*; 2 original specimens BMNH 1843.6.30.703–4, seen, both are *L. fabalis*; *nomen nudum*).

Littorina beanii MacGillivray, 1843: 139–140 (unjustified emendation of *Turbo fabalis* Turton, 1825).

Litorina obtusata [var.] *picta* Menke, 1845: 57 (Helgoland [Germany]; lectotype (here designated) Chemnitz, 1781: pl. 185, fig. 1854e, f).

Litorina limata Lovén, 1846: 154 (Boh.–Finm. [Bohuslan, Sweden, to Finnmark, Norway], here restricted to Finnmark; lectotype (here designated) SMNH 4591, seen; 13 paralectotypes, SMNH 4592, seen, 9 are *L. fabalis*, 4 are *L. obtusata*; 9 paralectotypes SMNH 4593, seen, 6 are *L. fabalis*, 3 are probable *L. obtusata*; 11 paralectotypes SMNH 3856, 4 are possible *L. fabalis*, 6 are *L. obtusata*, 1 is *L. saxatilis*; see remarks under Taxonomic History, p. 230). Philippi, 1848: **3**: 65; *Litorina* pl. 7, fig. 13. Küster, 1853: pl. 2, figs 10, 11; 1856: 13–14 (probably *L. fabalis*).

Littorina limata—Petit de la Saussaye, 1869: 122.

? *Nerita subtruncata* 'Chiereghini' Nardo, 1847: 95–96 ('Istria' [Italy; in error, not from Adriatic, Brusina, 1870]; types not traced; refers to Chiereghini MS figs 1157–1160; see remarks under Taxonomic History, p. 230).

? *Litorina subtruncata*—Nardo, 1847: 95–96.

Litorina vittata Philippi, 1848: **3**: 64; *Litorina* pl. 7, fig. 11 (Litus Britanniae gallicae; types not traced). Küster, 1853: pl. 2, figs 8, 9; 1856: 13.

Littorina (Neritrema) vittata—Récluz, 1869: 47 (*s.l.*).

Littorina obtusata littoralis var. ex forma *vittata*—Dautzenberg & Fischer, 1915: 115–116; pl. 2, figs 16–18.

Littorina (Littorina) obtusata var. *vittata*—Nordsieck, 1968: 40.

Littorina (Littorina) obtusata form *vittata*—Nordsieck, 1982: 61; pl. 26, fig. 23.00a.

Littorina rudis—Clark, 1850a: 362–364 (in part, includes *L. saxatilis* and *L. obtusata*; not *Turbo rudis* Maton, 1797=*L. saxatilis*).

? *Littorina obtusata* var. *compacta* Jeffreys, 1865: **3**: 357–358 (Loch Torridon and other parts of the Ross-shire coast [Scotland]; types not traced (Warén, 1980); not *Littorina compacta* Gabb, 1864).

Littorina bizonalis Récluz, 1869: 47 (Nos côtes de la Manche [French Channel coast]; in synonymy of *L. vittata*; *nomen nudum*).

Littorina fabula 'Turton' Kobelt, 1907: pl. 112, fig. 19 (error for *fabalis*; in text this figure is listed as *L. saxatilis*).

Littorina palliata—Kobelt, 1907: 62–63; pl. 111, fig. 8 (in part, other figs are *L. obtusata*; not *Turbo palliatus* Say, 1822=*L. obtusata*).

Littorina obtusata littoralis var. ex forma *minima* Dautzenberg & Fischer, 1915: 112; pl. 2, fig. 11 (Kérity, near Paimpol [France]; 1 spec., original of fig. 11, MNHNP, seen; infrasubspecific).

Littorina (Littorina) obtusata var. *minima*—Nordsieck, 1982: 61; pl. 26, fig. 23.00d (probably *L. fabalis*).

Littorina obtusata littoralis var. ex forma *pachychila* Dautzenberg & Fischer, 1915: 113; pl. 2, fig. 12 (Tour de Croy and pointe-aux-Oies, Wimereux; la Pointe de Rochebonne, near Paramé [both France]; Jersey; 1 spec., original of fig. 12, MNHNP, seen; infrasubspecific). Kaas & ten Broek, 1942: 147.

Littorina obtusata littoralis var. ex colore *fusca* 'Menke' Dautzenberg & Fischer, 1915: 118–119; pl. 2, fig. 17 (in part, other fig. is *L. obtusata*; infrasubspecific; not *L. fusca* Pfeiffer, 1840=*Nodilittorina mespillum* (Mühlfeldt, 1824)).

Littorina obtusata littoralis var. ex colore *inversicolor* Dautzenberg & Fischer, 1915: 122–123; pl. 2, figs 11, 18 (in part, other fig. is *L. obtusata*; infrasubspecific).

Littorina obtusata littoralis var. ex colore *inversicolorreticulata* Dautzenberg & Fischer, 1915: 123; pl. 3, fig. 7; pl. 4, fig. 10 (colour chart) (la Trinité-sur-Mer; La Pointe de Rochebonne; Chalibert, embouchure de la Rance; Rothéneuf; la Turballe [all France]; 1 spec., original of fig. 7, MNHNP, seen; as *inversicolor-reticulata*; infrasubspecific).

Littorina obtusata littoralis var. ex colore *reticulata* 'Lister' Dautzenberg & Fischer, 1915: 124–125; pl. 3, figs 5, 6; pl. 4, fig. 11 (colour chart) (Guéthary; Croisic; la Trinité-sur-Mer; Quiberon; Kerléon en Guipavas; Île de Bréhat; Val-André; Saint-Malo; Granville; Cherbourg; Dieppe [all France]; Jersey; Isle of Wight; Helgoland [Germany]; 2 specs, originals of figs 5 and 6, MNHNP, seen; *s.l.*; infrasubspecific; not *Litorina reticulata* Anton, 1838=*Nodilittorina* sp.).

Littorina obtusata littoralis var. ex col. *ziczac* Dautzenberg & Fischer, 1915: 125–126; pl. 3, fig. 9; pl. 4, fig. 12 (colour chart) (Le Croisic [France]; 1 spec., original of fig. 9, MNHNP, seen; in part, other fig. probably *L. obtusata*; infrasubspecific; not *Trochus ziczac* Gmelin, 1791=*Nodilittorina ziczac*).

Littorina mariae Sacchi & Rastelli, 1966: 351–369; figs 1, 2D, D', 3, pl. 11, fig. 2 (penis) (Grèves de Roscoff, Finistère, Bretagne septentrionale [northern Brittany, France]; 48 syntypes MCSNM, Fig. 84G, seen). Sacchi, 1967b: 339–355. Sacchi, 1968a: 213–219. Sacchi, 1968b: 2483–2485. Gratton, 1969: 53–57; fig. 1 (penis). Sacchi, 1969a: fig. 6e, f (penis); fig. 7e, f. Sacchi, 1969b: 33–47. Sacchi, 1972: 105–137. Sacchi, 1974: 64–75. Goodwin & Fish, 1977: 241–254; fig. 5 (radula); fig. 7 nos 2, 5. Reimchen, 1979: 1070–1085. Sneli & van Marion, 1979: fig. 1e; fig. 1i (penis). Fretter & Graham, 1980: 261–262; fig. 205D, E. Reimchen, 1981: 341–350. Moyse, Thorpe & Al-Hamadani, 1982: 7–15. Reimchen, 1982: 687–695. Jardine, 1984: 5; fig. 2b. Rolán, 1984: 107–108; figs. Sacchi, 1984: 209–232. Janke, 1986: 10–11. Rolán & Templado: 1987: 71–85; fig. 2B–D (penis); fig. 3 (protoconch); fig. 4; fig. 6 (radula) (discussed as *L. mariae*, *L.* A, *L.* B). Watson & Norton, 1987: fig. 2B (radula). Graham, 1988: 172; fig. 62 E, F. Warmoes, Backeljau & de Bruyn, 1988: 51–70; fig. 4L, M (penis); fig. 5C, D, F. Reimchen, 1989: 97–109; fig. 1. Hayward, Wigham & Yonow, 1990: 660; fig. 12.9. Reid, 1990a: 113–114; fig. 1B (oviduct). Poppe & Goto, 1991: 92; pl. 11, figs 5–7 (in part, other figs are *L. obtusata*). Sacchi & Sconfietti, 1992: 105–118. Tatarenkov & Johannesson, 1994: 105–125. Rolán-Alvarez, Zapata & Alvarez, 1995a: 1–9.

Littorina (Neritrema) mariae—Reid, 1989a: 102. Reid, 1990b: 1–19; fig. 1p. Taylor & Reid, 1990: 199–215 (mineralogy). Smith & Heppell, 1991: 17, 81.

Littorina mariae beeensis S. M. Smith, in Fretter & Graham, 1980: 262 (type locality (Smith, 1981) Duart Lochan, Loch Nedd, 58°14.9'N, 05°10.5'W [Sutherland, Scotland]; numerous syntypes NMS 1979075.11013, seen).

Littorina vadorum Sacchi, 1984: 220 (*nomen nudum*).

Note: for references in which *L. fabalis* and *L. obtusata* were not clearly differentiated, and which may have included both species, see Synonymy of *L. obtusata* (p. 196).

Taxonomic history

The history of this species is intimately bound up with that of the other member of the '*obtusata* group', *L. obtusata* itself, since these two species share superficially similar shells which were often not correctly distinguished. (The Synonymy and Taxonomic History of *L. obtusata*, pp. 196–206, should also be consulted for a discussion of the confusion between these two species). However, now that the two can be distinguished by unequivocal anatomical criteria, and the range of intraspecific shell variation is known, it is clear that shells alone can usually be identified with ease. Type specimens and figures can thus be allocated to the correct species, showing that a number of the species-level taxa, and many varieties, named in the *obtusata* group were based wholly or in part on forms of *L. fabalis*.

The oldest of these names is *Turbo fabalis*, named in honour of W. Bean by Turton in 1825. The brown, tessellated shell was well illustrated and was said to be very small, of 'diameter about a line' (i.e. 2.11 mm). Some later authors have therefore assumed that Turton was describing juveniles of *L. obtusata*, or that the species was unidentifiable. In fact, of four syntypes in the BMNH the largest is 3.6 mm tall by 4.3 mm in diameter; this shell contained a dried male animal, which has been rehydrated and the penis examined. The penis has six penial glands and a long filament (about 40% of the total penial length), confirming that this is indeed the species to which the name is here applied. This specimen is here designated the lectotype. Some authors followed Turton in recognizing *L. fabalis* as a species distinct from *L. obtusata* (Brown, 1827, 1837a, 1844a; Thorpe, 1844; Philippi, 1846–1848; Forbes & Hanley, 1850; Reeve, 1857; Sowerby, 1859; Récluz, 1869), but like him they were all using the name in a restricted sense to apply only to small, obtuse, brown, tessellated shells. MacGillivray (1843) also accepted this as a species, but objected to Turton's 'playful' Latinization of the name of the dedicatee and so renamed it *L. beanii*. Although Forbes & Hanley (1850) explicitly differentiated *L. fabalis* from juveniles of *L. obtusata*, they nevertheless suggested that the two would prove to be conspecific. Meanwhile, Clarke (1850, 1855) synonymized *L. fabalis* with *L. rudis* (=*L. saxatilis*), Sowerby (1859) suggested *L. fabalis* was the young of *L. obtusata*, and Jeffreys (1859) reduced it to a variety of *L. obtusata*, of which it was said to be the young males (Jeffreys, 1865). Thereafter, it was generally mentioned, if at all, only as a variety of *L. obtusata* (Mørch, 1869; Tryon, 1887; Dautzenberg, 1893; Marshall, 1898; Dautzenberg & Durouchoux, 1900, 1913) or as a synonym of it (Weinkauff, 1878; Kobelt, 1888; Dautzenberg & Fischer, 1915; Hidalgo, 1917). There followed a long gap before the name was again mentioned, by Smith (1982), as a valid senior synonym of *L. mariae*. Since then both Warén (1983) and Sacchi (1984) have considered it a synonym of *L. obtusata*.

Several other available names have also been based on examples of the present species. *Turbo expansus* Brown, in J. Smith, 1839, was established for a Pleistocene fossil from Scotland and, although types are probably lost and the original figure is somewhat stylized, the patulous shape, thick columella and apparent zigzag pattern support the identification. No types have been found for *Litorina obtusata picta* Menke, 1845, and the figures cited are a mixture of possible *L. obtusata* and certain *L. fabalis*; since at the type locality, Helgoland, the tessellated colour form described is usually *L. fabalis* (large samples in ZMB, BMNH), this concept of the name is here fixed by designation of a lectotype figure. The original description of *Litorina limata* Lovén, 1846, was inadequate for recognition of the species as either *L. obtusata* or *L. fabalis* but, of 34 syntypes in SMNH, 20 are *L. fabalis*, 13 are *L. obtusata* and one is *L. saxatilis*. The name *limata* has been little used subsequently, but both Philippi (1848) and Küster (1853) used it for the relatively tall-spired northern form of *L. fabalis*, and designation of a lectotype establishes this concept. *Nerita subtruncata* 'Chiereghini' Nardo, 1847, is tentatively included in the synonymy of *L. fabalis*, on account of the description of the shell as whitish with red-brown speckling (Nardo, 1847) and Brusina's (1870) observation that it is simply a small example of *L. obtusata*, 5 mm long and 6 mm wide. The determination of *Litorina vittata* Philippi, 1848, is quite certain, for the shell with enlarged aperture and thick columella is of a type seen only in *L. fabalis* (although the bizoned colour pattern which the author regarded as distinctive is in fact also seen in examples of *L. obtusata*). Philippi's species was accepted by Küster (1853) and by Récluz (1869), but was subsequently placed in the synonymy of *L. obtusata* by Weinkauff (1878) and Tryon (1887), and since then has been used only twice as a varietal name (Nordsieck, 1968, 1982). *Littorina obtusata* var. *compacta* Jeffreys, 1865, is tentatively included under *L. fabalis*, because it was described as smaller than typical *L. obtusata* and thick-set. There is no doubt about the determination of two form varieties and three colour varieties named by Dautzenberg & Fischer (1915), since the figured specimens are preserved in MNHNP, but these taxa are infrasubspecific. Thus many nineteenth century and a few later authors gave taxonomic recognition to forms now identifiable as the present species. However, since none of them examined anatomical characters, they did not recognize the broad range of shell types in this species that is now possible.

The modern concept of this species was established by Sacchi & Rastelli, who in 1966 described a new species, *L. mariae*. The existence of this second species in the *obtusata* group was first suggested by the observation of a dimorphism in shell size ('dwarf' forms predominating at exposed localities) and correlated differences in the proportions of shell colour morphs (Sacchi, 1961, 1962, 1963, 1964, 1966b, c). Most importantly, 'dwarf' and 'normal' forms differed in the shape of the penis (Sacchi, 1966a), and this character above all was taken to be diagnostic of *L. mariae*. Unfortunately, Sacchi & Rastelli (1966) made almost no reference to previous taxonomic work and gave no reliable characters for separating the shells of the two species. Consequently, it was not possible to trace the nomenclatural history of the newly-described species, and the name *L. mariae* was for a while unchallenged. The recognition of this additional species within the supposedly well-known taxon *L. obtusata* at first gained slow acceptance, despite continuing studies of its ecology, physiology, morphometrics and colour polymorphism (Sacchi, 1967b, 1968a, b, 1969a, b, 1970, 1972a, b, 1974; Gratton, 1969). Wider recognition followed a more detailed morphological comparison of shells and radulae of the two species by Goodwin & Fish (1977) and systematic accounts by Sneli & van Marion (1979) and Fretter & Graham (1980). Since then electrophoresis of allozymes (Morris, 1979; Wilkins & O'Regan, 1980; Moyse, Thorpe & Al-Hamadani, 1982; Warmoes, 1986; Backeljau & Warmoes, 1992; Zaslavskaya, Sergievsky & Tatarenkov, 1992a, b; Rolán-Alvarez, Zapata & Alvarez, 1995a; Tatarenkov, 1995a) and studies of the oviduct (Reid, 1990a) have amply confirmed the specific status of the two. There have been numerous additional ecological studies (e.g. Reimchen, 1979, 1981, 1982; Nielsen, 1980; Sacchi, 1984, 1985; Watson & Norton, 1987; Sacchi & Sconfietti, 1988, 1992; Williams, 1990a, b, 1992a, b, 1994, 1995) and *L. mariae* is now described even in textbooks (Graham, 1988; Hayward, Wigham & Yonow, 1990; Poppe & Goto, 1991).

Although criteria for separating the shells of *L. mariae* from *L. obtusata* have been given by several authors (Goodwin & Fish, 1977; Fretter & Graham, 1980, Warmoes, Backeljau & de Bruyn, 1988), there has so far been little attempt to trace the synonymy of *L. mariae* in the older literature. Smith (1982) concluded from Turton's (1825) figure that his *Turbo fabalis* was a senior synonym of *L. mariae* and recommended adoption of the older name. Sacchi (1984) responded that the diagnosis and figure could apply to a young *L. obtusata*, but that conchological criteria alone were insufficient for certain identification. Later Smith & Heppell (1991) decided that current usage was to be preferred and employed the name *L. mariae*. Type specimens of *Turbo fabalis* have now been discovered (Warén, 1983; herein), and the penial shape of the lectotype shown to be the same as that of *L. mariae*. *Turbo fabalis* is therefore indeed a senior synonym, and it must be decided whether adoption of the older name is desirable. As mentioned earlier, *fabalis* has been used as a valid name only once since 1913 (by Smith, 1982). However, the Code of Zoological Nomenclature (ICZN, 1985) no longer contains any mandatory rules on 'forgotten names'. Meanwhile, following its description in 1966, the name *L. mariae* has appeared in the taxonomic literature only since about 1980. Although it is at present widely used by both taxonomists and ecologists, many workers on *Littorina* are aware of the name *fabalis* and of the previous uncertainty surrounding its identity. Formal adoption of *mariae* as the valid name for this taxon would require not only suppression of *fabalis* by the ICZN, but also of the other seven available names (see preceding discussion and Synonymy). Since *mariae* has not long been established in the literature and *fabalis* is neither dubious nor forgotten, it has been decided to employ the latter name for this taxon.

Littorina fabalis shows considerable intraspecific variation in shell form, which has led some workers to suggest that additional taxa could be involved, although without proposing formal names (Reimchen, 1981; Rolán & Templado, 1987). This is discussed in the section on Shell Variation (p. 235). A brackish-water form from northwestern Scotland was mentioned by S. M. Smith (1979b, 1981), but a formal description did not appear. However, Smith's manuscript name *L. mariae beeensis* was quoted and briefly described by Fretter & Graham (1980) and this appears to satisfy the criteria of availability. This was apparently overlooked by Smith & Heppell (1991), who listed *beeensis* as a *nomen nudum*. The name '*Littorina vadorum*', attributed to 'des vieux auteurs anglais', was mentioned by Sacchi (1984) in connection with a similar form from *Zostera* meadows in Galicia. No reference to this name has been found in the older literature, however, so it is treated as a *nomen nudum*.

Diagnosis

Mature shell height 3.0–14.9 mm. Shell low-spired, globular or turbinate, sometimes patulous, usually with thick anterior columellar lip; relative columellar width (C/LA) greater than 0.29. Mineralogy of external irregular-prismatic calcite and internal crossed-lamellar aragonite. Sculpture smooth or with periostracal microstriae over whole surface. Colour highly variable: white, yellow, orange or brown, sometimes with reticulate or zigzag pattern, or with two broad dark bands; columella white, sometimes tinged with purple. Penis with long, vermiform filament, 30–60% total length, 3–17 mamilliform penial glands. Pallial oviduct of single loop of albumen gland, single loop of capsule gland and swollen, septate jelly gland; bursa short, less than half length of jelly gland. Spawn a benthic mass of singly-encapsulated eggs in gelatinous matrix; development nonplanktotrophic with intracapsular metamorphosis. Distribution: White Sea to southern Portugal, Iceland, southwestern Greenland.

Anatomical description

Head-foot. Pigmentation of head, tentacles and sides of foot varies from unpigmented to grey to entirely black. In lightly pigmented animals head and tentacles are usually darker than foot. Tentacles usually unpatterned, but sometimes show a single dark longitudinal stripe. The degree of pigmentation shows some correlation with shell colour; white, yellow, orange and pale reticulated shells contain unpigmented or pale grey animals, dark brown reticulated shells usually contain grey or black animals (Sacchi, 1969a, 1974; pers. obs.). For illustrations of variability of pigmentation see Rolán & Templado (1987, as *L. mariae*, *Littorina* A and B). Opercular ratio 0.38–0.61.

Penis (Fig. 82A–M). Base wrinkled, unpigmented or sometimes with grey to black pigmentation on outer surface; 3–14 (up to 17, Goodwin & Fish, 1977) mamilliform penial glands in single row. Filament long, vermiform, tip rounded or pointed, 30–60% total length, unpigmented. Sperm groove open, ending terminally. No seasonal change in penis (Sacchi, 1969a).

Sperm (Fig. 3H). Euspermatozoa 39–51 μm. Paraspermatozoa 14–21 μm diameter, round; vesicles unusually large (to 9 μm diameter), distinct; nucleus not visible.

Oviduct (Fig. 82N–R). Single loop of albumen gland; followed by single loop of capsule gland of about half to equal area of albumen gland, translucent part of capsule gland small, not usually visible except in histological sections; finally a swollen, septate jelly gland, unpigmented or containing grey to blackish pigment. Bursa small, opening at anterior end of jelly gland and extending back no more than halfway to anterior limit of capsule gland (Reid, 1990a). Seminal receptacle often lobulate.

Spawn and development. Few observations have been made specifically on the spawn of *L. fabalis*, although some of the descriptions of *L. obtusata s.l.* may have included this species (p. 206). Individual spawn masses up to about 10 mm long are laid on the thallus of algae, especially *Fucus serratus*, and are oval or kidney-shaped (Goodwin & Fish, 1977). The mean number of eggs in each gelatinous spawn mass ranges from 62 to 120; eggs are 195–200 μm in diameter and are enclosed singly in thin, almost spherical capsules 520–600 μm in diameter Goodwin & Fish, 1977; Gallardo & Götting, 1985; Rolán & Templado, 1987). Development is nonplanktotrophic with intracapsular metamorphosis. The spawning season has been recorded as spring to autumn in Wales (Reimchen, 1979) and in Brittany and Galicia (Sacchi, 1969a), although others have found spawn throughout the year (Goodwin & Fish, 1977, in Wales, most abundant from January to August; Gallardo & Götting, 1985, Helgoland and Brittany).

Radula and foregut (Fig. 83). Relative radular length 0.87–1.65. Rachidian: length/width 0.87–1.07; central cusp blunt to bluntly rounded; one pointed cusp and one outer denticle on each side. Lateral: four cusps; from inside these are: small rounded to pointed cusp; broad, short, blunt cusp; large, square to bluntly rounded cusp; outermost small pointed cusp. Inner marginal: five (occasionally four) cusps; from inside these are: small rounded or pointed cusp; three blunt cusps, of which inner two sometimes represented by a single cusp, third always largest; outermost pointed cusp. Outer marginal: 5–7 blunt to rounded cusps; narrow neck, projecting flange on inside of base only. All cusps narrower and more pointed in smaller shells (Fig. 83D). Intraspecific variation in radular tooth cusps has been illustrated by Goodwin & Fish (1977). Salivary glands enlarged, but only partially enveloping mid-oesophagus and radular sac. Oesophageal pouches large, usually greater than one third length of buccal mass.

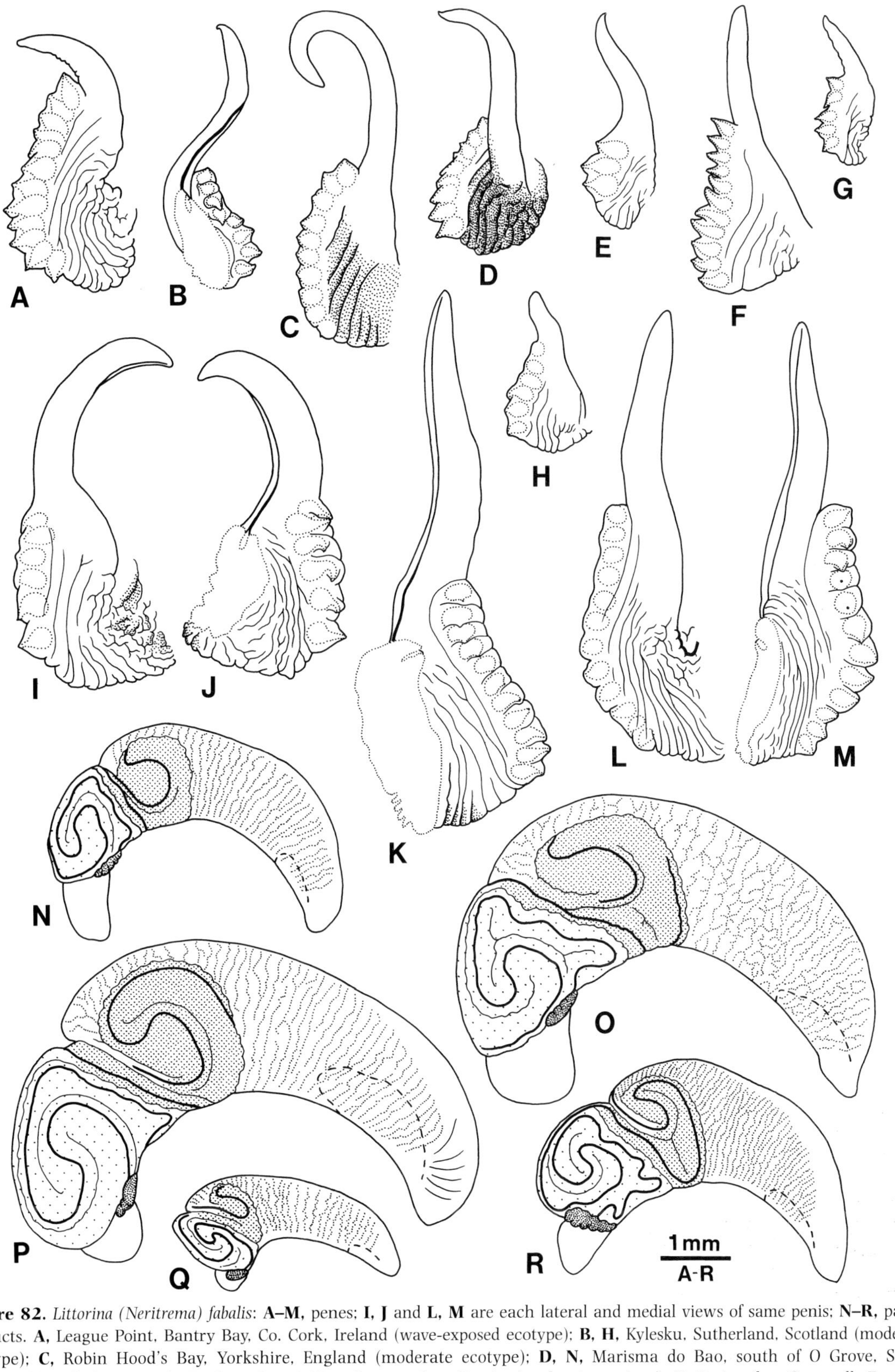

Figure 82. *Littorina (Neritrema) fabalis*: **A–M,** penes; **I, J** and **L, M** are each lateral and medial views of same penis; **N–R,** pallial oviducts. **A,** League Point, Bantry Bay, Co. Cork, Ireland (wave-exposed ecotype); **B, H,** Kylesku, Sutherland, Scotland (moderate ecotype); **C,** Robin Hood's Bay, Yorkshire, England (moderate ecotype); **D, N,** Marisma do Bao, south of O Grove, Spain (sheltered/brackish ecotype); **E,** Cape Svetoi Nos, Barents Sea, Russia (moderate ecotype); **F,** Portscatho, Cornwall, England (moderate ecotype); **G,** Rocamar, Baiona, Spain (wave-exposed ecotype); **I, J, P,** Flatey, Breidafjördur, Iceland (moderate ecotype); **K, O,** Ardivachar Point, South Uist, Outer Hebrides, Scotland (wave-exposed ecotype); **L, M,** Rubh' Aird-mhicheil, South Uist, Outer Hebrides, Scotland (moderate ecotype); **Q,** Ryashkov Island, White Sea, Russia (moderate ecotype); **R,** Tromsø, Norway (moderate ecotype). Shell heights: **A,** 7.3 mm; **B,** 5.5 mm; **C,** 9.7 mm; **D,** 7.1 mm; **E,** 6.9 mm; **F,** 6.7 mm; **G,** 3.5 mm; **H,** 4.9 mm; **I, J,** 10.5 mm; **K,** 11.0 mm; **L, M,** 10.0 mm; **N,** 7.9 mm; **O,** 12.4 mm; **P,** 13.2 mm; **Q,** 6.1 mm; **R,** 11.5 mm. For shading conventions of pallial oviducts see p. 6.

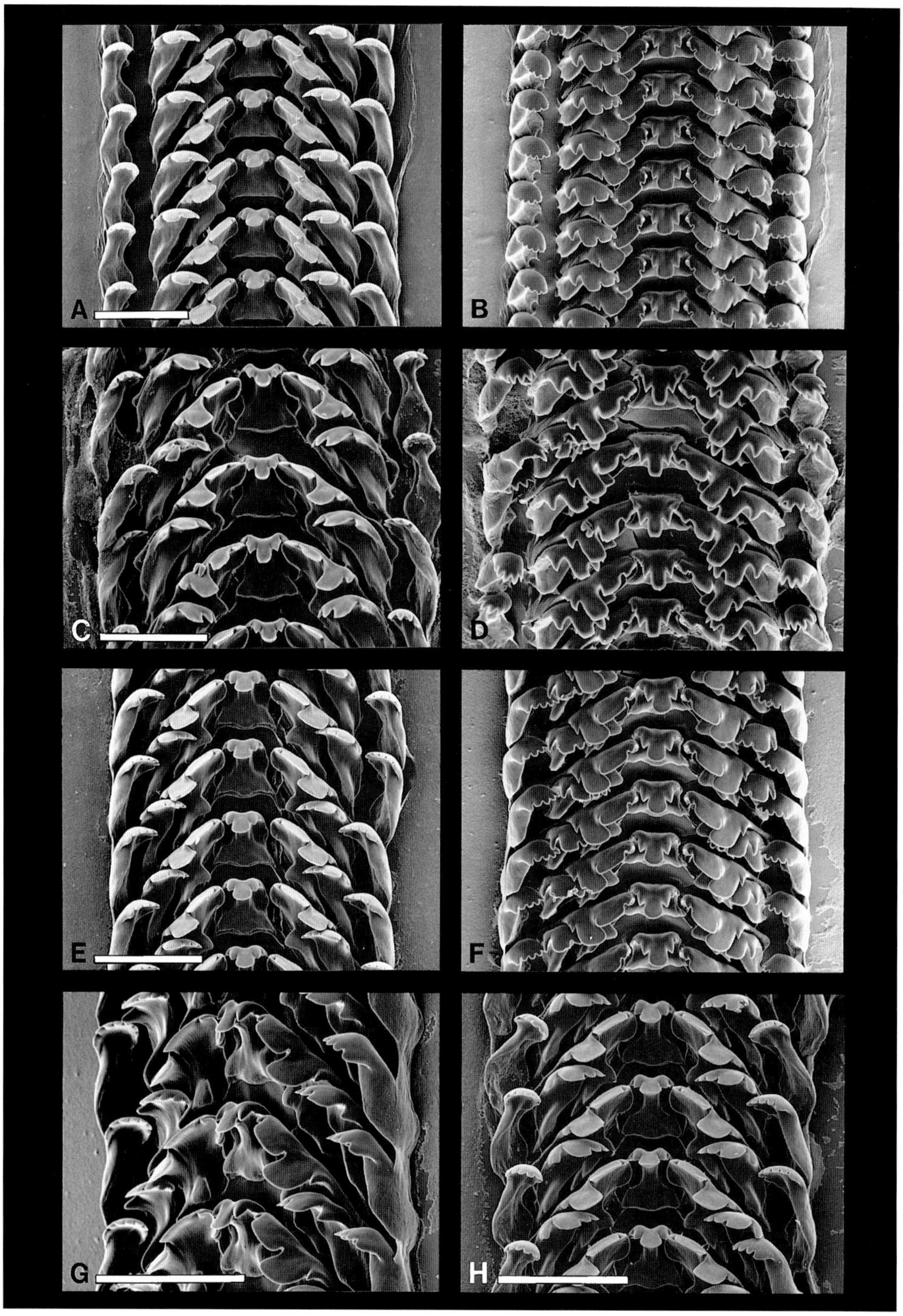

Figure 83. *Littorina (Neritrema) fabalis*: **A, B,** two views of radula (shell H=13.2 mm), flat (**A**) and at 45° (**B**); **C, D,** two views of radula (shell H=4.6 mm), flat (**C**) and at 45° (**D**); **E–G,** three views of radula (shell H=8.6 mm), flat (**E**), at 45° (**F**) and side view (**G**); **H,** radula (shell H=9.1 mm), flat. **A, B,** Flatey, Breidafjördur, Iceland (moderate ecotype); **C, D,** Rocamar, Baiona, Spain (wave-exposed ecotype); **E–G,** Robin Hood's Bay, Yorkshire, England (wave-exposed ecotype); **H,** Smelror, Vardø, Norway (wave-exposed ecotype). Scale bars: **A, B**=100 μm; **C, D**=50 μm; **E, F**=100 μm; **G**=100 μm; **H**=100 μm.

Shell variation (Figs 84–86)

Geographical variation. In *L. obtusata* (p. 208) a striking distinction exists between low-spired, thick-walled shells from the southern parts of the range (geog. var. *retusa*) and high-spired, thin-walled shells from northern areas (geog. var. *palliata*). A parallel distinction can be made in *L. fabalis*, with almost identical geographical boundaries between the two forms. The contrast between the two is, however, less striking in this species, because within the northern form there is considerable (apparently ecotypic) variation in spire height. The following diagnoses exclude the more extreme gerontic shells and ecotypes from estuarine and wave-exposed shores:

> Geog. var. *typica* (Figs 84A–P, 85A–F). H/B=0.76–1.02; SH=1.00–1.20. Depressed-globular outline, low to almost flat-spired; columella wide; relatively thick-walled. Range: Europe south of Lofoten Islands, southwestern Iceland.
>
> Geog. var. *limata* (Fig. 85G–N). H/B=0.96–1.28; SH=1.11–1.48. Outline turbinate; spire usually more projecting; columella relatively narrow; relatively thin-walled. Range: Lofoten Islands to White Sea, Iceland, Greenland.

Lovén's name *limata* has been adopted for the northern variety, since it was largely based on this shell form of *L. fabalis*, and has been used in this same sense by Philippi (1848) and Küster (1853) (see Taxonomic History, p. 230).

It is difficult to specify precisely the zone of overlap or the boundary between the two varieties, because of the considerable ecotypic variation that is superimposed on the geographical pattern; sheltered/brackish ecotypes of *typica* (Fig. 85F) may have a high spire and thin-walled shell similar to that of *limata*, while wave-exposed ecotypes of *limata* (Fig. 85G, N) are as low-spired as *typica*. Nevertheless, if shells from normal habitats are compared the geographical pattern is clear. In northern Europe geog. var. *limata* is found from the White Sea to Finnmark and Troms, as far south as Gryllefjord (TM, 1933) (no collections seen from Vesterålen), and intermediate forms at Narvik (DGR, 1989) and Flakstad, Lofoten Islands, whereas *typica* occurs elsewhere in Vestfjorden. In Iceland *limata* has been found from Reykjavík, many points around the north coast to Berufjördur in the south east; intermediates occur at Seltjarnarnes (DGR, 1989), and only *typica* at Útskálar, Grindavík (DGR, 1989) and the Vestmannaeyjar (DGR, 1989). All collections from western Greenland are of the *limata* form. The present study has not found any radular or anatomical differences between the two varieties, and it is concluded that the variation is intraspecific.

In the case of *L. obtusata* Seeley (1985, 1986) has shown that the thick-walled, low-spired geog. var. *retusa* is more resistant to crushing by the crab *Carcinus maenas*, and that the thin-walled, higher-spired *palliata* is found only where predatory crabs are rare or absent (p. 213). *Littorina fabalis* occurs in many of the same habitats as *L. obtusata* (although often extending lower on the shore) and the two are sympatric over much of the northeastern Atlantic. It seems likely that the two varieties of *L. fabalis* are analogous to those of *L. obtusata*, with which their distributions coincide, and that the same explanation applies to both species. The occurrence of some low-spired shells through much of the range of *limata* (in contrast to *palliata*) appears to be explained by the fact that *L. fabalis* can tolerate more wave-exposed conditions, under which low-spired, patulous shells are advantageous; where habitat descriptions are available, the low-spired shells of *limata* are all from wave-exposed sites.

Another geographical trend is the decreasing maximum size of shells in normal habitats from Britain to Brittany and Galicia (Sacchi & Rastelli, 1966; Sacchi, 1967b, 1969a), although this is often obscured by ecotypic variation.

Ecotypic variation. Three ecotypes can be recognized, occurring within the range of both geographical varieties:

> Moderate ecotype (Figs 84C, F–I, N–P; 85A, C, E, H–J, L, M): these are shells from intermediate habitats, which are neither very wave-exposed nor particularly sheltered and brackish; there is a wide range of adult size (H=7–14.9 mm); spire heights differ in the two geographical varieties.
>
> Wave-exposed ecotype (Figs 84A, B, D, E, K, J; 85B, G, N): *L. fabalis* is common on moderately exposed shores, where shells are usually larger than on nearby sheltered shores, but on shores

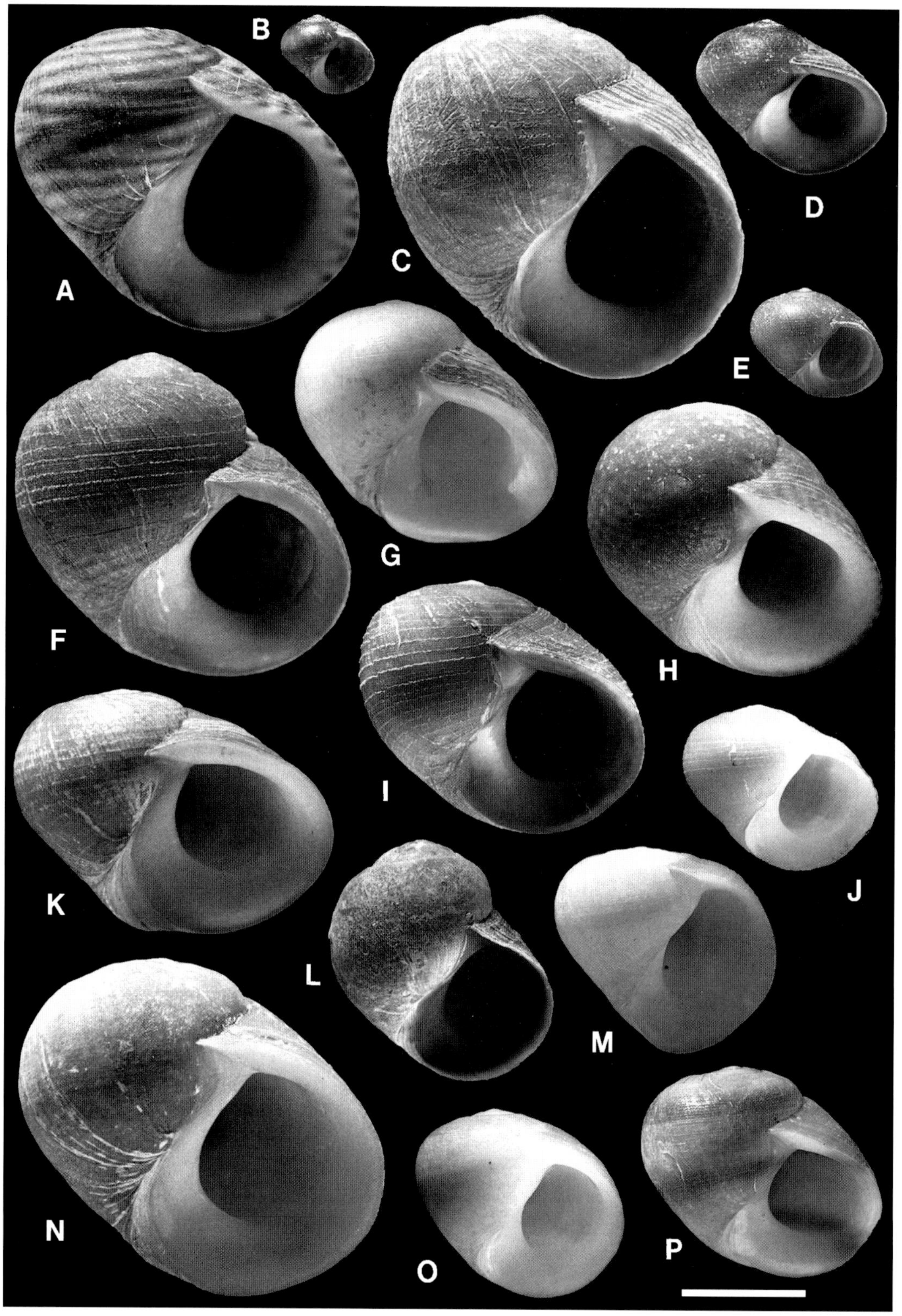

Figure 84. *Littorina (Neritrema) fabalis*: geog. var. *typica*; **A, B, D, E, K, J, M,** wave-exposed ecotype (**M,** juvenile); **C, F–I, N, P,** moderate ecotype; **L,** sheltered/brackish ecotype. **A, M,** Ardivachar Point, South Uist, Outer Hebrides, Scotland. **B,** *Turbo fabalis* Turton, 1825; lectotype, BMNH 1968356; Scarborough, Yorkshire, England. **C,** Cliftonville, Kent, England. **D,** Keem Strand, Achill Island, Co. Mayo, Ireland. **E,** Rocamar, Baiona, Spain. **F,** Binstead, Isle of Wight, England. **G,** *Littorina mariae* Sacchi & Rastelli, 1966; syntype, MCSNM; Grèves de Roscoff, Brittany, France. **H,** Finavarra, Co. Clare, Ireland. **I,** Carantec, Brittany, France. **J,** Baiona, Spain. **K,** League Point, Bantry Bay, Co. Cork, Ireland. **L,** Marisma do Bao, south of O Grove, Spain. **N,** Rubh' Aird-mhicheil, South Uist, Outer Hebrides, Scotland. **O,** Portscatho, Cornwall, England. **P,** Strome Ferry, Scotland. Scale bar = 5 mm.

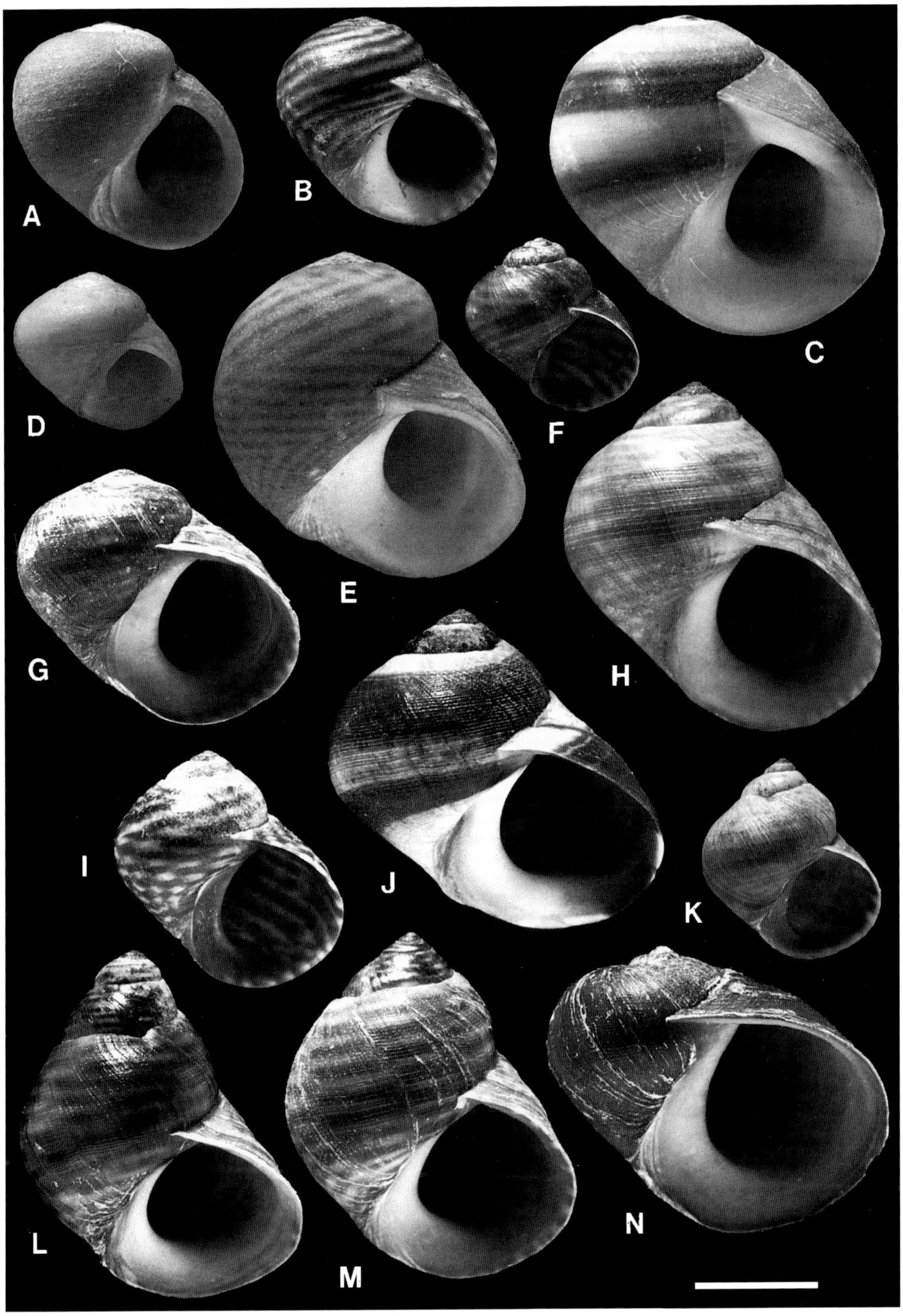

Figure 85. *Littorina (Neritrema) fabalis*: **A–F**, geog. var. *typica*; **G–N**, geog. var. *limata*; **A, C, E, H–J, L, M**, moderate ecotype; **B, G, N**, wave-exposed ecotype; **D, F, K**, sheltered/brackish ecotype; **L**, monstrosity of geog. var. *limata*. **A**, Hirtsholmene, Denmark; ZMK. **B**, Titran, Norway. **C**, Torness, Orkney Islands, Scotland; NMS 1974.90.11.002. **D**, North Locheynort, South Uist, Outer Hebrides, Scotland. **E**, Yesnaby, Orkney Islands, Scotland. **F**, W Saltholm, Øresund, Denmark; ZMK. **G**, 1 km SE of Valadalstorfa lighthouse, Tjörnes, Iceland. **H, J**, Dal'niye Zelentsy Bay, Barents Sea, Russia. **I**, shore by Tromsø Museum, Tromsø, Norway. **K**, S Flatey, Breidafjördur, Iceland. **L, M**, NW Flatey, Breidafjördur, Iceland. **N**, Smelror, near Vardø, Norway. Scale bar=5 mm.

of the most extreme exposure that can be tolerated, shells are dwarf (H as small as 3 mm); in either case the shell is low and broad (H/B=0.76–0.84), the spire flatter (SH=0.99–1.12), the aperture relatively larger, and the walls sometimes somewhat thinner, than in the moderate ecotype.

Sheltered/brackish ecotype (Figs 84L; 85D, F, K): shells are smaller in size (H=5–9 mm), often more thin-walled, the aperture is relatively smaller, and spire more produced (SH to 1.26 in geog. var. *typica*, to 1.44 in *limata*).

The wave-exposed ecotype is frequent in this species. Several authors have emphasized the striking differences between the moderate and wave-exposed ecotypes, not only in shell size and shape, but in colour frequencies, sexual dimorphism, ecology and life cycle (Sacchi, 1969a, 1974; Rolán & Templado, 1987; Williams, 1992a). It has often been observed that in passing from sheltered to moderately wave-exposed shores there is a marked increase in shell size (Sacchi, 1969a, 1974; Goodwin & Fish, 1977; Reimchen, 1982; Little & Williams, 1989; Williams, 1994), although Sacchi & Sconfietti (1992) believed increased size was correlated with stronger currents rather than exposure to waves. In Britain Reimchen (1982) found the mean shell diameter to be about 7 mm on sheltered shores, increasing to about 12 mm on moderately exposed ones. He explained this trend as the result of selection by crab predators; snails that attain maturity (and with it a thickened apertural lip) at small size are more resistant to crushing by small crabs than immature snails of the same size (all forms of *L. fabalis* are susceptible to large crabs). The size gradient is explained if small crabs are the major predators on sheltered shores. Small adult size on sheltered shores may also be an adaptation to prevent effective handling by crabs, and could in addition be related to the annual life cycle on such shores (Williams, 1992a). It has also been found that on sheltered shores the shells are relatively stronger, when adjusted for the difference in size (Fletcher, 1995). Less well known is the phenomenon of dwarf shells on exposed shores (Sacchi, 1969a). The most striking example was described by Rolán & Templado (1987) on exposed coasts in Galicia, northwestern Spain, where snails with a mean diameter of about 5 mm (but maturing as small as 1.6 mm) occur among tufts of *Mastocarpus stellatus* (Fig. 84E). Figure 84D is representative of a similar population from northwestern Ireland, living on *Fucus* and *Palmeria palmata*. A genetic study of the Spanish populations has confirmed that these dwarf forms are conspecific with typical examples from more sheltered shores (Rolán-Alvarez, Zapata & Alvarez, 1995a). Shells exactly comparable with Turton's syntypes have not been found among recent collections from Scarborough, the type locality, but the lectotype is a mature male only 3.6 mm in height (Fig. 84B) and is probably of this dwarf form from an exposed shore. As in other *Neritrema* species, the enlarged aperture of this ecotype is probably adaptive in relation to resisting dislodgement by waves, and under the most exposed conditions small size will allow animals to find refuge in sheltered microhabitats, among small weeds and barnacles.

On the most sheltered shores, which are usually also estuarine, the shells are not only small, but have more produced spires. S. M. Smith (1978, 1979b, 1981) first noticed this form in brackish lochs of the Outer Hebrides and the west of Scotland (and her manuscript name *L. mariae beeensis* was inadvertently published by Fretter & Graham, 1980). As Smith remarked (1982), it bears the same relationship to *L. fabalis* as the brackish ecotype *aestuarii* does to *L. obtusata*. It cannot, however, be regarded as a subspecies, because it is not geographically separated from typical *L. fabalis*. In these Scottish localities the shells are of normal thickness (Fig. 85D). In the brackish inlets and lagoons of Denmark this ecotype is also common and here is often thin-walled (Nielsen, 1980; Fig. 85F; see also Fig. 85K, from Iceland). A form similar to those brackish ecotypes that are of normal thickness is found in the peculiar habitat of *Zostera* meadows in enclosed, muddy bays in the rias of Galicia (Sacchi, 1984; '*Littorina* B' of Rolán & Templado, 1987; Sacchi & Sconfietti, 1992; Fig. 84L). Although the habitat is certainly sheltered, it is not clear whether it is also brackish, but this form is included here on the basis of shell resemblance. In a detailed study of shell and animal coloration, penis and ecology, Rolán & Templado (1987) suggested that this *Zostera*-dwelling form might be taxonomically distinct, but this appears unlikely in view of the known degree of differentiation between populations in this species (see Discussion, p. 245).

Morphological differentiation can be maintained even over a distance of a few metres. Working at sites on Anglesey, Wales, Reimchen (1981) found a 'dwarf' form on clumps of weed in sandy tidal channels, which differed not only in size, but also colour frequency and microsculpture, from shells on adjacent rock platforms that were more wave-exposed. Differences were so striking that Reimchen

Figure 86. *Littorina (Neritrema) fabalis*: juvenile shells; **A, B,** geog. var. *typica*; **C, D,** geog. var. *limata*. **A,** North Locheynort, South Uist, Outer Hebrides, Scotland; **B,** Kylesku, Sutherland, Scotland; **C, D,** Flatey, Breidafjördur, Iceland. Scale bar=1.0 mm.

considered the possibility of taxonomic subdivision. These two forms probably correspond with the moderate and wave-exposed ecotypes distinguished here, and similar selection pressures may be involved. Reimchen (1981, 1982) concluded that they were the result either of extremely localized adaptation to crab predation in the channels, or of a direct environmental effect on the phenotype. Fletcher (1995) has reported that shells are relatively stronger at lower levels on the shore, but again the explanation is not known.

Sculptural variation (Fig. 86). This species is never carinate or ribbed. The strength of the microstriae (which are purely periostracal) is quite variable, even between microhabitats at a single locality (Reimchen, 1981). Juveniles are more strongly sculptured (Fig. 86A), the microstriae often becoming obsolete in adults. Shells of geog. var. *limata* do not develop strong microstriae and may be almost smooth, even as juveniles (Fig. 86C, D).

Ontogenetic variation. Littorina fabalis shares with *L. obtusata* (p. 214) a more fully determinate course of shell growth than in other *Littorina* species. Juvenile shells have a thin growing edge and a more prominent anterior lip to the aperture; they also appear relatively more flat-spired, owing to the phenomenon of doming (Figs 84M, 86A, B). As adult size is reached, the suture of the final whorl descends towards the base and the lip becomes thickened. Sometimes the columella separates from the axis of coiling to give a pseudo-umbilical area (Fig. 84F, H, K). In extreme cases of gerontism the shell becomes slightly cylindrical (Fig. 85E). Allometric growth of shell shape and thickness (and the effects on shell strength) in geog. var. *typica* have been measured by Lowell, Fletcher, Grahame & Mill (1994; see also Fletcher, 1995). Doming is slightly less marked in geog. var. *limata*.

Strong growth lines, probably indicating annual interruptions to shell growth during winter, can sometimes be seen on shells from northern localities, but are less marked than in *L. obtusata*. Up to three such lines have been counted on shells of *L. fabalis* from the White Sea, implying that the species is not an annual in this area, in contrast to a population from Wales studied by Williams (1990a, 1992a).

Sexual dimorphism. This is pronounced in almost all populations that have been investigated, females being larger and more globose, males smaller and more patulous (Sacchi, 1968a, b, 1969b, Gallardo & Götting, 1985; Sacchi & Sconfietti, 1992; Williams, 1994). The shells of males are also significantly stronger at a given size than those of females, because they are more massive (Fletcher, 1995). Dimorphism is said to be greatest in populations from sheltered shores (Goodwin & Fish, 1977; Williams, 1994). There is some evidence for sexual selection, favouring small males and large females (Rolán-Alvarez, Zapata & Alvarez, 1995b).

Monstrosities. No malformations have been reported in this species, even in the few areas where epibionts are common (Sacchi, 1985). Infection by parasitic trematodes is not known to influence size or shape. Some of the most tall-spired forms of geog. var. *limata* (Fig. 85L) could be regarded as comparable with the 'Arctic monstrosities' of *L. obtusata* (p. 216), but shapes are less extreme.

Colour variation. The polymorphism of *L. fabalis* is closely similar to that of *L. obtusata* and many of the same considerations apply (p. 216).

Almost all populations in the southern parts of the range are polymorphic (Sacchi, 1974; Reimchen, 1979). To describe the colour phenotypes, Sacchi (1967b, 1968a, 1969a, 1972a, 1974) used the descriptive 'var. ex colore' system devised by Dautzenberg & Fischer (1915) for *L. obtusata s.l.* Since polymorphism in the strict genetic sense is based on discrete variation, the following non-intergrading categories can be recognized: reticulata (reticulate or zigzag pattern; divided by Sacchi and by Reimchen, 1979, into light reticulata, with orange reticulation on yellow to orange background, usually a pale animal; and dark reticulata, with dark pattern on brown background, usually a black animal; these categories intergrade; Figs 84A, H; 85B, E–I, K–M), citrina (yellow grading to white, the albescens of Sacchi; white interior), aurantia (orange), rubescens (brick red, not seen in present study), fusca (dark chestnut-brown), olivacea (olive-green to olive-brown), inversicolor (two broad dark bands; Figs 84P, 85C, J) and zonata (single pale peripheral band). Of these, only the first two are common in southern Europe; combining samples from around the British Isles, Reimchen (1979) recorded 47% citrina and 52% reticulata. The yellow phenotype has been recorded only as far north as Skarberget, in northern Norway (80 km southwest of Narvik, close to the boundary region between geog. vars *typica* and *limata*) and in the White Sea; this phenotype is apparently absent from Arctic Norway, the Barents Sea, Iceland and Greenland, where populations are almost entirely of reticulata, with occasional fusca and inversicolor. Distinctions between the morphs are not always clear; for example, in some southern samples faint reticulation is visible even on pale yellow or almost white shells. Once the structural and genetic basis of the polymorphism is known, a genetic classification similar to that proposed by Sergievsky (1983a, 1984, 1992) for *L. obtusata* will be more satisfactory. Reimchen (1974, quoted in Reimchen, 1979, 1989) has carried out breeding experiments that confirm the genetic basis of the polymorphism, dark reticulata behaving as a single allele dominant to citrina, for example.

In areas where the species is polymorphic there are some systematic patterns in the occurrence of phenotypes, implying that the polymorphism is not selectively neutral. Throughout Britain, Brittany, Galicia and Belgium the reticulata morph (and especially dark reticulata) predominates on wave-exposed shores, and citrina on more sheltered ones (Gratton, 1969; Sacchi, 1969a, 1974; Goodwin & Fish, 1977; Reimchen, 1979, 1981; Warmoes *et al.*, 1988), with few exceptions (Sacchi, 1969a). In very sheltered or brackish habitats in Denmark Nielsen (1980) found only brown or reticulate shells. Since these patterns were not obviously adaptive in terms of the visual properties of the morphs, Sacchi sought physiological and behavioural differences between them, which might by pleiotropy or linkage be connected with shell colour. He found that pale shells (yellow, white, orange and pale reticulata, all of which usually have pale or unpigmented animals) were relatively more abundant on the surface of algae when exposed during low tide, because of a greater tolerance of illumination (Sacchi, 1969a; 1974). There was some suggestion that darker colours were more resistant to desiccation, but in general the results of comparisons of thermal and desiccation tolerance among morphs were inconclusive and failed to account for their distribution on the shore (Sacchi, 1969a, 1972a). There were no significant differences in the sex ratios of the morphs (Sacchi, 1967b, 1984), and mating was not influenced by shell colour (Sacchi, 1974).

However, one case of apparently cryptic coloration (and therefore possible visual selection) was noted by Sacchi. The olivacea morph is rare or absent throughout most of the range of *L. fabalis* (Sacchi, 1967b, 1968a, 1969a, 1974; Reimchen, 1979), but in two enclosed bays (O Grove and Ría de Aldán) in Galicia it reaches high frequency or populations are even monomorphic (Sacchi, 1984; Rolán & Templado, 1987; Sacchi & Sconfietti, 1992). These populations occur in the unusual habitat of *Zostera* meadows, and the olive shells appear cryptic on the leaves of the seagrass. Another example is of the monomorphic chestnut brown (fusca) populations found in low turf of the red alga *Mastocarpus stellatus* on exposed coasts near Vigo (Rolán & Templado, 1987). Similar coloration has been observed on red algae at Achill Island (Ireland) and near Vardø (Arctic Norway) (pers. obs.).

There is thus some weak circumstantial evidence for visual selection in *L. fabalis*, but sound experimental evidence has been obtained by Reimchen (1979). The blenny *Lipophrys pholis* is an important predator on juvenile (less than 8 mm) *L. fabalis*. Dark reticulata shells are cryptic on the dark stem and holdfast of the principal algal host plant, *Fucus serratus*, while citrina morphs, although conspicuous when seen by reflected light on top of the alga, are also cryptic when viewed from below against the yellow lamina, illuminated by transmitted light. This was confirmed by laboratory feeding trials using blennies. In the field, on sheltered shores juvenile snails occur mainly on the lamina, but on wave-exposed shores more occur on the stem, to avoid dislodgement. Where visual predation by blennies or other aquatic predators is significant, selection will favour dark reticulate shells in wave-exposed habitats and yellow shells in sheltered sites, in agreement with the observed pattern. At sites with anomalous colour frequencies Reimchen was sometimes able to account for them by peculiarities of the colour or texture of the weed, light reticulata being common on an exposed shore with leathery, orange *Fucus*, for example. Visual selection may be strong, for clines of colour frequency occur between areas of contrasting exposure only 25–50 m apart (Reimchen, 1979, 1981). In a detailed study of the distribution of citrina and dark reticulata morphs on algal plants in Sweden, Ekendahl (1994) found the expected distribution of more yellow shells on *Fucus serratus* plants with relatively large areas of lamina (compared to stem), and more yellow shells on this alga in sheltered than exposed habitats. However, opposite results were obtained on *Fucus vesiculosus*, suggesting that overall background colour may be more significant than the lamina-to-stem ratio. She also noted that the blenny studied by Reimchen was absent from Swedish shores so that, if visual selection is occurring, it may be performed by other fish or crab predators.

Although Reimchen's (1979) results clearly show that selection for shell colour by a predator is possible, such selection will be directional in the absence of some balancing mechanism (e.g. frequency-dependent selection, or substrate choice in a heterogeneous environment) to maintain morph frequencies within populations. The nature of such mechanisms is not yet understood. In Wales Reimchen (1979) found only a small, non-significant tendency for citrina and reticulata morphs to occur at higher frequencies on backgrounds against which they were apparently cryptic (lamina and stipe respectively). In Sweden a significant difference in distribution of this kind was found only among adults during autumn, but in one of several laboratory trials a difference in substrate choice by the morphs was observed, reticulata morphs preferring the stipe of *Fucus serratus* (Ekendahl, 1995). There may also be seasonal differences in the appearance of the algae or behaviour of the snails, leading to temporal differences in selection, another possible balancing mechanism (Smith, 1976; Ekendahl, 1994).

Reimchen (1989) has also described the interesting case of mimicry of the polychaete *Spirorbis* by juveniles of *L. fabalis*. On shores in Wales where the small (3.5 mm diameter) white tubes of *Spirorbis* are common on *Fucus serratus*, both reticulata and citrina shells show white apices. This represents a distinct juvenile colour phase, of opaline shell with opaque white bands, which lasts only up to a diameter of 2–4 mm and bears a close resemblance to the worm tubes. Experiments with blennies demonstrated that white juveniles were at a selective advantage on weed with spirorbid tubes, and breeding experiments confirmed the genetic basis of the character.

In conclusion, the colour polymorphism of *L. fabalis* is clearly of some adaptive significance in relation to visual predation, at least in the southern parts of the range where the species is conspicuously polymorphic, although the balancing mechanisms maintaining morph frequencies are not yet known. In contrast, in the northern parts of the range and in some brackish habitats the species is almost entirely monomorphic (except in the White Sea). This might indicate that visual predation differs or is less intense in these areas, or that (as in *L. obtusata*) the physiological properties of the morphs are also significant. The more diverse polymorphism in southern populations is a parallel between this pair of sister-species.

Distribution

Habitat. Littorina fabalis is typically found on macrophytes, usually fucoid algae, in the mid to low eulittoral zone of sheltered to moderately exposed rocky shores. In Scandinavia, Germany, Britain, France and Galicia it occurs in the lower part of the *Fucus vesiculosus* and *Ascophyllum nodosum* zone, reaches maximum abundance on *F. serratus* and is only sporadic in the upper *Laminaria* zone (Sacchi, 1969a; Nielsen, 1980; Rolán & Templado, 1987; Watson & Norton, 1987; Warmoes *et al.*, 1988; Little & Williams, 1989; Janke, 1990; Williams, 1990a, b, 1992a, 1994; 1995; Fletcher, 1992). The greatest recorded depth is 10–12 m in the Øresund (ZMK). At least in Britain and France, on sheltered shores it lives almost exclusively on *F. serratus* low on the shore (reaching an abundance of 440 per m^2, Williams, 1992a), but where wave action is stronger it occupies a wider vertical range and can also be found on *F. vesiculosus* and even *F. spiralis* on the upper shore (Sacchi, 1969a; Williams, 1990b). On the atidal shores of western Sweden *L. fabalis* occurs throughout the narrow fucoid belt, on the species of *Fucus* and on *Ascophyllum* (Ekendahl, 1994). It is reported to become less common on steeply shelving shores (Fletcher, 1992). On exposed shores it can sometimes be found even in the absence of fucoids, for example among the red algae *Mastocarpus stellatus* in Galicia (Rolán & Templado, 1987), *Palmeria palmata* at Achill Island, Ireland, *Devaleraea ramentacea* at Grindavík, Iceland, and on red algal crusts or clinging to bare rock near Vardø, northern Norway (pers. obs.). At the opposite extreme, under very sheltered and muddy conditions it has been reported on the leaves of the seagrass *Zostera* in Galicia (Sacchi, 1984; Rolán & Templado, 1987; Sacchi & Sconfietti, 1992; pers. obs.) and Denmark (Nielsen, 1980), and is occasionally seen on green filamentous algae and *Ulva* (Rolán & Templado, 1987; pers. obs.). In brackish lochs in western Scotland and the Outer Hebrides it can be found just below the lowest tides, on *F. serratus* and *F. vesiculosus* (S. M. Smith, 1978, 1979b, 1981). It is relatively tolerant of brackish conditions and usually penetrates further into estuaries than *L. obtusata* (Nielsen, 1980, in Denmark; though the reverse was found by Little & Smith, 1980, and Smith & Little, 1980, in the Severn estuary in Britain). Little is known about possible breeding, seasonal or tidal migrations of *L. fabalis*. On tidal shores snails may spend their lives on algae (Williams, 1990a, b), but on exposed, atidal shores in Sweden Ekendahl (1994) often found juveniles higher on the shore than adults, sometimes on rocks above the algal zone.

The only other *Littorina* that is closely associated with intertidal macrophytes is *L. obtusata* but, although there is some overlap between the two species, at least on tidal shores *L. fabalis* reaches maximum abundance at a lower level, this distinction being most clear on sheltered shores (Sacchi, 1969a; Williams, 1990b, 1995). On atidal shores in Sweden there is no differential vertical zonation of the two species, but there is some separation according to wave-exposure, *L. fabalis* being the more tolerant of exposed conditions (Ekendahl, 1994). At an atypical sheltered locality near Brest, where *L. obtusata* was absent, Sacchi (1968a) found that *L. fabalis* extended much further up the shore, to the level of *Fucus spiralis*. Nevertheless, competition for food between the two species is perhaps limited, for *L. fabalis* is mainly a grazer upon epiphytes. Some thallus tissue may be ingested, but the radular and buccal apparatus is reported to be inadequate for deep excavation of the tough macroalgal tissues, so that the main components of the diet are surface microalgae and detritus (Watson & Norton, 1987; Williams, 1990a, b). Similarly, a population from *Zostera* ingested only diatoms and not the seagrass tissues (Sacchi, 1984). Grazing on diatoms is apparently non-selective and has been studied in detail by Williams (1992b).

Range (Fig. 87). The reported range of *L. fabalis* includes the White Sea (Sergievsky & Granovitch, 1987), Norway (Sneli & van Marion, 1979), Denmark (Nielsen, 1980), the British Isles (Seaward, 1990), The Netherlands (Fortuin, 1979), Belgium (Warmoes *et al.*, 1988), the southern Channel coast (Warmoes, Dumoulin & Reid, 1992), Galicia and as far south as 41.5°S in northern Portugal (Sacchi & Rastelli, 1966), and Iceland (Sannia & James, 1977; Sigurdsson, 1978; Alerstam, Gudmundsson & Johannesson, 1992). This range can now be confirmed in detail. In the White Sea it appears to be

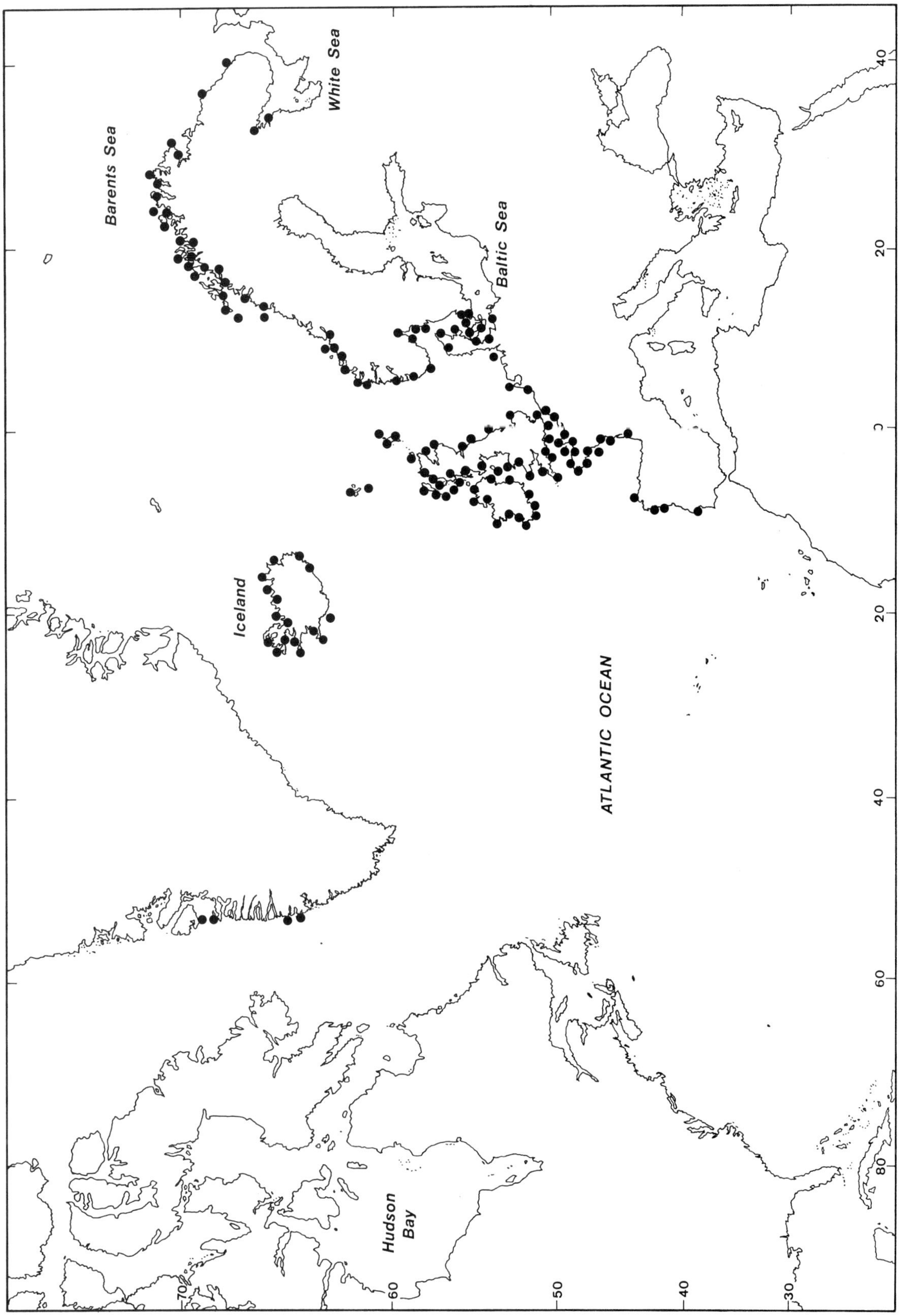

Figure 87. *Littorina (Neritrema) fabalis*: distribution map. Key: circles, specimens examined.

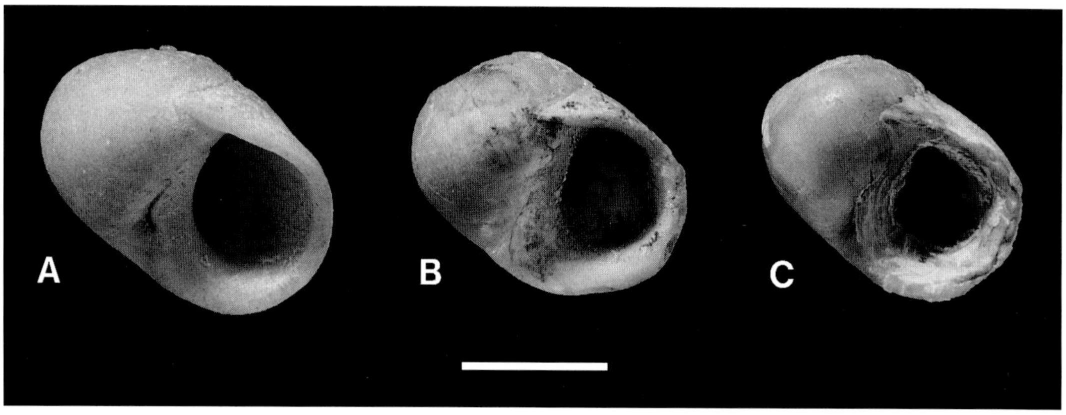

Figure 88. *Littorina (Neritrema) fabalis*: fossils. **A,** Affoud Taddert 25, Morocco; Ouljien, Pleistocene; MNHNP. **B,** bed G_1, Sidi Abd er Rahmane, near Casablanca, Morocco; between Maarifian and Anfatian, Pleistocene; MNHNP. **C,** bed J, near Casablanca, Morocco; between Maarifian and Anfatian, Pleistocene; MNHNP. Scale bar = 5 mm.

rare, and specimens have only been seen from Kandalakshsky Bay. It occurs on the Barents Sea coast and throughout Norway. It just penetrates the mouth of the Baltic, to Saltholm on the Øresund (ZMK) and Warnemünde in Mecklenburger Bucht (SMNH). There are few records from The Netherlands and Belgium, but it is abundant all around the French coast to Royan at the mouth of the Gironde (MNHNP) and thereafter occurs only in the Arcachon Lagoon (DGR, 1987). Changes in the distribution of *L. obtusata s.l.* in the Bay of Biscay during the twentieth century have been documented by Fischer-Piette (p. 221). However, in the present study no specimens have been seen from the Spanish Biscay coast. The species is common in Galicia, with a record from O Vicedo (DGR, 1987) and many in the region of Vigo. It is rare in Portugal, with a single modern record from Porto (BMNH, 1984) and an old one from Setúbal south of Lisbon (BMNH, before 1911). *Nerita subtruncata* Nardo, 1847, doubtfully synonymized with this species, was reported from Istria, but this was an error or human introduction, for there are no Recent records of *L. obtusata s.l.* from the Adriatic (Brusina, 1870). To the northwest *L. fabalis* is abundant all round the British Isles and in the Faroes. It occurs throughout Iceland, except on the inhospitably sandy and exposed southeastern coast between Höfn and the Vestmannaeyjar. It is here recorded from Greenland for the first time, on the west coast between Godhavn and Atangmik (ZMK), but is much less common there than *L. obtusata*. Ingólfsson (1992) listed this species from the Canadian Maritimes, but with no details; a very large quantity of *Littorina* material has been examined from North America without finding *L. fabalis*, so this record is as yet unsubstantiated. It may be significant that *Fucus serratus*, so frequently the host plant of *L. fabalis*, is of restricted distribution and perhaps recently introduced in the Canadian Maritimes (I. Tittley, pers. comm.).

Fossil record (Fig. 88). The earliest occurrences of *L. obtusata s.l.* are from Morocco, where there is a long record from the Upper Moghrebian-Messaoudian at Casablanca (about 2 Ma), from the Maarifian between Cape Rhir and Agadir (0.5–1 Ma) and through to the Ouljian of the Upper Pleistocene (Neuville & Ruhlmann, 1941; Lecointre, 1950, 1952, 1963; Brebion, 1979; ages from Biberson, 1977). The specimens on which the oldest record (from the Messaoudian at Sidi Messaoud, Neuville & Ruhlmann, 1941) was based have not been traced in MNHNP (Brebion, pers. comm.), but material is available from levels G_0, G_1 and J at Sidi Abd er Rahmane near Casablanca, between the Maarifian and Anfatian stages (about 0.5 Ma). These shells (Fig. 88) are not exactly comparable to any modern shells, but at least in the rather large aperture, thick columella, small size and only slight angulation at the shoulder, they resemble *L. fabalis* rather than *L. obtusata* (Fig. 88A does, however, approach some forms of *L. obtusata*). The records from Morocco are far to the south of the present southern limit and date from cold climatic intervals. Cataliotti-Valdina (1984) listed eight occurrences of *L. obtusata s.l.* at archaeological Palaeolithic and Neolithic cave sites in southern France, six of these close to or on the Mediterranean coast. Material from the Grotte du Lazaret, near Nice (Riss III glaciation, about 150 000 BP) has been examined, and the specimens are closely similar to those from Morocco and therefore also classified as *L. fabalis*. Cataliotti-Valdina (1984) interpreted these occurrences as evidence that the species inhabited the Mediterranean in the middle Upper Pleistocene, arguing that the shells were carried into the caves with algal and seagrass vegetation. This, however, is unlikely since the fossil assemblage includes rock-dwelling *Melarhaphe neritoides* from the littoral fringe, although there remains the probability that the littorinids were intentionally collected by man. Whether they were collected locally in the Mediterranean, or could have been carried from the Atlantic coast, is unknown; transport from the Atlantic at such an early date is believed to be unlikely (D. Kadolsky, pers. comm.), but until fossils are discovered in natural deposits around the Mediterranean the historical occurrence of *L. fabalis* there remains unproven. Late Pleistocene records of *L. obtusata s.l.* (p. 223) may have included *L. fabalis*; samples from Minchin Hole on the Gower Peninsula in Wales (BMNH; 120 000 BP, Jones & Keen, 1993) have been examined and found to contain both species. The type figure of *Turbo expansus* J. Smith, 1839, resembles this species and was described from the late glacial Clyde Beds of Scotland (11000–13000 BP, Jones & Keen, 1993).

Discussion

Intraspecific variation. Like its sister-species, *L. obtusata*, *L. fabalis* is highly variable in both shell shape and colour and is therefore a suitable subject for studies of microevolution (see remarks on *L. obtusata*, p. 224). In comparison with *L. obtusata*, *L. fabalis* shows lower heterozygosity and greater genetic subdivision, perhaps related to its lower effective population size, which is a consequence of its annual life cycle (Rolán-Alvarez, Zapata & Alvarez, 1995a; Tatarenkov, 1995a). The demonstration of visual selection on *L. fabalis* by a fish predator (Reimchen, 1979) is significant, since direct observations of this kind have seldom been made on polymorphic marine molluscs. As in *L. obtusata*, patterns of both local and geographical variation in *L. fabalis* suggest that crab predation is an important influence on shell shape, probably by natural selection.

The degree of intraspecific variation is greater in *L. fabalis* than in *L. obtusata*, corresponding to its wider habitat range, and several authors have suggested that further taxonomic division is warranted. Smith (1982) considered a small, brackish-water form to be a subspecies (named in Fretter & Graham, 1980). Reimchen (1981) suggested that a 'dwarf' form could be taxonomically distinct from larger shells on the same shores, but later considered that selection by crab predators could be responsible for the maintenance of small-scale clines (Reimchen, 1982). In Galicia Rolán & Templado (1987) described two unnamed forms from the extreme habitats of *Zostera* beds and exposed rocky shores, and at least in the latter case proposed that speciation had occurred, since differences from typical *L. fabalis* appeared to be maintained in areas of overlap. Genetic results have demonstrated differentiation of local populations in this nonplanktotrophic species, but have not supported species status for the dwarf wave-exposed ecotype in Galicia (Rolán-Alvarez, Zapata & Alvarez, 1995a). The relative influence of genetic and environmental effects on the phenotype are not yet known, but it is clear that striking morphological differences can be maintained over small distances, even on the same shore.

A detailed study of allozyme variation in *L. fabalis* on a group of small islands in Sweden (Tatarenkov & Johannesson, 1994) has also demonstrated significant differentiation on a small scale. For presumed selectively neutral loci, this could be explained by genetic drift. One locus (arginine kinase) showed much higher levels of variation, closely correlated with wave-exposure of the habitat, and in one case a cline from dominance of the sheltered alleles to the exposed alleles was found over a distance of only 17 m. This, however, was considered more likely due to selection than to the presence of two reproductively isolated taxa.

Epiphytic habit. As discussed in the remarks on *L. obtusata* (p. 224), the strong association of these two species with algae and other marine plants is remarkable in an otherwise rock-dwelling genus. The shell shape and colour, and in some locations the annual life cycle (Williams, 1992a), can be regarded as adaptations to the algal habitat, and it has been suggested that the algae also benefit from the association by removal of potentially harmful epiphytic growths by the grazing of the snails (Williams, 1990a, 1992b). There is, however, no clear evidence for coevolution in this relationship.

Phylogenetic relationships. Phylogenetic analyses of morphology, allozyme frequencies and mitochondrial DNA sequences all indicate that the sister-species of *L. fabalis* is *L. obtusata* (p. 225).

Similar species. For discrimination from *L. obtusata* see p. 225 and Table 6.

Littorina (Neritrema) islandica new species

(Fig. 89)

Synonymy

Littorina sp. Bárdarson, 1925: 66.
Littorina littorea—Gladenkov, Norton & Spaink, 1980: 65–66, 113; pl. 13, figs 10, 11 (in part, includes *L. littorea* and *L. squalida*).
Littorina aff. *saxatilis*—Gladenkov, Norton & Spaink, 1980: 66, 113; pl. 13, fig. 12.

Types

Holotype IMNH 515, Fig. 89A; 5 paratypes IMNH 3977, Fig. 89B, C; type locality: Bárdarson's (1925) horizon 25, Pliocene of Tjörnes Peninsula, Iceland.

Taxonomic history

Several authors have noted a similarity between this species and *L. saxatilis*. Bárdarson (1925) remarked that his *Littorina* sp. was 'possibly a variety of *L. rudis*' (a synonym of *L. saxatilis*). The *L. saxatilis* and *L. rudis* mentioned by Norton (1975) from Tjörnes, but without specifying the horizon, were probably this species, but he admitted to doubt about the identification of carinate *Littorina* from Iceland (Norton, 1976). Curiously, examples of this species were figured as both *L. littorea* and as *Littorina* aff. *saxatilis* by Gladenkov, Norton & Spaink (1980). Apparently they were deceived by the variability of *L. littorea* in the Norwich Crag of East Anglia (Norton, 1976; see p. 108), believing it to be sufficiently variable in sculpture to embrace shells with the coarse ribs of the present species, while they justified separation of their shells into two species on the basis of differing spire height and roundness of the whorls. In fact, shell sculpture appears to be a better aid to identification than shell shape in this case.

Diagnosis

Shell height up to 18.5 mm (estimated). Shell turbinate; whorls rounded, sutures distinct. Mineralogy of external calcite and internal aragonite. Sculpture of rounded ribs, of which up to 3 may become more prominent and carinate; narrow grooves between ribs. Distribution: late Pliocene of Iceland.

Shell description

The following description is based on seven specimens (holotype and 5 paratypes; one from Hoskuldsvík in RGD). All are somewhat damaged and heights range from 11 to 16 mm; the intact height of the largest is estimated as 18.5 mm. The shell shape is moderately variable, from specimens with a taller spire (Fig. 89A; estimated H/B=1.26, SH=1.71) to others broader and more depressed (Fig. 89B; estimated H/B=1.07, SH=1.62). Although shells have undergone some diagenesis, the two-layered structure is still clearly visible, corresponding to a thin internal aragonitic layer and a thick outer calcitic one. The sculpture consists of rounded ribs, with narrow grooves between them. In the paratypes there are three most prominent, somewhat carinate, ribs (at and above the periphery) with one or two smaller ribs between each; the enlarged ribs impart an angular outline. In the holotype there are no carinate ribs, but their widths are still uneven, with three larger than the rest; here the grooves between the ribs are little more than impressed lines. The paratypes are somewhat worn, and the apparently wider grooves seen near the periphery in Figure 89B, C are an artefact of wear and staining; careful examination shows that the grooves are narrow, similar to those in Figure 89A. On the base are finer ribs, about 7–10 in number. Fine spiral microsculpture is preserved in the RGD specimen, covering ribs and grooves.

Distribution

Fossil record. This species is known only from the Tjörnes Peninsula in northern Iceland. Bárdarson (1925) collected it frequently in his horizon 25, a 3 m thickness of rather coarse sandstone, at the top

Figure 89. *Littorina (Neritrema) islandica* new species: holotype (**A**), IMNH 515; paratypes (**B**, **C**) IMNH 3977; Tjörnes Peninsula, Iceland; Bárdarson's (1925) horizon 25, late Pliocene. Scale bar=10 mm.

of the *Serripes* zone. He also noted a *Littorina* in horizon 24 below, but this was not described. Norton (1976) mentioned carinate *Littorina* in Bárdarson's horizons 23 and 25. Gladenkov *et al.* (1980) described their own material only from horizon 25, but in a stratigraphic diagram also indicated an older occurrence in the *Mactra* zone below, with the provenance 'data of Hopkins & MacNeil'; this older record requires verification. The palaeoecology of the Tjörnes sequence has been discussed by Norton (1975), who described horizon 25 as a deposit from a tidal flat or river mouth. Palaeomagnetic dating suggests an age of about 2.0 Ma for the top of the *Serripes* zone (Einarsson, Hopkins & Doell, 1967; Gladenkov, 1981). This has been judged contemporaneous with the top of the Red Crag or lower Norwich Crag of East Anglia and assigned to the latest Pliocene (Gladenkov, 1981; but assigned to the Pleistocene by Norton (1975) following the convention for the North Sea).

Discussion

Phylogenetic relationships. The overall shape of this species is generalized for the genus and gives little indication of possible relationships. The distinctive feature is the shell sculpture which, in the rather irregular sizes of the ribs, is reminiscent of some forms of *L. saxatilis* and *L. arcana*, although in those species the grooves are usually relatively wider. Such narrow grooves and rounded ribs are often seen in *L. compressa*, although there the ribs are rather uniform in development over the mid part of the whorl. The sculpture of *L. islandica* is not exactly reproduced in any Recent species. Its occurrence in the northern Atlantic invites comparison with the Recent species in the same region, and again only the *L. saxatilis* group appears close. At an age of 2 Ma, *L. islandica* considerably predates estimates of the age of divergence of the members of the *L. saxatilis* group based on allozyme-frequency data (0.21–1.19 Ma) and is also older than the mean age of estimates based on mitochondrial DNA sequences (1.73 Ma, range 0–3.55 Ma, p. 365). For these reasons it is suggested that it is related (or possibly even ancestral) to the members of the *L. saxatilis* group, but since it cannot be clearly connected to any one of them, it has been described as new.

Similar species. In view of the extreme variability of shell shape and sculpture in members of the *L. saxatilis* group, an unambiguous diagnosis of *L. islandica* cannot be given. As mentioned above, the narrow grooves between the rounded ribs separate it from most Recent examples of *L. saxatilis* and *L. arcana*, whereas the carinate development of three of the ribs distinguish it from *L. compressa*. The only other *Littorina* species recorded as a Pliocene fossil from Iceland, but from the older *Mactra* zone, is *L. squalida* (Fig. 30B–D). In that species the fossils are often worn smooth or, if well-preserved, show fine sculpture of narrow ribs and incised spiral lines. A single larger shell (H=23.0 mm, but damaged, and estimated H=25 mm) has also been seen from Bárdarsson's horizon 25 at Hoskuldsvík (RGD, in lot together with one typical *L. islandica*). This shell is so worn that all traces of sculpture have been lost. It might be a large example of this species or alternatively a specimen of *L. littorea*, of which the shape is somewhat reminiscent. All modern examples of *L. littorea* are distinguished from *L. islandica* by their fine and regular spiral ribs, as well as by their less impressed sutures. Nevertheless, it may be noted that a few fossils of juvenile *L. littorea* (from the Pleistocene Tiglian stage in The Netherlands, Fig. 37N) do show coarser ribs, similar to the sculpture of *L. islandica*. Additional material of this species is desirable, to assess its range of variation and to confirm that, as concluded here, it is indeed related to the *L. saxatilis* group.

Littorina (Neritrema) compressa Jeffreys, 1865

(Figs 1, 4, 5, 90–95)

Synonymy

Litorina nigrolineata—Philippi, 1846: **2**: 104; *Litorina* pl. 1, figs 17–19 (not *Littorina nigrolineata* Gray, 1839= *L. saxatilis*).

Litorina rudis var. *nigrolineata*—Forbes & Hanley, 1850: pl. 83, figs 1, 2 (not *L. nigrolineata* Gray, 1839; determination based on original of fig. in BMNH).

Littorina nigrolineata—Troschel, 1858: 133; pl. 10, fig. 19 (radula) (not Gray, 1839). Deyglun, 1955 (not Gray, 1839). Crisp & Fischer-Piette, 1959: 364–365 (not Gray, 1839). Sacchi, 1971: 77 (not Gray, 1839). Sacchi, 1974: 84–87; fig. 5 (oviduct) (not Gray, 1839). Heller, 1975a: 140–143, 146–149; fig. 3C (penis); figs 4C, 5C (radula); pl. 1, figs 9–15, 20 (not Gray, 1839). Heller, 1975b: 153–170 (not Gray, 1839). Sacchi, 1975: 111–120; fig. 1 (radula); fig. 2 (oviduct); fig. 6 (penis); pl. 1 (not Gray, 1839). Heller, 1976: 201–213 (not Gray, 1839). Caugant & Bergerard, 1979: 247–262 (not Gray, 1839). Hannaford Ellis, 1979: 43–56; fig. 2B (ciliated field) (not Gray, 1839). Raffaelli, 1979a: 219–232; fig. 5 (penis); figs 8, 10A, B (radula) (not Gray, 1839). Elner & Raffaelli, 1980: 151–160 (not Gray, 1839). Fretter & Graham, 1980: 266–268; fig. 208 (not Gray, 1839). Wilkins & O'Regan, 1980: 355–359 (not Gray, 1839). Naylor & Begon, 1982: 17–30 (not Gray, 1839). Knight & Ward, 1986: 405–413 (not Gray, 1839). Matthews & Montgomery, 1987: 290–296 (not Gray, 1839). Graham, 1988: 174; figs 59D, 63 (not Gray, 1839). Hayward, Wigham & Yonow, 1990: 660; fig. 12.9. Knight & Ward, 1991: 81–91 (not Gray, 1839). Poppe & Goto, 1991: 92; pl. 11, fig. 10 (not Gray, 1839). Reid, 1993: 51–62; fig. 2; fig. 5A–F (penis); fig. 5G–I (oviduct) (not Gray, 1839). Caley, Grahame & Mill, 1995: 181–193. Crossland, Coates, Grahame & Mill, 1995: 205–209 (not Gray, 1839).

Littorina (Littorina) nigrolineata—Récluz, 1869: 44 (not Gray, 1839; *s.l.*).

Littorina (Littorina) rudis var. *nigrolineata*—Tryon, 1887: 241; pl. 41, fig. 95 (as *nigrilineata*; not *L. nigrolineata* Gray, 1839).

Littorina saxatilis nigrolineata—Dautzenberg & Fischer, 1912: 196–197; pl. 9, figs 28–32; pl. 10, figs 29, 30 (in part, includes refs to *L. saxatilis*; not Gray, 1839). Dautzenberg & Durouchoux, 1913: 22 (not Gray, 1839). Dautzenberg & Fischer, 1925: 54 (not Gray, 1839). McMillan, 1939: 173–174 (not Gray, 1839). McMillan, 1944: 100–103; figs (not Gray, 1839).

Littorina (Littorivaga) saxatilis nigrolineata—Winckworth, 1932: 221 (not Gray, 1839).

Littorina saxatilis var. *nigrolineata*—Fischer-Piette, Gaillard & Delmas, 1967: 891–896 (not Gray, 1839).

Littorina saxatilis rudis var. *nigrolineata*—James, 1968a: 152; pl. 1, figs 4c, d, f, 5e–l, 6g, h, 7d–f (not Gray, 1839). James, 1968b: fig. 33 (not Gray, 1839).

Littorina saxatilis var. ex col. *nigrolineata*—Fischer-Piette & Gaillard, 1971: 44–45 (in part, includes *L. saxatilis*; not Gray, 1839).

Littorina saxatilis form *nigrolineata*—Gaillard, 1977: 11 (not Gray, 1839).

Littorina (?Littorivaga) nigrolineata—Smith, 1982: 535–539 (not Gray, 1839).

Littorina (Neritrema) nigrolineata—Reid, 1989a: 102 (not Gray, 1839). Reid, 1990b: 1–19; fig. 1q; fig. 2h (oviduct); fig. 4h (radula) (not Gray, 1839). Taylor & Reid, 1990: 199–215 (mineralogy) (not Gray, 1839). Smith & Heppell, 1991: 17 (not Gray, 1839).

Litorina merwinii 'Thompson' Philippi, 1846: **2**: 104; *Litorina* pl. 1, fig. 18 (Belfast; original specimen not in BMNH; in synonymy of *Litorina nigrolineata*, therefore unavailable).

Litorina macerwinii 'Thompson' Philippi, 1847: **2**: index [236] (unjustified emendation of *Litorina merwinii* 'Thompson' Philippi, 1846; unavailable).

Turbo sulcatus Leach, 1852: 185, 187; pl. 9, fig. 6 (in mari Cornubiense [Cornwall, England]; Scilly Islands [England]; lectotype (here designated) Leach, 1852: pl. 9, fig. 6; not Gmelin, 1791).

Littorina rudis var. *sulcata*—Jeffreys, 1865: **3**: 365–366. Daniel, 1883: 342. Smart & Cooke, 1885: 296. Marshall, 1898: 121–122.

Littorina rudis var. ex forma *sulcata*—Dautzenberg & Durouchoux, 1900: 45.

Littorina rudis—Reeve, 1857: sp. 33; pl. 7, fig. 33a (in part, other figs are *L. saxatilis*; not *Turbo rudis* Maton, 1797=*L. saxatilis*). Sowerby, 1859: pl. 12, fig. 13 (in part, other fig. is *L. saxatilis*; not Maton, 1797). Sowerby, 1887: pl. 12, fig. 13 (in part, other figs are *L. saxatilis*).

Littorina (saxatilis var.) *rudis*—Kobelt, 1907: 59; pl. 110, figs 9, 10 (in part, other fig. is *L. saxatilis*; not Maton, 1797).
Littorina rudis var. *compressa* Jeffreys, 1865: **3**: 366 ('various parts of our shores' [British Isles; here restricted to Bordeaux Harbour, Guernsey, Channel Islands]; lectotype (here designated, 16.9 mm, Fig. 92A)+5 paralectotypes USNM 185682, seen; type collection figured by McMillan, 1944). Marshall, 1898: 122.
Littorina saxatilis nigrolineata var. *compressa*—Dautzenberg & Fischer, 1912: 197; pl. 9, figs 30, 31. Dautzenberg & Durouchoux, 1913: 22. McMillan, 1944: 100–103; fig.
Littorina saxatilis var. *compressa*—Fischer-Piette, Gaillard & Delmas, 1967: 891–896 (in part, includes *L. saxatilis*).
Littorina rudis var. ex colore *lutea* Dautzenberg & Durouchoux, 1900: 46 (St Malo [France]; in part, includes *L. saxatilis s.l.*; infrasubspecific; not *Litorina scabra* var. *lutea* Philippi, 1847=*Littoraria lutea*).
Littorina saxatilis nigrolineata var. *matoni* Dautzenberg & Fischer, 1912: 197; pl. 9, fig. 32 (Baie de Saint-Aubin, Jersey [Channel Islands]; original of fig., MNHNP, seen; infrasubspecific).
Littorina saxatilis var. ex col. *miniata* Dautzenberg & Fischer, 1912: 195, 197 (in part, includes *L. saxatilis s.l.*; infrasubspecific; not *L. littorea* var. ex col. *miniata* Dautzenberg & Durouchoux, 1900=*L. littorea*).
Littorina saxatilis nigrolineata var. *compressa* monstr. *canaliculatum* Dautzenberg & Fischer, 1912: 200–201; pl. 10, figs 29, 30 (St Servan [France]; original of figs, MNHNP, seen; monstrosity, therefore unavailable).
Littorina saxatilis rudis var. *rudissima*—James, 1968a: 152; pl. 1, fig. 8c, d; pl. 2, fig. 1a–c (in part; other figs *L. saxatilis*; not *Turbo rudissimus* Johnston, 1842=*L. saxatilis*).
Littorina saxatilis rudis var. *jugosoides* James, 1968a: 153; pl. 1, fig. 5m–o (St Mary, Channel Islands; in part; other figs are *L. saxatilis*; infrasubspecific).
Littorina (Littorivaga) saxatilis—Nordsieck, 1968: 40; pl. 6, fig. 23.20 (in part, includes *L. saxatilis*; not *Turbo saxatilis* Olivi, 1792). Nordsieck, 1982: 61; pl. 26, fig. 23.20a (in part, includes *L. saxatilis*, not Olivi, 1792).
Littorina saxatilis var. *cinerea* Fischer-Piette & Gaillard, 1971: 42–43 (Saint-Quay [France]; original specimen, MNHNP, seen; in part, includes *L. saxatilis s.l.*; infrasubspecific; not *L. cinerea* Pease, 1869=*Nodilittorina cinerea*).

Note: for references in which this species was not clearly differentiated from *L. saxatilis*, and may have been included with it, see Synonymy of *L. saxatilis* (p. 278). In the Synonymy and following text '*L. saxatilis s.l.*' (*sensu lato*) is used where it is likely, but not certain, that *L. compressa* and *L. arcana* may have been included with that species. The '*L. saxatilis* group' refers to all three species.

Taxonomic history

The history of this species is closely connected with that of *L. saxatilis*, a species so variable in shell form that *L. compressa* has been confused with it by many authors. Although unequivocal separation of the two depends upon anatomical characters (Deyglun, 1955; Sacchi, 1974, 1975; Heller, 1975a), the shell of *L. compressa* is sufficiently distinctive that good figures can usually be identified, allowing earlier references to be included in the synonymy. The more limited geographical distribution of *L. compressa* is also useful in this respect.

Philippi (1846) was the first author to separate this species from other forms in the *L. saxatilis* group, his description and figures indicating a clear concept of the species based on the strongly ribbed shell with black grooves. Unfortunately, he identified it as *L. nigrolineata*, a species that had been described by Gray in 1839 without a figure or a type locality. Examination of the types of Gray's species (Fig. 106P) has shown that they are in fact a nigrolineate form of *L. saxatilis* that is common on the northwestern coast of the Iberian Peninsula. Philippi's incorrect usage has been followed by the majority of subsequent authors who have recognized this species (a usage indicated as *L. 'nigrolineata'* in the following text) but, although it is now widely familiar by this name, it cannot be maintained and is here corrected. Of other names based on this species, *Litorina merwinii* 'Thompson' Philippi, 1846, was proposed in synonymy, and *Turbo sulcatus* Leach, 1852, is a junior primary homonym, so neither can be used. The earliest available name is *L. rudis* var. *compressa* Jeffreys, 1865, which was only briefly described and not figured in the original publication, but is supported by a type series (Fig. 92A). Although described as a variety, the name can be regarded as available at subspecific rank (ICZN, 1985: art. 45g). Since its description the name *compressa* has occasionally appeared in the literature as a variety (Marshall, 1898; Dautzenberg & Fischer, 1912; Dautzenberg & Durouchoux, 1913; McMillan, 1944; Fischer-Piette, Gaillard & Delmas, 1967; Sacchi, 1975).

After Philippi (1846) the species was briefly mentioned as *L. 'nigrolineata'* by Troschel (1858) and Récluz (1869), and described as *Turbo sulcatus* by Leach (1852), but was not again treated as a distinct species for almost a century (until reinstated by Deyglun, 1955). In the latter half of the nineteenth century a broad concept of the species *L. rudis* prevailed, embracing all three species of the

L. saxatilis group, and *L. compressa* was seldom noticed. Both Forbes & Hanley (1850) and Tryon (1887) reduced *'nigrolineata'* to a variety of *L. rudis*. Jeffreys (1865) recognized *L. rudis* vars *sulcata* and *compressa*, which were both forms of the present species, and Daniel (1883) even suggested that var. *sulcata* might be a different species. Most others included the present species in *L. rudis* (Reeve, 1857; Sowerby, 1859, 1887; Kobelt, 1907).

In the early twentieth century interest in shell variation increased. In a detailed study of the *L. saxatilis* group Dautzenberg & Fischer (1912) developed an elaborate system of subspecies, form varieties, colour varieties and monstrosities, to describe shell variation. Their *L. saxatilis 'nigrolineata'* corresponded largely with the present species but, from the cited references to published figures, evidently included the nigrolineate, Iberian form of *L. saxatilis*. They used *compressa* as a varietal name for uniformly yellow or brown shells and described a new variety, var. *matoni*, with finer sculpture. Dautzenberg & Fischer's classification of the present species as *L. saxatilis 'nigrolineata'* was followed by a few others (Dautzenberg & Durouchoux, 1913; Dautzenberg & Fischer, 1925; Winckworth, 1932), but in general the species was overlooked. In Britain McMillan (1939, 1944) pointed out the characteristic shell form, restricted geographical distribution and eulittoral zonation of *L. saxatilis 'nigrolineata'*, but still did not recognize it as specifically distinct.

The specific status of *L. 'nigrolineata'* was re-established in 1955 when, in an unpublished thesis, Deyglun (quoted in Crisp & Fischer-Piette, 1959, and Sacchi, 1975) distinguished it from *L. saxatilis* by its shell, radula, ecology and especially by its oviparous reproduction. This was accepted by Crisp & Fischer-Piette (1959), who confirmed the different intertidal zonation and geographical distribution of the two species. Nevertheless, this opinion was later rejected on conchological grounds, without any consideration of the new anatomical information, and *'nigrolineata'* again reduced to varietal status (Fischer-Piette, Gaillard & Delmas, 1967; Fischer-Piette & Gaillard, 1971). Gaillard (1977) reiterated this view, arguing that the difference between the jelly gland of *L. 'nigrolineata'* and the brood pouch of *L. saxatilis* was of minor anatomical significance, possibly a hormonal effect, which did not justify their separation. Rather surprisingly, even James (1968a, b), who had examined penial and radular characters as well as shells and ecology, failed to recognize the present species as distinct, including it with *L. saxatilis* as two varieties of *L. saxatilis rudis*.

Widespread recognition of *L. 'nigrolineata'* as a species distinct from *L. saxatilis* came after Sacchi (1971, 1974, 1975) publicized the earlier work of Deyglun (1955). Sacchi (1975) published a complete redescription, including shell, radular, penial and oviducal characters, together with a discussion of ecology and geographical distribution. He also made it quite clear that the nigrolineate Iberian form was merely a colour variety of *L. saxatilis* (nevertheless, confusion on this point has continued, e.g. Rolán, 1984; Burnay, 1986; and only now has the identity of Gray's type material been established). At about the same time Heller (1975a) also recognized the importance of anatomical characters in the systematics of the *L. saxatilis* group. He redescribed *L. 'nigrolineata'* from Wales, using the same characters of ridged shell, penis, radula and oviparous reproduction, and again noted its eulittoral zonation pattern. Heller (1975a) also used the evidence of characteristic esterase banding patterns revealed by gel electrophoresis, although these did not provide unequivocal separation from *L. saxatilis*. Similar results were obtained from studies of other polymorphic enzymes, which showed that sympatric populations of the two species maintained significantly different allele frequencies, suggesting a degree of reproductive isolation, although no diagnostic alleles were found (Caugant & Bergerard, 1979; Wilkins & O'Regan, 1980; Knight & Ward, 1986, 1991; Backeljau & Warmoes, 1992). Further work on the penis and oviduct of the British *Littorina* species (Hannaford Ellis, 1979; Raffaelli, 1979a; Fretter, 1980) supported the morphological characterization of the species by Sacchi (1974, 1975) and Heller (1975a). Thus by the end of the decade the status of *L. 'nigrolineata'* was virtually unchallenged (Fretter & Graham, 1980), except by a few conchological taxonomists (e.g. Rosewater, 1981), and it is now illustrated even in textbooks (Hayward, Wigham & Yonow, 1990; Poppe & Goto, 1991). Since this species is relatively easily recognized in the field, it has been the subject of a number of ecological studies, which have further emphasized its distinctness from the better known *L. saxatilis* (Heller, 1975b, 1976; Raffaelli, 1977; Elner & Raffaelli, 1980; Hughes, 1980; Hughes & Roberts, 1980, 1981; Naylor & Begon, 1982; Hannaford Ellis, 1983; Matthews & Montgomery, 1987; Reid, 1993). Recently, clear separation of all analysed individuals of *L. 'nigrolineata'* from other species in the *L. saxatilis* group has been achieved using the banding patterns of random amplified polymorphic DNAs (RAPDs) (Crossland, Coates, Grahame & Mill, 1995) although, as in allozyme studies, no individual bands were diagnostic.

Diagnosis

Mature shell height 3.6–22.9 mm. Shell turbinate, usually thick-walled, last whorl sometimes slightly compressed and cylindrical. Mineralogy of external irregular-prismatic calcite and internal crossed-lamellar aragonite. Sculpture usually of strong, regular, rounded or flattened ribs and fine spiral microstriae over whole surface. Colour variable: white, cream, yellow, fawn or brown, sometimes with brown to black grooves between the sculptural ribs; columella and aperture white or brown. Penis with very short filament, less than 5% total length, 1–8 mamilliform penial glands. Pallial oviduct of single loop of albumen gland, single loop of capsule gland and swollen, septate jelly gland; albumen and capsule glands together occupy about half total length of pallial oviduct; bursa usually extends only half length of jelly gland, rarely reaching back to capsule gland. Spawn a benthic mass of singly-encapsulated eggs in gelatinous matrix; development nonplanktotrophic with intracapsular metamorphosis. Distribution: Barents Sea, northern Norway, British Isles, Normandy and Brittany.

Anatomical description

Head-foot (Figs 1A, B; 90I). Head black to grey, with unpigmented tip of snout; usually pale around eye and across tentacle base; tentacles pale with two longitudinal black stripes, sometimes also with indistinct transverse bands; head and tentacles may be entirely black; sides of foot black to mottled grey or unpigmented. Sacchi (1974) reported pale or unpigmented animals to be more common in light-coloured shells. Opercular ratio 0.48–0.59.

Penis (Figs 1A, 90A-J). Base wrinkled, usually unpigmented, but sometimes with a little grey to black pigment on lower part of outer side; 3–8 (rarely 1–2) mamilliform penial glands of relatively large size in single row near distal end. Populations vary in the mean numbers of penial glands, but there is no correlation with either wave-exposure (Naylor & Begon, 1982) or with shell size (Heller, 1975a). Filament very short, less than 5% total length. Sperm groove open, ending terminally. Subepithelial glandular tissue around sperm groove often visible even in preserved specimens. Unparasitized animals of adult size are sometimes found with small penes, suggesting that reduction (or shedding) may occur outside the breeding season.

Sperm. Euspermatozoa 46–51 µm. Paraspermatozoa 11–16 µm diameter, round to slightly oval; vesicles large and distinct; nucleus not visible.

Oviduct (Figs 1B, 90K–O). Single loop of albumen gland; followed by single loop of capsule gland of about the same area as albumen gland, no translucent part of capsule gland visible (histological sections not seen); albumen and capsule glands together occupy about half total length of pallial oviduct; finally a swollen, septate jelly gland. In all 10 specimens seen from the Barents Sea (5 collections, all from Dal'niye Zelentsy Bay) the capsule gland is reduced in size, usually less than half the area of the albumen gland, and is a simple loop only slightly or not at all twisted into a spiral (Fig. 90M, N). Bursa opens at anterior end of jelly gland and usually (80% of individuals examined) does not extend back as far as capsule gland, commonly being about half total length of jelly gland. Seminal receptacle simple.

Spawn and development. Females deposit gelatinous egg masses in humid sites on the shore, such as in crevices in the barnacle zone, under stones and at the base of *Fucus* plants. Individual egg masses are variable in shape, usually reniform to circular, about 5 by 10 mm in area and 3 mm thick, and contain 100–200 pink or cream eggs, each singly enclosed in an egg capsule; several egg masses may be laid contiguously (Fretter & Graham, 1980; Hughes, 1980; Moreteau, 1981; Hannaford Ellis, 1983). Development nonplanktotrophic with intracapsular metamorphosis. On hatching, juvenile shell measures 480–600 µm in length (Fretter & Graham, 1980; Hughes, 1980; Hughes & Roberts, 1980; Moreteau, 1981). Protoconch 600 µm diameter and of 1.7 whorls in a shell from the Barents Sea. Spawning continues throughout the year, although at a low level during the warm summer months; in north Wales peak spawning has been recorded in either May to June (Hughes, 1978, 1980) or in autumn (Hannaford Ellis, 1983).

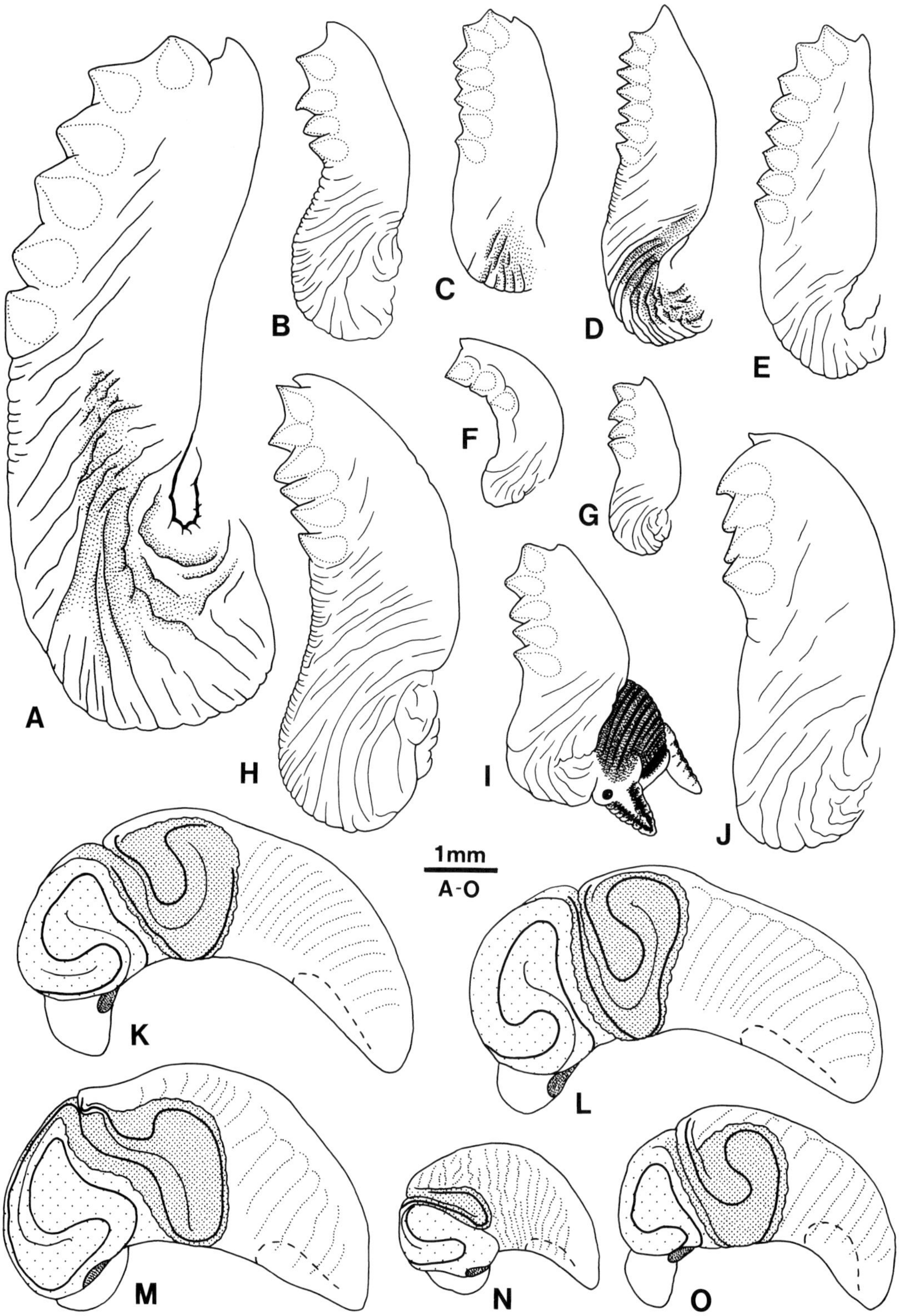

Figure 90. *Littorina (Neritrema) compressa*: **A–J,** penes; **I,** head pigmentation; **K–O,** pallial oviducts. **A,** Careg Hanner Trai, Anglesey, Wales (moderate ecotype); **B,** Lizard Point, Cornwall, England (moderate ecotype); **C, N,** Oscar Bay, Dal'niye Zelentsy Bay, Barents Sea, Russia (moderate ecotype); **D,** Loch Ainort, Skye, Scotland (moderate ecotype); **E, G,** 1 km W of St Govan's Head, Pembrokeshire, Wales (**E,** moderate ecotype; **G,** barnacle ecotype, shell illustrated Fig. 93H); **F, L,** Golam Head, Co. Galway, Ireland (**F,** barnacle ecotype, shell illustrated Fig. 93K; **L,** moderate ecotype); **H,** Carantec, Brittany, France (moderate ecotype); **I,** Coosfadda Bay, Co. Kerry, Ireland (barnacle ecotype); **J,** Rubh' Aird-mhicheil, South Uist, Outer Hebrides, Scotland (wave-exposed ecotype); **K,** Lezardrieux, near Paimpol, Brittany, France (moderate ecotype); **M,** Yarnishnaya Inlet, Barents Sea, Russia (moderate ecotype); **O,** Caerfai Bay, Pembrokeshire, Wales (moderate ecotype). Shell heights: **A,** 20.6 mm; **B,** 6.8 mm; **C,** 7.6 mm; **D,** 14.5 mm; **E,** 9.0 mm; **F,** 3.6 mm; **G,** 3.7 mm; **H,** 13.0 mm; **I,** 5.1 mm; **J,** 12.3 mm; **K,** 13.0 mm; **L,** 12.9 mm; **M,** 10.7 mm; **N,** 7.8 mm; **O,** 6.2 mm. For shading conventions of pallial oviducts see p. 6.

Radula and foregut (Fig. 91A–H). Relative radular length 1.10–2.77. Rachidian: length/width 1.00–1.40; central cusp blunt to bluntly rounded; one pointed cusp and one outer denticle on each side. Lateral: four cusps; second from inside bluntly rounded; third from inside largest and blunt to bluntly rounded; others small and pointed. Inner marginal: four cusps; second from inside rounded; third from inside largest and blunt to bluntly rounded; others small and pointed. Outer marginal: 4–6 rounded to blunt cusps (up to 7 in juveniles); narrow neck, projecting flange on inside of base only. All cusps more elongate and pointed in juveniles and small adults (Fig. 91B). Salivary glands enlarged.

Shell variation (Figs 4, 5, 92–94)

Geographical variation. Until recently, the geographical range of *L. compressa* was thought to be restricted to France and the British Isles, and within this region the morphological diversity of shell types is similar throughout. The species has now been recorded from northern Norway and the Barents Sea, and here the shells are conspicuously different, so that two geographical varieties can usefully be distinguished:

> Geog. var. *typica* (Figs 92A–J, 93E–S): H/B=1.06–1.33; SH=1.23–1.57. Mature shell height 3.6–22.9 mm; shape turbinate; often with laterally compressed and slightly shouldered last whorl; relatively thick-walled; sculpture of broad, regular, flattened or rounded ribs with deep but narrow grooves between them; colour variable. Range: northern France and British Isles.
>
> Geog. var. *barentica* (Fig. 93A–D): H/B=1.14–1.43; SH=1.41–1.80. Mature shell height 5.7–10.7 mm; shape turbinate to high-turbinate; whorls rounded; relatively thin-walled; sculpture of relatively poorly developed, sharper, narrower and more numerous ribs, becoming obsolete towards end of last whorl; almost all shells uniform brown. Range: Narvik (Norway) to Yarnishnaya (Russia).

So far, geog. var. *barentica* is known from only five localities (Vassvika near Narvik; Sørøy; Smelror near Vardø; Yarnishnaya and Dal'niye Zelentsy Bay, Russian Barents Sea), and only about 100 specimens have been examined. There is a large gap separating the known distribution of the two varieties, between the Barents Sea and the British Isles, and as discussed later (p. 261) this disjunction is probably real. Despite the striking differences in their shells, the varieties are believed to be conspecific. Neither penes (Fig. 90C) nor radulae (Fig. 91F) show significant differences; however, in samples from Yarnishnaya and Dal'niye Zelentsy Bay the capsule gland is always relatively small and consists of an almost uncoiled loop (Fig. 90M, N), differing from the typical form of the pallial oviduct (Fig. 90K, L, O). In a sample from Vassvika the capsule gland is of normal shape, although the shells are of geog. var. *barentica*, so the variation in the capsule gland is thought to be intraspecific. It would be interesting to examine variation in this character at intermediate localities, and to measure the genetic distance between the two geographical varieties.

The existence of a northern variety with a thin-walled, tall-spired shell is reminiscent of the geographical variation seen in *L. obtusata* and *L. fabalis*, and the explanation may be the same in each case (see discussion p. 213). The southern variety of *L. compressa* is preyed upon by crabs such as *Carcinus maenas* (Elner & Raffaelli, 1980) and the thicker and lower-spired shell has probably been selected for a degree of resistance to predation. In the range of the northern variety *Carcinus* is rare or absent (Vader, 1979), and other factors such as energetic considerations may constrain shell thickness and shape. The known southern limit of geog. var. *barentica* at Vassvika near Narvik is similar to that of *L. obtusata* geog. var. *palliata*, but not identical, for at this locality the shells of *L. obtusata* are of the thicker southern variety *retusa*. Since the microhabitats of these two species are not identical, an exact equivalence might not be expected. It should also be noted that both crab distribution and average shell morphology in local populations can change rapidly (over as little as 17 snail generations in *L. obtusata*, Seeley, 1986). At the time when the collection was made (DGR, 1989) *L. compressa* may not then have fully adapted to a recently increased level of crab predation, although spire height was in fact lower than in most of the shells from Russian localities. It would be interesting to search to the south of Narvik for populations of *L. compressa* which might coexist with crabs and might show the shell characters of the typical southern variety. Alternatively, perhaps in Norway *L. compressa* is unable to withstand heavy crab predation, possibly accounting for the southern limit near Narvik.

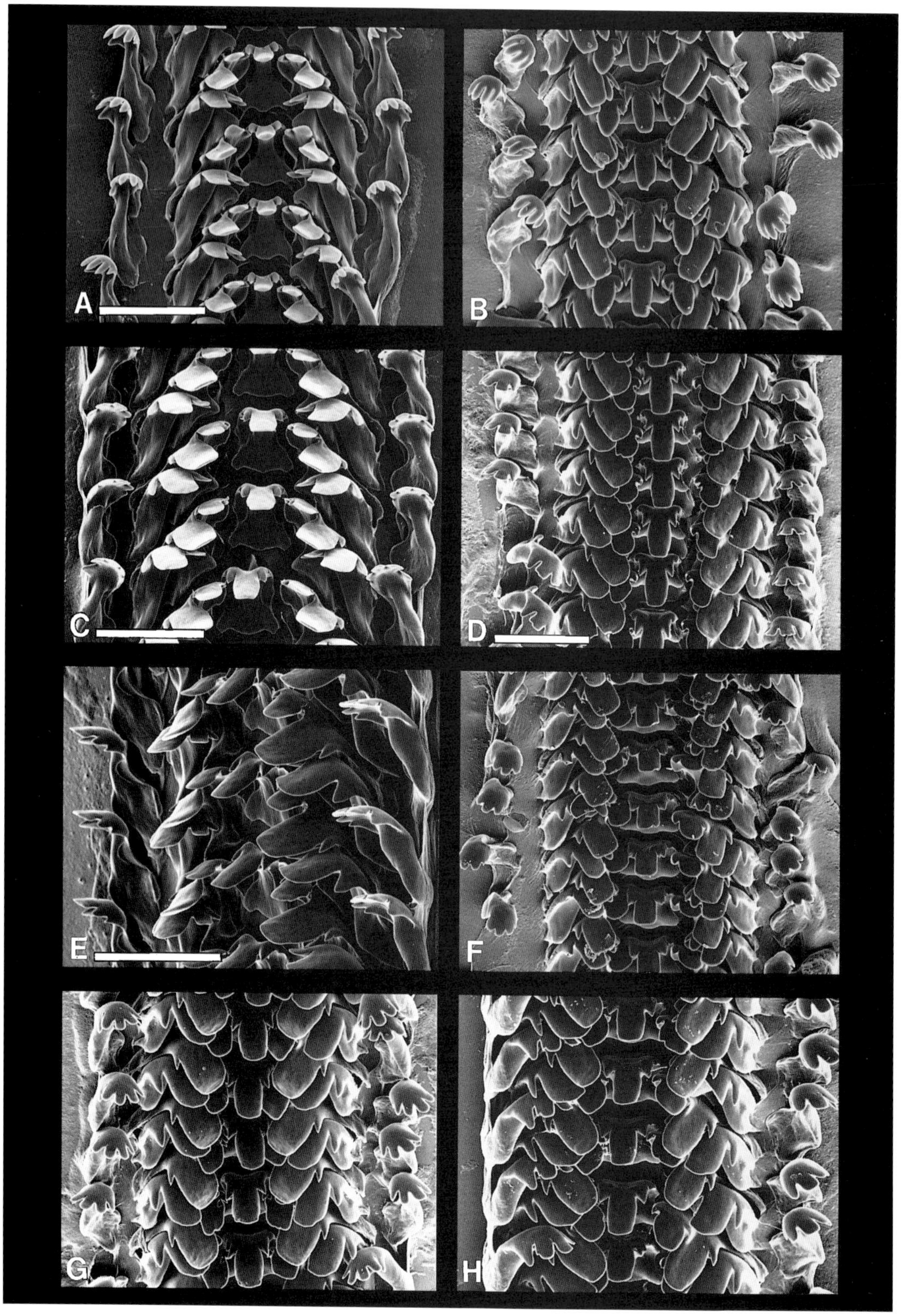

Figure 91. *Littorina (Neritrema) compressa*: **A, B,** two views of radula (shell H=5.1 mm), flat (**A**) and at 45° (**B**); **C–E,** three views of radula (shell H=14.7 mm), flat (**C**), at 45° (**D**) and side view (**E**); **F,** radula (shell H=9.5 mm), at 45°; **G,** radula (shell H=12.3 mm), at 45°; **H,** radula (shell H=16.6 mm), at 45°. **A, B,** Coosfadda Bay, Dingle Bay, Co. Kerry, Ireland (barnacle ecotype); **C–E,** Finavarra, Co. Galway, Ireland (moderate ecotype); **F,** Yarnishnaya Inlet, Barents Sea, Russia (moderate ecotype); **G,** Rubh' Aird-mhicheil, South Uist, Outer Hebrides, Scotland (wave-exposed ecotype); **H,** Barfleur, Normandy, France (moderate ecotype). Scale bars: **A, B**=50 μm; **C**=100 μm; **D**=100 μm; **E–H**=100 μm.

Ecotypic variation. Three ecotypes can be recognized:

> Moderate ecotype (Figs 92A–J, 93A–E, L–N, P, Q): these shells inhabit moderately exposed to sheltered shores, reach large size, have taller spires than the other ecotypes, and (only in geog. var. *typica*) a more thick-walled shell often with a compressed and shouldered final whorl.
>
> Wave-exposed ecotype (Fig. 93F, G, J, O, R, S): *L. compressa* is not common on extremely exposed shores, but near the exposure limit of the species the shells are smaller (to about 12 mm), thinner-walled, with a larger and rounder aperture, and lower spire (H/B=1.06–1.15; SH=1.23–1.37).
>
> Barnacle ecotype (Fig. 93H, I, K): shells are similar in shape to the wave-exposed ecotype, with a relatively large, round aperture, but are less than 5–6 mm in maximum height (maturing as small as 3.6 mm), with poorly developed narrow ribs or completely smooth, and are black, white, or tessellated or banded with black and white; they occur on exposed shores among barnacles and mussels.

Although nothing is known about the heritability of shell morphology in this species, the differences between the ecotypes have been interpreted as adaptive traits that are the result of selection, as in *L. saxatilis* which shows analogous ecotypes (p. 305; Fig. 4). On wave-exposed shores small size may be related to the ability to shelter from wind and waves in rock crevices, while shells need not be especially thick in the absence of crab predators (Heller, 1976; Naylor & Begon, 1982). An enlarged, rounder aperture (which produces a correspondingly more globose and usually lower-spired shell) was recorded on more exposed shores by Naylor & Begon (1982) and might be expected to accommodate a larger foot for adherence under exposed conditions; however, Heller (1976) failed to find correlations between either apertural roundness or relative spire height and exposure. On sheltered and boulder shores, where the risk of crushing by either crab predators or by mobile boulders is high, large and thick-walled shells may be selectively advantageous (Elner & Raffaelli, 1980; Naylor & Begon, 1982), and indeed Heller (1976) found a correlation between shell size and crab size on shores of similar wave-exposure. Significant differences between populations from contrasting microhabitats (e.g. upper-shore boulders and lower-shore crevices) can be maintained on the same shore (Naylor & Begon, 1982). The species is uncommon on exposed shores, so the wave-exposed ecotype is relatively infrequent. Since it lacks the characteristically compressed and shouldered outline of the moderate ecotype, it may be difficult to distinguish from *L. saxatilis* and *L. arcana* without anatomical examination. Although geog. var. *barentica* is typically rather tall-spired, at one relatively exposed locality (Smelror near Vardø) spire height was lower and the aperture enlarged (Fig. 93C), and a few shells fell within the shape range given above for the wave-exposed ecotype.

The barnacle ecotype is the rarest of the three and has so far been recorded only from localities in southwestern Wales and western Ireland (Reid, 1993). Its rarity may be due to the fact that few populations can survive the extreme wave-exposure that is apparently necessary to select for the small adult size that allows the animals to shelter within empty barnacle shells and in the interstices of mussel beds. In shell shape this ecotype resembles the wave-exposed ecotype, and its enlarged, rounded aperture may be similarly adaptive on the exposed shores where it is found. Some examples are completely smooth (Fig. 93K). The predominantly black and white coloration may be cryptic (see Colour Variation, p. 258). These barnacle ecotypes are so similar to the analogous ecotypes of *L. saxatilis* and *L. arcana* that identification must be confirmed anatomically.

This species does not occur in salt marshes and is not tolerant of brackish conditions (p. 260), so no shell types analogous to the sheltered/brackish ecotypes of other *Littorina (Neritrema)* species occur.

Sculptural variation. Typically (i.e. in geog. var. *typica* from moderate habitats) shell sculpture is of rather broad, distinct, flattened ribs, separated by deep and narrow grooves (e.g. Figs 92C, 94A), and this type of sculpture is a useful identification character for the species. In more wave-exposed habitats the ribs are often more prominent and rounded (e.g. Fig. 93F, S) although, as mentioned above, in the barnacle ecotype the shell is sometimes entirely smooth (Fig. 93K). In geog. var. *barentica* the sculpture is quite unlike the typical form, the ribs being narrower and more numerous, and sometimes becoming obsolete on the final whorl (Fig. 93A–D).

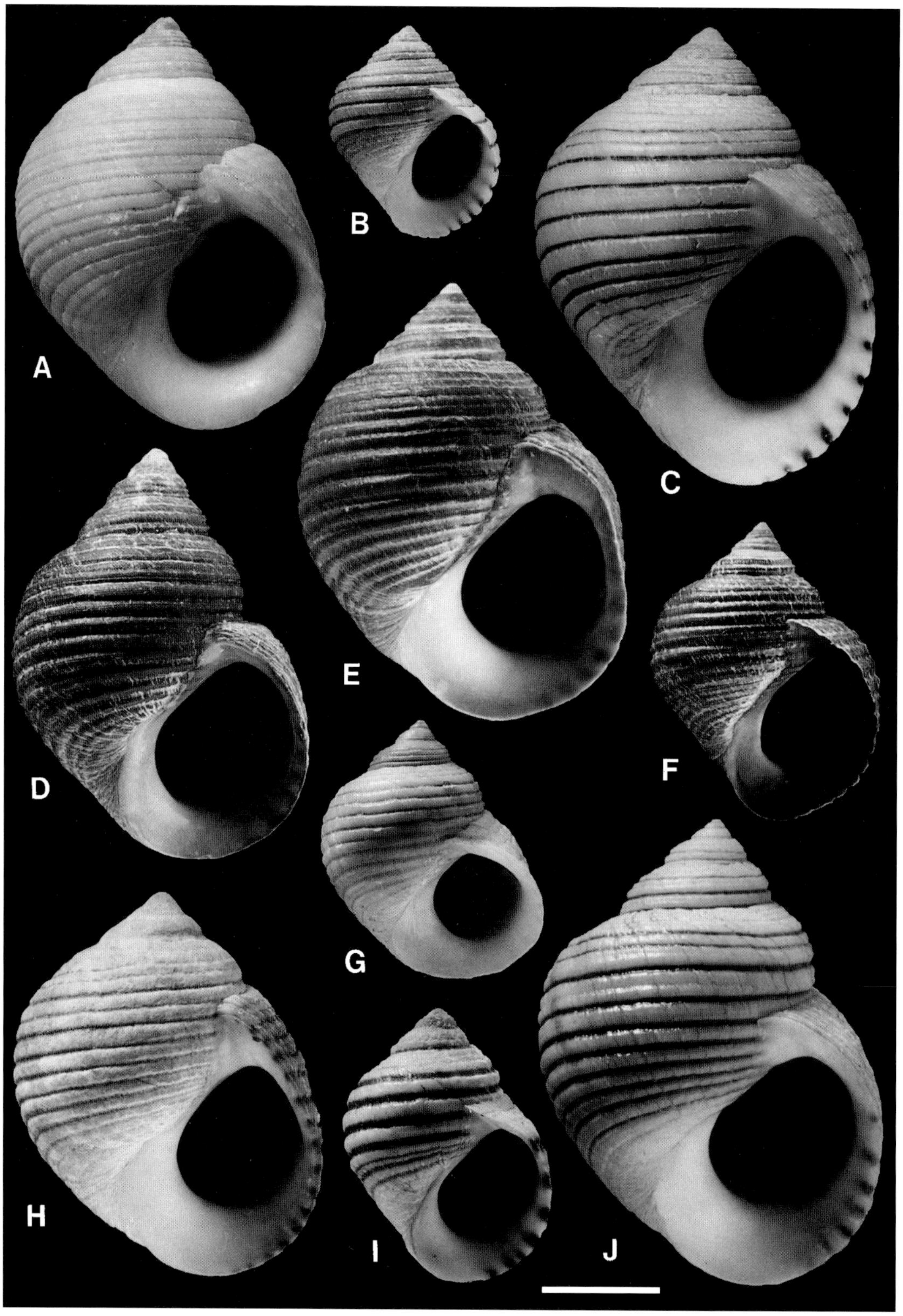

Figure 92. *Littorina (Neritrema) compressa*: geog. var. *typica*; moderate ecotype (**B**, **F**, juveniles). **A**, *Littorina rudis* var. *compressa* Jeffreys, 1865; lectotype, USNM 185682; Bordeaux Harbour, Guernsey, Channel Islands. **B**, **C**, 1 km NW of Trearddur, Anglesey, Wales. **D**, Lezardrieux, Brittany, France. **E**, **F**, Killyleagh, Co. Down, Ireland. **G**, Ynys-y-Fydlyn, Anglesey, Wales. **H**, Finavarra, Co. Clare, Ireland. **I**, Cow Strand, Sherkin Island, Co. Cork, Ireland. **J**, St Mary's, Scilly Isles, England. Scale bar=5 mm.

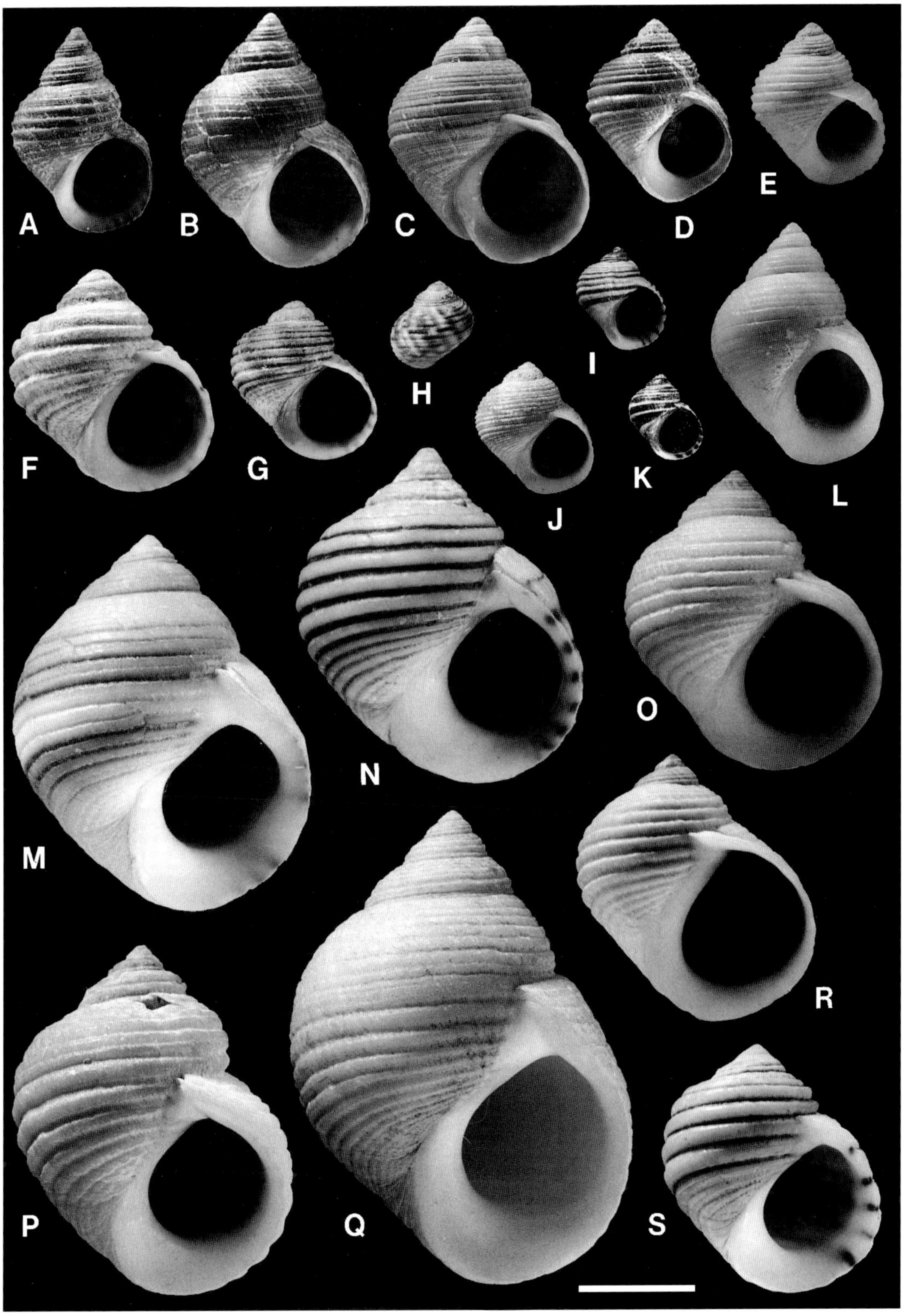

Figure 93. *Littorina (Neritrema) compressa*: **A–D,** geog. var. *barentica*; **E–S,** geog. var. *typica*; **A–E, L–N, P, Q,** moderate ecotype; **F, G, J, O, R, S,** wave-exposed ecotype; **H, I, K,** barnacle ecotype. **A, B,** Yarnishnaya, Barents Sea, Russia. **C,** Smelror, near Vardø, Norway. **D,** Vassvika, Narvik, Norway. **E,** Whitesand Bay, Pembrokeshire, Wales. **F, G,** Old Head of Kinsale, Co. Cork, Ireland. **H, J,** 1 km W of St Govan's Head, Pembrokeshire, Wales. **I, K,** Golam Head, Co. Galway, Ireland. **L,** Lochskipport, South Uist, Outer Hebrides, Scotland. **M,** Pointe de l'Arcouest, Brittany, France. **N,** 1 km NW of Trearddur, Anglesey, Wales. **O,** Rubh' Aird-mhicheil, South Uist, Outer Hebrides, Scotland. **P,** St Enogat, Dinard, Brittany, France. **Q,** Barfleur, Normandy, France. **R,** Islay, Scotland; NMS 1979087.11053. **S,** Sennen Cove, Cornwall, England. Scale bar=5 mm.

Ontogenetic variation. In juvenile shells the ribs are relatively more sharply pointed in section (Fig. 94B–G) and only become flattened at larger size (Fig. 94A) (also noted by Raffaelli, 1982). Actively growing juvenile shells show a thin apertural margin and a produced apertural lip (Fig. 92B, F). In the moderate ecotype of geog. var. *typica*, as growth slows and maximum size is approached, the aperture is strongly thickened within and its outline becomes laterally compressed and slightly shouldered (hence the specific name); the suture with the penultimate whorl also descends slightly towards the base (Figs 92A, E, H; 93N). In these examples growth therefore appears to be determinate. However, Moreteau (1981) reported alternations between active and static growth stages even in large animals with a thickened peristome.

Sexual dimorphism. Sacchi (1974) noted modest sexual dimorphism, presumably in size, and Naylor & Begon (1982) reported that females were squatter and broader.

Monstrosities. Dautzenberg & Fischer (1912) figured a shell with a sharp keel close to the suture on the final whorl, probably a consequence of earlier damage to the shell.

Colour variation (Fig. 5). In comparison with *L. saxatilis*, this species shows a reduced level of polymorphism. Strictly, the variation is not usually polymorphism at all, since the majority of populations are monomorphic, and where mixed populations are found intermediates may occur (Naylor & Begon, 1982). Nevertheless, across the species as a whole there is a range of shell colours, and the principal colour forms have long been known (Dautzenberg & Fischer, 1912; McMillan, 1944; Sacchi, 1974; Heller, 1975a). Shell ground colour may be white, grey, cream to yellow, or (less commonly) red-brown to blackish; superimposed on the pale ground colours may be a pattern of dark brown to purplish black lines that are restricted to the grooves of the shell. In the past, the names var. *compressa* and var. *nigrolineata* have been applied to the unlined and lined forms respectively (Dautzenberg & Fischer, 1912; McMillan, 1944; Naylor & Begon, 1982). Occasionally the dark lines may be present in one or two broad bands only (Fig. 93M). Tessellated shells are extremely rare; only five shells have been seen with a flecked pattern of brown and white (Fig. 93H).

So far there is no information on the heritability of shell colour in this species, although comparison with *L. obtusata* and *L. saxatilis* suggests that genetic determination is likely. The polymorphism appears not to be neutral, for some predictability can be found in the occurrence of the shell colours. The adaptive significance of this variation has been considered only in relation to possible visual selection. Working in Wales, Heller (1975b) found a strong association of red-brown shells with localities of reddish sandstone rock; in Pembrokeshire up to 70% of shells on Old Red Sandstone were red-brown, at localities mainly within the shelter of Milford Haven. This association of red-brown shells with Old Red Sandstone on sheltered shores has also been recorded in Northern Ireland (Matthews & Montgomery, 1987). On more exposed shores, even on reddish rock, the frequency of white shells increased, and here the predominant background was of whitish barnacles (Heller, 1975b). The yellow morph was found mainly on sheltered shores, particularly where these were thickly covered with fucoid algae, but also on shores with barnacles yellowed by fungal growth. These correlations certainly suggest visual selection, and visual predators might include birds, fish and possibly crabs. The distribution of the lined pattern was less clear; this form was regionally common in Pembrokeshire but rare in Anglesey. It was suggested that the dark lines added to the cryptic effect by breaking up the shell outline, or by resembling barnacle sutures against barnacle-covered backgrounds. In the Outer Hebrides Smith (1978) found both white and lined shells on exposed shores, but mostly yellow shells on sheltered (and probably algal-covered) shores.

A rather less clear picture emerged from a study of local variation around the northwestern coast of Anglesey (Naylor & Begon, 1982). Here, populations were mostly either lined or unlined, and occupied stretches of coastline of varying exposure, separated by obvious barriers to dispersal such as sandflats or exposed cliffs. There was some suggestion that the unlined forms were more common on sheltered shores, but it was not clear whether this was the result of local area effects or of selection. In contrast to Heller's (1975b) results, white shells were not found on exposed shores devoid of algae. This emphasizes the problem of the complexity of the maintenance and adaptive significance of visual polymorphism; different forms of selection may operate in different localities, and it is difficult to make generalizations, particularly when background colour is difficult to quantify. Nevertheless, the evidence for visual selection in some areas seems persuasive. Additional evidence is provided by the coloration

Figure 94. *Littorina (Neritrema) compressa*: juvenile shells of moderate ecotype. **A,** Lochskipport, South Uist, Outer Hebrides, Scotland. **B, E, G,** Mousehole, Cornwall, England. **C,** Caerfai Bay, St Davids, Wales. **D,** Porth Swtan, Anglesey, Wales. **F,** Loch Ainort, Skye, Scotland. Scale bar=1.0 mm.

of the rare barnacle ecotype, shells of which may be white, black, banded with black and white (presumably homologous with the lined morph) or occasionally tessellated, thus closely resembling the colour range found in the barnacle ecotype of *L. saxatilis*, which likewise appears to be visually cryptic against a background of white barnacles with dark sutures and crevices (Fish & Sharp, 1985; Reid, 1993).

The southern geographical variety *typica* shows the complete range of colour morphs. Almost all known examples of the northern geog. var. *barentica* are uniformly brown, very occasionally fawn or pale grey; against the dark ground colour a lined pattern cannot be clearly seen, although a very few shells have pale flecks in a tessellated pattern. This trend of reduced polymorphism in northern populations is also found in *L. obtusata* and *L. fabalis*.

Distribution

Habitat. Throughout its range in Brittany and the British Isles, *L. compressa* is characteristically a species of the mid to upper eulittoral zone. On moderately sheltered shores this corresponds to the zones of *Pelvetia* and *Fucus spiralis*, but it often extends further down to the level of *Fucus vesiculosus* and *Ascophyllum* (McMillan, 1939; Bakker, 1959; Heller, 1975a, b; Finet, 1978; Smith, 1979b; Hughes, 1980; Moreteau, 1981). Although common in the fucoid zones, it rarely occurs on algal fronds, but shelters in damp microhabitats between and among boulders (Sacchi, 1971, 1974; Hughes, 1980). On more wave-exposed shores it occurs throughout most of the barnacle zone, usually sheltering in crevices (Heller, 1975a, b; Naylor & Begon, 1982; Reid, 1993). The northern geographical variety *barentica* occurs somewhat lower, having been found only in the *Ascophyllum* zone on a sheltered shore (near Narvik, DGR), in the mid to low eulittoral on an exposed rock platform (Smelror near Vardø, DGR) and under stones in the lower eulittoral (Yarnishnaya, ZISP). In most parts of the range of *L. compressa* it is most common on sheltered and moderately sheltered shores, and is absent on wave-exposed coasts (Fischer-Piette & Gaillard, 1960; Fischer-Piette, Gaillard & Delmas, 1967; Sacchi, 1975; Moreteau, 1981; Matthews & Montgomery, 1987; Little & Williams, 1989). However, it can be found, usually in low numbers, on fully exposed coasts, for example in Cornwall, Wales and western Ireland (Heller, 1975a, b; Hughes, 1980; Reid, 1993), and then the shells are of the wave-exposed or barnacle ecotypes. At the opposite extreme of the exposure range it can be found at the mouths of estuaries (Fischer-Piette & Gaillard, 1960), sometimes on muddy rocks (pers. obs.), but does not appear to be very tolerant of brackish conditions. The chemical nature of the rock substrate does not seem to affect the occurrence of *L. compressa* (it can be found for example on granite or chalk), but a hard rock seems essential, probably for the successful attachment of egg masses (Mill & Grahame, 1990b; Warmoes, Dumoulin & Reid, 1992).

It is not found naturally in salt marshes, although transplanted adults will survive for up to 2 years, suggesting that conditions there are unfavourable for reproduction or early development (Hughes & Roberts, 1981). There is no apparent segregation of adults and juveniles, although juveniles are especially common in the shelter of barnacles (Hughes, 1980), and large animals have a wider vertical distribution (Moreteau, 1981). Several authors have suggested that the habitat range of this species is limited by its vulnerable benthic egg masses, for it is reportedly absent where tides, insolation, predatory fish and crabs, or soft sedimentary rock are unfavourable for their survival (Sacchi, 1971, 1974, 1975; Hughes & Roberts, 1981; Little & Williams, 1989; Mill & Grahame, 1990b; Warmoes *et al.*, 1992). On most shores it does not occur in large numbers, densities in the range 5–30 per m^2 having been recorded on shores of moderate exposure in Wales and Ireland (Hughes, 1980; Hughes & Roberts, 1981; Little & Williams, 1989). A much higher maximum density of 20 per dm^2 (2000 per m^2) was recorded by Moreteau (1981) in Brittany. The northern geographical variety *barentica* is uncommon at its known localities.

Throughout its range the species coexists with the five other northern Atlantic *Littorina* species. There is some overlap with both *L. saxatilis* and *L. arcana*, but these extend to higher tidal levels, and indeed this difference in zonation was used as early support for the specific distinction of *L.* '*nigrolineata*' (Crisp & Fischer-Piette, 1959; Sacchi, 1971, 1974, 1975); *L. saxatilis* and *L. arcana* are also more tolerant of wave-exposure. The zone of *L. compressa* may overlap those of *L. obtusata* and *L. fabalis*, but those species are found principally on fucoid algae. *Littorina littorea* may be found with *L. compressa*, but extends to lower levels on the shore.

Littorina compressa is an opportunistic grazer; on shores with abundant fucoids it eats mainly these larger algae (Sacchi, Ambrogi & Voltolina, 1981), whereas on more exposed shores with sparse algae its diet consists mainly of diatoms (Sacchi, Testard & Voltolina, 1977).

Range (Fig. 95). This species has a relatively restricted distribution in the northeastern Atlantic, in southern and western Ireland, westernmost Cornwall and Wales, most of Scotland, Brittany and on the shores of the Barents Sea. It has been included in several recent accounts of the distribution of littorinids in the British Isles and English Channel (Seaward, 1982, 1990; Mill & Grahame, 1990b, 1992b; Warmoes *et al.*, 1992), but with additional information the distribution can now be specified more precisely. Since at least the southern geographical variety of *L. compressa* is easily recognized from shells alone, and so much museum material and published accounts of the *L. saxatilis* group are now available, the known distribution is probably reasonably accurate and its patchiness does not simply reflect lack of collecting effort. In Britain specimens have been seen from the southwest of Cornwall (between Veryan Bay and Perranporth, including the Scilly Isles), Pembrokeshire (from Tenby to St David's), north Wales (northern Anglesey and Conway), the southern tip of the Isle of Man (all records BMNH), most of the west coast of Scotland (from the Sound of Jura to Skye, including the Western Isles and the Outer Hebrides), from scattered localities in the north and northwest of Scotland (between Scourie and Gardenstown, including the Orkney Islands), and from southeastern Scotland (between Carnoustie and Berwick-on-Tweed) (Scottish records BMNH, NMS, S. M. Smith Collection). To this may be added literature records from Lundy Island in the Bristol Channel and the Lleyn Peninsula in north Wales (Mill & Grahame, 1990b; Knight & Ward, 1991). A record from Portland in Dorset (Seaward, 1990) is considered doubtful. In BMNH there is an early nineteenth century collection from Northumberland and Durham (figured by Philippi, 1946a) and also a sample from Tynemouth collected in 1917, but a recent search failed to locate the species there (Mill & Grahame, 1990b). In Ireland specimens have been seen from a number of localities in the south (between Old Head of Kinsale and the Dingle Peninsula) and west (between Galway Bay and Achill Island) (BMNH). A record from Dublin (Seaward, 1982, 1990) is unconfirmed. In Northern Ireland there is a single isolated area of occurrence in Strangford Lough (mapped in detail by Matthews & Montgomery, 1987). On the southern side of the English Channel the easternmost records are from Yport (near Étretat; one 1988, BMNH, another older but undated, MNHNP). There is then a gap before occurrence on the Cotentin Peninsula (between Saint-Marcouff and Cherbourg), one record from the east side of the Cotentin Peninsula (Portbail), many from the Channel Islands, and a virtually continuous distribution in Brittany from Cancale to Saint-Guénolé (Pointe de Penmarch) in the extreme southwest. This most southerly collection was made in 1955 (MNHNP; Crisp & Fischer-Piette, 1959). Records of *L. nigrolineata* from Spain (Rolán, 1984) and Portugal (Burnay, 1986) are based on the Iberian black-lined form of *L. saxatilis*; *L. compressa* is absent. In the southern North Sea there is a single literature record from Yerseke, Zeeland, which may be explained by import of specimens with oysters from France (record by Stoker, discussed in Warmoes *et al.*, 1992). The littorinid faunas of Belgium (Warmoes, Backeljau & de Bruyn, 1988), Denmark (Nielsen, 1980), Sweden (Janson, 1982a) and southern Norway (Sneli & van Marion, 1979) are now well known, so the absence of records of *L. compressa* from these regions must be genuine. In addition, this species is usually easily recognized from shells, and the numerous dry museum samples seen from these areas in the present study have not revealed any examples. However, north of Trondheim the fauna has been less well collected, so the distribution of the northern geographical variety *barentica* is as yet poorly known. The occurrence of the species in the Barents Sea has previously been noted only briefly in Russian parasitological studies (Galaktionov, 1985; Galaktionov & Marasaev, 1986) and therefore has generally been overlooked. In the present study the northern variety has been seen from Vassvika near Narvik (DGR, 1989), Sørøy (NMS, 1982), Smelror near Vardø (DGR, 1989) and Yarnishnaya and Dal'niye Zelentsy Bay (ZISP, BMNH, 1987 to 1989).

The curiously patchy distribution of this species has been described in detail, because it is now probably known with considerable accuracy and provides clues about the factors that may limit the distribution. At least in the British Isles, its prevalence on western shores might suggest requirements for exposed conditions and clear oceanic water; however, this is clearly not the case, for it is often absent from the most exposed shores and can sometimes be found under turbid conditions near the mouths of estuaries. A restriction to the southwestern and western regions is well known in many littoral organisms that reach their northern limits in the British Isles, and can be correlated with winter sea temperatures which are too low for survival or reproduction in the eastern basin of the

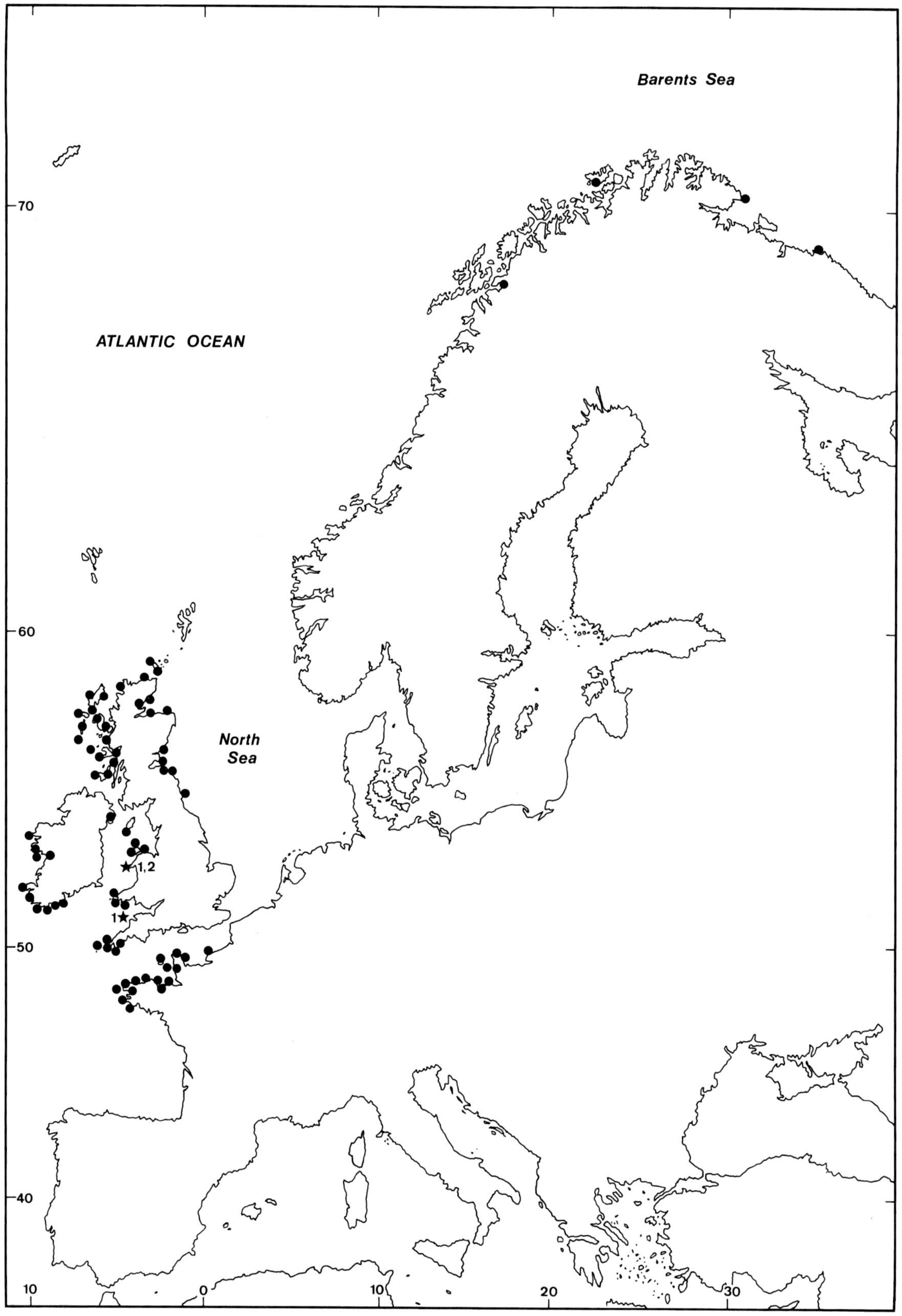

Figure 95. *Littorina (Neritrema) compressa*: distribution map. Key: circles, specimens examined; asterisks, records from literature (1, Mill & Grahame, 1990b; 2, Knight & Ward, 1991).

English Channel and in the North Sea (Southward & Crisp, 1954; Crisp & Southward, 1958; Lewis, 1964). However, when the distribution of *L. compressa* is considered in detail, it is seen that its range in Cornwall is more limited, and in eastern Scotland much more extensive, than expected if sea temperatures were a controlling factor. Moreover, *L. compressa* is not, as was first thought (Mill & Grahame, 1990b), at the northern extreme of its range in Scotland; populations on the Barents Sea coast exist under much colder temperature regimes. Mill & Grahame (1990b) first suggested that a firm rock substrate with sheltered crevices suitable for attachment and development of egg masses was the crucial requirement. This was supported by the correspondence between the distribution of *L. compressa* and of hard sedimentary and crystalline rocks on the French coast (Warmoes *et al.*, 1992). Clearly, suitable conditions for recruitment from local populations are essential for such a species without a dispersive larval stage, and the egg masses may be especially vulnerable. Desiccation of egg masses on the predominantly south-facing shores of southern Brittany may impose the southern limit of this species (Warmoes *et al.*, 1992). It has also been suggested that the type of development limits the dispersal capability of *L. compressa*, so that it has not yet reached the full extent of its potential range (Mill & Grahame, 1990b; Warmoes *et al.*, 1992). This seems improbable; Johannesson (1988) has argued that nonplanktotrophic developers can disperse effectively over long distances, because the progeny of a single immigrant fertilized female remain on the same shore, increasing the chance of founding a colony.

Nevertheless, on a wider scale there are still puzzling aspects to this distribution, implying that other local factors are also important. For example, the species is more restricted on both sides of the English Channel, on the east coast of England, in eastern Ireland and perhaps in Norway than is *L. arcana*; this is unexpected if dispersal capacity and requirements for egg survival determine distribution, since egg masses and type of development are the same in both species. The absence from most of Scandinavia cannot be accounted for by substrate or climatic factors. Further collecting in Norway may extend the known range of *L. compressa* south of Narvik; if, however, this is indeed the southern limit on this coast, the coincidence with the northern limit of the common occurrence of the crab *Carcinus maenas* (Vader, 1979) is notable, and possible limitation by crab predation could be investigated in this region.

Fossil record. No fossil records have been traced.

Discussion

Shell morphology. Littorina compressa has hitherto been considered less variable in shell form and consequently easier to recognize than other members of the *L. saxatilis* group. While this is true for the moderate ecotype of the southern geographical variety *typica*, the present survey has shown that, over its entire geographical and habitat range, shell form is almost as variable as in *L. saxatilis* itself. In particular, the wave-exposed and barnacle ecotypes, and the northern geographical variety *barentica*, have not previously been widely noticed.

On shores of moderate wave-exposure, the shells of geog. var. *typica* are more thick-walled and sometimes also larger than those of *L. saxatilis* at the same localities. This appears to be adaptive in relation to crab predation, for *L. compressa* occupies a lower tidal level than *L. saxatilis*, where predation is more severe (Heller, 1976; Elner & Raffaelli, 1980; Faller-Fritsch & Emson, 1985).

Genetics. Several studies of population genetics have found that *L. compressa* has a lower average heterozygosity than *L. saxatilis* or *L. arcana* (Beardmore & Morris, 1978; Wilkins & O'Regan, 1980; Ward, 1990; Knight & Ward, 1991; Backeljau & Warmoes, 1992). This cannot easily be explained by neutral or deterministic models (see Janson, 1987a, Ward, 1990, for reviews), since dispersal and habitat range of *L. compressa* and *L. arcana* appear to be similar. Perhaps, as in *L. littorea* (Janson, 1987a), a population bottleneck in the past may account for the relatively low genetic variability of *L. compressa*.

Distribution. The patchy geographical distribution is one of the interesting features of this species and cannot yet be adequately explained. At least in part, the specific requirements for sheltered and humid crevices in a hard rock for the successful attachment and development of the egg masses appears to limit both the geographical and habitat range. Whether it also limits colonizing ability is doubtful.

Phylogenetic relationships. Data from both allozyme and DNA studies support the monophyly of the three species of the *L. saxatilis* group, although no morphological synapomorphies have been recognized. There is conflicting evidence about relationships between the three species *L. compressa*, *L. arcana* and *L. saxatilis*, as discussed in the remarks on *L. saxatilis* (p. 336).

Similar species. Confusion is only likely with the two other members of the *L. saxatilis* group, and differentiation of all three is explained in the Discussion of *L. saxatilis* (p. 337; Table 7).

Littorina (Neritrema) arcana Hannaford Ellis, 1978

(Figs 96–100)

Synonymy

? *Turbo labiatus* Brown, 1827: pl. 46, figs 20, 21 (Penzance [Cornwall, England]; types lost; *nomen dubium*, see Taxonomic History, p. 265; not *Turbo labiata* Montagu, 1803). Brown, 1837a: 16; pl. 10, figs 20, 21.

? *Littorina labiata*—Brown, 1844a: 128; pl. 10, figs 20, 21.

? *Littorina patula* 'Jeffreys' Thorpe, 1844: 259 (Eddystone Rock [Cornwall, England]; *nomen dubium*, see Taxonomic History, p. 265).

Littorina patula—Heller, 1975a: 138–140, 146–149 (in part, includes *L. saxatilis*). Raffaelli, 1979a: 219–232; fig. 8 (radula).

Littorina rudis var. *patula*—Jeffreys, 1865: **3**: 365–366 (*s.l.*). Harmer, 1921: 660; pl. 53, fig. 16 (*s.l.*).

Littorina saxatilis patula—Dautzenberg & Fischer, 1912: 196; pl. 9, figs 24, 25 (in part, includes refs to *L. saxatilis*).

Littorina saxatilis tenebrosa var. *patula*—James, 1968a: 160; pl. 2, fig. 7d–f.

Littorina saxatilis patula var. *attenuata* Dautzenberg & Fischer, 1912: 196; pl. 9, figs 24, 25 (Scarborough [Yorkshire, England]; original of figs MNHNP, seen; infrasubspecific).

Littorina saxatilis jugosa var. *attenuata*—James, 1968a: 155; pl. 2, fig. 5c, d.

Littorina saxatilis—Seshappa, 1947: 335–336 (in part, includes *L. saxatilis*; not *Turbo saxatilis* Olivi, 1792). Seshappa, 1948: 702–703 (in part, includes *L. saxatilis*; not Olivi, 1792). Caugant & Bergerard, 1980: 107–112 (in part, includes *L. saxatilis*; not Olivi, 1792).

Littorina saxatilis tenebrosa var. *tenebrosa*—James, 1968a: 160; pl. 1, fig. 7a–c; pl. 2, fig. 5a, b (probable determination; in part; other figs *L. saxatilis*; not *Turbo tenebrosus* Montagu, 1803=*L. saxatilis*).

Littorina saxatilis rudissima—Seshappa, 1976: 222–229; figs 1–10 (spawn, embryos); figs 14, 15; fig. 16 (oviduct); fig. 19 (radula) (in part, includes *L. saxatilis*; not *Turbo rudissima* Johnston, 1842=*L. saxatilis*).

Littorina arcana Hannaford Ellis, 1978a: 304 (Porth Swtan, Anglesey, Wales; holotype (Fig. 99A) BMNH 197823W+21 paratypes BMNH 197824W, seen). Hannaford Ellis, 1979: 43–56; figs 1B, 2C, 3, 4B, 8B, D, 9A, B; pl. 1, figs 2, 4; pl. 2, figs 1, 2 (all reproductive anatomy); pl. 3, figs 5–8. Sneli & van Marion, 1979: 4–8; fig. 1c. Fretter, 1980: 148–153; fig. 1A (oviduct). Fretter & Graham, 1980: 270–272, 275–276. Ward & Warwick, 1980: 417–428. Atkinson & Warwick, 1983: 137–151. Jardine, 1984: 5–6; fig. 2e, f; fig. 3c (oviduct). Ward & Janson, 1985: 86–94. Grahame & Mill, 1986: 229–236. Knight & Ward, 1986: 405–413. Ward, Warwick & Knight, 1986: 233–241. Knight, Hughes & Ward, 1987: 417–426. Graham, 1988: 164, 180. Mill & Grahame, 1988: 347–353. Warmoes, Backeljau & de Bruyn, 1988: 51–70; fig. 4C (penis). Grahame & Mill, 1989: 837–855; fig. 4. Dytham, Grahame & Mill, 1990: 233–240. Grahame, Mill & Brown, 1990: 223–231. Hayward, Wigham & Yonow, 1990: 662. Sundberg, Knight, Ward & Janson, 1990: 29–40. Warwick, Knight & Ward, 1990: 109–116. Knight & Ward, 1991: 81–91. Dytham, Grahame & Mill, 1992: 39–44. Dytham, Mill, Grahame & O'Higgins, 1992: 127–133. Grahame & Mill, 1992: 99–106. Grahame, Mill, Double & Hull, 1992: 499–502. Mill & Grahame, 1992a: 31–37. Crossland, Coates, Grahame & Mill, 1993: 301–305. Reid, 1993: 51–62; fig. 3; fig. 6A–D (penis); fig. 6E–G (oviduct). Caley, Grahame & Mill, 1995: 181–193. Crossland, Coates, Grahame & Mill, 1995: 205–209. Grahame, Mill, Hull & Caley, 1995: 887–899.

Littorina (?Littorivaga) arcana—Smith, 1982: 535–539.

Littorina (Neritrema) arcana—Reid, 1989a: 102. Reid, 1990b: 1–19; fig. 1r. Smith & Heppell, 1991: 17, 81.

Littorina rudis—Raffaelli, 1979a: 219–232; fig. 4A; fig. 5 (penis) (in part, includes *L. saxatilis*; not *Turbo rudis* Maton, 1797=*L. saxatilis*). Hughes & Roberts, 1981: 251–268 (in part, includes *L. saxatilis*; not Maton, 1797).

Note: for references in which this species was not clearly differentiated from *L. saxatilis*, and may have been included with it, see Synonymy of *L. saxatilis* (p. 278). In the Synonymy '*s.l.*' (*sensu lato*) after a reference indicates that the author's concept of a name was probably based on *L. arcana*, but without giving clear diagnostic criteria. In the text '*L. saxatilis s.l.*' is used where it is likely, but not certain, that *L. arcana* (and in some cases *L. compressa* also) may have been included with that species. The '*L. saxatilis* group' refers to all three species.

Taxonomic history

This species was described in 1978 by Hannaford Ellis (1978a), who distinguished it from the closely related *L. saxatilis* principally by its oviparous reproduction and other anatomical details. In shell characters *L. arcana* is often indistinguishable from *L. saxatilis*, so it is sometimes difficult to determine whether earlier authors, who did not describe anatomical features, may have included *L. arcana* in their concepts of *L. saxatilis* and its synonyms. Nevertheless, knowledge of geographical distribution, habitat preferences and of shell variation in the two species at particular localities has allowed some early references to be assigned with some confidence to the synonymy of *L. arcana*. Where there is doubt the references are listed only in the synonymy of *L. saxatilis* (indicated by '*s.l.*').

A number of early authors distinguished a patulous form of *L. saxatilis s.l.* with large aperture, relatively low spire and thin-walled shell. Brown (1827) named one such form *Turbo labiatus*, but no types are known to exist. From his detailed shell description (in particular of the colour pattern 'pale chestnut or fawn, beautifully clouded or blotched with purple') and the type locality Penzance in Cornwall, it is very likely that this refers to *L. arcana*, because the marbled shells of this shape that have been seen from Cornish localities are all of this species. Nevertheless, because the modern definition of this species depends heavily upon anatomical characters, the concept of Brown's species could only be unequivocally established by designation of a neotype. This is not necessary, since the name *Turbo labiatus* is a junior primary homonym of *Turbo labiata* Montagu, 1803, and therefore must be rejected. A similarly patulous shell was described as *Littorina patula* by Thorpe (1844), from the Eddystone Rock off the south Cornish coast, and is again without known type specimens. Hannaford Ellis (1979) obtained a small sample of *L. saxatilis s.l.* from this locality and found that both *L. saxatilis* and *L. arcana* were represented, although neither fitted Thorpe's original description exactly. Her decision was to synonymize *L. patula* with *L. saxatilis*, describing *L. arcana* as new. In the interests of nomenclatural stability this decision is followed here, and *L. patula* is regarded as a dubious name with inadequate diagnosis. The only other new name likely to have been based on *L. arcana* is *L. saxatilis patula* var. *attenuata* of Dautzenberg & Fischer (1912), described from Scarborough in Yorkshire, where shells of this thin, patulous form are of *L. arcana*. However, this name must rejected because it was proposed as an addition to a trinomen and is therefore clearly of infrasubspecific rank (ICZN, 1985: art. 45g).

The name *patula* has appeared quite frequently in the literature, usually as a variety or subspecies of *L. saxatilis*, and some authors must have included *L. arcana* under that name (Jeffreys, 1865; Dautzenberg & Fischer, 1912; Harmer, 1921). The first undoubted indication of the present species was the observation by Seshappa (1947) in Northumberland of small, thin, patulous shells of variable sculpture which laid benthic egg masses, but coexisted in microsympatry with more typical larger, taller-spired, shells of *L. saxatilis* which were ovoviviparous. At first he suggested that these oviparous forms should be regarded as taxonomically distinct, comparing them with *L. rudissima* Bean, 1844 (i.e. *L. rudissima* (Thorpe, 1842)), but subsequently he found that some small, patulous shells were in fact ovoviviparous and therefore proposed that the single species, *L. saxatilis*, could exhibit both modes of reproduction (Seshappa, 1948). This was reiterated in a third paper (Seshappa, 1976) in which the patulous and predominantly oviparous forms were recognized as a subspecies, *L. saxatilis rudissima*.

Meanwhile, other workers were also appreciating the importance of anatomical characters in the classification of the *L. saxatilis* group. Another report of oviparity was made by Deyglun (1955) in Roscoff, Brittany, in a species that she called *L. 'nigrolineata'* (i.e. the species here identified as *L. compressa*). In redescribing *L. 'nigrolineata'* more fully, both Sacchi (1975) and Heller (1975a) assumed that Seshappa's record of oviparity applied to the same species. The influential work of Heller (1975a) defined four species in the *L. saxatilis* group in Wales. Of these, *L. patula* was separated from *L. rudis* (=*L. saxatilis*) principally by its more numerous penial glands, often arranged in two rows, and by the conchological features of low spire and large aperture; both species were said to be ovoviviparous, and the banding patterns of their polymorphic esterases did not differ significantly. Raffaelli (1979a) synonymized these two, suggesting that *L. patula* was a form of *L. rudis* adapted to wave-exposed conditions, and that there was continuous variation in penial characters. Neither Heller (1975a) nor Raffaelli (1979a) compared the reproductive anatomy of females. This, however, was the principal character of the new oviparous species, *L. arcana*, described by Hannaford Ellis (1978a), in which the brood pouch of the ovoviviparous *L. saxatilis* was replaced by an enlarged jelly gland. In a fuller description Hannaford Ellis (1979) showed characteristic, but not entirely diagnostic, differences in the penes of the two species. She suggested that Heller's *L. patula* was a mixture of the two, thus explaining his description of penial characters like those of *L. arcana* together with the ovoviviparity of *L. saxatilis*. The shells of the new species were extremely similar to those of *L. saxatilis*, although at some localities differences in shape or colour could be identified after anatomical separation. In particular, *L. arcana* was more often of a patulous shape (Hannaford Ellis, 1979), although in the most wave-exposed habitats *L. saxatilis* can sometimes be equally patulous.

At about the same time, *L. arcana* was also recognized in Norway (van Marion, 1978a; Sneli & van Marion, 1979). At first, some authors were reluctant to accept that the newly-described oviparous form was a distinct species, suggesting that a brood pouch was only a minor modification of a jelly gland, so that alternative modes of reproduction were possible within a single species (Caugant & Bergerard, 1980; Bergerard & Caugant, 1981; Hughes & Roberts, 1981; S. M. Smith, 1982); no doubt they were also influenced by the conchological similarity of the two, for by that time the oviparous *L. compressa* was widely accepted as distinct. Others simply ignored the newly-recognized anatomical characters; J. E. Smith (1981), for example, separated normal and patulous forms of *L. saxatilis s.l.* (as *L. rudis* and *L. saxatilis*) purely on shell shape, and must undoubtedly have included *L. arcana* in the patulous category. A redescription of the oviduct of *L. arcana* by Fretter (1980) added a long copulatory bursa and presence of an ovipositor to the known structural differences, and the species was therefore accepted as distinct by Fretter & Graham (1980). Subsequent studies have confirmed the usefulness of oviducal and penial characters in separating *L. arcana* from *L. saxatilis* (Jardine, 1984; Warmoes, Backeljau & de Bruyn, 1988; Reid, 1993).

The technique of gel electrophoresis of enzymes was also brought to bear on this problem. Caugant & Bergerard (1980) had been unable to detect any differences between oviparous and ovoviviparous forms of *L. saxatilis s.l.* at three polymorphic enzyme loci, in apparent agreement with their contention that they were conspecific. However, it was soon found that at a number of localities in Britain microsympatric populations of *L. arcana* and *L. saxatilis* maintained significantly different allele frequencies at several loci (Ward & Warwick, 1980; Ward & Janson, 1985; Knight & Ward, 1986, 1991; Warwick, Knight & Ward, 1990), suggesting a high degree of reproductive isolation between the taxa and thus supporting the specific status of *L. arcana*. The two species were clearly very closely related, with estimates of Nei's (1972) genetic distance ranging from 0.035 to 0.072 (Ward, 1990; Knight & Ward, 1991), or exceptionally up to 0.301 (Nei's, 1978, genetic distance, Backeljau & Warmoes, 1992). None of these genetic studies discovered any entirely diagnostic alleles (although phenotypes at one esterase locus were almost so, Ward & Janson, 1985), so that unequivocal identification of individuals by allozyme electrophoresis was not possible. *Littorina saxatilis* was found to be genetically more variable than *L. arcana*, and differences among populations within *L. saxatilis* (especially on a large geographical scale) were often greater than interspecific differences. As a result, cluster analysis of gene-frequency data usually produced only partial separation of populations into their respective species (Ward & Janson, 1985; Knight, Hughes & Ward, 1987; Ward, 1990; Knight & Ward, 1991). To some extent, however, the degree of separation depended upon the method of clustering used, with maximum likelihood performing better than the unweighted pair group method (Sundberg, Knight, Ward & Janson, 1990; Knight & Ward, 1991). Application of a different electrophoretic technique, isoelectric focusing, to the esterase system of the two species produced similar results, showing incomplete separation of populations and failing to detect any diagnostic bands (Mill & Grahame, 1988, 1992a; Dytham, Grahame & Mill, 1992). More recently the banding patterns of random amplified polymorphic DNAs (RAPDs) of these species have been examined, and this technique has proved to be more powerful, achieving unequivocal clustering of individuals from Cornwall into the two species (Crossland, Coates, Grahame & Mill, 1993). This has been repeated in samples from Wales, where one band was found to be present in all *L. arcana* examined (and some *L. compressa*), but absent in *L. saxatilis* (Crossland, Coates, Grahame & Mill, 1995). A final piece of evidence for the separate gene pools of *L. arcana* and *L. saxatilis* is the maintenance of significantly different shell colour frequencies (assumed to be under genetic control) in many sympatric populations (Atkinson & Warwick, 1983; Dytham, Grahame & Mill, 1990). Nevertheless, it has been found that the two species will produce fertile hybrids in the laboratory, although hybrids have not yet been detected in naturally sympatric populations (Ward, Warwick & Knight, 1986; Warwick, Knight & Ward, 1990).

In summary, therefore, the detailed genetic analysis of *L. arcana* and *L. saxatilis* has entirely confirmed the initial separation of the two on anatomical grounds. The species are still difficult (and in some cases impossible) to separate in the field using shell characters, although multivariate analyses of shell shape have confirmed the earlier observations that at many localities there are significant differences (Ward & Janson, 1985; Grahame, Mill & Brown, 1990; Grahame & Mill, 1989, 1992; Caley, Grahame & Mill, 1995; Grahame, Mill, Hull & Caley, 1995). This has hindered ecological studies but, where these have been carried out, further differences between the two taxa have emerged, in breeding season (Hannaford Ellis, 1983), behaviour (Hannaford Ellis, 1985), habitat (Grahame & Mill, 1986; Dytham, Grahame & Mill, 1990) and geographical distribution (Mill & Grahame, 1990b, 1992b; Warmoes, Dumoulin & Reid, 1992).

Diagnosis

Mature shell height 2.8–16.8 mm. Shell turbinate to slightly patulous. Mineralogy of external irregular-prismatic calcite and internal crossed-lamellar aragonite. Sculpture variable, occasionally almost smooth, usually narrow ribs of irregular width, sometimes as few as 7 strong, rounded ribs with interspaces equal to rib width; fine spiral microstriae over whole surface. Colour highly variable, e.g. black, brown, yellow, white or orange, sometimes banded or tessellated. Penis with triangular filament, 10–30% total length, 6–68 mamilliform penial glands usually in more than one row. Pallial oviduct of single loop of albumen gland, single loop of capsule gland and swollen, septate jelly gland; albumen and capsule glands together occupy about half total length of pallial oviduct; bursa long, extending back to capsule gland. Spawn a benthic mass of singly-encapsulated eggs in gelatinous matrix; development nonplanktotrophic with intracapsular metamorphosis. Distribution: northern Norway, British Isles, northern France and Brittany.

Anatomical description

Head-foot (Fig. 96A). Head black to grey with unpigmented tip of snout; usually pale around eye and across tentacle base, and tentacles pale with two longitudinal black stripes; head and tentacles may be entirely black; sides of foot speckled pale grey to black. Opercular ratio 0.48–0.64.

Penis (Fig. 96). Base wrinkled, usually unpigmented, but sometimes with a little grey to black pigment on lower part of outer side (or in sperm groove, Hannaford Ellis, 1979); 6–68 mamilliform penial glands, usually in more than one row or in a close-packed mass. Filament triangular, tip not usually mucronate; 10–30% total length. Sperm groove open, ending terminally. Subepithelial glandular tissue around sperm groove often visible even in preserved specimens. Hannaford Ellis (1979) described penial variation at sites in Wales and Scotland in detail. Since apparently unparasitized animals of adult size can sometimes be found with small penes, it is possible that penial reduction (or shedding) might occur outside the breeding season.

Sperm. Euspermatozoa 43–46 μm. Paraspermatozoa 12–16 μm diameter, round or slightly oval; vesicles large and distinct; nucleus not visible.

Oviduct (Fig. 97). Single loop of albumen gland; followed by single loop of capsule gland equal to about half or the same area as the albumen gland, no translucent part of capsule gland visible (histological sections not seen); albumen and capsule glands together occupy about half total length of pallial oviduct; finally a swollen, septate jelly gland. Bursa long, opening at anterior end of jelly gland and extending back at least to anterior part of capsule gland. Seminal receptacle simple. There is some variation in the relative size of the capsule gland, but only rarely is it as small as that shown in Figure 97G. Figure 97A–E illustrates the ontogeny of the pallial oviduct in juveniles, showing development of the loop of the renal oviduct and the loops of the egg groove around which the albumen and capsule glands form. Note that from the stage illustrated in Figure 97D the capsule gland is more than half the size of the albumen gland and the bursa is long, distinguishing the oviduct from that of *L. saxatilis* (p. 293). The pallial oviduct has also been described by Hannaford Ellis (1979) and by Fretter (1980), but the latter author did not recognize the spiral structure and used a different terminology for the glandular components.

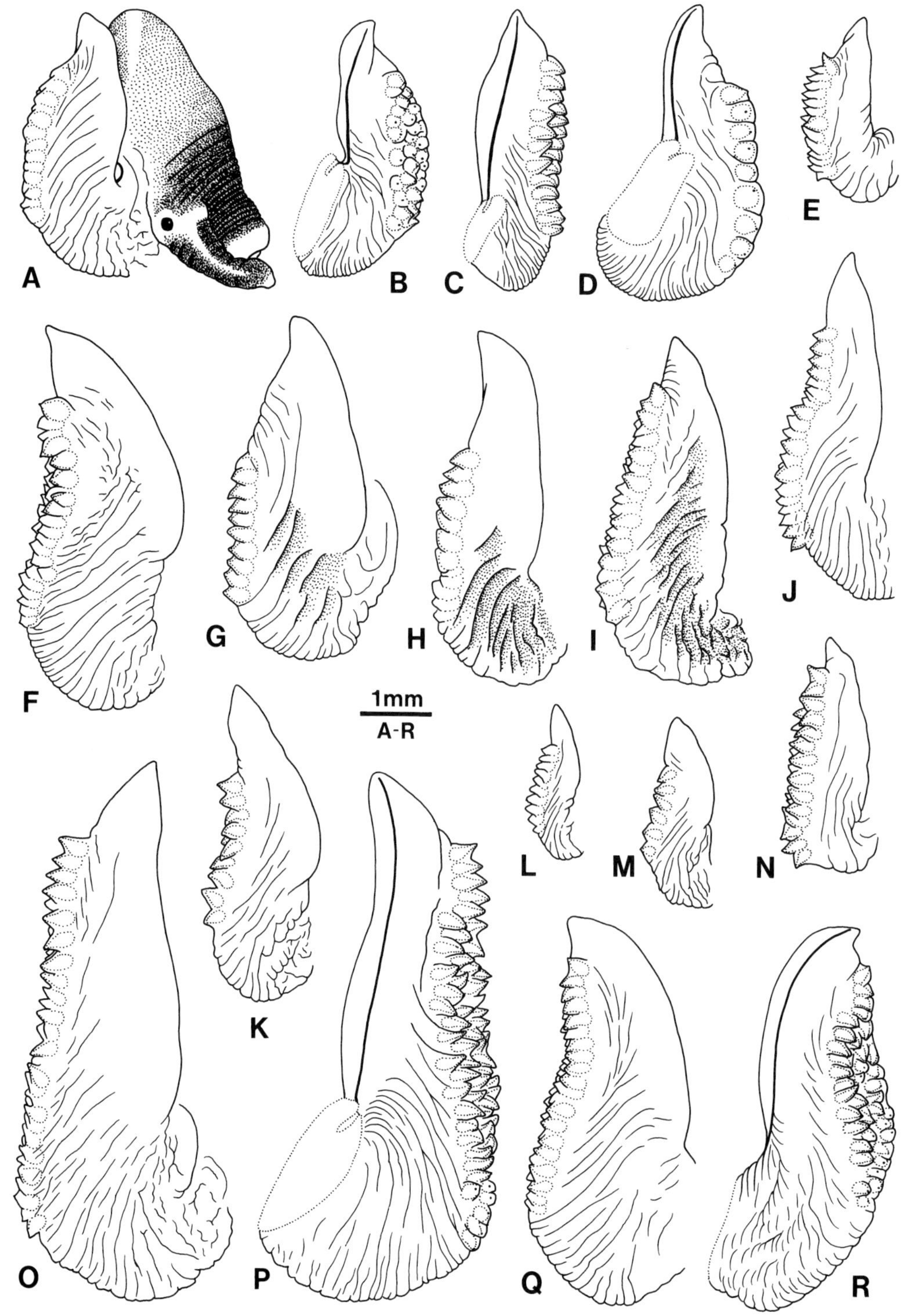

Figure 96. *Littorina (Neritrema) arcana*: **A**, head pigmentation; **A–R**, penes; **A**, **B** and **O**, **P** and **Q**, **R** are each lateral and medial views of same penis. **A**, **B**, Porth Swtan, Anglesey, Wales (paratype, BMNH 197824W; moderate ecotype); **C**, **L**, 1 km W of St Govan's Head, Pembrokeshire, Wales (**C**, wave-exposed ecotype; **L**, barnacle ecotype); **D**, Bell Rock, Anglesey, Wales (moderate ecotype); **E**, Laukhaugen, W of Trondheim, Norway (moderate ecotype); **F**, Pointe de l'Arcouest, Brittany, France (moderate ecotype); **G**, **H**, Smelror, near Vardø, Norway (moderate ecotype); **I**, Hartland Point, Devon, England (moderate ecotype); **J**, Dunbar, Scotland (moderate ecotype); **K**, Keem Strand, Achill Island, Co. Mayo, Ireland (moderate ecotype); **M**, Golam Head, Co. Galway, Ireland (wave-exposed ecotype); **N**, Dooagh, Achill Island, Co. Mayo, Ireland (wave-exposed ecotype); **O**, **P**, Carantec, Brittany, France (moderate ecotype); **Q**, **R**, Barfleur, Normandy, France (moderate ecotype). Shell heights (with number of mamilliform glands in parentheses, when these are not all clearly visible in view shown): **A**, **B**, 9.3 mm; **C**, 7.8 mm; **D**, 10.5 mm; **E**, 13.3 mm; **F**, 16.0 mm (30 glands); **G**, 14.7 mm (11 glands); **H**, 12.9 mm (11 glands); **I**, 8.4 mm (32 glands); **J**, 10.9 mm (24 glands); **K**, 7.9 mm (10 glands); **L**, 3.4 mm; **M**, 4.3 mm; **N**, 6.1 mm (23 glands); **O**, **P**, 14.0 mm (57 glands); **Q**, **R**, 12.3 mm (37 glands).

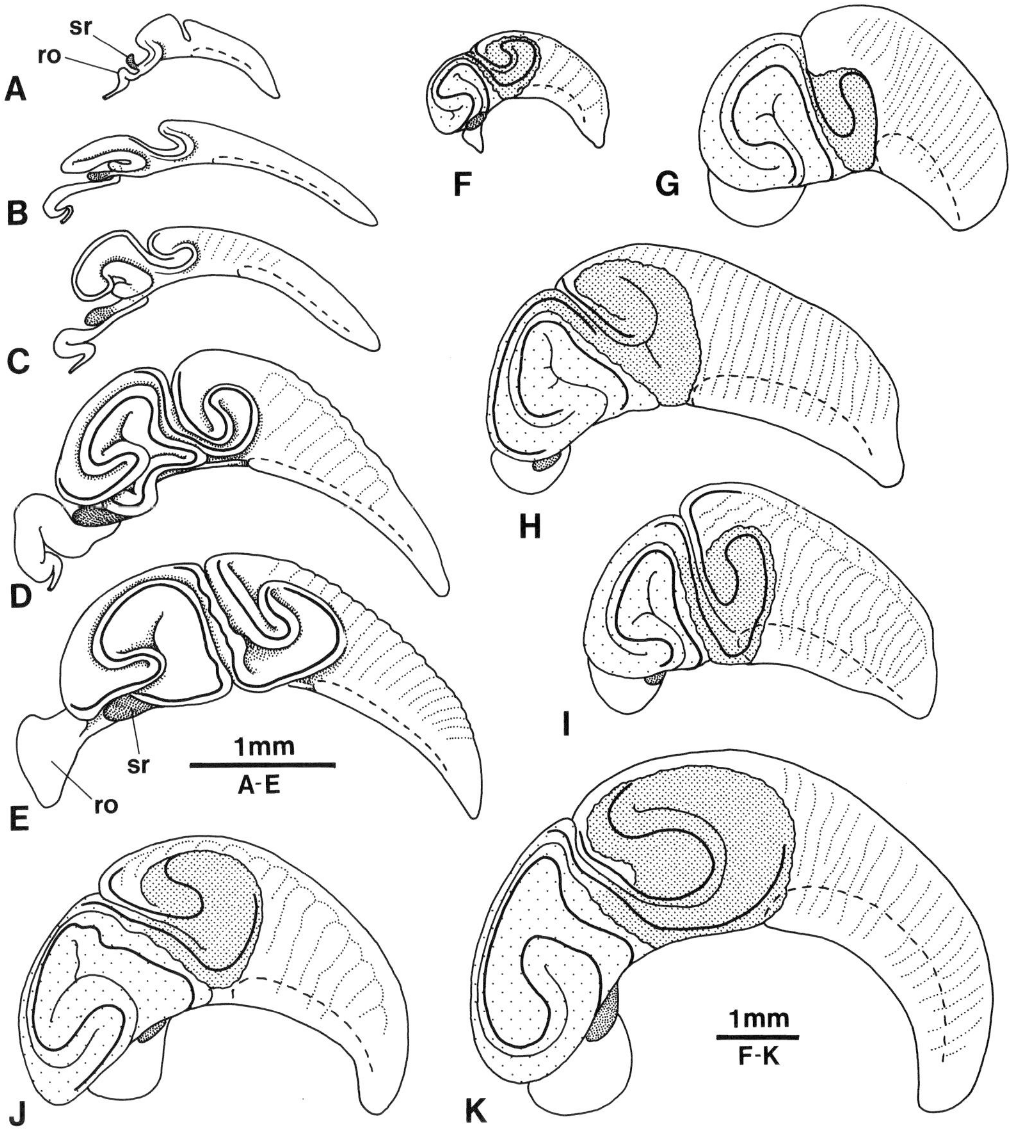

Figure 97. *Littorina (Neritrema) arcana*: pallial oviducts; **A–E,** ontogenetic series (note larger scale); **F–K,** mature pallial oviducts. **A–E,** Bell Rock, Anglesey, Wales (moderate ecotype); **F,** 1 km W of St Govan's Head, Pembrokeshire, Wales (barnacle ecotype); **G,** Marloes Bay, Pembrokeshire, Wales (moderate ecotype); **H,** Laukhaugen, W of Trondheim, Norway (moderate ecotype); **I,** Rhoscolyn, Holy Island, Anglesey, Wales (moderate ecotype); **J,** Dunbar, Scotland (moderate ecotype); **K,** Pointe de l'Arcouest, Brittany, France (moderate ecotype). Shell heights: **A,** 6.0 mm; **B,** 6.9 mm; **C,** 8.0 mm; **D,** 9.0 mm; **E,** 11.2 mm; **F,** 4.4 mm; **G,** 10.9 mm; **H,** 12.7 mm; **I,** 10.8 mm; **J,** 10.0 mm; **K,** 14.4 mm. Abbreviations: ro, developing renal oviduct; sr, developing seminal receptacle. Shading conventions as described on p. 6, except that in **A–E** albumen and capsule glands are unshaded, since they are not at these stages visibly differentiated.

Spawn and development. Benthic gelatinous egg masses have been reported in damp crevices in the upper barnacle zone (Hannaford Ellis, 1979) or over a wider tidal range, from high water of neaps to high water of spring tides, in crevices and on open rock (Seshappa, 1976). Communal spawning is frequent, but individual egg masses are oval, 5–8.5 mm in length, and contain 129–250 pink eggs; ova measure 285 by 235 μm and are individually enclosed in oval egg capsules 530–620 by 470–550 μm and 20 μm thick (Seshappa, 1976). Development nonplanktotrophic with intracapsular metamorphosis (Seshappa, 1976). In Northumberland some spawning occurs throughout the year, except in September and October (Seshappa, 1976); in north Wales there is maximum spawning from September to March, followed by a decline and then a lull in midsummer when high temperatures might impair egg survival (Hannaford Ellis, 1983). Juveniles hatch at a shell diameter of 500 μm (Jardine, 1984).

Radula and foregut (Fig. 98). Relative radular length 1.54–3.21 (to 6.47 in a mature shell of the barnacle ecotype 3.4 mm in height). Rachidian: length/width 1.00–1.21; central cusp bluntly rounded to narrowly rounded or slightly leaf-shaped; one pointed cusp and one outer denticle on each side. Lateral: four cusps; second and third from inside bluntly rounded or rounded, third largest; others small and pointed. Inner marginal: four cusps; second from inside rounded; third from inside largest and blunt to bluntly rounded; others small and pointed. Outer marginal: 6–7 rounded or bluntly rounded cusps; narrow neck, projecting flange on inside of base only. All cusps more elongate and pointed in small specimens (Fig. 98H). Salivary glands enlarged.

Shell variation (Fig. 99)

Geographical variation. This is analogous to that seen in all the other *Littorina (Neritrema)* species of the northeastern Atlantic, with northern and southern varieties:

Geog. var. *typica* (Fig. 99A–L, N, P, R): H/B=0.97–1.24; SH=1.16–1.47. Shape low-spired to turbinate; relatively thick-walled, especially when found on shores of moderate wave-exposure. Range: Brittany, northern France, British Isles, Trondheimfjord to Skarberget (Norway).

Geog. var. *finmarchica* (Fig. 99M, O, Q): H/B=1.21–1.31; SH=1.42–1.58. Shape turbinate to high-turbinate; relatively thin-walled, even when found on shores of moderate wave-exposure. Range: Norway north of Narvik.

The distinction between the two varieties is less striking than is seen for example in *L. compressa*, because of the superimposed ecotypic variation; the common wave-exposed ecotype of the southern variety has a relatively thin shell, comparable in thickness (but not in shape, since it is more low-spired) with the northern variety. The geographical variation is seen most clearly when shells from shores of moderate wave-exposure are compared.

The parallel geographical variation of the five species of *Littorina (Neritrema)* that inhabit the northern coast of Norway and Russia is believed to reflect the distribution of the predatory crab *Carcinus maenas*. This crab is rare or absent in Finnmark, north of Narvik (Vader, 1979), where each of the five species shows a thinner-walled, more tall-spired, northern variety. *Littorina arcana* is often microsympatric with *L. saxatilis* and, like *L. saxatilis* (Raffaelli, 1978b; Faller-Fritsch & Emson, 1985), it is probably preyed upon by *Carcinus*, particularly in habitats of moderate wave-exposure where the crabs are common, which may impose selection for thick-walled and somewhat lower-spired shells in populations of the southern geog. var. *typica*. So far, this selective mechanism has been studied only in *L. obtusata* (p. 213). The distribution of *L. arcana* in Norway is poorly known and there are no records of the species between Trondheim and Skarberget (60 km southwest of Narvik), but at both these localities shells were of the thick-walled southern variety, whereas at Vassvika (on the north side of the city of Narvik) on a fairly sheltered boulder shore the shells were thin-walled (DGR, 1989), so the southern limit of geog. var. *finmarchica* probably lies between Narvik and Skarberget. This agrees with the distributions of the northern varieties of the other *Neritrema* species, which all show southern limits close to Narvik. The limits may not, however, be identical for all the species; for example, at Vassvika *L. obtusata* was present in its southern variety, and *L. arcana* and *L. compressa* in their northern varieties. Selection pressures and morphological responses to them (whether genetic or ecophenotypic) may differ among the species. In addition, the distribution of the crab predator and the morphological response of the prey may change over a few snail generations (as shown in *L. obtusata*; Seeley, 1986). No anatomical differences have been noticed between northern and southern varieties in *L. arcana*.

Ecotypic variation. Three ecotypes can be recognized:

Moderate ecotype (Fig. 99A, B, G, K, L, M, O, P–R): these shells are found on moderately wave-exposed to sheltered shores, attain large size, have taller spires and (only in geog. var. *typica*) a thicker-walled shell.

Wave-exposed ecotype (Fig. 99C, D, F, H, I): the species is common on exposed shores and here the shells are somewhat smaller, with thinner walls, lower spires and larger apertures (H/B=0.97–1.07; SH=1.16–1.34); shell sculpture is often of strong ribs.

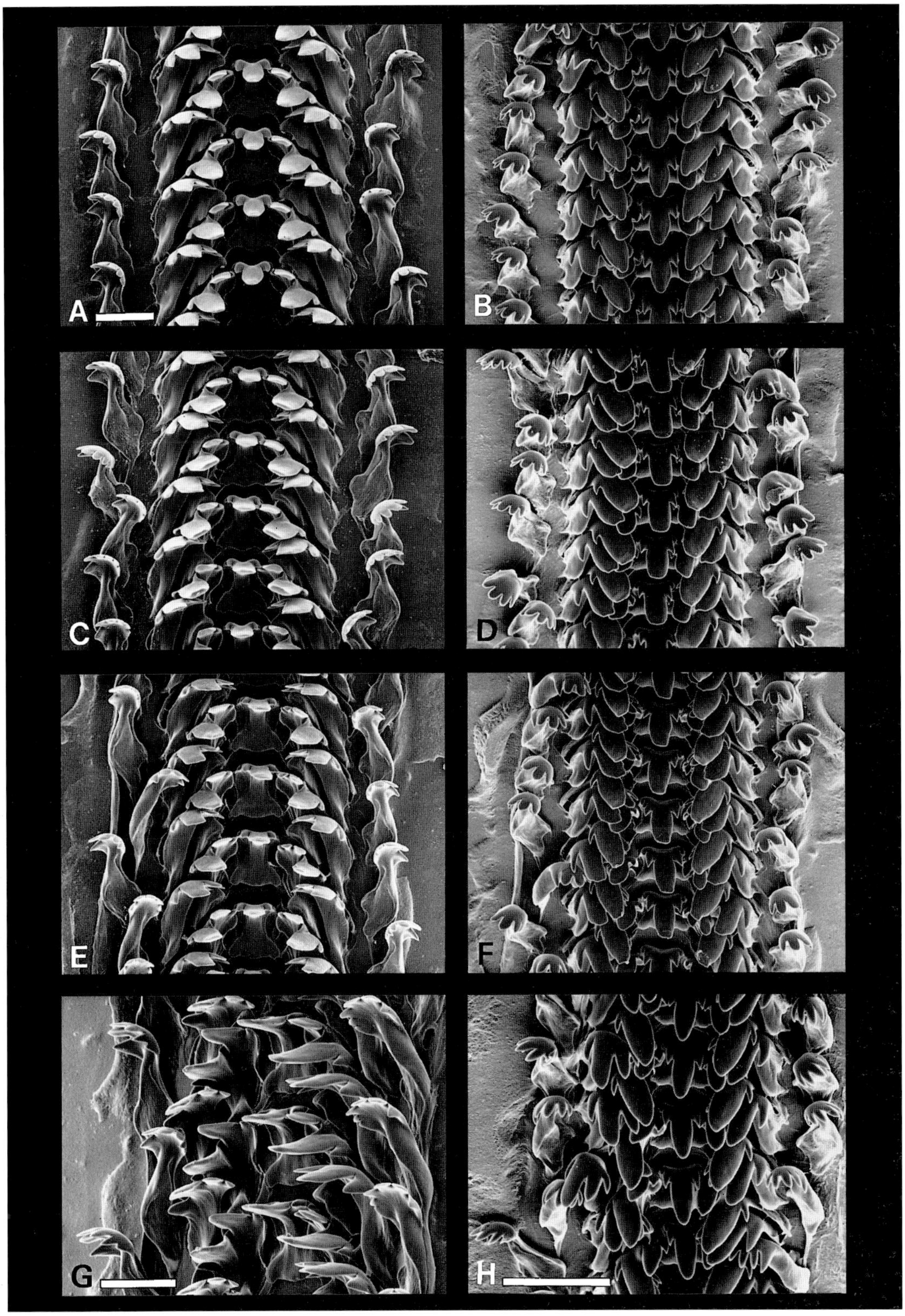

Figure 98. *Littorina (Neritrema) arcana*: **A, B,** two views of radula (shell H=8.4 mm), flat (**A**) and at 45° (**B**); **C, D,** two views of radula (shell H=13.0 mm), flat (**C**) and at 45° (**D**); **E–G,** three views of radula (shell H=9.5 mm), flat (**E**), at 45° (**F**) and side view (**G**); **H,** radula (shell H=3.4 mm), flat. **A, B,** Laukhaugen, Norway (moderate ecotype); **C, D,** Barfleur, Normandy, France (moderate ecotype); **E–G,** Robin Hood's Bay, Yorkshire, England (wave-exposed ecotype); **H,** St Govan's Head, Pembrokeshire, Wales (barnacle ecotype). Scale bars: **A–F**=50 μm; **G**=50 μm; **H**=50 μm.

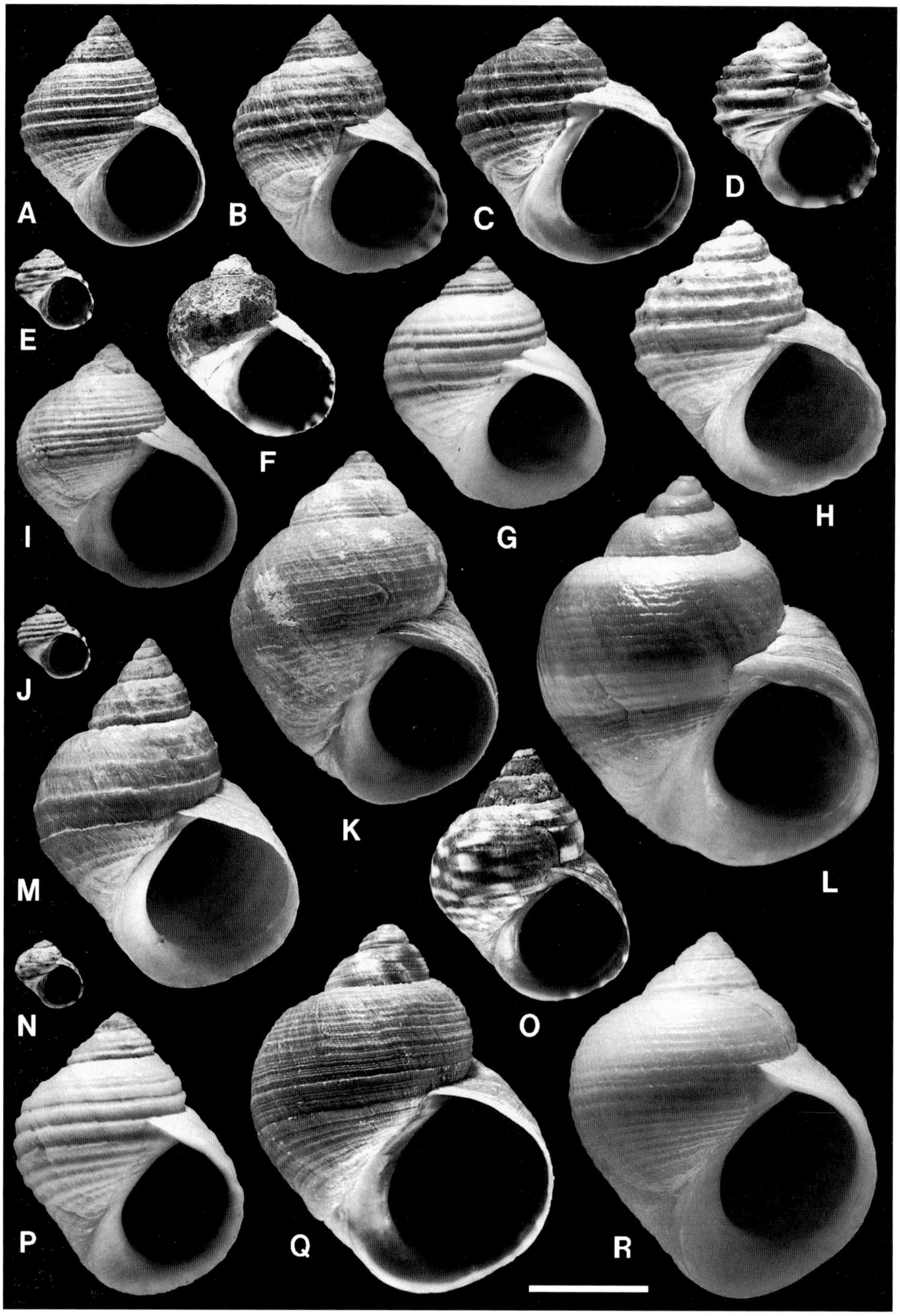

Figure 99. *Littorina (Neritrema) arcana*: **A–L, N, P, R,** geog. var. *typica*; **M, O, Q,** geog. var. *finmarchica*; **A, B, G, K, L, M, O, P–R,** moderate ecotype; **C, D, F, H, I,** wave-exposed ecotype; **E, J, N,** barnacle ecotype. **A,** *Littorina arcana* Hannaford Ellis, 1978; holotype, BMNH 197823W; Porth Swtan, Anglesey, Wales. **B, C, G,** Porth Ledden, Cornwall, England. **D,** Westdale Bay, Pembrokeshire, Wales. **E, J, N,** 1 km W of St Govan's Head, Pembrokeshire, Wales. **F,** Dooagh, Achill Island, Co. Mayo, Ireland. **H,** Ynys-y-Fydlyn, Anglesey, Wales. **I,** Lyme Regis, Dorset, England. **K, L,** Pointe de l'Arcouest, Brittany, France. **M, Q,** Smelror, near Vardø, Norway. **O,** Tromsø, Norway. **P,** Marloes Bay, Pembrokeshire, Wales. **R,** Carantec, Brittany, France. Scale bar = 5 mm. All shells from females of which identification was confirmed anatomically, except **L, M** from males with penes typical of *L. arcana*.

Barnacle ecotype (Fig. 99E, J, N): this is similar in shape to the wave-exposed ecotype and is likewise found on exposed shores, but within the shelter of barnacles and mussel beds; the shells are small, less than about 5 mm in height (maturing as small as 2.8 mm), and are usually tessellated or banded in black and white.

There has been little work on ecotypic variation in this species, but the pattern is analogous to that seen in *L. saxatilis* and *L. compressa* and similar adaptive explanations may apply (see discussion p. 305). On wave-exposed shores a larger aperture accommodates a larger foot (as shown by Grahame & Mill, 1986) for adherence to the substrate, and smaller size permits snails to shelter in crevices. On more sheltered shores there is an increased risk of crushing by crab predators and mobile boulders, which may impose selection for larger size and thicker shells.

The barnacle ecotype has only recently been noticed (Reid, 1993), perhaps because it had previously been confused with the almost identical barnacle ecotype of *L. saxatilis* (often recorded as *L. neglecta*). On an exposed shore in Pembrokeshire, Wales, the only conchological differences between the two were the tendency to stronger sculpture in *L. arcana* and its frequent lack of the basal black band that is usual in the barnacle ecotype of *L. saxatilis*, and identifications had always to be confirmed anatomically (Reid, 1993). At this locality there was a continuum between the wave-exposed and barnacle ecotypes, whereas in *L. saxatilis* these two ecotypes were quite distinct; it was suggested that since *L. arcana* seems generally more tolerant of wave-exposure than *L. saxatilis*, selection for an ecotype which shelters within barnacles may be less strong (see p. 309 for further discussion). The black and white tessellation and banding of the barnacle ecotype appear to be cryptic in this microhabitat.

This species is not found in salt marshes or brackish habitats, so no shell form analogous to the sheltered/brackish ecotype of *L. saxatilis* occurs.

Sculptural variation. Occasional examples are almost smooth, but most show at least some development of spiral ribs. Spiral sculpture is often more strongly developed on wave-exposed shores.

Ontogenetic variation. The suture of the last whorl descends slightly towards the base as maximum size is approached in some shells (Fig. 99B, K, L).

Sexual dimorphism. Grahame & Mill (1989) used multivariate morphometric analysis to compare the shapes of males and females in a sample from Wales and found the difference to be only just outside the range of significance (probability just greater than 0.05).

Colour variation. This species shows an extreme diversity of shell colours, comparable with that of *L. saxatilis*, and together they are the most variable of all *Littorina* species in this respect. The colour may be white, grey, yellow, fawn, orange, brown, purplish or blackish, sometimes with a superimposed darker pattern of a small number of broad bands, tessellations, or fine spiral lines in the shell grooves. The only detailed study of the polymorphism of this species has been that of Atkinson & Warwick (1983), who used the scoring system devised by Pettitt (1973a) for the *L. saxatilis* group. This scores shells for colour and pattern 'alleles' at a number of imaginary 'loci' (e.g. ground colour, bands, tessellations, lines). Although no details were given, they reported that preliminary breeding experiments had demonstrated that shell coloration does indeed have a genetic basis. Atkinson & Warwick (1983) worked mainly with samples from cliff habitats, all around Britain, and found that in the species as a whole about 80% of shells were unpatterned and in the continuously variable category of fawn to brown. Most of the populations were polymorphic and often more so than samples of *L. saxatilis* from the same shores. There were significant differences between the colour frequencies of the two species at most localities, but these were not always in the same direction. (On a few shores some colour patterns are restricted to one or other species and are then a useful aid to identification, e.g. Dytham *et al.*, 1990). In general, *L. arcana* showed a higher frequency of paler colours; tessellated morphs were more frequent (8.6%, cf. 1.9% in *L. saxatilis*), and white-banded, fawn tessellated and lined morphs were always more common in *L. arcana*. For both the fawn and brown morphs (actually part of a continuum) significant correlations of their frequencies in the two species were found, and this was interpreted as evidence that colour variation between shores is not random, but influenced by selection. In addition, a regression of the frequency of the brown morph in *L. arcana* on that in *L. saxatilis* had a slope significantly less than 1, suggesting the possibility of some form of frequency-dependent selection, such as apostatic selection. There was some evidence for positive correlations in the rarer morphs also, and between the overall diversity of the polymorphism in each species. Hannaford Ellis (1985) has noted that identical colour frequencies should not be expected in these two species, since they differ in their vertical zonation on the shore and may occupy microhabitats differing in visual properties or in accessibility to predators. Accepting the genetic basis of the polymorphism, and that these two species are reproductively isolated, these correlations provide persuasive evidence for selection of some kind. The nature of the selective agents, and indeed whether the selection is directly related to the visual appearance of the shell, are not yet known. Atkinson & Warwick (1983) suggested that apostatic selection by visual predators was likely. One apparent example of cryptic resemblance to the background is the barnacle ecotype which, like the corresponding ecotypes of *L. compressa* and *L. saxatilis*, is predominantly white with brown to black tessellations and bands.

It is notable that the level of colour polymorphism within populations of the northern geog. var. *finmarchica* appears no less than that in southern populations, in contrast to *L. obtusata*, *L. fabalis* and *L. compressa*, which all show reduced polymorphism in Arctic regions.

Distribution

Habitat. This species occurs on rocky shores in the upper eulittoral and littoral fringe. In Britain it has often been observed that it is most common on wave-exposed shores with steep cliffs, on sea walls or stable rocks, where it shelters in crevices (Hannaford Ellis, 1979; Hart & Begon, 1982; Grahame & Mill, 1986). On adjacent boulder shores and sheltered shores it may be less frequent or absent (Atkinson & Warwick, 1983; Ward & Janson, 1985; Grahame & Mill, 1989; Dytham *et al.*, 1990). In some regions it occupies a wider range of exposure conditions, including sheltered shores; at a locality in southwestern Ireland, for example, it occurred on open coasts regardless of exposure, although it was absent from an enclosed sea loch (Little & Williams, 1989). In northern Norway *L. arcana* was also found on both exposed and sheltered shores; at Tromsø it was abundant on dead algae at the top of the *Ascophyllum* zone on a protected shore (pers. obs.). It has only once been found on muddy rocks at the mouth of an estuary (pers. obs. at Lezardrieux, Brittany) and has not been found in salt marshes (Roberts & Hughes, 1980; Bergerard & Caugant, 1981). Its vertical zonation extends from the black *Verrucaria* zone of the littoral fringe down to the upper half of the barnacle zone (Hannaford Ellis, 1979; Reid, 1993), with the highest maximum tidal levels on exposed shores (Little & Williams, 1989). A downward breeding migration was first noted by James (1968a, as *L. saxatilis tenebrosa*, but this was probably mainly *L. arcana*) and said to occur in January to February and July to August. This was documented on Anglesey in Wales by Hannaford Ellis (1985), where adults moved down into the barnacle belt between September and October, and retreated upshore in June and July; this was related to the need to lay egg masses in damp sites. The requirement for a stable substrate for the egg masses appears to restrict the species to hard, non-friable rock types (Mill & Grahame, 1990a). The diet of this species has not been studied, but like the other members of the *L. saxatilis* group it is likely to be an opportunistic grazer on diatoms, green algae, fucoids and detritus.

In its high-shore habitat *L. arcana* usually coexists with both *Melarhaphe neritoides* and *L. saxatilis*. It appears to be more tolerant of extreme conditions of wave-exposure than is *L. saxatilis*, and the only localities at which it has been reported in the virtual or complete absence of *L. saxatilis* are on the most exposed natural rocky shores (Hart & Begon, 1982; Faller-Fritsch & Emson, 1985; Grahame & Mill, 1986) and on an exposed sea wall (Dytham *et al.*, 1990). It has been reported to occupy a vertical level above or identical to that of *L. saxatilis* (Faller-Fritsch & Emson, 1985; Mill & Graham, 1990a; Cole, 1991), but their relative zonation is affected by the downshore breeding migration which only takes place in *L. arcana* (Hannaford Ellis, 1979, 1985). The third member of the *L. saxatilis* group, *L. compressa*, is found at somewhat lower levels, in the mid to upper eulittoral zone.

Range (Fig. 100). *Littorina arcana* is known only from Brittany and northern France, the British Isles, and Norway north of Trondheim. Since the initial records of the then newly described species by Hannaford Ellis (1979), further British records have been added by Atkinson & Warwick (1983), Faller-Fritsch & Emson (1985), Ward & Janson (1985), Nunn & Smith (1987) and Knight & Ward (1991); maps of British localities have been given by Mill & Grahame (1990b, 1992b) and on a coarser scale by Seaward (1982, 1990). The distribution on the French Channel coast has been

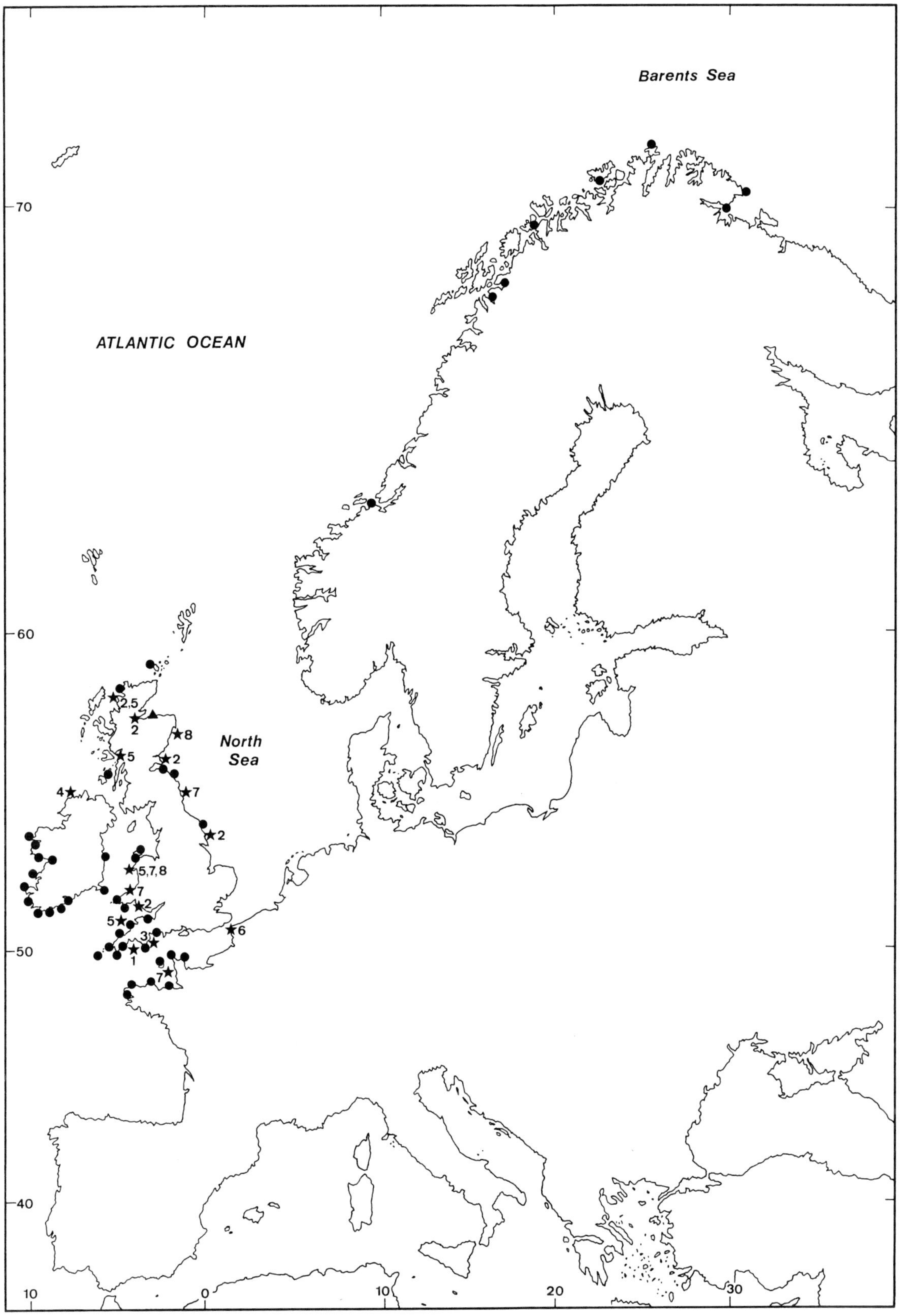

Figure 100. *Littorina (Neritrema) arcana*: distribution map. Key: circles, anatomical material examined; triangle, probable determination based on examination of shells only; asterisks, records from literature (1, Hannaford Ellis, 1979; 2, Atkinson & Warwick, 1983; 3, Faller-Fritsch & Emson, 1985; 4, Nunn & Smith, 1987; 5, Mill & Grahame, 1990b; 6, Warmoes, Dumoulin & Reid, 1990; 7, Knight & Ward, 1991; 8, Mill & Grahame, 1992b).

mapped by Warmoes *et al.* (1992). With this considerable collecting effort in recent years, and the addition of further records in the present study, the distribution in much of this geographical region is reasonably well known. However, since anatomical examination is always required for unequivocal determination of this species, dry museum collections are inadequate, and where records are sparse (as in Northern Ireland, Scotland and Norway) further collections are necessary.

In France the most southerly record is 5 km west of Brest (BMNH). It is common on the north coast of Brittany, the Channel Islands and the north side of the Cotentin Peninsula (BMNH). It appears to be absent further east, until the northernmost French records in the vicinity of Boulogne (Warmoes *et al.*, 1992). In Britain the easternmost limit on the Channel coast has been precisely established at Lyme Regis, Dorset (Mill & Grahame, 1990a). Further west it is found around the coast of southwestern England, penetrating the Bristol Channel as far as Porlock (BMNH) and Swansea (Atkinson & Warwick, 1983). It is common in Pembrokeshire; there is a single record from Cardigan Bay (Knight & Ward, 1991), others from the Lleyn Peninsula (Mill & Grahame, 1990b, 1992b; Knight & Ward, 1991) and many from Anglesey. To the north there is a large gap until the island of Jura (NMS) and Oban (Mill & Grahame, 1990b). Although the west coast and islands of Scotland have been reasonably well collected (there are numerous records for the other two members of the *L. saxatilis* group), there are no further records until the vicinity of Scourie (Atkinson & Warwick, 1983; Mill & Grahame, 1990b; NMS). Several authors have remarked upon this absence (Atkinson & Warwick, 1983; Mill & Grahame, 1992b) and it appears to be genuine (a note of '*arcana*-like shells' on the Uists by Smith, 1979b, lacked anatomical confirmation; it was not found there by DGR, 1986). In part, however, it could be due to a tendency to collect preferentially from the more sheltered and readily accessible shores of sea lochs, avoiding exposed coasts where *L. arcana* may be more abundant. Elsewhere in Scotland occurrence is patchy: Orkney Islands (NMS), Moray Firth (Atkinson & Warwick, 1983), Aberdeen (Mill & Grahame, 1992b) to Berwick-on-Tweed (BMNH). In northeastern England there are records from Northumberland (Seshappa, 1947; Knight & Ward, 1991) and Yorkshire as far south as Bridlington (Mill & Grahame, 1990a). From Ireland specimens are available from near Dublin, Carnsore Point, many localities on the south and west coasts (BMNH), and there is one published record from the north coast (Nunn & Smith, 1987), although none from Northern Ireland itself (Matthews & Montgomery, 1987). Elsewhere in Europe *L. arcana* is absent from Belgium (Warmoes, Backeljau & de Bruyn, 1988), and records from The Netherlands are considered unreliable (Warmoes *et al.*, 1988, 1992). In southern Scandinavia it has not been found in Denmark (Nielsen, 1980) or Sweden (Janson, 1982a). Sneli & van Marion (1979) reviewed the littorinid fauna of Norway and recorded *L. arcana* from Oslofjord; these specimens have been reexamined and found to be immature *L. saxatilis*; their record from the mouth of Trondheimfjord has been confirmed (DGR, 1989). The southern Atlantic coast of Norway should be searched for the presence of this species. North of Trondheim the central Norwegian coast has been poorly collected and, although no records are available, the species may well occur there. In the north of Norway there are seven records between Skarberget and Vadsø (DGR, 1989). Studies in the White Sea have not recorded this species, but further collections are required from the Russian Barents Sea coast.

As in the case of *L. compressa*, it is believed that the southern limit of *L. arcana* in Brittany may be determined by desiccation of the vulnerable egg masses on the south-facing coast (Warmoes *et al.*, 1992). Penetration up the English Channel cannot be accounted for by low winter temperatures, as is possible for some other invertebrates with predominantly western distributions (Crisp & Southward, 1958; Lewis, 1964), nor by intolerance of turbidity, for *L. arcana* is found in the colder and more turbid waters of the North Sea (Mill & Grahame, 1990a, b). However, on both sides of the Channel there is a close correlation between the occurrence of this species and the distribution of hard sedimentary and crystalline rock, leading to the suggestion that a firm substrate is necessary for the attachment and survival of egg masses (Mill & Grahame, 1990a, b; Warmoes *et al.*, 1992). This cannot explain the absence of *L. arcana* from Portland Bill, nor from most of the west of Scotland. Mill & Grahame (1990a, b) suggested that absence from apparently suitable areas might be due to the poor dispersal capacity of this oviparous species, but against this is the suggestion that non-planktotrophic developers can sometimes disperse effectively over long distances (Johannesson, 1988). The distribution of this species is quite similar to that of *L. compressa*, and similar factors have been invoked to account for both (Mill & Grahame, 1990b; Warmoes *et al.*, 1992), but in most parts of the range the distribution of *L. arcana* is the more extensive. The virtual absence of *L. arcana* from western Scotland is therefore especially puzzling, for *L. compressa* is frequent there.

Fossil record. No fossil records have been traced. Since identification depends heavily on anatomical examination, fossil material would be difficult to identify and morphometric comparison with Recent material would be necessary.

Discussion

Species status and population genetics. For a few years after the first description of *L. arcana* in 1978 there were some who argued that it was not a distinct species, but merely a reproductive morph of *L. saxatilis*. However, the species status of *L. arcana* is now no longer questioned (see Taxonomic History, p. 265). Sympatric populations of the two species maintain significant differences in allele frequencies at a majority of the polymorphic loci that have been analysed, indicating a high degree of reproductive isolation between them (Ward & Warwick, 1980; Ward & Janson, 1985; Knight & Ward, 1986, 1991). Furthermore, when allozyme frequency data are analysed on a geographical scale, populations of *L. arcana* do mostly cluster together (rather than with the geographically closest populations of *L. saxatilis*) (Ward & Janson, 1985; Knight *et al.*, 1987; Ward, 1990) and in fact some clustering techniques support the monophyly of *L. arcana* (Sundberg *et al.*, 1990; Knight & Ward, 1991). Genetic similarity, and inferred gene flow, are therefore greater between geographically separated populations of *L. arcana* than between *L. arcana* and *L. saxatilis* on the same shores.

Classically, biological species have been defined as non-interbreeding, reproductively isolated groups, and with this criterion in mind the possibility of hybridization between *L. arcana* and *L. saxatilis* has been examined. Laboratory breeding experiments have succeeded in crossing female *L. arcana* with male *L. saxatilis* to produce fertile hybrids, which in turn were back-crossed with male *L. saxatilis* (other reciprocal crosses were unsuccessful). Nevertheless, no hybrids were detected in naturally sympatric populations and, if they occurred at all, their incidence was estimated at less than 1–2% (Ward *et al.*, 1986; Warwick *et al.*, 1990). Prezygotic isolating mechanisms therefore appear to operate between these two species in the field, but these have yet to be investigated. It is now known that *L. arcana* and *L. saxatilis* are not so strictly micro-sympatric as was once believed, for *L. arcana* occupies a lower tidal level during its more restricted breeding season (Hannaford Ellis, 1985), which could lead to a degree of reproductive isolation. Presumably there is also assortative mating, as has recently been shown to operate even between ecotypes within the species *L. saxatilis* (Johannesson, Rolán-Alvarez & Ekendahl, 1995). The new DNA technology presents possibilities for further investigation of the genetic structure of these two sympatric species and for searching for possible cases of introgression in natural populations.

In comparison with the other two members of the *L. saxatilis* group, *L. arcana* has been found to have a higher average heterozygosity than *L. compressa*, but a slightly (Knight & Ward, 1986, 1991) or considerably lower value than *L. saxatilis* (Backeljau & Warmoes, 1992).

Shell morphology. In comparison with *L. saxatilis* from the same shores, *L. arcana* frequently has a relatively larger aperture, which accommodates a larger foot; this presumably improves adhesion ability and may be important in permitting *L. arcana* to be relatively more abundant than *L. saxatilis* on the most wave-exposed shores, as discussed by Grahame & Mill (1986). These authors suggested that there was a probable associated cost in terms of decreased resistance to desiccation and predation. However, since the species does seem able to produce shell forms with apertures that appear similar to those of *L. saxatilis* from sheltered shores (Fig. 99K, L), it is not clear whether the shell form of *L. arcana* is a cause or an effect of its distribution in relation to exposure.

In shell form this species is almost as variable as *L. saxatilis*, lacking only the sheltered/brackish ecotype of that species. The existence of the relatively thinner-walled northern geographical variety has not previously been noticed, and has parallels in the other Atlantic *Littorina (Neritrema)* species.

Distribution. As in the case of *L. compressa* (p. 261), it seems likely that the specific requirements for a hard rock substrate and a sheltered and humid microhabitat for successful attachment and development of egg masses limits the geographical and habitat range.

Phylogenetic relationships. The three species of the *L. saxatilis* group appear from both DNA and allozyme studies to form a monophyletic clade, although no morphological synapomorphies are known. The conflicting evidence about relationships between the three species *L. arcana*, *L. compressa* and *L. saxatilis* is discussed in the remarks on *L. saxatilis* (p. 336).

Similar species. Confusion should only arise with the two other members of the *L. saxatilis* group, and differentiation of all three is described in the remarks on *L. saxatilis* (p. 337, Table 7).

Littorina (Neritrema) saxatilis (Olivi, 1792)

(Figs 3I, 4, 5, 101–111)

Synonymy

Turbo littoreus—da Costa, 1778: 98–100 (in part, includes *L. littorea*; not Linnaeus, 1758=*L. littorea*; *s.l.*). Gmelin, 1791: 3588 (in part, includes *L. littorea*; not Linnaeus, 1758). Donovan, 1800: pl. 33, fig. 2 (in part, other figs are *L. littorea*; not Linnaeus, 1758). Maton & Rackett, 1807: 158–159; pl. 4, figs 8, 10, 11 (in part, other fig. is *L. littorea*; not Linnaeus, 1758). Pulteney, 1813: 49; pl. 19, fig. 2 (in part, other fig. is *L. littorea*; not Linnaeus, 1758). Dillwyn, 1817: 817 (in part; includes *L. littorea*; not Linnaeus, 1758; *s.l.*). Brown, 1827: pl. 46, figs 4–9 (in part, other figs are *L. littorea*; not Linnaeus, 1758). Blainville, 1830: 298–299 (in part; includes *L. littorea*; not Linnaeus, 1758). Brown, 1837a: 15; pl. 10, figs 4–9 (in part, other figs are *L. littorea*; not Linnaeus, 1758).

Nerita littorea—Fabricius, 1780: 403–404 (in part, includes refs to *L. littorea*; not *Turbo littoreus* Linnaeus, 1758).

Littorina littorea—Brown, 1844a: 128; pl. 10, figs 4–9 (in part, other figs are *L. littorea*; not *Turbo littoreus* Linnaeus, 1758).

Turbo saxatilis Olivi, 1792: 172; pl. 5, fig. 3a–c, A (Lagune nostre [i.e. Venice Lagoon, Italy]; no type material mentioned by Ghisotti, 1978, and presumed lost; lectotype (Rosewater, 1978) Olivi, 1792: pl. 5, fig. 3A, Fig. 101 herein). Blainville, 1828: 111.

Littorina saxatilis—Brusina, 1870: 190–193. Brusina, 1886: 206. Kobelt, 1907: 57–62; pl. 110, figs 6–8, 11–13; pl. 112, figs 1–4, 22, 23 (in part, other figs are *L. compressa*, *L. fabalis*; *s.l.*). Dautzenberg & Fischer, 1912: 187–201; pl. 9, figs 1–23, 26, 27; pl. 10, figs 1–28 (in part, other figs are *L. compressa*, *L. arcana*). Dautzenberg & Durouchoux, 1913: 21–22 (*s.l.*). Hidalgo, 1917: 391, 395–396. Bavay, 1920: 254 (*s.l.*). Dautzenberg & Fischer, 1925: 52–55 (in part, includes *L. compressa*; *s.l.*). Baden-Powell, 1930: 71, 74–76 (Pleistocene). van Benthem Jutting, 1933: 66–68; fig. 30. Ankel, 1936b: 47; fig. 117 (mantle); fig. 136 (penis); fig. 178 (oviduct); fig. 206D. Thorson, 1941: 26–30. Thorson, 1944: 31–32. Thorson, 1946: 170–173. Seshappa, 1947: 335–336 (in part; includes *L. arcana*). Seshappa, 1948: 702–703 (in part; includes *L. arcana*). Ankel, 1950: 19–27 (radula). Thorson, 1951: 27–28. Abbott, 1954a: 133; pl. 19, fig. d. Bousfield, 1960: 16; pl. 2, fig. 18. Fischer-Piette & Gaillard, 1960: 10–37 (*s.l.*). Fischer-Piette & Gaillard, 1961: 163–172 (*s.l.*). Fischer-Piette, Gaillard & Jouin, 1961: 320–328; text-figs 1–4. Fretter & Graham, 1962: figs 26A, 87A, 178, 222 (anatomy, embryo) (*s.l.*). Richards, 1962: 77; pl. 15, fig. 30 (Pleistocene). Fischer-Piette, Gaillard & James, 1963: 1–22 (*s.l.*). Fischer-Piette, Gaillard & James, 1964: 125–171 (*s.l.*). Fischer-Piette & Gaillard, 1966: 375–382 (*s.l.*). Ziegelmeier, 1966: 41; pl. 10, fig. 2a, b. Fischer-Piette, Gaillard & Delmas, 1967: 874–897 (*s.l.*). Muus, 1967: 137–139. Fischer-Piette & Gaillard, 1968: 103–118 (*s.l.*). James, 1968a: 143–165 (in part, includes *L. compressa*, *L. arcana*). Davis, 1971: 2–5; fig. 4: 3–10. Fischer-Piette & Gaillard, 1971: 1–90 (in part, includes refs to *L. compressa* and probably *L. arcana*). Fischer-Piette, Gaillard, Panouse & Bellon-Humbert, 1971: 970–974; fig. 1. Macpherson, 1971: 32–33; pl. 1, fig. 19. Bellon-Humbert, 1973: 20; pl. 3, fig. 15. Davis, 1973: 61–90; fig. 2; fig. 3 (head); fig. 6 (penis); fig. 7 (embryonic shell); fig. 10. Pettitt, 1973a: 531–538 (*s.l.*). Pettitt, 1973b: 339–343 (*s.l.*). Rasmussen, 1973: 241–242. Sacchi & Torelli, 1973: 181–188. Abbott, 1974: 67; text-fig. 551. Bandel, 1974: 105–106; figs 18–21, 26, 27 (radula). Matveeva, 1974: 105–116; fig. 14 (female animal). Sacchi, 1974: 75–84; fig. 5 (oviduct) (*s.l.*). Gofas, 1975: 95–99; figs 1–6. Hubendick & Warén, 1975: 10; figs 267–269. Newkirk & Doyle, 1975: 227–237. Sacchi, 1975: 111–120; fig. 1 (radula); fig. 2 (oviduct); figs 3–5 (penis); pl. 1. Daguzan, 1977: 17–40 (*s.l.*). Gaillard, 1977: 11–14 (in part, includes *L. compressa*; *s.l.*). Moore, 1977: 194; fig. 1. Ghisotti, 1978: 24. Torelli, 1978: 91–98. Caugant & Bergerard, 1979: 247–262 (*s.l.*). Sacchi, 1979: 7–21. Sacchi & Torelli, 1979: 125–126. Fretter, 1980: 148–153; fig. 1C, D (oviduct). Fretter & Graham, 1980: 270–272, 274–276; fig. 213. Sacchi, 1981: 51–67. J. E. Smith, 1981: 215–241; fig. 7A-D, J, K (embryonic shell); fig. 8A1, A2, C, F (*s.l.*). Janson, 1982a: 167–173. Smith, 1982: 535–539. Torelli, 1982a: 133–141. Torelli, 1982b: 73–74; fig. 41. Janson, 1983a: 224–227. Janson & Sundberg, 1983: 49–53. Sacchi & Malcevschi, 1983: 7–21. Cataliotti-Valdina, 1984: 211–219; fig. 1, nos 5, 6, 11, 12 (Pleistocene). Janson & Ward, 1984: 289–307. Rolán, 1984: 111, 114; figs. Fish & Sharp, 1985: 143–156; fig. 10.2B; fig. 10.2D (embryonic shell). Janson, 1985a: 64–68; fig. 2 (embryonic shell). Janson, 1985b: 51–59. Janson & Ward, 1985: 9–15. Ward & Janson, 1985: 86–94. Johannesson, 1986: 183–195. Knight & Ward, 1986: 405–413. Ward, Warwick & Knight, 1986: 233–241. Jagnow & Gosselck, 1987: 211–213; pl. 2, fig. 11. Knight, Hughes & Ward, 1987: 417–426. Graham, 1988: 164, 178; figs 59B, C, 65. Mill & Grahame, 1988: 347–353. Seshappa, 1988: 218–223 (*s.l.*). Sundberg, 1988: 169–184. Warmoes, Backeljau & de Bruyn, 1988: 51–70; fig. 4B, H, I–K (penis). Grahame & Mill, 1989: 837–855; fig. 4. Warmoes, 1989: 125–128. Byers, 1990: 3–10 (*s.l.*). Dytham, Grahame & Mill, 1990: 233–240. Grahame, Mill & Brown,

1990: 223–231. Hayward, Wigham & Yonow, 1990: 660; fig. 12.9. B. & K. Johannesson, 1990: 71–87. K. & B. Johannesson, 1990: 89–97. Sundberg, Knight, Ward & Johannesson, 1990: 29–40. Warwick, Knight & Ward, 1990: 109–116. Finet, Wüest & Mareda, 1991: 36; pl. 11a, b; pl. 11c, d (radula). Knight & Ward, 1991: 81–91. Poppe & Goto, 1991: 94; pl. 11, figs 16–21 (*s.l.*). Rózycki, 1991: 388–389; fig. 152. Buckland-Nicks & Worthen, 1992: 217–225 (penis). Dytham, Grahame & Mill, 1992: 39–44. Dytham, Mill, Grahame & O'Higgins, 1992: 127–133. Egorov, 1992: 170; fig. 1F. Grahame & Mill, 1992: 99–106. Grahame, Mill, Double & Hull, 1992: 499–502. Sergievsky, 1992: 235–245. Barnes, 1993: 477–495. Crossland, Coates, Grahame & Mill, 1993: 301–305. K. Johannesson, B. Johannesson & Rolán-Alvarez, 1993: 1770–1787; fig. 1. Reid, 1993: 51–62; fig. 4; fig. 7A–D (penis); fig. 7E, F (oviduct). Caley, 1994 (penis, oviduct, embryonic shell, radula). Caley, Grahame & Mill, 1995: 181–193. Crossland, Coates, Grahame & Mill, 1995: 205–209. Grahame, Mill, Hull & Caley, 1995: 887–899; fig. 3A. Mill & Grahame, 1995: 61–71. Rolán-Alvarez & Rolán, 1995: 1–8; figs 1, 3–5; fig. 6 (embryonic shell); fig. 7 (radula). Hull, Grahame & Mill, 1996: fig. 2. B. & K. Johannesson, 1996. Rolán-Alvarez, Rolán & Johannesson, 1996.

Littorina saxatilis forma typica [subsp. *saxatilis*]—Dautzenberg & Fischer, 1912: 192–193; pl. 9, figs 1–6.

Littorina rudis var. *saxatilis*—Harmer, 1921: 655; 1925: 874; pl. 65, fig. 28.

Littorina (Littorivaga) saxatilis—Winckworth, 1932: 221 (*s.l.*). Lebour, 1937: 127–128. Bequaert, 1943: 8–11; pl. 3, figs 1–10. Nordsieck, 1968: 40 (in part, fig. is *L. compressa*; *s.l.*). Nordsieck, 1982: 61; pl. 26, fig. 23.20 (in part, includes *L. compressa*; *s.l.*). Cataliotti-Valdina, 1984: 211–219; fig. 1 (Pleistocene). Sabelli, Giannuzzi-Savelli & Bedulli, 1990: 19, 146.

Littorina (Littorivaga) saxatilis saxatilis—Winckworth, 1932: 221.

Littorina saxatilis saxatilis—van Benthem Jutting, 1933: 67–68; fig. 31. Kaas & ten Broek, 1942: 144; pl. 11, fig. 6. James, 1968a: 146–147; pl. 1, fig. 1a–f. James, 1968b: 633–634; fig. 15; fig. 28 (head); fig. 48. Graham, 1988: 164, 178; figs 59B, 65A. de Bruyne, 1991: 101; fig. 65.

Littorina (Littorina) saxatilis—Rosewater, 1970: 423 (*s.l.*). Kilburn, 1972: 403–405; fig. 5 (also Pleistocene). Golikov & Kusakin, 1978: 86–87; fig. 52 (*s.l.*). Rosewater, 1981: 15–18; pl. 1A; pl. 2A–C; pl. 3A (operculum); pl. 4G (penis). Golikov, 1987: 76–77; fig. 28.

Littorina (Neritrema) saxatilis—Reid, 1989a: 102; pl. 3, fig. 3b (embryonic shell); fig. 5l (head). Reid, 1990b: 1–19; fig. 1s; fig. 2l (oviduct). Taylor & Reid, 1990: 199–215 (mineralogy). Smith & Heppell, 1991: 17, 81.

? *Turbo ruber* von Salis-Marschlins, 1793: appendix (Taranto [Italy]; refers to Chemnitz, 1781: pl. 185, fig. 1855; see Taxonomic History, p. 287).

Turbo rudis Maton, 1797: 277 (from p. 275: 1½ miles from Bere Alston, estuary of River Tavey [Tavy], Devonshire [England]; lectotype (here designated) Donovan, 1800: **1**: pl. 33, fig. 3; see Fig. 106S for topotype). Donovan, 1800: pl. 33, fig. 3. Montagu, 1803: 304–306. Maton & Rackett, 1807: 159; pl. 4, figs 12, 13. Pennant, 1812: 294. Pulteney, 1813: 49; pl. 18, figs 6, 7. Dillwyn, 1817: 818. Turton, 1819: 197. Lamarck, 1822: 49 (*s.l.*). Brown, 1827: pl. 46, figs 10–14, 25 (*s.l.*). Blainville, 1828: 108. Blainville, 1830: 297–298 (in part, fig. is *L. littorea*). Brown, 1837a: 15–16; pl. 10, figs 10–14, 25 (*s.l.*). Delessert, 1841: pl. 37, fig. 5a, b. Thorpe, 1844: 166. Leach, 1852: 187–188 (in part, includes ref. to *L. obtusata*).

Litorina rudis—Menke, 1830: 45. Anton, 1838: 52. Menke, 1845: 53–55. Lovén, 1846: 154 (*s.l.*). Philippi, 1846a: **2**: 103–104; *Litorina* pl. 1, figs 14–16. Küster, 1856: 7; pl. 1, figs 9, 10 (1853). Weinkauff, 1878: 27, 28, 30. Weinkauff, 1883: 215 (*s.l.*). Johnson, 1915: 121.

Littorina rudis—Lyell, 1835: 34 (Pleistocene). Gould, 1841: 257–259; fig. 175* (not 165 as cited). Johnston, 1842: 267–268 (*s.l.*). Deshayes & Milne Edwards, 1843: 216 (in part; includes ref. to *L. obtusata*). de Kay, 1843: 104; pl. 5, fig. 103a, b. MacGillivray, 1843: 137–138 (*s.l.*). Brown, 1844a: 128; pl. 10, figs 10–14, 25 (*s.l.*). Alder, 1848: 149 (*s.l.*). Middendorff, 1849: 383–384 (*s.l.*). Clark, 1850a: 362–364 (in part, includes *L. fabalis*; *s.l.*). Clark, 1850b: 32–33 (*s.l.*). Forbes & Hanley, 1850: 32–35, 53; pl. 83, figs 3, 5–7; pl. 86, fig. 1, 1a, 1b (in part, other figs are *L. compressa*; *s.l.*). Petit de la Saussaye, 1852: 182 (*s.l.*). Clark, 1855: 342–345 (in part; includes refs to *L. fabalis*, *L. obtusata*; *s.l.*). Reeve, 1857: sp. 33; pl. 7, fig. 33b, c (in part, other fig. is *L. compressa*). Jeffreys, 1859: 195. Sowerby, 1859: pl. 12, fig. 12 (in part, other fig. is *L. compressa*). Smith, 1860: 9. Jeffreys, 1865: **3**: 364–368; 1869: **5**: 206; pl. 65, figs 3, 3a, 3b (in part, includes *L. compressa*, *L. arcana*, *L. sitkana*). Petit de la Saussaye, 1869: 121 (*s.l.*). Gould, 1870: 304–306; fig. 575. Smith & Prime, 1870: 392. Mørch, 1871a: 186. Meyer & Möbius, 1872: 17–19; figs (opposite p. 20) 1–6 (animal, operculum, radula). Tryon, 1873–1874: 79; fig. 145. Sars, 1878: 165 (*s.l.*). Jeffreys, 1883: 112–113. Collin, 1884: 57–60. Herzenstein, 1885: 678–680. Locard, 1886: 283, 578, 580 (*s.l.*). Nobre, 1886: 20. Sparre-Schneider, 1886: 5–10 (*s.l.*). Sowerby, 1887: pl. 12, figs 12, 16–19 (in part, other fig. is *L. compressa*). Kobelt, 1888: 167–168 (*s.l.*). Hidalgo, 1890: pl. 89, figs 1–14. Locard, 1891: 188; fig. 164. Provancher, 1891: 92. Sykes, 1892: 191–198. Nobre, 1898: 79. Marshall, 1898: 121–123 (in part, includes *L. compressa*). Teilman-Friis, 1899: 430–431; fig. 2 (radula). Dautzenberg & Durouchoux, 1900: 45–46 (in part, includes *L. compressa*; *s.l.*). Brøgger, 1901: pl. 9, fig. 13; pl. 12, fig. 9a, b (Pleistocene). Johansen, 1901: 298–302; figs 4–6 (radula). Marshall, 1903: 53–54. Pelseneer, 1911: 9–11, 12–15; pl. 1, figs 1, 2; figs 2–25 (embryology, embryonic shell). Marshall, 1915: 329. Schlesch, 1916: 280. Bárdarson, 1919: 63. Bárdarson, 1920: 89. Harmer, 1921, 1925: 653–661, 874; pl. 53, figs 2, 3, 5, 6, 10, 12, 17, 19, 20, 28, 30; pl. 65, fig. 28 (in part, other figs are

L. littorea; includes *L. sitkana s.l.*; includes Pleistocene). Pelseneer, 1923: 72; fig. 4 (penis). Nobre, 1931: 145; pl. 33, fig. 8. Linke, 1933: 1–60 (anatomy); pl. 2, fig. 21 (penis); pl. 4, fig. 38, pl. 6, fig. 63 (oviduct). Spärck & Thorson, 1933: 13. Nobre, 1936: 101; pl. 82, fig. 11. Johansson, 1939: 344–396 (anatomy, histology). Richards, 1941: 10; pl. 1, fig. 11 (Pleistocene). Wium-Andersen, 1970: 267–273. Heller, 1975a: 135–138, 146–149; fig. 3A (penis); figs 4A, 5A (radula); pl. 1, figs 1–4, 18, 19; pl. 2, figs 21–26 (*s.l.*). Heller, 1975b: 153–170 (*s.l.*). Davis, 1976: 1. Heller, 1976: 201–213 (*s.l.*). Raffaelli, 1978b: 166–170 (*s.l.*). Clyne & Duffus, 1979: 178–185; figs 1, 4–6 (embryonic shell). Hannaford Ellis, 1979: 43–56; figs 1A, 2A, 4A, 8A, C, 9C, D; pl. 1, figs 1, 3; pl. 2, figs 3, 4 (reproductive anatomy); pl. 3, figs 1–4. Hughes, 1979: 119–126; fig. 1; fig. 2 (radula); fig. 3 (penis). Raffaelli, 1979a: 219–232; fig. 4B, C; figs 8, 9A–C (radula) (in part, includes *L. arcana*). Raffaelli, 1979b: 65–74 (*s.l.*). Sneli & van Marion, 1979: 4–8; figs 1a, b. Elner & Raffaelli, 1980: 151–160. Fretter & Graham, 1980: 270–276; figs 210–212. Ward & Warwick, 1980: 417–428. Wilkins & O'Regan, 1980: 355–359 (*s.l.*). van Marion, 1981: 99–108. J. E. Smith, 1981: 215–241; fig. 7E–H (embryonic shell); fig. 8A3, A4, D, E. Atkinson & Warwick, 1983: 137–151. Brandwood, 1983: 165–167. Atkinson & Newbury, 1984: 93–105. Jardine, 1984: 5–6; fig. 2d; fig. 3a (penis); fig. 3b (oviduct). Rolán, 1984: 113; figs. Brandwood, 1985: 551–565. Faller-Fritsch & Emson, 1985: 157–177. Grahame & Mill, 1986: 229–236. Hayward, Wigham & Yonow, 1990: 662; fig. 12.9.

Littorina vulgaris var. *rudis*—Maitland, 1854: 82.

Littorina (Littorina) rudis—Récluz, 1869: 44–45 (*s.l.*). Tryon, 1887: 241; pl. 41, figs 92–94, 96–100, 1–4 (in part, other fig. is *L. compressa*).

Litorina (Litorina) rudis—Kobelt, 1876–1878: 124; pl. 46, fig. 13 (*s.l.*).

Littorina (saxatilis var.) *rudis*—Kobelt, 1907: 59; pl. 110, fig. 6 (in part, other figs are *L. compressa*).

Littorina saxatilis rudis—Dautzenberg & Fischer, 1912: 197–198; pl. 10, figs 1–20. Dautzenberg & Durouchoux, 1913: 22–23 (*s.l.*). Dautzenberg & Fischer, 1925: 53–54 (*s.l.*). Kaas & ten Broek, 1942: 145; pl. 11, fig. 7. James, 1963: 30–31 (*s.l.*). James, in Fischer-Piette, Gaillard & James, 1964: 146–151; pl. 2, figs 1–3; text-figs 5 (head), 9 (penis), 16 (radula) (*s.l.*). James, 1968a: 147–153; pl. 1, figs 2d–f, 3c, d, 5a–d; pl. 2, figs 2a–h, 3d–f, 4a–c, 5e–f (in part, other figs are *L. compressa*). James, 1968b: 633–634; figs 11, 18; fig. 24 (head); figs 32, 34–36. Bandel, 1975: 6–7; figs 9–11 (embryonic shell). Seshappa, 1976: 222–229; figs 11–13; figs 18, 20, 21 (radula). Hoeksema, 1985: 22–28; figs 1–8. Graham, 1988: 164, 178; figs 59C, 65B. de Bruyne, 1991: 103; fig. 67.

Littorivaga rudis—Winckworth, 1922: 96–97 (*s.l.*).

Littorina (Littorivaga) rudis—Oldroyd, 1924: 149 (in part, includes *L. sitkana*). Oldroyd, 1927: 61 (in part, includes *L. sitkana*).

Littorina (Littorivaga) saxatilis rudis—Winckworth, 1932: 221 (*s.l.*). Nordsieck, 1968: 40; pl. 6, fig. 23.21. Nordsieck, 1982: 61; pl. 26, fig. 23.21.

Littorina saxatilis var. *rudis*—Thorson, 1941: 27.

Littorina saxatilis rudis var. *rudis*—James, 1968a: 152; pl. 1, figs 2d–f, 3c, d.

Littorina (Neritrema) saxatilis var. *rudis*—Smith & Heppell, 1991: 17, 81.

Turbo davidis Röding, 1798: **2**: 88 (lectotype (here designated, from original citation) Chemnitz, 1781: **5**: pl. 185, fig. 1855a, b, from Faroe Islands).

Turbo groenlandicus Röding, 1798: **2**: 88 (lectotype (here designated, from original citation) Chemnitz, 1781: **5**: pl. 185, fig. 1855f, g, from Greenland).

Turbo obtusatus—Pulteney, 1799: 45 (in part, includes ref. to *L. obtusata*; not Linnaeus, 1758=*L. obtusata*; *s.l.*).

Turbo tenebrosus Montagu, 1803: 303–304 (Devonshire; Kent [both England]; lectotype (here designated; Fig. 107C)+14 paralectotypes BMNH 1989177, seen; possible paralectotype RAMM Montagu Colln 4200,+1 unregistered spec., seen). Maton & Rackett, 1807: 160. Montagu, 1808: 182; pl. 20, fig. 4. Pennant, 1812: 295. Pulteney, 1813: 49; pl. 18, fig. 15. Dillwyn, 1817: 817–818. Turton, 1819: 197–198; pl. 10, figs 36, 37. Brown, 1827: pl. 46, figs 18, 19. Blainville, 1828: 108. Brown, 1837a: 16; pl. 10, figs 18, 19. Thorpe, 1844: 166. Leach, 1852: 188.

Littorina tenebrosa—Gould, 1841: 259–260; fig. 176* (not 166 as cited). de Kay, 1843: 105; pl. 6, fig. 106. MacGillivray, 1843: 138–139. Brown, 1844a: 128; pl. 10, figs 18, 19. Middendorff, 1849: 388–391; pl. 8, figs 6–12, 19–21 (in part, other figs are *L. obtusata*; *s.l.*). Forbes & Hanley, 1850: 39–42; pl. 84, figs 11, 12; pl. 85, figs 1–5. Sars, 1851: 180. Reeve, 1857: sp. 30; pl. 7, fig. 30. Sowerby, 1859: pl. 12, figs 16, 17. Gould, 1870: 306–307; fig. 576. Tryon, 1873–1874: 79–80; fig. 146. Locard, 1886: 284, 579, 580. Locard, 1891: 188. Teilman-Friis, 1899: 430–431; fig. 3 (radula). Fretter & Graham, 1980: 270–272, 275–276; fig. 214. Smith, 1982: 535–539. Backeljau, de Meyer, Janssens & Proesmans, 1984: 3. Rolán, 1984: 114–115; figs. Graham, 1988: 164, 181. Hayward, Wigham & Yonow, 1990: 662; fig. 12.9. Poppe & Goto, 1991: 94; pl. 11, fig. 23.

Litorina tenebrosa—Lovén, 1846: 154. Philippi, 1848: **3**: 67; *Litorina* pl. 7, figs 17, 20–22. Küster, 1856: 16–17; pl. 2, figs 18–20 (1853).

Medoria tenebrosa—Leach, 1852: 197.

Littorina rudis var. *tenebrosa*—Jeffreys, 1865: **3**: 365–366; 1869: **5**: pl. 65, fig. 3b. Sowerby, 1887: pl. 12, figs 16, 17. Marshall, 1898: 122. Brøgger, 1901: pl. 17, fig. 1a, b (Pleistocene). Dautzenberg & Durouchoux, 1906: 45. Harmer, 1921: 660–661; pl. 53, fig. 17 (in part, other fig. is *L. littorea*).

Littorina (Littorina) tenebrosa—Récluz, 1869: 45 (*s.l.*).
Littorina (Littorina) rudis var. *tenebrosa*—Tryon, 1887: 242; pl. 41, figs 1–4.
Littorina (saxatilis var.) *tenebrosa*—Kobelt, 1907: 61; pl. 110, figs 7, 8; pl. 112, fig. 22.
Littorina saxatilis tenebrosa—Dautzenberg & Fischer, 1912: 194–195; pl. 9, figs 13–17. Dautzenberg & Durouchoux, 1913: 22. Dautzenberg & Fischer, 1925: 55. Kaas & ten Broek, 1942: 144–145; pl. 11, fig. 11. James, in Fischer-Piette, Gaillard & James, 1963: 11–19; pl. 1, fig. 4; text-figs 5 (head), 6 (penis), 7 (radula) (*s.l.*). James, 1963: 31 (*s.l.*). James, in Fischer-Piette, Gaillard & James, 1964: 156–157; pl. 2, fig. 7; pl. 3, fig. 3; text-fig. 7 (head); text-fig. 11 (penis, perhaps *L. arcana*); text-fig. 14 (radula) (*s.l.*). James, 1968a: 156–160; pl. 1, fig. 6a, b; pl. 2, figs 1d–f, 3a–c, 4d–f, 6a–g, 7a–c (in part, other figs *L. arcana*). James, 1968b: 633–634; figs 13, 20; fig. 26 (head); figs 43–46 (*s.l.*). de Bruyne, 1991: 103; fig. 66.
Litorina rudis tenebrosa—Johnson, 1915: 121.
Littorina (Littorivaga) saxatilis tenebrosa—Winckworth, 1932: 221. Nordsieck, 1968: 41.
Littorina saxatilis var. *tenebrosa*—Thorson, 1941: 27–29. Thorson, 1946: 170–173; fig. 94 (embryo).
Littorina (Littorivaga) saxatile tenebrosa—La Roque, 1953: 168.
Littorina saxatilis form *tenebrosa*—Muus, 1967: 137–139; fig. 64 (animal, penis).
Littorina saxatilis tenebrosa var. *tenebrosa*—James, 1968a: 160; pl. 1, fig. 6a, b; pl. 2, fig. 6a–d (in part, other figs are probably *L. arcana*). Davis, 1973: 78.
Littorina (Littorivaga) saxatilis form *tenebrosa*—Nordsieck, 1982: 61; pl. 26, fig. 23.20d.
Littorina (Neritrema) saxatilis ?subsp. *tenebrosa*—Smith & Heppell, 1991: 17, 81.
Turbo jugosus Montagu, 1803: 586–588 (Dorset; St Ives, Cornwall [both England]; lectotype (here designated, 6.4 mm)+2 paralectotypes RAMM Montagu Colln unregistered, seen; possible paralectotype BMNH 1979185, seen). Maton & Rackett, 1807: 158; pl. 4, fig. 7. Montagu, 1808: 182; pl. 20, fig. 2. Pennant, 1812: 293 (*s.l.*). Pulteney, 1813: 49; pl. 19, fig. 1. Dillwyn, 1817: 820 (*s.l.*). Turton, 1819: 196–197. Brown, 1827: pl. 46, figs 15, 16 (*s.l.*). Blainville, 1828: 108. Blainville, 1830: 298. Brown, 1837a: 16; pl. 10, figs 15, 16 (*s.l.*).
Littorina jugosa—Brown, 1844a: 128; pl. 10, figs 15, 16 (*s.l.*). Thorpe, 1844: 258. Locard, 1886: 284, 579, 580. Locard, 1891: 189.
Litorina jugosa—Philippi, 1848: **3**: 67–68; *Litorina* pl. 7, fig. 23. Küster, 1856: 14–15; pl. 2, figs 12, 13 (1853).
Littorina rudis var. *jugosa*—Jeffreys, 1865: **3**: 365–366; 1869: **5**: pl. 65, fig. 3a.
Littorina (Littorina) jugosa—Récluz, 1869: 45 (*s.l.*).
Littorina (Littorina) rudis var. *jugosa*—Tryon, 1887: 241. Harmer, 1921: 656; pl. 53, fig. 5 (probably includes *L. littorea*).
Littorina rudis var. ex forma *jugosa*—Dautzenberg & Durouchoux, 1900: 45 (*s.l.*).
Littorina saxatilis jugosa—Dautzenberg & Fischer, 1912: 195; pl. 9, figs 18–27. Dautzenberg & Durouchoux, 1913: 22. Dautzenberg & Fischer, 1925: 54 (*s.l.*). James, 1963: 31 (*s.l.*). James, in Fischer-Piette, Gaillard & James, 1964: 151–155; pl. 2, figs 4–6; pl. 3, figs 1, 2; text-figs 6 (head); text-fig. 10 (penis); text-fig. 15 (radula). James, 1968a: 153–156; pl. 1, figs 2a–c, 3a, b, 4a, b, e, 6c, d, 8a, b (in part, other figs probably *L. arcana*). James, 1968b: 633–634; figs 12, 19; fig. 25 (head); figs 37–40.
Littorina (Littorivaga) saxatilis jugosa—Winckworth, 1932: 221.
Littorina saxatilis jugosa var. *jugosa*—James, 1968a: 155; pl. 1, figs 4a, b, e, 8a, b (*s.l.*).
Littorina (Littorivaga) saxatilis rudis form *jugosa*—Nordsieck, 1982: 61; pl. 26, fig. 23.21d, e.
Turbo diminutus Link, 1807: 134 (lectotype (here designated, from original citation) Chemnitz, 1781: **5**: pl. 185, fig. 1855a, b, from Faroe Islands).
Turbo albocinctus Link, 1807: 134 (lectotype (here designated, from original citation) Chemnitz, 1781: **5**: pl. 185, fig. 1855f, g, from Greenland).
Turbo litoreus Turton, 1819: 196 (in part, includes *L. littorea*; not *Turbo littoreus* Linnaeus, 1758=*L. littorea*, of which this was an unjustified emendation, p. 97).
Turbo obligatus Say, 1822: 241 (Portland, Maine; lectotype (here designated, 9.5 mm)+3 paralectotypes ANSP 18387, seen, 1 is *L. obtusata*).
Litorina obligata—Philippi, 1848: **3**: 67; *Litorina* pl. 7, fig. 19.
Littorina obligata—Reeve, 1858: sp. 98; pl. 17, fig. 98.
Littorina (Littorina) rudis var. *obligata*—Tryon, 1887: 241; pl. 41, fig. 94.
Littorina (Littorivaga) saxatilis obligata—Johnson, 1934: 102.
Turbo vestita Say, 1822: 241–242 (Maine; holotype ANSP 18403, seen).
Littorina vestita—Packard, 1867: 284.
Litorina groenlandica Menke, 1830: 45 (lectotype (here designated, from original citation) Chemnitz, 1781: **5**: pl. 185, fig. 1855a, b, from Faroe Islands; secondary homonym and subjective synonym of *Turbo groenlandicus* Röding, 1798). Møller, 1842: 82. Lovén, 1846: 154 (*s.l.*). Philippi, 1846a: **2**: 103; *Litorina* pl. 1, figs 11–13. Küster, 1856: 8–9; pl. 1, figs 11–13 (1853). Dall, 1886a: 203.
Littorina groenlandica (Menke)—Sars, 1851: 180. Mørch, 1857: 80. Reeve, 1857: sp. 52; pl. 10, fig. 52. Troschel, 1858: 133; pl. 10, fig. 20 (radula). Mørch, 1868: 78–79. Mørch, 1869: 207. Mørch, 1870: 14. Teilman-Friis, 1899: 430–431; fig. 1 (radula).

Littorina rudis var. *groenlandica* (Menke)—Sars, 1878: 165, 447; pl. 9, fig. 10a, b; anatomical pl. 6, fig. 4 (radula) (*s.l.*). Sparre-Schneider, 1886: 7–10 (*s.l.*). Posselt & Jensen, 1898: 231–232 (in part, includes *L. sitkana*). Norman, 1902: 355 (*s.l.*). Schlesch, 1916: 280. Harmer, 1921: 658; pl. 53, fig. 12 (in part, other fig. is *L. littorea*).

Littorina (saxatilis var.) *groenlandica* (Menke)—Kobelt, 1907: 62; pl. 110, figs 19, 20.

Littorina saxatilis groenlandica (Menke)—Dautzenberg & Fischer, 1912: 198; pl. 10, figs 21–28 (*s.l.*). James, 1968a: 161; pl. 2, fig. 8a–d. James, 1968b: 633–634; fig. 16; fig. 24 (head); fig. 42. Raffaelli, 1979a: 219–232; figs 8, 9D (radula).

Littorina (Littorivaga) groenlandica (Menke)—Oldroyd, 1927: 61 (in part, includes *L. sitkana*).

Littorina (Littorivaga) saxatilis groenlandica (Menke)—Winckworth, 1932: 221.

Littorina saxatilis var. *groenlandica* (Menke)—Thorson, 1941: 27. Laursen, 1944: 68–69 (Pleistocene). Rendall, 1956: 157.

Littorina (Littorivaga) saxatilis var. *groenlandica* (Menke)—Nordsieck, 1968: 40.

Littorina rudis groenlandica (Menke)—Wium-Andersen, 1970: 267–273. Backeljau, de Meyer, Janssens & Proesmans, 1984: 3 (*s.l.*).

Littorina (Littorivaga) saxatilis rudis form *groenlandica* (Menke)—Nordsieck, 1982: 61; pl. 26, fig. 23.21a, b.

Littorina (Neritrema) saxatilis var. *groenlandica* (Menke)—Smith & Heppell, 1991: 17, 81.

Litorina sulcata Menke, 1830: 45 (lectotype (here designated, from original citation) Chemnitz, 1781: **5**: pl. 185, fig. 1855f, g, from Greenland). Anton, 1838: 52.

? *Turbo ventricosus* Brown, 1837a: 16 (Clew Bay, Co. Mayo, Ireland; types not traced; not Woodward, 1833= *L. littorea*; *s.l.*).

Littorina ventricosa—Brown, 1844a: 128.

Littorina nigrolineata Gray, 1839: 140 (no loc.; type locality here designated: Porto, Portugal; lectotype (here designated, 14.7 mm, Fig. 106P)+17 paralectotypes BMNH 1968363, seen). Rolán, 1984: 112–113; figs. Burnay, 1986: 20–21.

Littorina saxatilis nigrolineata—Dautzenberg & Fischer, 1912: 196–197 (in part, includes *L. compressa*).

Littorina saxatilis var. *nigrolineata*—Fischer-Piette, Gaillard & Delmas, 1967: 891–896 (in part, includes *L. compressa*).

Littorina saxatilis var. ex col. *nigrolineata*—Fischer-Piette & Gaillard, 1971: 44–45 (in part, includes *L. compressa*).

Litorina marmorata L. Pfeiffer, 1839: 82 (Travemünde [near Lübeck, Germany]; types not traced).

Turbo rudissimus Johnston, 1842: 268 (in synonymy of *L. rudis*, without authority, but presumably a manuscript name from Bean, who subsequently published the name as new (Bean, 1844), thereby making it available (ICZN, 1985: art. 11e); authorship must be credited to Johnston (ICZN, 1985: art. 50g); type locality (here designated) Scarborough, Yorkshire, England; lectotype (here designated, 11.5 mm; material from Bean; ICZN, 1985: art. 72b(iii), recommendation 72B)+8 paralectotypes SMNH 3884, seen, some of the smaller specimens could be *L. arcana*; paralectotypes SM (Seshappa, 1976), not seen).

Littorina rudissima—Bean, 1844: 266–267 (although published as a new name, authorship is credited to Johnston, 1842, see above). Alder, 1848: 149–150 (*s.l.*).

Littorina saxatilis rudis var. *rudissima*—Dautzenberg & Fischer, 1912: 198; pl. 10, figs 4–20 (*s.l.*). Kaas & ten Broek, 1942: 145–146; pl. 11, figs 8, 9. James, 1968a: 152; pl. 1, fig. 5a–d; pl. 2, figs 2a–h, 3d–f (in part, other figs are *L. compressa*).

Littorina rudis var. *rudissima*—Harmer, 1921: 657–658; 1925: 874; pl. 53, fig. 10; pl. 65, fig. 30 (in part, other figs are *L. littorea*).

Littorina saxatilis var. *rudissima*—Lecointre, 1950: 241; fig. 17 (Pleistocene). Lecointre, 1952: 98; pl. 27, fig. 4 (Pleistocene). Lecointre, 1963: 59 (Pleistocene).

Littorina saxatilis rudissima—Seshappa, 1976: 222–229 (in part, includes *L. arcana*).

Littorina saxatilis Johnston, 1842: 268 (Berwickshire [Scotland]; types not traced; secondary homonym and subjective synonym of *Turbo saxatilis* Olivi, 1792; the authorship of this name is sometimes credited to Bean, who is cited at the end of the description). MacGillivray, 1843: 138. Brown, 1844a: 128. Thorpe, 1844: 258–259. Forbes & Hanley, 1850: 43–45; pl. 86, figs 4, 5. Reeve, 1857: sp. 91; pl. 16, fig. 91. Sowerby, 1859: pl. 12, fig. 19. Petit de la Saussaye, 1869: 122. Locard, 1886: 284–285, 579, 580. Kobelt, 1888: 169 (*s.l.*). Locard, 1891: 189.

Litorina saxatilis (Johnston)—Lovén, 1846: 154. Philippi, 1848: **3**: 66; *Litorina* pl. 7, fig. 16.

Littorina rudis var. *saxatilis* Johnston—Jeffreys, 1865: **3**: 365–366. Sowerby, 1887: pl. 12, fig. 19. Nobre, 1898: 79. Marshall, 1898: 121. Nobre, 1931: 146. Nobre, 1936: 102; pl. 82, fig. 12.

Littorina (Littorina) rudis var. *saxatilis* Johnston—Tryon, 1887: 242; pl. 41, fig. 97.

Littorina saxatilis (Olivi) *saxatilis* Johnston—James, 1963: 31. James, in Fischer-Piette, Gaillard & James, 1964: 157; pl. 2, fig. 9; pl. 3, fig. 4; text-figs 8 (head), 12 (penis), 13 (radula) (*s.l.*). Seshappa, 1976: 222–229; fig. 17 (radula).

Littorina castanea Deshayes, in Deshayes & Milne Edwards, 1843: 206–207 (replacement name for *Litorina groenlandica* Menke, 1830, restoring the non-binominal name of Schröter, 1779). Deshayes, 1850: 42; pl. 69, figs 8, 9.

Littorina zonaria Bean, 1844: 266 (Robin Hood Bay [Yorkshire, England]; lectotype (here designated, 15.1 mm)+14 paralectotypes BMNH 1849.2.2.48–53 and 1850.10.7.145–152, from Scarborough, Yorkshire, seen; 7 paralectotypes SMNH 3883, from Scarborough, seen).

Littorina saxatilis var. ex col. *zonaria*—Dautzenberg & Fischer, 1912: 199–200; pl. 10, figs 17, 24–26 (*s.l.*). Dautzenberg & Fischer, 1925: 54 (*s.l.*). Fischer-Piette & Gaillard, 1971: 43 (*s.l.*).

Littorina neglecta Bean, 1844: 266 (north shore of Scarborough [Yorkshire, England]; lectotype (here designated, 3.3 mm, Fig. 106G)+4 paralectotypes BMNH 1849.2.2.301–305, seen; 10 possible paralectotypes BMNH 1850.10.7.169–178, seen; 8 possible paralectotypes SMNH 3882, seen). Alder, 1848: 150. Kobelt, 1907: 66–67; pl. 111, figs 16–19; pl. 12, fig. 20. Heller, 1975a: 144–149; fig. 3D (penis); fig. 4D (radula); pl. 1, figs 16, 17. Raffaelli, 1979a: 219–232; fig. 5 (penis); figs 8, 10C, D (radula). Sneli & van Marion, 1979: 4–8; fig. 1h; fig. 1k (penis). Fretter & Graham, 1980: 268–270; fig. 209. Wilkins & O'Regan, 1980: 355–359. Smith, 1982: 535–539. Hannaford Ellis, 1984: 343–347; fig. 1. Jardine, 1984: 5. Fish & Sharp, 1985: 143–156; figs 10.1, 10.2A; fig. 10.2C (embryonic shell). Knight & Ward, 1986: 405–413. Graham, 1988: 176; fig. 64. Warmoes, Backeljau & de Bruyn, 1988: 51–70; fig. 4D–G (penis); fig. 5A, B. Warmoes, 1989: 125–128. Hayward, Wigham & Yonow, 1990: 660; fig. 12.9. Poppe & Goto, 1991: 94; pl. 11, fig. 14. Dytham, Mill, Grahame & O'Higgins, 1992: 127–133. Grahame, Mill, Double & Hull, 1992: 499–502. Caley, 1994 (penis, oviduct, embryonic shell, radula). Caley, Grahame & Mill, 1995: 181–193. Crossland, Coates, Grahame & Mill, 1995: 205–209. Grahame, Mill, Hull & Caley, 1995: 887–899; fig. 3B.

Littorina saxatilis neglecta—James, 1964: 26. James, 1968a: 161–162; pl. 1, fig. 6e, f. James, 1968b: 633–634; fig. 14; fig. 27 (head); figs 49, 50. Davis, 1973: 78.

Littorina (Neritrema) neglecta—Reid, 1989a: 102. Reid, 1990b: 1–19; fig. 1t; fig. 4i (radula). Smith & Heppell, 1991: 17, 81.

? *Littorina patula* 'Jeffreys' Thorpe, 1844: 259; fig. 7 (Eddystone Rock [Cornwall, England]; types not traced; *nomen dubium*; probably included *L. arcana*, see Taxonomic History, p. 288).

Littorina patula—Forbes & Hanley, 1850: 36–39; pl. 85, figs 6–10; pl. GG, fig. 2 (animal) (*s.l.*). Sowerby, 1859: pl. 12, fig. 18 (*s.l.*). Locard, 1886: 283–284, 579, 580 (*s.l.*). Locard, 1891: 188–189 (*s.l.*). Heller, 1975a: 138–140, 146–149; fig. 3B (penis); figs 4B, 5B (radula); pl. 1, figs 5–8 (*s.l.*).

Littorina rudis var. *patula*—Jeffreys, 1865: **3**: 365–366 (*s.l.*). Monterosato, 1872: 33. Sowerby, 1887: pl. 12, fig. 18 (*s.l.*). Harmer, 1921: 660 (*s.l.*). Nobre, 1931: pl. 33, fig. 9.

Littorina (Littorina) patula—Récluz, 1869: 45 (*s.l.*).

Littorina (Littorina) rudis var. *patula*—Tryon, 1887: 242; pl. 41, figs 98–100 (*s.l.*).

Littorina (saxatilis var.) *patula*—Kobelt, 1907: 59–61; pl. 110, figs 23–26; pl. 112, fig. 23 (*s.l.*).

Littorina (Littorivaga) saxatilis patula—Winckworth, 1932: 221 (*s.l.*). Nordsieck, 1968: 41 (*s.l.*).

Littorina saxatilis patula—Graham, 1988: 178.

Litorina incarnata 'Lovén' Philippi, 1846a: **2**: 103 (Greenland; in synonymy of *Litorina groenlandica* Menke, therefore unavailable).

Litorina arctica—Philippi, 1848: **3**: 68; *Litorina* pl. 7, fig. 26 (in part, other figs are *L. obtusata*; not Møller, 1842=*L. obtusata*).

Littorina arctica—Reeve, 1857: sp. 69 (in part, fig. is *L. sitkana*; not Møller, 1842=*L. obtusata*).

Nerita rustica 'Chiereghini' Nardo, 1847: 97–98 (Quarnero [Gulf of Venice, Italy]; in synonymy of *Littorina saxoides*, therefore unavailable).

Littorina saxoides 'Chiereghini' Nardo, 1847: 97–98 (Quarnero [Gulf of Venice, Italy]; refers to Chiereghini MS sp. 11, figs 1165, 1166; not seen).

Littorina rudis forma *elatior* Middendorff, 1849: 383–384 (infrasubspecific).

Littorina tenebrosa forma *elatior* Middendorff, 1849: 388–389 (infrasubspecific; not *L. rudis* forma *elatior* Middendorff, 1849).

Littorina tenebrosa [var.] *costulata* Middendorff, 1849: 389–390; pl. 8, figs 6–8 (infrasubspecific).

Littorina tenebrosa [var. ex col.] *rubidus* Middendorff, 1849: 390 (infrasubspecific).

Littorina tenebrosa [var. ex col.] *griseolacteus* Middendorff, 1849: 390 (infrasubspecific).

Littorina tenebrosa [var. ex col.] *tessellatus* Middendorff, 1849: 390 (infrasubspecific; not *Litorina tessellata* Philippi, 1847=*Littoraria tessellata*).

Littorina tenebrosa [var. ex col.] *zonatus* Middendorff, 1849: 390 (infrasubspecific; not *Litorina obtusata* var. *zonata* Menke, 1845=*L. obtusata*).

Littorina tenebrosa [var.] *densecostulata* Middendorff, 1849: 390 (infrasubspecific).

Litorina tenebrosa var. *intermedia* Forbes & Hanley, 1850: 41; pl. 83, fig. 4 (types not traced; not *L. intermedia* Philippi, 1846=*Littoraria intermedia*).

? *Littorina vulgaris* var. *albocincta* Maitland, 1854: 82 (Netherlands; *nomen nudum*).

Littorina simplex Reeve, 1857: sp. 63; pl. 12, fig. 63a, b (St Vincent, Cape de Verds [locality in error; probably from an Arctic locality]; lectotype (here designated, 15.5 mm)+3 paralectotypes (1 is *L. obtusata* geog. var. *palliata*), BMNH 1968313, seen).

Litorina simplex—Weinkauff, 1882: 83; pl. 11, figs 6, 7. Weinkauff, 1883: 215.

Littorina rudis var. *globosa* Jeffreys, 1865: **3**: 365–366 (Dublin Bay [Ireland]; Oban [Scotland]; 1 syntype USNM 337203, seen; not *Litorina globosa* Dunker, 1845=*L. striata*. Harmer, 1921: 656–657; pl. 53, fig. 6 (in part, other figs are *L. littorea*).

Littorina saxatilis rudis var. *globosa*—Dautzenberg & Fischer, 1912: 197; pl. 10, fig. 3.

Littorina rudis var. *similis* Jeffreys, 1865: **3**: 365–366 (Cornwall; South Wales; Aberdeenshire; Shetland [all British Isles]; types not traced). Harmer, 1921: 659; pl. 53, fig. 20.

Littorina saxatilis tenebrosa var. *similis*—Dautzenberg & Fischer, 1912: 194; pl. 9, figs 16, 17. James, 1968a: 160; pl. 2, fig. 1d–f, 3a–c, 4d–f, 6e–g (*s.l.*). Davis, 1973: 78.

Littorina rudis var. *laevis* Jeffreys, 1865: **3**: 365–366 (Sark [Channel Islands, British Isles]; Shetland [Scotland]; types not traced). Harmer, 1921: 655–656; pl. 53, fig. 3 (in part, other fig. is *L. littorea*).

Littorina rudis var. ex forma *laevis*—Dautzenberg & Durouchoux, 1900: 45 (*s.l.*).

Littorina saxatilis (Olivi) var. *adriatica* Brusina, 1870: 190–193, 236, 237 (Chioggia and Venice Lagoon [Italy]; types not traced; not in Jeffreys Collection, USNM).

Littorina rudis var. *rubescens* Monterosato, 1878: 83 (Venice Lagoon, Italy; *nomen nudum*).

Littorina saxatilis var. *rubescens*—Fischer-Piette & Gaillard, 1968: 103–118.

Littorina saxatilis Johnston var. *zonata* Daniel, 1883: 342 (Baie du Toulinguet [near Brest, France]; infrasubspecific; not *Litorina obtusata* var. *zonata* Menke, 1845=*L. obtusata*).

Littorina rudis var. *finmarchica* Herzenstein, 1885: 680 (Hammerfest and Vardø [Norway]; types not traced; *s.l.*).

Littorina danieli Locard, 1886: 283, 578–580 (Brest, Finistère [France] (here restricted); lectotype (here designated, 17.9 mm)+1 paralectotype MNHNP, seen). Locard, 1891: 188.

Littorina rudis var. ex forma *brevis* 'Jeffreys' Dautzenberg, 1887: 18 (error for *laevis* Jeffreys, see Dautzenberg & Durouchoux, 1900: 45; unavailable; *s.l.*).

Littorina rudis var. ex colore *albida* Dautzenberg, 1887: 18 (St-Lunaire (Ille-et-Vilaine) [France]; 10 original specimens IRSNB, seen; infrasubspecific; not *Litorina tenebrosa* var. *albida* Schrenck, 1867=*L. sitkana*). Dautzenberg & Durouchoux, 1900: 46 (*s.l.*).

Littorina saxatilis var. ex col. *albida*—Dautzenberg & Fischer, 1912: 199; pl. 10, figs 14, 23 (*s.l.*). Dautzenberg & Fischer, 1925: 53 (*s.l.*). Fischer-Piette & Gaillard, 1971: 40 (*s.l.*).

Littorina rudis var. ex colore *aurantia* Dautzenberg, 1887: 18 (St-Lunaire (Ille-et-Vilaine) [France]; infrasubspecific; *s.l.*). Dautzenberg & Durouchoux, 1900: 46 (*s.l.*).

Littorina saxatilis var. ex col. *aurantia*—Dautzenberg & Fischer, 1912: 199; pl. 10, figs 9–11 (*s.l.*). Dautzenberg & Fischer, 1925: 53 (*s.l.*). Fischer-Piette & Gaillard, 1971: 41 (*s.l.*).

Littorina rudis var. ex colore *fasciata* Dautzenberg, 1887: 18 (St-Lunaire (Ille-et-Vilaine) [France]; infrasubspecific; not *L. fasciata* Gray, 1839=*Littoraria fasciata*; *s.l.*). Dautzenberg & Durouchoux, 1900: 46 (*s.l.*).

Littorina rudis var. ex colore *tessellata* Dautzenberg, 1893a: 20 (Granville and St-Pair [France]; infrasubspecific; not *Litorina tessellata* Philippi, 1847=*Littoraria tessellata*; not *L. tenebrosa* var. ex col. *tessellatus* Middendorff, 1849=*L. saxatilis*; *s.l.*).

Littorina saxatilis var. ex col. *tessellata*—Dautzenberg & Fischer, 1912: 200; pl. 10, figs 18–20, 27, 28 (*s.l.*). Fischer-Piette & Gaillard, 1971: 43 (*s.l.*).

Littorina saxatilis tenebrosa var. *tessellata*—Kaas & ten Broek, 1942: 145.

Littorina nervillei Dautzenberg, 1893b: 35–36; pl. 1, figs 4, 4a, 4b (south of Sidi-Jamur, west coast of Djerba Island, Gulf of Gabès [Tunisia] (here restricted); 2 syntypes IRSNB, unregistered, seen). Seurat, 1929: 15; fig. 3, no. 7.

Littorina saxatilis forma typica [subsp. *saxatilis*] var. *nervillei*—Dautzenberg & Fischer, 1912: 193; pl. 9, figs 11, 12.

Littorina saxatilis var. *nervillei*—Ghisotti, 1972: 75; pl. 2, fig. 1.

Littorina (Littorivaga) saxatilis form *nervillei*—Nordsieck, 1982: 61; pl. 26, fig. 23.20b.

Littorina (Melaraphe) nervillei—Sabelli, Giannuzzi-Savelli & Bedulli, 1990: **1**: 19, 146; 1992: **2**: 368.

Littorina nervillei var. ex colore *livida* Dautzenberg, 1893b: 36; pl. 1, figs 5, 5a, 5b (south of Sidi-Jamur, west coast of Djerba Island, Gulf of Gabès [Tunisia] (here restricted); 2 syntypes (of *L. nervillei* and var. *livida*) IRSNB, unregistered, seen).

Littorina rudis var. *tenebrosapallida* L. E. Adams, 1896: 315 (creek in Brancaster Bay, near Hunstanton [Norfolk, England]; types not traced; as *tenebrosa-pallida*).

Littorina rudis var. ex. colore *lutea* Dautzenberg & Durouchoux, 1900: 46 (St Malo [France]; in part, includes *L. compressa*; infrasubspecific; *s.l.*; not *Litorina scabra* var. *lutea* Philippi, 1847=*Littoraria lutea*).

Littorina saxatilis var. ex col. *lutea*—Dautzenberg & Fischer, 1912: 199; pl. 10, fig. 13 (*s.l.*). Dautzenberg & Fischer, 1925: 53 (*s.l.*). Fischer-Piette & Gaillard, 1971: 40–41 (*s.l.*).

Littorina rudis var. ex. colore *sanguinea* Dautzenberg & Durouchoux, 1900: 46 (St Malo [France]; infrasubspecific; *s.l.*).

Littorina saxatilis var. ex col. *sanguinea*—Dautzenberg & Fischer, 1912: 199; pl. 9, fig. 20; pl. 10, fig. 7 (*s.l.*).

Littorina rudis var. *sulcata* monstr. *scalariformis* Martel, 1901: 111, 221; pl. 1, fig. 5 (Port-Mer, near Cancale [France]; original of fig. IRSNB, seen; monstrosity, therefore unavailable).

Littorina rudis var. *globosa* monstr. *elongata* Martel, 1901: 111, 222; pl. 1, figs 8, 9 (Cancale [France]; originals of figs IRSNB, seen; monstrosity, therefore unavailable; not *Turbo elongatus* Woodward, 1833=*L. littorea*).

Littorina rudis var. *globosa* monstr. *contabulata* Martel, 1901: 221; pl. 1, fig. 4 (Port-Mer [near Cancale, France]; original of fig. IRSNB, seen; monstrosity, therefore unavailable).

Littorina rudis var. *globosa* monstr. *scalariformis* Martel, 1901: 221; pl. 1, fig. 6 (Port-Mer [near Cancale, France]; original of fig. IRSNB, seen; monstrosity, therefore unavailable).

Littorina saxatilis forma typica [subsp. *saxatilis*] var. *lugubris* 'Recluz' Dautzenberg & Fischer, 1912: 193; pl. 9, figs 7–10 (Morbihan; l'embouchure de l'Escaut; Croisic; Pouliguen [all France]; originals of figs MNHNP, seen; infrasubspecific).

Littorina (Littorivaga) saxatilis form *lugubris*—Nordsieck, 1982: 61; pl. 26, fig. 23.20c.

Littorina saxatilis tenebrosa var. *elata* Dautzenberg & Fischer, 1912: 194; pl. 9, fig. 15 (Baie de St Malo [France]; original of fig. MNHNP, seen; infrasubspecific). Kaas & ten Broek, 1942: 145; pl. 11, fig. 12. James, 1968a: 160; pl. 2, fig. 7a–c.

Littorina rudis var. *tenebrosa elata*—Harmer, 1921: 661; pl. 53, fig. 19.

Littorina saxatilis jugosa var. *bynei* Dautzenberg & Fischer, 1912: 195; pl. 9, figs 22, 23, 26, 27 (Milford Haven and Bullslaughter Bay [Wales]; Morbihan [France]; Galicia; Grand-Vidé (Saint-Enogat) [France]; originals of figs MNHNP, seen; infrasubspecific).

Littorina saxatilis var. *bynei*—Fischer-Piette & Gaillard, 1971: 43–44 (*s.l.*).

Littorina saxatilis rudis var. *major* 'Kobelt' Dautzenberg & Fischer, 1912: 197 (Kobelt, 1888: 167, used 'major' only as an adjective, not a varietal name; in synonymy of *L. saxatilis rudis* var. *globosa*, therefore unavailable).

Littorina saxatilis forma *minor* Dautzenberg & Fischer, 1912: 199 (*nomen nudum*; *s.l.*).

Littorina saxatilis forma *abbreviata* Dautzenberg & Fischer, 1912: 199 (*nomen nudum*; *s.l.*).

Littorina saxatilis forma *conoidea* Dautzenberg & Fischer, 1912: 199 (*nomen nudum*; *s.l.*).

Littorina saxatilis forma *elongata* Dautzenberg & Fischer, 1912: 199 (*nomen nudum*; *s.l.*).

Littorina saxatilis var. ex col. *fusca* Dautzenberg & Fischer, 1912: 195, 198, 199; pl. 9, fig. 19; pl. 10, figs 5, 6, 21 (Embouchure de la Rance; La Trinité-sur-Mer [both France]; Vardø [Norway]; originals of figs MNHNP, seen; infrasubspecific; not *L. kurila* var. ex col. *fuscus* Middendorff, 1851=*L. sitkana*; *s.l.*). Dautzenberg & Fischer, 1925: 53–55 (*s.l.*). Fischer-Piette & Gaillard, 1971: 42 (*s.l.*).

Littorina saxatilis var. ex col. *miniata* Dautzenberg & Fischer, 1912: 195, 197–199; pl. 10, figs 8, 22 (La Trinité-sur-Mer [France]; north Iceland; originals of figs MNHNP, seen; infrasubspecific; in part, includes *L. compressa*; not *L. littorea* var. ex col. *miniata* Dautzenberg & Durouchoux, 1900=*L. littorea*; *s.l.*). Dautzenberg & Fischer, 1925: 54 (*s.l.*).

Littorina saxatilis var. ex col. *fulva* 'Monterosato' Dautzenberg & Fischer, 1912: 193, 198, 199; pl. 10, fig. 12 (Adriatic; St Pair [France]; original of fig. MNHNP, seen; infrasubspecific; *s.l.*). Fischer-Piette & Gaillard, 1971: 42 (*s.l.*).

Littorina saxatilis var. ex col. *trifasciata* Dautzenberg & Fischer, 1912: 198, 200; pl. 10, figs 15, 16 (Le Val André [France]; original of figs MNHNP, seen; infrasubspecific; *s.l.*). Fischer-Piette & Gaillard, 1971: 43 (*s.l.*).

Littorina saxatilis var. ex col. *flammulata* Dautzenberg & Fischer, 1912: 195, 200 (no loc.; infrasubspecific; *s.l.*). Fischer-Piette & Gaillard, 1961: 168–169; fig. 1.

Littorina saxatilis var. ex col. *lineata* Dautzenberg & Fischer, 1912: 195, 200; pl. 9, fig. 21 (La Corogne [La Coruña, Spain]; original of fig. MNHNP, seen; infrasubspecific; not *Litorina angulifera* var. *lineata* Philippi, 1847=*Littoraria angulifera*).

Littorina saxatilis rudis monstr. *sinistrorsum* Dautzenberg & Fischer, 1912: 200 (refers to Sykes, 1892: fig. 1, from Langton Herring, Fleet Backwater [Dorset, England]; monstrosity, therefore unavailable; not *L. litorea* var. *sinistrorsa* Jeffreys, 1865=*L. littorea*).

Littorina saxatilis rudis monstr. *contortum* Dautzenberg & Fischer, 1912: 200 (refers to Sykes, 1892: figs 2, 3, from Langton Herring, Fleet Backwater [Dorset, England]; 2 originals of figs BMNH 1990024/2 and 3, seen; monstrosity, therefore unavailable).

Littorina rudis var. *conoidea* Schlesch, 1916: 280 (Isafjördur, Iceland; types may be in Hull Museum, England).

Littorina palliata var. *turritella* Schlesch, 1916: 279 (Isafjördur, Iceland; syntype in TM, seen; further types may be in Hull Museum).

Littorina rudis var. *alticola* Dacie, 1917: 180 (Biggin Brough, Orkney [Scotland]; 15 syntypes BMNH 1915.7.16.1–15, seen).

Littorina neritoides—Harmer, 1921: 662–663; pl. 53, fig. 21 (in part, includes refs to *Melarhaphe neritoides*; not *Turbo neritoides* Linnaeus, 1758=*Melarhaphe neritoides*; Pleistocene).

Littorina nervillei var. *major* Pallary, in Seurat, 1924: 35 (Gightis, Golfe de Gabès [Tunisia]; infrasubspecific; not *L. littorea* var. ex f. *major* Dautzenberg & Durouchoux, 1913=*L. littorea*).

Littorina nervillei var. *major* var. ex col. *nigra* Pallary, in Seurat, 1924: 35 (Gightis, Golfe de Gabès [Tunisia]; infrasubspecific).

Littorina nervillei var. *major* var. ex col. *rufa* Pallary, in Seurat, 1924: 35 (Gightis, Golfe de Gabès [Tunisia]; infrasubspecific; not *Littorina rufa* 'Recluz' Reeve, 1857).

Littorina saxatilis groenlandica var. *sculpta* Schlesch, 1931: 134 (Húnaflói, north Iceland; fig. is probably *L. obtusata*; infrasubspecific).
Littorina saxatilis var. *sanguinea* Coen, 1933: 30–31 ([west coast of Adriatic]; *nomen nudum*; Mienis, 1975).
Littorina (Littorivaga) saxatile saxatile La Roque, 1953: 168 (unjustified emendation of *L. saxatilis*; in part, includes *L. sitkana*).
Littorina saxatilis var. *interrupta* Fischer-Piette, Gaillard & Jouin, 1961: 321; figs 1, 2 (Belle-Île (Castoul Plage) [France]; Santander [Spain]; originals of figs MNHNP, seen; infrasubspecific; not *Litorina ziczac* var. *interrupta* C. B. Adams, in Philippi, 1847=*Nodilittorina interrupta*).
Littorina saxatilis var. *tractibus* Fischer-Piette, Gaillard & Jouin, 1961: 321; fig. 3 (Barrica [Spain]; original of fig. MNHNP, seen; infrasubspecific).
Littorina saxatilis var. *hieroglyphica* Fischer-Piette, Gaillard & Jouin, 1961: 321; fig. 4 (Barrica [Spain]; original of fig. MNHNP, seen; infrasubspecific).
Littorina punctata—Barnard, 1963: 190 (not *Turbo punctatus* Gmelin, 1791=*Nodilittorina punctata*). Day, 1969: 162; text fig. (not Gmelin, 1791).
Littorina saxatilis tenebrosa var. ex col. *biinterrupta* Fischer-Piette & Gaillard, in Fischer-Piette, Gaillard & James, 1963: 7; pl. 1, fig. 1 (Castoul-Plage, Belle-Ile, France; original of fig. MNHNP, seen; infrasubspecific; as *bi-interrupta*).
Littorina saxatilis tenebrosa var. ex col. *maculata* Fischer-Piette & Gaillard, in Fischer-Piette, Gaillard & James, 1963: 7; pl. 1, fig. 2 (Castoul-Plage, Belle-Ile, France; original of fig. MNHNP, seen; infrasubspecific).
Littorina saxatilis tenebrosa var. ex col. *bizonaria* James, in Fischer-Piette, Gaillard & James, 1963: 14; pl. 1, fig. 4 (Rubna na Faing, Islay, Scotland; infrasubspecific).
Littorina saxatilis var. *bizonaria*—Fischer-Piette & Gaillard, 1971: 43 (*s.l.*).
Littorina saxatilis var. *nojensis* Fischer-Piette & Gaillard, in Fischer-Piette, Gaillard & James, 1964: 135; pl. 1, figs 1, 2 (Noja, province de Santander, Spain; original of fig. 1 MNHNP, seen; infrasubspecific). Fischer-Piette & Gaillard, 1971: 42 (*s.l.*).
Littorina saxatilis var. *sellensis* Fischer-Piette & Gaillard, in Fischer-Piette, Gaillard & James, 1964: 135; pl. 1, fig. 3 (Ribadesella (Asturias), Spain; original of fig. 3 MNHNP, seen; infrasubspecific).
Littorina saxatilis var. *rubrolineata* Fischer-Piette, Gaillard & Delmas, 1967: 874 (Bayona, Ría de Vigo, Spain; original specimen MNHNP, seen; infrasubspecific). Fischer-Piette & Gaillard, 1971: 44 (*s.l.*).
Littorina saxatilis var. *salvati* Fischer-Piette, Gaillard & Delmas, 1967: 874–875 (Île aux Oiseaux, bassin d'Arcachon, France; original specimen MNHNP, seen; infrasubspecific). Fischer-Piette & Gaillard, 1971: 43 (*s.l.*).
Littorina saxatilis nigrolineata var. *compressa*—Dautzenberg & Fischer, 1912: 197 (in part, includes *L. compressa*; not *L. rudis* var. *compressa* Jeffreys, 1865).
Littorina saxatilis var. *compressa*—Fischer-Piette, Gaillard & Delmas, 1967: 891–896 (in part, includes *L. compressa*; not *L. rudis* var. *compressa* Jeffreys, 1865).
Littorina saxatilis rudis var. *jugosoides* James, 1968a: 153; pl. 2, figs 4a–c, 5e, f (Twr Gwylanod, Aberystwyth; The Wick, Skomer I. [both Wales]; infrasubspecific; in part, other figs are *L. compressa*).
Littorina saxatilis rudis var. *rudissimoides* James, 1968a: 155; pl. 1, figs 2a–c, 3a, b (tidal inlet near Woodbridge [Suffolk, England]; mouth of River Leri, Ynyslas Estuary, near Aberystwyth [Wales]; infrasubspecific).
Littorina saxatilis jugosa var. *tenuis* James, 1968a: 155–156; pl. 1, fig. 6c, d (Thorny pit, Dale Peninsula, Wales; infrasubspecific; not *L. tenuis* Philippi, 1846=*Littoraria* sp.).
Littorina saxatilis var. *rubra* 'Philippi' Fischer-Piette & Gaillard, 1971: 41–42 (Philippi, 1846a, used 'rubra' as an adjective, not as a new varietal name; infrasubspecific; *s.l.*).
Littorina saxatilis var. *cinerea* Fischer-Piette & Gaillard, 1971: 42–43 (infrasubspecific; in part; original specimen is *L. compressa*; *s.l.*).
Littorina saxatilis var. *gascae* Fischer-Piette & Gaillard, 1971: 44 (Santa Maria del Mar [Spain]; infrasubspecific).
Littorina saxatilis var. *clarilineata* Fischer-Piette & Gaillard, 1971: 44 (Ria del Barquero, Xilloy Nord-Est [Spain]; original specimen MNHNP, seen; infrasubspecific).
Littorina saxatilis var. *nigra* Fischer-Piette & Gaillard, 1971: 72 (infrasubspecific; not *L. nervillei* var. *major* var. ex col. *nigra* Pallary, in Seurat, 1924=*L. saxatilis*).
Littorina (Littoraria) undulata (?)—Nordsieck & Garcia-Talavera, 1979: 57; pl. 11, fig. 8 (not *Littorina undulata* Gray, 1839=*Littoraria undulata*).
Littoraria undulata—Nordsieck, 1982: 62; pl. 26, fig. 23.45 (not Gray, 1839).
Littorina rudis var. *scotia* S. M. Smith, 1979a: 138–139 (Rockall; infrasubspecific).
Littorina [*neglecta* ecomorph] *scotia*—S. M. Smith, in Fretter & Graham, 1980: 270.
Littorina saxatilis scotia Graham, 1988: 178 (*nomen nudum* as a subspecific taxon).
Littorina (Neritrema) neglecta scotia—Smith & Heppell, 1991: 81.
Littorina saxatilis var. *lagunae* Barnes, 1993: 490–491 (infrasubspecific).

Note: In the synonymy, '*s.l.*' (*sensu lato*) following a reference indicates that an author probably included under a given name not only *L. saxatilis*, but also *L. arcana* and in some cases *L. compressa*,

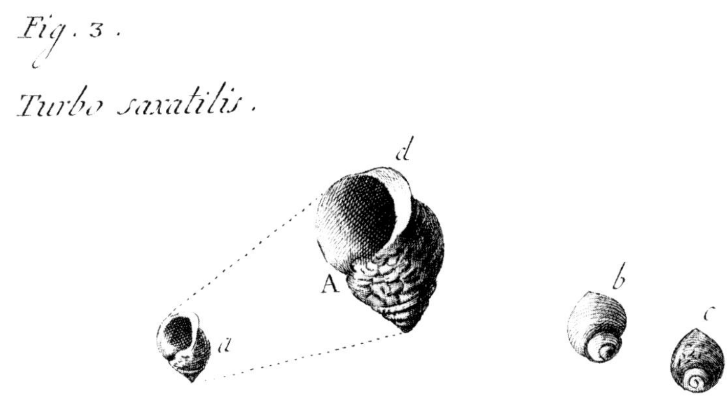

Figure 101. *Littorina (Neritrema) saxatilis*: lectotype (**A**) of *Turbo saxatilis* Olivi, 1792; Venice Lagoon, Italy (Olivi, 1792: pl. 5, fig. 3; actual length of fig. **a**=9 mm).

but without clear reference to, or figures of, these latter species. This may be so, for example, in particular habitats or localities at which more than one species is known to occur. In the text, '*L. saxatilis s.l.*' is used where it is not clear whether the other two species may have been included. The '*L. saxatilis* group' refers to all three species.

Taxonomic history

The synonymy of *L. saxatilis* must be among the most complex of any species of mollusc, including 28 specific names, 31 names proposed as varieties, forms or monstrosities (some of which are available if considered of subspecific rank), and 35 named colour varieties. This extensive synonymy is a reflection of the extreme intraspecific variability of this species, which is in turn a consequence of its nonplanktotrophic development, wide habitat range and large geographical distribution. The taxonomic history is further complicated by the existence of the conchologically almost identical species *L. arcana*, which was only clearly distinguished, on anatomical grounds, by Hannaford Ellis in 1978. Almost all those earlier authors who worked in areas of sympatry between these two species failed to discriminate between them. In addition, another closely related species, *L. compressa* (=*L. nigrolineata* of authors, not Gray, 1839), has frequently been classified with *L. saxatilis*, although in this case discrimination is usually possible from shells or good figures.

The species-level names that have been based on forms of *L. saxatilis* will first be reviewed, before considering the history of species concepts in this group. *Turbo saxatilis* was described by Olivi in 1792, from Venice. Elsewhere in the Mediterranean, this species occurs only near the Strait of Gibraltar and in the Gulf of Gabès, Tunisia. The only ubiquitous littorinid in the Mediterranean is *Melarhaphe neritoides*, and several authors therefore appear to have assumed that *Turbo saxatilis* was a synonym of this more common species (Philippi, 1847; Jeffreys, 1865; Récluz, 1869). Although no types are known to exist, Olivi's figures (Fig. 101) clearly show shells larger, broader and with more rounded whorls and more impressed sutures than *M. neritoides*, and with a maculated pattern that never occurs in that species; the figures are an adequate representation of a local form of *L. saxatilis* (Fig. 107A). Because of this misapprehension and the disjunt distribution, the synonymy of the Venetian species with the Atlantic form, then known as *L. rudis*, was not recognized until 1870 (Brusina, 1870, 1886; Monterosato, 1872; Jeffreys, 1883; Kobelt, 1907), and even then the name was not widely adopted until Dautzenberg & Fischer's monograph of 1912. Other available names for the Venetian form are *L. saxoides* Nardo, 1847, and *L. saxatilis* var. *adriatica* Brusina, 1870.

The identity of *Turbo ruber* von Salis-Marschlins, 1793, is uncertain. The author briefly described a small, smooth, shiny shell with round aperture from Taranto (where *L. saxatilis* is not known to occur), but referred to figures of *L. saxatilis* in Chemnitz (1781). Heppell (1995) suggested that a red form of *L. saxatilis* (as can be found in Venice) was intended.

In 1797 Maton described a large, solid, smooth and rather tall-spired shell from an estuary in southwestern England as *Turbo rudis*. Having been figured by Donovan (1800), this name was soon widely used in Britain (Maton & Rackett, 1807; Pennant, 1812), France (Lamarck, 1822; Blainville,

1830), Germany (Menke, 1830; Anton, 1838; Philippi, 1846), Scandinavia (Lovén, 1846) and North America (Gould, 1841; de Kay, 1843). Other specific names that have been based on similar large and relatively smooth shells include *L. zonaria* Bean, 1844, and *L. danieli* Locard, 1886.

From another part of the range of this species Chemnitz (1781: pl. 185, fig. 1855a–g) figured the large, colourful, inflated and coarsely ribbed northern Atlantic form, and these figures are the lectotypes of six taxa named by Röding (1798), Link (1807) and Menke (1830). Of these, only *Litorina groenlandica* Menke, 1830, became established in the literature, and was applied to shells from Greenland, Iceland and northern Norway (Møller, 1842; Lovén, 1846; Philippi, 1846; Sars, 1851; Mørch, 1857). This form was renamed *L. castanea* by Deshayes (Deshayes & Milne-Edwards, 1853), using Schröter's (1779) non-binominal name. Reeve's (1857) *L. simplex*, although said to originate from the Cape Verde Islands, is based upon incorrectly localized material of the same form.

Meanwhile, in the better-known parts of the range in western Europe and North America, several ecotypic forms and other distinctive shell types had also received specific names. In 1803 Montagu described the small, smooth, thin-shelled, brown form that is typical of sheltered and brackish localities as *Turbo tenebrosus*, and this name was soon familiar in the British literature and later in North America (Gould, 1841; de Kay, 1843), Russia (Middendorff, 1849) and Scandinavia (Sars, 1851). Similar shells were named *Turbo vestita* Say, 1822, and *Litorina marmorata* Pfeiffer, 1839, but neither name was used by other authors. Towards the end of the century Dautzenberg (1893b) described another small, delicate shell from salt marshes and sheltered habitats in the isolated Mediterranean locality of the Gulf of Gabès as *L. nervillei*, but this was soon reduced to a variety of *L. saxatilis* (Dautzenberg & Fischer, 1912).

To shells with a low spire and a few strong ribs, Montagu (1803) gave the name *Turbo jugosus*. This was widely used in Britain (e.g. Maton & Rackett, 1807; Pennant, 1812; Brown, 1827, 1844a; Thorpe, 1844), but less frequently mentioned elsewhere (Blainville, 1828, 1830; Philippi, 1848; Küster, 1856; Locard, 1886, 1891). A similar North American shell was named *Turbo obligatus* by Say (1822), and this specific name occasionally appeared in the literature (Philippi, 1848; Reeve, 1858). Johnston's (1842) *L. rudissima* was also based on sculptured shells.

On wave-exposed shores the shells of *L. saxatilis* are often patulous, with a low spire and relatively enlarged aperture. This shell type is also common in *L. arcana* and, since the two can only be unequivocally distinguished by anatomical characters, the determinations of some of the older names are uncertain. *Turbo labiatus* Brown, 1827, is more likely to have been based on *L. arcana*, but is here considered a *nomen dubium* (p. 265). *Turbo ventricosus* Brown, 1837, might apply to either species, since both occur at the type locality, but it is a junior primary homonym and therefore a rejected name. The identity of *Littorina patula* Thorpe, 1844, is also unclear, and again both species occur at the type locality (Hannaford Ellis, 1979); it too is considered a *nomen dubium* (p. 265).

Another distinctive ecotype is the very small, globose form with black and white shell that is typically found among barnacles on wave-exposed shores. This was named *L. saxatilis* by Johnston, 1842, who coincidentally chose the same specific name as that of Olivi (1792) for the Venetian form. Johnston's specific name was commonly used by nineteenth century authors, unaware of the secondary homonymy (e.g. Brown, 1844a; Thorpe, 1844; Lovén, 1846; Philippi, 1848; Forbes & Hanley, 1850). Johnston (1842) mentioned Bean at the end of the diagnosis of his species, perhaps indicating that he had obtained specimens from Bean, or crediting him with its discovery. In 1844 the same form was described by Bean himself, under the name *L. neglecta*, but this taxon was largely ignored (with the exceptions of Alder, 1848, and Kobelt, 1907) until reused as a subspecies by James (1964, 1968a, b) and as a species by Heller (1975a).

One other specific name from the synonymy of *L. saxatilis* must be mentioned. In 1839 Gray described *L. nigrolineata* in the *Zoology of Captain Beechey's Voyage*, without locality or figure. Thereafter, this name was erroneously used for the species here called *L. compressa* (which does frequently have a nigrolineate pattern), both by some nineteenth (Philippi, 1846a; Récluz, 1869) and many twentieth century authors. In the following text this incorrect usage of the name is indicated as *L. 'nigrolineata'*. Only Reeve (1857) correctly figured *L. nigrolineata* and synonymized it with *L. rudis*. Examination of the types of Gray's species (labelled in his hand; Fig. 106P) shows that they are unquestionably of the nigrolineate form of *L. saxatilis* that is common on exposed shores in northern Portugal and Galicia (see also Fig. 106O). (Curiously, Beechey's ship did not make landfall here or elsewhere in Europe (Beechey, 1831), so the source of Gray's specimens is unknown, and a type locality has here been designated). This completes the survey of all the synonyms of *L. saxatilis* that were first proposed at the rank of species. In addition, many of the varieties and forms described before 1961 could be deemed to be of subspecific rank and therefore available as specific names (ICZN, 1985: arts 45g, 46a), although none has been thus used. The history of concepts of this species can now be considered.

Since Linnaeus (1758) named three common northern European littorinids (*L. littorea, L. obtusata, Melarhaphe neritoides*), it is surprising that he did not mention the present species, when it is both more common and more easily collected than any of these three. It seems likely that Linnaeus included it in his concept of *Turbo littoreus*. Other early authors who failed to mention this common species may have done the same (e.g. Müller, 1776; Pennant, 1777), and some explicitly included it under *Turbo littoreus* (da Costa, 1778; Gmelin, 1791). However, both Chemnitz (1781) and Schröter (1779) clearly discriminated between the two in non-binominal works. Even after the present species had been given binominal names, it was often misidentified as *Turbo littoreus* (e.g. Donovan, 1800; Maton & Rackett, 1807; Pulteney, 1813; Brown, 1827, 1837a; Blainville, 1830). Pulteney (1799) apparently included it in *T. obtusatus*, since this was described as having a furrowed shell.

For the first half of the nineteenth century most authors divided *L. saxatilis s.l.* into between three and seven species. Following Montagu (1803), British conchologists accepted *Turbo rudis, T. tenebrosus* and *T. jugosus* (Maton & Rackett, 1807; Pennant, 1812; Pulteney, 1813; Dillwyn, 1817; Turton, 1819; Brown, 1827; also Blainville, 1828). To these were added *T. ventricosus* and *T. labiatus* by Brown (1837a), and *L. saxatilis* Johnston and *L. patula* by Thorpe (1844). Lovén (1846) accepted four of these species. The most elaborate subdivision was that of Philippi (1846–1848), who recognized *L. rudis, L. tenebrosa, L. jugosa, L. groenlandica, L. saxatilis* Johnston and *L. obligata*, and was also the first to distinguish (as *L. 'nigrolineata'*) the species now known as *L. compressa*. Meanwhile, some others had been more restrained; Lamarck (1822) had ignored Montagu's species and mentioned only *T. rudis*, and in their revision of his *Histoire Naturelle* Deshayes & Milne-Edwards (1843) added only *L. castanea* (i.e. *groenlandica* of other authors). Menke (1845) used only *L. rudis* for shells on the German coast. In North America Gould (1841) had synonymized Say's *T. obligatus* and *T. vestita* with *L. rudis* and *L. tenebrosa* respectively, and these two species were accepted by later American writers (de Kay, 1843; Tryon, 1873; Gould, 1870).

From the middle of the nineteenth century, with a growing understanding of variation in shell characters, workers began the process of synonymizing the previously recognized species. Thus Middendorff (1849) accepted only *L. rudis* and *L. tenebrosa* in Russia. More influential was the work of Forbes & Hanley (1850) on the British Mollusca; although they gave descriptions of four named species, they suggested that since they were anatomically similar (particularly in their shared ovoviviparity) they should be synonymized as *L. rudis*. Clarke (1850a) considered the anatomy in more detail and came to the same conclusion, that only one variable ovoviviparous species should be recognized. The existence of oviparous forms within the *L. saxatilis* group was not recognized until almost one hundred years later (Seshappa, 1947, 1948; Deyglun, 1955). During the remainder of the century a few conchologists continued to recognize between four and six species (Küster, 1856; Reeve, 1857–1858; Sowerby, 1859; Récluz, 1869; Locard, 1886), but the majority accepted only the single species *L. rudis* in western Europe (Petit de la Saussaye, 1852; Jeffreys, 1865, 1869, 1883; Weinkauff, 1878, 1882, 1883; Herzenstein, 1885; Dautzenberg, 1887, 1893; Sowerby, 1887; Tryon, 1887; Hidalgo, 1890). The northern form, *L. groenlandica*, continued to be treated as distinct during this period (M. Sars, 1851; Mørch, 1857, 1868, 1869, 1870; Dall, 1886b; Teilman-Friis, 1899), but was reduced to a variety of *L. rudis* by G.O. Sars (1878) and synonymized with it by Johansen (1901). In addition, the small barnacle-dwelling form was still occasionally accepted as a distinct species, *L. saxatilis* Johnston or *L. neglecta* (Petit de la Saussaye, 1869; Kobelt, 1888, 1907).

Although the synonymy of *L. rudis* with the older name *L. saxatilis* (Olivi) had been pointed out by earlier authors (Brusina, 1870, 1886; Monterosato, 1872; Jeffreys, 1883; Kobelt, 1907), it was the work of Dautzenberg & Fischer (1912) on the *L. saxatilis* group that brought the change of name to a wider audience. Even so, *L. rudis* continued in use for some time (e.g. Bárdarson, 1919, 1920; Harmer, 1920; Winckworth, 1922; Pelseneer, 1923; Nobre, 1931; Linke, 1933; Spärck & Thorson, 1933), before *L. saxatilis* was generally accepted (e.g. Colman, 1932; Winckworth, 1932; van Benthem Jutting, 1932; Bequaert, 1943; Thorson, 1944, 1946, 1951; Abbott, 1954a, 1974; Fretter & Graham, 1962; Davis, 1973; Rasmussen, 1973; Bandel, 1974).

Various systems of nomenclature were devised to describe the wide range of shell variation recognized in *L. saxatilis s.l.* Early authors often denoted varieties simply by letters followed by short

Latin descriptions (e.g. Philippi, 1846–1848). Middendorff (1849) developed a complicated system of named forms and colour varieties. Later, the taxa that had formerly been recognized as species were demoted to varieties, and new ones were added. Thus, for example, Jeffreys (1865, 1869) recognized nine named varieties of *L. rudis* (including some that can now be ascribed to *L. compressa* and in part to *L. arcana*). In 1912 Dautzenberg & Fischer published a detailed description of shell variation throughout the geographical range of the *L. saxatilis* group. They recognized seven of the most distinctive shell forms as subspecies (although these were not allopatric, in contrast to the modern usage of this category): *saxatilis, tenebrosa, jugosa, patula* (largely consisting of the species *L. arcana*), *nigrolineata* (mainly the species *L. compressa*), *rudis* and *groenlandica*. Some of these were further subdivided into a total of ten named varieties and five monstrosities, and in addition 12 colour varieties were defined in the species as a whole. This cumbersome system was followed by only a few authors (e.g. Schlesch, 1931; Winckworth, 1932; van Benthem Jutting, 1933; Kaas & ten Broek, 1942), but the named colour varieties provided the basis for a long series of papers on colour variation (e.g. Fischer-Piette & Gaillard, 1960, 1971; Fischer-Piette, Gaillard & James, 1963, 1964). The system of subspecies and varieties was also used, in slightly modified form, by James (1963, 1964, 1968a; James, in Fischer-Piette, Gaillard & James, 1964), who claimed that at least in Britain the 'subspecies' did not interbreed when sympatric, and he was followed by Davis (1973) in Nova Scotia. Soon afterwards this system was rendered obsolete by the new emphasis on anatomical characters and availability of new biochemical techniques, which together revolutionized the systematics of *Littorina*.

Use of anatomical information was not entirely new, however. More than a century before, both Clarke (1850a) and Forbes & Hanley (1850) had argued that *L. saxatilis s.l.* was a single species, because of its supposed anatomical uniformity and in particular its ovoviviparous development. In 1899 Teilman-Friis argued for the separation of *L. rudis, L. groenlandica* and *L. tenebrosa* on the basis of supposed differences in radular teeth, although larger samples showed intergradation in this character, which was used as evidence that they were conspecific (Johansen, 1901). Radular characters have also been examined more recently (James, in Fischer-Piette, Gaillard & James, 1964; James, 1968a; Heller, 1975a; Sacchi, 1975; Seshappa, 1976; Raffaelli, 1979a; Reid, 1990b), but in general have not been found to be useful in defining species within *L. saxatilis s.l.* Of much greater significance has been the investigation of reproductive anatomy and the type of development. The first hint that *L. saxatilis s.l.* was not always ovoviviparous was the discovery by Seshappa (1947, 1948) of sympatric oviparous and ovoviviparous forms in northeastern England. At the time, the systematic significance was not appreciated, and the egg-laying forms (now recognized as *L. arcana*) were regarded as reproductive variants (Seshappa, 1947, 1948; Fretter & Graham, 1962: 380) or as a 'subspecies' of *L. saxatilis* (Seshappa, 1976). Working in Brittany, Deyglun (1955; an unpublished thesis later quoted by Crisp & Fischer-Piette, 1959, and Sacchi, 1975) discovered another oviparous form and proposed that this was a distinct species, for which she adopted the name *L. 'nigrolineata'* (i.e. the species here called *L. compressa*). This was accepted by Crisp & Fischer-Piette (1959) and by Sacchi (1974, 1975), although Gaillard (1977) argued that the associated differences in the pallial oviducts of Deyglun's two forms were minor and that they were merely reproductive variants. The most influential paper of this period was the study by Heller (1975a) of the *L. saxatilis* group in Wales. Using a wide range of morphological characters of the penis, radula and shell, as well as developmental type, breeding season and zonation on the shore, he convincingly argued that four distinct species should be recognized: the ovoviviparous *L. rudis, L. patula* and *L. neglecta*, and the oviparous *L. 'nigrolineata'* (the name *L. saxatilis* (Olivi) was not used, because of uncertainty about the relationship between forms from Venice and Wales). Soon afterwards Raffaelli (1979a) reexamined variation in the shell, penis and radula, and suggested that the first two of these species were synonymous. Meanwhile, Hannaford Ellis (1978a, 1979) had emphasized developmental type and the associated morphology of the pallial oviduct, and described a new, oviparous species, *L. arcana*, of which the shell was often indistinguishable from that of *L. rudis*. Heller's *L. patula* was, she suggested, a mixture of oviparous *L. arcana* and ovoviviparous *L. rudis*. Once again there were some who argued that the new species was simply a reproductive morph of *L. saxatilis* or *L. rudis* (Caugant & Bergerard, 1980; Bergerard & Caugant, 1981; Hughes & Roberts, 1980, 1981). However, this possibility was discounted by Fretter (1980) in her study of the pallial oviducts of *Littorina* species. Thus, by 1980, in their monograph of British and Danish *Littorina* species, Fretter & Graham were able to recognize as distinct the oviparous species *L. 'nigrolineata'* and *L. arcana*, and the ovoviviparous *L. neglecta*. There was still some uncertainty about the remaining ovoviviparous forms; they suggested that *L. tenebrosa*, and possibly *L. groenlandica*, might also be specifically distinct, but that *L. saxatilis* and *L. rudis* were extremes of a cline in shell shape within a single species. In a study of shell variation in southwestern England, J. E. Smith (1981) ignored the recent anatomical advances and therefore undoubtedly included *L. arcana* in his samples, arriving at the erroneous conclusion that the ecotypes of *L. saxatilis s.l.* from exposed and sheltered shores were distinct species (which he called *L. saxatilis* and *L. rudis* respectively).

Meanwhile, a completely new class of characters had been made available for systematic work by advances in biochemistry and genetics. The first application of the new technology of gel electrophoresis to the study of the *L. saxatilis* group was by Wium-Anderson (1970), who from the electrophoretic banding patterns of haemoglobin and other proteins found *L. rudis* and *L. tenebrosa* to be inseparable, but suggested that *L. groenlandica* was perhaps distinct. Heller (1975a) also used electrophoretic banding patterns, this time of esterases, to support his classification of species in Wales. Caugant & Bergerard (1979) found significantly different allele frequencies at an enzyme locus in sympatric samples of *L. saxatilis* and *L. 'nigrolineata'*, which suggested that there was reproductive isolation between these two species that had recently been separated on morphological grounds. This result has since been amply confirmed at other loci, although no diagnostic alleles have been discovered (Wilkins & O'Regan, 1980; Knight & Ward, 1986, 1991; Backeljau & Warmoes, 1992). There is less genetic differentiation between *L. saxatilis* and the other oviparous species, *L. arcana*; at first no allozymic differences were found (Caugant & Bergerard, 1980), but frequency differences have now been detected in sympatric populations (Ward & Warwick, 1980; Ward & Janson, 1985; Knight & Ward, 1986, 1991; Backeljau & Warmoes, 1992), again supporting the initial morphological separation of the species. The technique of isoelectric focussing has also been used, demonstrating frequency differences in the banding patterns of esterases in *L. saxatilis* and *L. arcana* (Mill & Grahame, 1988, 1992a; Dytham, Grahame & Mill, 1992).

Within the ovoviviparous taxa, allozyme studies of exposed-shore and sheltered-shore ecotypes of *L. saxatilis* in Sweden found a high degree of genetic identity between them (Janson & Ward, 1984), supporting the synonymy of Heller's *L. patula* and *L. rudis*, which had earlier been suggested by clinal variation between them in shell shape and penial gland number (Janson, 1982a). Chromosome numbers were also found to be the same in these two ecotypes (Janson, 1983a). At about this time another powerful new technique, the multivariate analysis of shell shape, was first applied to the problem of variation in *L. saxatilis*. This permitted mathematical description and comparison of the subtle variations of shell form from one habitat to another (Janson & Sundberg, 1983). Using a combination of genetic and morphometric approaches, Janson & Ward (1985) demonstrated that the sheltered or brackish-water ecotype, often previously considered distinct under the name *L. tenebrosa*, was merely part of a cline which connected it with typical open-coast *L. saxatilis*. A successful laboratory cross between these two forms was also achieved (Ward, Warwick & Knight, 1986). Nevertheless, recognition of *L. tenebrosa* has persisted in the literature, especially in some textbooks (Rolán, 1984; Graham, 1988; Hayward, Wigham & Yonow, 1990; Poppe & Goto, 1991). Allozyme analysis was also applied to the vexed problem of the name of the species itself; from the high genetic identity of populations from Venice (the type locality of Olivi's *Turbo saxatilis*), Britain and Sweden, Janson (1985b) argued that they were probably conspecific, so that the name *L. saxatilis* should take priority over *L. rudis*. This result was also supported by successful laboratory hybridization between the two populations (Warwick, Knight & Ward, 1990).

By 1985 the problems surrounding the systematics of the *L. saxatilis* group therefore seemed largely to have been resolved. A certain consensus of opinion had been reached among researchers that four species were involved: the ovoviviparous *L. saxatilis* and *L. neglecta*, and the oviparous *L. 'nigrolineata'* (i.e. *L. compressa*) and *L. arcana*. Numerous ecological studies had been made of the newly-defined species, which appeared to confirm their classification (e.g. review by Raffaelli, 1982; also Naylor & Begon, 1982; Hannaford Ellis, 1983, 1984; Fish & Sharp, 1985). In fact those working with the species in the field were more ready to accept the new classification than some taxonomists, a few of whom still adhered to the concept of a single variable species (e.g. Rosewater, 1981; Nordsieck, 1982).

Some difficulties remained, however. Appropriately, the least known of the four species recognized at this time in the *L. saxatilis* group was the small, ovoviviparous *L. neglecta*. Ecologically, this seemed to be the most distinct of the group, being closely associated with barnacles in the mid eulittoral zone. Nevertheless, despite descriptions by Heller (1975a), Raffaelli (1979a), Hannaford Ellis (1984) and Fish & Sharp (1985), there were still no entirely reliable morphological criteria by which to distinguish it from *L. saxatilis*, and intermediates could be found in some habitats (e.g. Warmoes, Backeljau & de Bruyn, 1988). Examination of shell size, shape and colour over a wide geographical range also suggested that *L. neglecta* was not clearly separable by these characters from *L. saxatilis* found higher on the shore (B. & K. Johannesson, 1990), and no diagnostic anatomical characters could be found (Reid, 1990b, 1993). Furthermore, the pattern of genetic variation did not support their recognition as distinct species (K. & B. Johannesson, 1990, 1993). The existence of barnacle-dwelling ecotypes (almost identical in shell size, shape and colour to *L. neglecta*) in the species *L. arcana* and *L. compressa*, added support to the interpretation of *L. neglecta* as an ecotype of *L. saxatilis* (Reid, 1993). However, subsequent morphometric studies of shells have claimed that *L. neglecta* is at least morphologically distinct (Caley, Grahame & Mill, 1995; Grahame, Mill, Hull & Caley, 1995) and, on one shore in Yorkshire, apparently reproductively isolated from *L. saxatilis* (Hull, Grahame & Mill, pers. comm.). Recently, new DNA technology has been applied for the first time to this problem. Patterns of random amplified polymorphic DNAs (RAPDs) have been found to provide unambiguous separation of individuals of *L. saxatilis* from *L. arcana* (Crossland, Coates, Grahame & Mill, 1993) and in this respect provides a more powerful tool than the study of enzyme polymorphisms, from which no loci diagnostic of the species in the *L. saxatilis* group have so far been discovered. In an extension of the RAPDs technique to all four supposed species of the group in a small area in Wales, individuals of *L. arcana* and *L. compressa* formed discrete clusters, but *L. saxatilis* and *L. neglecta* were mixed in a single cluster, providing no evidence that *L. neglecta* is a distinct species (Crossland, Coates, Grahame & Mill, 1995). A small, smooth, low-shore ecotype comparable with *L. neglecta* occurs among mussels on shores in Galicia and is strikingly different from large, ridged shells of *L. saxatilis* found higher on the shore, but here again genetic analysis of polymorphic enzyme loci did not support specific separation of the two forms (K. Johannesson, B. Johannesson & Rolán-Alvarez, 1993; Rolán-Alvarez, Rolán & Johannesson, 1996), despite evidence of partial reproductive isolation between them. Work in Yorkshire has recently produced additional examples of morphological differentiation and reproductive isolation between forms that occupy different microhabitats on the same shore, and it has even been suggested that these deserve recognition as species (Grahame *et al.*, 1995; Hull, Grahame & Mill, 1996).

To summarize the systematics of this complex group, the two oviparous species *L. compressa* (i.e. *L. 'nigrolineata'* of authors, not Gray, 1839) and *L. arcana* are clearly distinct and their taxonomic status is no longer questioned. The history of their recognition is described in more detail in the Taxonomic History sections on each (pp. 249, 265). The evidence at present available does not support the further taxonomic division of the third species, *L. saxatilis*, to which all ovoviviparous *Littorina* are considered to belong. Nevertheless, this assertion is controversial, and the arguments for regarding *L. saxatilis* as a single species are examined in detail in the discussions of the distinctive ecotypes which can be recognized within this highly polymorphic taxon. The morphological characters most useful for the identification of the three species are summarized in Table 7 (p. 338).

Since the confusion with *L. littorea* in the eighteenth century, there have been few misidentifications of *L. saxatilis* with species outside the *L. saxatilis* group. Confusion with *L. obtusata* geog. var. *palliata* has been mentioned in the remarks on that species (p. 206). The rather tall-spired, spotted shells of the outlying populations of *L. saxatilis* in South Africa have been misidentified as *L. punctata* (Barnard, 1963; Day, 1969), a species that is now placed in *Nodilittorina*; similar shells from the Gulf of Gabès have also been compared with *N. punctata* (Sabelli, Gianuzzi-Savelli & Bedulli, 1992). In the Canary Islands spotted shells of *L. saxatilis* have been misidentified as *L. undulata* (Nordsieck & Garcia-Talavera, 1979), which is now in the genus *Littoraria*.

Diagnosis

Mature shell height 1.2–25.8 mm. Shell turbinate to high-turbinate, aperture often slightly constricted relative to width of preceding whorl, but alternatively slightly patulous. Mineralogy of external irregular-prismatic calcite and internal crossed-lamellar aragonite. Sculpture variable, almost smooth, or of narrow ribs of irregular width, or with as few as 5 strong, rounded ribs with interspaces wider than rib width; fine spiral microstriae over whole surface. Colour highly variable, e.g. white, yellow, orange, red, brown or black, sometimes banded or tessellated. Penis with relatively stout filament and usually a mucronate tip, 10–35% total length, 3–25 mamilliform penial glands usually mostly in a single row. Pallial oviduct of single small loop of albumen gland, single very reduced loop of capsule gland (albumen and capsule glands together occupy less than one third total length of pallial oviduct) and enlarged, thin-walled, septate brood pouch containing eggs and developing embryos; bursa small, less than half length of brood pouch, or vestigial. Development nonplanktotrophic with intracapsular metamorphosis; ovoviviparous; young released from brood pouch as shelled, crawling juveniles. Distribution: Novaya Zemlya and Svalbard, Atlantic coasts of Scandinavia and Europe, Morocco, Canary Islands, Azores, southern Africa, Mediterranean (Venice and Gulf of Gabès), Iceland, Greenland, Hudson Bay, southern Baffin Island to Chesapeake Bay.

Anatomical description

Head-foot (Fig. 103O, P). Head black to grey with unpigmented tip to snout; usually unpigmented around eye and across tentacle base; tentacles often unpigmented with two longitudinal black stripes, sometimes also with indistinct transverse bands; head and tentacles may be entirely black; sides of foot black, mottled grey or unpigmented. James (1968a) observed that newly-emerged juveniles were unpigmented and reported gradual darkening of head pigmentation with increasing size. However, Davis (1973) found no evidence of darkening with age in larger shells, but noted wide variation in pigmentation within populations and some differences between habitats. Opercular ratio 0.44–0.56.

Penis (Fig. 102). Base wrinkled, usually unpigmented, but outer side may be grey to blackish (black penes more common in geog. var. *groenlandica*); number of mamilliform penial glands ranges from 0–43, but usually 3–25 (97% of individuals examined); if less than 15–20 these are mostly in a single row. Filament usually unpigmented, occasionally grey; 10–35% total length, relatively stout, often wrinkled, usually with a mucronate tip. Sperm groove open, ending terminally. Subepithelial glandular tissue around sperm groove. There have been several studies of variation in the numbers of mamilliform glands. The number of glands increases with shell size (Linke, 1933; Heller, 1975a), although the correlation is not a close one (Raffaelli, 1979a; Janson, 1982a). Both James (1968a) and Heller (1975a) reported more numerous glands in the wave-exposed ecotype, but this result could have been caused by inclusion of *L. arcana* (in which glands are more numerous) in samples from exposed shores. An opposite trend, with fewer glands in the wave-exposed ecotype, have been found in Sweden (Janson, 1982a). In Nova Scotia (where *L. arcana* does not occur) samples from exposed shores were more likely to have glands arranged in more than one row (Davis, 1973). The ultrastructure of mamilliform glands has been described by Buckland-Nicks & Worthen (1992). There is evidence for a seasonal regression in penial size at some localities (in autumn in Helgoland, Linke, 1933; in summer in Kent, Berry, 1961; in summer in the barnacle ecotype in Wales, Fish & Sharp, 1985), although not in others (in lagoonal populations in both Sweden, Muus, 1967; and Venice, Sacchi & Torelli, 1973). A report of seasonal regression only among larger animals (at Roscoff, Bergerard, 1975) is probably explained by the inclusion of *L. arcana*, with more strongly seasonal reproduction. There is, however, some evidence of reproductive senescence among the largest males (Inagaki, 1972). Whether the penis is sometimes shed under natural conditions is unclear; Pettitt (1973a) observed this in an aquarium, but this may have been a response to stress (as suggested by Deutsch & Fioroni, 1992, for *L. littorea*). The abnormal development of a penis in females (pseudohermaphroditism), a condition often connected with high levels of the anti-fouling chemical tributyltin, has been recorded by Fioroni, Oehlmann & Stroben (1990).

Sperm (Fig. 3I). Euspermatozoa 46–53 µm. Paraspermatozoa 11–18 µm diameter, round or slightly oval; vesicles large and distinct; nucleus not visible.

Oviduct (Fig. 103A–N). Single relatively small spiral loop of albumen gland; followed by very reduced loop of capsule gland which shows only a slight spiral twist (only rarely as noticeable as in Fig. 103N); then a large, thin-walled, brood pouch with delicate septa (the modified jelly gland), which for most of the year is packed with eggs and developing embryos. Albumen and capsule glands together occupy one third or less of total length of pallial oviduct, even in spent females with empty brood pouch (Fig. 103C). Bursa short, opening near anterior end of brood pouch and extending back no more than half length of brood pouch and usually much less; in very small adults (less than 5 mm shell height) bursa is sometimes vestigial or absent. Descriptions have also been given by Linke (1933) and Fretter (1980), although using a terminology for the glands of the oviduct different from that used here. Since Losse & Greven (1993) found that there was no egg capsule in this species, a true

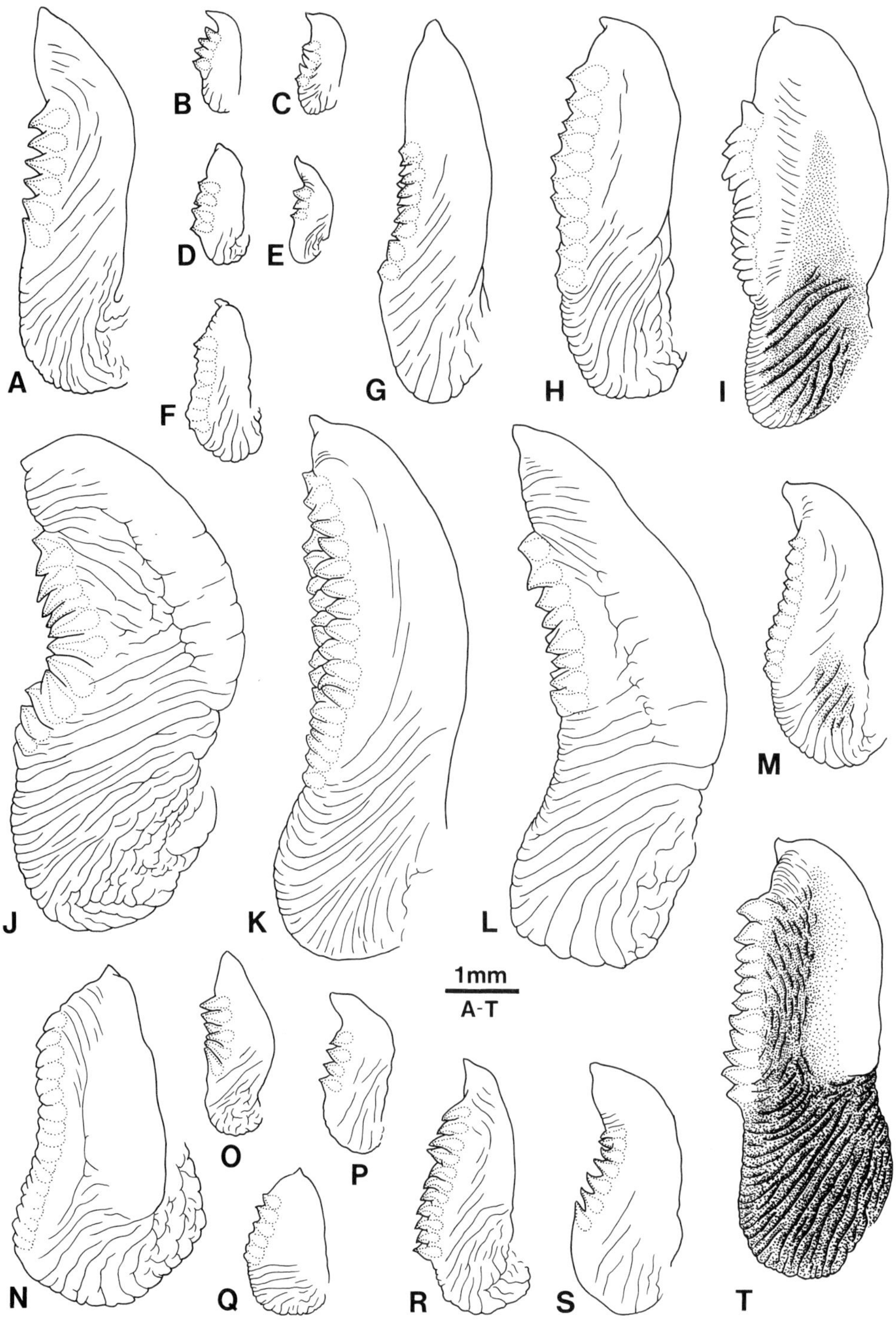

Figure 102. *Littorina (Neritrema) saxatilis*: penes. **A, D,** 1 km W of St Govan's Head, Pembrokeshire, Wales (**A,** moderate ecotype; **D,** barnacle ecotype); **B, C,** Ven, Sweden (ZMK; sheltered/brackish ecotype); **E,** 2 km SE of Malahide, Co. Dublin, Ireland (barnacle ecotype); **F,** Rockall (wave-exposed ecotype); **G,** Rubh' Aird-mhicheil, South Uist, Outer Hebrides, Scotland (moderate ecotype); **H,** Carantec, Brittany, France (moderate ecotype); **I,** Smelror, near Vardø, Norway (moderate ecotype); **J,** Barfleur, Normandy, France (moderate ecotype); **K,** 1 km S of Killough, Co. Down, Ireland (moderate ecotype); **L, Q,** Cabo Silleiro, Baiona, Galicia, Spain (**L,** moderate ecotype; **Q,** barnacle ecotype); **M,** Quoddy Head, near Lubec, Maine (moderate ecotype); **N,** Grindavik, Iceland (moderate ecotype); **O,** El Kantara, Djerba, Tunisia (sheltered/brackish ecotype); **P,** Dooagh, Achill Island, Co. Mayo, Ireland (barnacle ecotype); **R,** Alberoni, Venice Lagoon, Italy (moderate ecotype); **S,** Langebaan Lagoon, South Africa (sheltered/brackish ecotype); **T,** Lilleelv River mouth, Kangerdluarssuk Fjord, Narssaq, Greenland (moderate ecotype). Shell heights (with number of mamilliform glands in parentheses, when these are not all clearly visible in lateral view shown): **A,** 7.1 mm; **B,** 4.1 mm; **C,** 4.1 mm; **D,** 2.8 mm; **E,** 2.1 mm; **F,** 4.0 mm; **G,** 12.3 mm (12 glands); **H,** 11.0 mm; **I,** 13.9 mm (16 glands); **J,** 15.0 mm; **K,** 17.5 mm (28 glands); **L,** 12.3 mm (10 glands); **M,** 7.8 mm; **N,** 9.7 mm (21 glands); **O,** 4.8 mm; **P,** 3.4 mm; **Q,** 3.5 mm; **R,** 6.5 mm; **S,** 7.0 mm; **T,** 13.3 mm.

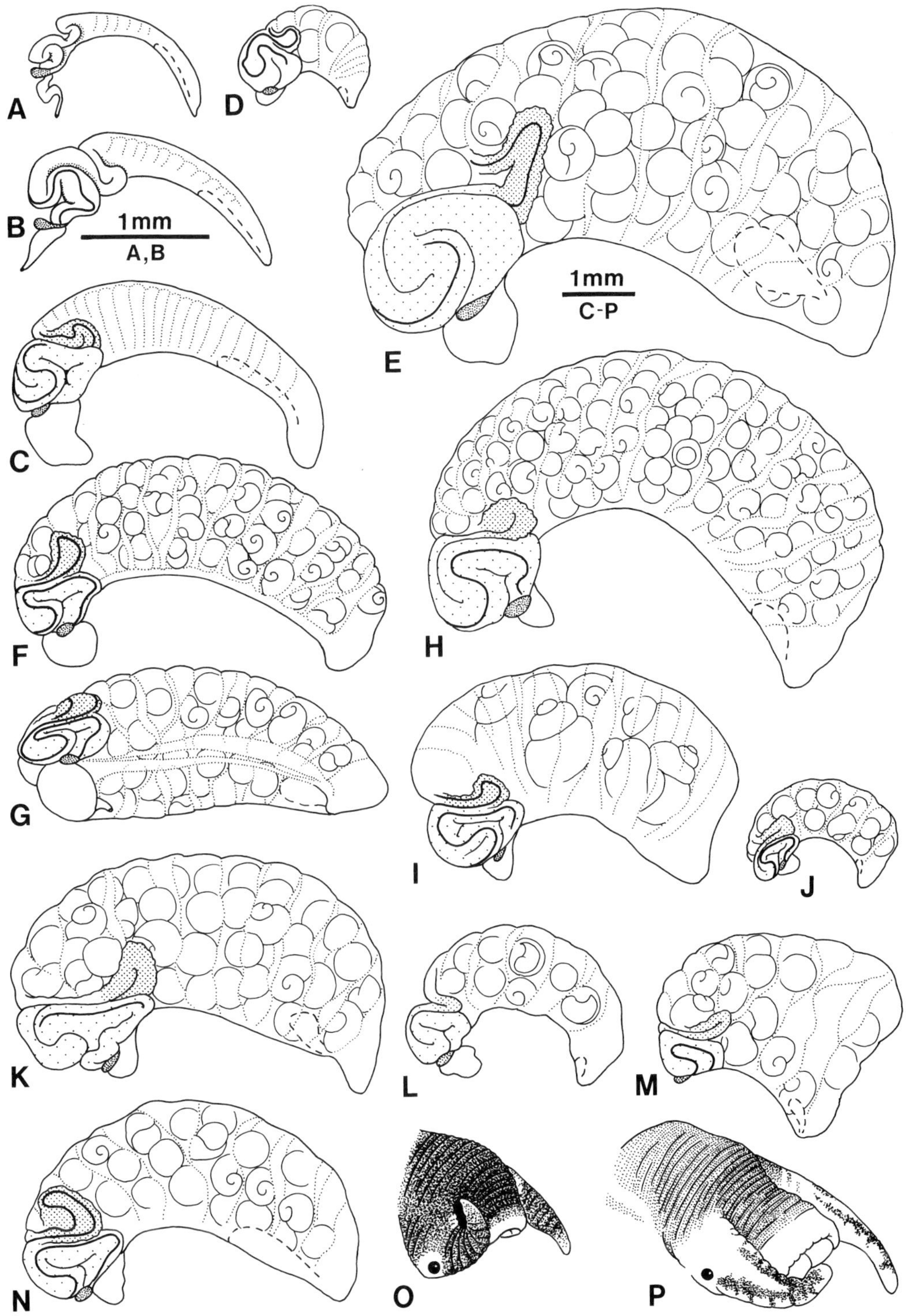

Figure 103. *Littorina (Neritrema) saxatilis*: **A–N**, pallial oviducts; **O**, **P**, head pigmentation. **A**, **B**, early ontogenetic stages (note larger scale); **C**, spent adult with empty brood pouch; **F**, **G**, lateral and ventral views of same oviduct; **I**, oviduct with giant embryos. **A**, **J**, 1 km W of St Govan's Head, Pembrokeshire, Wales (**A**, moderate ecotype; **J**, barnacle ecotype); **B**, Yesnaby, Orkney, Scotland (moderate ecotype); **C**, Dale, Pembrokeshire, Wales (moderate ecotype); **D**, Ven, Sweden (ZMK; sheltered/brackish ecotype); **E**, Vík, Iceland (wave-exposed ecotype); **F**, **G**, Carantec, Brittany, France (moderate ecotype); **H**, **P**, Alberoni, Venice Lagoon, Italy (moderate ecotype); **I**, Flókalundur, Iceland (moderate ecotype); **K**, Quoddy Head, near Lubec, Maine (moderate ecotype); **L**, El Kantara, Djerba, Tunisia (sheltered/brackish ecotype); **M**, Rockall (wave-exposed ecotype); **N**, Kyle of Durness, Scotland (moderate ecotype); **O**, Beauly Firth, Inverness, Scotland (sheltered/brackish ecotype). Shell heights: **A**, 4.4 mm; **B**, 6.3 mm; **C**, 10.0 mm; **D**, 4.2 mm; **E**, 13.5 mm; **F**, **G**, 12.0 mm; **H**, **P**, 10.2 mm; **I**, 11.5 mm; **J**, 3.2 mm; **K**, 10.6 mm; **L**, 5.1 mm; **M**, 8.0 mm; **N**, 10.5 mm; **O**, 8.7 mm. Shading conventions as described on p. 6, except that in **A**, **B** albumen and capsule glands are unshaded, since they are not at these stages visibly differentiated.

capsule gland may in fact also be absent; the gland here termed the capsule gland may consist of the covering gland alone (p. 17). This gland is somewhat enlarged just before ovulation (Fretter, 1980). Immature stages in the development of the renal and pallial oviduct are illustrated in Figure 103A, B. Reproductive senescence may occur in the largest females (Moreteau, 1978).

Embryos and development (Fig. 104E–J). It has been known since the early nineteenth century that *L. saxatilis* is ovoviviparous, retaining embryos within the brood pouch and releasing miniature snails at the crawling stage (Blainville, 1828; Johnston, 1842; Clarke, 1850a; Jeffreys, 1865). Release of earlier stages may occasionally occur, perhaps in response to disturbance (Hannaford Ellis, 1978b). When they enter the brood pouch, ova measure 180–211 µm in diameter (Delsman, 1914; Linke, 1933; Gallardo & Götting, 1985) and are enclosed singly in a spherical egg covering variously quoted as 420 µm (Linke, 1933) and 373–536 µm (Seshappa, 1976) in diameter; this covering is only 1.5–1.9 µm in thickness and it appears that no egg capsule is present (Losse & Greven, 1993). The course of development has been described and illustrated by Pelseneer (1911), Thorson (1946) and Fretter & Graham (1962). Since eggs continue to enter the brood pouch throughout the breeding season, all stages of development can be found together in this part of the pallial oviduct (Linke, 1933; Berry, 1961; Fretter, 1980; Stroud & Knudsen, 1982; Fish & Sharp, 1985). Within the brood pouch the osmotic pressure of the fluid bathing the embryos is regulated by the parent (Berger, 1975), but it has generally been believed that no maternal nutrition is supplied to the embryos, since they will develop to maturity (but sometimes with poor success) if cultured in seawater (Linke, 1933; Hannaford Ellis, 1978; Roberts & Hughes, 1980; Ross & Berry, 1990). However, there is now some indirect evidence that limited maternal nutrition may be possible, because there is no correlation between initial egg size and embryo size at hatching, and the egg covering has been shown to be permeable to proteins (Losse & Greven, 1993). The number of embryos in a single brood pouch is highly variable, but is clearly correlated with maternal size, ranging from 10 in a specimen of the barnacle ecotype only 2 mm in shell height (Fish & Sharp, 1985) to 749 (Gallardo & Götting, 1985) or even 900 (Linke, 1933) in the largest shells. Embryo numbers also vary significantly with season (p. 297) and between habitats (e.g. Faller-Fritsch, 1977; Hughes, 1979; Roberts & Hughes, 1980; Gallardo & Götting, 1985; Ross & Berry, 1990).

Many authors have recorded the sizes of juveniles on hatching from their egg coverings and emerging from the brood pouch. There is much variation, but in most habitats and under moderate climatic conditions shell diameters range between 400–650 µm and shell heights between 450–750 µm. The shell has 1.6–2.1 whorls and appears smooth, although at high magnification spiral rows of minute tubercles (1–2 µm diameter) can be seen over the surface (Rolán-Alvarez *et al.*, 1996), or only near the suture. Several trends in hatching size have been claimed, but the results are sometimes conflicting. On wave-exposed cliffs hatching size has often been reported to be larger than on more sheltered boulder shores nearby (Roberts & Hughes, 1980; J. E. Smith, 1981; Janson, 1982b; Hart & Begon, 1982; Hughes, 1995). Assuming the variation to be of genotypic origin (as argued by Rolán-Alvarez *et al.*, 1996), this trend has been explained either by a possibly superior adhering capacity of larger juveniles, or by their greater resistance to desiccation on open rock surfaces. The opposite result has also been found, suggesting that selective pressures such as predation, crushing by mobile boulders and desiccation may sometimes favour larger embryos on sheltered shores (Faller-Fritsch, 1977; Atkinson & Newbury, 1984); in particular, hatching sizes are larger in estuarine and salt-marsh populations (Hughes, 1978, 1995; Roberts & Hughes, 1980; Ross & Berry, 1990). Differences in embryo sizes may be found even in contrasting microhabitats on the same shore; on wave-exposed shores in Galicia the embryos of a small morph from mussel beds are larger than those of a larger, moderate ecotype from a relatively sheltered boulder zone higher on the shore (Rolán-Alvarez *et al.*, 1996); this situation is reversed on wave-exposed coasts in Yorkshire, where embryos are smaller in the mid-shore than in the high-shore form (Hull, Grahame & Mill, 1996), although here the topography of the shore is such that the latter appears to occupy the more exposed microhabitat. In the barnacle ecotype (*L. neglecta* of authors) hatching sizes have been found to be either smaller (350–400 µm shell height, James, 1968a; Fish & Sharp, 1985) or larger (600 µm diameter, Hannaford Ellis, 1983) than in more typical examples of *L. saxatilis*. In ecotypes of small adult size an additional consideration is the restricted brooding capacity, and embryo sizes may be a compromise between the demands of embryo survival and a minimum level of fecundity. The theoretical aspects of fecundity and reproductive investment in the ecotypes of this species have been discussed by Faller-

Fritsch (1977), Hughes & Roberts (1980), Roberts & Hughes (1980), Hart & Begon (1982), Atkinson & Newbury (1984), Faller-Fritsch & Emson (1985), Gallardo & Götting (1985) and Hughes (1995).

One possible trend in egg and embryo size that has yet to be accurately documented concerns the larger size of both in arctic populations (Stroud & Knudsen, 1982; Fig. 103D–N). The shell height of hatching embryos in northeastern Greenland was given as 720 µm by Stroud & Knudsen (1982), and James (1968a) reported sizes of 950–1200 µm in his 'subspecies *groenlandica*'. In the present study maximum shell diameters of unencapsulated embryos in the brood pouch have been found to be in the range 450–550 µm in Venice, Tunisia, Morocco and Portugal, 500–650 µm in France and the British Isles, and 600–800 µm in New England, Iceland and the Arctic. In some northern localities a proportion of the embryos that hatch at the normal size remain in the brood pouch and continue to grow, reaching a size of up to 1400 µm in height by 1200 µm in width, with up to 3 whorls. Such 'giant embryos' were first noticed by Stroud & Knudsen (1982) in Greenland. In this study they have been found commonly at localities in Newfoundland, Greenland, Iceland, Spitsbergen and the Barents Sea, and a single example has been seen from the Outer Hebrides. The size at which hatching from the egg covering occurs is clearly marked by the sharp transition between the smooth embryonic shell and the teleoconch with axial growth lines (Fig. 104F–J). The source of nutrition for these giant embryos is unknown. However, there is a striking negative correlation between the number of such embryos and their size; where the brood pouch is full of a mixture of normal encapsulated embryos and giant embryos, the latter show only slight growth of the teleoconch (Fig. 104G, I), but the very largest giant embryos (Fig. 104F, H, J) are found in small numbers (1–9) in pouches without any earlier developmental stages. This might suggest that cannibalism of siblings has occurred, although no partly-eaten embryos have ever been found in brood pouches. Alternatively, these giant embryos could represent embryos that have simply been retained for long periods within the brood pouch, perhaps while conditions are unfavourable for release, and have grown using nutrients from the parent. Matveeva (1974) reported that of the embryonic stages only shelled juveniles are able to survive in the brood pouch during the severe winters of the White Sea. The occurrence of giant embryos is clearly widespread and requires investigation.

Abnormal development of embryos has often been observed, in some cases producing planispiral, uncoiled or sinistral larval shells (Pelseneer, 1911; Thorson, 1946; Davis, 1973), although abnormalities are most common at the earliest stages of development of the eggs (Sokolova, 1995). Within many populations there is considerable variation between individuals in the occurrence of abnormal embryos, from 0 to 100% (Dixon & Pollard, 1985; Janson, 1985a; Sokolova, 1995). Contrary to an early suggestion by Cépède (1910), this is not correlated with infection of the brood pouch by a peritrich ciliate protozoan, nor with age of the female (Janson, 1985a; Sokolova, 1995), but disease may be a factor (Dixon & Pollard, 1985). It has been suggested that genetic factors may be involved, such as inbreeding, or hybrid breakdown in the hybrid zones between genetically differentiated ecotypes (Janson, 1985a; Hull, Grahame & Mill, 1996). There is also an environmental component to the variation in the occurrence of abnormalities between populations. Those living at very low salinities (5–8‰), close to the limit of tolerance, show the highest proportions of abnormal embryos, particularly at the later stages of their development (Thorson, 1946; Sokolova, 1995), and low temperatures may also have an adverse effect on development (Sokolova, 1995). Attempts have been made to correlate the frequency of abnormalities with environmental pollution, but with only limited success (Clyne & Duffus, 1979; Dixon & Pollard, 1985). Even in the absence of obvious environmental stress, developmental abnormalities may be common; in the White Sea from 20 to 90% of the females living at normal and moderate salinities contained broods entirely composed of abnormal eggs (Sokolova, 1995).

In *L. saxatilis* the spawning season and state of maturity of the females has usually been estimated by counts of the numbers of embryos in brood pouches, or by the percentages of females bearing embryos. In almost all populations most females contain some embryos throughout the year, at localities as diverse as Helgoland (Linke, 1933), Kent (Berry, 1961), Denmark (Rasmussen, 1973), Venice (Sacchi & Torelli, 1973; Torelli, 1974), The Netherlands (Bandel, 1974), Nova Scotia (Beland, 1974), New England (Pohley, 1976), Wales (Hannaford Ellis, 1983), France (Gallardo & Götting, 1985), Scotland (Berry & Smith, 1987), the Baltic Sea (Jagnow & Gosselck, 1987) and the White and Barents Seas (Kuznetsov, 1960; Matveeva, 1974). There is nevertheless often a seasonal fluctuation in embryo numbers, or in the percentage of brooding females, with reduced levels in late summer and winter (Berry, 1961; James, 1968d; Beland, 1974; Roberts & Hughes, 1980; Torelli, 1984; Gallardo &

Figure 104. *Littorina (Neritrema) saxatilis*: radulae, and shells of unencapsulated embryos from brood pouches. **A–C**, three views of radula of mature adult (shell H=9.5 mm), flat (**A**), at 45° (**B**) and side view (**C**); **D**, radula of juvenile (shell H=2.7 mm), at 45°; **E**, three embryos from single adult; **F, H, J**, giant embryos from single adult; **G, I**, giant embryos. **A–E**, Robin Hood's Bay, Yorkshire, England (moderate ecotype); **F, H, J**, Flókalundur, Iceland (moderate ecotype); **G**, Tjörnes Peninsula, Iceland (moderate ecotype); **I**, Smelror, Vardø, Norway (moderate ecotype). Scale bars: **A, B**=50 µm; **C**=50 µm; **D**=50 µm; **E–J**=200 µm.

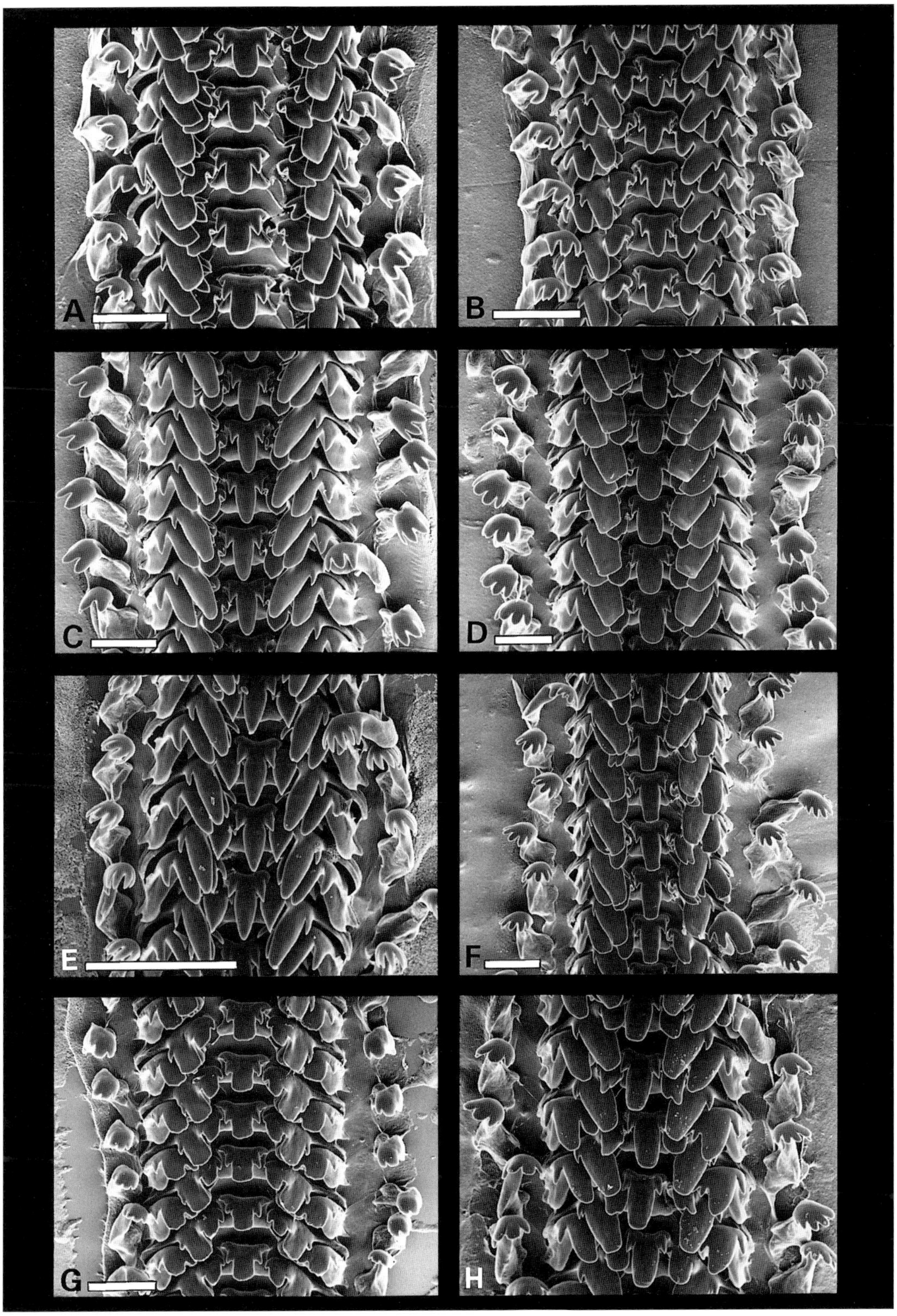

Figure 105. *Littorina (Neritrema) saxatilis*: radulae of mature adults, viewed at 45°. Shell heights: **A,** 10.2 mm; **B,** 8.9 mm; **C,** 9 mm; **D,** 13 mm; **E,** 3.4 mm; **F,** 8.8 mm; **G,** 9.8 mm; **H,** 10 mm. **A,** Alberoni, Venice Lagoon, Italy (moderate ecotype); **B,** Knysna, South Africa (sheltered/brackish ecotype); **C,** Punta Cires, Morocco (wave-exposed ecotype); **D,** Cabo Silleiro, Baiona, Galicia, Spain (moderate ecotype); **E,** Robin Hood's Bay, Yorkshire, England (barnacle ecotype); **F,** Dunbeath, Caithness, Scotland (moderate ecotype); **G,** Sredny Island, Chupa Bay, White Sea, Russia (moderate ecotype); **H,** Loud's Island, Maine (moderate ecotype). Scale bars: **A**=50 μm; **B**=50 μm; **C**=50 μm; **D**=50 μm; **E**=50 μm; **F**=50 μm; **G, H**=50 μm.

Götting, 1985) or in winter alone (Rasmussen, 1973; Caugant & Bergerard, 1980; Ross & Berry, 1991). The ovaries may be reduced at these times of climatic extremes (Berry, 1961), and in the coldest parts of the range eggs enter the brood pouch only during May to August (Matveeva, 1974). The small barnacle ecotype has a mainly annual life cycle, and females are mature during spring and summer in Wales (Hannaford Ellis, 1983, as *L. neglecta*). A report of a seasonal reproductive cycle among larger females only (Bergerard, 1975) can probably be explained by inclusion of *L. arcana*.

Radula and foregut (Figs 104A–D, 105). Relative radular length 1.34–2.63 (extremes of range 0.85–5.00). Shape of radular cusps is highly variable within this species and cannot yet be correlated with either habitat or geographical distribution, although in small (less than about 5 mm in shell height) adults (Fig. 105E) and immature specimens (Fig. 104D) all cusps are more elongate and pointed (see also Seshappa, 1976; Raffaelli, 1979a). Rachidian: length/width 0.92–1.46 (to 1.77 in a juvenile); central cusp blunt, rounded or pointed, varying in length; one pointed cusp and one outer denticle on each side. Lateral: four cusps; second and third from inside rounded or blunt, third largest; others small and pointed. Inner marginal: four cusps; second from inside pointed or rounded; third from inside largest and blunt to rounded; others small and pointed. Outer marginal: 4–6 rounded, pointed or blunt cusps; narrow neck, projecting flange on inside of base only. Cusps of outer marginal are more numerous, up to 7 or 8, in small adults and immature specimens (Raffaelli, 1979a; Rolán-Alvarez *et al.*, 1996; Fig. 104D). Salivary glands enlarged.

Shell variation (Figs 4, 5, 106–108)

The degree of shell variation shown by *L. saxatilis* is extraordinary, exceeding that in any other *Littorina* species, and has been a subject of much interest. Early conchologists described many of the shell varieties as distinct species (see Taxonomic History, p. 287), but in the twentieth century the first systematic approach to the subject was that of Dautzenberg & Fischer (1912). Recognizing that shell size and spire height were widely variable, they emphasized shell sculpture in their definitions of 'subspecies'. Thereafter, named 'subspecies' and 'varieties' were widely employed (e.g. James, 1968a). A most useful advance was the introduction of the 'ecotype' (or 'ecomorph') concept (Fretter & Graham, 1980; Janson & Ward, 1985) to describe similar phenotypes occurring in the same microhabitats (p. 26). Various mathematical descriptions of shell form have been employed. So far only Newkirk & Doyle (1975) have measured Raup's parameters of shell coiling in *L. saxatilis*. Others have used simple comparison of the ratios of linear measurements (e.g. Heller, 1976), or regressions of linear measurements and of their ratios on shell height (taken as a measure of size) (e.g. Raffaelli, 1979a; J. E. Smith, 1981; Janson, 1982a). More recently multivariate analyses of shell shape have been carried out (e.g. Janson & Sundberg, 1983; Sundberg, 1988; Grahame, Mill & Brown, 1990; Grahame & Mill, 1992), and one study has described shell shape using Fourier analysis (Dytham, Mill, Grahame & O'Higgins, 1992).

Although the consistent association of certain shell forms with particular habitats and geographical regions has long been known, its possible adaptive significance has been discussed only relatively recently, following the early work of James (1968a), Davis (1973), Heller (1976) and Raffaelli (1978b, 1979a). In examining ecotypic and geographical variation, the operation of various selective forces must be considered. Strong wave action, and perhaps wind, on exposed coasts may select for features that reduce the likelihood of dislodgement. A small, low-spired and globular shell may present less resistance to waves. Strong attachment to the substrate will be facilitated both by a large foot area, accommodated by an enlarged and more circular aperture (external apertural size may be the important variable) and also by reducing shell-wall thickness and consequent weight. Small size may also be important, allowing animals to shelter in protected microhabitats such as rock crevices and barnacle shells. On sheltered shores crabs are more common and may impose selection for traits that reduce susceptibility to predation. These include a large, thick-walled shell, and a small or narrow aperture with a small inner dimension to prevent insertion of crab chelae. On moderately sheltered shores the possibility of crushing between mobile boulders may also select for features such as thickness that increase shell strength. On shores with limited wave action, and at higher tidal levels, the danger of desiccation may be reduced by a smaller, more rounded aperture. Energetic considerations may also be important; production of shell material may be energetically costly, so that in the absence of opposing selection the shell may be as thin as possible. Less obvious are demographic factors, such as the relationship between size, fecundity and the lifetime mortality rate. For example, where there is a high rate of mortality throughout life it may be advantageous to divert energetic resources from somatic growth to early maturation, resulting in small adult size. It must be emphasized that discussions of adaptive shell variation are necessarily speculative if based on interpretation of correlations between shell form and environmental variables, rather than on experimental evidence. Final shell form must be the result of the interaction of a number of selective forces, which may sometimes conflict. It should also be noted that although shell shape parameters may be discussed separately, they cannot always vary independently, because of the constraints of shell geometry; for example, a relatively larger aperture necessarily produces a more globose and lower-spired shell. Experimental evidence for the action of natural selection on shell form is still limited. A few laboratory studies have shown that the susceptibility of snail morphs to crab predation is as predicted (Elner & Raffaelli, 1980; Johannesson, 1986), and creation of artificial microhabitats on the shore has demonstrated the importance of crevices as size-limiting refuges (Emson & Faller-Fritsch, 1976; Raffaelli & Hughes, 1978). Field transfer experiments have also been carried out to test some of the predictions of selection models (Atkinson & Newbury, 1984), and one of these has shown differential fitness of ecotypes (Janson, 1983b).

Implicit in the interpretation of shell morphological variation in terms of natural selection, is the assumption that the variation has a largely genetic origin. In fact there is only limited evidence that this is so. One approach has been the measurement of embryonic shells, which show significant regressions upon the shape of the mother (Newkirk & Doyle, 1975). Embryos from populations of contrasting shape (from wave-exposed and sheltered shores) lie on the same regression line of apertural area upon shell height as the adults from the same population (J. E. Smith, 1981), again suggesting a heritable component of shape variation. Multivariate analyses have confirmed the correlations of shape between embryonic and adult shells, from which a coefficient of determination of 20% has been estimated (Grahame & Mill, 1993). Further evidence comes from transplant experiments. Janson (1982b) showed by measuring growth rates of wave-exposed and sheltered ecotypes, transplanted between habitats and in the laboratory, that the size difference between them had a larger genetic and lesser environmental component. Behrens Yamada (1987) also found that differences in growth rate were maintained when individuals from geographically separated populations were transplanted. Despite the fact that the species can be bred in the laboratory (Warwick, 1982), few breeding experiments have yet been published. Juveniles of contrasting ecotypes reared in the laboratory are said to retain the shell characteristics of their parental populations (Janson, 1982a; Johannesson *et al.*, 1993). In a more detailed multivariate study, B. & K. Johannesson (1996) found that morphological (and behavioural) differences among three ecotypes persisted in laboratory-reared progeny, although they were less pronounced than among the parental populations. While it thus seems likely that variation between populations in shell shape and size has a large genetic component, the possibility of direct ecophenotypic effects has not yet been fully examined. Such effects are known in other species, for example the influence of growth rate on shell shape in *L. littorea* (Kemp & Bertness, 1984) and on sculpture in *L. sitkana* (Boulding, Buckland-Nicks & van Alstyne, 1993), and the adaptive thickening of the shell of *Nucella lapillus* in the presence of crabs (Palmer, 1990).

Geographical variation. The ecotypic variation of *L. saxatilis* is so great that it partly masks the broad-scale geographical trend in shell form. The contrast between northern and southern varieties is most clear on shores of moderate wave-exposure, for which the following definitions apply:

> Geog. var. *typica* (Figs 106, 107A–H, J, K): shape and sculpture variable, but in moderate ecotype the shell is relatively thick-walled; in large adults the suture of last part of final whorl may descend towards base, and aperture may be constricted relative to penultimate whorl. Range: western Europe south of Narvik (Norway); Maine and southwards in eastern North America.

> Geog. var. *groenlandica* (Fig. 107I, L–Q): shape and sculpture variable, but in moderate ecotype the shell is relatively thin-walled; whorls are more rounded, expansion rate more uniform; often relatively tall-spired (as a result of a smaller apical angle throughout growth, rather than constriction and descent of final whorl); annual growth checks may be conspicuous. Range: Norway north of Narvik, White Sea, Iceland, Greenland, Canada.

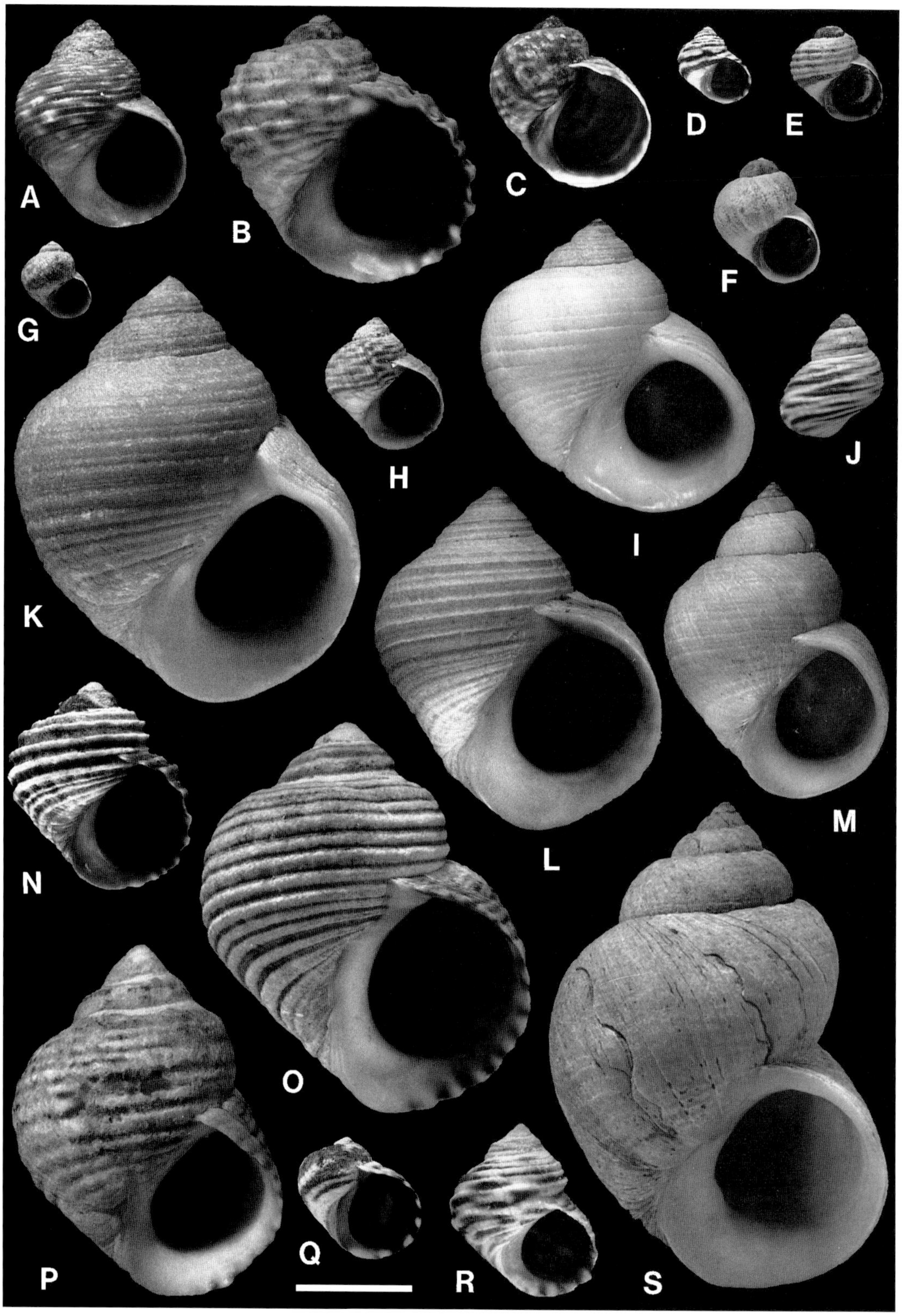

Figure 106. *Littorina (Neritrema) saxatilis*: geog var. *typica*: **A, H, I, K–P, R, S,** moderate ecotype (**H,** juvenile); **B, C,** wave-exposed ecotype; **D–G, J, Q,** barnacle ecotype. **A, H,** Cooden Beach, Bexhill, Sussex, England. **B,** Punta Cires, 40 km NE of Tangier, Morocco. **C,** Rockall. **D, J,** 1 km W of St Govan's Head, Pembrokeshire, Wales. **E, F,** Appledore Island, Isles of Shoals, Maine. **G,** *Littorina neglecta* Bean, 1844; lectotype, BMNH 1849.2.2.301; Scarborough, Yorkshire, England. **I, L,** Rubh' Aird-mhicheil, South Uist, Outer Hebrides, Scotland. **K,** Barfleur, Normandy, France. **M,** Narragansett, Rhode Island. **N,** Westdale Bay, Pembrokeshire, Wales. **O,** Porto, Portugal. **P,** *Littorina nigrolineata* Gray, 1839; lectotype, BMNH 1968363; no locality, but almost certainly from western Iberian Peninsula. **Q,** Cabo Silleiro, Baiona, Galicia, Spain. **R,** San Pedro do Estoril, Portugal. **S,** topotype of *Turbo rudis* Maton, 1797; Cargreen, River Tamar, Cornwall, England. Scale bar=5 mm.

Figure 107. *Littorina (Neritrema) saxatilis*: **A–H, J, K,** geog. var. *typica*; **I, L–Q,** geog. var. *groenlandica*; **A, F–K, M–Q,** moderate ecotype (**K,** juvenile); **B–E,** sheltered/brackish ecotype; **L,** barnacle ecotype. **A,** Alberoni, Venice Lagoon, Italy. **B,** Langebaan, South Africa. **C,** *Turbo tenebrosus* Montagu, 1803; lectotype, BMNH 1989177; no locality. **D,** Walton-on-the-Naze, Essex, England. **E,** El Kantara, Djerba, Tunisia. **F,** Prawl Point, Devon, England. **G,** Appledore Island, Isles of Shoals, Maine. **H,** Dooagh, Achill Island, Co. Mayo, Ireland. **I,** Shad Bay, SW of Halifax, Nova Scotia. **J,** Porlock Bay, Somerset, England. **K,** Kingsgate, Kent, England. **L, O,** Smelror, near Vardø, Norway. **M,** Diabasodden, Spitsbergen. **N,** Tuktu River mouth, Northwest Territory, Canada; NMC 35039. **P,** Höfdakaupstadur, Iceland. **Q,** Thingeyri, Iceland. Scale bar = 5 mm.

Both varieties display the full range of ecotypes, and only examples of the moderate ecotype are easily distinguishable. The other ecotypes converge closely; for example, Figure 107L shows a barnacle ecotype from within the geographical range of var. *groenlandica*, which is indistinguishable from barnacle ecotypes of var. *typica* (Fig. 106D–G, J). It has been suggested that the northern variety is larger (Sars, 1878), and in some parts of the range the northern variety may indeed reach large size, particularly in Iceland and northern Norway, where maximum sizes of local populations often exceed 20 mm. However, specimens of this size can also be found in the British Isles, and size variation is so great in each variety that no consistent trend is evident. The extreme variability of *L. saxatilis* also makes comparisons of relative spire height difficult in the two varieties. Simple measures of shape, such as H/B and SH, show broad overlap between the two varieties, even when restricted to the moderate ecotypes of both. Nevertheless, careful comparison shows that in geog. var. *typica* the apical angle of moderate ecotypes is usually somewhat larger, and that relatively tall, narrow shells are generally produced by the constriction of the aperture and descending suture of the final whorl. In contrast, in geog. var. *groenlandica* the apical angle of moderate ecotypes is smaller throughout growth, without a marked change in shape at the end of the final whorl, thus giving rise to relatively tall, narrow shells with less pronounced doming (compare Figs 106S and 107Q for extreme examples of this contrast).

It is difficult to specify the geographical ranges of the two varieties precisely, because of the superimposed ecotypic variation. Shells resembling the northern variety (thin-walled, tall-spired, with rounded whorls) can be found within the range of the typical variety, for example a form sometimes encountered at the highest tidal levels in crevices on wave-exposed rocky shores in the northern and western British Isles (Fig. 107H). These specimens probably account for the occasional records of var. *groenlandica* in the Orkney (Rendall, 1956) and Shetland Islands (James, 1968a). This overlap in shell characters has led several authors to deny that *groenlandica* is a recognizable variety (Johansen, 1901; Dautzenberg & Fischer, 1912; Bequaert, 1943). However, the thick-walled, moderate ecotype with constricted aperture and descending suture occurs only in the range of the southern variety and is here used to define its extent. The northernmost records of this shell form are Vassvika (just north of Narvik, Norway; DGR, 1989), Rockland (Maine; DGR, 1985) and there is a single collection from the harbour at Heimaey in the Vestmannaeyjar off southwestern Iceland (DGR, 1989). Detailed morphometric studies of moderate ecotypes of both varieties from comparable shores would be desirable to test these trends in shell shape.

This pattern of distribution is closely similar to those of northern and southern varieties of all the other *Littorina (Neritrema)* species in the northern Atlantic, and the same explanation may apply to each of them. The shore crab *Carcinus maenas* is a major predator of *L. saxatilis* on shores of moderate wave-exposure and imposes selection for thick-walled shells with constricted or narrow apertures (Elner & Raffaelli, 1980; Faller-Fritsch & Emson, 1985; Johannesson, 1986) and perhaps with lower spires (by analogy with *L. obtusata*, Seeley, 1986). *Carcinus* occurs throughout most of the southern range of *L. saxatilis*, but reaches only as far north as Finnmark (Vader, 1979), southwestern Iceland (Stephensen, 1939) and northern Maine (Seeley, 1986), thus coinciding with the distribution of crab-resistant shell forms which defines the extent of geog. var. *typica*. Distributions of the crab-resistant southern varieties of the *Neritrema* species are not quite identical, and in view of their differing habitats this is not surprising; for example, in southwestern Iceland the thick-walled variety of *L. obtusata* is known from a wider area than is that of *L. saxatilis*, perhaps because the former occupies a lower position on the shore where it is exposed to heavier predation. Detailed study of *L. obtusata* in New England has shown a close correlation between occurrence of crabs and of the resistant shell variety, and selection appears to be so strong that where the crab has extended its range snails show morphological adaptation within a few generations (Seeley, 1986; p. 213). Similar studies are necessary to investigate this hypothesis in *L. saxatilis*.

Many nineteenth century authors regarded *L. groenlandica* as a distinct species, but following Sars (1878) it has generally been considered as a variety or subspecies (see Synonymy, p. 278, and Taxonomic History, p. 287). Teilman-Friis (1899) justified its separation as a species by counts of cusps and denticles of the radular teeth, but as discussed by Johansen (1901) these characters are variable and do not show consistent geographical trends. Heller (1975a) found broad overlap in the anatomical characters of *groenlandica* and typical *L. saxatilis*, and did not separate them. In the present study some small differences have been noted; embryos are larger, giant embryos more frequent, and penial pigmentation sometimes darker in var. *groenlandica*. There are also some genetic differences, as expected in a species with limited dispersal; Wium-Andersen (1970) found differences in protein banding patterns between a sample from Greenland and others from Europe, and on this basis suggested that *groenlandica* was specifically distinct. Samples from Iceland were included in a study of allozyme variation by K. & B. Johannesson (1990) and in comparisons with Norwegian and Swedish samples showed relatively high genetic identities of between 0.75 and 0.9, which were not taken to indicate specific difference. In addition, karyological studies have not found any significant differences between samples from the Barents Sea, Britain, Sweden and Venice (Janson, 1983a; Birstein & Mikhailova, 1990; Vitturi, Libertini, Panozzo & Mezzapelle, 1995).

Geographical variation on a smaller scale has been investigated on exposed shores in southern and western Britain, with claims of clines in shell shape (Grahame & Mill, 1992; Mill & Grahame, 1995). However, variation between local populations differing in exposure to waves, and even between levels on the same shore, frequently exceeds the geographical differences reported. Such geographical clines are therefore difficult to substantiate, and their possible cause and significance remain unknown.

Ecotypic variation. Four ecotypes can be recognized:

Moderate ecotype (Figs 106A, H, I, K–P, R, S; 107A, F–K, M–Q): occurs on moderately wave-exposed to sheltered shores; shell height up to 25 mm; spire height moderate to tall (H/B=1.12–1.39; SH=1.34–1.76); in geog. var. *typica* shell is thick-walled, aperture may be constricted or narrowed, and suture may descend towards base at end of last whorl; sculpture and colour variable.

Wave-exposed ecotype (Fig. 106B, C): the species is often less common on extremely exposed shores; this ecotype occurs in microhabitats exposed to strong wave action; shell height up to 12 mm; low-spired and patulous (H/B=1.00–1.11; SH=1.18–1.33); thin-walled; aperture enlarged and rounded; sculpture and colour variable.

Barnacle ecotype (Figs 106D–G, J, Q; 107L): restricted to small crevices in beds of barnacles or mussels, mainly on wave-exposed shores; shell height less than 5 mm; spire height moderate to low (H/B=0.96–1.18; SH=1.19–1.63); thin-walled; aperture rounded, not constricted; sculpture smooth or with poorly-developed spiral ribs; colour usually pale grey or whitish with dark brown to black bands or tessellation, or unicolored whitish or dark brown.

Sheltered/brackish ecotype (Fig. 107B–E): occurs on macrophytes in salt marshes and lagoons, and on estuarine rocky shores; shell height up to 14 mm; spire height moderate to tall (H/B=1.17–1.47; SH=1.58–1.84); usually thin-walled; aperture rounded, not constricted; sculpture smooth or almost so; colour variable, but usually unicolored brown, or pale brown with darker bands and tessellation.

It must be emphasized that the scale of ecotypic variation can be small, distinctive ecotypes sometimes being separated by as little as 1 to 5 m. Where the typical microhabitats of different ecotypes adjoin, intermediate shell forms will be found, as shown by many detailed morphometric studies (e.g. Janson & Sundberg, 1983; Janson & Ward, 1985; B. & K. Johannesson, 1990; Johannesson *et al.*, 1993).

Moderate ecotype. The shells originally described from Venice by Olivi (1792) as *Turbo saxatilis* were of this type (p. 316). This ecotype also includes the form described as *Turbo rudis* by Maton (1797), which for much of its history has been a familiar name for the present species (see Synonymy, p. 278, and Taxonomic History, p. 287). Following the rediscovery of Olivi's earlier name (Brusina, 1870), the species gradually became known as *L. saxatilis*, but until quite recently some authors still retained the name *L. rudis* for the Atlantic populations, while their relationship with the Mediterranean *L. saxatilis* was unclear (e.g. Heller, 1975a; Faller-Fritsch & Emson, 1985). Atlantic and Venetian populations have since been shown to be conspecific by allozyme similarity (Janson, 1985b) and by successful laboratory hybridization between individuals from the type localities of *saxatilis* and *rudis* (Warwick *et al.*, 1990). The name *L. rudis* has, however, also been used in another sense, being applied to the moderate ecotype to distinguish it from the wave-exposed ecotype. This usage originated with Kobelt (1907, as a variety) and Dautzenberg & Fischer (1912, as a subspecies; see also James, 1968a). J. E. Smith (1981) went further, arguing (purely from comparisons of shell shape) that *L. rudis* was specifically distinct from the wave-exposed ecotype to which he restricted the name *L. saxatilis*. Fretter & Graham (1980) retained these two specific names, but suggested that they were merely extremes of a cline, and Graham (1988) reduced them to subspecies. The conspecificity of the moderate and wave-exposed ecotypes has been confirmed by morphometric analyses showing intermediates in habitats of intermediate wave-exposure, and by their high genetic identities (Janson, 1982a, b; Janson & Sundberg, 1983; Janson & Ward, 1984). Janson (1982a) named the moderate ecotype on Swedish shores the 'S morph' (sheltered morph).

It appears that crabs exert a strong influence on the shell morphology of the moderate ecotype and, as discussed earlier (p. 301), this is probably largely a result of natural selection. Throughout most of the geographical range of *L. saxatilis* the only molluscivorous crab that occurs on the upper shore is *Carcinus maenas*, and it is more common on shores of moderate exposure than on wave-exposed shores or in salt marshes. Evidence that this is a major predator of *L. saxatilis* comes from several sources, including field observations (review by Pettitt, 1975), the fact that *Carcinus* is a secondary intermediate host of trematode parasites of the snail (James, 1969), and examination of stomach contents (Brandwood, 1985). In laboratory feeding trials crabs adopt various techniques: completely crushing small shells; breaking the spire of larger shells; or in the largest shells inserting a chela into the aperture to either pull the animal out, to crush the columella, or to chip away the outer aperture (Elner & Raffaelli, 1980; Johannesson, 1986). Further evidence comes from breakage scars on shells, which record nonlethal damage that has been repaired by further shell growth. In some populations as many as 32 to 56% of shells have such scars (Raffaelli, 1978b; Elner & Raffaelli, 1980; Hart & Begon, 1982; Faller-Fritsch & Emson, 1985). Scar frequencies must be interpreted with care, however, for they are not necessarily an indication of lethal predation, and damage may also be caused by other agents, such as crushing between mobile boulders. Crushing both by stones and by crabs has been suggested to be more likely on boulder shores of moderate wave-exposure (where crabs are most common) and to exert similar selection for thick-walled shells (Raffaelli, 1978b, 1982); nevertheless, damage by stones is probably insignificant on most shores, because on a geographical scale thick-walled shells are found only in conjunction with *Carcinus* (see Geographical Variation, p. 301).

Several shell parameters have been correlated with the inferred intensity of crab predation and interpreted as adaptive. The larger size of shells on sheltered shores has been explained as the result of diversion of growth into somatic production to outgrow the sizes most vulnerable to predation (Faller-Fritsch, 1977; Hart & Begon, 1982; Janson, 1982b), and Heller (1976) reported a correlation between mean shell height and the size of local crabs on shores in Wales. Many authors have recorded greater thickness (or weight) of shells in moderately sheltered habitats where crabs are common (Heller, 1976; Raffaelli, 1978b; Elner & Raffaelli, 1980; Janson, 1982a; Faller-Fritsch & Emson, 1985; Johannesson, 1986), and laboratory trials have shown that thicker shells are often more resistant to *Carcinus* (Elner & Raffaelli, 1980; Johannesson, 1986), although it is difficult to compare shells that differ only in thickness. Some methods of attack by crabs involve insertion of a chela into the aperture, explaining the narrower, less circular, aperture of populations exposed to heavy crab predation (Heller, 1976; Janson, 1982a); Johannesson (1986) emphasized that a small inner apertural area is the important variable. In another method of attack the spire is crushed, so that a low spire might be predicted (as is the case in *L. obtusata* exposed to crab predation; Seeley, 1986). In comparison with shells from salt-marsh habitats (where predation may be low; p. 317), spires are somewhat lower on moderately sheltered shores. However, in comparison with the wave-exposed ecotype (also exposed to low predation, but with selection for an enlarged aperture; p. 307), spire height is greater. This demonstrates the difficulties of interpreting selection from correlations alone. It also illustrates that shell parameters cannot be considered independently; the constraints of shell geometry in *L. saxatilis* appear to be such that a low spire and narrow (or small) aperture cannot be achieved simultaneously (although this can be achieved in *L. obtusata* and *L. fabalis*, with much wider columellae). A possible solution to this conflict is suggested by the form of the largest adults of the moderate ecotype. Here the spire of juveniles may be relatively low (Fig. 107K), but at the end of the final whorl the size of the aperture is constricted by a combination of a reduction in apical angle and descent of the suture (Figs 106P, S; 107G, J); this has the result that relative spire height then increases, but the very largest shells are attacked by apertural chipping rather than by crushing of the spire, as shown by Elner & Raffaelli (1980). This interpretation is supported by the absence of marked constriction of the final whorl in geog. var. *groenlandica*, which is not exposed to selection by crabs (see Geographical Variation, p. 301).

It has also been observed that in proceeding from exposed towards more sheltered boulder shores, populations show a more pronounced anterior columellar lip (Dytham, Grahame & Mill, 1990). In part this is a feature of juvenile shells (van Marion, 1981; see Fig. 107K and Ontogenetic Variation, p. 318), and may also be emphasized by the relatively narrower aperture of the moderate ecotype.

Wave-exposed ecotype. The name *L. patula* was sometimes used for this shell form during the nineteenth century, later reduced to a variety or subspecies, but again recognized as a species by Heller (1975a) (see Synonymy, p. 278, and Taxonomic History, p. 287). In northeastern England the patulous form was identified as *L. rudissima* by Bean (1844), later used as a subspecific name by Seshappa (1948, 1976). James (in Fischer-Piette *et al.*, 1964; James, 1968a) used the subspecific name *tenebrosa* for this form. Closer investigation of clinal variation between the wave-exposed and moderate ecotypes of *L. saxatilis* indicated that they were conspecific (Raffaelli, 1979a). Meanwhile, description of the new species *L. arcana* suggested that many previous authors had included it with the conchologically similar wave-exposed ecotype of *L. saxatilis* (Hannaford Ellis, 1978, 1979). J. E. Smith (1981) segregated shells from exposed and sheltered coasts in southwestern England on the basis of shell shape alone (and therefore undoubtedly included *L. arcana*), distinguishing them as *L. saxatilis* and *L. rudis* respectively. Fretter & Graham (1980) followed this nomenclature for the wave-exposed and moderate ecotypes of *L. saxatilis*, but noted that they were probably extremes of a cline, and Graham (1988) retained these names at subspecific level.

The most detailed investigation of the relationship between these two ecotypes has been by Janson (1982a, b, 1983a, b) on the Swedish west coast. Here a continuous population of *L. saxatilis* occupied a 2 km stretch of coastline with alternating exposed cliffs and sheltered boulder shores, each with the appropriate ecotype. The shell of the wave-exposed ecotype (termed the 'E morph', or exposed morph) was significantly smaller, thinner-walled, lower-spired and with a larger aperture; its lower growth rate and possibly its shell shape were in part genetically determined; and there were fewer penial glands relative to shell size. Both physiological (Sundell, 1985) and behavioural (B. & K. Johannesson, 1996) differences have also been demonstrated. Significant morphological and genetic differences were found between subpopulations from contrasting habitats less than 60 m apart (Janson, 1982b; Janson & Sundberg, 1983; Janson & Ward, 1984). Nevertheless, the two ecotypes were considered to be conspecific, because morphological intermediates were found in intermediate habitats (Janson, 1982a; Janson & Sundberg, 1983), because allozyme analysis revealed high genetic identities (Nei's, 1972, *I* greater than 0.98) between the ecotypes, and since no correlation was found between genetic and morphological differentiation (Janson & Ward, 1984). There was, however, some evidence for reduced fecundity in hybrids between the moderate and wave-exposed ecotypes, since the morphologically variable intermediate forms showed the highest proportions of abnormal embryos in the brood pouches, suggesting hybrid breakdown (Janson, 1985a). Transfer of wave-exposed, intermediate and moderate ecotypes between habitats showed that each survived better than the others in its habitat of origin, supporting the explanation of the morphological differences in terms of natural selection (Janson, 1983b).

It has long been known that on wave-exposed shores the shells of *L. saxatilis* are usually smaller, thinner-walled, with lower spires and larger, more rounded apertures. This has been observed on shores in Britain (James, in Fischer-Piette *et al.*, 1964; James, 1968a; Heller, 1976; Moore, 1977; Raffaelli, 1978b, 1979a; S. M. Smith, 1979a; J. E. Smith, 1981), Iberia (Fischer-Piette, Gaillard & Delmas, 1967; Fischer-Piette & Gaillard, 1971; Sacchi, 1981), Nova Scotia (Davis, 1973; Newkirk & Doyle, 1975), Venice (Sacchi & Torelli, 1973), Morocco (Gofas, 1975), as well as Sweden (see above; Johannesson, 1986; Sundberg, 1988). The major selective agent in this case is the force of waves (and perhaps wind), because on steeply shelving shores dislodgement is likely to be lethal. The patulous shell form is adaptive because it may accommodate a larger foot for attachment to the rock (James, 1968a; Davis, 1973; Heller, 1976; Raffaelli, 1979a; Janson, 1982a; Atkinson & Newbury, 1984), as confirmed by measurements of foot area (J. E. Smith, 1981; Grahame & Mill, 1986). The low spire is a geometrical consequence of the enlarged, rounded aperture, but has also been considered adaptive, for a globular shell may reduce drag (James, 1968a; Newkirk & Doyle, 1975; Denny, 1988). Other agents, which elsewhere select against an enlarged aperture, are less significant on wave-exposed shores, because crabs are uncommon (Heller, 1976; Raffaelli, 1978b; Elner & Raffaelli, 1980), while constant wave-splash and availability of crevices reduce the risk of desiccation (Raffaelli, 1979a; Faller-Fritsch & Emson, 1985). Sometimes selection pressures may conflict, as on wave-exposed shores where crabs are common; Johannesson (1986) has argued that under these circumstances shells show an enlarged outer apertural area (to shelter a large foot) and a reduced inner apertural area (to avoid insertion of chelae). Although a thin-walled, light shell may be adaptive in reducing dislodgement on wave-exposed shores (Heller, 1976), the energetic cost of shell production in relation to the productivity of the habitat may also be significant; selection may favour the minimum expenditure of energy on shell growth (Raffaelli, 1978b, 1982; Elner & Raffaelli, 1980).

The adaptive significance of shell size is a complex issue; size is related to both survival and fecundity (Hughes & Roberts, 1980; Hart & Begon, 1982; Faller-Fritsch & Emson, 1985; Hughes, 1995), and there may be an obligate relationship between size and shape (Sundberg, 1988). Size must have a strong environmental component, for example relating to food availability, but shell size (or at least growth rate) has also been shown to have a significant genetic component (Janson, 1982b). In general, shells are smaller on strongly wave-exposed shores (e.g. H less than 6 mm in Wales, Heller, 1976; 3 mm in Sweden, Janson, 1982b). In littorinid snails, with relatively poor powers of adhesion to the substrate, selection for small size is unlikely to be imposed directly by the mechanical force of waves, but rather by the ability to shelter from them (Denny, Daniel & Koehl, 1985). On exposed coasts, therefore, shell size is strongly influenced by microtopography of the substrate; smaller shells are able to shelter from waves and wind in narrow rock crevices, and selection may favour early maturation and a maximum size less than the width of available refuges (Heller, 1976; Hart & Begon, 1982). It has been shown experimentally that on exposed shores small snails have a higher survival rate than large ones (Janson, 1983b). A strong correlation between the size of crevices and of shells has been found in several studies; on a boulder shore in Kent juveniles sheltered in empty barnacles, and larger shells in depressions in the rock (Emson & Faller-Fritsch, 1976); in Wales both shell size and crevice size increased at higher tidal levels on a cliff face (Raffaelli & Hughes, 1978), and in both examples provision of artificial crevices increased snail abundance in the predicted size ranges. Crevices may provide protection from other agents of mortality, such as predators and desiccation, but shelter from wave action is arguably the most important, for dependence of shell size on crevice availability is only pronounced on wave-exposed shores (Emson & Faller-Fritsch, 1976; Raffaelli & Hughes, 1978). Not only size-specific mortality (as when shells grow too large for available crevices), but also lifetime mortality may influence shell size. For example, on a shore of moderate exposure Atkinson & Newbury (1984) found a population of small shells inhabiting empty barnacles, but since they found that large shells survived equally well in this habitat, they argued that selection was simply for early maturation as a result of heavy mortality throughout life (animals died on a mudflat below if they were dislodged). This demonstrates the complex interdependence of selection for size and for fecundity.

Shells from habitats differing in wave-exposure usually differ in size and, in order to compare aspects of shape and to remove allometric effects, most authors have attempted to remove the effects of size when making mathematical comparisons of shell form. Various methods have been used, including ratios and regressions on shell height, analysis of covariance and several multivariate techniques. Sundberg (1988) suggested that such procedures are not strictly adequate and that the differences in form (even between wave-exposed and moderate ecotypes, which look so different in shape) are mainly connected with a general multidimensional size factor, one that cannot be represented simply as a linear measurement of shell height. As a result of allometric effects, such as the changing apical angle with growth, the wave-exposed ecotype is similar in shape to a small moderate ecotype. This may imply that modifications of shell form are achieved by developmental heterochrony.

The influence of crevice size also illustrates that marked differences in shape and size can be found between subpopulations of *L. saxatilis* inhabiting different microhabitats at successive tidal levels on a wave-exposed shore. Several studies have reported examples of clinal change in relation to tidal level (B. & K. Johannesson, 1990; Grahame & Mill, 1992). In contrast, sheltered shores show little differentiation of microhabitats, or of shells, over the generally narrow tidal range occupied by the species on those shores. On exposed shores *L. saxatilis* inhabiting barnacle and mussel beds may be so different from crevice-dwelling populations at higher levels that they are here discussed as a separate ecotype; in such cases, selection pressures are believed to be sufficiently strong that morphological and genetic differences are maintained over vertical distances of only a few metres (see Barnacle Ecotype, p. 309). Meanwhile, at the upper tidal limit of *L. saxatilis* on exposed shores, it may live beyond the reach of direct wave action, or in deep and sheltered crevices. Here, shell form may resemble that characteristic of more sheltered shores, and such examples cannot be classified as wave-exposed ecotypes. For example, Figure 107H illustrates a shell from a subpopulation inhabiting deep crevices near the top of the littoral fringe on an extremely exposed shore in western Ireland; this form is thin-walled and tall-spired, thus resembling both the sheltered/brackish ecotype and geog. var *groenlandica*.

A striking example of this kind of differentiation at successive tidal levels on exposed shores is presently under investigation on several shores in Yorkshire, northeastern England (Hull *et al.*, 1996). Here, a form that could be assigned to the moderate ecotype occurs among boulders in the mid eulittoral zone. On larger boulders at the top of the shore the shells are smaller, thin-walled, more globular and with a relatively larger aperture, thus corresponding to the wave-exposed ecotype. Conceivably, in the latter microhabitat the animals may be more exposed to wave action or less subject to crab predation. The investigators have gone further, noting that morphological intermediates between the two forms are rare (only about 2% of the population). In the high-shore form the eggs are larger (larger embryos are a recognized character of the wave-exposed morph, see p. 296). Interestingly, the rare intermediates showed a bimodal distribution of egg sizes and a high incidence of aborting eggs, implying that they were indeed hybrids and suggesting partial reproductive isolation between the two forms. Allozyme data from one shore have supported the presence of a partial barrier to gene flow (Hull, Grahame & Mill, pers. comm.). The authors have speculated that these two forms could be recognized as distinct species, but the evaluation of this claim must await additional information on gene flow over a wider geographical range (see Discussion, p. 335). In a comparable example of shore-level differentiation, in Galicia it has been clearly shown that morphological differences (including both shell characters and size of embryos) and partial reproductive isolation (by assortative mating) do not preclude some gene flow (Johannesson *et al.*, 1993; Johannesson, Rolán-Alvarez & Ekendahl, 1995; Rolán-Alvarez *et al.*, 1996; p. 314).

Some influences on shell shape are less well understood. *Littorina saxatilis* is sometimes absent in situations of the strongest wave exposure, but where it does occur in such conditions it shows a patulous shell form of extreme type. An example is the form that occurs on the cliffs of Rockall (Moore, 1977; Smith, 1979a; Fig. 106C). In some parts of its range it is replaced on strongly exposed shores by *L. arcana*, which may in some way be more suited to this habitat, perhaps because of a generally more patulous shell shape and larger foot (Grahame & Mill, 1986). In southern Britain, at sites where both species occur on the same exposed shores, Grahame & Mill (1989, 1992) have made the interesting observation that *L. saxatilis* is then less patulous in shape than on similarly exposed shores where it occurs alone; that is, where *L. saxatilis* occurs allopatrically on exposed shores it is more like *L. arcana* in shape than is the case when it occurs sympatrically. These authors interpreted this as a case of character displacement. The most significant difference between sympatric and allopatric *L. saxatilis* was the wider penultimate whorl and smaller aperture of the former; it was suggested that this might relate to the accommodation of a larger brood (Grahame & Mill, 1989), although subsequent studies either did not support this (Grahame *et al.*, 1990) or provided equivocal evidence (Grahame & Mill, 1992). In view of the strong connection between microhabitat and shell shape in *L. saxatilis*, an alternative interpretation might be that in sympatry the two species occupy subtly different microhabitats, perhaps as an outcome of some form of interspecific competition. It is already known that intertidal zonation and migratory behaviour differ in *L. saxatilis* and *L. arcana* (Hannaford Ellis, 1979, 1985); the less patulous shell of the former in sympatry suggests that it might occupy a microhabitat more sheltered from the full force of waves, perhaps at a higher tidal level (as shown by Hannaford Ellis, 1985, for part of the year) or in deeper crevices. This remains to be investigated. More recently Mill & Grahame (1995) have suggested that the apparent character displacement might merely be an artefact of a geographical cline in the shape of *L. saxatilis*.

In both the wave-exposed and moderate ecotypes shell sculpture shows wide variation (see Sculptural Variation, p. 318).

Barnacle ecotype. The small ovoviviparous form typically found among barnacles on exposed shores is the most distinctive of the ecotypes of *L. saxatilis*, and there is a continuing controversy about its possible status as a distinct species. Confusingly, it was first named *L. saxatilis* by Johnston (1842), unaware of the secondary homonymy with the species of Olivi (1792), and as such it was accepted as a distinct species by a number of nineteenth century authors (see Synonymy, p. 278, and Taxonomic History, p. 287). It was reduced to a variety by Jeffreys (1865) and thereafter occasionally appeared in the literature as *L. rudis* var. *saxatilis* Johnston. The form was redescribed as *L. neglecta* by Bean (1844), but this name was largely overlooked in the early literature (exceptions were Alder, 1848; and Kobelt, 1907). Dautzenberg & Fischer (1912) included the barnacle ecotype in the typical subspecies of *L. saxatilis* (Olivi). This ecotype became more widely known as a result of the work of James, first as *L. saxatilis saxatilis* Johnston (James, 1963; James, in Fischer-Piette *et al.*, 1964) and then as *L. saxatilis neglecta* (James, 1964, 1968a, b), although he evidently included some examples of the sheltered/brackish ecotype from salt marshes in his concept of the taxon. In Heller's (1975a) influential revision of the *L. saxatilis* group, *L. neglecta* was recognized as a full species. Since then it has been the subject of studies of morphology (Raffaelli, 1979a; Fish & Sharp, 1985; Dytham, Mill, Grahame & O'Higgins, 1992; Caley *et al.*, 1995; Grahame *et al.*, 1995), shell coloration (Hannaford Ellis, 1984), ecology (Raffaelli, 1978a; Hannaford Ellis, 1983), genetics (Wilkins & O'Regan, 1980; Knight & Ward, 1986), physiological tolerance (Cannon & Hughes, 1992) and enzyme activity (Hull, Mill & Grahame, 1994; Grahame *et al.*, 1995); it has also been treated in reviews and monographs (Sneli & van Marion, 1979; Fretter & Graham, 1980; Smith, 1982; Graham, 1988). Although difficulties of identification were sometimes noted in these studies, none seriously questioned the specific status of *L. neglecta*. Doubts began to appear when no unequivocal morphological characters could be found for the separation of *L. neglecta* from *L. saxatilis* (Warmoes *et al.*, 1988; B. & K. Johannesson, 1990). A combined study of shell morphometrics and allozymes, using samples from a wide area of northwestern Europe, failed to find any consistent differences, and as a result the two taxa were synonymized (B. & K. Johannesson, 1990; K. & B. Johannesson, 1990, 1993; but see Grahame, Mill, Double & Hull, 1992, 1993). This was supported by anatomical examination, which also failed to find significant qualitative differences (Reid, 1990b, 1993). Furthermore, it was discovered that the other two species in the *L. saxatilis* group also produced barnacle ecotypes, almost identical in shell form to *L. neglecta*, supporting its status as an ecotype (Reid, 1993). More recently it has been suggested, from morphometrics and measurements of enzyme activity, that *L. neglecta* exists as a species microsympatrically with a barnacle ecotype of *L. saxatilis* (Grahame *et al.*, 1995). The application of new DNA technology has not so far supported the specific status of *L. neglecta* (Crossland *et al.*, 1995).

Some significant work has been carried out in Galicia, in northwestern Spain, where, on wave-exposed shores, a small, smooth-shelled form of *L. saxatilis* occurs among mussels on the lower shore, in striking contrast to a larger, ridged, moderate ecotype found at higher levels (Johannesson *et al.*, 1993; Johannesson, Rolán-Alvarez & Ekendahl, 1995; Rolán-Alvarez, Johannesson & Ekendahl, 1995; Rolán-Alvarez & Rolán, 1995; Rolán-Alvarez *et al.*, 1996). Detailed studies have revealed that these two forms are conspecific, because there is some gene flow between them, but that there is a degree of assortative mating in their narrow zone of overlap. The mussel-dwelling form on Galician shores is morphologically analogous to the barnacle ecotype in northwestern Europe, but for the sake of clarity, and since it is not yet clear if the same genetic interpretations can be made in each case, it will be discussed separately (p. 314). Although most of the studies of the barnacle ecotype have been carried out in northwestern Europe, and particularly in the British Isles, a corresponding ecotype also occurs in the western Atlantic (Fretter & Graham, 1980; pers. obs.; Fig. 106E, F).

In the following review of the status of the barnacle ecotype the terms 'barnacle ecotype' and 'typical *L. saxatilis*' (including the moderate and probably also the wave-exposed ecotypes) will be used for uniformity, irrespective of the nomenclature adopted by the authors quoted. Exceptions are necessary when discussing results which suggest differentiation between *L. neglecta* and an additional barnacle ecotype of *L. saxatilis*.

All authors have noted the smaller size of the barnacle ecotype, which may mature at a shell height as small as 1.2 to 1.5 mm (James, in Fischer-Piette *et al.*, 1964; Fish & Sharp, 1985) and reaches a maximum size of about 5 mm (Heller, 1975a; Fish & Sharp, 1985). However, mature individuals of typical *L. saxatilis* can also be found in this size range (Warmoes *et al.*, 1988; Caley *et al.*, 1995; Grahame *et al.*, 1995). Shell shape has also usually been mentioned, the barnacle ecotype having a moderate to low spire, rounded whorls, and rounded aperture lacking an anterior lip (at least when mature), in contrast to typical *L. saxatilis* from higher tidal levels on the same shores. Shape has been regarded as diagnostic in some qualitative studies, with few intermediates being found (Fish & Sharp, 1985), while others have found difficulty in discriminating the two forms by shape (Warmoes *et al.*, 1988; Reid, 1993), as illustrated in Figures 108E-J.

Multivariate morphometric studies have also been carried out, but the results have been contradictory, depending upon the sampling strategy. B. & K. Johannesson (1990) analysed material from geographically distant sites in northwestern Europe and compared samples from the barnacle zone with those collected higher on the shore. Significant differences in shape were found between these microhabitats on some shores, but were not consistent between shores. In addition, geographical variation between samples from barnacles was more pronounced than variation among samples from successive levels within shores. This approach is susceptible to the objection that real differences in shape between two taxa might be obscured if there is overlap in their intertidal zonation, as suggested by Caley *et al.* (1995). These authors therefore first classified their samples according to qualitative characters of shell sculpture and coloration, before multivariate analysis. As a result, it was claimed that a category identified as *L. neglecta* could be recognized at a number of sites in the northern Atlantic. Furthermore, this was more distinct morphologically from *L. saxatilis* than was *L. arcana*, as shown by different principal components of shape variation in *L. neglecta*; it was also less variable, as shown by its higher scores in cross-validation of discriminant analysis. These differences were not simply a consequence of size, for the *L. neglecta* category differed significantly from *L. saxatilis* in the same size range. A similar study of material from a single site in Yorkshire supported these results (Grahame *et al.*, 1995). However, this approach can also be criticised, for the criteria used for sorting the samples before analysis were not clearly defined, and a subjective assessment of shell shape may have been inadvertently included. In addition, the method of analysis was designed to confirm differences between groups, rather than to detect intermediates. In fact, the study of Grahame *et al.* (1995) in Yorkshire did detect an additional shell type, similar to *L. saxatilis* but maturing at small size (5 mm), inhabiting the barnacle zone together with *L. neglecta*. These small shells accounted for only about 10% of the snails in the barnacle zone and in size and shape appear to be intermediates, and therefore possible hybrids, between the categories identified as *L. saxatilis* and *L. neglecta*. Nevertheless, the authors suggested that these small snails corresponded to a barnacle ecotype of *L. saxatilis*, distinct from *L. neglecta* itself, because of their microsympatry in the barnacle zone and differences in enzyme activity (p. 313). Yet another technique applied to the description of shell shape has been Fourier analysis (Dytham, Grahame, Mill & O'Higgins, 1992). From these various qualitative and quantitative comparisons of shape it emerges that, on shores where ecotypic differentiation is pronounced, the barnacle ecotype is a highly distinctive form.

Another aspect of shell form is the degree of sculpture. Most authors have observed a range of sculpture, from smooth to ridged, in the barnacle ecotype (James, 1968a; Heller, 1975a; Fish & Sharp, 1985; see Fig. 108H, I), although at some sites barnacle ecotypes are more often smooth-shelled than *L. saxatilis* at higher levels (B. & K. Johannesson, 1990). Some have gone further, using the absence of ridged sculpture as the single most important character of the shell by which to distinguish the barnacle ecotype, and identifying a strong correlation with shell shape (Caley *et al.*, 1995; Grahame *et al.*, 1995).

Shell coloration has also been used to characterize the barnacle ecotype, which is usually whitish to grey, often with brown to blackish lines breaking up into tessellation (especially at larger size), or entirely blackish brown. Nevertheless, while coloration may separate the ecotypes on individual shores, it is clear from Figures 106 and 107 that the colour patterns of the barnacle ecotype are not unique in the group as a whole. One frequently-mentioned trait is the presence of a brown or black band on the base, running into the aperture (James, in Fischer-Piette *et al.*, 1964; James, 1968a, b; Heller, 1975a), but this is not consistently present in the barnacle ecotype (Hannaford Ellis, 1984; B. & K. Johannesson, 1990; Reid, 1993). Furthermore, juveniles of typical *L. saxatilis* sometimes show a similar band (Raffaelli, 1979a; Fish & Sharp, 1985; compare juvenile and adult in Fig. 106H and A).

In the interpretation of the morphological evidence discussed so far, it is clear that the presence of numerous intermediates between the ecotypes counts against their recognition as distinct species. However, the converse, where the ecotypes are significantly different and intermediates are few, should not be accepted uncritically as evidence for specific difference. Studies have repeatedly shown that the scale of ecotypic variation in *L. saxatilis* can be very small indeed, over distances of 2 to 3 m or less (e.g. Janson & Ward, 1985; B. & K. Johannesson, 1990; Johannesson *et al.*, 1993). Even on a microsympatric scale (within 1 m^2) it is believed that a degree of assortative mating, combined with strong disruptive selection, can maintain significant morphological differences between ecotypes in the presence of some gene flow (Johannesson *et al.*, 1993; Johannesson, Rolán-Alvarez & Ekendahl, 1995). The design of the sampling strategy on a fine scale, and the relation of samples to their microenvironment, have not been adequately considered in the studies so far described. In addition, it is necessary to consider possible intermediates between ecotypes, and their micro-distribution. Nevertheless, morphological evidence alone will remain inadequate to resolve the relationship between the barnacle ecotype and typical *L. saxatilis*, and can only be assessed in combination with genetic and behavioural studies. These points are clearly demonstrated in the fine studies of Galician *L. saxatilis* (p. 314).

The anatomical characters that distinguish some other closely-related *Littorina* species have also been examined. The cusps of the radular teeth of the barnacle ecotype are all more pointed than in adult *L. saxatilis* of larger size, the cusps of the outer marginal tooth are more numerous (7–8, cf. 4–6), and the rachidian tooth is more elongate (Fig. 105E); however, all these characters are found in juvenile *L. saxatilis* of comparable size (Fig. 104D) (Seshappa, 1976; Raffaelli, 1979a; Reid, 1990b; Caley, 1994; p. 300). A 'figure of 8' pattern of black pigment is generally visible through the operculum of the barnacle ecotype and has been considered reliable for its identification by some authors (Hannaford Ellis, 1984), but not others (Fish & Sharp, 1985; B. & K. Johannesson, 1990; Caley, 1994). In fact a similar pattern is found not only in some darkly pigmented *L. saxatilis*, but in many other *Littorina* species also. Strangely, penial characters, so useful for identification in some other species of the genus, have received little attention. Fish & Sharp (1985) recorded 1–9 mamilliform glands in the barnacle ecotype and described them as relatively larger than in *L. saxatilis*, but no qualitative differences in penial shape were noticed by Reid (1993) or in the present study (Fig. 102D, E, P, Q). The numbers of mamilliform glands of this ecotype do fall in the lower part of the range in *L. saxatilis*; however, the numbers of glands are closely correlated with shell height in this group (Heller, 1975a; Raffaelli, 1979a, but note that both these authors may have included *L. arcana*; Janson, 1982a), so that this difference can be accounted for as a function of body size (Caley, 1994). Furthermore, gland number does sometimes show variation without indicating specific difference; for example, significant differences in the regression of penial gland number on shell height have been shown between the moderate and wave-exposed ecotypes of *L. saxatilis* (Janson, 1982a).

Oviducts have also been examined, but without noting obvious qualitative differences (Reid, 1993; Fig. 103J). In the present study it has been found that the bursa is sometimes vestigial or absent in very small adults (less than 5 mm), but this does not seem to be correlated with ecotype. In a quantitative comparison of the pallial oviducts of the barnacle ecotype and typical *L. saxatilis* on the same shores, Caley (1994) found some small, but statistically significant, differences; there were differences in the allometric relationships between the developing glands, and the barnacle ecotype showed a slightly larger albumen gland than *L. saxatilis* of comparable size. The embryos of the barnacle ecotype have been said to be either smaller (Fish & Sharp, 1985) or larger (Hannaford Ellis, 1983) than in typical *L. saxatilis*. Fecundity in relation to shell height has been found to be greater in the barnacle ecotype, and slight differences in embryo shape have been reported (Caley, 1994). Embryo number and size are also known to show significant variation between populations of typical *L. saxatilis* (p. 296).

It is perhaps the characteristic habitat, combined with the distinctive appearance, of the barnacle ecotype that has contributed most to its acceptance as a species by so many field workers. At least on British shores, *L. saxatilis* often appears to be restricted to the littoral fringe, whereas the barnacle ecotype occurs in the mid and upper eulittoral zone (Heller, 1975a; Raffaelli, 1978a). Elsewhere this pattern is not maintained; for example, typical *L. saxatilis* can be found throughout the eulittoral zone, and even subtidally, in other parts of its range (see Habitat, p. 324). Nor is the barnacle ecotype always so specific in its habitat, analogous forms being found in other sheltered microhabitats in the eulittoral zone of exposed shores, for example among mussels (Johannesson *et al.*, 1993), in *Laminaria* holdfasts and among *Corallina* (Fretter & Graham, 1980). Shells morphologically identical to the barnacle ecotype have been reported in rock crevices above the level of the barnacle belt (Caley, 1994; Caley *et al.*, 1995). Habitat details alone can provide no evidence about specific or ecotypic status; strong disruptive selection might produce shells resembling barnacle ecotypes in a variety of interstitial microhabitats on wave-exposed coasts. Nevertheless, the microdistribution of these forms in relation to that of supposed conspecifics is of considerable interest. For example, at a site in Yorkshire the occurrence of small, reproductively mature *L. saxatilis* in the barnacle zone together with morphologically distinct *L. neglecta* has been cited as evidence that these are separate species (Caley, 1994; Caley *et al.*, 1995). Additional studies, of gene flow and behaviour (possible hybridization, assortative mating, and migration), will be of great interest in this case. So far, there have been no detailed studies of mating behaviour (of the kind carried out in Galicia, see p. 314) in the barnacle ecotype. Hybridization between the barnacle ecotype and typical *L. saxatilis* was attempted by Warwick *et al.* (1990), but without success. However, since crosses within the barnacle ecotype were not successful either, this negative result may not be significant. In further work T. Warwick (quoted in Reid, 1993) has achieved successful hybridization between the two forms.

Different demographic patterns have also been found; in Wales *L. saxatilis* breeds almost continuously, whereas the barnacle ecotype breeds mainly in the spring and behaves as an annual (Hannaford Ellis, 1983; Fish & Sharp, 1985). The annual life cycle could, however, be a consequence of the small size, early maturity and perhaps high mortality rate of the barnacle ecotype, and would not appear to preclude the possibility of interbreeding.

A few investigations of the barnacle ecotype have involved physiological comparisons. Tolerance of high temperature, low salinity and starvation have been examined by Cannon & Hughes (1992), but although some differences between the two forms were indicated, these could have been correlated with size. In a study of the activity and heat sensitivity of enzymes (notably aspartate aminotransferase), significant differences have been detected, even between *L. neglecta* and the small forms of typical *L. saxatilis* with which they were microsympatric among barnacles (Hull *et al.*, 1994; Grahame *et al.*, 1995).

Since biological species are defined in terms of interbreeding and gene flow, these lines of evidence must be the most significant in assessing the status of the barnacle ecotype. It seems likely that much of the morphological difference between the ecotypes of *L. saxatilis* may have a genetic origin (p. 301). The question is whether the pattern of genetic differentiation between the barnacle ecotype and typical *L. saxatilis* is such as to consider them distinct species. Several studies have examined allozyme patterns at single localities. Heller (1975a) reported different esterase banding patterns in Wales. In Ireland, the barnacle ecotype was significantly different from typical *L. saxatilis* in allele frequencies (at the phosphoglucomutase locus) at three localities (Wilkins & O'Regan, 1980). Both these studies are open to the criticism that *L. arcana* may have been inadvertently included with the 'typical' form. This is not so in the work of Knight & Ward (1986), who at a site in Yorkshire found a significant difference at the nucleoside phosphorylase locus. Nevertheless, in view of the microgeographic scale of genetic differentiation in *L. saxatilis*, the interpretation is unclear. Recently, on a single well-studied shore in Yorkshire, significant differences in allele frequencies at five loci have been found between *L. neglecta* and a small, barnacle-dwelling form of *L. saxatilis* with which it is microsympatric, which imply a barrier to gene flow between them (Hull, Grahame & Mill, pers. comm.).

Such results are extremely interesting, but their taxonomic implications can only be assessed in a wider, geographical, context. In samples from several localities in northwestern Europe, K. & B. Johannesson (1990) were able to show that at four loci 83–95% of between-sample variation was due to geographical variation, and only less than 4% to differences between samples from the barnacle zone and littoral fringe. They concluded that there was more gene flow between the barnacle ecotype and typical *L. saxatilis* on the same shore than between snails of similar morphology on geographically separated shores, and that ecotypic status was therefore appropriate. This is the test that distinguishes ecotypes from 'sibling' species (see Species Concepts, p. 375). However, as noted earlier, advocates of *L. neglecta* as a distinct species have questioned the validity of the initial sorting of samples in this study, based on tidal level rather than morphology (Caley *et al.*, 1995). One caveat in the use of allozymes to estimate gene flow is that the alleles are assumed to be selectively neutral. In this case there was evidence for selection at a fifth locus (aspartate aminotransferase), which did show marked and consistent differentiation over the tidal range, but regardless of whether barnacle ecotypes were present or not. This locus was therefore excluded when considering gene flow between ecotypes (K. & B. Johannesson, 1990, 1993; misinterpreted by Grahame, Mill, Double & Hull, 1992). (Further evidence for selection on this locus is that the same pattern of allelic differentiation over the tidal range appears in *L. arcana*; Grahame, Mill, Double & Hull, 1992, 1993). A preliminary investigation using the new technique of random-amplified polymorphic DNAs (RAPDs) has been carried out at several sites on one headland in Wales (Crossland *et al.*, 1995). Phenetic cluster analysis of the resulting banding patterns failed to show consistent differentiation of the barnacle ecotype from the typical one, although on one shore there was slightly greater similarity within than between ecotypes. The genetic study of barnacle-dwelling forms in Yorkshire by Hull, Grahame & Mill has not yet been extended to other shores in the vicinity; if, on a scale of tens of kilometres, it can be demonstrated that there is more gene flow among populations of *L. neglecta* than between it and *L. saxatilis* within shores, it may well deserve recognition as a distinct species.

In summary, on individual shores the barnacle ecotype (*L. neglecta* of authors) may show striking morphological differences from typical *L. saxatilis* on the same shore, with few intermediates, and may even show significant genetic differentiation. As yet, no geographically consistent morphological character has been discovered with which to characterize *L. neglecta*, making objective sampling strategies difficult. Furthermore, many of the earlier studies were carried out before the fine scale of microgeographical variation within the variable taxon *L. saxatilis* was fully appreciated, so that the sampling procedure was inadequate. From the sometimes conflicting accounts and interpretations of comparisons between *L. saxatilis* and *L. neglecta*, it appears that there may perhaps be regional variation in the relationship between the two. In some areas, notably Yorkshire, there may be a considerable degree of reproductive isolation between the two, and at least on individual shores *L. neglecta* is distinct morphologically, physiologically and genetically from *L. saxatilis*, leading some authors to recognize it as a separate species (Caley *et al.*, 1995; Grahame *et al.*, 1995). In other areas, such as Wales, Iceland, Scandinavia and Belgium, workers have noted morphological intermediates between the two (Fish & Sharp, 1985; Warmoes *et al.*, 1988; B. & K. Johannesson, 1990; Reid, 1993), and have found some genetic evidence to support the interpretation of *L. neglecta* as a barnacle ecotype of *L. saxatilis* (K. & B. Johannesson, 1990; Crossland *et al.*, 1995). Another consideration is that so far almost all comparative studies have been carried out on wave-exposed shores, where differentiation between the two forms is most pronounced, perhaps because disruptive selection is then most intense (p. 315). The limited genetic evidence so far available suggests greater gene flow between forms on the same shore than among morphologically similar ecotypes on geographically distant shores, and is therefore consistent with the ecotype status of the barnacle form. However, there is a clear need for detailed genetic and behavioural studies, of the kind described in the following example.

On wave-exposed shores in Galicia there is a similar case of two contrasting ecotypes that occupy different tidal levels (Fischer-Piette, Gaillard & Delmas, 1967; Johannesson *et al.*, 1993; pers. obs.). Here a small (less than 5 mm), greyish white, often tessellated, but unbanded, smooth-shelled form occurs among beds of mussels (*Mytilus*) low on the shore (Fig. 106Q). This is the 'smooth and unbanded morph' of Johannesson *et al.* (1993) and, although rather larger, is otherwise morphologically similar to the barnacle ecotype in other geographical areas. Its habit of nestling in crevices, but among mussels rather than barnacles, at relatively low tidal levels on strongly wave-exposed rocks, is also similar. At higher levels a larger form is found on barnacle-covered rocks and boulders, with a ridged shell, banded in black on a whitish ground (Fig. 106O, P). This microhabitat is somewhat sheltered from the full force of the waves and, although found in a zone of abundant barnacles, the size and shape of this form show that it can be included in the category of the moderate ecotype. This has been named the 'ridged and banded morph' (Johannesson *et al.*, 1993). The two forms are largely restricted to their respective habitats, but their distributions overlap in a narrow belt 2–3 m wide, where the barnacle and mussel zones meet. Even in this region of overlap the typical forms predominate, but intermediates also occur at a low frequency (8–29%). These appear to be viable and fertile hybrids, of which the females carry developing embryos. Other morphological differences have also been described. The smooth, unbanded form has larger embryos and more numerous cusps on the outer marginal radular teeth (Rolán-Alvarez *et al.*, 1996).

A genetic study of this interesting case (Johannesson *et al.*, 1993) has shown that on a geographical scale of tens of kilometres of shoreline, morphological and genetic distances are not correlated. This suggests that the two forms are conspecific, sharing a common gene pool (see also Rolán-Alvarez & Rolán, 1995). However, on a microscale on single shores, the genetic distance between the smooth, unbanded form and the ridged, banded form just 10–20 m away is greater than between samples taken from similar distances within the habitat of each morph (mean Nei's genetic distances of 0.046 and 0.007 respectively, based on five polymorphic enzyme loci; Johannesson, Rolán-Alvarez & Ekendahl, 1995), indicating a partial barrier to gene flow between the two forms. Laboratory rearing has shown that the conchological differences are probably genetically determined (Johannesson *et al.*, 1993). Furthermore, gene flow between the two forms is estimated to be sufficient to counteract random genetic drift (Rolán-Alvarez *et al.*, 1996). The morphological and genetic patterns on these Galician shores are therefore believed to be maintained by strong disruptive selection between low-shore and high-shore habitats, together with a partial barrier to gene flow between the two forms.

The nature of this barrier has been investigated in detailed studies of mating behaviour (Johannesson, Rolán-Alvarez & Ekendahl, 1995; Rolán-Alvarez, Johannesson & Ekendahl, 1995). Analysis of mating pairs in the field confirmed partial reproductive isolation between high-shore and low-shore forms. This was maintained by three distinct mechanisms: microhabitat choice in the zone of overlap covering patches of barnacles and mussels; assortative mating even within patches (perhaps partly mediated through size selection); and a degree of sexual selection, hybrid females having reduced sexual fitness. Frequency-dependent and density-dependent components of sexual selection have been identified (Rolán-Alvarez, Johannesson & Ekendahl, 1995). This presents an important case study showing how incipient reproductive isolation may arise in sympatry, perhaps leading to speciation. However, it must be emphasized that reproductive isolation is not yet complete, and that since gene flow between the two forms in sympatry exceeds that within the forms on a geographical scale, they must be still regarded as members of a single biological species (Johannesson *et al.*, 1993; Rolán-Alvarez *et al.*, 1996). There is a clear parallel with the case of the barnacle ecotype known as *L. neglecta*, but until comparable genetic and behavioural investigations are carried out on northern European shores it remains uncertain that the genetic interpretation and systematic conclusion are the same. The Galician example does emphasize that extreme morphological differentiation, even in microsympatry and combined with partial reproductive isolation, must not be accepted as evidence of species-level differentiation without detailed study of patterns of gene flow.

In considering the status of the barnacle ecotype, it is also pertinent to consider patterns of variation in related species. The two other species of the *L. saxatilis* group, *L. arcana* and *L. compressa*, also sometimes occur in the barnacle zone on exposed shores. Since they too have nonplanktotrophic development and therefore restricted dispersal, they can be expected to respond in a similar way to the selective regime of this microhabitat. Indeed, both have been found to produce shells almost identical to those of the barnacle ecotype of *L. saxatilis* under these conditions (Reid, 1993; pp. 255, 273). While this is not direct evidence for the status of the latter, it does emphasize that it is no more distinct morphologically from the typical form of the species than is the case for ecotypes found in closely-related species.

There is at present little evidence for the possible selective forces acting on the barnacle ecotype. Assuming that its shell characters are determined genetically, as suggested by laboratory rearing of the Galician snails (Johannesson *et al.*, 1993), some appear to be adaptive. In size and shape it approaches the wave-exposed ecotype (although usually smaller and less patulous), and similar arguments apply. Small size allows it to shelter from strong wave action in the eulittoral zone inside crevices, whether among barnacles, mussels or weed holdfasts. It is a common observation that the barnacle ecotype is most well developed and distinctive on strongly wave-exposed coasts where barnacles are abundant, and most authors who have claimed species status for this ecotype have worked on such shores. In contrast, barnacle ecotypes are difficult to differentiate, or cannot be found, on moderately exposed shores in the English Channel (Warmoes *et al.*, 1988), and typical barnacle ecotypes are absent on Swedish shores, where the barnacle belt is narrow and subject to severe ice-scour in winter (B. & K. Johannesson, 1990). Reid (1993) suggested that *L. saxatilis* is intolerant of extreme wave-exposure (see Habitat, p. 324), and on strongly exposed coasts is able to survive only in the shelter of crevices; these are usually available mainly in the small interstices among barnacles and mussels in the eulittoral zone, and in larger rock crevices or among boulders higher up the shore in the littoral fringe, above the full force of the waves. The dependence of *L. saxatilis* upon crevices of suitable size is well known (Emson & Faller-Fritsch, 1976; Raffaelli & Hughes, 1978) and applies equally to the barnacle ecotype (Raffaelli, 1978a). The combination of wave action with the availability of crevices in two microhabitats results in strong disruptive selection, producing the conchologically distinct lower-shore barnacle ecotype on exposed shores. Where wave action is more moderate, disruptive selection may be weaker and other factors (such as crab predation, desiccation and fecundity) may be important in determining shell form, so that a distinct barnacle ecotype does not develop. On shores of certain rock types suitable crevices might be available throughout the tidal range, and there more gradual clinal change with tidal level would be predicted. The outcome of selection therefore depends upon local characteristics of the shore, including topography, rock type, exposure, and zonation of barnacles and mussels. Since such characters may be similar on comparable shores within geographical regions, the repeated differentiation of morphologically similar barnacle ecotypes on a geographical scale (as appears to have occurred in the Galician example) is not unexpected. Visual selection by predators may account for the cryptic coloration of the barnacle ecotype (Raffaelli, 1982; see Colour Variation, p. 320).

In several characteristics (shell size and shape, radular teeth, number of penial glands) the barnacle ecotype resembles juvenile *L. saxatilis* of more typical shell form. This has led several authors to suggest that, in the barnacle zone, selection for maturity at small size has resulted in a paedomorphic form (Emson & Faller-Fritsch, 1976; Raffaelli, 1979a; Fish & Sharp, 1985). The larger number of cusps on the outer marginal teeth of the low-shore Galician form (Rolán-Alvarez *et al.*, 1996) is also a juvenile character.

Sheltered/brackish ecotype. Historically, this form was first named by Montagu (1803) as *Turbo tenebrosus*. There has since been some uncertainty about the interpretation of this name (James, 1968a; Barnes, 1993), but the original description mentions chocolate-brown colour, 'ventricose spires' (i.e. rounded whorls), suborbicular aperture, thin outer lip, diameter of one quarter of an inch (6.3 mm), and it was said to be found on mud, rocks and 'even in ditches subject to the daily flux of the tide' (suggesting brackish creeks). '*Turbo tenebrosus* var.' was figured by the author (Montagu, 1808), showing a tall-spired shell with swollen whorls. The evidence thus gives a reasonable picture of what is here considered the sheltered/brackish ecotype, and this interpretation of *Turbo tenebrosus* is fixed by designation of a lectotype (Fig. 107C). Under this name, or as *L. tenebrosa*, it was recognized by many authors in the first half of the nineteenth century, before being reduced to a variety or subspecies (see Synonymy, p. 278, and Taxonomic History, p. 287). However, doubts about its status persisted; Teilman-Friis (1899) considered that small differences in denticles and cusps of radular teeth justified its separation as a species, but Johansen (1901) showed that these characters were not constant. James (1968a) did not recognize one purely brackish-water subspecies in his classification of British *L. saxatilis s.l.*, but considered that most of his subspecies could extend into brackish habitats; inexplicably, he applied the subspecific name *tenebrosa* to the wave-exposed ecotype. Further confusion was introduced by Davis (1971, 1973) who attempted to follow the descriptions given by James (1968a) and identified shells from salt marshes and eelgrass beds in Nova Scotia as *L. saxatilis neglecta*. Robertson & Mann (1982) also used the name *L. neglecta*, as a full species, for this same Nova Scotian form. Working in lagoons in Denmark, Muus (1967) found such a contrast between the habits and distributions of form *tenebrosa* and the fully marine *L. saxatilis* that he found it hard to accept their specific identity. Fretter & Graham (1980; also Graham, 1988) provisionally accorded it full specific status, and Smith (1982) argued that this was appropriate in view of the habitat segregation between *tenebrosa* submerged on weeds and *L. saxatilis* on the shores of the same lagoons. However, closer investigation has revealed clinal change between the moderate ecotype of *L. saxatilis* and the sheltered/brackish ecotype referred to as *L. tenebrosa*, for example in the Isefjord in Denmark (Rasmussen, 1973), along a brackish inlet in Dorset (Brandwood, 1983, 1985) and over a distance of only 5 m between sheltered rocks and submerged *Zostera* in Sweden (Janson & Ward, 1985). In addition, allozyme analysis showed high genetic identities (Nei's, 1972, *I* of about 0.99) between the ecotypes, and no correlation between morphological and genetic differentiation, when samples from Scotland and Sweden were analysed together (Janson & Ward, 1985). As a final piece of evidence for conspecificity, the two ecotypes have been successfully hybridized (Ward *et al.*, 1986). Recently, Barnes (1993) has argued that snails living permanently submerged on macrophytes in lagoons are morphologically and behaviourally different from those inhabiting lagoonal and estuarine shores, and are effectively reproductively isolated from them by virtue of their habitat, although laboratory crosses were successful; consequently he distinguished the submerged type as var. *lagunae*. This is regarded as an extreme form of the sheltered/brackish ecotype here, but is discussed separately (p. 317).

A further taxonomic issue concerns the nature of *Turbo saxatilis* itself, described from Venice (Olivi, 1792). Much of the shore of the Venice Lagoon consists of salt-marsh vegetation, where shells of the sheltered/brackish ecotype are abundant; however, on artificial rock substrates on less sheltered shores the shells can be classified as moderate ecotypes (Fig. 107A). Since Olivi (1792) mentioned occurrence in fissures of rocks, and illustrated shells of moderate spire height with spiral striations and a projecting anterior columellar lip (Fig. 101), it was evidently to these moderate ecotypes that his name referred (Sacchi & Torelli, 1973; Torelli, 1977, 1982b; Janson, 1985; see discussion by Barnes, 1993).

A number of authors have described the changes in shell morphology that are observed when *L. saxatilis* is sampled along an environmental gradient from an open coast into a sheltered inlet or estuary. In the absence of figures of shells or of salinity measurements, it can be difficult to interpret these results, but in many cases the changes described seem to apply only to the transition between wave-exposed and moderate ecotypes, involving increasing spire height, less pronounced sculpture and a smaller and narrower aperture (Fischer-Piette & Gaillard, 1960; Fischer-Piette *et al.*, 1964, 1967;

Fischer-Piette & Gaillard, 1971; Newkirk & Doyle, 1975; Berry & Smith, 1987). When observations are continued into salt marshes and into more brackish water, additional trends appear: shells become smaller, thin-walled, with small and rounded aperture, and often have a tessellated colour pattern, as well as being smooth and tall-spired (Jaeckel, 1951; Rasmussen, 1973; Brandwood, 1983, 1985). These belong to the typical sheltered/brackish ecotype, although they are connected with the moderate ecotype by a range of intermediates, so that no sharp distinction is possible.

Whether this shell form is the product of selection is difficult to determine, for little experimental work has been done. The habitats of this ecotype differ in several respects from those of the typical moderate ecotype, for example in extreme shelter from wave action, low salinity and sometimes in apparently low crab predation, and the relative influence of these variables is hard to assess. Since the same trends towards small, thin-walled, tall-spired shells are seen in sheltered/brackish forms of many other *Littorina* species, such as *L. littorea*, *L. sitkana*, *L. subrotundata*, *L. obtusata* and *L. fabalis*, and as these include both planktotrophic and nonplanktotrophic developers, and occupy various microhabitats, it is difficult to escape the possibility of some direct ecophenotypic effect. For example, it seems likely that low salinity may induce small, thin-walled shells. It is also possible that food may be abundant in the highly productive habitat of salt marshes, and that this may have an ecophenotypic effect on shell shape and thickness, by means of an increased rate of growth; for example, it has been shown that at higher rates of growth *L. subrotundata* produces shells that are larger, narrower, taller-spired and with thinner walls than in controls (Boulding & Hay, 1993). Nevertheless, there is evidently also a genetic component, as suggested by laboratory rearing of juveniles from a population of small, thin, tall-spired shells from a *Zostera* mud flat in Sweden, which largely retained the parental characteristics (B. & K. Johannesson, 1996).

It has been argued that the form of the shell with tall spire, small apical angle and correspondingly small aperture is an adaptive response to desiccation stress, which is expected to be more severe on protected shores with little wave splash (Newkirk & Doyle, 1975). A smaller aperture will reduce evaporative water loss, but its shape may also be significant; it has been found that in a sheltered inlet in Nova Scotia (Newkirk & Doyle, 1975) and in some Welsh salt marshes (Raffaelli, 1979a; Elner & Raffaelli, 1980) apertures become narrow relative to those on more exposed shores. This cannot be related to selection by crabs (in contrast to the moderate ecotype, p. 306), since *Carcinus* is absent in Nova Scotia and said to be rare in the salt-marsh sites, and was therefore interpreted as a device to reduce desiccation. However, if evaporative water loss occurs principally from around the margin of the operculum, a more circular aperture would be expected to be adaptive where desiccation exerts selection. This has been tested in snails from successive tidal levels on a wave-exposed shore in Wales, where a more rounded operculum was indeed associated with a lower rate of water loss (Grahame *et al.*, 1990). In fact the aperture of the sheltered/brackish ecotype is usually rather circular (Fig. 107B–E). Nevertheless, it is not clear why desiccation in, for example, a salt marsh should be any more severe than at high levels on a fairly sheltered rocky shore, and this rounded apertural form may simply be the norm in the absence of significant crab predation (as in geog. var. *groenlandica*). Desiccation may perhaps be particularly severe in the most southerly parts of the range, for example on sheltered shores in Tunisia and southern Morocco, and here the sheltered/brackish ecotype shows a notably tall spire and small aperture (Fischer-Piette, Gaillard, Panouse & Bellon-Humbert, 1971; Fig. 107E), and is reminiscent of the shape of tropical littorinids (Vermeij, 1973).

It is also uncertain whether crab predation is significant in salt marshes. In Wales low frequencies of shell scars suggest that it is not; shells may nevertheless be relatively thick, leading to the suggestion that other factors such as productivity may influence shell thickness (Raffaelli, 1978b; Elner & Raffaelli, 1980). However, since juvenile crabs are common in these marshes, the thick shells may be the product of selection on young snails (Raffaelli, 1982). Typically, however, the sheltered/brackish ecotype is relatively thin-shelled. This is so at two salt-marsh sites in South Africa, where local crabs appeared to be weak, and it was assumed that crab predation was low (Hughes, 1979; Fig. 107B). It has since been found that the common crab in this habitat (*Cyclograpsus punctatus*) can in fact successfully attack even the largest local *L. saxatilis* in laboratory trials, but that the snails escape by climbing the stems of *Spartina* (Knight *et al.*, 1987).

An extreme form of this ecotype occurs in brackish lagoons and seagrass beds, living permanently submerged on macrophytes. It is small (H=1.7–4.6 mm, Robertson & Mann, 1982), fragile, smooth, with round whorls and a moderate spire, brown or tessellated in colour (Muus, 1967; Davis, 1973; these shells closely resemble the lectotype of *Turbo tenebrosus*, Fig. 107C). This form has been discussed in detail by Barnes (1993), who showed that the spire was lower and the operculum relatively bigger than in the larger *L. saxatilis* found on the shores of similar lagoons. He argued that the submerged, macrophytic habitat of the 'lagoon winkle' effectively isolated it from the shore populations. Nevertheless, no species-level distinction is appropriate, because the two forms hybridized in the laboratory. Furthermore, Janson & Ward (1985) have shown clinal change in morphology between snails taken from a sheltered shore and from submerged *Zostera* 5 m away in an inlet in Sweden, with no significant genetic differentiation between them. Presumably neither crab predation nor desiccation are significant in this habitat.

Sculptural variation. There is pronounced sculptural variation in *L. saxatilis* as a whole, even within most of the ecotypes; the sheltered/brackish ecotype is the exception, always with weak or absent spiral sculpture. The range of shell sculpture is illustrated in Figures 106–108. Dautzenberg & Fischer (1912) emphasized shell sculpture as the most important character distinguishing their 'subspecies' within the *L. saxatilis* group. The same subspecific and varietal names were adopted by Fischer-Piette, Gaillard and their co-workers: *jugosa* for strongly ribbed shells, *rudissima* for those with fine ribs, and *rudis* for virtually smooth shells. In a number of descriptive studies on the coasts of France and Iberia they established a general rule that populations on wave-exposed shores were often strongly sculptured, and those on sheltered shores often smooth or weakly sculptured, although exceptions were noted (Fischer-Piette & Gaillard, 1960, 1961, 1971; Fischer-Piette, Gaillard & Jouin, 1961; Fischer-Piette *et al.*, 1964, 1967). A similar trend, but again with exceptions, has been observed in other regions, including Nova Scotia (Davis, 1973) and the British Isles (James, 1968a; Heller, 1975a; Faller-Fritsch & Emson, 1985).

The cause and possible adaptive significance of this variation are not yet understood. In another *Neritrema* species, *L. sitkana*, it has recently been shown that shell sculpture is connected with growth rate, ridges being produced under conditions of low food availability and slow growth, and smooth shells under more favourable conditions (Boulding *et al.*, 1993). There is some evidence for a similar ecophenotypic effect in *L. saxatilis*. Gaillard (1965) found no change in sculpture when individuals were transferred between shores, but when juveniles from sculptured parents were reared in the laboratory (presumably under favourable conditions of growth) sculpture was less marked. In contrast, the two shell types, smooth and ridged, from wave-exposed shores in Galicia have been reported to retain their sculptural features when their progeny are reared in the laboratory, suggesting genetic determination of the character in this case (Johannesson *et al.*, 1993).

If genetic control is assumed, adaptive explanations can be considered. A ridged shell might be expected to be stronger than a smooth one of the same size and mass, leading to the prediction of ridged shells in environments where shell material is energetically costly to produce, but where a strong shell is necessary. This hypothesis was rejected by Brandwood (1985), since he found no evidence that shell strength was significantly influenced by ridging, and in low-salinity environments (where production of shell material might be costly) shells were smooth. On wave-exposed shores, where crab predation is low, shells were weak but ridged. Brandwood therefore suggested that ridges served to resist dislodgement by gripping the sides of crevices, as earlier proposed by Emson & Faller-Fritsch (1976). The possible significance of hydrodynamic considerations is unknown; the relationship between surface roughness and drag is complex, but in general roughness increases drag, although it is doubtful that this is significant for snails on wave-swept shores (Denny, 1988). On tropical shores stronger sculpture is found in species from higher levels in the littoral fringe, and is believed to be adaptive in increasing convective heat loss (Vermeij, 1973). Such considerations probably do not apply to *L. saxatilis* on temperate shores, and sculpture does not become noticeably stronger in the south of the geographical range.

Ontogenetic variation. As in all *Littorina* species, there is a degree of allometric variation, connected with the reduction in apical angle as shell height increases (doming). This has been demonstrated by regressions of shell dimensions and their ratios on shell height (van Marion, 1981; Janson, 1982a). Other than this gradual change, there is little ontogenetic variation in many populations of *L. saxatilis*, and growth appears to be indeterminate. There is a more striking change in some populations of the moderate ecotype of geog. var. *typica*. Here juveniles with a thin growing edge may have a projecting lip at the base of the columella (Figs 107K, 108E), which becomes less pronounced as the aperture becomes thickened when maximum size is approached (van Marion, 1981). This juvenile apertural

Figure 108. *Littorina (Neritrema) saxatilis*: **A–D**, details of adult shell sculpture in sheltered/brackish ecotype (**A**) and moderate ecotypes (**B–D**); **E**, **J**, immature shells of moderate ecotype; **F**, **G**, mature shells intermediate between moderate and barnacle ecotypes; **H**, **I**, mature shells of barnacle ecotype. **A**, Walberswick marshes, Suffolk, England; **B**, Lochskipport, South Uist, Outer Hebrides, Scotland; **C**, Loch Ainort, Skye, Scotland; **D**, Great Castle Head, Pembrokeshire, Wales; **E–J**, Robin Hood's Bay, Yorkshire, England. Scale bar=1.0 mm.

form may be connected with selection for a narrow aperture in the presence of crab predators, or perhaps with opposing selection by wave-exposure (promoting a large external apertural area) and crabs (promoting a small inner apertural area) (Johannesson, 1986). It has earlier been noted (p. 306) that there is often a pronounced constriction of the aperture, achieved by a combination of reduction of the apical angle and descent of the final suture, at the end of the last whorl in large shells of the moderate ecotype of geog. var. *typica* (Figs 106P, S; 107G, J), possibly connected with resistance to apertural attack by crabs. In such forms the pattern of growth appears more determinate.

Moreteau (1976) argued that growth is not strictly determinate, a thickened peristome indicating only a temporary cessation of growth. Alternation of phases of growth and non-growth occur throughout life, the non-growing phases predominating as size increases. Secretion of shell material on the inner surface may continue during non-growth phases, accounting for the thicker shells of older individuals (van Marion, 1981). In arctic regions the overall growth rate is slower, and interruptions to growth take place during the winter, leading to conspicuous annual growth checks on the shell (Fig. 107M, N, Q). These can be used to estimate ages of individuals of geog. var. *groenlandica*, showing that they may live for up to 13 years (Stroud & Knudsen, 1982; Sergievsky, Granovitch & Mikhailova, 1991).

Sexual dimorphism. In *L. saxatilis* dimorphism is not usually conspicuous, and in some populations statistical tests have shown no significant dimorphism in size or shape (Davis, 1973; Newkirk & Doyle, 1975; Sacchi, 1981; van Marion, 1981; Grahame & Mill, 1989). In other cases females have been found to be slightly, but significantly, larger than males (Sacchi & Torelli, 1973, 1979; Daguzan, 1977; Hughes, 1979; Sacchi, 1981). Gallardo & Götting (1985) found size dimorphism to be most pronounced in populations of small adult size. The larger size of females has been explained by either greater longevity (Pelseneer, 1926) or faster growth (Daguzan, 1977), but laboratory rearing has supported the latter interpretation (B. & K. Johannesson, 1996). Differences in shape between the sexes were reported by several early authors, who suggested that male shells were not only smaller but also more tall-spired, with a proportionately smaller aperture (Jeffreys, 1865; Pelseneer, 1911; Bequaert, 1943). Modern studies have used a variety of measurements and analytical techniques, so are difficult to compare, but there appears to be some disagreement between their results. Some have confirmed the greater globosity (i.e. lower H/B) of females (Sacchi & Torelli, 1979; Sacchi, 1981), but Daguzan (1977) reported a relatively smaller aperture length (i.e. greater SH) in females. Grahame & Mill (1989, 1992; Mill & Grahame, 1995) used multivariate analyses and, while not investigating the details of the dimorphism, they did report that males resembled *L. arcana* more closely than did females, implying a narrower width across the penultimate whorl (and possibly a larger aperture) in males. Both Jeffreys (1865) and Grahame & Mill (1992) proposed that dimorphism has a functional significance, the broader shell of the female accommodating the large brood pouch.

Monstrosities. Although this is such a common and widely-distributed species, few monstrosities have been mentioned. There are occasional reports of shells with a keel at the shoulder (Jeffreys, 1865; Martel, 1901; Bequaert, 1943) and of scalariform shells (Martel, 1901; Reed, 1982). Sykes (1892) described cases of irregular coiling in the sheltered/brackish ecotype from The Fleet, Dorset, including one apparently sinistral and other hyperdextral examples, and suggested that interruption of growth by attached algae was responsible. Attachment of barnacles can also cause distortion (Hermans, 1978), but is surprisingly rare. Breakage of the aperture (as can be caused by unsuccessful attack by crabs) can also influence shell form in subtle ways, for the regrown shell may have a smaller apical angle and more rounded aperture (van Marion, 1981). Abnormalities of larval shells have already been discussed (p. 297).

Colour variation (Fig. 5). Of all *Littorina* species, *L. saxatilis* shows one of the widest ranges of shell colour and pattern, only matched in this respect by *L. arcana*. Colour variation has been the subject of many descriptive studies, but so far there has been surprisingly little experimental work so that, in contrast to the variation of shell morphology, that of shell colour is poorly understood.

Various systems of classification have been used to describe the polymorphism. Dautzenberg & Fischer (1912) formalized a Latin 'var. ex colore' system, recognizing a total of 12 varieties, and this was used and added to by some later workers (see Synonymy, p. 278). Most notable of these were Fischer-Piette, Gaillard and their co-authors, who in a series of nine papers described colour variation along the coastline of western Europe. This approach culminated in the adoption of 25 named varieties (Fischer-Piette & Gaillard, 1971), but was clearly unwieldy for such a variable species; more than 150 descriptive categories were necessary to describe the most variable samples (Fischer-Piette, Gaillard & James, 1963). A simpler system, based on the assumption of genetic control of the polymorphism, was devised by Pettitt (1973). Here, variation was described by six 'loci': ground colour (G: with 'alleles' white, fawn, grey, brown, orange, purple, black), band number (B: unbanded, 1, 2, or more bands), band colour (C: colours as in G), tessellation (T: present or absent), interrupted lines (I: present or absent) and lines in grooves (LG: present or absent). Although this 'genetic' system was imaginary, it reduced the diversity to a manageable number of categories (Pettitt found 63 of the possible phenotypes) and has been successfully used in some subsequent studies (e.g. Raffaelli, 1979b; Atkinson & Warwick, 1983). Both these systems were devised before widespread recognition of the three species within the *L. saxatilis* group, but this does not detract from their utility, for a similar range of coloration is shown by each of the three. Some recent workers have dismissed the earlier descriptive studies on the grounds that several species may have been confused under the name '*L. saxatilis*'. In fact, now that geographical ranges of the species are well known, it can be seen that this reservation applies to relatively few of the early studies, in particular to those carried out in Brittany, southwestern England, Wales and Scotland. Only those studies clearly referable to *L. saxatilis s.s.* will be quoted here, unless otherwise indicated.

Shell coloration can be viewed in terms of a ground colour with superimposed darker patterning of tessellation, bands or fine lines (Sacchi, 1974). Although the possible range of coloration is great in the species as a whole, local populations are often monomorphic, and highly diverse populations are infrequent (e.g. Belle-Île population described by Fischer-Piette *et al.*, 1963). Local and regional variation is so great that few generalizations can be made, but in the species as a whole a brown to fawn, unpatterned shell is most common, and the most visually diverse populations occur in moderate habitats of neither extreme shelter nor exposure. Some workers have found difficulty in classifying the range of morphs into discrete classes (Fischer-Piette & Gaillard, 1971) but, if intergrading classes are combined (Atkinson & Warwick, 1983) and ontogenetic variation taken into account (Hannaford Ellis, 1984), a workable classification can be achieved. The structural and chemical basis of shell coloration has not been investigated, although Sergievsky (1992) has noted that in the White Sea the white phenotype is produced by a superficial 'calcination' of the shell.

Before the origin, maintenance and adaptive significance of the colour polymorphism can be discussed, it must be established that it has a genetic basis. Unfortunately there is no detailed evidence of this, but only a brief note that it has been shown to be the case by preliminary breeding experiments (Atkinson & Warwick, 1983). The localized and unpredictable occurrence of some rare colour morphs has sometimes been taken as indirect evidence of genetic control, because it suggests the operation of 'founder effects' (Fischer-Piette *et al.*, 1963, 1964; Fischer-Piette & Gaillard, 1968). The restricted geographical ranges of some morphs also supports a genetic interpretation (Fischer-Piette, Gaillard & Jouin, 1961; Fischer-Piette & Gaillard, 1968; Torelli, 1978). Occasional reports of changes of coloration in single shells (Fischer-Piette *et al.*, 1963; Fischer-Piette & Gaillard, 1971; Davis, 1973) should not necessarily be taken as evidence against genetic control, but may be explained by ontogenetic change. In a well-documented example, banded forms of the barnacle ecotype become tessellated during growth (Hannaford Ellis, 1984). There is also a report of sudden change in colour following damage to the shell (Fischer-Piette & Gaillard, 1971).

Following the protocol of Cook & Garbett (1992), the maintenance of the polymorphism can now be considered. Such is the bewildering array of colour variation in this species, that some authors have been unable to discern any predictable patterns, and have implied that the polymorphism may be selectively neutral (James, in Fischer-Piette *et al.*, 1963; Fischer-Piette & Gaillard, 1968, 1971). However, there is now a body of evidence for nonrandom associations, so that the neutral hypothesis can be ruled out, at least as a general explanation. Unfortunately, as yet the only evidence for selection is rather weak, in the form of correlations of colour frequencies with environmental variables, and experimental work is needed to test the various selective hypotheses.

While recognizing that predictable associations between colour (if genetically determined) and habitat imply selection, some authors have concluded that this is a result of non-visual effects, connected with shell colour only through pleiotropy or linkage of genes. One frequently documented trend is the change between populations on wave-exposed and sheltered shores. In France there is a preponderance of yellow shells in sheltered inlets, whereas on exposed coasts the shells are more variable (Fischer-Piette & Gaillard, 1960, 1961). A more striking trend is found in Galicia, where on exposed coasts shells are mostly a lineated morph and in shelter mostly yellow and fawn, with high colour diversity and less predictable occurrence of morphs at intermediate localities (Fischer-Piette & Gaillard, 1960, 1961; Fischer-Piette *et al.*, 1961, 1964; Fischer-Piette, Gaillard & Delmas, 1967; Fischer-Piette & Gaillard, 1971; Sacchi, 1979; Sacchi & Malcevschi, 1983; Torelli, 1982a). In contrast, in Britain Raffaelli (1979b) could find no correlation between shell colour and exposure (although in this study *L. arcana* was included). These trends have been interpreted in terms of the physiological tolerance of the several morphs to wave-exposure and reduced salinity (Fischer-Piette & Gaillard, 1971; Gaillard, 1977; Torelli, 1982a). Differences in physiological tolerance of colour morphs have indeed been shown in the related species *L. obtusata* (p. 218), but in *L. saxatilis* this has yet to be verified experimentally, and hypotheses involving visual predation offer alternative explanations for the observations, as discussed below.

Several morphs are associated with habitats of reduced salinity and with salt marshes. In Spain Fischer-Piette & Gaillard (1971) found that orange shells extended furthest into brackish water and suggested that they might be most tolerant of reduced salinity. There are high proportions of tessellated shells in brackish habitats and marshlands in many parts of the range, including Denmark (Rasmussen, 1973), South Africa (Hughes, 1979; Fig. 107B) and Britain (Raffaelli, 1979b; J. E. Smith, 1981; Brandwood, 1983; Fig. 107D). In the Cies Islands of Galicia, however, the tessellated morph occurs in a warm and slightly hypersaline lagoon, leading to the suggestion that it is a temperature-tolerant form (Sacchi, 1974, 1979; Sacchi & Malcevschi, 1983), as supported by its occurrence in Venice (Fig. 107A), Tunisia (Fig. 107E), southern Morocco (Fischer-Piette *et al.*, 1971) and South Africa (Hughes, 1979). In the Venice Lagoon the situation is more complex, with tessellated shells predominating in the canals of the city and unicolored red, orange and fawn shells in the marshes (Torelli, 1978). Without experimental evidence, these correlations are open to alternative interpretations; for example, it has also been suggested that tessellated shells are visually cryptic against a background of *Spartina* stems (Knight *et al.*, 1987). Climatic factors have also been invoked for other morphs, as in the case of the lineated morph (Fig. 106O, P) that is common in Galicia, but absent from warmer regions to the south and east (Fischer-Piette *et al.*, 1961; Fischer-Piette & Gaillard, 1971).

One curious case concerns the observation by Fischer-Piette & Gaillard (1966) that at many fully marine localities in France and Spain the percentage of entirely black shells increased from near zero when their surveys began in 1959 to 10 to 20% a few years later. Again they were unable to account for this in terms of visual effects, finding no correlation with pollution by oil for instance, and suggested that climatic change might have played a part. Later they found high percentages of black shells at some localities in northern Norway and suggested that this colour morph may have originated there and spread rapidly in southern Europe (Fischer-Piette & Gaillard, 1971). Sacchi (1974) favoured a correlation with pollution, perhaps by metals.

Some workers have sought pleiotropic correlations of colour with other characters. For example, Torelli (1978) observed that in Venice sculptured shells were always fawn. Raffaelli (1979b) found some correlations between shell colours and both the frequency of scars on the shell and stability of the habitat, fawn shells having more scars and predominating on boulder shores of intermediate wave-exposure. Since he found no association between coloration and shell thickness, he suggested that selection on some other pleiotropic character (such as susceptibility to parasites) could explain his observations. Although there is no evidence for this in *L. saxatilis*, differential infection by parasites has been shown in *L. obtusata* (p. 218). In the White Sea it has been suggested that the peculiar shell structure of the white morph might be connected with calcium metabolism and the neutralization of acidic products of anaerobic respiration, leading to some selective advantage on open rock surfaces where this morph is frequent (Sergievsky, 1992). Another possible non-visual effect which could maintain polymorphism is heterosis, although again there is no evidence for this.

Many of the authors quoted above have dismissed visual selection as unimportant, because they have not found obvious correlations with the appearance of the habitat. Indeed, there are cases of populations which, to the human eye, seem conspicuously mismatched to their backgrounds (Davis, 1973; Raffaelli, 1979b). However, there are also examples of cryptic resemblance. In a study in Wales Heller (1975b; *L. arcana* included in samples) found a strong association of red plus brown shells with red sandstone (mean 62% red plus brown, cf. 10% on other substrates). This was confirmed by Byers (1990; also including *L. arcana*), who examined the small-scale distribution of grey and brown shells at geological junctions of grey and red rocks in Wales. Nevertheless, in this second study other factors besides visual selection were also implicated, for significant differences in colour frequencies occurred in samples from the same rock types, even as little as 5 m apart. Again in Wales, Raffaelli (1979b; also including *L. arcana*) did not find correlations between orange shells and red rocks. Cryptic resemblance is not apparent in some other parts of the range; in a survey of red shells throughout France and Spain no obvious environmental correlates were noticed, and founder events were invoked to explain the distribution (Fischer-Piette & Gaillard, 1968). Other examples of apparent crypsis on unusual rock types have been observed in Nova Scotia (Davis, 1973), South Africa (Knight *et al.*, 1987) and Yorkshire (Dytham *et al.*, 1990). Heller (1975b) also found the yellow morph to be common at sites where *L. saxatilis* occurred among the brown alga *Pelvetia*. Perhaps the most striking example of crypsis is shown by the barnacle ecotype, with greyish white shells tessellated or banded with blackish brown, or uniformly whitish or blackish (Hannaford Ellis, 1984; Fish & Sharp, 1985; Reid, 1993; Figs 106D–G, J; 107L). These colour forms may all be equally cryptic against the fine-grained background of white, striated barnacle plates and dark empty tests. In many areas these same colours are unusual in typical *L. saxatilis* on rock surfaces higher up the shore. A different situation occurs on exposed coasts in Galicia, where the low-shore equivalent of the barnacle ecotype inhabits crevices in mussel beds and has a whitish, weakly banded or tessellated shell, which erodes to give a black spire (Fig. 106Q), whereas the high-shore subpopulation has a whitish shell strikingly lineated with black (Fig. 106O, P). Significantly, this high-shore population is found together with barnacles on these extremely exposed coasts (Johannesson *et al.*, 1993; pers. obs.), suggesting that this colour pattern is indeed cryptic among barnacles. This may also explain the black and white banded pattern seen among juveniles in some typical populations of *L. saxatilis*, where the juveniles inhabit the sheltered crevices of barnacle beds (Fish & Sharp, 1985; partly visible in Fig. 106H). It must be stressed that the same selective processes may not operate throughout the range of the species; in the White Sea, for example, the proportion of white morphs is not correlated with the density of barnacles, and there is no evidence for visual selection upon it by predators (Sergievsky, 1992). Here, climatic selection might be significant (see below).

There is therefore some suggestion of visual selection acting on *L. saxatilis*, but the possible predators have not been identified. Pettitt (1975) reviewed the predators of this species, and many of the birds, crabs and fish involved do have high visual acuity and possible colour vision. Since *L. saxatilis* usually inhabits the littoral fringe, bird predators have been considered potentially important, including rock pipits in Wales (Heller, 1975b; Hughes & Roberts, 1981) and waders in Belgium (Warmoes *et al.*, 1988) and Iceland (Alerstam, Gudmundsson & Johannesson, 1992). If predators are small, selection may take place mainly on small juvenile snails (Faller-Fritsch & Emson, 1985). In the case of the barnacle ecotype lower on the shore aquatic predators may be significant, and it is known to be consumed in large numbers by the blenny, *Lipophrys pholis* (Raffaelli, 1982; Fish & Sharp, 1985). This fish has been implicated in the visual selection of *L. fabalis* in laboratory trials (Reimchen, 1979). It is interesting that whereas some other polymorphic *L. (Neritrema)* species in the Atlantic (*L. obtusata*, *L. fabalis*, *L. compressa*) show reduced polymorphism in the northern parts of their ranges, this is not the case for *L. saxatilis* (or for *L. arcana*) in which the northern geog. var. *groenlandica* can exhibit the same range of polymorphism as the typical southern form. If bird predation is indeed responsible for maintaining the polymorphism in this species, this observation is consistent with its generally higher tidal level, where bird predation may be significant throughout its range.

There remains the possibility of visual effects unconnected with predation, of which climatic selection is an example. This has been investigated in the White Sea, where white shells are more common on the exposed surfaces of boulders during low tide and also at higher levels up the shore. Laboratory experiments showed that white shells reached lower temperatures and had higher survival rates under strong illumination, but it is unclear whether colour significantly affects fitness in the field (Sergievsky & Granovitch, 1989, 1990; Sergievsky, 1992). There is no obvious increase in pale shells at more southerly latitudes, so climatic selection does not seem to be of general importance. Hannaford Ellis (1984) discussed the ontogenetic change from a banded to a tessellated shell in the barnacle ecotype in terms of temperature regulation, but this seems improbable.

If a genetically-based colour polymorphism is to be maintained within a population by selection (whether visual or otherwise), there must be a balancing or frequency-dependent component (see Reid, 1987, for review). This might be achieved by disruptive selection in a coarse-grained environment, with a balance between selection and migration of genotypes. The littoral environment may be heterogeneous, but on a small scale, so this hypothesis does not seem to be clearly applicable to *L. saxatilis*, in which migration rates are about 1 to 4 m in 3 months (Janson, 1983b). There is, however, one piece of evidence that colour morphs might select an appropriate matching background, which would promote polymorphism maintained by this means; the two morphs on wave-exposed shores in Galicia show microdistributions consistent with microhabitat choice (Johannesson, Rolán-Alvarez & Ekendahl, 1995). In addition, behavioural differences between morphs have been suggested by Sergievsky (1992). At least for the barnacle ecotype, the visual elements of the background would appear to present a very fine-grained environment; here the polymorphism might be stable if each of the black, white and banded morphs is an equally cryptic random sample of the background pattern. For a population exposed to visual predation, frequency-dependence could also be achieved by apostatic selection. This has been suggested by Heller (1975b), who noted that a low frequency of apparently conspicuous red shells was found even on shores without red rocks; however, at the present state of knowledge this could equally imply selectively neutral variation. There is nevertheless some slight evidence for frequency-dependent selection. Atkinson & Warwick (1983) separated the two species *L. saxatilis* and *L. arcana* in sympatric samples from exposed shores throughout Britain, and showed a significant correlation between the frequency of the commonest, brown to fawn, morph in the two (which in itself suggests that selection of some kind is responsible for the variation between shores). However, the slope of the regression differed significantly from 1; in other words, if the morph was common in *L. saxatilis*, it was less so in *L. arcana*, showing an element of frequency-dependence. These authors also suggested that apostatic selection was a possibility. One further mechanism that could effectively maintain polymorphism is assortative mating. This has not yet been investigated with respect to colour polymorphism, but has been found in the case of the two ecotypes on exposed shores in Galicia (Johannesson, Rolán-Alvarez & Ekendahl, 1995), where it may be mediated by selection for mate size (Erlandsson & Johannesson, 1994).

To summarize, the available evidence for the maintenance and adaptive significance of the shell colour polymorphism in *L. saxatilis* consists of little more than some suggestive correlations with environmental variables, which are often open to alternative, but equally speculative, interpretations. It appears that the same correlations do not apply throughout the wide geographical range of the species, and the wide range of habitat types also militates against general explanations. The need for experimental investigations, for example of environmental tolerances of morphs, and of visual selection by predators, is clearly apparent. As in the case of the variations in shell morphology, the small scale over which differences in morph frequencies can be maintained must be emphasized (Byers, 1990; Johannesson *et al.*, 1993; Reid, 1993).

Distribution

Habitat. Owing to the taxonomic confusion surrounding the three members of the *L. saxatilis* group, a number of early references to the ecology of *L. saxatilis* probably included more than one species. Only references which clearly apply to *L. saxatilis s.s.* (by virtue of explicit identification or known geographical and habitat ranges of the three species) will be quoted here, unless otherwise noted. *Littorina saxatilis* occurs in a wider range of habitats than any other *Littorina* species. It also occupies a very wide geographical range, and observations on habitats will be discussed separately for low-boreal regions (the geographical range of var. *typica* as far south as Portugal), high-boreal areas (the range of geog. var. *groenlandica*) and the southern limits (the Mediterranean and Africa).

Throughout its low-boreal range, *L. saxatilis* is most familiar as a species of the upper eulittoral zone and the littoral fringe, a dominant animal of the *Verrucaria* and upper barnacle zones on wave-exposed coasts, and of the corresponding *Verrucaria*, *Enteromorpha* and *Pelvetia* zones on coasts of moderate exposure, extending in lower numbers to the levels of *Fucus spiralis* and *Ascophyllum* (e.g. Heller, 1975a; Faller-Fritsch & Emson, 1985; Janke, 1986). The upper vertical limit is extended by spray on the most exposed coasts, reaching 20 to 40 m above the barnacle zone under extreme conditions (Hunter, 1953; James, in Fischer-Piette *et al.*, 1964; James, 1968a; all apply to *L. saxatilis s.l.*, but probably this species). An even higher record, of *L. 'littorea'* from 120 m above sea level on St Kilda, Scotland (Waterston & Taylor, 1906; Hewitt, 1907) probably applies to this species (Cooke, 1914; Dacie, 1917). In fact the species appears unable to withstand the full force of waves on some extremely exposed coasts (Fischer-Piette & Gaillard, 1971) and is there confined to sheltered microhabitats such as crevices and the spray zone, or may even be absent and replaced by *L. arcana*

(Grahame & Mill, 1986; Mill & Grahame, 1990b). On other highly exposed cliffs it may be present, but in low numbers, as on the islet of Rockall (Moore, 1977; Smith, 1979a). It attaches only to hard substrates, including wooden pilings (Warmoes *et al.*, 1988), is scarce on soft and friable rocks (Fisher-Piette & Gaillard, 1968; Mill & Grahame, 1990b) and does not often crawl on weed (Sumner, Osburn & Cole, 1913). Particularly on exposed shores, it is strongly dependent on the availability of suitably sized crevices, which may limit both its size and density (Emson & Faller-Fritsch, 1976; Raffaelli & Hughes, 1978). Juveniles occur together with adults, but show a greater tendency to congregate in small crevices, such as among barnacles (Emson & Faller-Fritsch, 1976; Faller-Fritsch & Emson, 1985) and in *Lichina* tufts (Colman, 1940; applies to *L. saxatilis s.l.*). No large-scale breeding or seasonal migrations seem to occur in low-boreal regions (Berry, 1961; Hannaford Ellis, 1985). In areas of sympatry with *L. arcana* the two species show considerable overlap, *L. arcana* sometimes extending a little above (Faller-Fritsch & Emson, 1985) or migrating below *L. saxatilis* (Hannaford Ellis, 1985). There is some indirect evidence of competition between these species, both from morphological 'character displacement' (p. 309) and from the apparent exclusion of *L. saxatilis* by *L. arcana* on cliff faces exposed to strong wave action (Grahame & Mill, 1986; Mill & Grahame, 1990b). Another littorinid, *Melarhaphe neritoides*, has its centre of distribution higher in the littoral fringe, but again there is overlap. At lower levels *L. saxatilis* overlaps with *L. compressa* and to a lesser extent with *L. littorea*.

On wave-exposed shores *L. saxatilis* may also extend to lower levels in beds of barnacles, where it nestles in small crevices (e.g. Heller, 1975a; Fish & Sharp, 1985; B. & K. Johannesson, 1990; Reid, 1993). Here it is present as the barnacle ecotype, and the association of this distinctive shell form with a characteristic habitat has led some to consider it a separate species, although it has been argued that this is simply the product of strong selection by wave action in this particular microhabitat (see Barnacle Ecotype, p. 309). This small ecotype can also be found in beds of *Mytilus* (Johannesson *et al.*, 1993), in *Laminaria* holdfasts and among *Corallina* (Fretter & Graham, 1980). Among barnacles it may reach densities of 1000 to 4000 per m^2 (Hawkins, 1981; Little & Williams, 1989) or even 100 per 10 cm^2 (equivalent to 100 000 per m^2; Fish & Sharp, 1985).

In low-boreal regions the optimal habitat for *L. saxatilis* (in terms of density, growth rate and size) appears to be moderately sheltered rocky and boulder shores (Roberts & Hughes, 1980). Here densities may reach 1200 to 3000 per m^2 (Moore, 1940; Faller-Fritsch, 1977; Berry & Smith, 1987). On such shores the upper limit of zonation is the mean high-water mark of spring tides, but the lower limit is rather variable; in general *L. saxatilis* is only abundant down to the upper edge of the *Ascophyllum* zone and continues at low density down to mean low-water of spring tides (Moore, 1940; Hughes, 1980; Roberts & Hughes, 1980). Predatory crabs (especially *Carcinus maenas*) also reach greatest abundance on such shores; although there are no precise data on the influence of crabs on the zonation of the snails, it is likely that they contribute to the low abundance throughout much of the eulittoral zone. On sedimentary shores the species is only found where there is a hard substrate for attachment, even if this is in the form of scattered stones, shells and debris (Spooner & Moore, 1940; Dexter, 1947; Jaekel, 1952; James, 1968a). It is also often abundant in salt marshes, where it attaches to the bases of *Spartina* and other vegetation, and to firm mudbanks; densities here may reach 600 to 1300 per m^2 (Dexter, 1945, 1947; Hughes & Roberts, 1981; Hardwick-Witman & Mathieson, 1983). The species is more tolerant of brackish conditions than other *Littorina* species, penetrating further into estuaries (Fischer-Piette, 1931; Spooner & Moore, 1940; Brattegard, 1966; Rasmussen, 1973; Little & Smith, 1980; Smith & Little, 1980) and surviving at salinities of 5–7‰ (Jaeckel, 1951, 1952; Muus, 1967; Sokolova, 1995). The similar species *L. arcana* does not penetrate into estuaries nor occur in salt marshes. In sheltered inlets and far up estuaries *L. saxatilis* extends lower on the shore, down to low water of spring tides (Little & Smith, 1980; Smith & Little, 1980; Little & Williams, 1989).

Continuing this trend, some populations of the sheltered/brackish ecotype live permanently submerged in sheltered, usually brackish, lagoons, generally attached to vegetation such as *Zostera* (Jaeckel, 1951, 1952; Rasmussen, 1973; Brandwood, 1985; Janson & Ward, 1985), *Fucus* (Janson & Ward, 1985), *Ruppia* and *Potamogeton* (Muus, 1967; Barnes, 1993). The animals are usually very small (p. 317) and may occur at extremely high densities, Muus (1967) having recorded maxima of 226 000 juveniles per m^2 and 26 000 adults per m^2 on submerged macrophytes. This form extends to depths of 13 m in the southwestern Baltic Sea (Jaeckel, 1952; Muus, 1967; Jagnow & Gosselck, 1987).

Throughout much of the range of geog. var. *groenlandica* the species occupies a similar spectrum of habitats to that of the southern geog. var. *typica*, although in more northerly areas it may be absent on open coasts that are severely affected by ice scour. On protected shores it can be found on rock with ephemeral green algae in the lower part of the littoral fringe (often rendered barren by ice action), extending down all across the eulittoral zone, among barnacles, *Ascophyllum* and *Fucus* (Gowanloch & Hayes, 1926; T. A. & A. Stephenson, 1954a, b; Barsanova, 1961; Ellis & Wilce, 1961; Petersen, 1962; Golikov & Kusakin, 1978; Stroud & Knudsen, 1982; Golikov, 1987). Even when found abundantly in the *Fucus* zone, few snails occur on the thallus of the weed (Matveeva, 1974). In contrast to the southern variety, var. *groenlandica* reaches maximum abundance in the middle eulittoral zone; here exceptional densities of up to 43000 per m^2 (9800 g per m^2) have been recorded on the Murman coast (Matveeva, 1974), but more common maximum densities are 2900 per m^2 in the White Sea (Matveeva, 1974) and 2000 per m^2 in Spitsbergen (Hansen & Haugen, 1989). This lower extent on the shore may be a consequence of the absence of the shore crab *Carcinus maenas* in the geographical range of geog. var. *groenlandica* (p. 304). On sedimentary shores the sheltered/brackish ecotype may be found in salt marshes (Davis, 1973) and in the shallow sublittoral on the fronds of *Zostera* (Zenkevitch, 1963; Davis, 1973; Robertson & Mann, 1982), where it reaches a density of 3750 per m^2 in the White Sea (Matveeva, 1974). It is tolerant of estuarine conditions in the White Sea (Berger, 1986b; Sokolova, 1995), but reportedly absent from brackish water in Greenland (Petersen, 1962; Stroud & Knudsen, 1982). Throughout its northern range *L. saxatilis* is principally an intertidal species, but there are also many records of its occurrence alive at considerable depths, from 47 to 62 m in Iceland (Bárdarson, 1919, 1920; Thorson, 1941), from 46 to 94 m in Greenland (Posselt & Jensen, 1898; Thorson, 1944), 90 m in Finnmark (SMNH), from 46 m and 50 m in Spitsbergen (Odhner, 1915; SMNH), 187 m on the Murman coast (ZISP) and even 290 m in the Gulf of St Lawrence (Whiteaves, 1901). Although some of these records may have been of individuals transported into deep water, the records are so numerous that it appears the species does survive there. Some of the museum collections from deep water are large samples, raising the possibility that populations may exist at such depths. Again this may perhaps be related to the absence of many predators in subarctic seas. However, in a shallow embayment up to 5 m deep in Newfoundland a subtidal population has been found to coexist with abundant crabs (*Cancer irroratus*) (Gilkinson & Methven, 1991). In shore populations there is a seasonal migration, the animals dispersing throughout the eulittoral zone and into the littoral fringe in summer, and congregating in crevices and under stones in the mid-eulittoral zone in autumn. They do not migrate sublittorally to avoid winter ice cover, but remain on the shore, where they become frozen into the icefoot (Pohley, 1976; Petersen, 1962; Matveeva, 1974). Migratory patterns differ among size-classes, smaller animals remaining lower on the shore during summer (Matveeva, 1974; Stroud & Knudsen, 1982). The species is remarkably tolerant of low temperatures; in Spitzbergen animals may be frozen into the ice for eight to nine months of the year (Aarefjord & Skreslet, 1969), and in New Hampshire they have been observed actively feeding at $-11^{\circ}C$ (Batchelder, 1915). There may also be some small-scale tidal migrations (Mikhailova, Granovitch & Sergievsky, 1988; Sergievsky, 1989).

In southern Portugal and further south *L. saxatilis* becomes increasingly rare and sporadic in its occurrence. Near the Strait of Gibraltar it is found principally in the upper eulittoral zone on fairly exposed rocky shores (wave-exposed ecotype) and also on sheltered, often muddy and lagoonal, rocky shores near Cadiz (Gofas, 1975; pers. obs.; where it approaches the sheltered/brackish ecotype). In the Azores it has been found in algal turf on wave-exposed rocky shores (S. J. Hawkins, pers. comm.). The other southern records all refer to lagoonal, salt-marsh or other sheltered habitats. In southern Morocco it has been found on firm mud under *Salicornia* plants (Fischer-Piette *et al.*, 1971). In two lagoons in South Africa it occurs on firm, muddy sandflats in the upper eulittoral zone, on *Zostera* debris, stones and the stems of the marsh grass *Spartina*, and is absent from open rocky coasts (Kilburn, 1972; Hughes, 1979; Knight *et al.*, 1987). The type locality, Venice, is also a lagoonal habitat, and the local ecology of *L. saxatilis* is well known (Sacchi & Torelli, 1973; Torelli, 1974, 1977, 1978). Here it is most abundant in the mid-eulittoral *Enteromorpha* zone on artificial hard substrates such as the walls of canals, where it may reach a density of 635 per m^2. On natural shores it is common on hard mud under salt-marsh vegetation, on the stems of *Spartina*, under wood debris, and among pebbles with the bivalve *Lasaea* and isopod *Sphaeroma*. It is absent from the open rocky coasts of the Gulf of Trieste (where competition with *Melarhaphe neritoides* has been suggested), and from the estuaries of Venetian rivers, although it tolerates a salinity range down to 9‰ in the lagoon.

In the Gulf of Gabès, Tunisia, the habitat and mid-eulittoral zonation are similar again, the species having been reported together with *Lasaea* and *Sphaeroma* under stones on muddy sand, among pebbles and seagrass debris, and also on firm mud under halophytes, but not on open rocky shores (Seurat, 1924, 1934; Molinier & Picard, 1954; pers. obs.). A density of 100 per m^2 has been recorded in this area (Zaouali, 1978). The reasons for this restriction of habitat in the southern parts of the range are unclear. The polarization on exposed and lagoonal shores around the Strait of Gibraltar is consistent with an hypothesis of exclusion from intermediate shores by crab predators, which at least in more northern parts of the range are less common on exposed and salt-marsh coasts. High temperatures and desiccation may also be unfavourable in southern regions, as suggested by the restriction to shaded and damp microhabitats in the African localities. The two Mediterranean localities are unique in that sea by virtue of their significant tidal amplitude; whether this is necessary for the survival of *L. saxatilis*, or simply for the development of suitable lagoonal marshland habitats, is unknown.

Most reports of the diet of *L. saxatilis* suggest that it is a relatively nonselective grazer on diatoms, blue-green algae, ephemeral filamentous and foliose algae, and the decaying tissues and detritus of macrophytes (Lami, 1937; Berry, 1961; Petersen, 1962; Matveeva, 1974; Torelli, 1974; Sacchi, Testard & Voltolina, 1977; Sacchi, Torelli & Voltolina, 1977; Sacchi, Ambrogi & Voltolina, 1981; Robertson & Mann, 1982; Hawkins & Hartnoll, 1983; Konan, Cardinal & Lamontagne, 1992). Whether or not it feeds on intact fucoids is disputed; it has been said to graze on *Pelvetia* (Clyne & Duffus, 1979) and to eat large amounts of fucoids on shores where these algae are available (Sacchi *et al.*, 1981). Others have reported that only decaying tissue of fucoids is eaten (Fretter & Graham, 1962) and that adult plants are consumed reluctantly (Chapman, 1989; Barker & Chapman, 1990). Occasionally small animals may be ingested (Berry, 1961; Sacchi, Torelli & Voltolina, 1977), perhaps inadvertently, and it has been said to consume barnacles (Smith, 1978) and their spat (Elmhirst, 1923). Nevertheless, a carnivorous diet, proposed for the barnacle ecotype (Smith, 1978), seems unlikely.

Range (Figs 109, 110). This species has an extremely wide geographical distribution, throughout the subarctic province of the northern Atlantic, extending south at least to Virginia in the western Atlantic and to the Strait of Gibraltar in the eastern Atlantic, with outlying populations in the Mediterranean, Morocco, Azores, Canary Islands and South Africa. This wide distribution can partly be explained by its equally remarkable range of habitats and environmental tolerance (see Habitat, p. 324). In addition, its ovoviviparous reproduction may actually be advantageous. Although absence of a planktonic dispersive stage may lead to reduced rates of dispersal on a local scale (Janson, 1987b), over medium (Johannesson & Warmoes, 1990) and large distances (Johannesson, 1988) colonization may still be effective, because of the high probability that even a single fertilized female immigrant will be able to found a new colony (see Ovoviviparity, Distribution and Dispersal, p. 335).

In the northern parts of the range the distribution follows closely the influence of warmer Atlantic water intruding into the Arctic Ocean, which defines the extent of the subarctic area (Ellis, 1955). The explanation is unclear, but the species is not limited by freezing in ice, nor by meltwater, so perhaps some minimum temperature is required for development. The most northeasterly records are from Varnek Bay on the Kara Peninsula (Strelnikov, 1929; ZISP) and southern Novaya Zemlya (BMNH, ZISP; and presumably the record by Lech, 1878, of *L. 'obtusata'* which is not known to occur there). There are several accounts of the species in Spitsbergen (Mørch, 1870; Odhner, 1915; Hansen & Haugen, 1989; Rózycki, 1991), where it reaches almost 80°N (SMNH). It is found all around Iceland, wherever the coastline is suitable (Thorson, 1941). On the Russian mainland *L. saxatilis* occurs in Cheshskaya Bay (BMNH, ZISP) and throughout the White Sea and Murman Coast. Continuing southwards, the species is continuously distributed around the coasts of Norway and western Sweden. It penetrates the Baltic as far as Svaneke on Bornholm (ZMB) and northern Rügen (SMNH; also Schlesch, 1937; Fretter & Graham, 1980; Jagnow & Gosselck, 1987). It inhabits the whole coast of Denmark, of the southern North Sea, the English Channel and the entire British Isles, including the isolated islet of Rockall, 400 km west of the Outer Hebrides (Crisp, 1956; Moore, 1977; Johannesson, 1988). In this area it has been mapped in some detail in the English Channel (Warmoes *et al.*, 1988; Warmoes, Dumoulin & Reid, 1992) and in Britain (Mill & Grahame, 1990b, 1992b).

In France the species is abundant along both northern and southern coasts of Brittany. In 1955 it was reported to be much less common south of the Loire, disappearing first from wave-exposed and estuarine habitats, and reaching its southern limit in France at the mouth of the Gironde (Crisp &

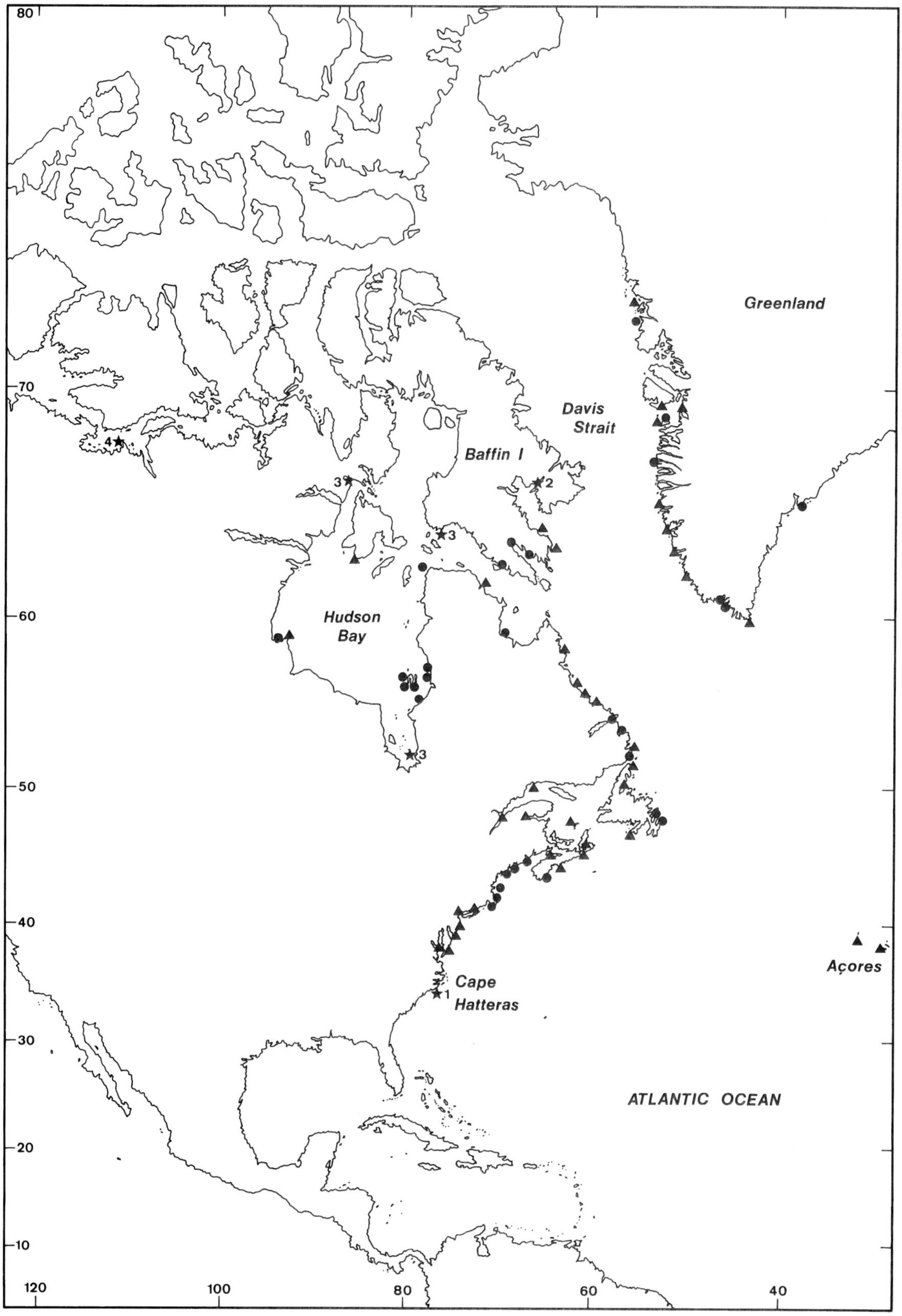

Figure 109. *Littorina (Neritrema) saxatilis*: distribution in western Atlantic. Key: circles, anatomical material examined; triangles, probable determination based on examination of shells only; asterisks, records from literature (1, T. A. & A. Stephenson, 1952; 2, Ellis, 1955; 3, Macpherson, 1971; 4, Dall, 1919).

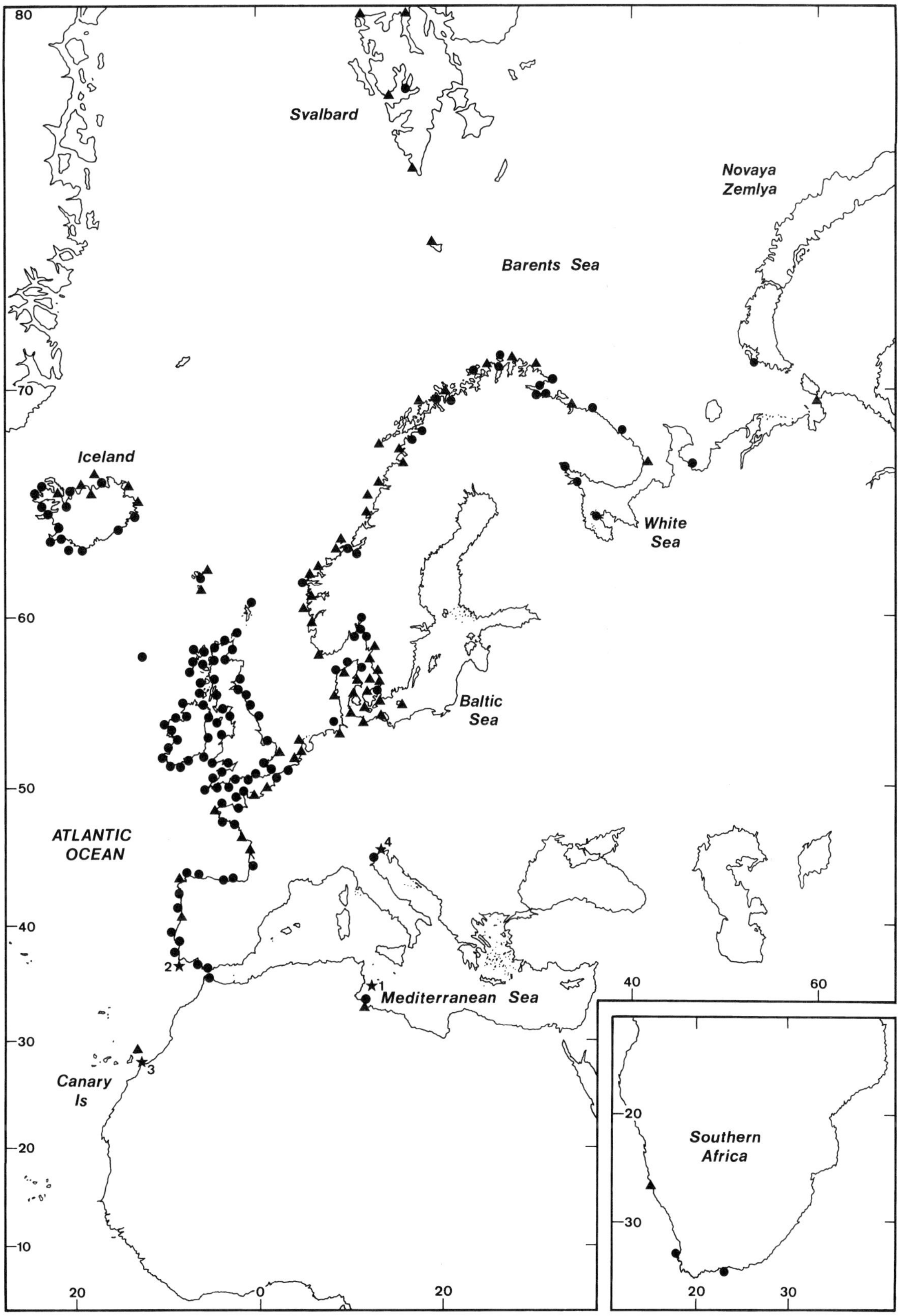

Figure 110. *Littorina (Neritrema) saxatilis*: distribution in eastern Atlantic and (inset) southern Africa. Key: circles, anatomical material examined; triangles, probable determination based on examination of shells only; asterisks, records from literature (1, Seurat, 1924; 2, Fischer-Piette, 1960; 3, Fischer-Piette, Gaillard, Panouse & Bellon-Humbert, 1971; 4, Torelli, 1980). There are in addition unconfirmed records from Senegal (Nordsieck & Garcia-Talavera, 1979; Nordsieck, 1982) and Ghana (R. N. Kilburn, pers. comm.).

Fischer-Piette, 1959). It was not then found in the Arcachon Lagoon, but has since been collected on concrete piers at Andernos (DGR, 1987). Elsewhere on this coast, between the Gironde and Biarritz, the coastline is sandy and unsuitable for *L. saxatilis*. Its recorded distribution in the Bay of Biscay has fluctuated during the twentieth century. At the beginning of the century it was present at St Jean-de-Luz and Guéthary close to the Spanish border, but in 1935 just two specimens were recorded from San Sebastián, and by 1955 it was absent in this angle of the Bay of Biscay (Crisp & Fischer-Piette, 1959). Similarly, in northern Spain there are many records from the early part of the century from San Sebastián and to the west along the coast (Hidalgo, 1917), but by 1949 it had apparently retreated westwards to Santoña (Fischer-Piette, 1953, 1955). Thereafter, there was an extension 65 km to the east (Fischer-Piette, 1956). In the present study it was collected in this same vicinity at Castro Urdiales (DGR, 1987).

The species occurs sporadically between Santoña and Salinas, and then abundantly in a wide range of habitats further west and throughout Galicia (Fischer-Piette, 1953, 1955). There are a number of records from Portugal, but the distribution is discontinuous because of the stretches of unsuitable sandy coastline. At the southern limit in Portugal there appears to have been a retreat followed by an advance during the twentieth century. Early authors recorded it from the Algarve on the south coast (Hidalgo, 1917; Nobre, 1931; Carvalho, 1945), by 1955 it was present only as far south as Vila Nova de Milfontes on the west coast (Fischer-Piette, 1957a, 1959), but by 1960 it had returned to Sagres and Portimao (Fischer-Piette, 1960). A similar pattern of retreat and advance (although on the west coast of Portugal) was recorded in *L. obtusata s.l.* (p. 223), and the coincidence suggests that climatic fluctuation may have been the cause. Continuing southwards, the coastline is again rocky in southern Spain, but occurrence here is sporadic, *L. saxatilis* being known only from Cadiz and Tarifa (Fischer-Piette, 1957a, 1959; Fischer-Piette & Gaillard, 1971; Gofas, 1975; DGR, 1987). Elsewhere in the Strait of Gibraltar it has been found on Alboran Island (Richard & Neuville, 1897; Hidalgo, 1917) and at several localities between Ceuta and Tangier (Gofas, 1975).

This completes the more or less contiguous distribution in the eastern Atlantic, but *L. saxatilis* also occurs in several outlying populations further south, and within the Mediterranean. In southern Morocco it has been found at Puerto Cansado (28°N) (Fischer-Piette *et al.*, 1971); it occurs on the islands of Flores and Faial in the Azores (BMNH) and on Lanzarote in the Canary Islands (Thorson, 1951; Nordsieck & Garcia-Talavera, 1979; Rosewater, 1981; Nordsieck, 1982; USNM). In addition, there are unconfirmed records in West Africa from Senegal (Nordsieck & Garcia-Talavera, 1979; Nordsieck, 1982) and Ghana (R. N. Kilburn, pers. comm.); further populations should be searched for in this region, particularly among halophytes. Further south still, there are living records from Lüderitz in Namibia (NM E5281), and in South Africa from Langebaan Lagoon and Knysna Lagoon (Day, 1969; Kilburn, 1972; Hughes, 1979; Knight *et al.*, 1987; BMNH). Within the Mediterranean this species occurs in only two areas, the first being the adjacent lagoons of Venice (Sacchi & Torelli, 1973; Torelli, 1974, 1977, 1978) and Grado-Marano (Torelli, 1980, 1982b); although formerly abundant on canal walls and piers within the city of Venice itself, the species is adversely affected by domestic and industrial pollution (Sacchi & Torelli, 1973; Torelli, 1977) and is now rare in the city, although still abundant elsewhere in the lagoon (pers. obs., 1992). The second Mediterranean area is the Gulf of Gabès in Tunisia; here it has been recorded from a number of localities near Sfax, Gightis, on the Kerkennah Islands and the island of Djerba (Dautzenberg, 1893b; Seurat, 1924, 1929, 1934; Zaouali, 1978; SMNH). However, it appears to have become scarce and may possibly be in danger of extinction, for a recent visit failed to find it at Gightis (abundant there according to Zaouali, 1978) and on Djerba it was found at only one locality, El Kantara (DGR, 1991). Two specimens have been reported alive from the Balearic Islands (Hidalgo, 1867), but if so this may have been a chance introduction, and it has not been collected there since. A record from the Crimea (Kutorga, quoted by Jeffreys, 1865; probably repeated by Dautzenberg & Fischer, 1912, and others) is considered unreliable.

There are several possible interpretations of these outlying Mediterranean and African occurrences, although in the absence of a good fossil record (p. 331) and of knowledge of possible limiting environmental factors in the southern parts of the range (p. 327) these interpretations are speculative. It is notable that the two Mediterranean localities are (together with the Strait of Gibraltar) the only parts of that sea where the spring tidal amplitude approaches one metre (Fischer, 1943; Maloney & Burns, 1958). Since *L. saxatilis* was more widespread during the Pleistocene, reaching the coast of Morocco near Casablanca where it does not now occur, this led to the hypothesis that the Mediterranean populations were relicts of a wider Pleistocene distribution, surviving only where tidal and climatic conditions are favourable (Crisp & Fischer-Piette, 1959; Torelli, 1980; Cataliotti-Valdina, 1984). Against this is the lack of any reliable fossil record from the Mediterranean (p. 332). An alternative view is that human introduction took place, either with imports of edible molluscs (Smith, 1982) or accidentally on ships; both might be considered likely in the case of Venice, long established as a centre for shipping. The hypothesis of human introduction has been extended to the two South African populations, since both are in natural harbours frequented by early ships (Hughes, 1979; Knight *et al.*, 1987). However, this view is rendered much less likely by the discovery that *L. saxatilis* was present near Cape Town during the Pleistocene (Kilburn, 1972; Schalke, 1973; p. 332). Although it has not been recorded alive from this particular locality, it is at least possible that the two known living populations elsewhere in South Africa predate European settlement. Genetic analyses of the Venetian (Janson, 1985b) and South African populations (Knight *et al.*, 1987; Sundberg *et al.*, 1990) have shown that both are genetically distant from northern European samples; they also have a reduced average heterozygosity, which suggests that they have passed through a population bottleneck at some stage, but cannot further illuminate the history of these populations. An alternative to the hypotheses of either relictual distribution or human introduction is offered here. As discussed earlier (p. 326), all Mediterranean, southern Moroccan and South African records from natural habitats are from salt marshes, lagoons, or sheltered and muddy shores, and this habitat restriction might be imposed by predation pressure and/or insolation. Such habitats are frequented by the wading birds which migrate between Europe and Africa, and the small, light shells of the sheltered/brackish ecotype could perhaps be passively transported by them. Dispersal by birds has been suggested in the case of a 3 mm living *L. obtusata s.l.* found 21 km from the sea in Scotland (Russell Hunter, Maitland & Yeoh, 1964; see Rees, 1965, for other molluscan examples). Although introduction by man cannot yet be ruled out, it is at least possible that not only the South African populations, but also the other outliers, are indigenous. If so, additional occurrences in similar habitats in Africa are to be expected.

Turning to the western Atlantic, the northern limit again follows closely that of the influence of warmer Atlantic water in the Arctic (Ellis, 1955). In eastern Greenland the northern limit was given as Angmagssalik by Thorson (1944; BMNH; ZMK), although Madsen (1936) suggested that it was further north, at 66–67°N. In western Greenland it occurs along the coast as far as Tasiusak at 73°N (Madsen, 1940; Laursen, 1944; Thorson, 1951; ZMK). On the western side of the Davis Strait the northern limit is believed to be Cumberland Sound on Baffin Island (Oughton, 1940; Ellis, 1955; map in Macpherson, 1971), because further north the shore is almost barren and the species has not been found on shores in the arctic regions of the Canadian Archipelago (Ellis & Wilce, 1961). Elsewhere in northern Canada it occurs in the Hudson Strait, Roes Welcome Sound (Repulse Bay, Macpherson, 1971), and throughout Hudson Bay and James Bay (Richards, 1936; Pelletier, Wagner & Grant, 1968; Macpherson, 1971; NMC). There is one record much further to the west, from Cape Barrow in Coronation Gulf (68°N, 111°W) (Dall, 1919); the material on which this was based has not been located in museum collections, and it remains unsubstantiated. The species is found in Ungava Bay and along the coast of Labrador and Newfoundland, penetrates the Gulf of St Lawrence at least as far as Rivière-du-Loup (USNM) and is abundant throughout New England. The southernmost records in the United States are from Cape May, New Jersey (Richards, 1929; Bequaert, 1943), Chincoteague Island, Virginia (Kraeuter, 1974; Rosewater, 1981; USNM 709282), and from Beaufort, North Carolina (T. A. & A. Stephenson, 1952).

In the past there has sometimes been confusion between *L. saxatilis* and *L. sitkana*, synonyms of the former being used for some examples of the latter (see p. 149). This accounts for erroneous statements that the range of *L. saxatilis* extends into the Pacific Ocean (e.g. Jeffreys, 1865, 1874; Harmer, 1921; Thorson, 1941, 1944, 1951; Macpherson, 1971; Abbott, 1974; Fretter & Graham, 1980). It has, however, recently been discovered that *L. saxatilis* has been accidentally introduced to San Francisco Bay (J. Geller, J. T. Carlton, pers. comm., 1995).

Fossil record (Fig. 111). Since shells of *L. saxatilis* can be indistinguishable from those of *L. arcana*, fossils should strictly be referred to as *L. saxatilis s.l.* However, because *L. arcana* is, at the present time, often absent from the more sheltered shores where fossil preservation is most likely, many of the fossil shells are probably of *L. saxatilis s.s.* Fossils are common in deposits from the last glacial and the post-glacial period throughout much of the present range of *L. saxatilis*, in Spitsbergen (Feyling-Hanssen,

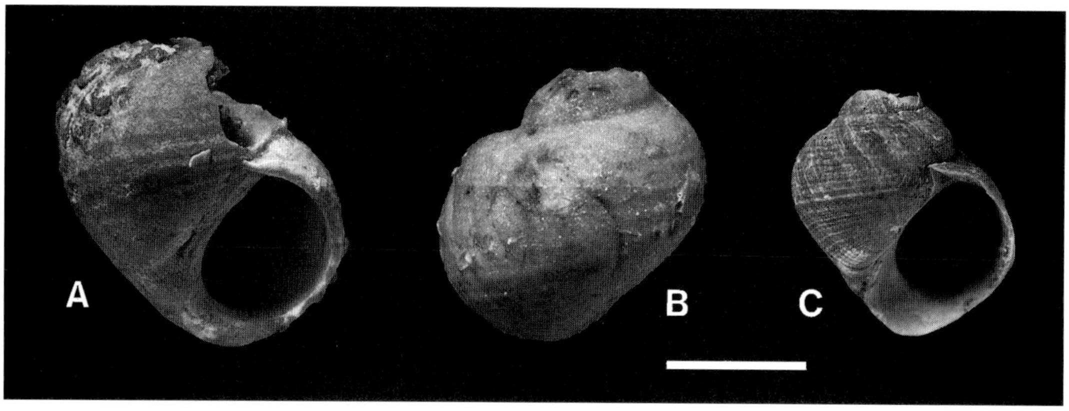

Figure 111. *Littorina (Neritrema) saxatilis*: fossils. **A, B,** bed J, Sidi Abd er Rahmane, near Casablanca, Morocco; between Maarifian and Anfatian stages, Pleistocene; MNHNP. **C,** BS 88–283, Boxgrove, Sussex, England; Pleistocene; ZMC. Scale bar=5 mm.

1955), Russia (Knipowitsch, 1900), Norway (Brøgger, 1901), Sweden (Jeffreys, 1865; Antevs, 1928), the Baltic (during the Littorina Sea stage; Munthe, 1931; Kessel & Raukas, 1967); Scotland (Harmer, 1921; Berry & Smith, 1987), Iceland (Thoroddsen, 1892; Bárdarson, 1923), Greenland (Laursen, 1944, 1950), Hudson Bay and James Bay (Richards, 1936, 1941, 1962; Pelletier *et al.*, 1968), Newfoundland (Richards, 1940), Quebec (Dawson, 1872) and New England (Knight, 1934). Of particular interest are records from the Cape Flats Formation to the north of Cape Town, where the species was found in deposits dating from the middle of the last glaciation (about 40000 BP) and from the post-glacial period (Schalke, 1973), which indicate that *L. saxatilis* may be indigenous to South Africa and not introduced by man (p. 331). Somewhat older are fossils from the raised beaches at Portland in Dorset (Harmer, 1921; Baden-Powell, 1930) and from the Gower Peninsula in Wales (Chatwin, 1932; George, 1932; BMNH), which date from about 100 000 to 200 000 BP (Jones & Keen, 1993).

The fossil record in the Mediterranean was summarized by Cataliotti-Valdina (1984). He listed Upper Palaeolithic records from two inhabited caves near the Ligurian coast and from another in the Dordogne. Of greater age was archaeological material from the Grotte du Lazaret, near Nice (Riss III glaciation, about 150 000 BP). The only Mediterranean record not associated with human occupation was from the Egades Archipelago in Sicily (Tyrrhenian II) (Malatesta, 1957). Cataliotti-Valdina (1984) considered that these records supported the presence of *L. saxatilis* in the Mediterranean at these times. However, as pointed out by D. Kadolsky (pers. comm.) the record from Sicily is unreliable, for it was not mentioned by Malatesta & Zarlenga (1986) in a review of northern species in the Mediterranean. It is considered unlikely that the archaeological material could have been transported by man from Atlantic shores at such an early date in human history (D. Kadolsky, pers. comm.). Nevertheless, until fossils are found in natural deposits, the Pleistocene occurrence of *L. saxatilis* in the Mediterranean remains unconfirmed.

Older records of *L. saxatilis* are notably scarce. Those from the Norwich Crag of East Anglia (Harmer, 1921) are based on misidentifications of *L. littorea* (p. 100). The latter species is abundant in this formation and, since both can today be found living together in sheltered, depositional environments, it appears that *L. saxatilis* was indeed absent from East Anglia at that time. Elsewhere in Britain specimens typical of the moderate ecotype have recently been collected at Boxgrove in Sussex (ZMC, Fig. 111C; R. Preece, in prep.), where associated mammalian fossils indicate oxygen isotope stage 13 (524 000–478 000 BP) (Roberts, Stringer & Parfitt, 1994). Similar in appearance, but less well preserved, are shells from levels G_0, G_1 and J at Sidi Abd er Rahmane near Casablanca (MNHNP; Fig. 111A, B) from between the Maarifian and Anfatian stages of the North African Pleistocene, dated at about 0.5 Ma (Lecointre, 1950, 1952, 1963; Brebion, 1979; age from Biberson, 1977). The species does not now occur on rocky shores in this part of Morocco, but these fossils are from a cold climatic period, when the contiguous distribution of *L. saxatilis* presumably stretched further south than at present.

Fossils from the latest Pliocene of Iceland have been referred to *L. saxatilis* and are apparently related to the *L. saxatilis* group, but are here considered specifically distinct (p. 246).

Discussion

Exceptional characteristics. Among *Littorina* species (and indeed among gastropods in general) *L. saxatilis* stands out as exceptional in several respects. It exhibits extreme morphological and genetic variation, it occurs in a very wide variety of habitats and over a large geographical range, and it is ovoviviparous. These characteristics are to a great extent interrelated, and in particular it can be argued that its ovoviviparity is central to its variability and versatility.

Littorina saxatilis is also exceptional as the most well-studied species of its genus and one of the most well-known of marine gastropods. This can partly be ascribed to the interest generated by its variability and by the relatively recent taxonomic revision of the *L. saxatilis* group, and also to its abundance, ecological importance and widespread distribution. Nevertheless, there are still conspicuous gaps in the knowledge of this species, for example in the genetic and ecophenotypic control of shell form, the maintenance and adaptive significance of the shell colour polymorphism, the action of natural selection on shell traits, the nature of species recognition or isolation mechanisms, patterns of gene flow on a geographical scale, and the physiology of ovoviviparity.

Morphological variation. The variation in shell shape, sculpture and mature size among populations is greater than in any other *Littorina* species, and the complete range of shell coloration is at least as great as in any other. Less well known is the variation in some anatomical features such as radular tooth form, penial shape and embryo size, which also appear to be unusual. Assuming, initially, that the variation is under genetic control, the reasons for it can be explored. The ovoviviparity, non-planktotrophic development and intracapsular metamorphosis of the species preclude the widespread dispersal (and consequent gene flow) usually associated with broadcast of spawn and with planktotrophic development. As an adult it is relatively sedentary, migrating only about 1 to 5 m per month (Berry, 1961; Janson, 1983b; K. & B. Johannesson, 1995), and there has also been a suggestion of homing behaviour (Janson, 1983b), which would further reduce mobility. The area of the panmictic unit and the population size are consequently small. Under such conditions both stochastic and selectionist models predict high levels of interpopulation variability. Stochastic processes such as founder effects and genetic drift are more likely to be significant in small populations with low dispersal and have sometimes been invoked to explain phenotypic variation (see Colour Variation, p. 320). On the other hand, if variation is adaptive and influenced by natural selection, low gene flow permits local adaptation. In a species that occupies such a wide geographical and habitat range, this will lead to differentiation of local populations. Where selection is strong, morphological clines can be maintained over distances of just a few metres (e.g. Janson & Sundberg, 1983; Byers, 1990; B. & K. Johannesson, 1990; Johannesson *et al.*, 1993; Grahame *et al.*, 1995; Hull *et al.*, 1996). There is circumstantial evidence that crab predation, wave action and visual predation are important agents of natural selection on shell characters, although experimental evidence is limited (p. 301). It is also likely that some aspects of shell form, particularly shell size, may be directly affected by the environment (e.g. Kemp & Bertness, 1984; Palmer, 1990; Boulding *et al.*, 1993), although the possibility has been little studied in *L. saxatilis*. If ecophenotypic effects are significant, a sedentary adult life in combination with a wide habitat range will again enhance variability between populations.

Although the mechanisms are incompletely known, the intraspecific morphological variation of *L. saxatilis* is clearly connected with its nonplanktotrophic development and wide habitat range. In comparison with planktotrophic species such as *L. littorea*, with the potential for pelagic dispersal as both eggs and veligers, *L. saxatilis* shows much greater interpopulation differentiation (Janson, 1987a). When compared with other nonplanktotrophic species (subgenus *Neritrema*), such as *L. arcana* and *L. compressa*, it shows similar ecotypic variation in the habitats shared with these species (i.e. moderate, wave-exposed and barnacle ecotypes), but in addition displays a sheltered/brackish ecotype, as a consequence of its wider habitat range.

Genetics and speciation. There have been numerous studies of the population genetics of *L. saxatilis*, mostly based on the allozymic variation detected by electrophoresis (review by Ward, 1990). At the level of variation between populations, the genetic results give the same picture of differentiation as the morphological results, both being connected with low dispersal and a wide habitat range. In common with other nonplanktotrophic *Littorina* species, *L. saxatilis* shows far greater differentiation on geographical and local scales than planktotrophic species such as *L. littorea* (E. M. Berger, 1973; Janson, 1987a; Ward, 1990). The scale of genetic differentiation can be very small, over one or two metres (Janson & Ward, 1984, 1985; K. & B. Johannesson, 1990; Johannesson *et al.*, 1993). Variation in allozymes is often assumed to be selectively neutral, in which case their differentiation simply reflects the low gene flow between populations and arises by random processes such as genetic drift and founder effects. On a local scale it has been shown that founder effects are in fact an insignificant cause of genetic differentiation (Janson, 1987b; Johannesson & Warmoes, 1990), although they may be more important in cases of long-distance colonization (Ward, 1990). However, there is evidence that some allozymes are under selection, for clines in their frequencies can occasionally be correlated with environmental variables such as wave-exposure (Janson & Ward, 1984) or shore level (K. & B. Johannesson, 1989, 1990, 1993; Grahame, Mill, Double & Hull, 1992, 1993; Johannesson *et al.*, 1993; K. Johannesson, B. Johannesson & Lundgren, 1995). In such cases genetic differentiation will be related to both low gene flow and habitat range.

Using data on rates of selection and migration in natural populations of *L. saxatilis* on adjacent wave-exposed and sheltered shores, Johannesson & Sundberg (1992) have constructed a model to investigate whether differentiation could proceed to complete reproductive isolation between the ecotypes adapted to each shore, in other words leading to speciation. Since intermediate forms (heterozygotes in the model) were more fit than either extreme ecotype in intermediate habitats, they concluded that speciation under these conditions was unlikely. However, studies on exposed shores in Galicia have since shown that genetic (and morphological) differentiation can be maintained in adjacent subpopulations occupying different microhabitats (mussel beds and barnacle-covered rocks) on the same shore (Johannesson *et al.*, 1993). Gene flow between the subpopulations does occur, but is restricted by a degree of assortative mating (Johannesson, Rolán-Alvarez & Ekendahl, 1995), which is one of the conditions required for parapatric or sympatric speciation. There is also the possibility that post-zygotic isolating mechanisms may have arisen between some ecotypes as a by-product of their genetic differentiation. This was suggested by the high incidence of abnormal embryonic development in subpopulations intermediate between wave-exposed and moderate ecotypes in Sweden (Janson, 1985a). A more extreme case of embryonic inviability in apparent hybrids between high-shore and low-shore ecotypes has recently been reported in northeastern England (Hull *et al.*, 1996). *Littorina saxatilis* seems to provide an excellent system in which to test models of speciation processes. Speciation in *Littorina* is discussed in more detail in Chapter 6.

Although the conditions necessary for sympatric speciation appear to be present in *L. saxatilis*, differentiation and dispersal of incipient species may be a relatively slow process. Cycles of glaciation and eustatic sea-level change during the Pleistocene no doubt caused considerable disruptions and shifts to the geographical range of this species. The last glaciation came to an end only about 10000 years ago, and much of the coast of northern Europe may only have been repopulated since then. This may partly account for the complex picture that is beginning to emerge.

Another aspect of genetic variation is the level of polymorphism within populations. In allozyme studies this is often measured by average heterozygosity per locus. It is difficult to compare figures from different studies in which various enzymes have been assayed, but nevertheless the values for *L. saxatilis* appear to be among the highest for *Littorina* species, perhaps exceeded by *L. scutulata* and *L. plena* (Janson, 1987a; Ward, 1990; Backeljau & Warmoes, 1992). This impression of high genetic variability within populations is supported by other measures, including the diversity of esterase bands revealed by isoelectric focussing (Mill & Grahame, 1988, 1992a; Dytham, Grahame & Mill, 1992) and the diversity of random-amplified polymorphic DNAs (Crossland *et al.*, 1995). The theoretical explanations are complex and poorly understood. It is notable that within littorinids as a whole there is no relationship between dispersal capability (as assessed by type of development) and heterozygosity, so that genetic variation within populations is apparently little influenced by gene flow between them, nor by effective population size (Janson, 1987a; Ward, 1990; Tatarenkov, 1992). Neither stochastic nor selectionist models entirely account for these observations, and Janson (1987a) has suggested that historical effects (such as population bottlenecks, in *L. saxatilis* seen only in outlying populations) may be more important than gene flow or selection. Nevertheless, one selectionist model predicts a correlation between niche width and genetic diversity, because the diversity is adaptive in a heterogeneous environment (Noy, Lavie & Nevo, 1987). Niche width is difficult to quantify, but on individual shores *L. saxatilis* does often appear to extend over a wider range of microhabitats and tidal levels than other *Littorina* species, and the possibility that this is related to its genetic diversity should be further explored.

Classification. In view of the morphological and genetic variability of this species, it is not surprising that there has been considerable disagreement about its classification. As reviewed earlier, its nomenclature has had an unusually complicated history (see Synonymy, p. 278, and Taxonomic History, p. 287). However, it has been argued (see Geographical Variation, p. 301, and Ecotypic Variation, p. 305) that there is as yet no compelling evidence, from either morphology, allozymes or DNA, for the further subdivision of this species. At present, the picture is of a single highly polymorphic species in which (in the heterogeneous environment of rocky shores) steep morphological clines can be maintained by disruptive selection over distances as small as one or two metres. The resulting morphs are here regarded as ecotypes.

Nevertheless, faced with evidence of morphological distinctness together with a degree of reproductive isolation (both behavioural and post-zygotic), some authors have suggested that the morphs should be recognized as separate species (Grahame *et al.*, 1995; Hull *et al.*, 1996, pers. comm.). Such a conclusion cannot, however, be sustained by observations on a single shore. As emphasized by K. & B. Johannesson (1990), if gene flow between different morphs on the same shore is more extensive than between similar morphs on geographically separated shores, they are appropriately regarded as ecotypes, but if the converse is true, the forms are to be regarded as separate species (often referred to as 'sibling species' if morphological differences are slight) (see also Rolán-Alvarez & Rolán, 1995). So far, this test has only been applied in one example (Johannesson *et al.*, 1993), which illustrates the importance of repeating observations on a number of separate shores.

Further study of gene flow within and between the ecotypes of *L. saxatilis* is clearly necessary. As more information becomes available, it may perhaps be discovered that they behave as isolated biological species in some geographical regions, but not in others. Such an outcome would present problems for conventional classification, and the solution would then depend on the operational definition of a species that might be adopted (p. 373).

Ovoviviparity, distribution and dispersal. In the family Littorinidae ovoviviparity involving brooding throughout the larval stage, with release following metamorphosis, is rare. Besides *L. saxatilis*, it is known only in two tropical littorinids, *Tectarius viviparus* (Rosewater, 1982) and *Littoraria aberrans* (Philippi, 1846), both of which brood embryos in the mantle cavity and not within the oviduct (Reid, 1989a). Among *Littorina* species the ovoviviparity of *L. saxatilis* is clearly an autapomorphy derived from an ancestral condition represented by *L. arcana* and *L. compressa*, in which gelatinous egg masses are deposited on the rock substrate (Reid, 1990b). In anatomical terms, the evolution of ovoviviparity represents only a small modification of the pallial oviduct: eggs are retained in the jelly gland, which serves as a brood pouch; the secretory epithelium of the jelly gland becomes low (Linke, 1933); and the albumen and capsule glands are reduced in size (or the capsule gland in a strict sense may be lost, Losse & Greven, 1993; p. 293). In energetic terms, this evolutionary step has taken place without significant change in the allocation of resources to reproductive effort (Hughes, 1995). The anatomical changes might easily be achieved by regulatory genes (Ward & Warwick, 1980), for example by retarding development of the glands of the oviduct. Nevertheless, this minor reproductive modification appears to have had a profound effect on the biology of the species.

The most obvious advantages of ovoviviparity lie in the protection of the eggs and embryos during their development, and in the consequent broadening of the range of habitats and latitudes over which development is possible. In *L. arcana* and *L. compressa* the egg masses are believed to be susceptible to desiccation, predation and mechanical damage (although this has yet to be observed or demonstrated), and must be laid in suitably damp, protected sites on hard rock types. The adults are limited in their distribution to such sites and also do not extend into warm climatic zones (pp. 263, 276). In contrast, *L. saxatilis* has a far wider habitat range, extending both higher and lower on the shore, into sheltered inlets and salt marshes, and also reaching higher and lower latitudes (see Distribution, p. 324). The physiology of ovoviviparity is still poorly known, but involves osmotic regulation within the brood pouch and possible maternal nutrition (see Embryos and Development, p. 296). At high tidal levels the young of *L. saxatilis* are, however, still at risk from desiccation after they are released from the parent (Berry & Hunt, 1980) and generally shelter in rock crevices.

Ovoviviparity may increase juvenile survivorship, but at the cost of reduced fecundity, because of the increased parental energetic investment per offspring. The fecundity of *L. saxatilis* may thus be an order of magnitude lower than that of, for example, *L. compressa* (Hughes, 1978; Hughes & Roberts, 1980, 1981), so that it has a lower potential rate of increase. The higher fecundity of oviparous species may explain why this strategy has persisted in the relatively benign habitat of the mid shore (Hughes, 1995). Nevertheless, the benefit of protection of the young would seem to be of overriding importance in the upper eulittoral zone and littoral fringe inhabited by most littorinids, so that the rarity of ovoviviparity in this family is surprising. It has often been considered that another cost of nonplanktotrophic development in general (including ovoviviparity through metamorphosis) lies in the absence of a pelagic larval stage through which widespread dispersal can be achieved; in consequence, nonplanktotrophic species are often evolutionarily short-lived, with an increased probability of extinction and speciation (Jablonski & Lutz, 1983; Jablonski, 1986). This argument has been used to explain the rarity of ovoviviparity through metamorphosis in littorinids (Reid, 1989a). While this may well apply to tropical littorinids, it does not appear to be the case in *L. saxatilis*, in which it has been suggested that its type of development may actually enhance long-distance colonization (Johannesson, 1988).

On a small scale (distances of the order of 1 km or less) the dispersal potential has been estimated from the rate of colonization of new habitats, such as artificial breakwaters and barren skerries. It has been found that the colonization rate is less than that of the planktotrophic *L. littorea*, but is nevertheless still quite high, new habitats being populated within about two to ten years (Janson, 1987b; Johannesson & Warmoes, 1990; K. & B. Johannesson, 1995). The mechanisms of dispersal are not known, but may include rolling on sand, rafting on weed, or accidental release of encapsulated embryos (Hannaford Ellis, 1978b; Johannesson & Warmoes, 1990). Over much greater distances, however, nonplanktotrophic species may be at an advantage. Dispersal to a distant site is a rare event for planktotrophs and nonplanktotrophs alike, but the former can establish a colony only if a large number of larvae or adults arrive together, whereas for the latter a single fertilized female may found a colony. Johannesson (1988) illustrated this argument by the presence of *L. saxatilis* alone among *Littorina* species on the isolated islet of Rockall, 400 km west of the Outer Hebrides. However, this advantage of nonplanktotrophic development could apply equally to oviparous species such as *L. arcana* and *L. compressa*, yet neither has the extensive geographical range of *L. saxatilis*. Once again the ovoviviparity and associated environmental tolerance of *L. saxatilis* may be significant, perhaps enhancing survival of the young from founder females. Other considerations are that females of *L. saxatilis* often contain embryos throughout the year (in contrast to the other two members of the *L. saxatilis* group, which have shorter breeding seasons), and that the wider habitat range, including weeds and marshlands, presents more opportunities for rafting of adults and (as argued for the foundation of African colonies, p. 331) carriage by birds. *Littorina saxatilis* has been recorded on floating clumps of algae 10 km off the coast of Iceland (Ingólfsson, 1995).

Littorina saxatilis occupies the widest range of habitats and shows by far the most extensive geographical distribution of any *Littorina* species. Ovoviviparity is not necessarily the only factor permitting this wide distribution, for physiological, dietary and behavioural adaptations might also be involved. It does, however, seem likely that ovoviviparity has been a key adaptation in the evolution of this species. This is suggested by comparisons with *L. arcana* and *L. compressa*, which are genetically and phylogenetically very closely related to *L. saxatilis*, yet both occupy a narrower range of habitats (and consequently do not show the full range of ecotypic variation), have a restricted geographical distribution, and show lower genetic variation within populations.

Phylogenetic relationships. The three species of the *L. saxatilis* group, *L. saxatilis*, *L. arcana* and *L. compressa*, would appear from their morphological similarities to be closely related. These similarities include shell form, the single spirals of both albumen and capsule glands in the pallial oviduct, penial shape and radular resemblances. However, none of these are synapomorphies, and therefore the cladistic analysis (p. 351) does not resolve them as a monophyletic group.

Genetic data from enzyme electrophoresis have also been used to reconstruct phylogeny, and most studies of these species have employed phenetic clustering, using the UPGMA algorithm (see Sundberg *et al.*, 1990, for review of methodology). Where several populations of each species have been analysed together, *L. compressa* has emerged as a discrete cluster (Ward, 1990; Knight & Ward, 1991), as have the other two species in some studies (Ward & Warwick, 1980; Ward & Janson, 1985). Nevertheless, some intermixing of populations of *L. saxatilis* and *L. arcana* has often been found (Knight *et al.*, 1987; Ward, 1990; Knight & Ward, 1991). This, however, appears to be an artefact of the phenetic clustering method, which assumes equal rates of divergence in lineages; when other methods have been used, improved and sometimes complete separation has been achieved (Sundberg *et al.*, 1990;

Knight & Ward, 1991). The three species do therefore each appear to be monophyletic units. Where more species have been included, the three members of the *L. saxatilis* group have always emerged as a monophyletic group, but within this there are two competing topologies. The more common, supported by UPGMA clustering of data derived from starch-gel electrophoresis, shows a closer relationship between *L. saxatilis* and *L. arcana* than either shares with *L. compressa* (Ward, 1990; Knight & Ward, 1991; Zaslavskaya, Sergievsky & Tatarenkov, 1992a, b). In contrast, clustering of gene-frequency data from polyacrylamide-gel electrophoresis, by both UPGMA and distance-Wagner methods, has linked *L. arcana* and *L. compressa* most closely among the three (Backeljau & Warmoes, 1992).

Use of the technique of random amplification of polymorphic DNAs (RAPDs) has supported the monophyly of *L. saxatilis* and of *L. arcana* (Crossland *et al.*, 1993). Inclusion of the third species and UPGMA clustering of Nei & Li's F-percent similarity index of the banding pattern has again linked *L. saxatilis* more closely with *L. arcana*, while *L. compressa* appeared more distantly related to both (Crossland *et al.*, 1995). Analysis of DNA sequence data (from the 12S, 16S and cytochrome-*b* mitochondrial genes) has supported the alternative topology, connecting *L. arcana* and *L. compressa* most closely, although since the numbers of base differences were small (only 12 to 15 out of 1469, among pairs of the three species) and in parsimony analyses the alternative topologies differed in length by only one step, little confidence can be placed in this result (Reid, Rumbak & Thomas, 1996; p. 363).

From measures of allozyme frequencies in *L. saxatilis* and *L. arcana*, in combination with estimates of the rates of divergence in lineages of other animal groups, the age of their separation has been calculated as 0.22 to 0.79 Ma (Ward & Warwick, 1980), or 0.34 to 0.55 Ma (Knight *et al.*, 1987). This compares with a date of 1.73 Ma (range 0–3.55 Ma) for the separation of the three species in the *L. saxatilis* group, calculated from divergence of their mitochondrial DNA sequences (Reid *et al.*, 1996). The oldest fossils of *L. saxatilis s.l.* are about 0.5 Ma (p. 332), and are considerably predated by fossils of a related species from Iceland, here described as a new taxon, *L. islandica* (p. 246).

Similar species. Care is needed when identifying the three members of the *L. saxatilis* group, but with experience the majority of mature animals can be confidently assigned using morphological characters (summarized in Table 7). Using characters of the shell, *L. compressa* is the most distinctive of the three; its moderate ecotype of the southern geog. var. *typica* is easily recognized by its compressed, slightly cylindrical and shouldered final whorl, and regularly-spaced, flattened ribs with narrow grooves between them. The shells of juveniles and of wave-exposed and barnacle ecotypes can be difficult or impossible to identify, although the regular sculpture is sometimes a useful guide; in the northern geog. var. *barentica*, however, the sculpture becomes finer and irregular, and identification must always be confirmed anatomically. The shells of *L. saxatilis* and *L. arcana* are extremely similar to each other. Morphometric analyses have often shown significant differences in shape between sympatric samples of their southern geographical varieties (Hannaford Ellis, 1979; Ward & Janson, 1985; Grahame & Mill, 1989, 1992; Grahame *et al.*, 1990, 1995; Dytham, Mill, Grahame & O'Higgins, 1992; Caley *et al.*, 1995), emphasizing the more patulous form of *L. arcana* (i.e. a shorter spire and enlarged, more rounded aperture) and its relatively narrower penultimate whorl. However, over the entire geographical and habitat range, these characters are not constant; in particular, the wave-exposed and barnacle ecotypes, and the northern geographical varieties, are often impossible to distinguish using shell characters. Marked constriction of the aperture in adult examples of the moderate ecotype is rarely seen in *L. arcana*, but common in *L. saxatilis*. In addition, the sheltered/brackish ecotype with a small, thin, smooth, relatively tall-spired shell is only found in *L. saxatilis* among the three species. Shell coloration is also a poor guide to identification. In *L. compressa* the tessellated pattern is very rare, but its common 'nigrolineate' pattern can also be developed in the others. Over the British distribution of *L. saxatilis* and *L. arcana* there are significant differences in the frequencies of some of their minor colour morphs (Atkinson & Warwick, 1983), but none is diagnostic. However, on particular shores an initial anatomical separation of these two species may reveal that a particular shape, sculptural type or shell colour is associated only with one or other, permitting local identification (Hannaford Ellis, 1979; Faller-Fritsch & Emson, 1985; Dytham *et al.*, 1990).

Of the anatomical characters, the most useful is the mature pallial oviduct and this was used in the first clear distinctions of *L. compressa* and *L. arcana* from *L. saxatilis* (Sacchi, 1975; Hannaford Ellis, 1979). Both *L. compressa* and *L. arcana* are oviparous, laying benthic gelatinous egg masses, and the

Table 7. Comparison of *Littorina saxatilis*, *L. arcana* and *L. compressa*.

Character	*L. saxatilis*	*L. arcana*	*L. compressa*
Penis	Fig. 102	Fig. 96	Fig. 90A–J
Filament	10–35% total length, stout with mucronate tip	10–30% total length, triangular	<5%, triangular
Mamilliform glands	Usually 3–25 (range 0–43), mostly in one row	6–68, usually in more than one row	1–8, relatively large, in one row
Development	Ovoviviparous, brooding embryos through metamorphosis in brood pouch of pallial oviduct	Benthic gelatinous egg mass; nonplanktotrophic	Benthic gelatinous egg mass; nonplanktotrophic
Pallial oviduct	Fig. 103A–N	Fig. 97	Fig. 90K–O
Albumen and capsule glands	Together one third or less of total length of pallial oviduct	Together at least one half of total length of pallial oviduct	Together at least one half of total length of pallial oviduct
Capsule gland	Small, inconspicuous loop	Conspicuous spiral loop	Conspicuous spiral loop
Jelly gland	Modified to form thin-walled, septate brood pouch filled with eggs and embryos	Swollen with fleshy septa when mature	Swollen with fleshy septa when mature
Copulatory bursa	Less than half length of brood pouch	Full length of jelly gland	Usually only half length of jelly gland
Ciliated field (Hannaford Ellis, 1979)	Often large	Small	Small
Shell	Figs 106–108	Fig. 99	Figs 92–94
Range of ecotypes	Moderate, wave-exposed, barnacle and sheltered/brackish	Moderate, wave-exposed and barnacle	Moderate, wave-exposed and barnacle
Moderate ecotype of southern geog. var.	Sculpture variable, but ribs usually narrow and irregular; aperture often constricted	Sculpture variable, but ribs usually narrow and irregular; aperture not usually constricted	Sculpture of strong, wide, flattened ribs; last whorl laterally compressed and slightly shouldered
Coloration	Variable	Variable	Tessellation is rare
Distribution			
Habitat	Diverse; includes salt marshes and littoral fringe	More common on wave-exposed shores; absent from salt marshes; upper eulittoral zone and littoral fringe	More common on sheltered shores; absent from salt marshes; mid to upper eulittoral zone
Range	Figs 110, 111. Sub-arctic region of N Atlantic, S to Chesapeake Bay and Morocco, Mediterranean, Azores, Canary Is, South Africa	Fig. 100. N Norway, British Isles, N France	Fig. 95. Barents Sea, N Norway, British Isles, N France

glands of their pallial oviducts are similar: a single spiral loop of albumen gland followed by a single spiral loop of capsule gland (which in area is equivalent to or only half that of the preceding albumen gland) and finally an enlarged jelly gland, which is swollen with fleshy septa in mature individuals. To distinguish between these two species, the relative length of the copulatory bursa is useful; in *L. arcana* it always extends back along the jelly gland to the capsule gland, but in *L. compressa* it usually (80% of individuals examined) does not extend back to the capsule gland and is commonly only half the length of the jelly gland (Fretter, 1980; Reid, 1993). In the far north of the range of *L. compressa* its capsule gland is reduced in size to a simple (not spiral) loop. In contrast to the two other species, *L. saxatilis* is ovoviviparous and mature females usually contain embryos in the brood pouch (which is a modified jelly gland, thin-walled and with delicate septa) throughout the year. Even if the brood pouch is empty, adult females of *L. saxatilis* can be recognized by the small size of the albumen and capsule glands in relation to the length of the entire pallial oviduct (these two glands together occupying only one third or less of the total length, compared with about one half in *L. arcana* and *L. compressa*), and by the fact that the capsule gland is a small and inconspicuous loop which lacks spiral structure or shows only a small spiral twist (compare Fig. 103C with oviducts of *L. arcana*, Fig. 97D–K, and *L. compressa*, Fig. 90K–M, O). Another useful character is the small size of the copulatory bursa, which in adults of *L. saxatilis* is less than half the length of the brood pouch and in the smallest individuals sometimes vestigial or absent. Immature and juvenile females of the three species are more difficult to identify, for then the jelly glands of the oviparous species are not so swollen, and the distinctive adult proportions of the bursae are not established. Nevertheless, beyond a certain stage, at least juveniles of *L. saxatilis* can be distinguished from the other two by the proportions of the three glands of the pallial oviduct as described (compare Figs 97D, E and 103B, C, although the stages shown in Figs 97A–C and 103A are not distinguishable).

The second useful anatomical character is the shape of the penis. In *L. compressa* this is diagnostic, with only 1–8 relatively large mamilliform glands and a small filament less than 5% of the total length (Sacchi, 1975; Heller, 1975a). For the remaining two species there are statistical differences in penial shape and gland number, but these are not diagnostic in every case (Hannaford Ellis, 1979). In *L. arcana* the penis has a triangular filament and only rarely is the tip mucronate; there are 6–68 mamilliform penial glands, usually arranged in more than one row or in a closely-packed mass. In *L. saxatilis* the filament is often conspicuously mucronate, the penial glands are less numerous, usually less than 25, and they are most commonly arranged in a single row, which does not extend as far down towards the base as in *L. arcana*. These characters show some regional variations; for example in Wales the shape of the tip is particularly useful (Hannaford Ellis, 1979; pers. obs.), but less so in Brittany, where instead the numbers of glands show little overlap (pers. obs.); penial characters are said to be unreliable in Yorkshire (Jardine, 1984). Identification of males of *L. saxatilis* and *L. arcana* is thus often less certain than of females, but at some sites can be assisted by locally diagnostic shell features. It has also been noted that the mature penis of *L. arcana* is smaller relative to the shell height than in *L. saxatilis* (Hannaford Ellis, 1979). As in the case of females, the males of *L. saxatilis* are often mature throughout the year, whereas the breeding season of *L. arcana* is more restricted; since penes may be reduced in length outside the breeding season, this may further contribute to the difference in penial size in these two species.

In view of the difficulties that can be experienced in identification, particularly in the males of *L. saxatilis* and *L. arcana*, other anatomical characters have been sought. Hannaford Ellis (1978a, 1979) described the 'ciliated field', an area of pinkish, wrinkled epithelium lying between the prostate gland (or pallial oviduct) and the columellar muscle. Among British specimens this was said to be of largest extent in *L. saxatilis* (in which the prostate was correspondingly lengthened) and inconspicuous in the others. However, this feature can be difficult to use, since it is not readily seen in preserved specimens and in addition has not proved to be constant throughout the geographical range (Hannaford Ellis, 1979; Atkinson & Warwick, 1983; Grahame & Mill, 1986; Mill & Grahame, 1990b). The radula has been examined in several studies, but despite early claims (James, 1968a; Sacchi, 1975) no consistent differences have been found among the three species (Heller, 1975a; Seshappa, 1976; Raffaelli, 1979a; pers. obs.).

Ecological preferences and geographical distribution can also aid identification. For example, neither *L. arcana* nor *L. compressa* is known to occur in salt-marsh habitats, and *L. arcana* becomes relatively more common on wave-exposed shores. The geographical ranges of the three species now seem to be known with some precision, and only between the Barents Sea and southern Brittany, including the British Isles, can all three be found; elsewhere *L. saxatilis* occurs alone (see Ranges of each species, pp. 261, 274, 327, for details).

Finally, there is the possibility of using molecular characters for identification. So far, no diagnostic alleles have been discovered by allozyme electrophoresis, although one esterase system was almost specific in *L. arcana* and *L. saxatilis* at a locality in Yorkshire (Ward & Janson, 1985). The technique of RAPDs has been shown to be effective for the unequivocal clustering of individuals of these two species from Cornwall, although no diagnostic bands were found (Crossland *et al.*, 1993). This technique has been extended to samples of all three species from Wales (Crossland *et al.*, 1995) and, although no single band was diagnostic of any species, a combination of two bands appeared to provide unequivocal separation of all three.

There is much less likelihood of confusion with other littorinids that are sympatric with members of the *L. saxatilis* group. *Littorina littorea* has finer sculpture, less impressed sutures and both penis and oviduct are diagnostic. In the ranges of the southern geographical varieties of the Atlantic *L. (Neritrema)* species (i.e. approximately south of Narvik, Norway, in the eastern Atlantic and south of

Maine in the western Atlantic) no confusion should arise with moderate or wave-exposed ecotypes of *L. obtusata* or *L. fabalis*, which have globular or almost flat-spired, smooth shells, often with strong microstriae. However, the sheltered/brackish ecotypes and the northern geographical varieties of both these species are more tall-spired and can approach smooth forms of the moderate and sheltered/brackish ecotypes of *L. saxatilis*. In these cases, useful characters of the shell of *L. saxatilis* are the more rounded whorls and more impressed sutures, and the absence of strong microstriae; characters of the oviduct, penis and radula all distinguish *L. saxatilis* from *L. obtusata* and *L. fabalis*.

Some Pacific species, particularly *L. sitkana*, *L. subrotundata* and *L. natica*, can resemble *L. saxatilis* and *L. arcana*. However, their distributions do not overlap, and both oviduct and radula distinguish them, although penes are sometimes similar.

Doubtful species

Several fossil species are known from so few examples, sometimes poorly preserved, that their affinities within *Littorina*, and in some cases even their membership of the genus, are doubtful. These are discussed here.

Littorina sp. Woods & Saul, 1986

(Fig. 112A)

Synonymy

Littorina sp. Woods & Saul, 1986: 636; fig. 3.1 (Punta Rosarito, SW Baja California Norte, Mexico, UCLA loc. 7083; Palaeocene; UCLA 59432, now in LACM, lost, Fig. 112A).

Remarks

This fossil was figured, but not described or named, in a report on a collection of shallow-water and often wave-worn gastropod molluscs (largely Neritidae) from a locality on the western side of Baja California (Woods & Saul, 1986). The deposit was included in the Sepultura formation, said to be of Palaeocene or early Eocene age, although from comparison of gastropod fossils the authors considered the former to be more likely. Unfortunately the four specimens collected have since been lost (L. Saul, pers. comm.), but the original figure of one of them is here reproduced (Fig. 112A). The shell is 7.7 mm in height, of turbinate form, with five broad and shallow grooves on the base, becoming obsolete or eroded towards the periphery. No sculpture is clearly visible above the periphery. The overall shape, aperture and the form of the columella are typical of members of the Littorininae, and on this basis assignment to *Littorina* is possible. However, some of the more generalized shell shapes of species of the related genera *Nodilittorina* and *Littoraria* are extremely similar, and the generic assignment is therefore uncertain.

This species is of particular importance because of its great age. The oldest certain *Littorina* species, *L. sookensis*, is only of Upper Oligocene age, and no undoubted fossils of the Littorinidae are yet known that are older than the present species (Reid, 1989a). For comparison, the age of the divergence of the sister-genera *Littorina* and *Nodilittorina* has been estimated at 43 to 67 Ma (Middle Eocene to latest Cretaceous) from mitochondrial DNA sequence data (Reid, Rumbak & Thomas, 1996), similar to the age of this fossil. If this is indeed a member of a relatively derived littorinid genus, such as *Littorina*, *Nodilittorina* or *Littoraria*, the diversification of the family must have occurred earlier, and this record pushes back the likely origin of the family. New material would be of great interest.

Littorina subobesa Cooper, 1896

Synonymy

Littorina subobesa Cooper, 1896: 331; pl. 47, figs 3, 4 (Morley, Shasta Co., California, Cretaceous A [probably Santonian (Member V), Saul, quoted by Coan, 1981]; Marysville Buttes [Sutter Co.], Cretaceous B or Eocene [Capay Stage, Middle Eocene, Coan, 1981]; False Bay, San Diego [Mission Bay], Cretaceous B or Eocene [probably Rose Canyon shale, Domengine Eocene, Coan, 1981]; types in CAS probably destroyed in San Francisco fire, Keen & Bentson, 1944). Keen & Bentson, 1944: 168. Coan, 1981: 173.

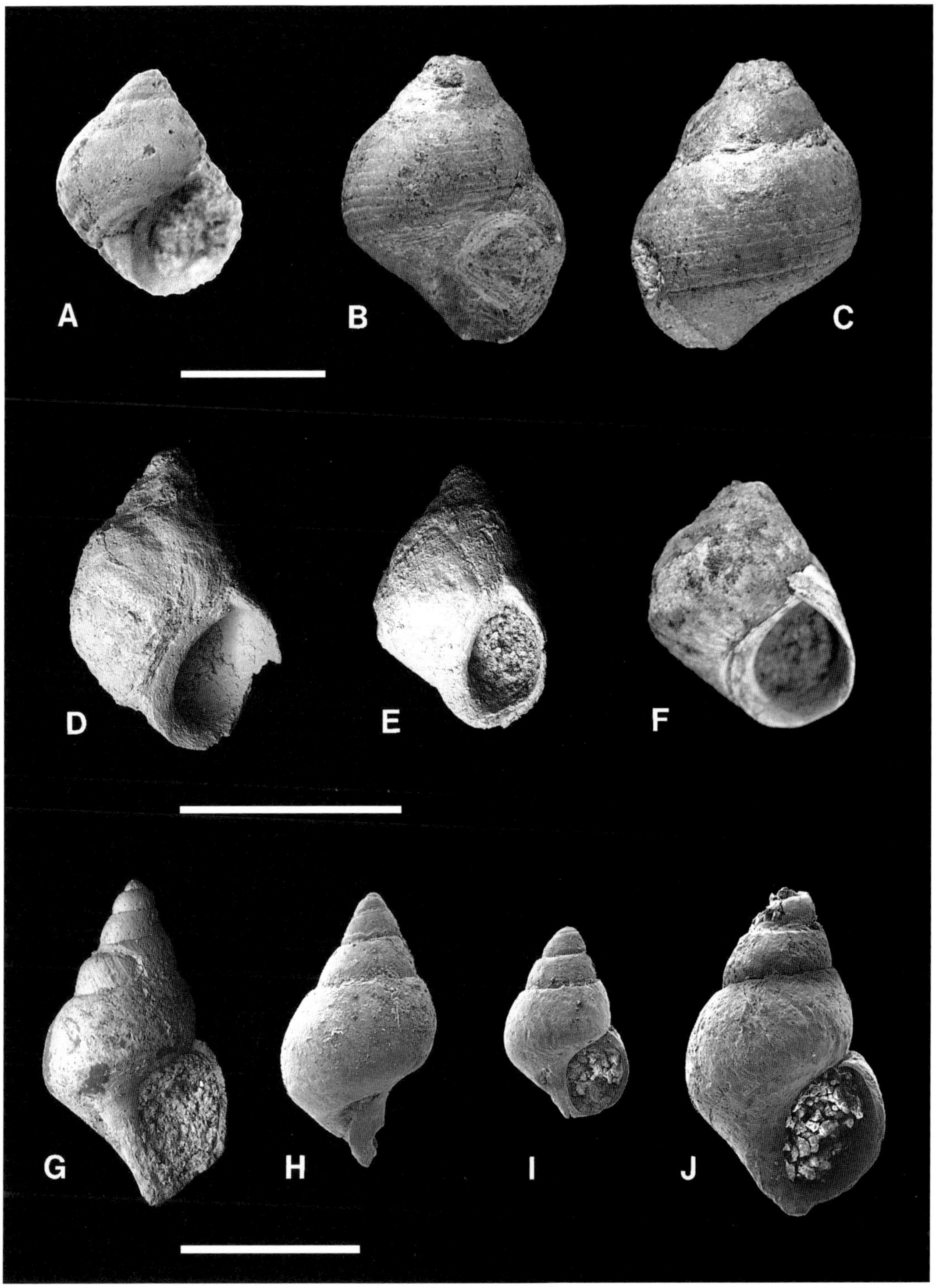

Figure 112. Doubtful fossil species of *Littorina*. **A,** *Littorina* sp. Woods & Saul, 1986; copy of Woods & Saul, 1986: fig. 3.1; Punta Rosarito, SW Baja California Norte, Mexico; Palaeocene; UCLA 59432 (now LACM). **B, C,** *Littorina kozaiensis* Nomura & Onisi, 1940; holotype, SHKM 21762; Yôsuibori, Simizu, Kozai-mura, Sennan district, Miyagi Prefecture, Japan; Middle Miocene. **D–F,** *Littorina* cf. *mountsoloensis* Weaver & Palmer, 1922; quarry on NW side of Oakville, Grays Harbor County, Washington; basal Lincoln Creek formation, Upper Eocene; BMNH PI.TG.2013. **G,** *Littorina oligocenica* Dickerson, 1917; holotype, CAS 181.01; probably a member of the genus *Lacuna*; Gries Ranch, Lewis County, Washington; basal Lincoln Creek formation, Upper Eocene. **H–J,** *'Littorina'* (probably *Lacuna*) *oligocenica*; Gries Ranch, Lewis County, Washington; basal Lincoln Creek formation, Upper Eocene; BMNH PI.TG.2014. Scale bar (**A–C**)=5 mm; scale bar (**D–F**)=5 mm; scale bar (**G–J**)=2 mm.

Remarks

These shells were described as up to 18x11 mm, ovate, with rounded whorls, and microscopic spiral impressed lines. The form of the columella illustrated, and the basal angle, appear to exclude this species from *Littorina*. Cooper himself compared his species to *Littorina compacta* Gabb, 1864, which was placed in the Ataphridae by Wenz (1938). Since the types are lost (Keen & Bentson, 1944), this is best considered a *nomen dubium* (Coan, 1981).

Littorina mountsoloensis Weaver & Palmer, 1922

(cf. Fig. 112D–F)

Synonymy

Littorina mountsoloensis Weaver & Palmer, 1922: 30–31; pl. 11, fig. 12 (Coal Creek, Cowlitz County, about half a mile above old dam, Section 11, Township 8 North, Range 3 West, Washington; UWP loc. 337; Middle Eocene; holotype UWP 190, lost (Weaver, 1942); plaster cast UCMP 32371, seen). Weaver, 1942: 330 (giving a different type locality, perhaps in error: McClarity Ranch on south bank of Stillwater Creek, one mile west of Vader, Lewis County, Section 30, Township 11 North, Range 2 West; UWP loc. 319).

Remarks

This species was described from the Cowlitz formation of Washington, of Middle Eocene age (about 41–44 Ma; Armentrout, Hull, Beaulieu & Rau, 1983). The original description was as follows: 'shell small, smooth; apex acute; spire about one-half the length of the body whorl; whorls six, sides straight; suture linear and only minutely excavated; aperture oval; anterior end of outer lip not completely entire', and the dimensions were given as 7 by 4 mm (Weaver & Palmer, 1922). Unfortunately the type is lost; an existing plaster cast of it is poor, but measures about 6.3 by 4.0 mm. The published photograph shows a shell reminiscent of *L. scutulata* and *L. plena* in its high spire and indistinct sutures. Details of the aperture and columella are not clear in this illustration; the final phrase of the description could be interpreted to suggest an anterior sulcus, which would exclude this species from the Littorinidae.

The following species, *L. oligocenica*, was described from the slightly later Lincoln Creek formation, in the adjacent Lewis County of southern Washington. In comparing the two, Effinger (1938, who presumably had access to the type of *L. mountsoloensis*) noted that *L. mountsoloensis* was more tumid, indicating that it was relatively lower spired. From the available specimens of *L. oligocenica*, which do not exceed 3.6 mm in height, it also appears that *L. mountsoloensis* was larger. Weaver (1942) also compared these two species but, confusingly, wrote 'This species [*L. oligocenica*] . . . differs from the Cowlitz species *L. mountsoloensis* in its larger size, wider aperture, less deeply incised sutures, and wider axial angle'; presumably it was intended thus to describe *L. mountsoloensis*.

It has not been possible to locate specimens, among collections in major museums, agreeing with the description of this species from its type locality or formation. Specialists on the Cowlitz fauna have also been unable to trace any additional records of *L. mountsoloensis* from this formation (E. Nesbitt, pers. comm.). However, a sample of 15 shells has been examined from a quarry on the northwestern side of Oakville in Grays Harbor County, also in southern Washington (BMNH PI.TG.2013). These were part of an intertidal fauna in a basaltic sandstone at the base of the Lincoln Creek formation (J. L. Goedert, pers. comm.), of Upper Eocene age (about 40 Ma; Armentrout *et al.*, 1983). This fauna is considered to be equivalent to that of the 'Gries Ranch bed' from which *L. oligocenica* was described. Nevertheless, these specimens are clearly not conspecific with *L. oligocenica* (a smaller, thin-walled, narrower, slightly umbilicate shell, which may be a *Lacuna* species, see p. 343 and Fig. 112G–J). The largest measures 6.7 mm in height, and the shape is rather variable (Fig. 112D–F), from turbinate to high-turbinate (H/B=1.13–1.37). The spire is flat-sided, with weakly impressed sutures, and there is a distinct keel at the periphery of the last whorl. The columellar pillar is short, with no excavation or umbilicus, and the anterior end of the aperture is rounded. The shell is solid, its surface is smooth, and no apex is preserved. Electron microscopy of fractured surfaces shows that the shell consists of two layers of similar thickness, both of crossed-lamellar structure and therefore aragonitic; there is also an indistinct outermost layer about 20–30 μm in thickness with finer granular structure (which might be a result of diagenetic alteration); there is no outermost banded crossed-lamellar structure (as described, for example, in *L. keenae* by Taylor & Reid, 1990). Until material from the type locality of *L. mountsoloensis* is available for comparison, the identification of these shells must remain uncertain. They do not correspond with any other littorinid so far described from this region, and at least in their general characters of size, shape and age, they do resemble the description of *L. mountsoloensis*. The poor published figure of that species is, however, somewhat more tall-spired (H/B=1.57) as also suggested by the dimensions of the plaster cast of the holotype (6.3x4.0 mm; H/B=1.58).

Unfortunately the generic assignment of these shells from Oakville is also uncertain. The overall shell shape is comparable with that of littorinine genera such as *Melarhaphe*, *Nodilittorina* and *Littorina*, and species lacking spiral sculpture are known in all three. *Melarhaphe* usually has a distinctively long and pinched columella, and is known only from the Atlantic (Kadolsky, 1973), so this assignment is unlikely. There are no entirely diagnostic conchological characters by which to distinguish the sister-genera *Nodilittorina* and *Littorina*. The former usually has distinct and regular spiral sculpture, and frequently an eroded parietal area adjacent to the columella, but not all species show these traits. These two characters are also seen in the basal *Littorina* species *L. striata* and *L. keenae*, suggesting that they are plesiomorphic in this genus. In contrast, the Oakville shells are smooth and the parietal area is not eroded; in these characters, as well as in shell outline and columellar shape, they resemble *L. scutulata*. However, that species is a member of the large clade of *Littorina* with the synapomorphy of an outer shell layer of calcite, a character which (from molecular data, Fig. 120) appeared more recently than about 26 Ma. This apomorphic character is absent in the Oakville shells. In short, these shells might belong to an early and atypically smooth-shelled lineage of *Littorina*, or to *Nodilittorina*. They date from a time close to the age of separation of these two genera, as estimated from their molecular divergence (43–67 Ma, p. 365).

Littorina oligocenica Dickerson, 1917

(Fig. 112G–J)

Synonymy

Littorina oligocenica Dickerson, 1917: 178; pl. 31, fig. 1 (just back of the Greeco [Gries] Ranch house, 4 miles E of Vader, on E bank of Cowlitz River, Lewis County, Washington, CAS loc. 426; Upper Eocene; holotype CAS 181.01, seen, Fig. 112G). Effinger, 1938: 379. Weaver, 1942: 328; pl. 63, fig. 22; pl. 67, fig. 19.

Remarks

This species was described from the Gries Ranch beds of Washington, occurring together with genera that indicate a nearshore or littoral, subtropical environment (Dickerson, 1917; Effinger, 1938). This fauna was included in the Toutle formation, of Upper Eocene or Lower Oligocene age, by Roberts (1958). More recently, Armentrout *et al.* (1983) included it in the basal, basaltic sandstone member of the Lincoln Creek formation (J. L. Goedert, pers. comm.), of Upper Eocene age (about 40 Ma). Although abundant at the type locality (Effinger, 1938), this species could not be located among available museum collections. The following description is based on the well-preserved holotype and on an additional sample of 24 shells recently collected at the type locality by J. L. Goedert (BMNH PI.TG.2014).

The holotype measures 3.6 by 2.1 mm, and no larger specimens have been seen. Effinger (1938) noted that 'the numerous specimens obtained show a great deal of variation in shape from conic to more elongate', although the available shells are mostly relatively narrow and tall-spired (H/B=1.43–1.71). The whorls are slightly rounded, with clear sutures, and only a slight peripheral angulation. Of particular significance is the columella, which in well-preserved specimens is entire, with sometimes a suggestion of an umbilical groove. However, the columellar callus over the umbilical region appears to be delicate and easily eroded to reveal a distinct umbilical groove between the pillar and the base, which is filled with rock matrix (Fig. 112H, I). The shell is relatively thin-walled and its surface entirely smooth. The protoconch is smooth, relatively elongate and slightly bulbous; although its limits are unclear, it is about 0.5 mm in diameter and of 2–2.5 whorls. The shell microstructure is entirely of crossed-lamellar aragonite, with no obvious layers.

In considering the likely generic assignment of this species, the umbilical groove is considered important. An umbilicus partly covered by a thin callus is not known in *Littorina* (except in a few atypical or gerontic specimens), but is present in some other genera. A similar arrangement is seen in *Lacuna*; in *L. vincta* (Montagu, 1803) the umbilical chink is variable in width and sometimes virtually obscured by the columella (see Fretter & Graham, 1980, for description of columellar area of *Lacuna*). The protoconch is unlike that of any *Littorina*; it is within the size range for nonplanktotrophic *Littorina* species (Table 15), and yet the number of whorls suggests planktotrophic development. In comparison with *Lacuna* species, the protoconch is more projecting, but is otherwise not unlike the smooth protoconch of about 2.5 whorls illustrated for the planktotrophic *Lacuna vincta* by Reid (1989a), which can reach 0.47 mm in diameter (Thorson, 1946; as *Lacuna divaricata*). The smooth, elongate and delicate shell are also reminiscent of some *Lacuna* species. Placement in the genus *Littorina* is therefore considered unlikely, and it is suggested that this species should be transferred to *Lacuna*.

Littorina kozaiensis Nomura & Onisi, 1940

(Fig. 112B, C)

Synonymy

Littorina kozaiensis Nomura & Onisi, 1940: 191–192; pl. 19, fig. 6a, 6b (Yôsuibori, Simizu, Kozai-mura, Sennan district, Miyagi Prefecture, Japan; Lower [Middle] Miocene; holotype SHKM 21762, seen, Fig. 112B, C).
Littorina (Littoraria) kozaiensis—Rosewater, 1970: 453; pl. 349, figs 6, 7. Masuda & Noda, 1976: 221.
Littorina (Littorina) kozaiensis—Higo & Goto, 1993: 73.

Remarks

This species was described from a single specimen collected in northeastern Honshu. Accompanying molluscan fossils are members of the early Middle Miocene Kadonosawa fauna (15–16.5 Ma) and indicate subtropical conditions (Chinzei, 1986). The holotype measures 9.8 by 7.7 mm and has slightly flattened whorls but with clear sutures; the periphery is a little angled and marked by a barely enlarged rib; below this there are 6 incised lines on the base, and above it about 14 incised lines between periphery and suture. The shell is recrystallized, so that no details of mineralogy are evident, but it appears relatively thick and solid. Although clearly a littorinine, there are few details to suggest a generic allocation. The shell lacks the eroded parietal area and long columellar pillar characteristic of many *Nodilittorina* species. It does not recall any Recent *Littorina* species, although membership of this genus cannot be entirely excluded. Rosewater (1970) gave no explanation for his assigment to *Littoraria* (then considered a subgenus of *Littorina*), but he may have been correct. The slightly flattened whorl profile, peripheral angulation and incised, regular sculpture are similar to some Recent species such as *Littoraria articulata* (Philippi, 1846), a tropical and subtropical species which occurs in the Indo-Pacific region, including southern Japan where it is found mainly on intertidal rocks (Reid, 1986a). The presence of *L. kozaiensis* together with a subtropical assemblage of fossils also supports assignment to *Littoraria*, which is an exclusively warm-water genus.

Littorina iwakiana Nomura & Hatai, 1936

Synonymy

Littorina iwakiana Nomura & Hatai, 1936: 144; pl. 16, figs 16a, 16b (Okada, Yamaoka-mura, Higasi-sirakawa-gun, Tanagura, Iwaki Prefecture, Japan; Middle Miocene; holotype SHKM 6895, seen).
Nodilittorina (Granulilittorina) iwakiana—Rosewater, 1970: 506; pl. 386, figs 19, 20.
Littorina (Granulilittorina) iwakiana—Masuda & Noda, 1976: 221.

Remarks

A single example of this species was described from the Tanagura beds in northeastern Honshu, in a molluscan association of the Shiobara type, indicating temperate climatic conditions (Chinzei, 1986) and an age of 10–12 Ma (T. Ozawa, pers. comm.). The specimen is now crushed, but the original description was as follows: 'Shell small, globose, consisting of about four convex whorls. Last whorl very large with subangulate periphery. Surface marked by spiral threads, separated by narrower grooves. Base convex, spirally, finely threaded. Aperture ovate, inner lip covered with a moderately thick callus' (Nomura & Hatai, 1936). The dimensions given were 3.8 by 3.2 mm. No details of shape or sculpture are visible from the remaining fragments of this specimen. However, the shell fragments have a thin inner chalky layer in which crossed lamellae are visible, and a thick outer translucent layer, corresponding to aragonitic and calcitic shell layers respectively which, together with the description, suggests that the generic placement is correct. *Littorina squalida* has been recorded from the Middle Miocene of Kamchatka (p. 89), and the description above might possibly apply to a small specimen of that species.

Chapter 5

PHYLOGENY

It is now widely recognized that phylogenetic analysis is an essential prerequisite for many evolutionary hypotheses, for example those relating to biogeography, speciation and adaptation. Increasingly, phylogenetic analysis is also being used as a basis for classification. A number of methods are available for the reconstruction of phylogenetic relationships, utilizing data from morphology, allele frequencies, DNA and the fossil record. The phylogenetic analysis of morphological data proceeds by the selection and numerical coding of characters, whose distribution is most commonly analysed by cladistic methods (see Forey, Humphries, Kitching, Scotland, Siebert & Williams, 1992, for an introduction to cladistics). Genetic data usually take the form of frequencies of polymorphic enzymes, determined by electrophoresis, from which the frequencies of alleles of the genes specifying the enzymes are inferred. The allele frequencies can be used to reconstruct the phylogeny of closely-related species, using a variety of distance measures and clustering techniques (e.g. Sundberg, Knight, Ward & Johannesson, 1990; Backeljau, de Wolf, van Dongen & Brito, 1994). With advances in molecular techniques, it is now possible to examine the DNA directly, leading to another class of characters for analysis. A relatively coarse and nonspecific technique is the random amplification of polymorphic DNA fragments, leading to electrophoretic banding patterns (e.g. Crossland, Coates, Grahame & Mill, 1995). It is also possible to determine directly the sequence of nucleotide bases in known stretches of DNA, which provides a very large number of characters from which to construct gene phylogenies (e.g. Hillis & Moritz, 1990). Under the controversial assumption of a 'molecular clock', molecular data may be used to estimate the age of branching points on the phylogeny. In addition, the fossil record is an important source of evidence, to be used in conjunction with morphological and molecular data from living species (Smith & Littlewood, 1994). In this chapter the morphological characters discussed in the Systematic Descriptions (Chapter 4) will first be used in a cladistic analysis of living *Littorina* species. The resulting phylogeny will be integrated with information from the fossil record and from geological events to provide an evolutionary time scale, and comparisons will then be made with molecular phylogenies and estimates of the ages of clades derived from enzyme and DNA data. Lastly, the implications for the phylogenetic classification of the species will be considered.

Morphological Phylogenetic Analysis

Previous analyses

The first attempt at a cladistic analysis of morphological characters among the genera of the Littorinidae was that of Reid (1986a), which identified the arrangement of consecutive spiral loops of the egg groove through albumen and capsule glands of the pallial oviduct as a synapomorphy of the genus *Littorina*. This was refined by including one representative of each littorinid subgenus (Reid, 1989a). As a result, *Nodilittorina* was identified as the sister-genus of *Littorina*, while *Littorina* itself was defined as a monophyletic clade by two synapomorphies, the first describing the two consecutive loops of the egg groove and the second the absence of rod-pieces in the paraspermatozoa. The problematic species *L. striata* was clearly included in *Littorina* by this analysis. However, neither of the two synapomorphies of the genus was unique within the family, and as a result there was uncertainty about the possible inclusion of the genus *Mainwaringia* within the clade *Littorina*. A cladistic analysis of all the *Littorina* species then recognized (including the two species of *Mainwaringia*) was made by Reid (1990b), using three representatives of *Nodilittorina* as an outgroup.

During the present study of new material, the morphological characters used in these previous analyses have been reexamined; some have been found to be unreliable or wrongly interpreted, while other new characters have been identified. Furthermore, three species have since been added to the genus (*L. kasatka*, *L. horikawai*, *L. natica*), the two species of *Mainwaringia* have been removed, and two of the species included by Reid (1990b) have been recognized as synonyms of others (*L. 'kurila'* of *L. subrotundata*; *L. neglecta* of *L. saxatilis*). The cladistic analysis of living *Littorina* species has therefore been redone and is described below (a summary of this same analysis has been reported by Reid, Rumbak & Thomas, 1996).

Choice and coding of characters

Shell. Such is the variability of shell form, sculpture and coloration within individual species, that it is difficult to recognize any of these shell characters as phylogenetically significant. Reid (1990b) defined two, one a measure of spire height, the other the presence or absence of nodules, but this is unsatisfactory. Whether or not shell shape is primarily under genetic control, it is highly plastic within species, sometimes covering the complete range from nearly flat to tall (e.g. *L. littorea*, *L. obtusata*, *L. saxatilis*). Nodulose sculpture is present in only two *Littorina* species (*L. striata*, *L. aleutica*), but not in all individuals of these, and although common in the outgroup (*Nodilittorina*), is not constant in its development there either. Nodulosity is apparently not a stable character among these species. Nevertheless, shell form is clearly not without some content of phylogenetic information, for some obvious groupings of species can be recognized. The solution has been to take four distinct groups as states of a single character. These are not readily defined by quantitative criteria, but are conchologically easy to distinguish. '*Nodilittorina*-like' describes those species which, without anatomical information, could not with confidence be assigned to one or other of the two sister-genera; their shells are usually relatively narrow and tall-spired; nodules may be developed; in non-nodulose species the sculpture is of regularly-spaced, incised lines; a crescentic, eroded area may be present on the parietal wall adjacent to the columella; the columellar pillar is relatively long and straight; a pale spiral band is usually visible on the base and within the aperture. A second category is described as '*L. squalida*-like' and includes those species that have sculpture of numerous fine ribs with narrow grooves between them, a round aperture and uniformly rounded columellar pillar. The category '*L. obtusata*-like' includes only two species, of which many examples are low-spired or almost flat-spired and which always lack spiral ribs, the surface being marked only by microstriae. The remaining shells are '*Neritrema*-like', very variable in shape from patulous to high-turbinate, with an ear-shaped or oval aperture; development of sculpture is variable, but often of a few carinate ribs with wide grooves between them. A second character of the shell is its mineralogy, whether the entire thickness is of crossed-lamellar aragonite, or an external layer of irregular-prismatic calcite is present.

Penis. The number of mamilliform penial glands has been coded as four states: absent, 0 or 1 in the same species, 2 glands, and many (3 or more) glands. This coding has been adopted for the following reasons. In some species (*L. striata*, *L. scutulata*, *L. plena*) mamilliform glands are invariably absent. In *L. kasatka* some populations lack mamilliform glands, while others have only a single gland. In *L. keenae* the great majority of individuals have a pair of glands (although occasionally there may be 1 or 3). The remaining species show a wide range in the number of glands. While some typically have high numbers (10–100 in *L. squalida*, 6–68 in *L. arcana*) and others low (*L. brevicula* 0–7, *L. mandshurica* 1–5), there is much overlap in the numbers in different species, while within species there is often a correlation with shell size. Clearly, in these species the number of penial glands is labile, and it is unlikely to be meaningful to attempt to subdivide this category further. In another modification of the data used by Reid (1990b), it is now recognized that the projection on the filament of *L. scutulata* and *L. plena* is not in fact a mamilliform gland, but simply an undifferentiated protruberance of subepithelial glandular tissue; this feature has therefore been coded as a separate character.

Sperm. The absence of rod-pieces in the paraspermatozoa is a synapomorphy of the genus *Littorina*. These structures are present in genera such as *Nodilittorina* and *Littoraria*, which are related to *Littorina* (Reid, 1989a).

Pallial oviduct. This is a source of three characters. The first relates to the arrangement of the spiral loops followed by the egg groove through the albumen and capsule glands. Although difficult, the attempt has been made to code these complex patterns in a way that reflects the likely homologies of the loops. Seven states of differing complexity have been defined (Fig. 113). In state 0, found in the

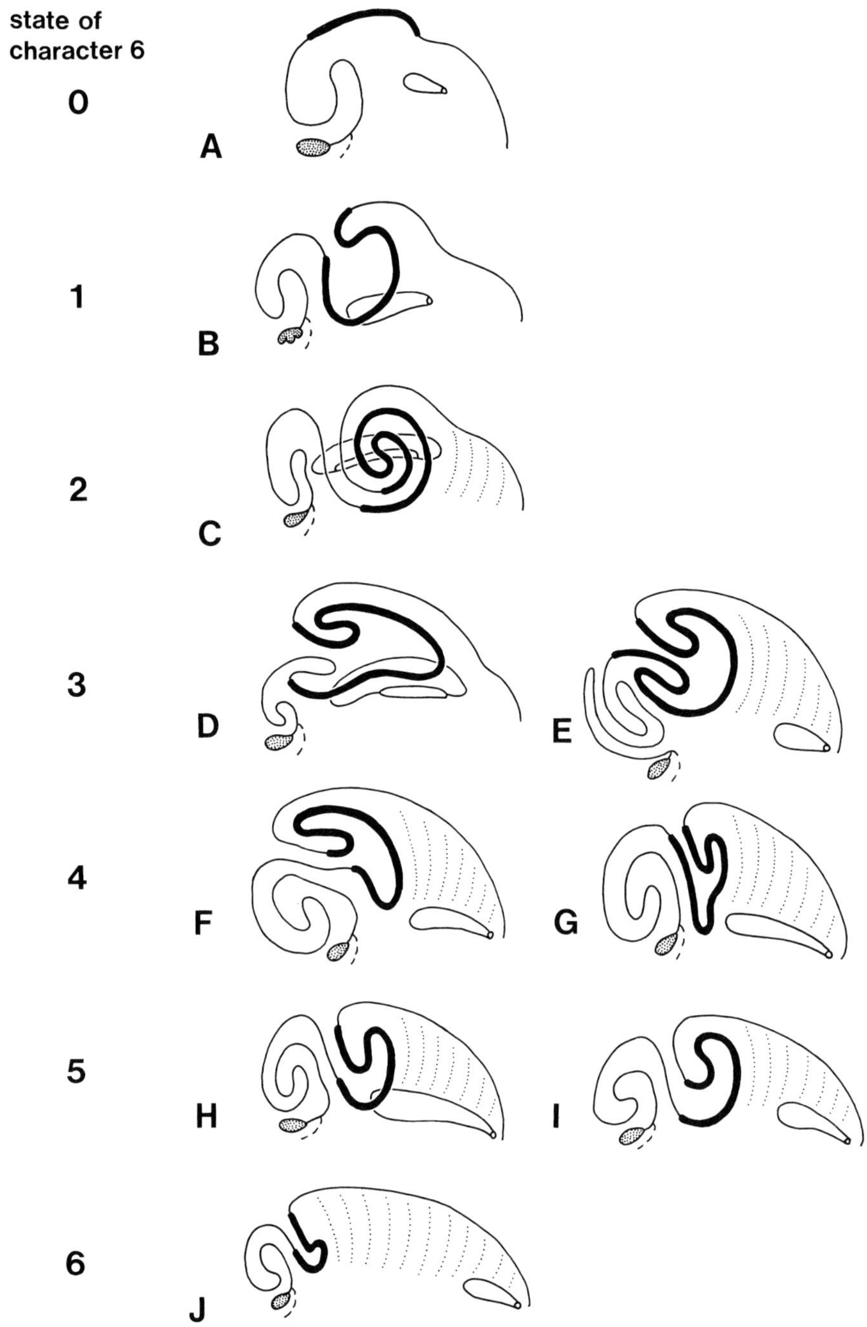

Figure 113. Diagrammatic representations of pallial oviducts of *Nodilittorina* and *Littorina* species, demonstrating the states of character 6 (see Table 8) used for the morphological cladistic analysis. The solid line represents the route of the ova along the egg groove from the junction of the seminal receptacle (stippled) with the renal oviduct (dashed line) on the left, to the anterior opening on the right; the capsule gland is indicated by the thickened line. The septa of the enlarged jelly gland (or, in **J**, of the brood pouch) are shown as dotted lines. The copulatory bursa is also shown. The sperm groove is not shown, but runs directly from the mouth of the bursa back to the seminal receptacle. See text for details. **A**, *Nodilittorina meleagris*; **B**, *L. striata*; **C**, *L. keenae*; **D**, *L. littorea*; **E**, *L. natica*; **F**, *L. sitkana*; **G**, *L. subrotundata*; **H**, *L. obtusata*; **I**, *L. compressa*; **J**, *L. saxatilis*.

outgroup of *Nodilittorina* species (Fig. 113A), there is a simple initial loop through the albumen gland, but the pathway through the capsule gland is straight. In state 1, found in *L. striata* (Fig. 113B), the initial loop is the same, but there is a large invagination in the path through the capsule gland, and in state 2 this second loop is elaborated into a spiral, as seen in *L. keenae* (Fig. 113C). In state 3, an additional loop appears between the other two and is lined either by albumen (e.g. *L. littorea*, Fig. 113D) or capsule gland (e.g. *L. natica*, Fig. 113E). This additional loop is represented only by an inflexion in state 4 (e.g. *L. sitkana*, Fig. 113F; *L. subrotundata*, Fig. 113G), whereas in state 5 there are merely two sequential and rather simple loops through albumen and capsule glands (e.g. *L. obtusata*, Fig. 113H; *L. compressa*, Fig. 113I). Finally, state 6 describes the reduced capsule gland of *L. saxatilis* (Fig. 113J). Classification of most species is straightforward, but *L. subrotundata* is unusually variable in this respect and has here been placed in the same category as *L. sitkana*. A second character describes the additional first loop of the egg groove present in the albumen gland of *L. aleutica* and *L. natica* (Fig. 113E), which appears to be homologous in these two. A third character relates to the condition of the jelly gland, which may be small (in those species producing pelagic spawn without an enveloping gelatinous matrix), enlarged with thick glandular walls (in *L. keenae* and all oviparous members of the subgenus *Neritrema*, in which egg capsules are surrounded by copious jelly), or enlarged but thin-walled and serving as a brood pouch for developing embryos (*L. saxatilis* alone). One of the characters used by Reid (1990b), the position of the copulatory bursa, has been omitted. This is because in some species with a small jelly gland, in a short straight section of the oviduct, the relative position of the bursa is difficult to describe (e.g. *L. scutulata*, *L. kasatka*, *L. littorea*) or is variable in position (*L. squalida*).

Spawn and development. The form of the egg capsule has been classified in five states: a pelagic, sculptured cupola (*Nodilittorina* species, *L. striata*); a pelagic, biconvex disc (the majority of planktotrophic species); a large capsule with a double rim (*L. scutulata*, *L. plena*); an ovoid, thin-walled capsule within a benthic gelatinous mass (oviparous *Neritrema* species); and the state in *L. saxatilis*, in which the capsule is apparently absent, the embryos within the brood pouch being enclosed only in thin, spherical egg coverings. The egg capsules of *Nodilittorina dilatata* and *N. meleagris* have been described (Borkowski, 1971; Bandel, 1974). That of *N. pyramidalis* is unknown, but is assumed to be of the sculptured cupola type, since this is present in the closely-related species *N. trochoides* (Berry, 1986, as *N. pyramidalis*). Reid (1990b) was incorrect in classifying, without first-hand observation, the egg capsules of *L. sitkana* and *L. subrotundata* as biconvex, and in supposing that egg capsules were absent in *L. obtusata* and *L. fabalis*. The type of development (planktotrophic, nonplanktotrophic or ovoviviparous) has been included as an additional character since, although it is correlated with the type of spawn, there is no *a priori* reason why this should be so; in other littorinid genera planktotrophic veligers may hatch from benthic egg masses, for instance (Reid, 1989a). The form of the protoconch is dependent upon the type of development (p. 19), and has therefore not been included as an additional character.

Radula and foregut. The radula is a particularly difficult structure to classify for analysis, owing to interspecific, intraspecific and ontogenetic variation in tooth shapes. Examination of large numbers of radulae with the scanning electron microscope has shown that some of the assessments of Reid (1990b), based on light microscopy, were incorrect. One clear character relates to the outer marginal teeth, which may show a definite central rachis from the neck to the base of the tooth, with a prominent flange or projection on its outer side, and cusps of the tooth head that are long and finger-like. This condition is seen, for example, in *L. keenae* (Fig. 14E). Alternatively, the rachis and outer flange of the base may be less well developed, and (in adults) the tooth cusps are shorter and rounded or blunt, as exemplified by *L. littorea* (Fig. 34C, E) and *L. saxatilis* (Fig. 104A, C). The features of the five central radular teeth have been considered together as a single character, for the shapes of the largest cusp on each of these teeth vary in the same way. These cusps may be long, narrow and mainly rectangular in shape (at least twice as long as broad), as in *L. keenae* (Fig. 14F). Both *L. scutulata* and *L. plena* are included in this category for, despite considerable intraspecific variation, some individuals of each do show long cusps (Figs 19F, 23E). A second state of this character is exemplified by *L. littorea* (Fig. 34B), in which the major cusps are pointed, and triangular to leaf-shaped. There are problems with the classification of *L. squalida* (Fig. 27D, F, H), in which the cusps are generally rounded (but similar to some examples of *L. littorea*, Fig. 34F, which clearly

Table 8. Characters used in the morphological phylogenetic analysis of *Littorina* species.

Character no.	Character	Character state	Description of character state
1	Shell shape and sculpture	0	*Nodilittorina*-like
		1	*L. squalida*-like
		2	*Neritrema*-like
		3	*L. obtusata*-like
2	Shell mineralogy	0	aragonite only
		1	outer calcite, inner aragonite layer
3	Mamilliform penial glands	0	absent
		1	0 or 1 gland in same species
		2	2 glands
		3	many (3 or more) glands
4	Glandular projection of penial filament	0	absent
		1	present
5	Paraspermatozoa	0	rod-pieces present
		1	rod-pieces absent
6	Egg groove of pallial oviduct	0	*Nodilittorina*-like
		1	*L. striata*-like
		2	*L. keenae*-like
		3	*L. littorea*-like
		4	*L. sitkana*-like
		5	*L. obtusata*-like
		6	*L. saxatilis*-like
7	Albumen gland of pallial oviduct	0	single loop
		1	additional first loop
8	Jelly gland of pallial oviduct	0	small
		1	enlarged, with thick glandular septa
		2	modified as brood pouch
9	Egg capsules	0	pelagic, sculptured cupola
		1	pelagic, biconvex
		2	pelagic, two rims
		3	within benthic gelatinous mass, ovoid
		4	retained in brood pouch, thin egg covering only
10	Development	0	planktotrophic
		1	nonplanktotrophic
		2	ovoviviparous
11	Outer marginal radular teeth	0	outer basal projection, long cusps
		1	basal projection reduced, cusps shorter
12	Cusps of five central radular teeth	0	long, rectangular
		1	short, pointed
		2	short, blunt
13	Salivary glands	0	small
		1	enlarged

belongs to this category), and with *L. kasatka* (Fig. 40) and *L. horikawai* (Fig. 60C–H), which both show rather lengthened cusps (probably a paedomorphic feature connected with their small size). Some of the species included in this group show extreme intraspecific variation, from pointed to almost blunt cusps (e.g. *L. horikawai*; *L. subrotundata*, Fig. 64), but in these cases some individuals still fall within the typical range of the group. The remaining species fall into a category with predominantly square to rectangular (less than twice as long as broad), blunt to rounded cusps, as

Table 9. Character state matrix for the morphological phylogenetic analysis of *Littorina* species. For description of characters and character states see Table 8. The outgroup comprises three species of *Nodilittorina*. Unknown character states are indicated by a query.

	Character number												
Species	1	2	3	4	5	6	7	8	9	10	11	12	13
N. meleagris	0	0	0	0	0	0	0	0	0	0	0	0	0
N. dilatata	0	0	1	0	0	0	0	0	0	0	0	0	0
N. pyramidalis	0	0	1	0	0	0	0	0	0	0	0	0	0
L. striata	0	0	0	0	1	1	0	0	0	0	0	0	0
L. keenae	0	0	2	0	1	2	0	1	1	0	0	0	0
L. scutulata	0	1	0	1	1	3	0	0	2	0	0	0	1
L. plena	0	1	0	1	1	3	0	0	2	0	0	0	1
L. squalida	1	1	3	0	1	3	0	0	1	0	1	1	1
L. littorea	1	1	3	0	1	3	0	0	1	0	1	1	1
L. kasatka	2	1	1	0	1	3	0	0	?	0	1	1	1
L. brevicula	2	1	3	0	1	3	0	0	1	0	1	1	1
L. mandshurica	2	1	3	0	1	3	0	0	1	0	1	1	1
L. sitkana	2	1	3	0	1	4	0	1	3	1	1	1	1
L. horikawai	2	1	3	0	1	4	0	1	?	1	1	1	1
L. subrotundata	2	1	3	0	1	4	0	1	3	1	1	1	1
L. aleutica	2	1	3	0	1	3	1	1	3	1	1	2	1
L. natica	2	1	3	0	1	3	1	1	3	1	1	2	1
L. obtusata	3	1	3	0	1	5	0	1	3	1	1	2	1
L. fabalis	3	1	3	0	1	5	0	1	3	1	1	2	1
L. compressa	2	1	3	0	1	5	0	1	3	1	1	2	1
L. arcana	2	1	3	0	1	5	0	1	3	1	1	2	1
L. saxatilis	2	1	3	0	1	6	0	2	4	2	1	2	1

exemplified by *L. obtusata* (Fig. 77F) and *L. compressa* (Fig. 91H). The species of this group are less variable as adults, but show pronounced ontogenetic variation, with longer and more pointed cusps in small individuals (Figs 83D, 104D). In the foregut the salivary glands provide another character; they may be either small and narrow, lying concealed beneath the mid-oesophagus, or alternatively much enlarged and widened, extending up the sides of the mid-oesophagus and sometimes even covering the radular sac lying above it.

Cladistic analysis

The morphological characters are summarized in Table 8, and in Table 9 their states are tabulated for all 19 living species of *Littorina* and for the outgroup. The latter comprises the same three species of *Nodilittorina* used by Reid (1990b): *N. meleagris* (Potiez & Michaud, 1838), *N. dilatata* (d'Orbigny, 1842) and *N. pyramidalis* (Quoy & Gaimard, 1833). A Wagner parsimony analysis was performed using the program PAUP, version 3.1.1 (Swofford, 1993). All characters were specified as unordered and were unweighted. The 'heuristic' search option was employed, with 10 repeats of a random addition sequence of taxa, and with the 'mulpars' option in effect. 'Acctran' optimization and 'outgroup' rooting were used.

The result was 21 equally parsimonious trees, each of 30 steps (consistency index=0.933; retention index=0.973; rescaled consistency index=0.908). A strict consensus tree of these is shown in Figure 114, with only the unambiguous character state changes indicated. Ordering of those multi-state characters for which there was some justification for recognizing a transformation sequence (characters 3, 6, 8, 10 in Table 8) had only a small effect on the consensus tree. The number of steps was then increased to 33, removing *L. kasatka* as the first branch of the central polychotomy, making *L. squalida* and *L. littorea* sister-species, and reducing all the *Neritrema* species to a polychotomy, within which *L. aleutica* and *L. natica* were sister-species and the Atlantic species (*L. obtusata*, *L. fabalis*, *L. compressa*, *L. arcana*, *L. saxatilis*) formed a clade.

In spite of the addition and removal of taxa, and the many modifications to the data set, the topology of this cladogram is broadly similar to that of Reid (1990b). Both present and former

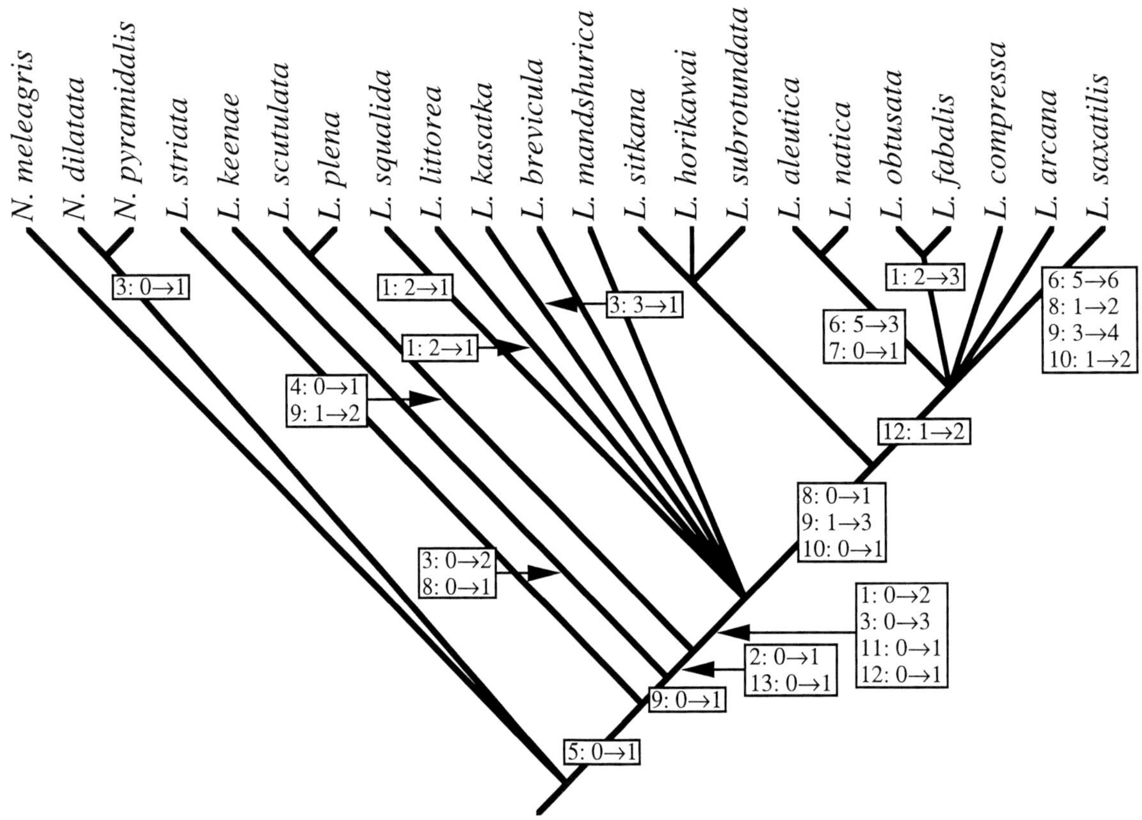

Figure 114. Morphological cladogram for the 19 living species of *Littorina*. This is the strict consensus tree of 21 equally parsimonious trees derived from characters and character states given in Tables 8 and 9. Boxes indicate only unambiguous character state changes (in the format 'character number: state change') and refer to the branches on which they are superimposed or otherwise indicated by large arrows. (From Reid, Rumbak & Thomas, 1996).

analyses agree in the basal position of *L. striata*, followed by *L. keenae* and then the pair *L. scutulata* and *L. plena*, and also agree in the monophyly of the subgenus *Neritrema*. Notable changes in the revised analysis include the collapse of *L. brevicula*, *L. mandshurica*, *L. kasatka*, *L. squalida* and *L. littorea* to a polychotomy, and the appearance of *L. aleutica* and *L. natica* together as the sister-clade of the Atlantic *Neritrema* species, although the monophyly of the latter is only supported if character number 6 (the egg groove of the pallial oviduct) is specified as ordered.

Fossil record

The first attempt to reconstruct the phylogenetic history of *Littorina* was by Golikov & Tzvetkova (1972). Although it was claimed that the tree diagram of six Pacific and three Atlantic species was based on analysis of palaeontological and comparative-morphological data, no details were given. The implied relationships appear to owe more to the idea that the genus evolved from tropical ancestors in the northwestern Pacific, under the influence of climatic cooling. It was suggested that *Littorina* arose from the tropical genus *Nodilittorina*, between the Eocene and Oligocene, and that the diversification of the genus in the northern Pacific took place during the Miocene, as a consequence of climatic change. The suggested origin of the Atlantic species *L. littorea*, *L. obtusata s.l.* and *L. saxatilis s.l.* from *L. squalida*, *L. kurila* (i.e. smooth-shelled *L. sitkana* and similar species) and *L. sitkana* respectively, was apparently based on shell similarity.

Unfortunately the fossil record of *Littorina* is rather poor. Most of the living species inhabit intertidal rocky shores, which are often exposed to moderate or strong waves, and under these conditions preservation of fossils is rare. Those that can survive on sheltered sedimentary shores and at subtidal depths (such as *L. littorea* and *L. squalida*) are more common in the fossil record. Another difficulty in tracing the fossil history of the group is that there is no synapomorphic character of the shell by which all members of the genus can be recognized. The outer calcite layer of the shell is a unique synapomorphy of the more derived *Littorina* species (Taylor & Reid, 1990), separating them from other littorinids of similar turbinate form. However, the more basal members are conchologically indistinguishable from species of the sister-genus *Nodilittorina*. When attempting to trace the history of individual species, and to classify extinct species, the considerable intraspecific variation in shell characters is an added difficulty. There have been no previous attempts to review the fossil history of *Littorina* in detail. Early works (e.g. Cossmann, 1915) are unreliable, predating the separation of conchologically similar genera such as *Littoraria* and *Nodilittorina* as monophyletic units by anatomical criteria (Bandel & Kadolsky, 1982; Reid, 1986a).

The stratigraphic record of living and extinct *Littorina* species is summarized in Table 10. It has not been possible to include extinct species in the cladistic analysis, since this depends so heavily on anatomical characters. Although in this case the fossil record can provide little direct evidence for phylogenetic relationships, it is nevertheless important for estimating the evolutionary time scale. In Figure 115 an attempt has been made to integrate the fossil record with the morphological phylogeny. There are two caveats. First, there is no objective means of distinguishing between ancestors and sister-species in the fossil record, and extinct species are represented as ancestors on the lineages of living species only for simplicity. Second, the earliest occurrence of a particular species or lineage is used to provide a minimum age for the branch. However, since the clades concerned are not defined by any unambiguous synapomorphies of the shell, this convention involves the implicit assumption that the subjective features of shell resemblance, which are used to assign fossils to species and lineages, are synapomorphies. For the clade comprising *L. squalida* and *L. littorea*, there is some weak evidence that this assumption may be correct for, if certain characters are specified as ordered in the cladistic analysis (p. 351), these two appear as sister-species, with the *L. squalida*-like state of the shell-form character as a synapomorphy.

In the absence of any known conchological synapomorphies for the entire genus, there is no reliable palaeontological evidence for the first appearance of the group. The oldest possible member of the genus is the undescribed *Littorina* from the Palaeocene or Lower Eocene of Baja California (Woods & Saul, 1986; p. 340). Another possible member is *L. mountsoloensis* from the Middle or Upper Eocene of Washington (p. 342), but other *Littorina* species from this period are more doubtful (pp. 340, 343). Two extinct taxa of Miocene or greater age are included in Figure 115. *Littorina sookensis* from the Upper Oligocene is so similar in appearance to the Recent *L. keenae*, that it can be placed in *Littorina* with some confidence. Although the precise affinities of *L. remondii* from the Upper Miocene are uncertain, it is clearly a member of the genus, as shown by the outer calcite layer of its shell. Two other extinct species of younger age are *L. petricola* and *L. islandica*; the former is believed to be related to *L. squalida* and *L. littorea*, based on its shell form and sculpture, and for similar reasons the latter is connected with the *L. saxatilis* group.

Figure 115 differs in several respects from the morphological cladogram (Fig. 114). *Littorina squalida* and *L. littorea* are shown as sister-species; this is based on the occurrence of *L. squalida* in the Upper Pliocene of northern Alaska and Iceland, the resemblance of the oldest *L. littorea* to *L. squalida*, and the assumption that the divergence of these species involved allopatric speciation in the two oceans. This increases the resolution of the phylogeny and does not conflict with the morphological evidence. Furthermore, the geological event of the opening of the Bering Strait at 3.5 to 4 Ma (Hopkins, 1967b) gives a maximum age for the separation of these species under this particular speciation hypothesis; the actual age of separation (brought about by Pleistocene cooling of the Arctic) may have been more recent, between the opening of the Bering Strait and the appearance of *L. littorea* in the Red Crag at about 2.4 Ma. Another difference from the morphological cladogram is that the Pacific *L. aleutica* and *L. natica* are together shown as the sister-group of the five Atlantic *Neritrema* species. The monophyly of the Atlantic *Neritrema* species is not unequivocally supported by the morphological data, but only by alternative optimizations of the character describing the looping of the egg groove. The reconstruction shown is preferred on the grounds that it requires only one migration of an ancestral *Neritrema* from the Pacific to the Atlantic. Once again, the opening of the Bering Strait provides a maximum age for this event. Finally, *L. islandica* is shown as an ancestor of the *L. saxatilis* group.

Table 10. Summary of classification, geographical distribution and fossil record of *Littorina* species. † indicates extinct species. Geographical distribution is the modern range for Recent species (except as indicated), and the entire geographical range of fossil records for extinct species.

Species	Geographical distribution	Fossil record
? *Littorina* sp. Woods & Saul, 1986	Baja California	Palaeocene
Littorina (Liralittorina) striata King & Broderip, 1832	Azores, Madeira, Canary Is, Cape Verde Is	Pleistocene to Recent
† *Littorina (Planilittorina) sookensis* Clark & Arnold, 1923	Vancouver I.	Upper Oligocene
Littorina (Planilittorina) keenae Rosewater, 1978	Oregon to Baja California	Pleistocene to Recent
† *Littorina (Littorina) remondii* Gabb, 1866	Central California	Upper Miocene
Littorina (Littorina) scutulata Gould, 1849	SE Alaska to Baja California	Pliocene to Recent
Littorina (Littorina) plena Gould, 1849	Gulf of Alaska to California	Not known
Littorina (Littorina) squalida Broderip & Sowerby, 1829	N Sea of Japan to Bering Strait (N Alaska and Iceland in Upper Pliocene)	Middle Miocene to Recent
† *Littorina (Littorina) petricola* Arnold, 1908	California and Oregon	Pliocene
Littorina (Littorina) littorea (Linnaeus, 1758)	White Sea to Portugal, Newfoundland to Virginia	Upper Pliocene to Recent
Littorina (Littorina) kasatka Reid, Zaslavskaya & Sergievsky, 1991	Kamchatka to Hokkaido	Not known
Littorina (Littorina) brevicula (Philippi, 1844)	Japan, Korea, China	Middle Miocene to Recent
Littorina (Littorina) mandshurica (Schrenck, 1861)	N Sea of Japan	Pleistocene to Recent
Littorina (Neritrema) sitkana Philippi, 1846	N Pacific, from Oregon to Bering Sea to N Sea of Japan	Not known
Littorina (Neritrema) horikawai Matsubayashi & Habe, in Habe, 1979	Islands off Kyushu	Not known
Littorina (Neritrema) subrotundata (Carpenter, 1864)	N Pacific from N California to Aleutian Is to Kurile Is	Not known
Littorina (Neritrema) aleutica Dall, 1872	Aleutian Is, Bering Sea	Not known
Littorina (Neritrema) natica Reid, new name	Kurile Is, Kamchatka, NW Bering Sea	Not known
Littorina (Neritrema) obtusata (Linnaeus, 1758)	White Sea to Portugal, Iceland, Greenland, Labrador to Long I.	Pleistocene to Recent
Littorina (Neritrema) fabalis (Turton, 1825)	White Sea to Portugal, Iceland, SW Greenland	Upper Pliocene to Recent
† *Littorina (Neritrema) islandica* Reid, n. sp.	Iceland	Upper Pliocene
Littorina (Neritrema) compressa Jeffreys, 1865	Barents Sea, British Isles, N France	Not known
Littorina (Neritrema) arcana Hannaford Ellis, 1978	N Norway, British Isles, N France	Not known
Littorina (Neritrema) saxatilis (Olivi, 1792)	Novaya Zemlya to Morocco, Venice, Tunisia, Canary Is, Azores, S Africa, Iceland, Greenland, Baffin I. to Hudson Bay and Chesapeake Bay	Pleistocene to Recent

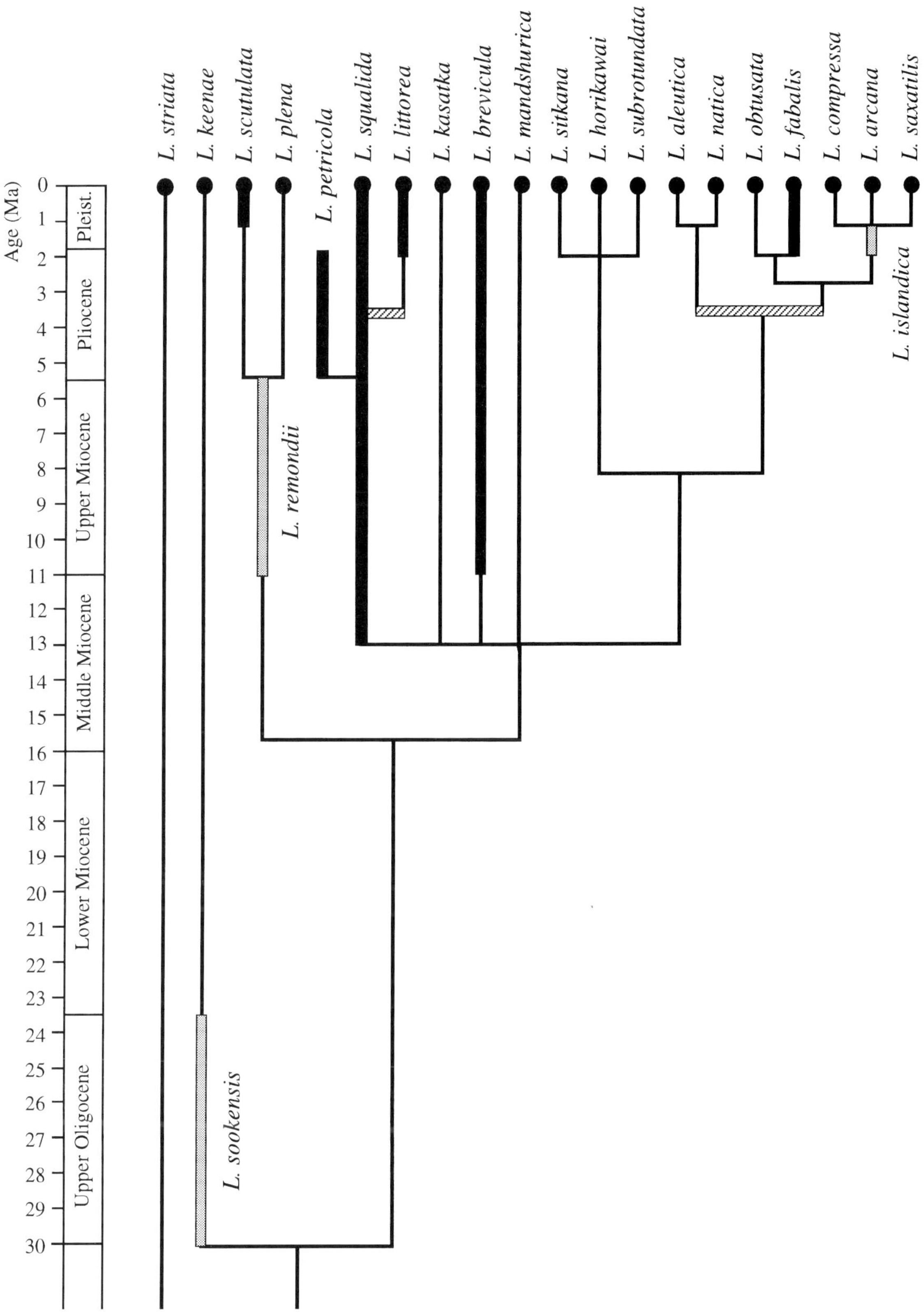

Figure 115. Phylogenetic hypothesis for the 19 living and 4 extinct species of *Littorina*. Topology is based on the morphological cladogram (Fig. 114), with addition of extinct species as suggested by resemblances of shells (see text). Time scale is based on the fossil record. Thick, solid bars span the known stratigraphic ranges of species (Table 10); thick, stippled bars indicate possible ancestral or sister-species (precise stratigraphic ranges of *L. sookensis* and *L. remondii* are uncertain). Horizontal hatched bars indicate separation of lineages datable by the opening of the Bering Strait about 3.5 Ma. Thin lines indicate tree topology only, in those parts where time scale is unknown.

Analysis of allozyme frequencies

Allozyme electrophoresis is a widely used method for the estimation of gene frequencies, which in turn can be used to derive a measure of genetic differentiation between species. In general, genetic similarity must bear some relation to phylogenetic history but, as in the case of morphological data, rigorous models are required for the precise analysis of genetic information. Some of the early electrophoretic studies of *Littorina* species hinted at patterns of genetic similarity, but without any phylogenetic analysis. Thus, for example, *L. saxatilis* and *L. littorea* were found to share two haemoglobin alleles, both differing from those in *L. striata* (Wium-Andersen, 1970). In two other studies *L. saxatilis s.l.* shared more allozymes with *L. obtusata s.l.* than either did with *L. littorea* (Wilkins, O'Regan & Moynihan, 1978; Warmoes, 1986).

The phylogenetic analysis of allozyme-frequency data requires first the computation of a matrix of some measure of genetic distance between the species and second the cluster analysis of these data. There are several available methods for each of these steps, based on different theoretical assumptions, and conflicting trees may result when different methods are used on the same set of data (Sundberg *et al.*, 1990; Backeljau & Warmoes, 1992; Backeljau *et al.*, 1994). One of the most common approaches is phenetic clustering (UPGMA algorithm) of a matrix of Nei's (1972) genetic distances; this clustering method is based upon the assumption that the degree of protein differentiation is proportional to the time since divergence, or in other words that rates of evolution are equal in all branches of the tree (i.e. a 'molecular clock'). This method was used by Ward (1990) for the analysis of 16 to 25 loci of the six northern Atlantic *Littorina* species, showing a close relationship among members of the subgenus *Neritrema* (within which the *L. obtusata* and *L. saxatilis* groups were differentiated) and a more distant relationship of *L. littorea* to the rest (Fig. 116A). A similar analysis of 22 loci of just the three members of the *L. saxatilis* group produced the same topology, with *L. arcana* and *L. saxatilis* showing a closer relationship with each other than to *L. compressa* (Knight & Ward, 1991).

Some Pacific and Atlantic species were included together in the same analysis by Zaslavskaya, Sergievsky & Tatarenkov (1992a, b). Again using UPGMA clustering of Nei's genetic distance data (derived from 17 loci), they produced the topology shown in Figure 116B. This places all the Pacific species near the base of the tree (together with *L. littorea* as the sister-species of *L. squalida*) and shows *L. sitkana* as the sister-species of three Atlantic *Neritrema* species. The topology of the Atlantic *Neritrema* species differs only slightly from that of Ward (1990), in the position of *L. obtusata*. A similar analysis of a different set of 13 loci for the same four Pacific species and the three species of the *L. saxatilis* group, produced the same topology among the Pacific species and agreed with Ward (1990) in the topology of the *L. saxatilis* group, with *L. sitkana* as its sister-species (Zaslavskaya *et al.*, 1992a, b). Some other clustering methods do not depend upon the assumption of equal rates of evolution. These include the Fitch-Margoliash and distance-Wagner procedures, which in this case gave the same topology of the Pacific species and *L. littorea*, differing only in the arrangement of the terminal cluster of *L. saxatilis*, *L. obtusata* and *L. fabalis*; one of the alternatives agreed with the tree of Ward (1990). Four Pacific species similar in appearance to *L. sitkana* were investigated at 17 loci by Zaslavskaya (1995), again using UPGMA clustering of Nei's genetic distances, with the result that *L. sitkana*, a possible unnamed sibling species and *L. subrotundata* together formed a group, to which *L. kasatka* was only distantly related (Fig. 116C).

A different selection of three Atlantic and three Pacific species (including a number of populations here regarded as *L. subrotundata*, see p. 176) was analysed by Boulding, Buckland-Nicks & van Alstyne (1993) and the distance-Wagner method used to cluster Cavalli-Sforza & Edwards' chord distances. Only ten loci were assayed, and of these six were invariant or showed low levels of variability, so that little confidence should be placed in the results. The unrooted tree obtained had generally low bootstrap values. If branches with less than 50% support were to be collapsed, and the tree rerooted with *L. scutulata*, the resulting topology would show *L. littorea* as the first branch, followed by *L. sitkana* and then, as a trichotomy, *L. subrotundata*, *L. obtusata* and *L. saxatilis*. Such a tree is not inconsistent with other phylogenetic analyses of the genus. However, UPGMA clustering of Nei's genetic distances from these data gives a substantially different tree (E. G. Boulding, pers. comm.), with *L. obtusata* at the base, followed by *L. littorea*, and is not in agreement with the results of any other authors.

A wider range of species, all from the Atlantic, and including all seven *Littorina* species and single members of three other genera (*Melarhaphe*, *Littoraria*, *Nodilittorina*), was analysed at nine loci by

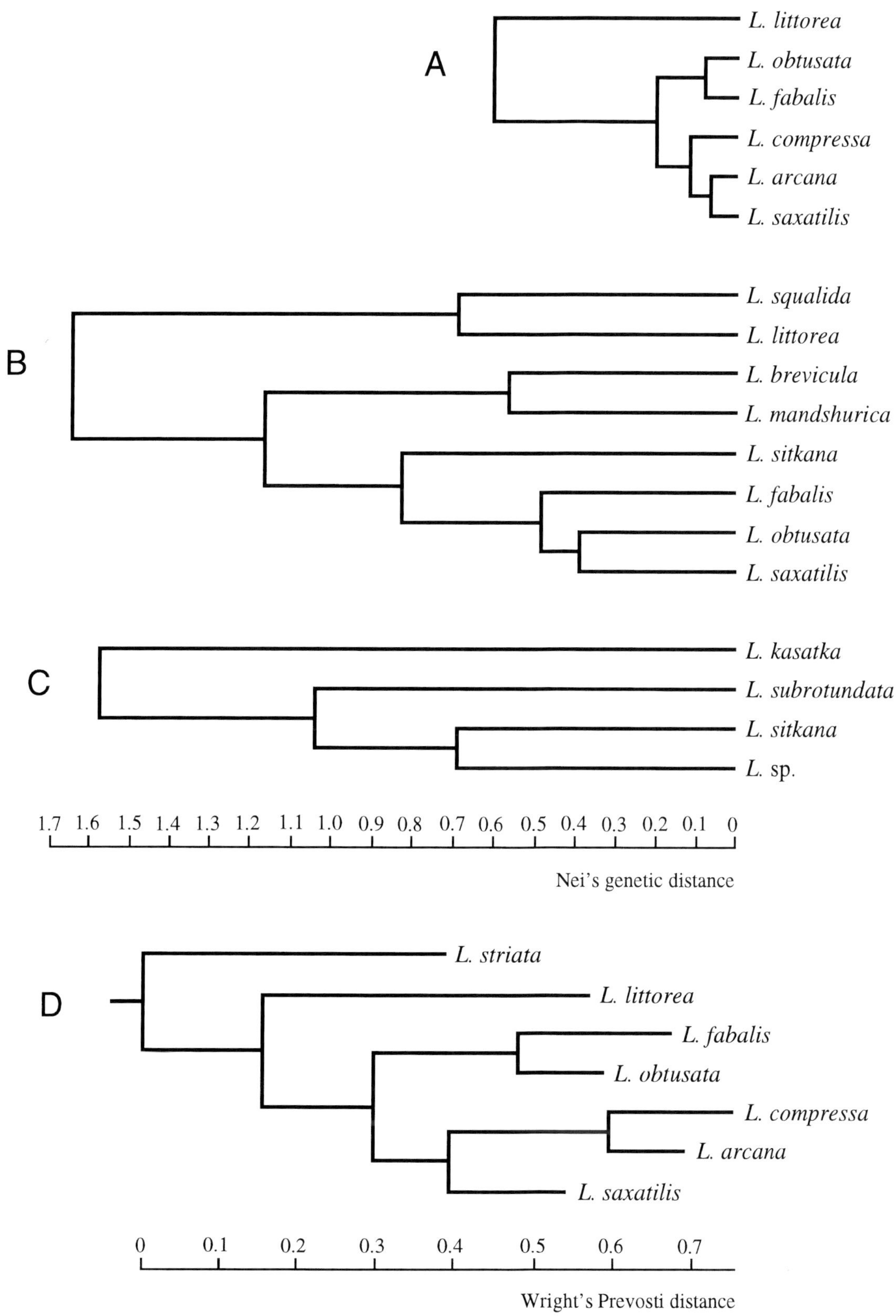

Figure 116. Genetic relationships among *Littorina* species, derived from allozyme-frequency data. Trees **A–C** are UPGMA trees constructed from Nei's (1972) genetic distances; tree **D** is a part of a distance-Wagner tree constructed from Wright's Prevosti distance, rooted using a more distant outgroup (*Melarhaphe neritoides*, not shown). (**A**, after Ward, 1990; **B**, after Zaslavskaya, Sergievsky & Tatarenkov, 1992a; **C**, after Zaslavskaya, 1995; **D**, after Backeljau & Warmoes, 1992).

Backeljau & Warmoes (1992). Here, UPGMA clustering of Nei's genetic identities gave the usual relationships among the *L. saxatilis* group, *L. obtusata* group and *L. littorea*, but in the basal parts of the tree *L. striata* was clustered with *Melarhaphe neritoides*, so that *Littorina* did not appear as a monophyletic group. However, for a group including such distantly-related species, the assumption of constant evolutionary rates throughout the tree is probably not realistic, so that the authors preferred to use the rate-independent distance-Wagner algorithm to cluster Wright's Prevosti distances. Rooting the resulting tree at *M. neritoides* produced the topology shown in Figure 116D, in which the seven *Littorina* species do form a monophyletic group. The relationships among these species are consistent with previous results in all but one respect, the resolution of the *L. saxatilis* group, in which *L. arcana* and *L. compressa* are clustered as sister-species. It should, however, be noted that the number of loci used in this study, only nine, was rather small. Furthermore, polyacrylamide-gel electrophoresis was used (rather than the common starch-gel technique of most other studies), which sometimes detects more allelic variation; this may account for the rather high genetic distances reported in this study and may also have contributed to the different topology of the *L. saxatilis* group (Backeljau & Warmoes, 1992). In a further analysis of this same data set Backeljau *et al.* (1994) investigated the effects on the tree topology of the use of different genetic distance measures and rate-independent clustering methods. The topologies obtained were sufficiently inconsistent that the authors were unable to confirm the clustering of *L. striata* with the other members of the genus *Littorina*; allozyme data are evidently unsuitable for resolution of such distant relationships. However, most clustering methods confirmed the topology of the *L. saxatilis* group found in the earlier analysis; only the parsimony method produced the alternative relationship, clustering *L. saxatilis* and *L. arcana* together (the topology found by Ward, 1990, and Knight & Ward, 1991).

DNA analysis

Two methods have been employed to investigate phylogenetic relationships among *Littorina* species from analyses of their DNA. One is the technique of random amplification of polymorphic DNAs (RAPDs) by the polymerase chain reaction, which produces a banding pattern characteristic of the genome, from which dendrograms of phenetic similarity among genomes can be produced by various clustering methods. Among *Littorina* species this method has so far been used only in the differentiation of the closely related species *L. saxatilis* and *L. arcana* (Crossland, Coates, Grahame & Mill, 1993), and in the investigation of relationships among the members of the *L. saxatilis* group (Crossland *et al.*, 1995). In the latter study one phenetic clustering method showed *L. saxatilis* and *L. arcana* to be more similar to each other than to the third species, *L. compressa*, and this could be interpreted as indicative of phylogenetic relationships.

The direct sequencing of DNA is a more widely applicable and sensitive technique, which can provide an enormous amount of data for phylogenetic analysis, because each point mutation carries potential phylogenetic information. Mitochondrial DNA is particularly useful for the investigation of phylogenetic relationships, because it is haploid, does not undergo recombination, and usually shows simple matrilineal transmission (Avise, Arnold, Ball, Bermingham, Lamb, Neigel, Reeb & Saunders, 1987; Avise, 1994). Furthermore, different mitochondrial genes or parts of genes are known to evolve at different rates; rapidly evolving sequences are suitable for the investigation of recently diverged taxa, and slowly evolving sequences for distantly related taxa (Simon, 1991). This technique was first applied to the phylogeny of *Littorina* by Rumbak, Reid & Thomas (1994), who employed sequences of the 12S ribosomal RNA gene from 11 *Littorina* species, with two *Nodilittorina* species as an outgroup. Since this gene evolves relatively slowly, the resulting phylogenetic trees were poorly resolved, but did support the monophyly of the genus *Littorina*, with *L. striata* as its basal member, followed by *L. keenae*.

This work has since been extended to cover all 19 recognized living species of *Littorina*, for most of which DNA sequences were obtained from fragments of three mitochondrial genes, the 12S and 16S ribosomal RNA genes and the cytochrome-*b* gene (Reid, Rumbak & Thomas, 1996). For the 16S gene, sequences were available not only for all the *Littorina* species, but also for six *Nodilittorina* species, three *Littoraria* species, and for *Mainwaringia rhizophila*. This provided the opportunity for a more rigorous test of the monophyly of *Littorina* than had hitherto been possible. The results of three different analytical methods (maximum parsimony, neighbour-joining clustering of Kimura's corrected pairwise distances, and maximum likelihood) showed a similar overall pattern: *L. striata* clustered with the *Littoraria* species, this clade joined with a monophyletic group comprising all the remaining

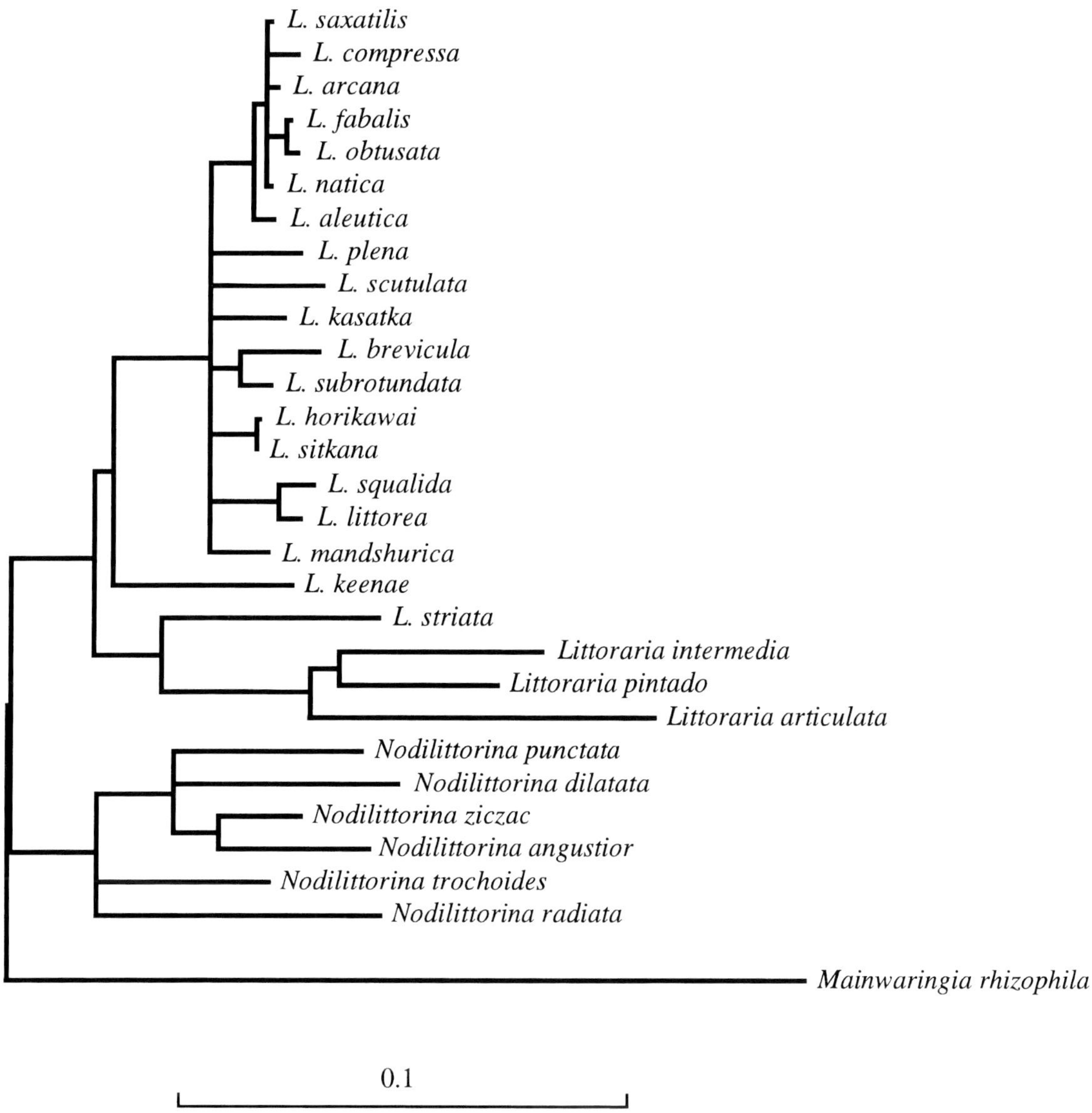

Figure 117. Phylogenetic tree from DNA sequence of a 444 base-pair region of the 16S ribosomal RNA gene, for *Mainwaringia rhizophila*, 6 *Nodilittorina* species, 3 *Littoraria* species and all species of *Littorina*. This is a neighbour-joining tree constructed with Kimura's corrected pairwise distances. Branch lengths are scaled in terms of the expected numbers of substitutions (scale bar indicates a proportion of one substitution out of ten positions). In this bootstrapped 50% majority-rule consensus tree (100 replicates), groups with less than 50% support have been collapsed to polychotomies. *Mainwaringia rhizophila* has been used as an outgroup to root the tree. Note the anomalous position of the genus *Littoraria*. (From Reid, Rumbak & Thomas, 1996).

Littorina species, the *Nodilittorina* species were the sister-group to all the preceding taxa, and finally *M. rhizophila* was the most distant taxon (Fig. 117). This clearly shows that the enigmatic genus *Mainwaringia* does not belong to the genus *Littorina*. According to Reid's (1989a) morphological phylogeny, *Littoraria* is the sister-group of *Littorina* and *Nodilittorina* together, so the appearance of *Littoraria* within the expected monophyletic clade *Littorina* was puzzling. However, this was explained as an artefact of bias in the base-compositions of the DNA sequences and therefore may not reflect the actual phylogenetic relationships of the taxa. Inspection of the branch lengths of the *Littoraria* species (Fig. 117) and tests for rate heterogeneity supported this interpretation, suggesting marked heterogeneity if this topology was accepted. This heterogeneity was considerably reduced if the relationship indicated by the morphological analysis was imposed. In all analyses of the 16S data the six *Nodilittorina* species formed a monophyletic group, and *L. striata* never clustered with them, even if the *Littoraria* species were removed from the analysis. In summary, although the monophyly of the genus *Littorina* is not unequivocally supported, there is no strong reason for rejecting it.

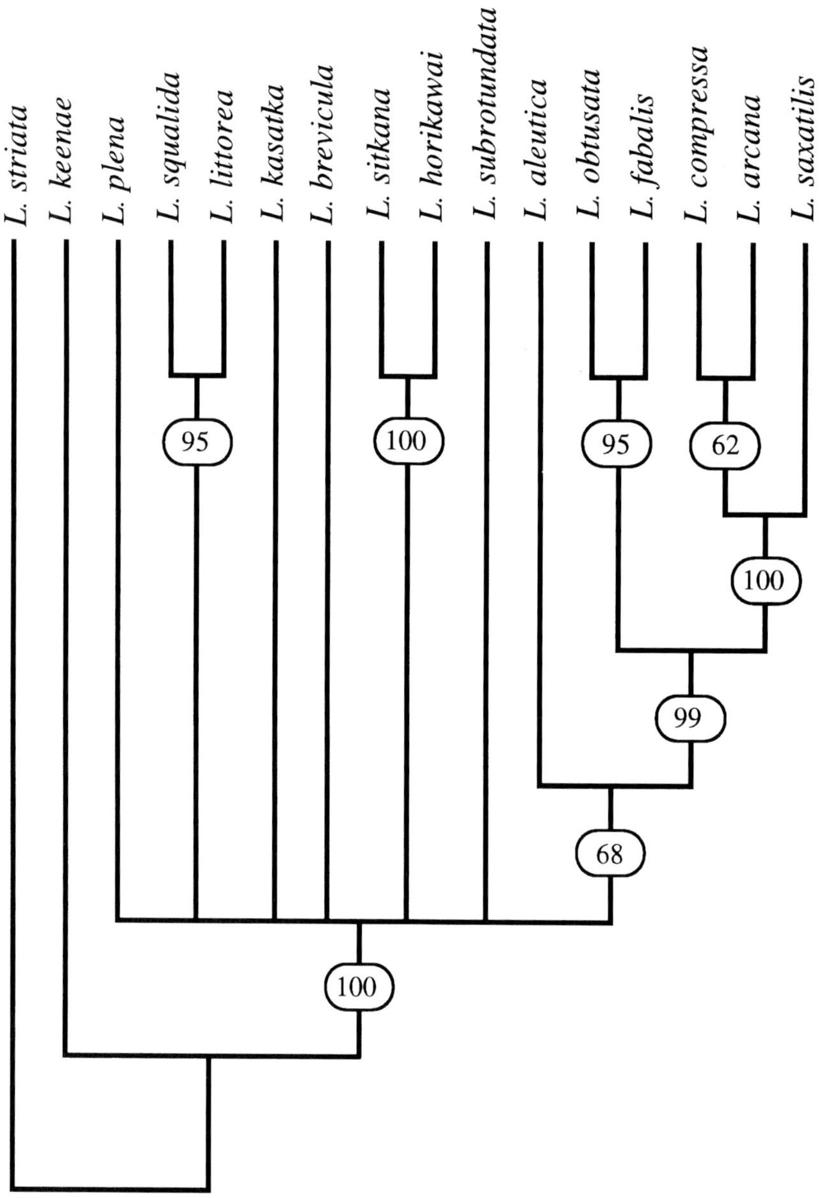

Figure 118. Maximum-parsimony tree based on DNA base sequences of fragments of the 12S, 16S and cytochrome-*b* mitochondrial genes (a total of 1469 base-pairs with 395 informative sites) for 16 *Littorina* species. This is a bootstrapped 50% majority-rule consensus tree, in which clades with less than 50% support have been collapsed to polychotomies. Numbers below nodes are the percentage of 100 replicates in which the taxa originating from the node occur together. *Littorina striata* has been used as the outgroup to root the tree. (From Reid, Rumbak & Thomas, 1996).

Regarding phylogenetic relationships within the genus *Littorina*, sequences from all three genes were available for only 16 species, and Figure 118 shows a maximum-parsimony tree constructed from this combined data set (Reid *et al.*, 1996). This tree was rooted with *L. striata*, as indicated by the 16S analysis described earlier and by the 12S analysis of Rumbak *et al.* (1994). There is strong support for the basal position of *L. keenae*, for the monophyly of the *L. saxatilis* group, for the monophyly of the five Atlantic *Neritrema* species, and for the terminal sister-species pairings of *L. squalida* with *L. littorea*, *L. obtusata* with *L. fabalis*, and *L. sitkana* with *L. horikawai*. There is weaker support for *L. aleutica* as the sister-taxon of the Atlantic *Neritrema* species, but the large polychotomy of all species except *L. striata* and *L. keenae* remains unresolved. Separate analyses of the sequences from each of the three mitochondrial genes produced slightly less well resolved trees with lower bootstrap support, but the topologies were almost all consistent with each other. To summarize the mitochondrial gene phylogeny, the three additional species not represented in Figure 118 have been added to the tree as indicated by the results of the separate analyses of each gene, producing the topology of Figure 120. To the sister-species pairs listed earlier, are added *L. scutulata* with *L. plena*, and *L. brevicula* with *L. mandshurica*. In Figure 120 the *L. saxatilis* group is shown as a trichotomy, as is the relationship between *L. natica*, *L. aleutica* and the Atlantic *Neritrema* species; for each of these trichotomies the alternative dichotomous resolutions differ by only a single step and are therefore not adequately supported.

Consensus phylogeny

The four independent approaches to the reconstruction of phylogenetic relationships among *Littorina* species have produced a large measure of agreement. Although the topologies of the several phylogenetic hypotheses vary in their resolution, they are almost entirely consistent with each other, so that some kind of consensus can be attempted. There has been recent debate about the most appropriate method for combining phylogenetic information from different sources of data. Some authors recommend the analysis of a single, combined, data set, which maximizes descriptive and explanatory power (e.g. Eernisse & Kluge, 1993). Others advocate separate analyses, arguing that congruent topologies from independent data sets provide the best estimate of the true phylogeny (e.g. Miyamoto & Fitch, 1995). The latter view is followed here, and in Figure 119 the independent topologies have been combined as a consensus tree. This is the preferred phylogenetic hypothesis with which to consider the macroevolution of the group (Chapter 6). It must be emphasized, however, that this remains an hypothesis, to be tested and modified as further data become available.

The preferred hypothesis (Fig. 119) is essentially a combinable-components consensus of the morphological (Fig. 114) and DNA (Fig. 120) trees. Clades supported by either data set are recognized, if not contradicted by the other tree (and in this case there are no contradictions), thus retaining more information than a strict consensus tree (Forey *et al.*, 1992). Several slight modifications to the consensus tree, and additional evidence from allozyme analyses, will be mentioned.

The basal positions of *L. striata* followed by *L. keenae* (nodes 2 and 3 in Fig. 119) are strongly supported by both morphological and DNA data. In the molecular analysis a large polychotomy unites all the remaining *Littorina* species. This polychotomy is partly resolved by the morphological analysis, which separates *L. scutulata* and *L. plena* together as the first branch (clade 16 at node 4), on the basis of shell, penial and radular characters. Of these characters the shape of the outer marginal teeth is the most convincing, for there is some subjectivity in the interpretation and coding of shell form and radular cusp shapes, while the optimization of the number of penial glands is uncertain (p. 399). This partial resolution of the polychotomy is consistent with a bootstrapped consensus tree of the molecular data (Fig. 118) and furthermore does not increase the number of steps in the most parsimonious trees derived from the combined molecular data (Reid *et al.*, 1996). The fact that the remaining species form an unresolved polychotomy (at node 5) in both morphological and DNA analyses suggests that this represents a genuine episode of rapid cladogenesis during the evolution of the genus, and is not simply a result of insufficient data. Additional molecular data may ultimately resolve the relationships at nodes 4 and 5 more convincingly, but relationships are likely to be very close. Some analyses of allozyme frequencies have clustered *L. brevicula* and *L. mandshurica* more closely with *Neritrema* species than with *L. squalida* and *L. littorea* (Zaslavskaya *et al.*, 1992a, b; Fig. 116B), but this resolution cannot yet be incorporated, because *L. kasatka* was not included.

It is interesting that the DNA data provide little information on the possible monophyly of the nonplanktotrophic subgenus *Neritrema* (clade 6 in Fig. 119). This group is well defined by three unequivocal morphological synapomorphies (Fig. 114). However, one of these is the character of nonplanktotrophic development itself, while the other two relate to the production of a benthic spawn mass (an enlarged jelly gland which produces the spawn matrix, and simplification of the egg capsules within the spawn mass) and are presumably functionally correlated with developmental strategy. There is therefore no independent morphological evidence for the unique origin of nonplanktotrophic development in *Littorina*. Of the various analyses of mitochondrial DNA sequences reported by Reid *et al.* (1996), only one (distance analysis of cytochrome-*b* gene) gave weak support (bootstrap value 54%) to a monophyletic clade of *Neritrema* species, two analyses weakly contradicted it (bootstrap values 51% and 54%), while the majority of analyses (including those of the combined molecular data

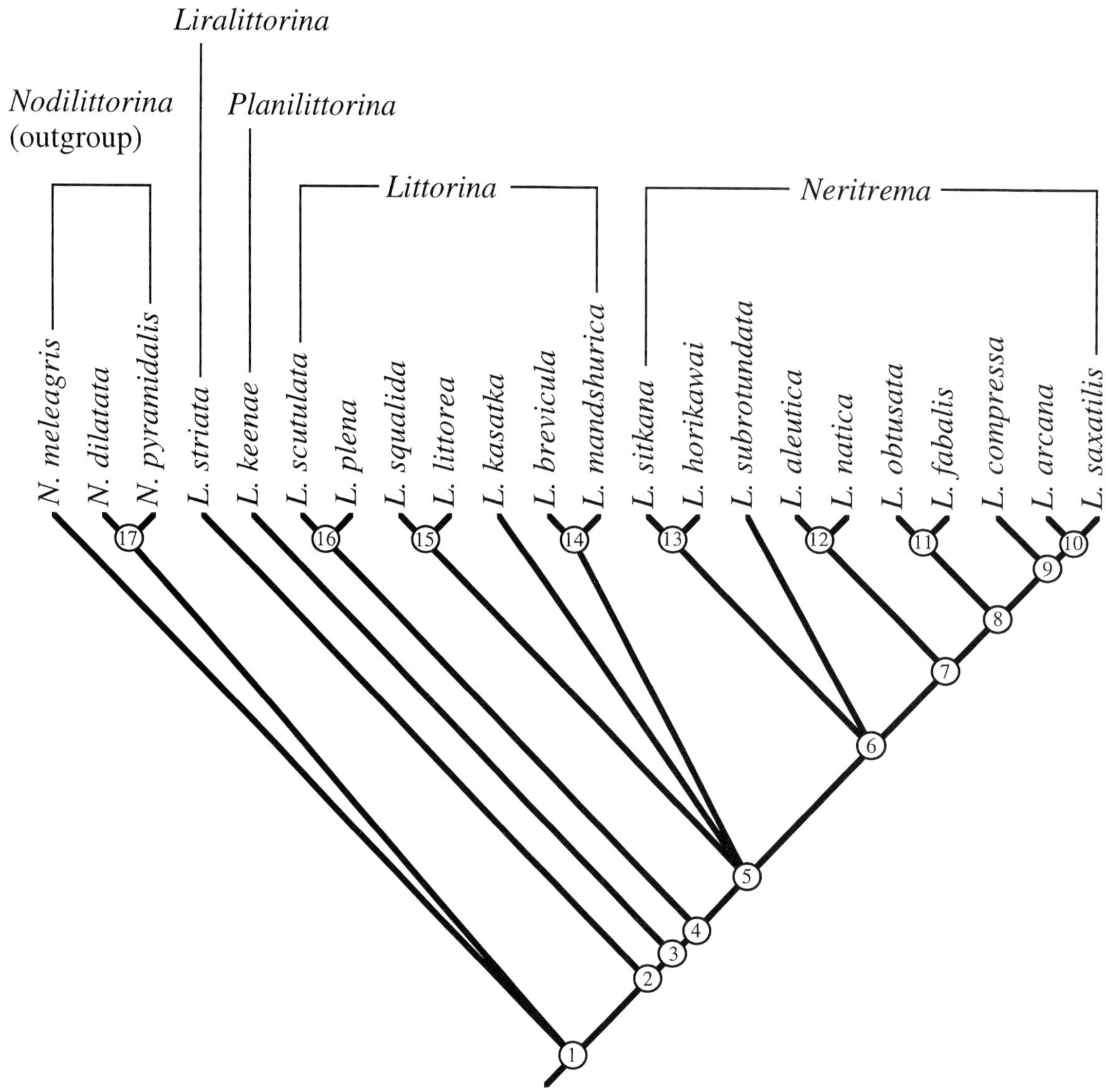

Figure 119. Summary of phylogenetic relationships of *Littorina* species, based on morphological, DNA and allozyme data. The outgroup consists of three species of the probable sister-genus, *Nodilittorina*. This is essentially a combinable-components consensus tree of the morphological (Fig. 114) and molecular (Fig. 120) trees, with several slight modifications (see text for details). The nodes (clades) of the phylogeny are numbered arbitrarily for ease of reference. The four recognized subgenera of *Littorina* are superimposed.

set, Fig. 118) failed to resolve it. Imposition of this resolution of a monophyletic *Neritrema*, based on the morphological tree, did not increase the length of the most parsimonious trees found by the parsimony analysis of the complete molecular data set. The molecular data are therefore equivocal on this question, and further information is desirable. Available allozyme data lend some support to the monophyly of *Neritrema*, for in two separate analyses *L. sitkana* clustered with three Atlantic members of the *Neritrema* group (Zaslavskaya *et al.*, 1992a).

For the more recent divergences the morphological and molecular trees are sometimes in agreement (clades 7, 11 and 16, in Fig. 119). In several cases the molecular data offer improved resolution with strong support (clades 8, 9, 13, 14 and 15). Within *Neritrema* there are some uncertainties. The monophyly of *L. sitkana*, *L. horikawai* and *L. subrotundata* is supported only by a morphological character (the egg groove of the pallial oviduct) with ambiguous optimization, and has therefore not been accepted. The morphological resolution of *L. aleutica* and *L. natica* as sister-species (supported by two unequivocal synapomorphies) has been accepted. This was in fact weakly contradicted by the analysis of the 16S gene (the only molecular analysis to include both these species), which placed *L. natica* as the sister-species of clade 8; a parsimony tree with *L. aleutica* and *L. natica* as sister-species was, however, only a single step longer.

In the case of the three species of the *L. saxatilis* group (clade 9 in Fig. 119) no morphological synapomorphies were discovered from which to determine their relationships. Although the DNA analysis resolved *L. compressa* and *L. arcana* as sister-species (Fig. 118, bootstrap value of 62%), an alternative resolution of *L. arcana* and *L. saxatilis* as sister-species was just one step longer in a parsimony tree. A majority of allozyme studies (p. 356) have produced this latter resolution, and this has been adopted in Figure 119.

Evolutionary time scale

As discussed earlier (p. 352), palaeontological and geological evidence can be used to estimate the dates of origin of clades, if certain assumptions are made. If some clades of a molecular phylogeny can be dated in this way, there is also the possibility of estimating the ages of clades for which there is no fossil record, if it is assumed that there is a constant rate of molecular divergence over time. Such estimation of average evolutionary rates is subject to various sources of error and is therefore controversial. Rates may vary within and between lineages (although within *Littorina* tests show uniformity of rates, Reid *et al.*, 1996). The fossil record may be incomplete; ages of fossils are often imprecise and provide only minimum estimates of the ages of clades. Despite such problems, the potential interest of a time scale for phylogenetic trees is such that it is necessary to attempt to estimate rates of evolutionary change. For *Littorina* this has been done by Reid *et al.* (1996), from which the following account is taken.

Mutations accumulate at various rates in different genes and in different locations within genes; for example, in protein-coding genes mutations are more frequent in the third positions of codons, because these do not always entail a change in the amino acid encoded. Furthermore, some kinds of mutations occur more frequently than others; for example, transition substitutions of one purine or pyrimidine base by another of the same class are more frequent than transversion substitutions of a purine by a pyrimidine or vice versa. Given sufficient time, all the potentially variable sites within a gene may be subject to mutation, leading to eventual saturation and to loss of discernible phylogenetic information content. Nevertheless, within specified time intervals the accumulation of mutations is approximately linear.

For *Littorina* species there are five possible points for calibration: the age of *L. sookensis* (23.5–30 Ma, dating the separation of *L. keenae* from the more derived members of the genus), the earliest occurrence of *L. squalida* (12–15 Ma, dating the unresolved polychotomy), the period between the opening of the Bering Strait and the effective closure of the trans-Arctic migration route by climatic cooling (2.4–4 Ma, providing two calibration points, of the separation of *L. littorea* from *L. squalida*, and of the Atlantic *Neritrema* species from their Pacific sister-taxa, *L. aleutica* and *L. natica*), and lastly the oldest fossils of *L. obtusata s.l.* and of *L. islandica* (each 2 Ma, together estimating the separation of the *L. obtusata* group from the *L. saxatilis* group). Using these dated points on the molecular phylogeny, Reid *et al.* (1996) calculated rates of accumulation of transitions and of transversions for the 12S, 16S and cytochrome-*b* (third positions only) genes, giving six linear regressions with which to estimate unknown ages. When estimating molecular ages for clades for which fossil evidence of age was available, regressions of age on divergence were recalculated without that particular point, to avoid circularity. To illustrate the different rates of divergence, transversions were found to accumulate at the rate of 1% every 11.9 Ma in the 16S gene over the range 0 to about 30 Ma, whereas third position transitions in the cytochrome-*b* gene accumulated much more rapidly, at the rate of 1% every 0.13 Ma over the range 0 to about 3 Ma.

The resulting molecular estimates of the ages of clades are shown in Figure 120 and listed in Table 12 (p. 377). The ranges of these estimates are sometimes quite wide, which is unsurprising in view of the potential sources of error in their calculation. Nevertheless, for the five points used for the calibration of evolutionary rates the independent molecular estimates of age do overlap those based on palaeontological and geological evidence, which gives some confidence in the estimates of otherwise unknown ages. The ranges of estimates are widest in the cases of sister-species pairs, because for these there is only a single value of divergence from which to calculate age from each regression, whereas

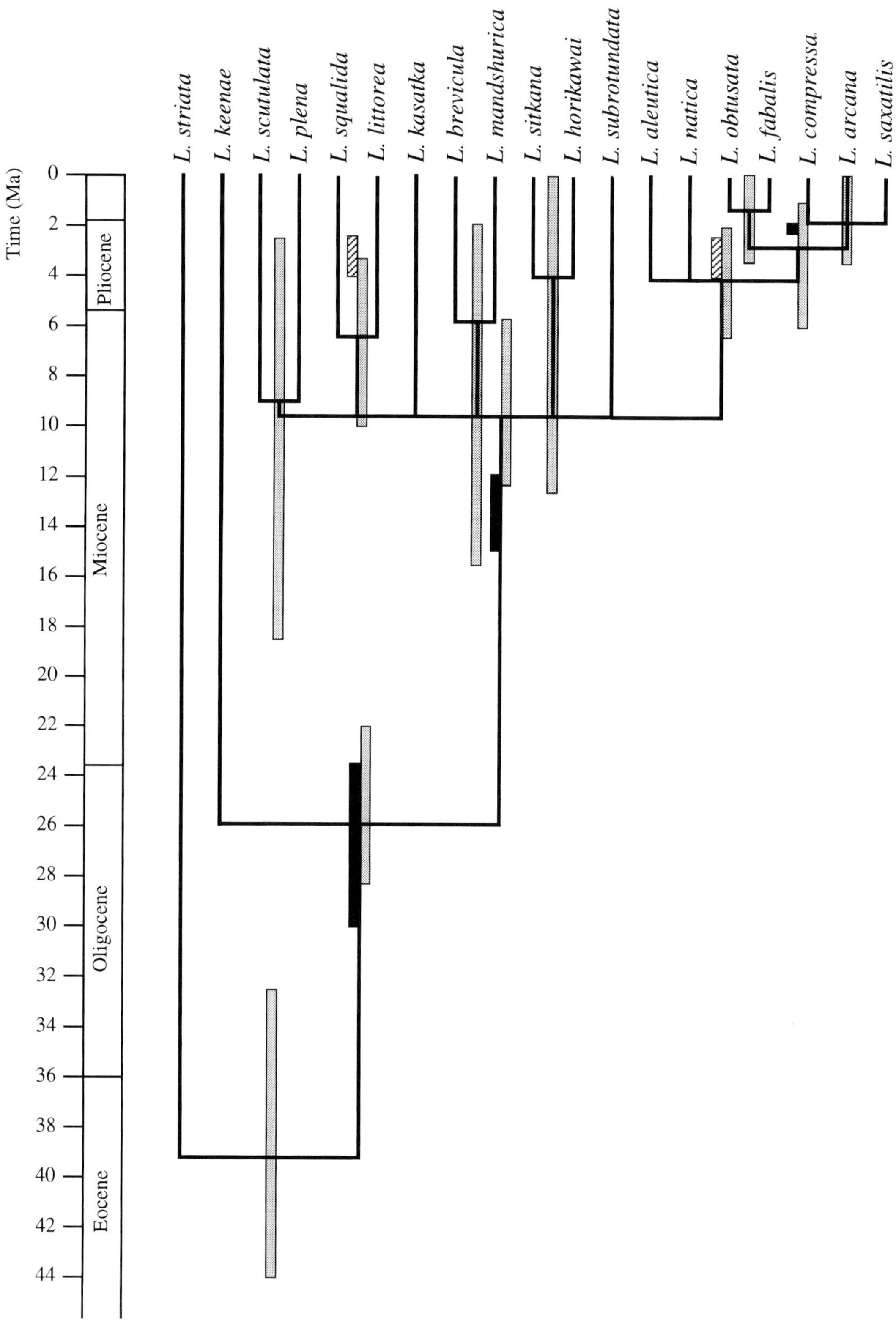

Figure 120. Phylogenetic tree of *Littorina* species, summarizing analyses of DNA base sequences of the 12S, 16S and cytochrome-*b* mitochondrial genes, with a time scale calibrated from fossil and geological evidence. Solid error bars indicate palaeontological estimates of age (the oldest fossil member of the clade). Hatched error bars indicate geological estimates of age (the estimated age of the opening of the Bering Strait; Hopkins, 1967b). These palaeontological and geological estimates of age have been used to calibrate rates of molecular divergence. Stippled error bars indicate the resulting range of estimates of ages of each clade, based on rates of accumulation of transitions or transversions for each of the three genes. The origins of the clades are shown at the means of the molecular estimates of age (except for the pair *L. scutulata* and *L. plena*, where the mean age of their divergence was older than that of the polychotomy from which they arose). (From Reid, Rumbak & Thomas, 1996).

for polychotomies each age was estimated from the averaged divergence among the members of the clade.

Alternatively, estimates of ages of divergence between sister-species can be made from the genetic distance between them, as assessed from allozyme studies. This has been done in two studies of *Littorina*, in which Nei's (1987) calibration of the genetic distance *D* (Nei, 1972) was used. According to this calibration, *D* equal to 1.0 corresponds to a divergence time of 5 Ma. A different calibration, derived from the pulmonate snail *Partula* (*D* equal to 1.0 corresponds to 7.7 Ma; Johnson, Murray & Clarke, 1986) was applied by Knight, Hughes & Ward (1987) to estimate the age of divergence between *L. saxatilis* and *L. arcana*.

The ages of clades can now be considered from these sources of evidence. The divergences of species within the *L. obtusata* and *L. saxatilis* groups are evidently the most recent. For the *L. saxatilis* group the estimated age of separation of the three species based on DNA-sequence divergences ranges from 0 to 3.55 Ma (mean 1.73 Ma). This can be compared with allozyme-based estimates for the separation of *L. saxatilis* and *L. arcana*, of 0.22–0.79 Ma (Ward & Warwick, 1980) and 0.34–0.55 Ma (Knight *et al.*, 1987). For the pair *L. saxatilis* and *L. compressa* an age of 0.21–1.19 Ma can be calculated from genetic distances given by Ward (1990; using Nei's calibration of *D*). For the pair *L. obtusata* and *L. fabalis* the corresponding figures are 0–3.49 Ma (mean 1.32 Ma) from DNA sequences, and 0.79–1.73 (mean 1.25 Ma) from allozyme frequencies (Tatarenkov, 1995a), showing closer agreement between the two methods. From DNA sequences the divergence between the two clades of *Neritrema* in the Atlantic is estimated at 1.31–5.99 Ma (mean 2.83 Ma).

The range of molecular dates for the separation of the Atlantic *Neritrema* species from their probable Pacific sister-taxa, *L. aleutica* and *L. natica*, is 1.95 to 6.36 Ma (mean 4.12 Ma), which is close to the estimated time during which trans-Arctic migration was possible (2.4–4 Ma). In comparison, the molecular ages for the divergence of the other pair of Pacific and Atlantic sister-taxa, *L. squalida* and *L. littorea*, are mostly greater, ranging from 3.28 to 10.07 Ma (mean 6.42 Ma). Several explanations could be considered, and further evidence will be necessary to evaluate them. The rate of molecular evolution of the three mitochondrial genes in this pair of species might have been higher than the average for the genus, giving a falsely high estimate of their age. Indeed, using Nei's (1987) calibration of genetic distance, Zaslavskaya *et al.* (1992a) calculated an age of 3.45 Ma for the separation of this species pair. Alternatively, their separation could perhaps have occurred in the Pacific, before trans-Arctic migration began. (A possible explanation is that it was *L. petricola*, from the Lower Pliocene of the northeastern Pacific, that was the ancestor of *L. littorea*, although morphological comparison among the few fossil specimens of all three species do not at present support this idea). There is also the possibility that the geological estimate of the age of the Bering Strait of 3.5–4 Ma (Hopkins, 1967b) is too young; a more recent estimate is a little older, 3.8–4.4 Ma (Basilyan, Barinov, Oreshkina & Trubichin, 1991).

Within the Pacific the molecular estimates for the divergence of the pair *L. brevicula* and *L. mandshurica* cover a wide range, 1.99–15.62 Ma (mean 5.84 Ma), compared with an allozyme estimate of 2.85 Ma (applying Nei's calibration to the value of *D* given by Zaslavskaya *et al.*, 1992a). For *L. sitkana* and *L. horikawai* the molecular estimates are again variable, 0–12.69 Ma (mean 3.99 Ma). This is also the case for *L. scutulata* and *L. plena*, 2.48–18.63 Ma (mean 12.81 Ma); here the allozyme estimate is much younger, 1.38–2.39 Ma (*D* from data of Mastro, Chow & Hedgecock, 1982, recalculated by Ward, 1990; applying calibration of Nei, 1987). The age of the major polychotomy on the molecular phylogeny is estimated at 5.74–12.39 Ma (mean 9.60 Ma). This molecular estimate is a little younger than the oldest fossils of this group (*L. squalida* 12–15 Ma, *L. brevicula* 10–12 Ma), although the ranges overlap. For the separation of the *L. keenae* lineage, the molecular estimates of 22.00–28.11 (mean 25.92 Ma) agree closely with the Upper Oligocene age (23.5–30 Ma) of its possible ancestor, *L. sookensis*. By extrapolation of the rates of molecular evolution, the basal species *L. striata* may have diverged about 32.47–43.93 Ma (mean 39.23 Ma), in the Upper Eocene.

As discussed earlier, the time of origin of the genus *Littorina* is unlikely ever to be known from the fossil record. It can, however, be estimated by further extrapolation. Using the evolutionary rates for transversions in the 16S and 12S mitochondrial genes, Reid *et al.* (1996) estimated the date of divergence between *Littorina* and its presumed sister-group, *Nodilittorina*, as 43 or 67 Ma, in the Middle Eocene or latest Cretaceous. Such figures are very approximate, but do suggest that the earliest possible *Littorina* species, from the Palaeocene or Lower Eocene (Woods & Saul, 1986), could date from close to the time of origin of the genus.

Formal classification

It is possible to construct a formal classification from a cladogram or phylogenetic tree, from which (with knowledge of the conventions adopted) the branching pattern of the tree can be recovered (see discussion of the several alternative systems of phylogenetic classification by Wiley, 1981; Forey *et al.*, 1992). However, such systems of rigorously phylogenetic classification, which require exclusively monophyletic groups, tend to be cumbersome and unstable. Difficulties arise with the inclusion of fossils together with Recent taxa in the same classification, and in cases where relationships are incompletely known. Traditional 'evolutionary' classifications, on the other hand, suffer from the disadvantage of attempting to express both general resemblance and genealogy, which may be incompatible. In practice, many taxonomists prefer to use a tree diagram to indicate phylogenetic relationships, while retaining a pragmatic approach to formal classification, in which both monophyletic and paraphyletic groups are used. This approach is adopted here, and the formal classification of *Littorina* is summarized in Table 10 and Figure 119.

The genus *Littorina* is believed to be monophyletic. This is supported by the cladistic analysis of morphological characters presented here, but a more rigorous demonstration requires the inclusion of additional outgroups. This was done by Reid (1989a), who, in an analysis of all subgeneric groups within the family Littorinidae, identified the lack of rod-pieces in the paraspermatozoa, and the presence of at least two spiral loops in the pathway of the egg groove through the pallial oviduct, as synapomorphies of the genus, although neither of these characters was unique in the family. The molecular evidence from the 16S mitochondrial gene does not support the monophyly of *Littorina*, but as discussed earlier (p. 359) this result (Fig. 117) is probably unreliable, and there is therefore no strong evidence for questioning the monophyly of this genus (Reid *et al.*, 1996).

Within the genus, four subgenera are accepted here. *Liralittorina* is monotypic, containing only *L. striata*; in the morphological analysis it has no unambiguous autapomorphies. *Planilittorina* contains one living species, *L. keenae*, of which the enlarged jelly gland is an autapomorphy (although not a unique one in *Littorina*). The two mamilliform glands of the penis appear as an autapomorphy of *L. keenae* in the cladistic analysis (Fig. 114), but this disappears if the character is ordered in the analysis. The subgenus *Littorina* contains seven living species. According to the morphological phylogeny (Fig. 114), this is a paraphyletic group, a grade of organization defined by a combination of apomorphic (outer calcite shell layer, enlarged salivary glands) and plesiomorphic characters (planktotrophic development, pelagic egg capsules, small jelly gland). The molecular phylogeny, however, does not preclude the possibility that this may be a monophyletic group. The remaining ten recognized Recent species are placed in the subgenus *Neritrema*. Morphologically, this is a well-defined monophyletic clade, with three unambiguous morphological synapomorphies (Fig. 114), all connected with spawn and larval development (enlarged jelly gland or brood pouch; benthic gelatinous egg masses with ovoid capsules or, if ovoviviparous, no capsules; nonplanktotrophic larval development). Nevertheless, the DNA analysis offers only limited support for the monophyly of *Neritrema*, as does that of incomplete allozyme data. In the absence of synapomorphies of the shell, fossil species have been assigned to these subgenera purely on the basis of overall similarity to living species.

This subgeneric classification scheme does not differ from that of Reid (1989a, 1990b). However, whereas the affinities of the two species of *Mainwaringia* were uncertain in these earlier accounts, these have now been unequivocally excluded from *Littorina* (see DNA Analysis, p. 358; and p. 11).

Chapter 6

MACROEVOLUTIONARY HISTORY

The phylogenetic hypothesis derived in the preceding chapter can now be used as a framework within which to consider various aspects of the macroevolution of *Littorina*. Branching points of lineages within the phylogeny can be interpreted as speciation events, and the processes by which species are formed is a central theme of this chapter. This in turn is closely connected with the concepts and definitions of species, which influence both the view of speciation processes and the practical decisions made by taxonomists attempting to classify species. Since speciation processes frequently involve shifts and disjuntions of geographical distributions, it is first necessary to examine the geographical ranges of *Littorina* species, the factors that limit them and the processes that extend them. On a geological time scale, information about speciation and distribution can be combined with evidence from fossils, and of geological and climatic events, to attempt a synthesis of the historical biogeography of the group. Phylogenetic trees are also a source of hypotheses about the evolutionary history of the attributes of their constituent species. By superimposing the characters of the terminal taxa onto the phylogeny, the character states of hypothetical ancestors can be reconstructed. This provides insights into the process of adaptation, by which morphological characters have responded to the forces of natural selection.

Geographical distribution

Geographical ranges and their limits

The 19 living species of *Littorina* are distributed almost entirely in the subarctic and temperate regions of the Northern Hemisphere. In the Atlantic Ocean the limits of the range of the genus are defined by its most widely-distributed species, *L. saxatilis*, and extend from Novaya Zemlya and Svalbard (80°N) in the northeast to southern Morocco (20°S) in the south (with outlying populations in South Africa and the Mediterranean), and in the northwestern Atlantic from western Greenland (73°N) to North Carolina (35°S). Turning to the Pacific Ocean, the northern limit is at or north of the Bering Strait (perhaps to 70°N) and the southern limits are in southern Baja California (23°S) in the east and at Hong Kong (22°S) in the west. The diversity of strictly sympatric species, that can be found on a single shore, is highest between Brittany and northern Norway, where it reaches six species; in Spain and Portugal four species occur, in North America only three (one, *L. littorea*, perhaps introduced), and in subarctic regions also three. In the Pacific Ocean regional diversity ranges from three species in California to four in British Columbia, five in the Bering Sea, five in the Okhotsk Sea and five in the Sea of Japan, although the number of strictly sympatric species is not known to exceed four.

The majority of the species have widespread distributions (Table 11); in only two cases is the maximum width of the range less than 2000 km (1400 km in *L. mandshurica*, 450 km in *L. horikawai*). In some other gastropod groups the sizes of ranges of species are correlated with the type of larval development, those species with planktotrophic larvae (and therefore the potential for widespread dispersal) having the larger ranges (Hansen, 1980; Jablonski, 1982; Kohn & Perron, 1994). Such a correlation is not found in *Littorina*, however (Table 11); of the two most restricted species listed above, *L. horikawai* has nonplanktotrophic development and *L. mandshurica* planktotrophic, while in both oceans the species with the widest distributions are nonplanktotrophic (*L. saxatilis* in the Atlantic, *L. sitkana* in the Pacific). Explanations for this are explored later (see Dispersal and Colonization, p. 370).

It is of course difficult to generalize about factors that limit geographical ranges, when species differ in their physiological tolerances, ecological requirements and dispersal capabilities. Nevertheless, the question is important, particularly for consideration of the historical expansion, contraction or vicariance of ranges, which have been involved in allopatric speciation events.

Table 11. Geographical range, oceanic or continental distribution pattern, and type of development of *Littorina* species. Ranges are indicated as the shortest great circle distances between the most distant recorded localities. For *L. littorea* the range in North America (where the species may have been introduced by man) has been excluded. For *L. saxatilis* the South African localities have been included.

Species	Geographical range (km)	Distribution pattern	Spawn; larval development
L. striata	2600	oceanic	pelagic; planktotrophic
L. keenae	2000	oceanic	pelagic; planktotrophic
L. scutulata	4400	intermediate	pelagic; planktotrophic
L. plena	4100	continental	pelagic; planktotrophic
L. squalida	4800	continental	pelagic; planktotrophic
L. littorea	4900	continental	pelagic; planktotrophic
L. kasatka	2300	oceanic	pelagic; planktotrophic
L. brevicula	3700	continental	pelagic; planktotrophic
L. mandshurica	1400	continental	pelagic; planktotrophic
L. sitkana	8300	both	benthic; nonplanktotrophic
L. horikawai	450	oceanic	benthic; nonplanktotrophic
L. subrotundata	7000	both	benthic; nonplanktotrophic
L. aleutica	2500	oceanic	benthic; nonplanktotrophic
L. natica	2900	continental	benthic; nonplanktotrophic
L. obtusata	7400	continental	benthic; nonplanktotrophic
L. fabalis	4400	continental	benthic; nonplanktotrophic
L. compressa	3300	continental	benthic; nonplanktotrophic
L. arcana	3100	intermediate	benthic; nonplanktotrophic
L. saxatilis	12700	both	ovoviviparous; nonplanktotrophic

It is well known that the distributions of marine invertebrates are frequently limited by the direct or indirect effects of temperature (Kinne, 1970). In *Littorina* species this is indicated by correlations between distributions and sea temperatures near the limits of the ranges of some species. For example, limitation by low temperatures at the northern extremes of ranges is suggested by the close correlation between the occurrence of *L. saxatilis* in the northernmost Atlantic and the penetration of warm Atlantic water into the Arctic region (Ellis, 1955). Another case is the restriction of *L. brevicula* to warm bays and its absence from open coasts on the southeastern side of Hokkaido, which is under the influence of a cold southward current (Ohgaki, 1983). At southern extremes the opposite effects can sometimes be seen. For example, at the limit of *L. sitkana* in the western Sea of Japan it is absent from enclosed bays that are warm in summer (Golikov & Skarlato, 1967; Golikov & Kusakin, 1978). The absence of *Littorina* species from the head of the Bay of Biscay is usually ascribed to the localized warm climate of the area (Crisp & Fischer-Piette, 1959). Another example is the restriction of *L. kasatka* to the cold eastern side of Hokkaido. If ranges are indeed limited by temperature, climatic change should produce corresponding shifts in distribution (under the assumption that genetic determinants of physiological tolerance and of ecological requirements evolve more slowly). Evidence for this can sometimes be found in the fossil record. During cold climatic intervals of the Pleistocene both *L. littorea* and *L. fabalis* occurred on the coast of Morocco (Lecointre, 1950, 1952), respectively 850 km and 950 km south of their present limits. During warm interglacial periods ranges were extended northwards, in the case of *L. littorea* to Svalbard and the Taymyr Peninsula in Siberia, more than 2000 km beyond its modern limit. Significant changes in distribution may also occur on a shorter timescale. In a remarkable series of observations Fischer-Piette (1953, 1955, 1956, 1957a, b, 1958, 1960) documented the contraction of the ranges of *L. littorea*, *L. saxatilis* and *L. obtusata s.l.* along the northern and southern coasts of the Iberian Peninsula during the first half of the twentieth century, and in the case of *L. saxatilis* and *L. obtusata s.l.* their subsequent small advance. In this example no correlation with temperature has yet been attempted but, since many molluscs were affected in the same way, some climatic effect is likely.

The mechanisms by which temperature may control range limits are poorly understood. However, direct effects upon the survival of adults appear to be unlikely. In Britain heat coma temperatures of

L. littorea, *L. obtusata s.l.* and *L. saxatilis s.l.* have been measured as 32°C, 33–34°C and 35°C respectively, and their thermal death points as 46°C, 44°C and 45°C (Evans, 1948); these are not only higher than environmental temperatures on British shores (Evans, 1948), but the death points are about 10°C higher than tissue temperatures recorded for tropical littorinids on dry rocks (Vermeij, 1971) and comparable with the highest temperatures recorded in splash pools on tropical shores (Britton, 1992). There is also little evidence for mortality of adults at low temperatures. For example, a severe winter in Britain had no obvious effects on *Littorina* species (Crisp, 1964). In more extreme northern climates some species migrate to low-tide level or below to avoid low air temperatures; an example is *L. littorea* (Gowanloch & Hayes, 1926; Lambert & Farley, 1968; Matveeva, 1974), but this species can still tolerate freezing (Kanwisher, 1966), although it dies at −8°C (Matveeva, 1974). Of exceptional tolerance is *L. saxatilis*, which shows no winter migration and in Svalbard survives freezing into the icefoot on the shore for eight to nine months of the year (Aarefjord & Skreslet, 1969). Temperature limitation is much more likely to act at other stages of the life cycle. At southerly latitudes benthic egg masses and juveniles exposed on the shore at low tide are likely to be susceptible to desiccation, as shown near the southern limit of *L. sitkana* (Behrens, 1972). This has also been suggested as a limiting factor at the southern extreme of the distributions of *L. arcana* and *L. compressa* in Brittany, both of which lay benthic egg masses in damp crevices on rocks (Warmoes, Dumoulin & Reid, 1992). Planktotrophic larvae may also be adversely affected by high temperatures; at the southern limit of *L. littorea* in North America the sporadic settlements can be correlated with cooler than average seasonal temperatures, although whether this is a consequence of ocean current patterns (Kraeuter, 1974) or a direct effect of temperature upon settlement (Wells, 1965) is not clear. Towards the northern limits of their ranges the most widely-distributed species often show shortened breeding seasons, restricted to the summer months (e.g. *L. sitkana* in the Kurile Islands, Golikov & Kusakin, 1978; *L. littorea*, *L. obtusata s.l.*, *L. saxatilis* in the White Sea, Kuznetsov, 1960, Matveeva, 1974), suggesting that a minimum summer temperature may be necessary for spawning. In *L. saxatilis* it has been found that eggs and early larval stages are most susceptible to cold and cannot survive the severe winters of the White Sea (Matveeva, 1974; Sokolova, 1995). Other effects of temperature may be indirect, for example on food plants, predators or competitors.

More obviously, geographical ranges can also be circumscribed by stretches of unsuitable habitat. Clearly, depending upon the capacity for dispersal, the open sea may be a barrier to range expansion in this genus of almost exclusively intertidal species. The means and distances of dispersal, and possible influence upon geographical range, will be considered later (see Dispersal and Colonization, p. 370). Other aspects of the physical habitat can be more subtle. Low salinity limits distributions on a geographical scale in large estuarine regions such as the Baltic Sea, where the distance of penetration is determined by physiological tolerance of low salinity. The Atlantic species and their Pacific sister-taxa are among the more euryhaline members of the genus, suggesting that tolerance of low salinity may have been a requirement for those species involved in the trans-Arctic migration during the Pliocene (V. Ya. Berger, 1973, 1978a, b). Even where shorelines are present, the lack of a suitable rocky substrate may interrupt or limit the range; this is obvious on a local scale, but can also be seen on a geographical scale. For example, no littorinids occur on the French coast south of the Gironde, where the open coast is entirely of sand (Crisp & Fischer-Piette, 1959). In eastern North America the most southerly large area of natural rocky shore is the Connecticut coast, and the sporadic occurrences of *L. littorea*, *L. obtusata* and *L. saxatilis* further south are limited to isolated rock outcrops and artificial hard substrates. Similarly, sandy shores in Oregon may contribute to the southward limitation of *L. sitkana* (Behrens Yamada, 1977b). The composition of the rock may also be important. For example, the distributions of *L. compressa* and *L. arcana* appear to be limited in some parts of Britain and the English Channel by their requirement for a hard, crystalline rock for the attachment of their gelatinous egg masses (Mill & Grahame, 1990b; Warmoes *et al.*, 1992).

Another aspect of geographical distribution that is related to habitat requirement is the distinction that can be drawn between 'oceanic' and 'continental' distribution patterns. Species with oceanic distributions occur on high islands and sometimes on peninsulas of larger land masses, whereas those with continental distributions are mainly restricted to continental coastlines, including turbid and even estuarine habitats. This distinction has been recognized in various gastropods, including other littorinid genera (Reid, 1986a, 1989b, 1992), and is not a sharp division, but rather a gradient between extremes. Typical examples of the oceanic type include *L. striata* (found only on the eastern Atlantic islands and rarely, if at all, on the African mainland), *L. aleutica* (found on the Aleutian

Islands and on the Chukotsk Peninsula of the Siberian mainland) and *L. horikawai* (found on islands off Kyushu and rarely on mainland peninsulas). At the continental extreme are *L. littorea* (common on Atlantic shores, but rare or absent from some small islands off the European and British mainland), *L. squalida* (common on the mainland coasts of the Okhotsk and Bering Seas, but virtually absent from the Kurile and Aleutian Islands) and *L. natica* (largely restricted to continental shores of the Bering Sea and Kamchatka, although also occurring on some of the Kurile Islands). Other species appear to span the spectrum, for example *L. sitkana*, *L. subrotundata* and *L. saxatilis* occur on both mainland coasts and oceanic islands. In some cases subtle distinctions can be made between superficially similar distributions; for instance, of three *Littorina* species found in California, *L. keenae* and *L. scutulata* are the more oceanic in character (occurring on open oceanic shores and on the Channel Islands off the coast) and *L. plena* the more continental (occurring frequently in estuarine situations and not recorded from the Channel Islands). The distinction between oceanic and continental species is a purely empirical one and the basis for it is unknown. At least in part it is probably simply the manifestation on a geographical scale of habitat requirements and physiological tolerance. Those species tolerant of strong wave action, and unable to survive in turbid and estuarine environments, will fall in the oceanic region of the spectrum. Other oceanographic factors may also be involved, such as the high nutrient levels, primary productivity and turbidity near continental margins, and also oceanic circulation patterns, which may particularly influence pelagic eggs and planktotrophic larvae. In the case of *L. littorea* it has been suggested that its absence from islands near the coast is a consequence of its planktotrophic development, which results in poor powers of colonization and inability to maintain populations on islands where larvae are swept away by ocean currents (Johannesson, 1988). This, however, cannot be a general explanation for the phenomenon, because there is no correlation between oceanic or continental distribution patterns and type of development in the genus as a whole (Table 11). Whatever the cause, the distinction is a useful one, and the contrasting patterns of distribution described (contiguous distributions on linear continental margins as opposed to discontinuous occurrences on scattered oceanic islands) have important consequences for genetic divergence of populations and the likelihood of allopatric and parapatric speciation (see Modes of Speciation, p. 376).

Transfer of adults beyond their natural range has sometimes been attempted, to investigate range limitation. For *L. keenae* Behrens Yamada (1977b) showed that adults survived and produced spawn in Washington, north of their northern limit in Oregon, and suggested that unfavourable currents or larval mortality limited natural spread. It has also been suggested that some supposedly poorly-dispersing species may not yet have reached the full extent of their potential range on continental coastlines (Mill & Grahame, 1990a, b) but, in view of dispersal capabilities reviewed below, this may be unlikely.

Little is known about possible limitation of ranges by biotic factors. Predation by crabs is more severe at lower latitudes and is known to be an important selective force acting on *Littorina* species (e.g. Heller, 1976; Raffaelli, 1978; Johannesson, 1986; Seeley, 1986). On a local scale crab predation may exclude some species from certain habitats (see discussions of *L. subrotundata*, p. 178, and *L. saxatilis*, p. 327), and in the case of *L. sitkana* has been suggested to contribute to the limitation of southward spread (Behrens Yamada, 1977b, 1992). The only species known to have specialized requirements for a narrow range of food plants are *L. obtusata* and *L. fabalis*, both closely dependent upon fucoids and a few other algae, and where these plants are rare or absent (at the head of the Bay of Biscay, for example) these species cannot occur. There is no evidence of competitive exclusion on a geographical scale among *Littorina* species, although on a local scale interspecific competition for food can occur (e.g. Behrens Yamada & Mansour, 1987; Behrens Yamada, 1992).

In summary, where otherwise suitable coastlines exist, the direct or indirect influence of temperature seems to be one of the more general causes of range limitation in *Littorina* species, although the mechanisms involved are poorly known and probably diverse. Too little is known about possible biotic factors.

Dispersal and colonization

The dispersal of individuals is the process by which geographical ranges are maintained or enlarged. Unfortunately the word 'dispersal' is used in two senses: to cover those short-term processes by which individuals are distributed, and in a narrowly biogeographical sense to describe long-term changes to the geographical range. The concepts are closely related, for distributional processes are the proximate cause of range shifts, but some analytical biogeographers prefer to distinguish the former as 'dispersion', restricting 'dispersal' to the latter (e.g. Rosen, 1988). Here 'dispersal' will be used in both senses, and the meaning made clear by the context. The dispersal ability of a species is an attribute of profound evolutionary significance. Individual dispersal to a new area may result in colonization and extension of the range, whereas dispersal to a populated area, followed by interbreeding, results in gene flow. This is an important species cohesion mechanism (see Species Concepts, p. 373), and its rate and pattern influence effective population size, differentiation among populations, and ultimately the rate and pattern of speciation and extinction.

Dispersal of post-larval stages must generally be slow or infrequent. Release and recapture methods have been used to estimate the rate of dispersal of adults by crawling on the shore; results for *L. saxatilis* are 1–4 m in 3 months (Janson, 1983b) and up to 9 m in 2–3 months (Berry, 1961), and for *L. sitkana* and *L. subrotundata* a maximum of 15 m in 6 weeks (Boulding, 1990). Similarly, expansion of colonies of *L. saxatilis* into uninhabited stretches of shore proceeds at a rate of about 5 m per month (K. & B. Johannesson, 1995). Occasional long-distance transport must occur, and for species lacking pelagic eggs or larvae is the only means of colonizing islands and other isolated habitats, but the mechanisms are largely a matter of speculation. Some small gastropods are able to drift in the water column, attached to a long mucous thread. This behaviour has been described in detail in the littorinid *Lacuna*, in which animals raise the posterior end of the foot and secrete mucus to initiate drifting, as a response to water currents (Martel & Chia, 1995a). The early postlarval stages of *Littorina sitkana* are also reported to drift in this way and have been found in small numbers on mid-water collecting devices placed on the shore (Martel & Chia, 1995b). Other *Littorina* species may likewise possess this capacity when very small. It has been suggested that dislodged individuals of *L. saxatilis* might be rolled across sand to a nearby rocky shore by currents, and rafting on floating weeds and debris is a possibility (Johannesson & Warmoes, 1990). Rafting has also been proposed as an explanation of pre-European records of *L. littorea* in North America (Clarke, 1971; Johannesson, 1988). Two species, *L. obtusata* and *L. fabalis*, are always found in association with macroalgae, and for these rafting is likely to be relatively frequent; they have been found as a common component of the fauna on drifting clumps of weed more than 10 km off the coast of Iceland (Ingólfsson, 1995). Transport by migratory birds would appear to be the only likely explanation for the presence of *L. saxatilis* in southern Africa before European settlement, and may have been facilitated by the occurrence of this species in muddy, salt-marsh habitats where wading birds abound, and where its shell is often small and lightweight (p. 331). A record of *L. obtusata s.l.* 21 km from the sea in Scotland has also been ascribed to transport by birds (Russell Hunter, Maitland & Yeoh, 1964; see Rees, 1965, for other molluscan examples). Deliberate or accidental transport by man has also taken place, particularly in the case of *L. littorea*. This species has been widely used and transported for food in Europe, and this probably accounts for occasional records from the Mediterranean and Canary Islands outside its natural range (e.g. Aradas & Benoit, 1870; Moolenbeek, 1989) and for at least one temporary establishment of a breeding population in Italy (Barsotti & Campani, 1982; Johannesson, 1988). The most famous example of possible human introduction of this species is its appearance in North America in about 1840, followed by rapid spread in areas of the western Atlantic where it was previously unknown. Whether this was indeed a case of human introduction is still debated (p. 116). Inadvertent transport of *Littorina* species with imports of oysters for cultivation has also occurred in the cases of *L. littorea* (Carlton, 1969, 1992) and *L. brevicula* (Hanna, 1966) on the west coast of North America, and of *L. compressa* in The Netherlands (Stoker, discussed in Warmoes *et al.*, 1992), but have not resulted in establishment of populations. In addition, *L. saxatilis* has recently been introduced to San Francisco Bay, apparently among seaweed used to pack bait (J. Geller, J. T. Carlton, pers. comm.).

For those species with pelagic eggs and planktotrophic larvae, transport of larval stages by marine currents would appear to be the chief means of long-distance dispersal. Details of early life history are available for two *Littorina* species. In Denmark *L. littorea* hatched from the egg capsule after 6 days and developed to the stage of metamorphosis and settlement one month later, although under unfavourable laboratory conditions larval life could be prolonged for a further six weeks (Thorson, 1946; see also Williams, 1964), making a total possible pelagic life of more than 10 weeks. In Washington larvae of *L. scutulata* lived for up to 25 days in laboratory culture, but did not survive to metamorphosis (Buckland-Nicks, Chia & Behrens, 1973). For comparison, a planktotrophic tropical littorinid, *Nodilittorina hawaiiensis* Rosewater & Kadolsky, 1981, took an average of 24 days from spawning until metamorphosis at 25°C (Struhsaker & Costlow, 1968, as *Littorina picta*). The velocities of directional currents in the open ocean range from 10 to 75 km per day (Scheltema, 1971; Kraeuter, 1974), so that there is the potential for considerable long-distance transport of larval stages. However, this potential is not necessarily realized, because larval stages may remain close to the shore, where tidal waters are little influenced by directional currents. This appears to be the case for *L. littorea*, of which eggs and larvae are only common in inshore plankton samples (Fretter & Shale, 1973) and in the calm conditions of estuaries may even develop among benthic detritus (Fish, 1972). Nevertheless, medium-distance dispersal in the range of tens to hundreds of kilometres can be effective, as shown by the rapid spread of *L. littorea* in North America during the nineteenth century; after its appearance in the Gulf of St Lawrence, it extended its range for 1500 km to New York in only 40 years (p. 116), an average rate of 37 km per year. At the southern limit of its modern range in North America, settlements occasionally occur on the Delmarva Peninsula, for which the probable source of larvae is the closest large area of suitable habitat, Long Island Sound, some 300 km to the north (Kraeuter, 1974).

For *Littorina* species with benthic eggs and nonplanktotrophic development (subgenus *Neritrema*) pelagic dispersal of larval stages is limited to occasional accidental transport. Rafting on detached macroalgae may not be uncommon in a few species, especially *L. obtusata* and *L. fabalis* which typically deposit spawn on their fucoid host plants (Tatarenkov & Johannesson, 1994). *Littorina saxatilis* normally retains embryos in the brood pouch until the crawling stage, but may occasionally release earlier, encapsulated stages, perhaps as a result of disturbance; these develop normally and might conceivably undergo pelagic dispersal (Hannaford Ellis, 1978b).

The expectation of poor dispersal capacity in nonplanktotrophic *Littorina* species is largely supported by their genetic structure; for example, the nonplanktotrophic *L. saxatilis* shows a much higher degree of genetic differentiation between geographically separated populations than does the planktotrophic *L. littorea*, which (assuming that the genetic variation is selectively neutral) may be explained by genetic drift in the former and by the unifying effects of extensive gene flow in the latter (E. M. Berger, 1973; Janson, 1987a; Johannesson, 1992; see Ward, 1990, and Tatarenkov, 1995b, for reviews of population genetic structure in the genus as a whole). Dispersal capability has also been examined more directly by recording the rate of colonization of new or barren habitats. In Sweden *L. saxatilis* has colonized recently-emergent rock skerries up to 2.5 km from mainland populations over the course of about a century, but took only 2 years to colonize a denuded skerry 15 m from an established mainland population (Janson, 1987b). The rate of recolonization of skerries following local extinction caused by a bloom of a toxic flagellate has also been studied in Sweden (K. & B. Johannesson, 1995). New breakwaters in Belgium only a few hundred metres from established populations were colonized in two years; *L. littorea* colonized the same breakwaters after only one year, but in evolutionary terms this difference is small (Johannesson & Warmoes, 1990). Therefore, on this small scale (about 1 km or less) the rate of dispersal of *Littorina* can be sufficient to permit quite rapid colonization, irrespective of the type of larval development. The numbers of colonizing individuals or immigrants are, however, much greater for planktotrophic species, leading to their higher rates of gene flow.

Over medium and long distances, however, improved dispersal does not automatically enhance colonization potential. The probability of successful colonization of a distant site depends not only upon the chance of arrival of a single individual, but also upon the likelihood that it will establish a population, as emphasized in a model by Johannesson (1988). For species with pelagic development larval transport over medium distances will be relatively frequent, but to establish a new colony a sufficiently large number of larvae must settle and survive to find mates. Even then, the density of the resulting eggs and larvae will be low, and especially on isolated islands many will be swept away from the founding population. There will therefore be a maximum distance away from an established population at which pelagic larval transport will be likely to found new colonies, and the probability will be affected by hydrological conditions at the new site. Larvae may arrive at more distant sites, but their numbers will be too low to found new self-sustaining colonies. For nonplanktotrophic species the rate of transport of individuals over any medium or long distance may be extremely low, but even a single fertilized female or rafted egg mass will have a good chance of establishing a colony, because offspring remain on the shore close to the parent. Over long distances, perhaps several hundreds of kilometres or more, it can thus be argued that nonplanktotrophic species are, paradoxically, more likely to found new colonies and extend their range, despite their much lower rates of individual dispersal and gene flow (Johannesson, 1988).

The relationship between geographical range and individual dispersal is therefore not a simple one. In some gastropod groups, mainly in tropical regions, it has been reported that planktotrophic species show wider geographical ranges than nonplanktotrophic ones (e.g. Hansen, 1980; Jablonski, 1982; Scheltema, 1989; Kohn & Perron, 1994), and in some cases there is even a correlation between the length of the pelagic phase and the extent of the range (e.g. in *Conus*, Kohn & Perron, 1994). These results are as expected purely from the greater potential for long-distance dispersal provided by long-lived planktotrophic larvae, but the indirect effects of dispersal, such as gene flow and repopulation after local extinction, may also be important in promoting a wide geographical range (Emlet, 1995). However, among shallow-water gastropods of temperate and cold waters the simple relationship between geographical range and type of development disappears (Vermeij, Palmer & Lindberg, 1990). Here macroalgae are much more abundant, providing opportunities for dispersal of small species by rafting (Highsmith, 1986). This type of dispersal will be particularly effective for colonization by nonplanktotrophic species, which are more likely to found a population from a small number of immigrants, as discussed above. Some small, nonplanktotrophic species may thereby achieve even wider ranges than their planktotrophic congeners, as in the minute bivalve *Lasaea* (Ó Foighil, 1989). In *Littorina* there is no doubt that nonplanktotrophic species can indeed disperse, by whatever means, across wide expanses of open sea to extend their range. For example, during the last glaciation the fauna of rocky shores in the northwestern Atlantic, Greenland and Iceland became extinct, subsequently recolonizing the region by dispersal from northwestern Europe (Ingólfsson, 1992); it was three nonplanktotrophic species (*L. saxatilis*, *L. obtusata*, *L. fabalis*) that first spread successfully to the northwestern Atlantic, whereas the planktotrophic species *L. littorea* may only have become established in North America in the middle of the nineteenth century, and then perhaps only by human introduction. In the genus as a whole, it should therefore not be surprising that there is no clear correlation between the type of larval development and the width of the geographical range (Table 11). The mean width of the range of nonplanktotrophic species is 5205 km, which exceeds the 3355 km of planktotrophic ones, although such is the spread of values that this difference is not significant ($t=1.448$, $P>0.1$). The widest geographical range of all *Littorina* species is found in *L. saxatilis*, a nonplanktotrophic species that is also ovoviviparous, which may further enhance its colonizing ability (p. 335). Of course colonizing ability does not depend only upon aspects of dispersal and development, but also on a complex of factors that may include habitat range, physiological tolerance and genetic variability. On each of these counts *L. saxatilis* appears to be well suited as a successful colonizer.

Species and speciation

Species concepts

Many different definitions of species have been proposed in the evolutionary literature (see Ereshefsky, 1992, for a review). Before considering the processes of speciation in *Littorina*, it is necessary to consider briefly what is understood by a species in this group, and whether species as recognized by morphological criteria correspond with species defined by their biological attributes. Although genetic data are now available for some *Littorina* species, in the first instance species have so far always been delimited using morphological characters. The characters used by early taxonomists were primarily those of the shell, and the resulting groupings were often arbitrary and unstable, as shown by the complex taxonomic histories of some of the more conchologically variable species. Morphological characters are still widely, and often exclusively, used to define species, but are now carefully chosen in an attempt to reflect the underlying processes that are responsible for the genetic and phenetic cohesion of species.

A number of the biological species concepts have been defined solely in terms of reproductive attributes. Most influential has been that of Mayr (1963), often known as *the* biological species concept, which defines species as ‘groups of interbreeding natural populations that are reproductively isolated from other such groups’. Under this view, species are defined by the reproductive barriers, or isolating mechanisms, that mark their boundaries, and it is therefore more aptly known as the isolation species concept (Paterson, 1985). Gene flow caused by interbreeding, within and among populations, is seen as the mechanism responsible for the genetic integrity of species. The speciation process is suggested to begin with the spatial isolation of a population or part of the parent species, followed by genetic divergence, during which reproductive isolating mechanisms develop merely as an accidental byproduct of genetic reorganization. There can be active selection to enhance isolating mechanisms (reinforcement) only when the divergent populations are reunited and hybrids between them are at a disadvantage. The perceived importance of interbreeding accounts for the emphasis on reproductive characters for identifying species. Direct tests of interfertility are rare, but a comparison of appropriate reproductive characters may suggest whether interbreeding is possible. In *Littorina* the shape of the penis has been found to be one of the most useful characters for discriminating between superficially similar species (p. 14), with the implication that penis shape may be an isolating mechanism. The relative size of the copulatory bursa of females may also serve to isolate sympatric species (p. 18).

A different view of isolating mechanisms has been proposed by Paterson (1985) in his recognition species concept. He suggested that during the speciation process there is selection for characters that promote successful interbreeding between individuals. These constitute the so-called specific-mate recognition system (SMRS); they may function as isolating mechanisms if and when the divergent populations are reunited, but that was not their original function during the speciation process. In *Littorina* penis shapes can be as different in some allopatric pairs of sister-species (e.g. *L. squalida* and *L. littorea*) as in largely sympatric pairs (e.g. *L. scutulata* and *L. plena*; *L. obtusata* and *L. fabalis*), and furthermore there is no evidence for character displacement where sister-species show partial overlap of their distributions (p. 14). It therefore seems more appropriate to regard penis shape as part of the SMRS of species, rather than as a prezygotic isolating mechanism.

These purely reproductive concepts of species emphasize gene flow as the mechanism responsible for the cohesion of species. However, in asexual species and in sexual species with geographically isolated populations, interbreeding cannot occur, and in such cases cohesion is achieved through natural selection. This mechanism is emphasized in ecological concepts of species, in which niches or adaptive zones are regarded as distinct for each taxon. These two views have been reconciled in the cohesion species concept of Templeton (1989), defined as 'the most inclusive population of individuals having the potential for phenotypic cohesion through intrinsic cohesion mechanisms'. Such mechanisms include both those factors promoting genetic identity through gene flow, and ecological and demographic factors which control the spread of new genetic variants through natural selection and descent from a common ancestor. Since the reproductive and ecological aspects of species are usually correlated, gene flow and selection often act together to promote cohesion, but the role of each becomes clear in the rare cases where they are in conflict. For example, intense selection may maintain distinct populations even in the face of gene flow between them; such conditions may be rare, because there will be corresponding selection for mate choice, reducing gene flow.

The concepts discussed so far emphasize processes that are responsible for the maintenance of living taxa as cohesive units, and provide a framework for viewing the process of speciation, by which altered cohesion mechanisms are genetically assimilated. Some other species concepts consider the history of species lineages through time. An example is Wiley's (1981) modification of the evolutionary species concept, defined as 'a single lineage of ancestor-descendant populations which maintains its identity from other such lineages and which has its own evolutionary tendencies and historical fate'. This is a broad definition that does not specify the cohesion mechanisms responsible for evolutionary tendencies and is compatible with all models of speciation. It provides a theoretical basis for interpreting a branching diagram as a phylogenetic tree, in which each branch is the product of a speciation event. Furthermore, it allows for the survival of an ancestral species after a speciation event, in cases where one of the daughter species retains the evolutionary tendencies of the ancestor. It does, however, preclude the division of a single fossil lineage into arbitrary 'chronospecies'. Such considerations are of practical importance only in groups with an extensive fossil record. In *Littorina*, for example, this concept is the basis for the use of the name *L. squalida* for fossils that may be ancestral to both *L. petricola* and *L. littorea*, as well as to the living taxon of the same name. Some workers have adopted a more restrictive phylogenetic or cladistic species concept, in which species must (like higher categories in a cladistic hierarchy) be strictly monophyletic, defined by synapomorphies of their individuals, and cannot persist after a speciation event (Mishler & Donoghue, 1982; Baum, 1992). Under this view, a species is an ancestral-descendant lineage between two successive branch points (or a branch and a terminus) on a phylogenetic tree.

For several *Littorina* species there exists a considerable body of information on the genetic structure of populations, and it must now be considered how this relates to concepts and definitions of species in this group. Information about genetic structure has so far been derived largely from the analysis of allozyme frequencies, and in general the results have supported the genetic coherence of species as defined by morphological criteria. Measures of genetic distance between conspecific populations are in most cases much less than between populations of different species (e.g. Mastro, Chow & Hedgecock, 1982; Moyse, Thorpe & Al-Hamadani, 1982; Ward, 1990; Zaslavskaya, Sergievsky & Tatarenkov, 1992a, b; Tatarenkov, 1995a). Since littorinids are sexually-reproducing, mobile and generally occur in large and often contiguous populations, it is likely that gene flow is more important than selection in maintaining genetic cohesion, so that reproductive definitions of species are adequate. Indeed, in allozyme studies it is often assumed that variation is selectively neutral, so that similarity of allele frequencies within species is used as evidence of gene flow. However, such data should be interpreted cautiously, for there is now evidence that some enzyme loci are under selection (Janson & Ward, 1984; K. & B. Johannesson, 1989, 1990, 1993; Grahame, Mill, Double & Hull, 1992; Tatarenkov & Johannesson, 1994; K. Johannesson, B. Johannesson & Lundgren, 1995). The influence of gene flow is likely to be strongest in those species that are most widely dispersed, by pelagic eggs and planktotrophic larvae.

Where morphologically distinct taxa are sympatric, the occurrence of significantly different allele frequencies is taken as evidence of their reproductive isolation and supports their status as species. This, for example, has been repeatedly shown in the discrimination of *L. compressa* from *L. saxatilis* (e.g. Knight & Ward, 1986, 1991; p. 250). In this case the morphological evidence includes differences in both male and female reproductive structures, which are interpreted as isolation or recognition mechanisms, so that genetic and morphological evidence lead to the same conclusion. Genetic differences between allopatric samples are difficult to interpret, for gene flow may be reduced by geographical isolation and differences due to selection or drift may become significant. Such problems are demonstrated by the genetic differentiation on a geographical scale within *L. subrotundata*, which has been interpreted as evidence for specific differences by Boulding, Buckland-Nicks & van Alstyne (1993; p. 174); in this case anatomical characters provide no independent support for reproductive isolation.

Genetically, the most well-studied species are *L. saxatilis* and *L. arcana*. These are very closely related (p. 266); genetic distances between them are small and no alleles are known to be diagnostic, but differences in the pallial oviduct do provide an unequivocal character for morphological separation. This example illustrates the importance of sampling on a geographical scale. Under reproductive definitions, putative species should show genetic cohesion; in practical terms, there should be evidence of greater gene flow between geographically distant samples within species, than between species at single localities (K. & B. Johannesson, 1990). To test this, genetic data from *L. arcana* and *L. saxatilis* (separated by morphological criteria) have been analysed by various clustering techniques. Simple phenetic clustering of such data initially achieved only partial separation of the two species. However, methods appropriate for phylogenetic analyses, such as the distance-Wagner procedure, improved the separation (Knight & Ward, 1991), and indeed the maximum-likelihood technique separated the populations into two discrete clusters, corresponding to their morphological identification (Sundberg, Knight, Ward & Johannesson, 1990). These two species therefore show genetic cohesion and some evidence of monophyly. Interestingly, hybrids between these taxa have been produced in the laboratory, but not so far identified in the wild, suggesting that some kind of weak prezygotic isolating mechanism may operate where the two are sympatric. There is also partial postzygotic isolation for, of the four possible F1-to-parent backcrosses, only the hybrid female to *L. saxatilis* male cross has been successful (Warwick, Knight & Ward, 1990).

In the examples given so far, the definitions of species are not controversial, because the genetic evidence is clear and is in agreement with morphological evidence for reproductive isolation. However, this is not always the case. Recently, detailed investigation of some nonplanktotrophic *Littorina* species has revealed the existence of morphological and genetic differentiation on a fine microgeographic scale. This is most striking and well known in *L. saxatilis*, in which samples from microhabitats only one to two metres apart may show pronounced differences in shell characters. In the present study these forms are designated as ecotypes (p. 26). In some cases there are correlated differences in allele frequencies, but no consistent differences in reproductive anatomy have yet been reported. One controversial case concerns the barnacle ecotype, sometimes considered as a species (*L. neglecta*) distinct from typical *L. saxatilis* on the same shores (p. 309). In the absence of any consistent qualitative characters by which to define this ecotype, the morphometric evidence is disputed. Some have claimed that the barnacle ecotype is a phenetically coherent unit (Caley, Grahame & Mill, 1995;

Grahame, Mill, Hull & Caley, 1995), whereas others have found intergradation between the two (e.g. B. & K. Johannesson, 1990). Genetic studies of barnacle ecotypes in Britain, Scandinavia and Spain have used evidence from (presumed) selectively neutral allozymes to suggest that there is more gene flow between the barnacle ecotype and *L. saxatilis* on the same shore than between barnacle ecotypes on geographically distant shores (K. & B. Johannesson, 1990; K. Johannesson, B. Johannesson & Rolán-Alvarez, 1993), and DNA studies in Britain suggest a similar pattern (Crossland, Coates, Grahame & Mill, 1995). Unfortunately, information is still incomplete, for in northeastern England, where genetic differentiation may be most pronounced, the necessary geographical sampling has yet to be undertaken (Hull, Grahame & Mill, pers. comm.). In this case species definitions are controversial, because the boundaries of genetic cohesion (determined by gene flow) are larger than those of phenetic cohesion (determined by natural selection within each microhabitat). Species definitions that emphasize genetic cohesion and monophyly lead to recognition of a single polymorphic species, disrupted by hybrid zones which separate ecotypes (K. & B. Johannesson, 1990; Johannesson *et al.*, 1993; Rolán-Alvarez & Rolán, 1995; Rolán-Alvarez, Rolán & Johannesson, 1996). Those definitions that emphasize phenotypic cohesion and ecological characteristics may be taken to indicate recognition of a number of species, perhaps within a syngameon (the most inclusive unit of interbreeding) (Caley *et al.*, 1995; Grahame *et al.*, 1995; Hull, Grahame & Mill, pers. comm.).

Further information about possible reproductive isolation between ecotypes of *L. saxatilis* is available from detailed local studies of mating behaviour, gene flow and fertility. On shores in Galicia partial reproductive isolation between two ecotypes is maintained by a combination of microhabitat choice, assortative mating and sexual selection (Johannesson, Rolán-Alvarez & Ekendahl, 1995). In northeastern England allozyme data suggests that there may be a complete reproductive barrier between the barnacle ecotype (*L. neglecta*) and forms of *L. saxatilis* with which it is strictly microsympatric (Hull, Grahame & Mill, pers. comm.). In the same region a high degree of reproductive isolation between upper-shore and mid-shore ecotypes of *L. saxatilis* has been suggested by the low fertility of hybrids, and supported by allozyme data (Hull, Grahame & Mill, 1996, pers. comm.). Uncritical application of reproductive species concepts might, in these cases, suggest that distinct species should be recognized. However, the importance of a geographical perspective must again be emphasized. Although information is still meagre, one possible interpretation is that the strength of reproductive barriers varies geographically, being complete on some shores, but breaking down elsewhere. If so, it would not be desirable (or practical) to give taxonomic recognition to purely local cases of reproductive isolation. Broader geographical patterns might, however, require such recognition.

Such complicated cases, still poorly understood, present challenges to current species concepts and taxonomic practices. Taxonomic debates must not, however, overshadow the exceptional interest of these problematic cases for the study of the mechanisms by which species are generated. It must be emphasized that speciation is a process, and not an event, so that controversies will occasionally arise about the practical classification of lineages undergoing divergence.

So far, genetic studies have not revealed the existence of species that were not first recognized, or at least suggested, by morphological characters. Nevertheless, the possibility remains that such truly 'cryptic' or 'sibling' species may be found. In *L. littorea*, for example, there is a report of some anomalous gene frequencies in a small number of individuals from France and Belgium (Johannesson, 1992). Another example concerns the relatively large genetic distances between typical *L. sitkana* and sympatric, morphologically similar, animals at one locality in the Kurile Islands (Zaslavskaya, 1995). Both cases require further investigation.

Modes of speciation

For sexually reproducing organisms, in which gene flow is a major cohesive mechanism, speciation is the process of cladogenesis by which gene flow between sister-populations is sufficiently reduced to allow each to become irrevocably committed to a different evolutionary fate (Bush, 1994). Various classifications of speciation processes have been proposed, depending on the degree of geographical separation and relative sizes of the sister-populations (e.g. Bush, 1975, 1994; Wiley, 1981; Brooks & McLennan, 1991). In the allopatric mode geographical separation is the cause of the interruption to gene flow. Where the sister-populations are each of relatively large size, sufficient to exclude inbreeding as a factor in their divergence, this is described as vicariant (dichopatric or large-

Table 12. Speciation modes in *Littorina*, as suggested by relative geographical distributions of sister-taxa. See Figure 119 for phylogeny and for numbering of its nodes. The ages of divergence are the means (and ranges in parentheses) of the estimates based on rates of sequence divergence of up to three mitochondrial genes (Reid, Rumbak & Thomas, 1996). Abbreviations: NP, nonplanktotrophic development; P, planktotrophic development; PA, parapatric speciation; PI, peripheral isolates speciation; S, sympatric speciation; V, vicariant speciation.

Node number	Sister elements	Estimated age of divergence (Ma)	Relative distributional areas	Separation (km)	Overlap of areas (as % of smaller)	Type of development	Inferred speciation mode
(10)	*arcana+saxatilis*	1.73 (0–3.55)	1:6	—	100	NP	S
(9)	*compressa*+(10)	1.73 (0–3.55)	1:6	—	100	NP	S
(11)	*fabalis+obtusata*	1.32 (0–3.49)	1:1.3	—	100	NP	S
(8)	(11)+(9)	2.83 (1.31–5.99)	1:1.5	—	100	NP	S
(12)	*natica+aleutica*	< 4.12	1:1	250	0	NP	V or PA
(7)	(8)+(12)	4.12 (1.95–6.36)	1:3	4000	0	NP	V
(13)	*horikawai+sitkana*	3.99 (0–12.69)	1:30	1200	0	NP	PI or PA
(14)	*mandshurica+brevicula*	5.84 (1.99–15.62)	1:2.5	—	30	P	?
(15)	*squalida+littorea*	6.42 (3.28–10.07)	1:1	5500	0	P	V
(16)	*scutulata+plena*	12.81 (2.48–18.63)	1:1	—	80	P	? S

subdivision) speciation. In contrast, in the peripheral isolates model (peripatric or founder-effect speciation) one of the sister-populations is small, promoting its rapid genetic divergence by mechanisms that include inbreeding, drift, selection and genomic reorganization. In allo-parapatric (reinforcement) speciation initial divergence takes place in geographically isolated populations, but is completed in parapatry or sympatry when the sister-populations are reunited, followed by selection against hybrids. In the non-allopatric mode lineage independence is achieved without complete geographical isolation, and most models (excluding those involving hybridization or asexual reproduction) require segregation imposed by specialization to different habitats, or by timing of reproduction. Where the habitats of the sister-populations are contiguous, and limited hybridization occurs within a narrow zone of contact, the process is described as parapatric speciation. If speciation is sympatric, however, spatial segregation is on a finer scale, within the normal range of dispersal of the populations. The categories within both allopatric and non-allopatric modes may be parts of a continuum, rather than discrete processes.

Several lines of evidence can be used to determine the likelihood of these speciation mechanisms. First, it is possible to examine the results of speciation, after it has occurred, by investigating the geographical distributions of pairs of sister-taxa. Comparisons can then be made with the phylogenetic and biogeographic patterns predicted by the different speciation models (Wiley, 1981; Lynch, 1989; Brooks & McLennan, 1991; Frey, 1993). These predictions depend upon the assumptions that character change is fast enough to reconstruct the history of speciation and that no extinctions of taxa have occurred. They also require that no significant dispersal (i.e. change in geographical range) has taken place to obscure the pattern of speciation, so that the geographical distribution of an ancestral species can be reconstructed by summing the distributions of its descendants. This is not to deny that dispersal has occurred in some cases; nevertheless, this assumption is necessary for first-order explanations, because dispersal is a non-parsimonious explanation that can account for *any* pattern of distribution of sister-taxa and should be invoked only when there is clear evidence to support it.

For *Littorina* the phylogenetic analyses (Chapter 5) have identified ten pairs of sister-taxa, including not only sister-species, but also deeper branches in the phylogeny (Fig. 119; Table 12). For each pair the relative areas of their modern distributions have been estimated, and their area of overlap expressed as a percentage of the distribution of the more narrowly distributed of the two. The analysis has not been extended throughout the phylogeny, because for older branches it becomes increasingly likely that dispersal may have obscured speciation patterns. All the divergences listed are estimated to have occurred less than about 6 million years ago (with the exception of that of *L. scutulata* and *L. plena*, for which the estimated age range is particularly wide). There is no correlation between age and degree of overlap, which might question the assumption of no dispersal. Under the assumptions, the most parsimonious first-order explanations for the speciation events are that complete overlap implies sympatric speciation, whereas no overlap implies allopatric speciation. Within the allopatric category similar sizes of the areas of the sister-taxa suggest vicariant speciation, and highly dissimilar areas indicate peripheral isolates speciation. No clear conclusions can be drawn from partially overlapping distributions, and this method may not be able to distinguish between parapatric and allopatric processes, which will be confused by even small changes in distributions. The results (Table 12) suggest four (perhaps five) cases of possible sympatric speciation, and four of allopatric speciation (of which one may have been by peripheral isolation, and one possibly by parapatric speciation). A similar, although preliminary, analysis of the whole family Littorinidae has suggested that vicariant allopatric speciation is most common, but that there is some evidence of sympatric speciation (Reid, 1994). These results contrast with a view that has become almost dogmatic in the literature, that the peripheral isolates process is the most common and that sympatric speciation is rare (e.g. Mayr, 1963). However, in a phylogenetic survey of vertebrates, Lynch (1989) found that the vicariant mode was most common, and sympatric speciation least so. Individual cases of speciation in *Littorina* will later be discussed in detail.

A second source of evidence about the processes of speciation is to examine cases of intraspecific variation in living species, under the assumption that these patterns (if of genetic origin) might indicate processes of differentiation of populations that could ultimately lead to speciation. Variation on a geographical scale might indicate the early stages of allopatric or parapatric speciation, whereas ecotypic variation falls within the range of sympatric speciation processes. Examples of geographical variation in *Littorina* include the shell forms designated as geographical varieties; these involve clinal changes on a large scale rather than disjunct distributions, and may be relevant to parapatric speciation. Genetic studies have shown that isolated peripheral populations are divergent. Most of the nonplanktotrophic *Littorina* species (subgenus *Neritrema*) are known to show ecotypic variations in shell form, for which a genetic basis has sometimes been demonstrated. These have sometimes been used to estimate gene flow and natural selection in adjacent microhabitats, from which genetic models of speciation have been constructed. Other cases of ecotypic variation suggest how reproductive isolation may arise in sympatry. The methods of investigation described so far require inference about past or future speciation events, but a final and more direct source of evidence about actual speciations may sometimes be obtained from the fossil record. Although the fossil record is sparse in *Littorina*, in one case it does provide corroborating evidence.

Vicariant speciation. The analysis of distributions of sister-taxa (Table 12) has identified three possible cases of vicariant speciation. Of these, palaeontological evidence is available only for the case of *L. squalida* and *L. littorea* (pp. 89, 118). In the northern Pacific *L. squalida* has a fossil record extending back to the Middle Miocene. The opening of the Bering Strait during the Upper Pliocene (3.5 to 4 Ma) permitted the expansion of its range into the Arctic and northern Atlantic Oceans, as confirmed by fossils from northern Alaska (2.4–3 Ma) and Iceland (3–3.5 Ma). It is suggested that climatic cooling towards the end of the Pliocene, perhaps assisted by periodic closure of the Strait, brought about vicariance of this contiguous distribution, splitting it into two isolated parts in the northern Pacific and northern Atlantic. The earliest fossils of *L. littorea* in the Atlantic are from the southern North Sea (2.4 Ma), and these rare specimens show some resemblance to the Pacific sister-species, thus hinting at the anagenetic transformation of the Atlantic isolate into the modern *L. littorea*. However, estimates of the age of the divergence between these two species, based on mitochondrial DNA sequences, range from 3.28 to 10.07 Ma (mean 6.42 Ma), thus predating the geological estimates of the date of opening of the Bering Strait (Reid, Rumbak & Thomas, 1996). This throws some doubt on the proposal of vicariant speciation outlined above and suggests that the divergence could have occurred in the Pacific, before the opening of the Strait (p. 365).

A second lineage of *Littorina*, the ancestor of the monophyletic radiation of the Atlantic *Neritrema* species (clade 8, see Fig. 119), was also involved in this trans-Arctic migration and subsequent vicariance during the Upper Pliocene. In this case the molecular estimates of the age of the vicariant event, 1.95 to 6.36 Ma (mean 4.12), correspond more closely to the geological age of the opening of the Bering Strait, supporting the vicariant hypothesis. A third possible case involves *L. aleutica* and its probable sister-species *L. natica* but, since the separation is small and involves habitat specialization, this will be discussed as a case of possible parapatric speciation (p. 380).

If, as seems likely, ranges of *Littorina* species are often limited by environmental temperature (p. 369), climatic change may be a common cause of distributional shifts and perhaps of vicariance. As mentioned above, climatic cooling probably contributed to vicariance in the two cases of trans-Arctic migration. Climate-induced vicariance is also likely to have occurred within the Pacific. For example,

Reid (1990c) proposed that the ancestor of all members of the subgenera *Littorina* and *Neritrema* (i.e. the ancestor of clade 4, see Fig. 119) was widely distributed across the northern Pacific, but that the general decline in temperature during the Neogene caused a southward range shift and extinction in the far north, thus splitting the ancestral range into eastern (clade 16) and western (clade 5) parts. (If this did take place, some dispersal has occurred since then, because some members of clade 5 now occur in the northeastern Pacific). Climatic change has been specifically invoked in one model of speciation in marine invertebrates (Valentine & Jablonski, 1983; Valentine, 1984). As an example, consider a temperature-limited species inhabiting a north–south coastline in the northern hemisphere. Gradual climatic cooling will bring about a southward shift in the range. Since temperature gradients are rarely uniform, vicariance might occur, leaving a relictual population in a warm refuge north of the main receding range. Depending upon the relative sizes of the populations, this model could include cases of peripheral isolates speciation and, if adaptation to a lower temperature by the northern population occurred while still in contact with the southern parent population, could also include parapatric speciation. From an earlier phylogenetic analysis of morphological characters of *Littorina* species (Reid, 1990b, c), it was suggested that the species in the western Pacific showed a close correspondence between their phylogenetic position and geographical distribution, with successively more derived species occupying more northerly, but overlapping, ranges. This was explained by successive speciation events towards the north, as just described, during a period of prolonged climatic cooling. However, the new phylogenetic analyses (Chapter 5) have not supported this detail of the earlier topology, so that this suggestion is no longer tenable in its original form. Nevertheless, some connection between climatic change and speciation is still likely. Both the morphological (Fig. 114) and DNA analyses (Fig. 120) show a major polychotomy in the centre of the phylogeny and, although there are slight differences between the groupings (polychotomy at node 5 or 4 respectively), their lack of resolution in the same area suggests that the polychotomy may represent a real episode of sudden and rapid cladogenesis. The age of this polychotomy may be estimated from the oldest fossil occurrence of *L. squalida* (12–15 Ma) and from estimates of rates of mitochondrial DNA evolution (5.74–12.39 Ma, mean 9.60 Ma). This corresponds with a period of particularly rapid climatic change during the Middle Miocene (p. 385). At this time range shifts must have been accelerated, providing the opportunity for vicariance and consequent speciation. This episode of rapid cladogenesis can also be viewed as an adaptive radiation following the invasion of a new adaptive zone, including adaptation to cooler climatic conditions (p. 394).

Peripheral isolates speciation. There is a continuum between allopatric speciation by vicariance and by peripheral isolation. In its typical case the latter involves an isolate sufficiently small (from one fertilized female to a single deme) that small population size is a factor in the rapid differentiation of the isolate which, by processes including inbreeding, drift, genomic reorganization and selection, will become apomorphic relative to the parental species. An alternative explanation of the process (the centrifugal speciation model) is that genetic novelties are more likely to arise and become fixed in the large parental population, presumed to occupy the optimal habitat, leaving a relatively plesiomorphic peripheral isolate (Frey, 1993). Since effective population sizes are seldom known, Lynch (1989) selected an arbitrary ratio of distributional areas between sister-species of more than 20:1 as the criterion for indicating a possible case of peripheral isolates speciation. In *Littorina* only a single case is indicated (Table 12); *L. horikawai* is restricted to a small area off Kyushu, about 1200 km from the closest part of the very large northern Pacific distribution of its sister-species, *L. sitkana*. The cause of the isolation event in this case is not known. The classic cause of peripheral isolation is 'waif dispersal' of a small number of founding individuals, but this is unlikely over such a distance in a nonplanktotrophic species, without colonization of the intervening coastlines. No obvious physical vicariant event would appear to account for it, but the presence of the supposed isolate at the southern extremity of the range of *L. sitkana* does suggest a range retraction caused by climatic change (as could be envisaged, for example, during a short warming trend under the generally deteriorating conditions of the Upper Pliocene). An alternative view is that parapatric speciation (p. 380) was involved before the range retraction of *L. sitkana* occurred, because there is a habitat difference between these sister-species.

Evidence from intraspecific variation supports the classic view that peripheral populations are distinctive. The best examples are the South African and Mediterranean populations of *L. saxatilis*, geographically and genetically distant from the contiguous northern Atlantic population of this species; the histories of these populations are still unknown, but they may well have been founded by a small number of individuals transported over the large intervening distances (p. 330). There are no clear examples of peripheral populations with sufficiently distinctive shell types that they could be regarded as geographical varieties. Nevertheless, some peripheral populations show morphological peculiarities, for example the oviduct of *L. compressa* at the northern extremity of its range (p. 251) and the penis of *L. kasatka* in the south of its distribution (p. 121); in these cases distributions are not sufficiently well known to discern whether peripheral isolation or clinal change is involved.

Two factors probably militate against the importance of peripheral isolates speciation in *Littorina*. Firstly, the species are relatively widely dispersed, reducing the potential for isolation from the effects of gene flow. This is more obviously the case for the planktotrophic species, but also seems to be true for the nonplanktotrophic ones (see Dispersal and Colonization, p. 370). Secondly, the majority of species occupy continental coastlines. Especially in the northern Pacific, these provide more or less continuous stretches of habitat, along which isolation by waif dispersal is unlikely. Even in the northern Atlantic, where gaps between continental land masses are larger, they are apparently insufficient to prevent wide distributions of several *Neritrema* species. One famously disjunct distribution in the Atlantic is that of the planktotrophic *L. littorea*; the American population is certainly a post-glacial arrival, but whether by natural dispersal or human introduction is unclear (p. 116). Waif dispersal to isolated islands is considered a classic type of peripheral isolates speciation, but such opportunities are limited in *Littorina*. Species with distributions on oceanic islands include *L. striata* in the Atlantic, but here even the 1400 km separating the Cape Verde group from the Canary Islands seems insufficient to have resulted in any obvious degree of morphological divergence in this planktotrophic species; genetic comparisons have yet to be made. In the northeastern Pacific, *L. keenae* and perhaps *L. scutulata* occur in oceanic-type habitats on the mainland coast and on nearby islands, but there are no suitably distant islands in this part of the ocean on which isolation might occur. In the northern and northwestern Pacific the nonplanktotrophic species *L. aleutica* and *L. horikawai* both occupy island habitats, but here distances between the islands are small, and isolation consequently unlikely.

Parapatric speciation. Models of parapatric speciation require differentiation of two contiguous populations, between which interbreeding occurs in a narrow hybrid zone (Endler, 1977; Barton & Hewitt, 1985). Speciation ensues if hybrids are selected against, leading to reinforcement of isolating or recognition mechanisms. The divergence may be driven by adaptation to different, adjacent habitats, for example, and will be promoted by low dispersal. Hybrid zones are visible as stepped clines, but the reality of parapatric speciation is still controversial, because clines can also be interpreted as cases of primary allopatric divergence followed by secondary contact. Direct evidence for the occurrence of this process is difficult to obtain. Several *Littorina* species show clinal variation in shell form on a geographical scale (e.g. *L. plena*, *L. brevicula*, *L. sitkana*), but striking stepped clines are known only in the five species of *Neritrema* from the northern Atlantic. In these five cases there is convincing evidence that the thick-walled, low-spired shells of the southern varieties are more resistant to high levels of predation by a shore crab, which is absent in the north. Although direct evidence is lacking, the morphological differences may have a genetic basis. Nevertheless, northern and southern varieties are genetically similar as measured by allozyme frequencies (K. & B. Johannesson, 1990; Lewis & Williams, 1995), and there seems no evidence here of incipient speciation. Among the known pairs of sister-taxa there are none with shell features suggesting adaptation to greater and lesser degrees of crushing predation.

The simple comparison of geographical ranges of sister-taxa (Table 12) cannot distinguish between parapatric and allopatric modes of speciation. Nevertheless, it is worth pointing out that in one case there is only a small geographical separation, combined with a difference in habitat. This pair is *L. aleutica* (found only on islands and a peninsula of the Siberian mainland) and its sister-species *L. natica* (this accepts the morphological topology, p. 187; the same discussion applies if the DNA topology is accepted for, like *L. natica*, the Atlantic *Neritrema* species are all typical of continental shores). The two distributions are separated by a sea gap of 250 km between Kamchatka and Bering Island, and on the northern coast of the Bering Sea by 350 km, although in the latter case no collections are available from the intervening coast. These distances are small compared with the known dispersal distances of *Neritrema* species, and the two are not isolated by any obvious barrier. The possibility remains that they may have differentiated and adapted to their respective oceanic and continental habitats while in contact, although it is unlikely that this can ever be clearly shown.

Sympatric speciation. Where the spatial segregation imposed by adaptation to different habitats is on a finer scale, the resulting genetic isolation and divergence is described as sympatric speciation. This too is a controversial concept, but there is now good evidence for such a process in animals that show marked preferences for microhabitats, food plants or hosts, resulting in habitat-specific assortative mating (reviews by Bush, 1993, 1994). The classical view is that sympatric speciation does not occur, so that sympatry between sister-taxa must be the result of post-speciation dispersal. If this is so, a correlation between degree of range overlap and time since divergence would be expected. However, this is not the case in a variety of vertebrate groups (Lynch, 1989), nor in the small sample of sister-taxa of *Littorina* (Table 12). It can therefore be argued that substantial sympatry is not necessarily due to dispersal, and sympatric sister-taxa must at least be considered as possible candidates for sympatric speciation.

The intraspecific variation of some *Neritrema* species has provided insights into the conditions under which such a process might occur. The differences in shell form between wave-exposed and moderate ecotypes are well known, and there is at least some evidence for their genetic basis (e.g. Newkirk & Doyle, 1973; Janson, 1982b; Behrens Yamada, 1987; Boulding & Hay, 1993; B. & K. Johannesson, 1996). This situation was modelled by Boulding (1990), who assumed a polygenic trait with an environmental component. Field measurements of selection and migration rates were made using *L. sitkana* from a sheltered shore and *L. subrotundata* from an exposed one, and used to demonstrate that selection (probably by waves and crab predation) was sufficiently strong to account for the maintenance of wave-exposed and moderate ecotypes as a stable genetic polymorphism, at the low levels of gene flow shown by these nonplanktotrophic animals. This did not directly address the possibility of speciation, although Boulding did suggest that it may have been the first step in the separation of these two species. In this particular case this is unlikely; there is as yet no firm evidence that these two are sister-taxa, and more importantly *L. subrotundata* is not restricted to exposed shores (p. 181). Nevertheless, the model is applicable to the maintenance of ecotypic differentiation along microenvironmental gradients within *Neritrema* species.

A model explicitly concerned with the possibility of speciation was developed by Johannesson & Sundberg (1992). This was based on three subpopulations over a one-dimensional environmental cline and incorporated two genetic loci, one determining the phenotype, the other for assortative mating according to phenotype. This modelled the clinal ecotypic variation of *L. saxatilis* over a gradient of wave-exposure (which is a coarse-grained division of the habitat relative to the mobility of the animals) and investigated the conditions of selection and gene flow under which either stable polymorphism or speciation were predicted. Their conclusion was that under the rates of selection and migration observed in the field, polymorphism would be stable, but would not proceed to speciation (interruption of gene flow between homozygotes) even with the establishment of assortative mating between extreme phenotypes. This was the case as long as intermediate phenotypes (heterozygotes) were of superior fitness in the intermediate habitat, at least at low migration rates. However, their finding that high migration rates would *increase* the likelihood of speciation seems intuitively unlikely and is contradicted by some other models of sympatric speciation; this may suggest that some assumptions of the model are unrealistic. The point that selection against intermediate phenotypes in intermediate habitats promotes sympatric speciation is nevertheless important. Field observations suggest that in a few cases such selection may achieve almost complete microgeographic isolation between ecotypes, a striking example being the sheltered/brackish and wave-exposed ecotypes of *L. subrotundata* at the southeastern extremity of its range, where crab predation in intermediate habitats perhaps maintains the separation (p. 178).

Until recently, the possibility of assortative mating between ecotypes had been hypothetical. Two ecotypes of *L. saxatilis* on wave-exposed shores in Galicia are found respectively in shore-zones of barnacles or mussels, and intermediates are rare in the narrow belt of contact between the two zones (Johannesson *et al.*, 1993). Like other cases of ecotypic differentiation in this species, this appears to be a primary cline produced by disruptive selection, but on an unusually small spatial scale (fine-grained relative to the mobility of the animals). Contributing to the maintenance of this steeply stepped cline is a degree of microhabitat choice. Crucially, there is also assortative mating in the zone of overlap (Johannesson, Rolán-Alvarez & Ekendahl, 1995). One possible interpretation is that assortative mating may have evolved through reinforcement, as a consequence of the reduced fitness of hybrids. Alternatively, a pleiotropic effect of selection on size, in combination with size-assortative mating, could be suggested (Johannesson, Rolán-Alvarez & Ekendahl, 1995). Whatever the explanation, this example shows that incipient reproductive isolation can develop in sympatry. Some other examples from *L. saxatilis* suggest that hybrids between different ecotypes may indeed be of reduced fitness. This was hinted at by Janson (1985a) in her work on the wave-exposed and moderate ecotypes on Swedish shores; although intermediate phenotypes showed higher survivorship in intermediate habitats (Janson, 1983b), there was a higher than expected incidence of abnormalities in embryonic development in the morphologically variable populations in intermediate habitats. Another, more extreme, case of apparently reduced fecundity in probable hybrids has been discovered by Hull, Grahame & Mill (1996), in which female intermediates between high-shore and lower-shore forms of *L. saxatilis* in Yorkshire contained mainly abnormal embryos in the brood pouch. If due to post-zygotic isolation, this should lead to selection for assortative mating. A further example concerns the barnacle ecotype of *L. saxatilis*, which on some shores appears from genetic data to be reproductively isolated from other ecotypes (Hull, Grahame & Mill, pers. comm.). (Whether these forms are appropriately regarded as separate species depends upon the operational species concept that is adopted, and has been discussed earlier (p. 375).

These examples support the possibility of sympatric speciation in nonplanktotrophic *Littorina* species. However, one difficulty for sympatric speciation under these conditions must be mentioned. The intertidal environment is an essentially linear one, along which there is separation of the several habitat types to which the recognizable ecotypes are adapted. Assuming that long-distance dispersal is rare, this means that where two ecotypes are in contact, there will be greater migration (with associated possibilities of gene flow) between them than between similar ecotypes on distant shores. Differentiation of lineages is therefore likely to be a geographically localized event, with attendant likelihood of extinction. If speciation were to proceed in this way, dispersal would have to be invoked to account for the wide geographical distributions of the sympatric sister-species of *Neritrema* that are observed.

The four pairs of completely sympatric sister-taxa of *Littorina* are all members of the nonplanktotrophic subgenus *Neritrema*, from the northern Atlantic, which may have speciated less than 3 million years ago (Table 12). (The branching order among the three members of the *L. saxatilis* group is still unclear (p. 363) but, whichever order is assumed, the two pairs of sister-taxa show virtually complete overlap and are therefore candidates for sympatric speciation). These sister-taxon pairs do show some habitat segregation, for example by wave-exposure in the pair at node 10, tidal level at node 9, and tidal level, wave-exposure and weed substrate at node 11 (Table 12). If sympatric speciation, by microhabitat specialization as described earlier, did indeed produce this radiation, these modern ecological differences might indicate the ancestral microhabitats. Thus, for example, *L. arcana* and *L. saxatilis* might have diverged as wave-exposed and moderate ecotypes respectively of their common ancestor. However, the degree of overlap between the modern habitats of each of the pairs is considerable and perhaps too great to support such a simplistic interpretation. Furthermore, although adaptive differences in shell form between microhabitats are visually obvious, they may not represent the most significant aspects of habitat-related spatial and genetic segregation within species. Ecotypic variation should therefore be seen as a model, but not necessarily a cause, of sympatric speciation in *Littorina*.

A fifth possible case of sympatric speciation is the pair *L. scutulata* and *L. plena*, with 80% overlap of their distributions (Table 12). In contrast to the *Neritrema* species previously discussed, these are planktotrophic and therefore probably widely dispersed, with high levels of gene flow. This is partly supported by allozyme data, although some surprising examples of inter-population differences have been found (Mastro *et al.*, 1982; Ward, 1990; Correa Sandoval, 1994). Their divergence is unlikely to have occurred by the 'micro-parapatric' process discussed earlier, which requires relatively low levels of gene flow, often in a coarse-grained environment. Sympatric speciation could nevertheless still take place in planktotrophic species, for example if larvae exhibit genetically-determined habitat preferences during settlement, although this is hypothetical in littorinids. The DNA evidence suggests that the divergence between this pair may in fact be an ancient one (mean 12.81 Ma, although an estimate from allozyme divergence is much less, about 2 Ma; p. 365), so that allopatric speciation followed by dispersal may be a more likely explanation of their overlap.

Speciation, extinction and life history

A common theme throughout this discussion of species and speciation processes has been that gene flow is an important cohesion mechanism, and its reduction (or interruption) a precondition for speciation. It has previously been shown (see Dispersal and Colonization, p. 370) that the type of development has a profound influence on gene flow, for individual dispersal is greater in species with pelagic eggs and planktotrophic larvae. Those speciation processes in which geographical isolation is the cause of reduced gene flow, leading to differentiation, are thus more likely to occur in non-planktotrophic species, because isolation is possible over smaller distances. Even the non-allopatric processes of parapatric and sympatric speciation will mostly be enhanced by low gene flow. Speciation in marine invertebrates with high dispersal, large population sizes and extensive ranges presents a challenge to the classic models of allopatric speciation. Where allopatric subdivision of species does occur, it must often be on a large geographical scale, and the resulting isolates are expected to be of large size, rather than small peripheral populations (Palumbi, 1994). The sample of speciation events in *Littorina* is a small one, but does illustrate these points. The vicariant event following trans-Arctic migration was on a sufficiently large scale to affect both planktotrophic and nonplanktotrophic lineages (nodes 7 and 15, see Figs 119, 121 and Table 12), whereas the possible cases of peripheral isolates speciation and of parapatric speciation are found among nonplanktotrophic taxa (nodes 12 and 13). Four possible cases of sympatric speciation have been identified among nonplanktotrophic forms (nodes 8, 9, 10, 11), although the one possible planktotrophic case (node 16) suggests that low gene flow may not be essential for this process.

It has therefore often been suggested that rates of speciation in marine invertebrates will be greater in nonplanktotrophic than planktotrophic lineages (review by Jablonski & Lutz, 1983). However, the diversity of species through time is the product not only of their rate of origin, but also of their rate of removal by extinction. The widespread dispersal of planktotrophic groups may reduce their rate of speciation, but this may be partly offset by a reduced likelihood of extinction. In several gastropod groups planktotrophic species show significantly longer durations in the fossil record (Hansen, 1980; Jablonski, 1986; Gili & Martinell, 1994). The important consideration here is believed to be the extent of the geographical range, rather than dispersal *per se*, with more localized species being more susceptible to extinction. As discussed earlier (p. 373), a wider geographical range is correlated with planktotrophic development in some gastropod groups, although not in *Littorina*. Despite this possible influence of extinction, radiation rates do appear to be greater in nonplanktotrophic clades, leading to a process of 'species selection' (Jablonski, 1986; Lieberman, Allmon & Eldredge, 1993). In *Littorina* this is clearly shown by comparing the radiations of the two lineages that invaded the northern Atlantic following trans-Arctic migration. Both lineages presumably reached the Atlantic at about the same time, between the opening of the Bering Strait 3.5 to 4 Ma and the onset of widespread Arctic glaciation about 2.4 Ma. The planktotrophic isolate in the Atlantic is now represented by only the single species *L. littorea*, whereas during the same interval the nonplanktotrophic (*Neritrema*) lineage (clade 8) has radiated to produce five living species. Similarly, since the unresolved proliferation of clades in the Middle to Upper Miocene (clade 5), five living (and one extinct) planktotrophic species have sprung from the polychotomy, whereas ten living *Neritrema* species have appeared. This radiation of *Neritrema* species could be described as non-adaptive (Skelton, 1993), for it does not appear to have been initiated primarily by a new ecological opportunity or adaptive innovation (p. 394).

In addition to species selection, another mechanism contributes to the increasing ratio of nonplanktotrophic to planktotrophic species through time. As a result of developmental constraints, it appears that once the feeding structures of a planktotrophic larval stage have been lost, they are difficult to reacquire (Strathmann, 1978, 1985), and embryological processes may also restrict the possibility of developmental change in nonplanktotrophic species (Lieberman *et al.*, 1993). As a result, while planktotrophic species can give rise to nonplanktotrophic daughter-species, the reverse is very rarely the case (but see Reid, 1989a, for two possible examples in Littorinidae). In *Littorina* it is possible that nonplanktotrophic development has arisen only once (although the evidence for the monophyly of *Neritrema* is weak, p. 361), and no reversions are indicated.

Historical biogeography

With a sound knowledge of the phylogeny of *Littorina*, its fossil record and the modern geographical ranges of its constituent species, and against the background of likely means of dispersal, extent of gene flow and possible speciation mechanisms, it is now possible to attempt a synthesis of the historical biogeography of the group. Of the methods of biogeographic analysis that are based on cladistic phylogenetic analysis, there are two extreme schools. That of 'vicariance biogeography' seeks congruent patterns among the area cladograms derived from a number of different taxonomic groups, attempting to explain these in terms of common biogeographic history (e.g. Humphries & Parenti, 1986). Vicariant events are seen as the only processes sufficiently general to cause congruent patterns and so are taken as first-order explanations for them, whereas dispersal (range shift) is viewed as an untestable and often unnecessary assumption. In contrast, the method of 'phylogenetic biogeography' is applied to individual monophyletic groups and is based on the assumption of range expansion from the centre of origin, by a process of peripheral isolates speciation, the so-called 'progression rule' (e.g. Brundin, 1988). This leads to the expectation that the occurrence of primitive taxa marks the centre of origin of the group, whereas derived taxa occur further away. In practice, it is often impossible to choose between competing vicariant and dispersalist hypotheses that attempt to account for observed patterns of distribution, and resolution would require a complete fossil record. Reid (1990c) showed how the two approaches gave different, but complementary, interpretations of the historical biogeography of *Littorina*, and argued that both should be used. Vicariant events may explain congruent, large-scale patterns among a variety of taxonomic groups, whereas detailed consideration of the fossil record and of means of dispersal and speciation illuminate the history of an individual group.

An area cladogram is produced by superimposing areas of endemicity upon the phylogeny, as shown in Figure 121. Because of overlap of ranges, areas of endemicity are difficult to define precisely, but five broad areas have been recognized. These are the eastern Atlantic (the island groups of the Azores, Madeira, Canaries and Cape Verdes), the northeastern Pacific (Baja California to Oregon), the northern Pacific (Okhotsk Sea, Bering Sea and Alaska to Washington), the northwestern Pacific (China Sea and Japan Sea) and the northern Atlantic (temperate and subarctic regions). In the area cladogram the eastern Atlantic and then northeastern Pacific are basal, and there are two examples of sister-group relationships between the northern Atlantic and northern Pacific. This differs from the area cladogram of Reid (1990c) in the omission of Southeast Asia and in the sister-group relationship between the northeastern and northern Pacific. The only comparable phylogeny so far available for a marine mollusc is based on the DNA sequence of the mitochondrial cytochrome-*b* gene of seven species of the intertidal gastropod *Nucella* (Collins, Frazer, Palmer, Vermeij & Brown, in press). That radiation is believed to have taken place since the earliest Miocene, and there are some similarities with that of *Littorina*. The basal species of *Nucella* occurs in the northern and northwestern Pacific, and the single northern Atlantic species is the sister-taxon of a northern Pacific form; the more derived species are from the northern Pacific, but extend into the northeastern Pacific area. There is no extant basal species of *Nucella* in the northeastern Pacific, but interestingly the earliest fossil record is from the Lower Miocene of California and then from the Middle Miocene of Hokkaido and Kamchatka (Amano, Vermeij & Narita, 1993). Although no other rigorous phylogenetic analyses are available for molluscs in this region, several comparative systematic accounts suggest likely sister-group relationships between taxa in the northern Pacific and northern Atlantic. These include the gastropod *Neptunea* (Strauch, 1972; Nelson, 1978), and the bivalves *Mya* (MacNeil, 1965; Strauch, 1972) and *Chlamys* (Waller, 1991).

The time and area of origin of the genus *Littorina* will probably never be known with certainty, for the shells of its living basal members are so similar to those of the sister-genus *Nodilittorina* that fossils of suitable age cannot be assigned unequivocally. The earliest possible *Littorina* species in the fossil record is the unnamed species from the Palaeocene or Lower Eocene of California (Woods & Saul, 1986; p. 340). This is followed by another doubtful species, provisionally identified as *L. mountsoloensis*, from the Middle to Upper Eocene of Washington (p. 342). The ages of these doubtful fossils may be compared with the molecular estimates of the age of *Littorina*, of 43 or 67 Ma (Middle Eocene or latest Cretaceous; Reid, Rumbak & Thomas, 1996). However, strict application of the progression rule to the area cladogram would identify the eastern Atlantic as the area of origin of *Littorina*, where its least derived living species, *L. striata*, is found. This area is close to the Palaeogene Tethys Sea, where a number of Cenozoic molluscan radiations may have been initiated. Some of the Atlantic islands were then already in existence, for example the Canary Island group may be of Cretaceous age (Rothe, 1968). From there, westward dispersal to the eastern Pacific could have occurred at some time during the Palaeocene or Eocene, assisted by the westward circum-global equatorial current which existed at that time (Berggren & Hollister, 1977). Although the Atlantic Ocean may have been only about 75% of its present width in the early Palaeogene (Adams, 1981), this distance is still large relative to the known dispersal capabilities of planktotrophic *Littorina* species.

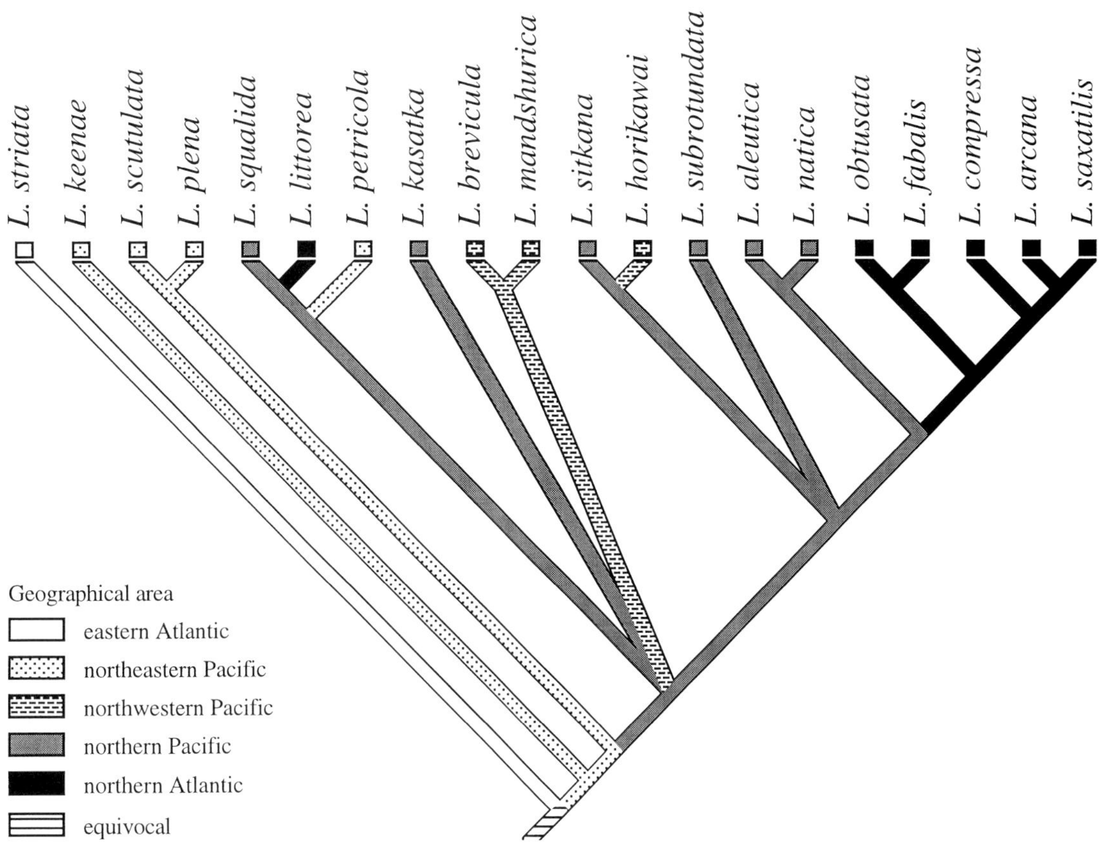

Figure 121. Area cladogram for *Littorina* species, obtained by superimposing areas of endemicity upon the phylogeny of Figure 119. All taxa are Recent except for the Pliocene *L. petricola*. See text for definitions of areas of endemicity. Note that the parsimonious reconstruction of ancestral areas as shown here assumes a dispersalist interpretation of distribution patterns.

However, surface currents across the Atlantic at that time are believed to have been stronger than at present, and Scheltema (1995) has estimated that during the Palaeocene to Eocene only two to four weeks were required for the passive transport of larvae from east to west. This is well within the likely duration of pelagic life in planktotrophic *Littorina* species (p. 371). Furthermore, since the earliest *Littorina* species were, like *L. striata* and *L. keenae*, presumably oceanic in habitat, they may more easily have been dispersed by directional oceanic current flows. As Reid (1990c) pointed out, some modern planktotrophic *Nodilittorina* and *Littoraria* species show amphi-Atlantic distributions, so trans-Atlantic dispersal is possible even at the present time. This episode in the biogeography of *Littorina* is, however, purely speculative. An alternative hypothesis could be proposed from a literal interpretation of the fossil record. Since no fossils resembling *L. striata* are known from the European Cenozoic, and the oldest *Littorina* fossils occur in western North America, the genus may have arisen in the northeastern Pacific. In that case, *L. striata* might have reached the eastern Atlantic by eastward dispersal through the open Panama Strait.

The palaeoenvironmental history of the northern Pacific during the Cenozoic is now well known, and the evolution of *Littorina* there can be considered against the background of climatic oscillation, eustatic sea level change and geological events. Following the warmest period during the Lower Eocene, there was a general trend of climatic cooling throughout the Cenozoic, but with periodic reversals (Flower & Kennett, 1994). Evidence from oxygen isotope analyses (Savin *et al.*, 1981), microfossils (McGowran, 1986; Barron & Baldauf, 1990) and fossil molluscs (Addicott, 1969; Itoigawa & Yamanoi, 1990) indicates that the climatic optimum of the Neogene occurred during the early Middle Miocene (15–16 Ma) and was followed by further cooling, interrupted by short warm intervals (10–11 Ma and 6–7 Ma). There were corresponding fluctuations in sea level, with high but unstable levels in the early Middle Miocene and lower levels during cool intervals (Flower & Kennett, 1994). These climatic events had pronounced effects on the distribution, extinction and speciation of marine fauna. During climatic optima warm-water molluscan assemblages migrated 10–15° to the north, a distance of more than 2000 km (Gladenkov, 1994). The fossil records of some warm-water groups show episodes of speciation correlated with climatic optima (e.g. *Umbonium*, Ozawa & Okamoto, 1993), whereas boreal groups originated or diversified during periods of climatic deterioration (e.g. Buccinoidea and Turritelloidea, Titova, 1994).

The oldest undoubted fossil member of *Littorina* is *L. sookensis* from the Upper Oligocene (23.5–30 Ma) of Vancouver Island (p. 51). Fossils of the genus next appear in the Middle Miocene as *L. squalida* (as early as 12–15 Ma) and *L. brevicula*, both in Kamchatka, and as *L. iwakiana* in Japan. Several other shallow-water gastropod genera (e.g. *Nucella*, *Lirabuccinum*) apparently have a similar history, with older fossils in the northeastern than in the northwestern Pacific, and literal interpretation of this fossil evidence suggests that they dispersed westward across the North Pacific before the early Middle Miocene (Amano *et al.*, 1993). Such long-distance dispersal is perhaps unlikely, and in the case of *Littorina* an alternative vicariant hypothesis has been proposed (Reid, 1990c). During the early Middle Miocene the relatively warm climate may have permitted an ancestral *Littorina* species (the ancestor of clade 4, Fig. 119) to extend its range from western North America around the coastline of the North Pacific. Morphological (and probably physiological) adaptations to the cooler northern climate, and a major ecological shift, also occurred in the ancestor of clade 4 (see Historical Ecology, p. 388, and Adaptation, p. 394), accompanying this northward expansion. Even at this time of climatic optimum, marine glaciation was beginning in the Gulf of Alaska (Marincovich, 1990) and this, together with the widespread climatic cooling following the optimum, may have caused *Littorina* to become extinct in the northernmost Pacific, dividing its distribution into eastern (clade 16) and western (clade 5) parts.

A major event in the evolution of *Littorina* was the episode of rapid speciation during the Middle Miocene, revealed as a polychotomy in phylogenetic hypotheses (Figs 115, 120). The oldest fossils of clade 5 (*L. squalida*) suggest a minimum age of 12–15 Ma for this event, whereas molecular estimates are somewhat younger, 5.74–12.39 Ma (mean 9.60 Ma) (p. 365). These estimates cover a rather wide range, but correspond broadly to the period of climatic cooling between 15 and 10 Ma. According to the area cladogram, this episode of speciation took place in the northern Pacific. Fossil evidence, from the early records of *L. squalida* and *L. brevicula* in Kamchatka, suggests a more precise localization in the western part of the region. This is also supported by the concentration of the modern descendants of this radiation in the western half of the Pacific. Furthermore, during this period molluscan faunas were similar from Kamchatka to Hokkaido, belonging to a single zoogeographical province with a mild-temperate to cool-temperate climate (Chinzei, 1986; Ogasawara, 1994). It is suggested that this remarkable episode of rapid speciation may have been promoted by the rapid climatic cooling and fluctuation during the Middle Miocene. Under the influence of changing sea levels at this time, the region between the Japanese archipelago and the Kamchatka peninsula displayed a complex and varying topography of islands, seaways and currents (see palaeogeographic reconstructions by Chinzei, 1986; Ogasawara, 1994). This may have provided suitable conditions for climatic vicariance and isolation of populations of these planktotrophic *Littorina* species.

The time of origin of nonplanktotrophic development in *Littorina* is not known from the fossil record. This type of development is interpreted as an adaptive response to a cool climate (p. 401). The molecular analysis does not distinguish the time of origin of the nonplanktotrophic subgenus *Neritrema* (clade 6) from the Middle Miocene speciation episode. It cannot yet be determined whether this clade arose during the same cooling climatic period, or perhaps the subsequent one. The area cladogram does imply an origin in the northern Pacific. Speciation within this lineage was promoted by the reduced gene flow and increased opportunities for isolation consequent upon the loss of a pelagic larval phase. Nevertheless, dispersal was not precluded, and at least the two species, *L. sitkana* and *L. subrotundata*, eventually achieved a wide geographical range, dispersing to the northeastern Pacific. One planktotrophic lineage apparently also dispersed to the east, for during the Pliocene *L. petricola*, closely related to *L. squalida*, occurred in California and Oregon. During cool periods dispersal of boreal groups from their origins in the western part of the northern Pacific to the northeastern Pacific, also occurred in other molluscan groups (e.g. Titova, 1994).

There are two examples of sister-group relationships between northern Pacific and northern Atlantic taxa in *Littorina*: the pair *L. squalida* and *L. littorea*, and clades 12 and 8 (Figs 119 and 121). This, together with the parallel occurrences in unrelated groups mentioned earlier (p. 384) suggests a large-scale vicariant explanation. This is provided by the trans-Arctic interchange of marine organisms, a major biogeographic phenomenon that involved many marine groups (Durham & MacNeil, 1967; Vermeij, 1991). Migration between the Pacific and Atlantic Oceans was made possible following the first opening of the Bering Strait during the Pliocene (3.5–4.0 Ma, Hopkins, 1967a, b; 3.8–4.4 Ma, Basilyan, Barinov, Oreshkina & Trubichin, 1991). The possibility of an earlier opening of the Strait is now discounted (Herman & Hopkins, 1980; Vermeij, 1991). The Arctic climate was then warmer than today (Repenning, Brouwers, Carter, Marincovich & Ager, 1987; Herman & Hopkins, 1980) and was no barrier to dispersal. A total of 295 molluscan species either took part in the trans-Arctic interchange, or are descended from those that did so, and of these 88% are of Pacific ancestry (Vermeij, 1991). The reasons for the asymmetry of the interchange are debated, but may include the greater biodiversity in the northern Pacific than northern Atlantic, impoverishment of the Atlantic by extinctions which created ecological opportunities for invading species, and prevailing current flows favouring Pacific migrants (Vermeij, 1991). In the case of *Littorina*, however, species were present at northern latitudes only on the Pacific side, and the corresponding ecological habitats they encountered in the northern Atlantic were apparently largely empty of native species. The migration route was perhaps more likely to have been eastwards through the Canadian Archipelago and Labrador Sea, rather than around the north coast of Siberia. The latter route is at higher latitude, with large river outflows that may have presented a barrier. In addition, ocean circulation patterns produce an eastward flow from the Bering Strait to the Atlantic (Vermeij, 1991) and may have assisted the dispersal of pelagic eggs and larvae and the rafting of egg masses and post-larval stages. However, the only fossil evidence for this eastward migration route is the occurrence of *L. squalida* on the north coast of Alaska about 2.4–3.0 Ma (Repenning *et al.*, 1987). This species was an early arrival in the Atlantic, as shown by its presence at the top of the *Mactra* zone (3–3.5 Ma) of the Tjörnes Beds in northern Iceland, just before the main influx of Pacific species is recorded in the overlying *Serripes* zone (Gladenkov, 1981). The opening of the Bering Strait simply permitted dispersal and range expansion. Speciation ensued following vicariance, which was probably caused by climatic cooling in the late Pliocene, about 2.4 Ma, when widespread glaciation began (Shackleton *et al.*, 1984; Loubere, 1988). Periodic closure of the Bering Strait with fluctuating sea levels (Hopkins, 1967a; Basilyan *et al.*, 1991) may have contributed to the isolation of the two oceans. As mentioned earlier (p. 378), molecular data suggest that the divergence of *L. littorea* and *L. squalida* may have preceded the trans-Arctic interchange, although the meagre fossil evidence does not support this interpretation. It is notable that both lineages of *Littorina* that took part in the trans-Arctic migration (the ancestor of clade 15 and the ancestor of clade 7) probably had distributions of the continental type (p. 389, Fig. 123B). Successful migrants were presumably those able to tolerate the continental conditions prevailing on the Arctic route. V. Ya. Berger (1973, 1978a, b; Berger & Kovaleva, 1974) suggested that the successful migrants were, like their Atlantic descendants, euryhaline, although he included *L. sitkana* in this category, which is not now believed to have been involved in the migration. Nevertheless, physiological tolerance and habitat range were clearly important attributes of successful migrant species.

Since arriving in the northern Atlantic the planktotrophic *L. littorea* has differentiated from its Pacific ancestor, *L. squalida*, but has not speciated further, whereas the nonplanktotrophic lineage has radiated to produce five modern species in the Atlantic. The Pleistocene glaciations have probably severely affected the biogeographic pattern in the Atlantic. For example, in the northwestern Atlantic glaciation resulted in the extinction of the fauna of rocky shores, for there were no southern refugia available (Ingólfsson, 1992). This may partly explain the higher diversity of species in the northeastern Atlantic. The modern pattern of extensive overlap of ranges of *Neritrema* species in the northeastern Atlantic suggests the possibility of sympatric speciation in this nonplanktotrophic clade, and this is to some extent supported by genetic studies (p. 381). Nevertheless, it is also possible that the extensive modern sympatry has been produced by the compression of geographical ranges during Pleistocene glaciations, followed by subsequent wide dispersal, which may have obscured the original patterns of allopatric speciation.

Two further speciation events in the northwestern Pacific, in the region of Japan, also deserve comment. These are the divergences between *L. brevicula* and *L. mandshurica*, and between *L. horikawai* and *L. sitkana*, of which the average DNA estimates of age are 5.84 and 3.99 Ma respectively. In each case the first-named member of the pair now shows a warm-temperate distribution, which is a derived condition different from the cold-temperate distribution of the common ancestor of each pair (p. 388, Fig. 123A). A possible common explanation for these two events is a mid-Pliocene warm period

(3.5–4 Ma, Gladenkov, 1994). This may have resulted in climatic vicariance of each ancestral range, giving rise to a warm-adapted southern isolate, while the main population retreated northwards. Another relatively recent speciation event in the northern Pacific is that between *L. aleutica* and *L. natica*, dated to about 4 Ma by molecular analysis. The former species is virtually limited to the Aleutian Island chain, and the date is close to the geological estimate of the emergence of these islands at about 3 Ma (Scholl, Buffington & Marlow, 1975).

During the late Pliocene and Pleistocene, both the northern Pacific and northern Atlantic were affected by pronounced climatic fluctuation and cycles of glaciation. None of the known speciation events in the Pacific appear to date from this period. However, during the same time interval there appear to have been at least four speciation events among the *Neritrema* species of the northern Atlantic. The nature of these speciations is still unclear, so that it is not known whether they were influenced by climatic change, and sympatric speciation is an alternative possibility. Using examples from marine ostracods, Cronin (1985) suggested that only sustained, unidirectional, climatic change causes speciation, whereas rapid cycles of glaciation may be too short to allow reproductive isolation to develop.

Historical ecology

Fossils of *Littorina* species are rare and seldom provide unambiguous information about the environments which they inhabited. However, ecological attributes can be treated in the same way as other biological characteristics and probable ancestral states reconstructed using the methods of phylogenetic analysis. Such reconstructions have wide application in macroevolutionary studies. They can be used to consider possible relationships between ecological shifts and phylogenetic diversification, as in the case of radiation following entry into a new adaptive zone. They are also central to the study of the evolutionary responses of organisms to selective pressures in the environment (see Adaptation, p. 394).

In order to reconstruct ancestral ecological characters, those of extant taxa are first coded as discrete states (Tables 13 and 14) and then superimposed on the phylogenetic tree (Fig. 122). Ancestral states of each character can then be reconstructed according to the principle of maximum parsimony (Figs 123, 124), a process known as character optimization (Swofford & Maddison, 1987). Optimization has here been performed using the program MacClade (Maddison & Maddison, 1992), with all characters specified as unordered, and all polychotomies treated as multiple speciation events (i.e. 'hard' polychotomies, as opposed to 'soft' polychotomies, which are uncertainties in resolution). For some characters more than one equally parsimonious reconstruction may be found, and in such cases the ancestral states are equivocal. The successful application of this method depends upon the assumption that lineages change their ecological characteristics slowly, relative to the rate at which lineage branching (cladogenesis) occurs. For many taxonomic groups this does appear to be the case (Baum & Larson, 1991; Brooks & McLennan, 1991), although the generality of the assumption has been questioned (Frumhoff & Reeve, 1994).

Based on the accounts of distributions and habitats of species in the Systematic Descriptions (Chapter 4), eight characteristics have been selected and coded into discrete states (Table 13). Clearly, ecological data do not lend themselves readily to this simplified approach, for species tolerate a range of environmental conditions, and habitats overlap to varying degrees. Furthermore, the characters are not all independent; for example, a species found in the littoral fringe does not have the opportunity to use macrophytic algae as a substrate; and oceanic species do not occur in estuarine conditions. Nevertheless, some useful conclusions can be drawn by considering each character in turn. For ease of reference the nodes of the phylogeny, and the clades they define, have been numbered arbitrarily (Figs 119, 122).

The first ecological character is described as 'climatic regime' and indicates the positions of species on a gradient from the tropics to the subarctic zone. A simple definition by geographical latitude is inappropriate, failing to take account of the influence of regional climatic effects such as oceanic currents. Environmental temperature can be measured in many ways, for example as annual mean, maximum, summer or winter, sea or air temperature, and it is not clear which of these may be most relevant to *Littorina* species. Latitudinal range may be influenced by numerous factors other than climate, including the nature of the shore, hydrographic conditions and biotic factors (see Geographical Ranges and Their Limits, p. 367), and individual species may differ in their sensitivity to each. It is therefore not possible at this stage to define this character precisely. The solution has been to categorize the species with respect to the global marine biogeographic zones defined by Briggs (1974). The distributions of most species fall clearly into one of the three categories, but where overlaps occur (e.g. *L. scutulata*, *L. brevicula*) they have been placed in the zone that includes the greater part of their range. Optimization produces the tree shown in Figure 123A. The ancestral state in *Littorina* (retained by *L. striata* and *L. keenae*) was warm-temperate, but between nodes 3 and 4 there was a shift to the cold-temperate and subarctic region. This typifies all the remaining species, with the exception of *L. brevicula* and *L. horikawai*, which have shifted back towards the ancestral warm-temperate region. It should, however, be noted that *L. brevicula* does also extend northwards into the cold-temperate region. Turning to the sister-genus, *Nodilittorina*, the great majority of the species have tropical distributions, although one of the three outgroup species has a warm-temperate distribution. Further information is required before it can be certain that a shift from tropical to warm-temperate distribution characterized the ancestor of all Recent *Littorina* species, but this is likely, since more distant outgroups are exclusively tropical (Reid, 1989a).

The concept of oceanic and continental distribution patterns has been discussed earlier (p. 369) and, although the basis for the distinction is unclear, it remains a useful one. When the data from Table 11 are superimposed on the phylogeny (Fig. 123B), it is seen that *L. striata* and *L. keenae* retain the ancestral oceanic pattern, as shown by the three outgroup species (however, a few other *Nodilittorina* species do have continental distributions). The ancestor of clade 4 is reconstructed as of continental distribution type (note that the state is equivocal if soft polychotomies are assumed), but within this clade the character appears to have been relatively labile, with no clear phylogenetic pattern, and some examples of sister-species with contrasting characters can be seen.

A related character is that describing tolerance of estuarine conditions. Only species with continental distributions encounter large river estuaries, but not all are able to inhabit such environments. Occurrence in estuaries may depend upon tolerance of turbidity, soft sediment, marsh plants as food or substrate, and of reduced and fluctuating salinity. Of these factors, salinity may perhaps be the most significant, and for a few species there have been detailed ecophysiological studies of salinity tolerance, or at least records of environmental salinity range. These, however, are not usually standardized or readily comparable. In Table 14 species have been coded as tolerant of estuarine conditions if they are regularly found in salt-marsh habitats within river estuaries, or in brackish lagoons. Optimization of this character suggests that estuarine tolerance may have appeared independently at least six times in *Littorina* (Fig. 123C) (or, if soft polychotomies are assumed, as few as twice, with four reversals).

A more easily coded ecological character is that describing the characteristic tidal level occupied by each species. This has been divided into three categories (Table 13) and, despite some variation of zonation pattern with wave exposure, season and geographical region, most species can easily be categorized. There is doubt only about *L. kasatka* and *L. natica*, for which little habitat information is available, but both appear to occupy the middle to upper eulittoral zone. Once again, however, this character undoubtedly reflects a variety of environmental influences, such as physiological tolerance of the physical environment, and the occurrence of predators, food and competitors. Optimization reveals a shift from the littoral fringe to the mid-shore in the ancestor of clade 4, while the high-level habitat of *L. striata* and *L. keenae* is primitive within the genus, shared with most members of the sister-genus (*N. meleagris* is atypical of *Nodilittorina* species in this respect) (Fig. 123D). Within clade 4 a shift to the low shore has occurred in the sister-species pair *L. squalida* and *L. littorea*, in *L. mandshurica* and in *L. fabalis*, whereas only *L. arcana* and *L. saxatilis* appear to have successfully reinvaded the littoral fringe. The majority of *Littorina* species are, like the outgroup, exclusively littoral in habit, any occasional sublittoral records probably representing accidental occurrences. Only the sister-species *L. squalida* and *L. littorea*, and also *L. saxatilis*, regularly inhabit the sublittoral zone (Fig. 124A), and then mainly in the northern parts of their range and in lagoonal habitats.

Two of the ecological characters refer to the substrate. Whereas most *Littorina* species, like the sister-genus *Nodilittorina*, are restricted to rock substrates, a few can sometimes be found also on sand and mud. Optimization suggests that this trait has appeared in as many as four independent lineages, one of which led to the pair of sister-species *L. squalida* and *L. littorea* (Fig. 124B) (but possibly only once, with three reversals, if soft polychotomies are assumed). In this case there is some support from the fossil record, for both *L. squalida* and *L. littorea*, together with the related, extinct, *L. petricola*, are relatively common as fossils and occur in assemblages suggesting sheltered, depositional environments. As a separate character, some species of *Littorina* have also been recorded crawling on macrophytes,

Table 13. Ecological characters of *Littorina* species, coded for phylogenetic analysis.

Character no.	Character	Character state	Description of character state
14	Climatic regime	0	tropical
		1	warm-temperate
		2	cold-temperate and subarctic
15	Oceanic or continental distribution	0	oceanic
		1	intermediate or both
		2	continental
16	Estuarine tolerance	0	intolerant
		1	occurs in estuarine conditions
17	Zonation level on shore	0	upper eulittoral and littoral fringe
		1	middle to upper eulittoral
		2	middle to lower eulittoral
18	Sublittoral occurrence	0	accidental only
		1	regular occurrence in some locations
19	Soft sediments	0	absent on soft sediments
		1	can occur on sand or mud
20	Association with macrophytes	0	not found with macrophytes
		1	sometimes crawl on macrophytes
		2	obligate association with macrophytes
21	Diet	0	epilithic and endolithic microalgae only
		1	macrophytes sometimes included in diet

Table 14. Character state matrix for ecological characters of *Littorina* species. For description of characters and character states see Table 13. The outgroup comprises three species of *Nodilittorina*. Unknown character states are indicated by a query.

	Character number							
Species	14	15	16	17	18	19	20	21
N. meleagris	0	0	0	1	0	0	0	0
N. dilatata	0	0	0	0	0	0	0	0
N. pyramidalis	1	0	0	0	0	0	0	0
L. striata	1	0	0	0	0	0	0	0
L. keenae	1	0	0	0	0	0	0	0
L. scutulata	2	1	0	1	0	0	1	1
L. plena	2	2	1	1	0	0	1	1
L. squalida	2	2	1	2	1	1	1	?
L. littorea	2	2	1	2	1	1	1	1
L. kasatka	2	0	0	1	0	0	0	?
L. brevicula	1	2	0	1	0	0	1	1
L. mandshurica	2	2	0	2	0	1	0	?
L. sitkana	2	1	0	1	0	1	1	1
L. horikawai	1	0	0	1	0	0	0	?
L. subrotundata	2	1	1	1	0	1	1	1
L. aleutica	2	0	0	1	0	0	1	?
L. natica	2	2	1	1	0	0	1	?
L. obtusata	2	2	1	1	0	0	2	1
L. fabalis	2	2	1	2	0	0	2	1
L. compressa	2	2	0	1	0	0	0	1
L. arcana	2	1	0	0	0	0	0	?
L. saxatilis	2	1	1	0	1	0	1	1

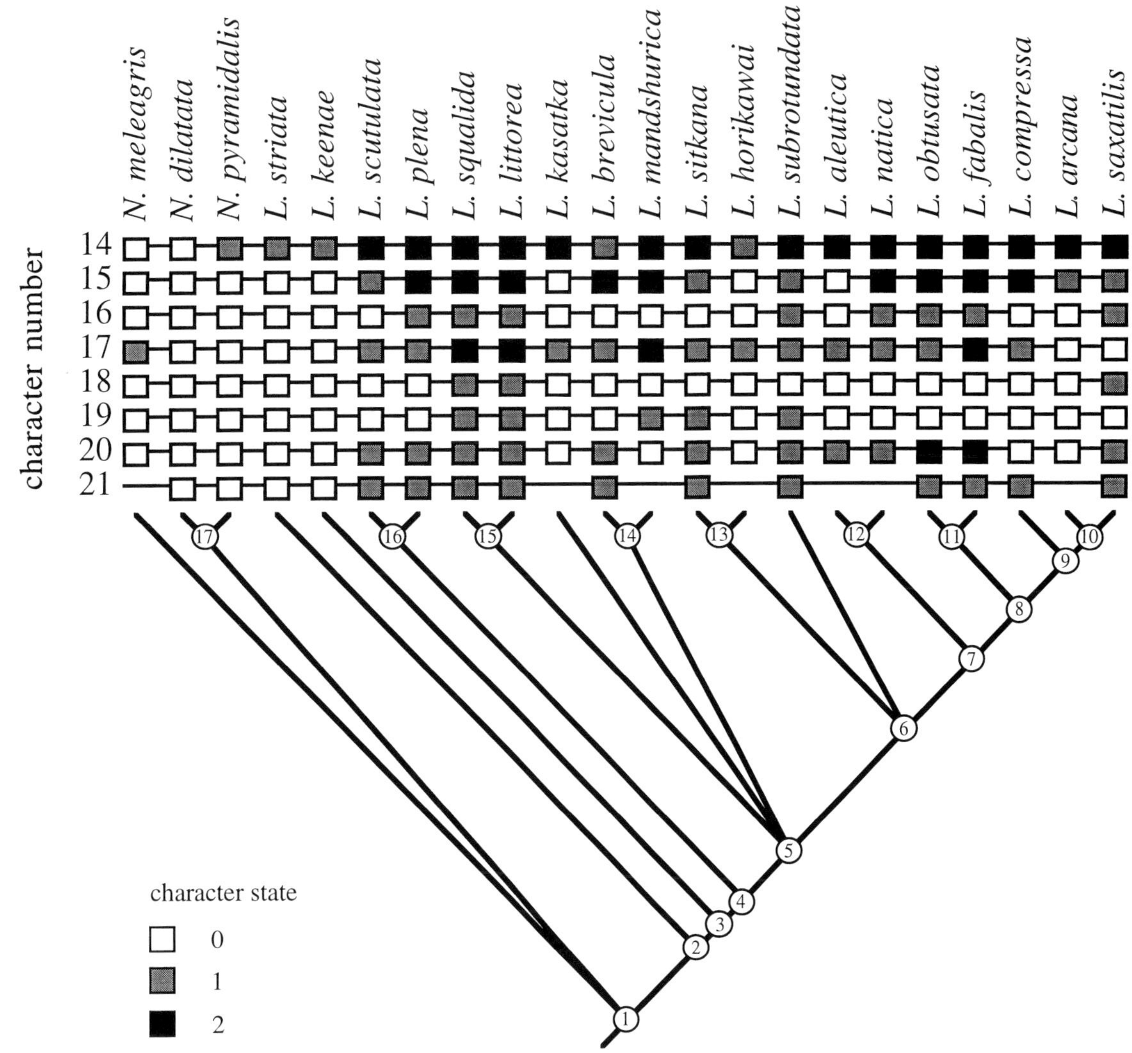

Figure 122. Ecological characters superimposed on the phylogeny of *Littorina* species, with three species of *Nodilittorina* as the outgroup. Character states are listed in Table 13. Nodes (clades) are numbered arbitrarily.

but this has developed into an obligate epiphytic habit only once, in the ancestor of the sister-species *L. obtusata* and *L. fabalis* (Fig. 124C; p. 224).

A character relating to diet has been included. However, this is difficult to code for analysis, since many *Littorina* species are dietary generalists, grazing on a variety of food sources, including microalgae, macroalgae and detritus, as available (Fig. 131). Nevertheless, some are known to show clear dietary preferences, and dietary components are considered in more detail later (p. 405). The diet of *Nodilittorina* species occupying the superficially bare rocks of the littoral fringe is believed to consist largely of epilithic and endolithic microalgae. Since the habitat of *L. striata* is similar, this species is assumed to have a predominantly microalgal diet also. In the case of *L. keenae*, likewise inhabiting the littoral fringe, it is known that microalgal films and black lichens (*Verrucaria*) are the main components of the diet, and that intact macroalgae cannot be consumed. Of the remaining 17 species of *Littorina*, diets have been studied in ten, and in each case macroalgae (i.e. corticated and leathery species) have been recorded as a dietary component. This leads to the prediction that the capacity to graze macroalgae is a character shared by the members of clade 4 (Fig. 124D).

To summarize, some ecological characters (climatic regime, zonation level, sublittoral occurrence, diet) seem relatively conservative, with ecological shifts characterizing large clades. Other ecological characters are more labile, with multiple independent shifts (and consequent uncertainties in the reconstruction of ancestral states). The polychotomy at node 5 is not obviously associated with any of

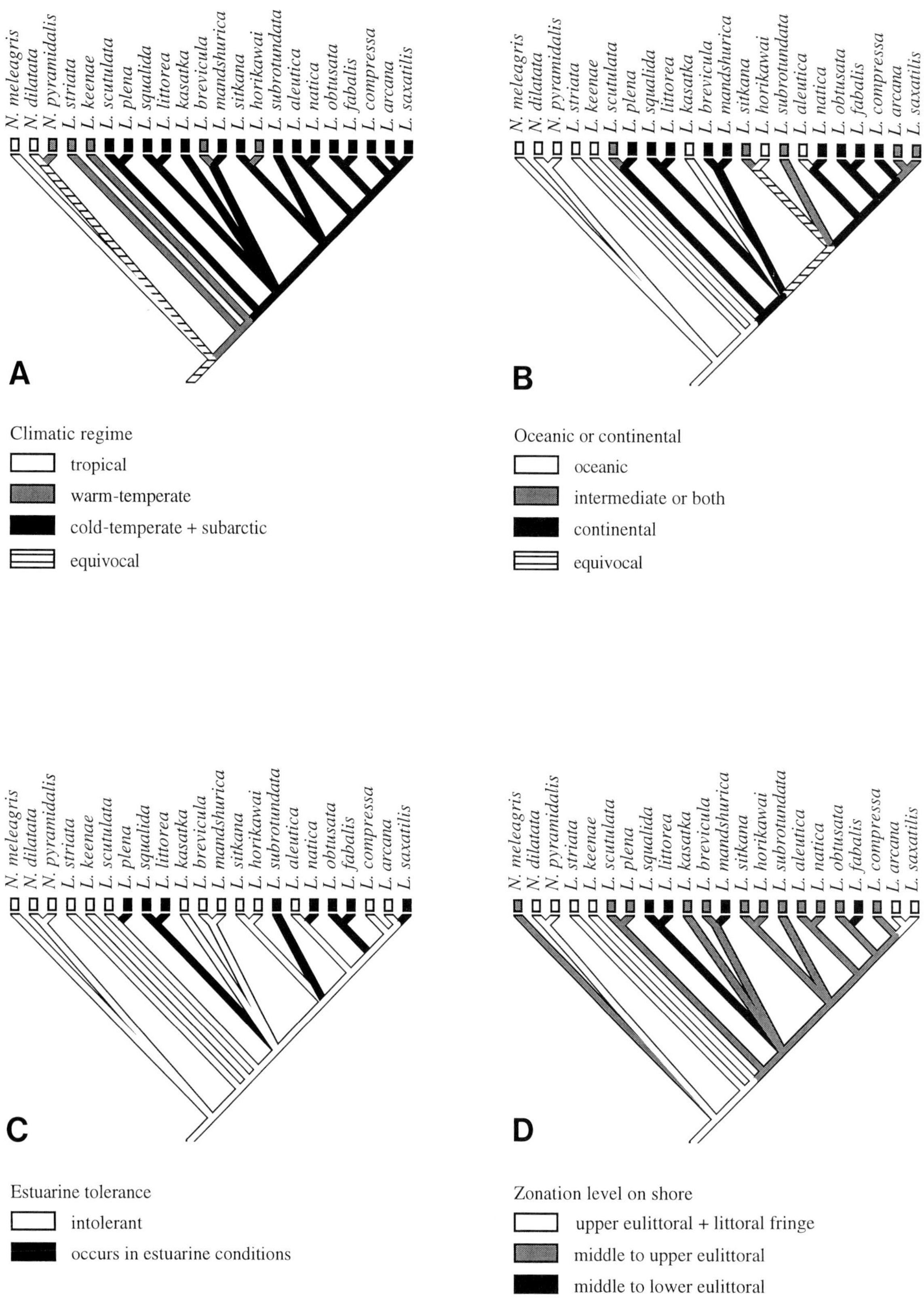

Figure 123. Optimization of ecological characters on the consensus phylogeny of *Littorina* species, for the reconstruction of the ecology of ancestral lineages.

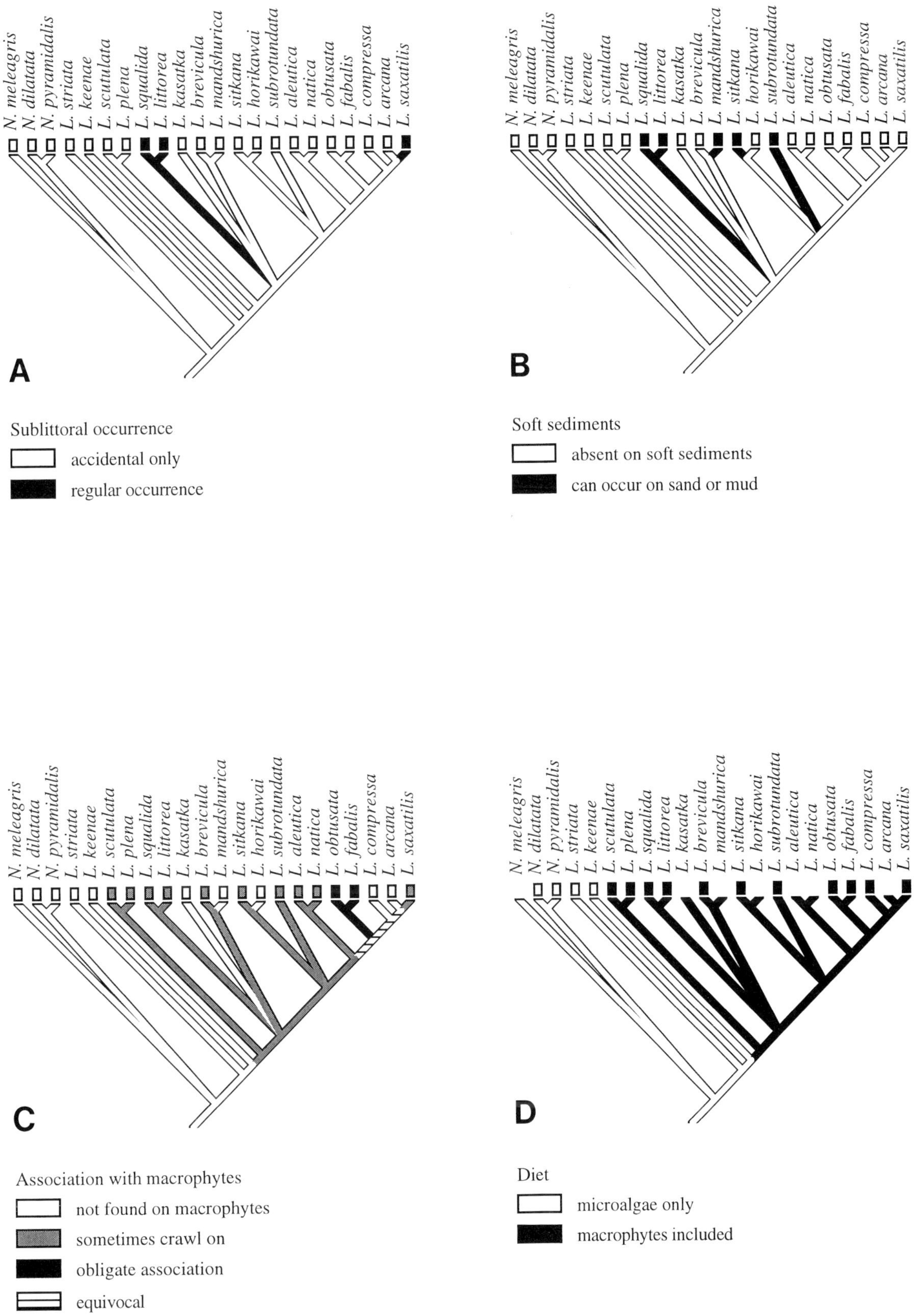

Figure 124. Optimization of ecological characters on the consensus phylogeny of *Littorina* species, for the reconstruction of the ecology of ancestral lineages (continuation of Fig. 123).

the conservative ecological shifts, at least in the form in which they have been coded here. However, the analysis demonstrates that a number of ecological shifts (in characters 14, 15, 17, 20, 21) have taken place concurrently between nodes 3 and 4. In terms of their ecology, the two basal *Littorina* species, *L. striata* and *L. keenae*, resemble the sister-genus *Nodilittorina* in most respects, having retained the ancestral characters of habitat in the littoral fringe, oceanic distribution and microalgal diet, although their warm-temperate, rather than tropical, distribution may represent a derived state. Less is known about life-history traits, but it is possible that this group is also characterized by long life span (up to 15 years), slow growth and low recruitment (as in *L. keenae*, Behrens Yamada, 1992; and in some tropical *Nodilittorina* species, Borkowski, 1974), which may be plesiomorphic features connected with the high-shore habitat. In contrast, clade 4 shows a complex of synapomorphic ecological characters: cold-temperate to subarctic range, continental distribution pattern, eulittoral zonation, inclusion of macroalgae in the diet (and sometimes as a substrate). These characteristics are clearly interrelated, for at high latitudes the shorelines are predominantly of the continental type, biotic interactions (such as predation and competition) may be less limiting in the eulittoral zone, and macroalgae provide an abundant food resource on the shore. The species have shorter life spans (mostly less than 5 years, except in Arctic populations), grow rapidly and show high recruitment, as expected in the more favourable eulittoral habitat (e.g. Moore, 1937; Moreteau, 1976; Hughes, 1980; Janson, 1982b; Behrens Yamada, 1992). Together, these ecological characters indicate a major shift to a new adaptive zone (*sensu* van Valen, 1971), the middle to upper eulittoral zone of cold-temperate and subarctic rocky shores. This shift may have been initiated by the range expansion of the ancestor of clade 4 into the northern Pacific, and subsequent climatic cooling, during the Middle Miocene (p. 386).

It has been suggested that entry into a new adaptive zone may bring about an increased diversity of lineages, through enhanced speciation or reduced extinction or both, producing what may be termed an adaptive radiation (*sensu* Skelton, 1993). Comparison with the time-calibrated molecular phylogeny (Fig. 120) indicates that a burst of rapid cladogenesis took place close to the origin of clade 4. While the time at which the adaptive shift took place in the long lineage leading to clade 4 cannot yet be specified, the ecological reconstruction indicates that it preceded, or was concurrent with, this episode of rapid cladogenesis. The initial rapid diversification of clade 4 could therefore be interpreted as an adaptive radiation in the new adaptive zone. In comparison, there is no evidence for contemporaneous diversification in its sister-lineage, leading to *L. keenae*, in which no ecological shift appears to have occurred. The sudden adaptive radiation of clade 4 may be contrasted with the pattern of sustained cladogenesis within clade 6 (the subgenus *Neritrema*). In the latter case no obvious ecological shifts can be identified, and here the increased rate of cladogenesis may have a non-adaptive explanation, the result of the high speciation rate of this nonplanktotrophic clade (see Speciation, Extinction and Life History, p. 382). The possible mechanisms by which the 'ecological opportunity' of a new adaptive zone might enhance speciation processes are poorly understood. For example, ecological segregation is a necessary component of some models of sympatric and parapatric speciation. However, in vicariant speciation models resource-based adaptive divergence might precede speciation (for example, following dispersal into a new habitat) or follow it (for example, resulting from competition following secondary sympatry) (Skelton, 1993). At present, therefore, the integration of this view of adaptive radiation with the models of speciation and with the biogeographic hypotheses discussed earlier is unclear. An alternative perspective on adaptive radiation is the search for 'key innovations', apomorphic traits believed to account for enhanced diversification of lineages (p. 408).

Adaptation

The concept of adaptation is central to evolutionary theory, for it is the predicted outcome of natural selection. There has been much discussion of the meaning of adaptation and the appropriate methodology for studying it (see, for example, reviews by Coddington, 1988, 1994; Baum & Larson, 1991; Brooks & McLennan, 1991; Harvey & Pagel, 1991). An important contribution was that of Gould & Vrba (1982), who defined an adaptation as any trait that promotes fitness and was built by natural selection for its current biological role. This emphasized that adaptation must be viewed in an historical context, in contrast to the ahistorical concept of biological function. To identify a trait as an adaptation for a particular function, it is not sufficient to demonstrate that the trait enhances organismal performance; it must also be shown that it evolved for this role by the process of natural selection. With the recognition of this historical component in the study of adaptation, the potential contribution of phylogenetic analysis became clear. Coddington (1988) redefined adaptation in explicitly cladistic terms, as 'apomorphic function promoted by natural selection, as compared with plesiomorphic function'. Like apomorphy, adaptation is therefore seen to be a relative term that must be considered at the appropriate level of the phylogenetic hierarchy. To qualify as an adaptation for a particular function, Coddington (1988) proposed that a trait must be shown (by reference to a phylogeny) to be an apomorphic state, to exhibit a superior performance of the derived function (relative to that of the plesiomorphic state of the trait), and thus to confer enhanced fitness. The historical action of natural selection can rarely be examined directly, so Coddington's test relies on the assumption that the selective advantage of the derived trait, as identified by the functional comparison, was also responsible for the initial origin and spread of the trait. This assumption has been made more explicit in an extension of this phylogenetic test of adaptation by Baum & Larson (1991). They used the term 'selective regime' to embrace the whole complex of interactions between the organism and its environment that determines how natural selection will act on heritable character variation. By superimposing critical aspects of the selective regimes of terminal taxa on to a phylogeny, they predicted the selective regime under which the character of interest evolved, thus testing whether or not natural selection may have been responsible for the initial fixation of the trait. To constitute an adaptation for a particular function, a trait must have enhanced utility relative to its antecedent state and must have evolved within the derived selective regime (Baum & Larson, 1991).

The methodology for recognizing adaptations is as follows (Baum & Larson, 1991). Where possible, all extant members of the clade should be included, to maximize the likelihood of accurate reconstruction. Any characters can be used to reconstruct the phylogeny; contrary to earlier suggestions (Coddington, 1988) it is not necessary to exclude those under study from this initial reconstruction, for the best estimate of phylogeny is obtained by including all available data (Deleporte, 1993). The principle of parsimony is used to locate character-state changes of the trait on the branches of the phylogeny. The identification and scoring of the selective regime depends on the hypothesis being tested; detailed study of the natural history of the taxon should indicate the relevant environmental and organismal features to be included in the selective regime. By superimposing the selective regimes on the phylogeny, and again optimizing using the principle of parsimony, the selective regimes of ancestral lineages are reconstructed (as done earlier, see Historical Ecology, p. 388). In appropriate cases palaeoecological information may aid in this reconstruction. Where optimization of traits or selective regimes is ambiguous, all possible alternatives must be considered in constructing the adaptive hypothesis. To assess the performance advantage of a derived trait, its current utility must be compared with that of its antecedent state. Strict comparison of fitness is possible only in populations polymorphic for the two states, but performance can be assessed in other ways, such as experimental manipulation of the organisms. A more general method is functional comparison with a sister-taxon showing the antecedent state, under the conditions of the derived selective regime. Within the confines of purely morphological studies such functional tests may be difficult, and suggested adaptive hypotheses are conditional upon their supposed outcome. Having demonstrated, or assumed, a performance advantage, the final stage is to compare the distribution of trait and selective regime on the phylogeny. If the trait originated in a lineage that already occupied the derived selective regime, it is an adaptation. If, on the other hand, the origin of the trait preceded that of the derived selective regime, the trait is an exaptation. This term was introduced by Gould & Vrba (1982) to describe a trait that confers performance advantage in its current biological role, but was not produced by natural selection directly for that role. If, as is often the case, the derived trait and the selective regime in which it confers a performance advantage appear simultaneously in a lineage, these possibilities, adaptation or exaptation, cannot be unequivocally distinguished. In practice, such cases are generally taken as evidence of adaptation, for natural selection will rapidly follow exposure to a new selective regime (Arnold, 1994).

There have been several criticisms of this phylogenetic method for testing adaptive hypotheses concerning the single, unique origins of derived traits. One has questioned the accuracy with which ancestral character states and selective regimes can be inferred; if characters show much homoplasy, or if there have been frequent changes of selective regime within lineages, the method will be unreliable (Frumhoff & Reeve, 1994). Its application requires that the rate of cladogenesis should be high relative to the rates of change of characters and of selective regimes. For this reason, the method is most suited to analyses at low taxonomic levels. It has also been emphasized that relationships among traits, performance and fitness are potentially so complex that the required phylogenetic correlation between derived selective regime and derived trait may not necessarily indicate cause and effect. For example, genetic correlations caused by pleiotropy or linkage might lead to incorrect conclusions about the relevant performance advantage. Furthermore, the components of the selective regime may also be complex and correlated, leading to doubt about the appropriate design of a performance test (Lauder, Leroi & Rose, 1993). The phylogenetic method can only provide a set of tests by which a given hypothesis of adaptation can potentially be falsified; alternative hypotheses cannot thereby be excluded (Arnold, 1994). Indeed, adaptation is itself only one of several alternative explanations for apomorphic traits; others, for example involving genetic drift, may not require the action of selection. Because of these limitations of hypotheses derived from single, unique, evolutionary events, an alternative approach to the phylogenetic study of adaptation has been developed. This involves the search for correlations between multiple, independent origins of a derived trait (convergences) and a particular aspect of the selective regime. If the statistical correlation exceeds that expected by chance, a general explanation is required and suggests a general pattern of adaptive change (e.g. Harvey & Pagel, 1991). Once again, a phylogeny is essential for the analysis, and tests of performance advantage strengthen the case. Such a comparative approach is clearly appropriate to studies across lineages at high taxonomic levels, in which a significant number of convergent appearances of a trait have occurred. These two approaches to the study of adaptation, the individual and the general (termed homology and convergence methods by Coddington, 1994), are complementary. The strength of the former lies in the detailed consideration of the natural history of a taxon or clade, in the understanding of the utility of homologous traits, and in the specification of precise and falsifiable hypotheses about ancestral selective regimes and current performance. On the other hand, the latter seeks general explanations for evolutionary patterns, albeit at the cost of imprecision in the understanding of the processes responsible (see discussions by Arnold, 1994; Coddington, 1994). One obvious difficulty of the convergence method is that there may be alternative adaptive responses in different lineages that are exposed to the same selective regime; conversely, a similar trait may be adaptive in more than one selective regime. Together, these two phylogenetic approaches are concerned with explanations for the origin and initial spread of adaptive traits; the implicit assumption is that traits are then maintained by stabilizing selection. The examination of the maintenance of heritable traits is the province of microevolutionary studies at the level of populations within species, and these can provide additional insights into the mechanisms of adaptation.

Against this background, hypotheses of the possible adaptive functions of certain traits in *Littorina* species can now be examined. These have been derived principally from the phylogenetic analysis of unique character states, using the methodology described, but with additional evidence from convergences in other groups and from microevolutionary studies, where appropriate. The principal traits considered are those morphological characters that have been used in the construction of the cladogram (p. 347) and have previously been coded for phylogenetic analysis (Tables 8, 9). Their character states have been optimized on the consensus phylogeny, which is the best hypothesis of relationships currently available for this genus. For this purpose the program MacClade (Maddison & Maddison, 1992) has been used, with all characters specified as unordered and polychotomies treated as multiple speciation events. Nodes have been referred to by arbitrary numbers as in Figure 122. Other traits, less well known in their occurrence or less amenable to discrete coding, are also discussed. Identification of likely selective regimes has been based upon the analysis and coding of ecological characters (Tables 13, 14; Figs 122–124). Tests of the hypotheses are here limited to determining the sequence (or concurrence) of derived traits and derived selective regimes on the phylogeny, since few functional studies have been made, and no controlled measurements of performance are available.

Shell shape, size and sculpture. These shell characters present difficulties for phylogenetic analysis and the development of hypotheses of adaptive function. As a consequence of their often considerable intraspecific variation, they are difficult to code as discrete character states for analysis (p. 347). However, one advantage of this variability is the scope that it provides for functional and microevolutionary studies, so that there is now much information about the likely influence of such selective agents as predatory crabs, wave action, microclimate and crevice size upon shell form (pp. 22–29). Problems arise because shell form is influenced by this wide range of selective factors, which are frequently confounded in natural habitats. Furthermore, several aspects of shell form (e.g. spire height and wall thickness) may respond together to the same selective factor (e.g. crab predation). Therefore, shell form cannot readily be analysed as discrete traits and correlated with independent aspects of the selective regime. A further complication arises from the poorly-understood ecophenotypic effects upon shell form (p. 23). While ecophenotypic variation may be functional, it cannot be said to be adaptive, since natural selection can only promote evolutionary change in genetically-based traits. On the other hand, the capacity to respond phenotypically to environmental factors is likely to be adaptive, although as yet insufficiently known to permit any phylogenetic test.

Examining the optimization of shell characters on the phylogeny (Fig. 125A) shows the derived states of *Neritrema*-like and *L. squalida*-like shells in clade 5, whereas the more basal branches of *Littorina* retain the plesiomorphic *Nodilittorina*-like shell. In the light of microevolutionary studies of shell form, various hypotheses for the adaptive origin of the *Neritrema*-like shell could be suggested. In comparison with the ancestral form, examples of this shell type usually show a relatively lower spire (and consequently enlarged aperture), which might confer some advantage in the eulittoral zone, where crab predation and wave action may be more severe and microhabitats are more humid than in the littoral fringe. As a partial test of this hypothesis, the *Neritrema*-like shell is seen to have originated within the selective regime of the middle to upper eulittoral zone (by comparison with the optimization of this ecological character, Fig. 125B). The additional necessary tests for performance advantage could be designed to compare fitness-related attributes of *Neritrema*-like species with those of a suitable outgroup (e.g. *L. scutulata* and *L. plena*). However, because of the difficulties of coding, analysis and experimental design mentioned earlier, it is unlikely that this type of phylogenetic analysis will be able to identify single shell traits (e.g. relative spire height) which serve an adaptive function in relation to defined aspects of the selective regime (e.g. crab predation). Microevolutionary studies of intraspecific variation will probably be more fruitful in this respect. However, only phylogenetic analysis can offer a broad, historical view. From this perspective, perhaps the key characteristic of all three derived shell types within *Littorina* is not a single shell trait, but rather their greater versatility of form (through genetic adaptation and/or ecophenotypic plasticity) than in the ancestral type, and this may be adaptive in relation to the microspatial and temporal variability of the selective regime in the cool-temperate eulittoral zone. This suggestion is at present impossible to test and requires much further information about the control of shell variation and about selective regimes. Since the intraspecific variability of the members of the subgenus *Neritrema* is also related to their nonplanktotrophic development (and consequent low gene flow, p. 24), they should be excluded from a phylogenetic test of this particular hypothesis.

Turning to other shell traits, the *L. squalida*-like state is confined to this taxon and its sister-species, *L. littorea* (Fig. 125A). This trait has been defined by reference to shell sculpture and apertural shape, of which the possible adaptive functions are unknown. It is, however, notable that these two species reach the largest sizes of any *Littorina*, and rapid growth to large size is one means of achieving some immunity to crab predation (e.g. Reimchen, 1982; Lowell, Fletcher, Grahame & Mill, 1994). The two species occupy the derived habitat of the lower eulittoral zone (Fig. 125B) and extend to sublittoral depths (Fig. 125C), where crab predation is likely to be more severe than higher on the shore. The distributions of trait and selective regime thus do not refute the hypothesis of adaptive function of large size. One other species, *L. saxatilis*, also occurs regularly in the sublittoral zone (Fig. 125C); in this case the shells are small and delicate, and may avoid crab predation by living on macrophytes (p. 325), illustrating an alternative adaptive response to a similar selection pressure.

The *L. obtusata*-like shell type, characterized both by the ability to produce a low-spired, globular shell and by lack of sculpture, is similarly restricted to a pair of sister-taxa, *L. obtusata* and *L. fabalis* (Fig. 125A). Within each species the low-spired, globular shell form occurs only in southern populations that are exposed to crab predators (pp. 208, 235) and is known (from microevolutionary studies in *L. obtusata*, Seeley, 1986) to be more resistant to crab predation. Although the genetic basis is unknown, at the microevolutionary level this appears to be a case of natural selection and morphological adaptation. The shell form of the common ancestor of the two species is not known, but the parsimonious assumption is that it too had the ability to produce the low-spired, globular form. The phylogenetic analysis suggests that the common ancestor occupied the middle to upper eulittoral zone (Fig. 125B), where crab predation is likely to be a significant selective pressure, and this may account for the adaptive origin of the trait. However, none of the other lineages in this zone developed a similar shell form. An alternative possibility is suggested by the correlation with the obligate association with macrophytes displayed only by *L. obtusata* and *L. fabalis* (Fig. 125D). If visual predation was significant, the low-spired, globular shape may initially have been favoured by its cryptic

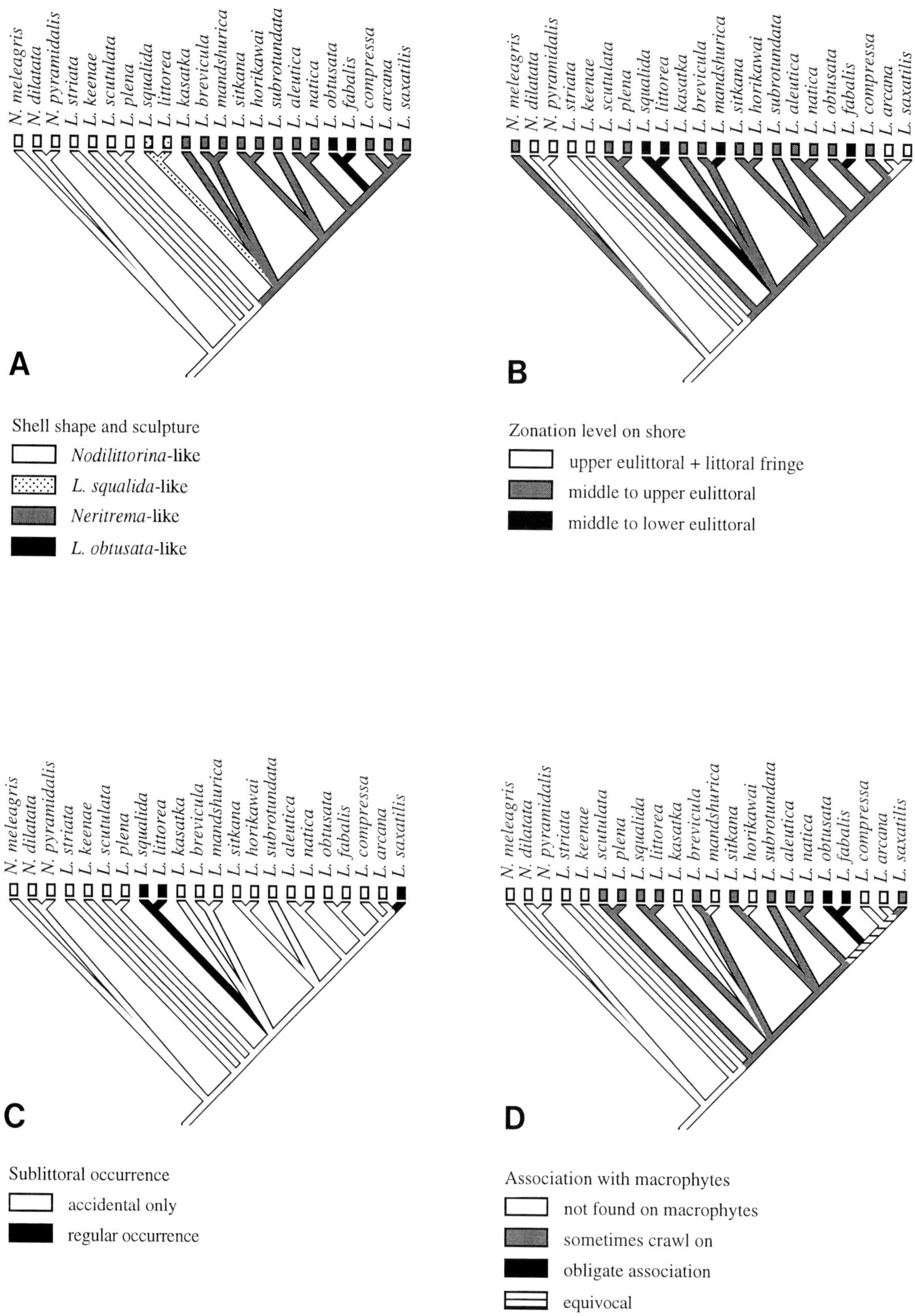

Figure 125. Hypotheses of the adaptive functions of shell characters (**A**) can be based on reconstruction of the selective regimes of ancestral lineages in *Littorina* (**B–D**). See text for details.

resemblance to the bladders of certain fucoid algae (Smith, 1976). These possibilities could be tested by appropriate experimental comparisons with outgroup taxa.

Shell colour polymorphism. This has not been included as a character in the morphological cladistic analysis, owing to difficulties of coding. The level of intrapopulation polymorphism varies widely between species, and the genetic basis (which might support the uncertain homology of the polymorphisms) is not understood. Polymorphisms with a probable genetic basis have been described in *L. squalida*, *L. littorea*, *L. brevicula* and in all *Neritrema* species with the exception of *L. natica*. Optimizing on the phylogeny suggests a single origin in the ancestor of clade 5, with reversals in *L. kasatka*, *L. mandshurica* and *L. natica*, and thus lends weak support to the homology of the trait. Microevolutionary studies have provided some information about the diverse mechanisms which maintain the polymorphism within species (p. 32). Comparison with the distributions of ecological characters supports a possible adaptive explanation for the origin of the trait. This occurred within the selective regime of the middle to upper eulittoral (Fig. 123D), which is probably a visually more diverse habitat, with macrophytes (Fig. 124C) and barnacles, than the ancestral habitat in the littoral fringe. A consistently high level of intrapopulation colour polymorphism is found only in the sister-taxa *L. obtusata* and *L. fabalis*; these alone show an obligate association with macrophytes (Fig. 124C), which provide the most visually diverse background. Various hypotheses of the origin and maintenance of colour polymorphism are related to the visual diversity of the background. At a higher phylogenetic level, striking colour polymorphism is known elsewhere in the Littorinidae only in two subgenera of *Littoraria*, in species that occur among the foliage of mangrove trees (Reid, 1986a, 1987). These two convergent origins of the trait in the family, both in association with a visually diverse background, offer further support for the adaptive hypothesis. However, it is notable that not all littorinids that have an association with macrophytes (e.g. *Lacuna* species) show striking colour polymorphism.

Shell mineralogy. This trait provides a particularly clear example of the use of the phylogenetic approach to adaptation. Coding of the presence or absence of the outer calcite layer is unambiguous. An adaptive hypothesis suggests that since calcite is less soluble than aragonite, an outer calcite layer may serve the adaptive function of reducing shell dissolution in cold water (Taylor & Reid, 1990; p. 36). Optimization on the phylogeny shows that the calcite layer originated in the ancestor of clade 4 (Fig. 126A), which corresponds with a shift in climatic regime from warm-temperate to cold-temperate and subarctic (Fig. 126B). The hypothesis therefore passes the initial phylogenetic test. Experimental tests of current performance advantage could be performed, to compare the rates of shell dissolution of members of clade 4 with those of *L. keenae* and *L. striata* under the same cold-temperate to subarctic conditions. The hypothesis predicts that the former group will show the lower rates of dissolution. The phylogenetic reconstruction shows that the two species of clade 4 that occupy a predominantly warm-temperate range (*L. brevicula* and *L. horikawai*) have been derived from cold-temperate ancestors and have retained the calcite layer. Without the phylogenetic information, it would not have been clear whether the calcite layer was a synapomorphy within *Littorina* (although outgroup comparison alone would have suggested this) and, if so, whether it had originated under a cold-temperate or warm-temperate regime.

A survey of the distribution of outer calcite layers throughout the Littorinidae shows that they have appeared independently in two other clades, the genera *Risellopsis* and *Pellilitorina* (Taylor & Reid, 1990; note that these authors included *Mainwaringia* in the clade *Littorina* in their phylogenetic analysis; this is now excluded, see p. 358, removing an anomaly and confirming the single origin of this trait in *Littorina*). *Risellopsis* occurs in cold-temperate waters in New Zealand, and *Pellilitorina* in the Antarctic, so that this convergence approach also supports the adaptive hypothesis. However, several other cold-water clades of littorinids have shells entirely composed of aragonite, and therefore the lack of an outer calcite layer does not appear to be a constraint precluding a cold-water habitat.

Anatomical characters. It has been suggested that the shape of the penis, and possibly the number of mamilliform penial glands, may be used in species recognition (p. 14). If this is its main function, no correlation with the selective regime of the physical environment can be expected. The number of glands has, however, been used in construction of the morphological cladogram. Optimization of this character on the phylogeny does reveal a certain pattern (Fig. 127). The basal members of *Littorina* show few or no glands, as in the outgroup, but in these lower branches the character state is labile. The trait is also

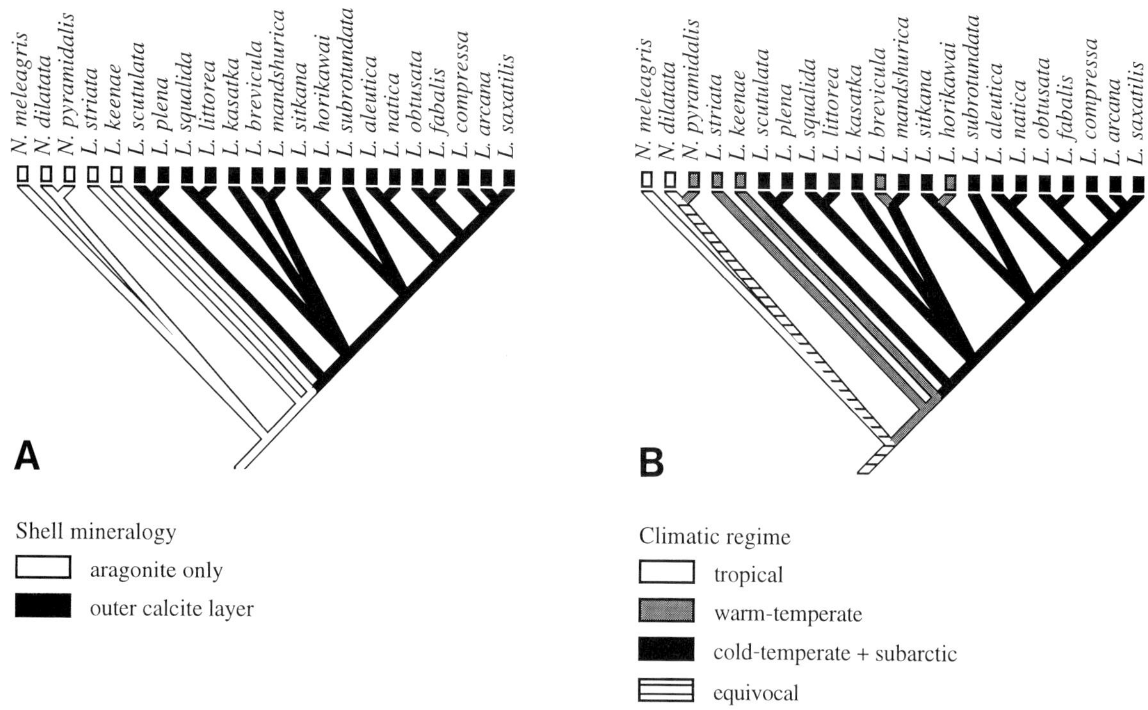

Figure 126. The origin of the outer calcite layer of the shell in *Littorina* (**A**) coincided with a shift to a colder climatic regime (**B**), supporting an hypothesis of its adaptive function. See text for details.

variable within species. The reconstruction shown implies three independent origins of mamilliform penial glands (as does an optimization assuming soft polychotomies); since these glands are complex and apparently homologous structures that are also present in other littorinid genera, multiple loss (a non-parsimonious reconstruction) is perhaps more likely (Reid, 1989a). Despite this uncertainty, there is an obvious phylogenetic pattern, for within clade 5 all species (with the exception of *L. kasatka*) show many glands, which is a derived state under all optimizations. At present, no adaptive explanation can be suggested. Within species, some correlations with shell size and with wave-exposure of the habitat have been found (p. 14), but do not account satisfactorily for the phylogenetic trend. Similarly, character states of the paraspermatozoa and of the arrangement of the egg groove are without known adaptive function. However, the characters of the jelly gland, egg capsule and type of development form an interrelated complex, of which the adaptive functions will be considered together.

Spawn and larval development. The extraordinary diversity of types of spawn and of larval development in the genus *Littorina* (summarized in Table 15) have long been of interest. In attempting to find functional or adaptive explanations, early authors sought correlations with zonation level on the shore (e.g. Woodward, 1909; Mileikowsky, 1975) and later with demography (Hughes & Roberts, 1981; Behrens Yamada, 1992), but these were largely unsuccessful. The concept of an optimal reproductive strategy in a particular habitat failed to explain the cases of *Littorina* species with contrasting types of spawn and development that could be found together on the shore. In the absence of a phylogeny for the genus, the polarity of the evolutionary sequence of spawn types was read in the wrong direction (Fretter, 1980). Only with the development of a preliminary phylogeny were testable adaptive hypotheses proposed (Reid, 1989a, 1990b) and these can now be refined.

There is a large literature devoted to theoretical aspects of the life histories of marine invertebrates, which have in the past often been considered as optimal strategies under different regimes of environmental instability, demography, isolation and patchiness of the habitat, latitude, depth and body size (e.g. Jablonski & Lutz, 1983; Grahame & Branch, 1985; Roff, 1992). Of particular interest in relation to the phylogenetic analysis of adaptation in *Littorina* are two explanations for a well-documented trend among prosobranch gastropods in general, which display a decreasing proportion of planktotrophic forms from the tropics towards the poles (Thorson, 1950, 1965; Spight, 1981). This

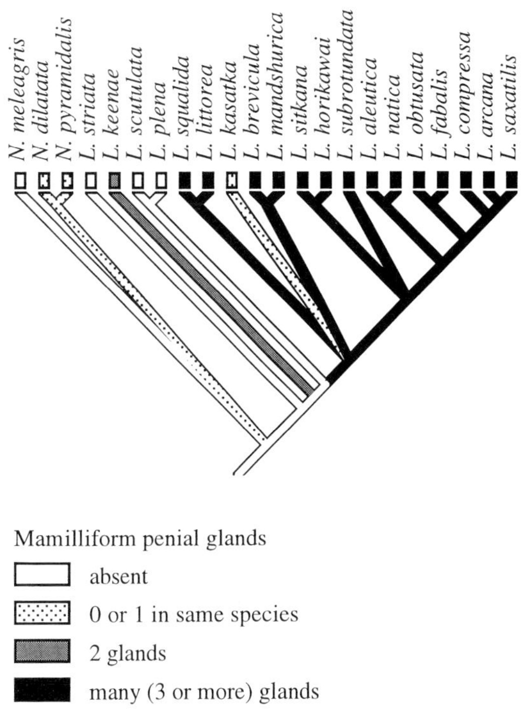

Figure 127. Optimization of the numbers of mamilliform penial glands on the phylogeny of *Littorina* species.

was explained by Thorson (1950) in terms of the energetics of larval growth; at high latitudes the brief period of availability of planktonic microalgal food, and the slow growth rate of larvae at low temperatures, were suggested to lead to suppression of planktotrophic larval stages. An alternative explanation, in terms of latitudinal gradients in predation, was proposed by Highsmith (1985); at high latitudes predation on pelagic larval stages, by pelagic predators and benthic filter feeders, is likely to be more severe, favouring the elimination of pelagic stages, whereas in the tropics intense benthic predation may favour pelagic development.

In Figure 128 the phylogenetic distributions of three morphological characters connected with spawn and development are compared with that of the latitudinal climatic regime. Although segregated for the purpose of the cladistic analysis (p. 347), the morphological characters are partly interrelated. For example, an enlarged jelly gland (Fig. 128A) produces the gelatinous matrix of the benthic spawn of *Neritrema* species (Fig. 128B) (although the convergently enlarged jelly gland of *L. keenae* produces an ephemeral pelagic egg mass). Furthermore, spawn type and development are exactly correlated in *Littorina*, in which all benthic eggs show nonplanktotrophic development (although this is not invariably so in other littorinids, Reid, 1989a). An adaptive hypothesis based on Thorson's idea is supported by the origin of nonplanktotrophic development (Fig. 128C) within the cold-temperate to subarctic selective regime (Fig. 128D), although at least six planktotrophic species also occur at high latitudes. Closer examination reveals an additional morphological pattern. If the size of the pelagic capsule is superimposed on the phylogeny (Fig. 129), a closer correlation with the climatic regime emerges: the two warm-temperate species *L. striata* and *L. keenae* have retained the ancestral small (180–400 μm diameter) capsules shared with tropical *Nodilittorina* species, whereas the pelagic capsules of cold-temperate and subarctic members of clade 4 are all considerably larger (600–1100 μm diameter). One possible adaptive hypothesis has been based on Highsmith's idea by Reid (1990b), who suggested that these large capsules might be less susceptible to the pelagic predators (e.g. hydromedusae, ctenophores, chaetognaths, euphausiids, decapod larvae, filter-feeding fish and fish larvae) and benthic filter feeders (e.g. barnacles, bivalves, polychaetes) that abound in colder waters (see Young & Chia, 1988, for a review of predators of pelagic larvae). Experimental tests could be designed to test whether this is in fact the case. However, limited data on the size of food particles of some filter feeders do not support this explanation; the mussel *Mytilus* consumes particles only up to 100 μm in

Table 15. Summary of characteristics of spawn and larval development in *Littorina* species and three members of the sister-genus, *Nodilittorina*. Data compiled from Systematic Descriptions in Chapter 4, and for *Nodilittorina* species from Lewis (1960), Borkowski (1971), Bandel (1974) and pers. obs. Values in parentheses are rare extremes of range.

Species	Climatic regime	Spawn type	Development	Capsule diameter (μm)	No. of ova per capsule	Ovum diameter (μm)	Protoconch diameter (μm)
N. meleagris	tropical	pelagic capsule	planktotrophic	180–240	1	55–80	?
N. dilatata	tropical	pelagic capsule	planktotrophic	200–250	1	42–80	?
N. pyramidalis	warm-temperate	pelagic capsule	planktotrophic	?	?	?	?
L. striata	warm-temperate	pelagic capsule	planktotrophic	240	1	70	260
L. keenae	warm-temperate	pelagic capsule	planktotrophic	300–400	1(3)	77–90	240
L. scutulata	cold-temp./subarc.	pelagic capsule	planktotrophic	700–1000	(1)3–10(14)	105	240
L. plena	cold-temp./subarc.	pelagic capsule	planktotrophic	1100	(4)17–32(41)	96	300
L. squalida	cold-temp./subarc.	pelagic capsule	planktotrophic	870–880	14–15	95	340
L. littorea	cold-temp./subarc.	pelagic capsule	planktotrophic	(300)600–1000(1100)	1–3(9)	120–160(205)	380–390
L. kasatka	cold-temp./subarc.	pelagic capsule	planktotrophic	?	?	?	?
L. brevicula	warm-temperate	pelagic capsule	planktotrophic	310–450(540)	1(5)	84–114	350–360
L. mandshurica	cold-temp./subarc.	pelagic capsule	planktotrophic	1000	9–12	120	?
L. sitkana	cold-temp./subarc.	benthic mass	nonplanktotrophic	550–1000	1(2)	175–200	450
L. horikawai	warm-temperate	benthic mass	nonplanktotrophic	?	?	?	390–440
L. subrotundata	cold-temp./subarc.	benthic mass	nonplanktotrophic	890–960	1	?	530
L. aleutica	cold-temp./subarc.	benthic mass	nonplanktotrophic	900	1	?	?
L. natica	cold-temp./subarc.	benthic mass	nonplanktotrophic	700–1000	1	?	?
L. obtusata	cold-temp./subarc.	benthic mass	nonplanktotrophic	630–680	1	210–255	420–500
L. fabalis	cold-temp./subarc.	benthic mass	nonplanktotrophic	520–600	1	195–200	?
L. compressa	cold-temp./subarc.	benthic mass	nonplanktotrophic	?	1	?	480–600
L. arcana	cold-temp./subarc.	benthic mass	nonplanktotrophic	530–620	1	235–285	?
L. saxatilis	cold-temp./subarc.	ovoviviparous	nonplanktotrophic	capsule absent	not applicable	180–211	400–650(800)

diameter, rejecting larger particles in pseudofaeces (Bayne, Thompson & Widdows, 1976), whereas barnacles are able to consume particles larger than 1 mm in diameter (Crisp & Southward, 1961). Further information on predation upon spawn is required, but perhaps there is some other, unknown, advantage of large propagule size in cold seas. It should be noted that the homology of these large capsules is not certain; the double-rimmed shape of the capsule shared by *L. scutulata* and *L. plena* might suggest a separate origin from a small, biconvex, ancestral type (Fig. 128B; coding of character based on shape and not size). In addition, it is not clear whether the benthic egg mass of *Neritrema* originated from an ancestor with a small or large pelagic capsule (this depends upon the interpretation of the polychotomy at the base of clade 5 as hard or soft, i.e. as a real multiple speciation event, or as an uncertain resolution). Depending upon this outcome, nonplanktotrophic development and large pelagic capsules could be interpreted as alternative adaptive solutions to intense pelagic predation (or some other selective factor), or otherwise an hypothesis concerning the possible energetic advantage of nonplanktotrophic development could be proposed. As well as being of large size, the pelagic capsules of cold-water *Littorina* species also typically contain many eggs, in contrast to the primitive type which contains only one, although this trait is evidently labile (Table 15) and any possible adaptive significance is unclear. If large capsules are indeed advantageous for some reason in cold water, then selection for increased size of hatching veligers might also be expected and (taking ovum diameter as a guide) there is some suggestion that this could be so (Table 15). The maximum size of veligers (i.e. protoconch size, Table 15) is, however, presumably limited by considerations of swimming and buoyancy.

Three species add interesting details to the adaptive explanations for the correlations between climate and spawn type. In *L. keenae* the capsules are small and the ova usually single, as expected from its warm-temperate Californian distribution. However, the capsules are initially released in an ephemeral gelatinous mass up to 15 mm long (Schmitt, 1979). Various adaptive interpretations are possible (p. 58), but one possibility is that this is another means of increasing the effective size of the propagule. A second species, *L. brevicula*, also conforms to the general correlation, having small capsules, generally single ova, and a (partly) warm-temperate distribution. However, some of the

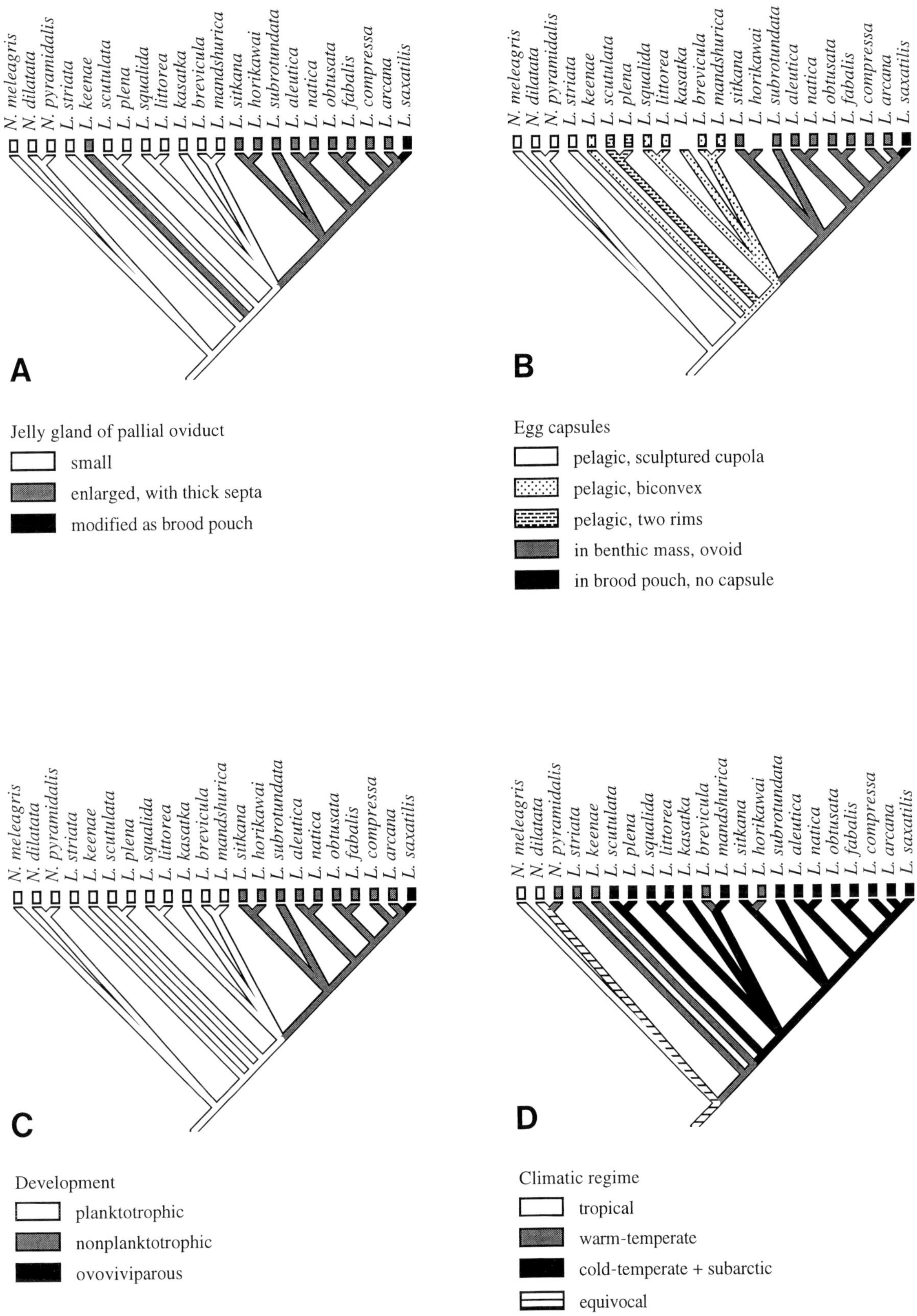

Figure 128. Three morphological characters of *Littorina* (**A–C**) form an interrelated complex connected with spawn type and larval development. Adaptive hypotheses can be based on the reconstruction of the climatic regimes of ancestral lineages (**D**). See text for details.

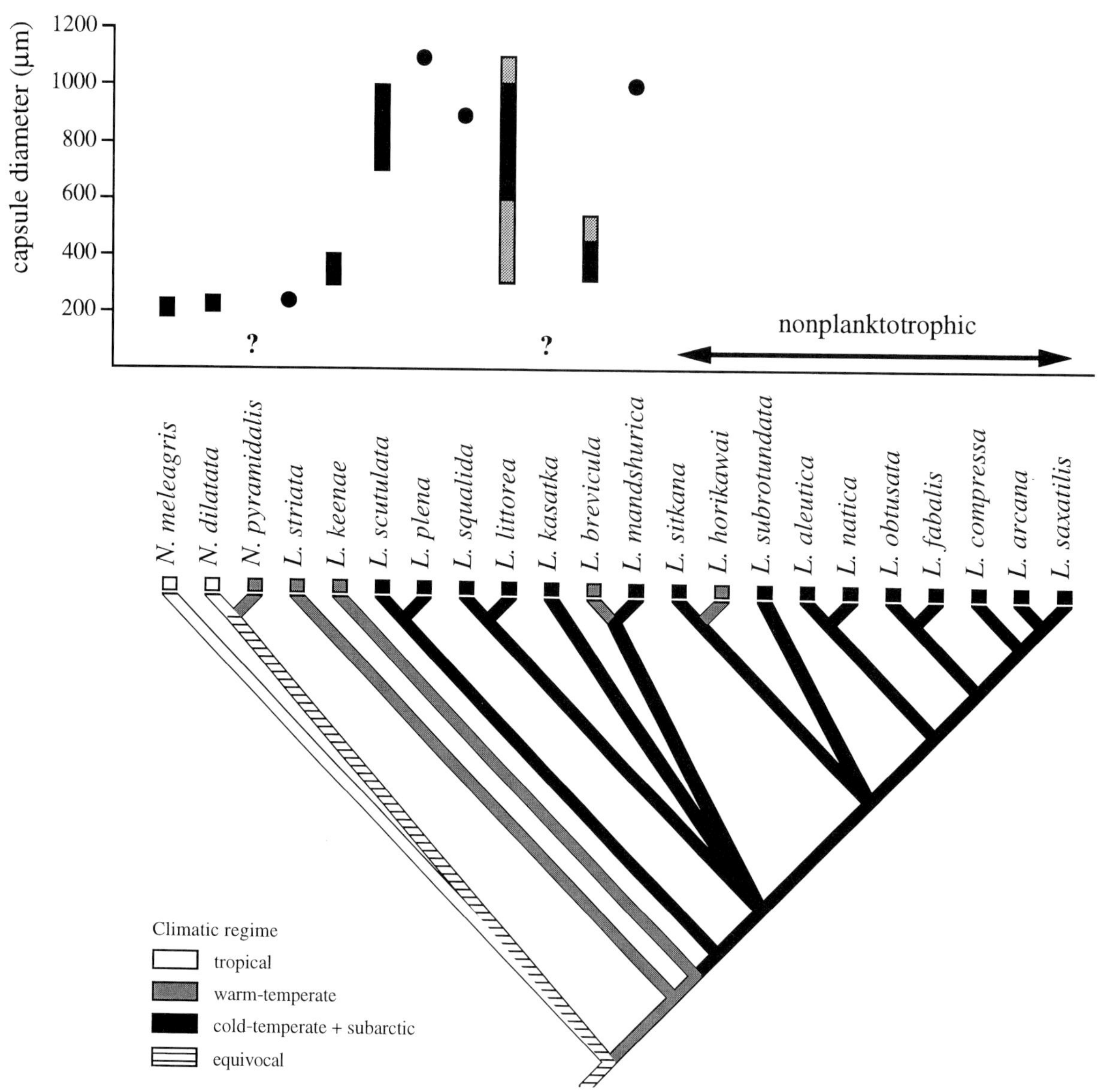

Figure 129. Diameters of pelagic egg capsules of planktotrophic *Littorina* species, and two *Nodilittorina* species, superimposed on the phylogeny, with reconstruction of ancestral climatic regimes based on those of Recent species. For capsule diameters solid bars indicate normal range, stippled bars indicate rare extremes, dots indicate single measurements; data for *Littorina* species from sources quoted in Chapter 4 and for *Nodilittorina* species from Lewis (1960), Borkowski (1971) and Bandel (1974).

possible phylogenetic reconstructions (e.g. assuming homology of large capsule types and a multiple speciation event at node 5; Fig. 129) show that it may have reverted to a small capsule size. If so, at this level of the phylogeny small capsules may be adaptive, which could be tested by functional comparison with the sister-species, *L. mandshurica*. *Littorina brevicula* might well repay a detailed study of the relationships between capsule size, ovum number and predation, since its spawn is variable in these characters, and it occupies a range of latitudes from subtropical to subarctic (p. 133). The third interesting species is *L. horikawai*, which has retained a benthic gelatinous egg mass despite a shift of climatic regime to the the warm-temperate zone. This is the most southerly known example of this spawn type in *Littorina*. At least in some other *Neritrema* species the benthic spawn appears to limit southward spread into warmer climatic regimes (e.g. *L. arcana* and *L. compressa*, pp. 263, 276), an example of phylogenetic constraint.

A third type of development, ovoviviparity, is unique to *L. saxatilis*. This type of development must retain any functional advantage of the ancestral benthic egg mass in relation to predation and energetics in cold water, and may well be superior in both respects. An overriding consideration, however, is likely to be improved protection of the embryos, permitting colonization of greatly increased ranges of habitat and latitude (p. 335). This adaptive hypothesis is supported by the optimization of zonation level on the phylogeny (Fig. 123D), which suggests that ovoviviparity arose in the selective regime of the upper eulittoral and littoral fringe, where the habit presumably conferred an advantage in comparison with the ancestral benthic egg masses, in which embryos were more susceptible to desiccation. Ovoviviparity may confer an additional advantage in terms of dispersal, and indeed this may also be a consideration when comparing benthic and pelagic spawn. However, the short-term advantage of wide dispersal of larvae is disputed (Palmer & Strathmann, 1981) and may therefore not be relevant to adaptive hypotheses (but see p. 383 for a discussion of the long-term evolutionary consequences).

Comparisons of convergences at higher phylogenetic levels can also be carried out for characters of spawn and development (Reid, 1989a). In the case of the type of spawn, benthic egg masses elsewhere in the Littorinidae are in fact plesiomorphic and do not therefore require adaptive explanation. Nonplanktotrophic development is a derived character in the two basal subfamilies, Lacuninae and Laevilitorininae, and in both is correlated with cold climatic regimes. Ovoviviparity through metamorphosis occurs elsewhere in only two tropical littorinids, and at least in one (*Littoraria aberrans*) may be adaptive in relation to an extremely high-tidal habitat.

At the microevolutionary level, measurements have been made of reproductive effort, fecundity and egg size in populations of nonplanktotrophic species from contrasting habitats, and differences have been related to demography and sources of mortality (Hughes & Roberts, 1980, 1981; Hart & Begon, 1982; Hughes, 1995). Such studies provide insights into the maintenance of some life-history traits by current selection, but cannot yet be related to the evolutionary origins of spawn and developmental types.

Radula and salivary glands. Like characters relating to shell shape, those of the radula do not lend themselves readily to simple phylogenetic analysis. Problems of coding radular teeth into discrete states, arising from ontogenetic and individual variation, have already been mentioned (pp. 20, 349). Some adaptive function of radular teeth in relation to diet is obviously to be expected, and phylogenetic correlations with diet must therefore be considered. However, different radular types may effectively exploit the same food source (e.g. Nicotri, 1977; Hawkins, Watson, Hill, Harding, Kyriakides, Hutchinson & Norton, 1989). Furthermore, the effectiveness of the radula as a tool depends not only upon the shape of the unused teeth that are commonly examined in morphological studies, but also on patterns of tooth wear, cusp hardness and effectiveness of the buccal musculature (Padilla, 1985). The recording of items in the diet also poses problems. Algal plants may differ in their palatability according to size, part of the plant and state of decomposition. *Littorina* species appear to be mainly opportunistic feeders on a variety of algal foods, so that the preferred diet may not be clear without experimental tests. The radula is therefore probably a generalized tool, a compromise among the requirements for a wide range of foods, which may obscure correlations of its morphology with diet.

With these reservations in mind, the optimization of radular traits on the phylogeny can be examined in relation to diet. Owing to difficulties of coding and limited information, the selective regime has been represented by only a single character, the inclusion of some macroalgae in the diet (Fig. 130D). The origins of the derived states of both radular characters (Fig. 130A, B) are seen to have occurred within the lineage that included macroalgae in the diet, supporting hypotheses of the adaptive function of these traits in relation to macroalgal grazing. The enlargement of the salivary glands (Fig. 130C) shows an exact correlation with the derived dietary regime; the physiological role of salivary secretions does not appear to have been investigated, but this correlation suggests a possible link with macroalgae in the diet. Enlargement of the salivary glands has not been noted elsewhere in the family (Reid, 1989a).

A more detailed analysis of dietary components in relation to phylogeny and radular form is shown in Figure 131. Here, the dietary components have been classified into eight categories, based partly on the functional groups of algae devised by Steneck & Watling (1982). As reviewed earlier (p. 21), knowledge of radular function is limited, but three structural (and possibly functional) types of radulae have been defined in *Littorina*, based on the shape of the cusps of the five central teeth. The phylogenetic information does not conflict with this interpretation. The two basal species, *L. striata* and *L. keenae*, retain the ancestral habitat of the littoral fringe in warm-temperate climates and share with the sister-group, *Nodilittorina*, the plesiomorphic radular type with long, fine, chisel-shaped cusps. This functions (and at some deeper level of the phylogeny, may well be adaptive) for grazing epilithic and endolithic

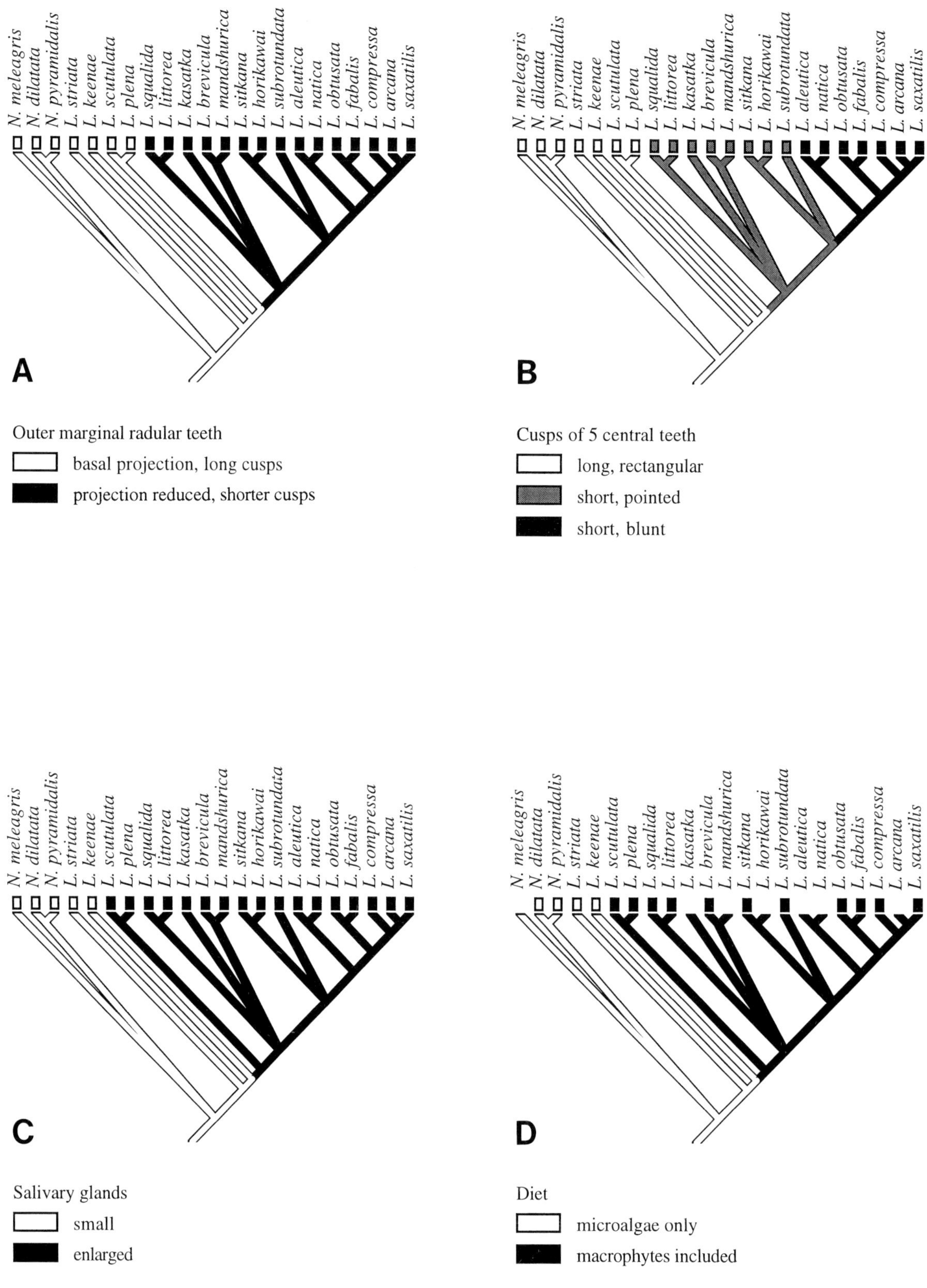

Figure 130. Phylogenetic correlations between derived traits of the radular teeth (**A**, **B**) and of salivary glands (**C**) with the presence of leathery and corticated macroalgae in the diet (**D**), suggest hypotheses of their adaptive function. See text for details.

	N. meleagris	N. dilatata	N. pyramidalis	L. striata	L. keenae	L. scutulata	L. plena	L. squalida	L. littorea	L. kasatka	L. brevicula	L. mandshurica	L. sitkana	L. horikawai	L. subrotundata	L. aleutica	L. natica	L. obtusata	L. fabalis	L. compressa	L. arcana	L. saxatilis
epi/endolithic microalgae 1	?+	++	?+	?+	+	++	?+															
black lichens 2			?+	?+	++	++	?+															
diatom films 3	?+				+	++	+		+				+		+			+	++	++		++
filamentous algae 4	++					+	+		++		+		+									+
foliose algae 5						+			++		+		+		+							+
macroalgae 6						+	+		+				+					++	+	++		++
calcareous red algae 7						+							+									
plant detritus						+			+				+						+	+		+

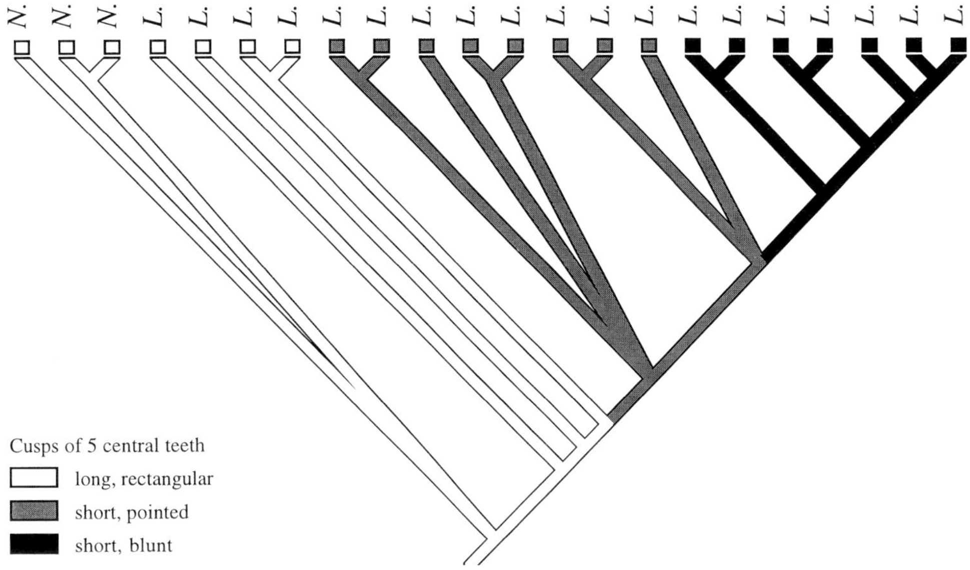

Figure 131. Diets of *Littorina* species and of three species of *Nodilittorina* superimposed on phylogeny showing optimization of cusp shape of central radular teeth. Symbols: +, recorded as a component of diet; ++, major dietary component; ?+, assumption of dietary component based on known habitat. Notes on dietary categories: 1, epilithic and endolithic flora of superficially bare rocks in littoral fringe, including high proportion of blue-green algae (rock particles also appear in gut contents); 2, encrusting black lichens (e.g. *Verrucaria*); 3, diatom films on damp rocks and epiphytic films on macroalgae; 4, includes sporelings of macroalgae; 5, e.g. *Ulva, Enteromorpha, Porphyra*; 6, corticated and leathery macroalgae (e.g. *Fucus, Laminaria, Chondrus, Mastocarpus*); 7, e.g. *Corallina, Lithothamnion*. Data from sources quoted in Chapter 4 and for *Nodilittorina* species from Bandel & Wedler (1987) and Chapman & Underwood (1994).

microalgae, algal crusts and lichens from the surface of bare, high-level rocks. Rock particles are frequent in the gut, and the radula is relatively long to compensate for rapid wear of the teeth. The diet of *L. striata* has not been studied, but in *L. keenae* it is noteworthy that macroalgae cannot be eaten unless sufficiently small or fragmented to be consumed whole (Dahl, 1964). *Littorina scutulata* (and, by analogy, also the less well-known *L. plena*) has shifted to a slightly lower tidal level in the upper eulittoral and extends into the cold-temperate climatic zone where macroalgae are abundant. The principal components of the diet are still epilithic and endolithic microalgae, lichens and diatom films, but in addition a wide variety of larger algae are consumed if available. The ancestral radular type is retained (although as noted by Behrens Yamada, 1992, and p. 21, it is in some respects intermediate between this and the following type), but evidently can utilize a range of food items. Short, pointed, radular cusps originated in the ancestor of clade 5 which, like *L. scutulata*, presumably had a wide dietary range. It has been suggested that this radular type is adapted for puncturing and tearing thin, foliose algal thalli (as preferred by *L. littorea*, Watson & Norton, 1985b), although it evidently also functions for grazing on filamentous algae, diatom films and detritus, and can even be used on macroalgae. Also in this morphological group, *L. sitkana* has a broad dietary range, but is unable to utilize epilithic or endolithic microalgae and no rock particles appear in its gut (Voltolina & Sacchi, 1990). The third functional type was derived in the ancestor of clade 7 and is characterized by short, blunt cusps. At least three species in this group include fucoids and other macroalgae as a major component of their diets, and the cusps may be adapted for excavation of leathery algal thalli. Surface diatom films are also effectively grazed from rocks and macroalgal fronds. Much additional information on dietary preference, radular function, tooth wear and feeding efficiency is required to test these preliminary adaptive hypotheses. Simple functional comparisons with suitable outgroup taxa are, however, unlikely to be suitable, being confounded by dietary preferences and tolerance of algal chemical defences.

Key innovations and phylogenetic constraints. Some authors have attempted to characterize those particular adaptations that open up a new adaptive zone as 'key innovations'. The selective regime may be greatly modified in the new adaptive zone, and an adaptive radiation may ensue. A test for a key innovation is the exact correlation between the origin of the trait and the shift to the new adaptive zone; no lineages without the trait are able to occupy the new adaptive zone (Baum & Larson, 1991). The corollary is that the ancestral trait was a 'phylogenetic constraint', preventing occupancy of the new adaptive zone (Harvey & Pagel, 1991). However, these concepts may be of limited explanatory value (Brooks & McLennan, 1991). Clades may be characterized by more than one apomorphic trait, so that a single key innovation cannot be identified. In addition, the link between the key innovation and enhanced speciation (or reduced extinction) is seldom clear. Other, undiscovered, traits (for example of behaviour or physiology) may have been equally or more important. Using the criteria of Baum & Larson (1991) the key morphological innovations, that permitted the major adaptive shift to the upper and middle eulittoral zone of cold-temperate and subarctic rocky shores, may be identified as the outer calcite layer of the shell, the large pelagic egg capsules (and the alternative or modified state of benthic egg masses), and the enlarged salivary glands. Without some or all of these, the extraordinary domination of *Littorina* species on the temperate shores of the northern hemisphere, for at least the last 10 million years, might not have been possible.

REFERENCES

AAREFJORD, F. & SKRESLET, S. 1969. Frozen up *Littorina saxatilis* Olivi in the icefoot in Spitsbergen. *Norsk Polarinstitutt Årbok*, **1967**: 240.

ABBOTT, D. P. & HADERLIE, E. C. 1980. Prosobranchia: marine snails. In: *Intertidal Invertebrates of California* (Morris, R. H., Abbott, D. P. & Haderlie, E. C., eds), 230–307. Stanford University Press, Stanford.

ABBOTT, R. T. 1954a. *American Seashells*. Van Nostrand, New York.

ABBOTT, R. T. 1954b. Review of the Atlantic periwinkles, *Nodilittorina*, *Echininus*, and *Tectarius*. *Proceedings of the United States National Museum*, **103**: 449–464.

ABBOTT, R. T. 1974. *American Seashells*. Edition 2. Van Nostrand Reinhold, New York.

ABE, N. 1935. The colony of the Littorina: *Littorivaga brevicula* (Philippi). *Science Reports of Tôhoku University. Biology*, **9**: 279–296.

ADAMS, A. & REEVE, L. 1850. Mollusca [parts 2, 3]. In: *The Zoology of the Voyage of H. M. S. Samarang; under the Command of Captain Sir Edward Belcher, C. B., F. R. A. S., F. G. S. during the years 1843–1846* (Adams, A., ed.). Reeve, Benham & Reeve, London.

ADAMS, C. G. 1981. An outline of Tertiary palaeogeography. In: *Chance, Change and Challenge: The Evolving Earth* (Cocks, L. R. M., ed.), 221–235. British Museum (Natural History), London.

ADAMS, H. & ADAMS, A. 1853–1858. *The Genera of Recent Mollusca; Arranged according to their Organization*, **1** and **3**. John van Voorst, London.

ADAMS, L. E. 1896. Albino forms of *Littorina rudis* var. *tenebrosa*. *Journal of Conchology*, **8**: 315.

ADDICOTT, W. O. 1969a. Late Pliocene mollusks from San Francisco Peninsula, California, and their paleogeographic significance. *Proceedings of the California Academy of Sciences. Series 4*, **37**: 57–93.

ADDICOTT, W. O. 1969b. Tertiary climatic change in the marginal northeast Pacific Ocean. *Science, New York*, **165**: 583–586.

ADDICOTT, W. O. 1970. Tertiary paleoclimatic trends in the San Joaquin basin, California. *United States Geological Survey Professional Papers*, **644–D**: 1–19.

ADDICOTT, W. O. 1976. Neogene molluscan stages of Oregon and Washington. In: *Neogene Symposium, Pacific Section, Annual Meeting, Society of Economic Paleontologists and Mineralogists, San Francisco*, 95–115.

ADEGOKE, O. S. 1969. Stratigraphy and paleontology of the marine Neogene formations of the Coalinga region, California. *University of California Publications in Geological Sciences*, **80**: 1–241.

ALDER, J. 1848. A catalogue of the Mollusca of Northumberland and Durham. *Transactions of the Tyneside Naturalists' Field Club*, **1**: 97–209.

ALERSTAM, T., GUDMUNDSSON, G. A. & JOHANNESSON, K. 1992. Resources for long-distance migration: intertidal exploitation of *Littorina* and *Mytilus* by knots *Calidris canutus* in Iceland. *Oikos*, **65**: 179–189.

ALEXANDER, R. C. 1947. *Littorina littorea* on the New Jersey coast. *Nautilus*, **60**: 73–76.

ALIFIERAKIS, N. S. & BERRY, A. J. 1980. Rhythmic egg-release in *Littorina littorea* (Mollusca: Gastropoda). *Journal of Zoology, London*, **190**: 297–307.

ALSTYNE, K. L. VAN 1988. Herbivore grazing increases polyphenolic defenses in the intertidal brown alga *Fucus distichus*. *Ecology, New York*, **69**: 655–663.

ALTENA, C. O. VAN R., BLOKLANDER, A. & POUDEROYEN, L. P. 1954. De fossiele schelpen van de Nederlandse stranden en zeegaten, 1. *Basteria*, **18**: 54–64.

ALTUKHOV, Y. P. & KALABUSHKIN, B. A. 1974. Stable polymorphism in present and fossil populations of the mollusc *Littorina squalida*. *Doklady Akademii Nauk S. S. S. R*, **215**: 1477–1480. (Russian. Translated in *Doklady Biological Sciences* **215**: 140–142).

AMANO, K., VERMEIJ, G. J. & NARITA, K. 1993. Early evolution and distribution of the gastropod genus *Nucella*, with special reference to Miocene species from Japan. *Transactions and Proceedings of the Palaeontological Society of Japan. New Series*, **171**: 237–248.

AMIO, M. 1963. A comparitive embryology of marine gastropods, with ecological consideration. *Journal of the Shimonoseki University of Fisheries*, **12**: 229–358.

ANKEL, W. E. 1930. Die atypische Spermatogenese von *Janthina*. *Zeitschrift für Zellforschung und Mikroskopische Anatomie*, **11**: 491–604.

ANKEL, W. E. 1936a. Die Frasspuren von *Helcion* und *Littorina* und die Funktion der Radula. *Verhandlungen der Deutschen Zoologischen Gesellschaft. Supplementband, Zoologischer Anzeiger*, **38**: 174–182.

ANKEL, W. E. 1936b. Prosobranchia. In: *Die Tierwelt der Nord- und Ostsee* (Grimpe, G. & Wagler, E., eds), lief. 29, pt. IX.b1. Akademische Verlagsgesellschaft, Leipzig.

ANKEL, W. E. 1938. Wie frisst *Littorina*? I. Radula–Bewegung und Fresspur. *Senckenbergiana*, **19**: 317–333.

ANKEL, W. E. 1950. Über die relative Grösse und die ökologische Bedeutung des Radula-Apparates bei *Littorina*-Arten und anderen Prosobranchiern. *Zoologischer Anzeiger, Ergänzungsband*, **145**: 19–27.

ANTEVS, E. 1928. Shell beds on the Skagerack. *Geologiska Föreningens i Stockholm Förhandlingar*, **50**: 479–750.

ANTON, H. E. 1838. *Verzeichniss der Conchylien welche sich in der Sammlung von Hermann Eduard Anton befinden*. Halle.

ANTONIO, C. D' 1985. Epiphytes on the rocky intertidal red alga *Rhodomela larix* (Turner) C. Agardh: negative effects on the host and food for herbivores? *Journal of Experimental Marine Biology and Ecology*, **86**: 197–218.

APPLETON, R. D. & PALMER, A. R. 1988. Water-borne stimuli released by predatory crabs and damaged prey induce more predator-resistant shells in a marine gastropod. *Proceedings of the National Academy of Sciences of the U.S.A.*, **85**: 4387–4391.

ARADAS, A. & BENOIT, L. 1870. *Conchigliologia Vivente Marina della Sicilia*. C. Galatola, Catania.

ARMENTROUT, J. M. 1975. Molluscan biostratigraphy of the Lincoln Creek Formation, southwest Washington. In: *Paleogene Symposium and Selected Technical Papers, American Association of Petroleum Geologists, Pacific Section Annual Meeting, 1975* (Weaver, D. W., ed.), 14–48. California.

ARMENTROUT, J. M., HULL, D. A., BEAULIEU, J. D. & RAU, W. W. 1983. Correlation of Cenozoic stratigraphical units of western Oregon and Washington. *Oregon Department of Geology and Mineral Industries, Oil and Gas Investigation*, **7**: 1–90.

ARNOLD, E. N. 1994. Investigating the origins of performance advantage: adaptation, exaptation and lineage effects. In: *Phylogenetics and Ecology. Linnean Society Symposium Series*, **17** (Eggleton, P. & Vane-Wright, R. I., eds), 123–168. Academic Press, London.

ARNOLD, R. 1903. The paleontology and stratigraphy of the marine Pliocene and Pleistocene of San Pedro, California. *Memoirs of the California Academy of Sciences*, **3**: 1–420.

ARNOLD, R. 1906. The Tertiary and Quaternary pectens of California. *United States Geological Survey Professional Papers*, **47**: 1–264.

ARNOLD, R. 1908. Descriptions of new Cretaceous and Tertiary fossils from the Santa Cruz Mountains, California. *Proceedings of the United States National Museum*, **34**: 345–390.

ARNOLD, R. 1910. Paleontology of the Coalinga District, Fresno and Kings Counties, California. *United States Geological Survey Bulletin*, **396**: 1–173.

ARNOLD, R. & ANDERSON, R. 1910. Geology and oil resources of the Coalinga District, California. *United States Geological Survey Bulletin*, **398**: 1–354.

ARNOLD, R. & HANNIBAL, H. 1913. The marine Tertiary stratigraphy of the north Pacific coast of America. *Proceedings of the American Philosophical Society*, **52**: 559–605.

ASH, V. B. 1989. Resistance to shell breaking in two intertidal snails. *Zoological Journal of the Linnean Society*, **96**: 167–184.

ATKINSON, W. D. & NEWBURY, S. F. 1984. The adaptations of the rough winkle, *Littorina rudis*, to desiccation and to dislodgement by wind and waves. *Journal of Animal Ecology*, **53**: 93–105.

ATKINSON, W. D. & WARWICK, T. 1983. The role of selection in the colour polymorphism of *Littorina rudis* Maton and *Littorina arcana* Hannaford-Ellis (Prosobranchia: Littorinidae). *Biological Journal of the Linnean Society*, **20**: 137–151.

AURIVILLIUS, C. W. S. 1885. Öfversigt öfver de af Vega-Expeditionen insamlade Arktiska hafsmollusker. II. Placophora och Gastropoda. In: *Vega-Expeditionens Vetenskapliga Iakttagelser* (Nordenskiöld, A. E., ed.), **4**: 311–383. Stockholm.

AVISE, J. C. 1994. *Molecular Markers, Natural History and Evolution*. Chapman & Hall, New York.

AVISE, J. C., ARNOLD, J., BALL, R. M., BERMINGHAM, E., LAMB, T., NEIGEL, J. E., REEB, C. A. & SAUNDERS, N. C. 1987. Intraspecific phylogeography: the mitochondrial DNA bridge between population genetics and systematics. *Annual Review of Ecology and Systematics*, **18**: 489–522.

BACKELJAU, T., BREUGELMANS, K., BRITO, C., BRUYN, L. DE, WOLF, H. DE & TIMMERMANS, J.-M. 1995. Macrogeographic genetic homogeneity in *Littorina striata* from the Azores (Mollusca, Prosobranchia). *Açoreana. Supplement*, **1995**: 159–171.

BACKELJAU, T., MEYER, M. DE, JANSSENS, L. & PROESMANS, R. 1984. Prosobranch and shelled opisthobranch molluscs from Store Ekkerøya, Varangerfjorden (northern Norway). *Fauna Norvegica. Series A*, **5**: 1–5.

BACKELJAU, T. & WARMOES, T. 1992. The phylogenetic relationships of ten Atlantic littorinids assessed by allozyme electrophoresis. In: *Proceedings of the Third International Symposium on Littorinid Biology* (Grahame, J., Mill, P. J. & Reid, D. G., eds), 9–24. Malacological Society of London, London.

BACKELJAU, T., WOLF, H. DE, DONGEN, S. VAN & BRITO, C. 1994. The phylogenetic relationships of *Littorina striata* as deduced from allozyme data. *Cahiers de Biologie Marine*, **35**: 239–240.

BADEN-POWELL, D. F. W. 1930. Notes on raised beach Mollusca from the Isle of Portland. *Proceedings of the Malacological Society of London*, **19**: 67–76.

BAGENAL, T. B. 1957. The vertical range of some littoral animals on St Kilda. *Scottish Naturalist*, **69**: 50–51.

BAILY, J. L. 1939. The taxonomic value of the radula. *Annual Report of the American Malacological Union*, **7**: 3.

BAKKER, K. 1959. Feeding habits and zonation in some intertidal snails. *Archives Néerlandaises de Zoologie*, **13**: 230–257.

BANDEL, K. 1974. Studies on Littorinidae from the Atlantic. *Veliger*, **17**: 92–114.

BANDEL, K. 1975. Embryonale und larvale Schale einiger Prosobranchier (Gastropoda, Mollusca) der Oosterschelde (Nordsee). *Hydrobiological Bulletin*, **9**: 3–22.

BANDEL, K. 1984. The radulae of Caribbean and other Mesogastropoda and Neogastropoda. *Zoologische Verhandelingen*, **214**: 1–188.

BANDEL, K. & KADOLSKY, D. 1982. Western Atlantic species of *Nodilittorina* (Gastropoda: Prosobranchia): comparative morphology and its functional, ecological, phylogenetic and taxonomic implications. *Veliger*, **25**: 1–42.

BANDEL, K. & WEDLER, E. 1987. Hydroid, amphineuran and gastropod zonation in the littoral of the Caribbean Sea, Colombia. *Senckenbergiana Maritima*, **19**: 1–129.

BÁRDARSON, G. G. 1919. Mollusca marina Islandiae. *Skyrsla um hid Íslenska Náttúrufraedisfjelag*, **1917–1918**: 45–75.

BÁRDARSON, G. G. 1920. Om den marine Molluskfauna ved Vestkysten af Island. *Kongelige Danske Videnskabernes Selskabs Skrifter. Biologiske Meddelelser*, **2** (3): 1–139.

BÁRDARSON, G. G. 1923. Old sea-deposits in Borgarfjördur and Hvalfjördur. *Rit Visindafjelags Islendinga*, **1**: 1–117.

BÁRDARSON, G. G. 1925. A stratigraphic survey of the Pliocene deposits at Tjörnes, in northern Iceland. *Biologiske Meddelelser, Kongelige Danske Videnskabernes Selskab*, **4** (5): 1–118.

BARKER, K. M. & CHAPMAN, A. R. O. 1990. Feeding preferences of periwinkles among four species of *Fucus*. *Marine Biology, Berlin*, **106**: 113–118.

BARKMAN, J. J. 1955. On the distribution and ecology of *Littorina obtusata* (L.) and its subspecific units. *Archives Néerlandaises de Zoologie*, **11**: 22–86.

BARNARD, K. H. 1963. Contributions to the knowledge of South African marine Mollusca. Part III. Gastropoda: Prosobranchiata: Taenioglossa. *Annals of the South African Museum*, **47**: 1–199.

BARNES, R. S. K. 1993. On the nature of the coastal lagoon winkles attributed to *Littorina tenebrosa* and *Littorina saxatilis*. *Cahiers de Biologie Marine*, **34**: 477–495.

BARRON, J. A. & BALDAUF, J. G. 1990. Development of biosiliceous sedimentation in the North Pacific during the Miocene and Early Pliocene. In: *Pacific Neogene Events: Their Timing, Nature and Interrelationships* (Tsuchi, R., ed.), 43–63. University of Tokyo Press, Tokyo.

BARSANOVA, N. G. 1961. Some comparative characteristics of the rocky littoral fauna in the Barents and White Seas. *Trudy Instituta Okeanologii Akademiya Nauk S. S. S. R.*, **46**: 140–146. (Russian).

BARSOTTI, G. & CAMPANI, E. 1982. Il Promontorio di Castiglioncello (LI): III. Rinvenimento di una popolazione di *Littorina littorea* (L.) — Moll. Gastropoda Prosobranchia. *Quaderni del Museo di Storia Naturale di Livorno*, **3**: 65–71.

BARTON, N. H. & HEWITT, G. M. 1985. Analysis of hybrid zones. *Annual Review of Ecology and Systematics*, **16**: 113–148.

BASILYAN, A. E., BARINOV, K. B., ORESHKINA, T. V. & TRUBICHIN, V. M. 1991. Pliocene transgressions of the Bering Sea. In: *Paleogeography and Biostratigraphy of the Pliocene and Quaternary. 13th Congress INQUA, Peking, 1991*, 5–24. Nauka, Moscow. (Russian).

BASTER, J. 1759–1762. *Opuscula Subseciva*, **1**. Harlem.

BATCHELDER, C. H. 1915. Migration of *Ilyanassa obsoleta*, *Litorina litorea* and *Litorina rudis*. *Nautilus*, **29**: 43–46.

BAUER, B., FIORONI, P., IDE, I., LIEBE, S., OEHLMANN, J., STROBEN, E. & WATERMANN, B. 1995. TBT effects on the female genital system of *Littorina littorea*: a possible indicator of tributyltin pollution. *Hydrobiologia*, **309**: 15–27.

BAUM, D. 1992. Phylogenetic species concepts. *Trends in Ecology and Evolution*, **7**: 1–2.

BAUM, D. A. & LARSON, A. 1991. Adaptation reviewed: a phylogenetic methodology for studying character macroevolution. *Systematic Zoology*, **40**: 1–18.

BAVAY, A. 1920. Notes sur les littorines (projet d'études). *Bulletin de la Société Zoologique de France*, **45**: 249–255.

BAXTER, R. 1987. *Mollusks of Alaska*. Shells and Sea Life, Bayside, California.

BAYNE, B. L., THOMPSON, R. J. & WIDDOWS, J. 1976. Physiology: I. In: *Marine Mussels: Their Ecology and Physiology* (Bayne, B. L., ed.), 121–206. Cambridge University Press, Cambridge.

BEAN, W. 1844. A supplement of new species. In: *British Marine Conchology* (Thorpe, C., ed.), 263–267. Edward Lumley, London.

BEARDMORE, J. A. & MORRIS, S. R. 1978. Genetic variation and species coexistence in *Littorina*. In: *Marine Organisms: Genetics, Ecology and Evolution* (Battaglia, B. & Beardmore, J. A., eds), 123–140. Plenum Press, New York.

BECHER, C. 1886. Mollusken von Jan Mayen. *Die Österreichische Polarstation Jan Mayen, Beobachtungs-Ergebnisse*, **3**: 67–82.

BEECHEY, F. W. 1831. *Narrative of a Voyage to the Pacific and Beering's Strait to Cooperate with the Polar Expeditions: Performed in His Majesty's Ship Blossom, under the Command of Captain F. W. Beechey, R. N. in the Years 1825, 26, 27, 28*. Colburn & Bentley, London.

BEETS, C. 1946a. The Pliocene and Lower Pleistocene gastropods in the collections of the Geological Foundation in The Netherlands (with some remarks on other Dutch collections). *Mededeelingen van de Geologische Stichting. Serie C-IV-1*, **6**: 1–166.

BEETS, C. 1946b. On a shell-deformation of *Littorina littorea* (Linné) caused by *Balanus*. *Basteria*, **10**: 20–25.

BEHRENS, S. 1972. The rôle of wave impact and desiccation on the distribution of *Littorina sitkana* Philippi, 1845. *Veliger*, **15**: 129–132.

BEHRENS YAMADA, S. 1977a. Range extension in *Littorina sitkana* Philippi, 1845, and range contraction in *Littorina planaxis* Philippi, 1847. *Veliger*, **19**: 368.

BEHRENS YAMADA, S. 1977b. Geographic range limitation of intertidal gastropods *Littorina sitkana* and *L. planaxis*. *Marine Biology, Berlin*, **39**: 61–65.

BEHRENS YAMADA, S. 1987. Geographic variation in the growth rates of *Littorina littorea* and *L. saxatilis*. *Marine Biology, Berlin*, **96**: 529–534.

BEHRENS YAMADA, S. 1989. Are direct developers more locally adapted than planktonic developers? *Marine Biology, Berlin*, **103**: 403–411.

BEHRENS YAMADA, S. 1992. Niche relationships of northeastern Pacific littorines. In: *Proceedings of the Third International Symposium on Littorinid Biology* (Grahame, J., Mill, P. J. & Reid, D. G., eds), 281–291. Malacological Society of London, London.

BEHRENS YAMADA, S. & MANSOUR, R. A. 1987. Growth inhibition of native *Littorina saxatilis* (Olivi) by introduced *L. littorea* (L.). *Journal of Experimental Marine Biology and Ecology*, **105**: 187–196.

BELAND, P. 1974. Studies on population structure of *Littorina saxatilis* and *L. obtusata*. *Littorinid Tidings*, **2**: 17.

BELLON-HUMBERT, C. 1973. Les mollusques marins testacés du Maroc. *Travaux de l'Institut Scientifique Chérifien, Série Zoologique*, **37**: 1–144.

BENTHEM JUTTING, T. VAN 1933. *Mollusca (I). A. Gastropoda Prosobranchia et Pulmonata. Fauna van Nederland*, **7**. A. W. Sijthoff, Leiden.

BEQUAERT, J. C. 1943. The genus *Littorina* in the Western Atlantic. *Johnsonia*, **1** (7): 1–27.

BERGER, E. M. 1973. Gene-enzyme variation in three sympatric species of *Littorina*. *Biological Bulletin of the Marine Biological Laboratory, Woods Hole*, **145**: 83–90.

BERGER, E. M. 1977. Gene-enzyme variation in three sympatric species of *Littorina*. II. The Roscoff population, with a note on the origin of North American *L. littorea*. *Biological Bulletin of the Marine Biological Laboratory, Woods Hole*, **153**: 255–264.

BERGER, V. YA. 1973. The adaptation of littorinids to changes in environmental salinity in the Sea of Japan, White Sea and Barents Sea, and its role in the evolution of the molluscan genus *Littorina*. *Tezisy Dokladov. Otchetnaya Nauchnaya Sessiya po Itogam Rabot, Zoologicheskii Institut*, **1973**: 5–6. (Russian).

BERGER, V. YA. 1975. The changes of euryhalinity in ontogenesis and adaptation connected with reproduction of the White Sea mollusc *Littorina saxatilis*. *Biologiya Morya*, **1**: 43–50. (Russian).

BERGER, V. YA. 1978a. A study of salinity adaptations in *Littorina sitchana*, with special reference to the evolution of the genus *Littorina*. *Zoologicheskii Zhurnal*, **57**: 1786–1789. (Russian).

BERGER, V. YA. 1978b. Euryhalinity and the evolution of *Littorina*. *Morphologiya Sistematika i Evolyutsiya Zhivotnikh, Leningrad*, 46–47. (Russian).

BERGER, V. YA. 1986a. Adaptations of marine molluscs to environmental salinity changes. *Issledovaniya Fauny Morei*, **32** (40): 1–216. (Russian).

BERGER, V. YA. 1986b. Features of the ecology of *Littorina saxatilis* (Olivi) from estuaries of the White Sea. In: *Ecological Investigations of the Benthic Organisms of the White Sea* (Skarlato, O. A., ed.), 5–16. Zoologicheski Institut, Academiya Nauk S. S. S. R., Leningrad. (Russian).

BERGER, V. YA. & KOVALEVA, N. M. 1974. A study of adaptations in the genus *Littorina* of the Sea of Japan to changes in salinity with respect to the evolution of the genus. *Zoologicheskii Zhurnal*, **53**: 1459–1465. (Russian).

BERGER, V. YA. & SERGIEVSKY, S. O. 1986. Differences in adaptive responses to changes in salinity between two shell-colour phenotypes of the periwinkle *Littorina obtusata*. *Biologiya Morya*, **1986** (1): 36–41. (Russian).

BERGERARD, J. 1975. Cycle saisonnier dans une population naturelle de *Littorina saxatilis* (Olivi) (Gastéropode, Prosobranche). *Bulletin de la Société Zoologique de France*, **100**: 133–145.

BERGERARD, J. & CAUGANT, D. 1981. Cycle sexuel saisonnier de *Littorina saxatilis* (Mollusque Prosobranche) et son déterminisme. *Bulletin de la Société Zoologique de France*, **106**: 277–283.

BERGGREN, W. A. & HOLLISTER, C. D. 1977. Plate tectonics and paleocirculation—commotion in the ocean. *Tectonophysics*, **38**: 11–48.

BERMAN, J. & CARLTON, J. T. 1991. Marine invasion processes: interactions between native and introduced marsh snails. *Journal of Experimental Marine Biology and Ecology*, **150**: 267–281.

BERRY, A. J. 1961. Some factors affecting the distribution of *Littorina saxatilis* (Olivi). *Journal of Animal Ecology*, **30**: 27–45.

BERRY, A. J. 1986. Semi-lunar and lunar spawning periodicity in some tropical littorinid gastropods. *Journal of Molluscan Studies*, **52**: 144–149.

BERRY, A. J. & HUNT, D. C. 1980. Behaviour and tolerance of salinity and temperature in new-born *Littorina rudis* (Maton) and the range of the species in the Forth estuary. *Journal of Molluscan Studies*, **46**: 55–65.

BERRY, A. J. & SMITH, S. M. 1987. Aspects of the molluscan fauna of the rocky shores of the Firth of Forth and Forth estuary, Scotland. *Proceedings of the Royal Society of Edinburgh. Section B, Biological Sciences*, **93**: 431–447.

BERTNESS, M. D. 1984. Habitat and community modification by an introduced herbivorous snail. *Ecology*, **65**: 370–381.

BERTNESS, M. D., YUND, P. O. & BROWN, A. F. 1983. Snail grazing and the abundance of algal crusts on a sheltered New England rocky beach. *Journal of Experimental Marine Biology and Ecology*, **71**: 147–164.

BIBERSON, P. J. 1977. Le Villafranchien du nord de l'Afrique. *Giornale di Geologia. Serie 2a*, **41**: 225–241.

BIGELOW, R. P. & RATHBUN, E. P. 1903. On the shell of *Littorina litorea* as material for the study of variation. *American Naturalist*, **37**: 171–184.

BINGHAM, F. O. 1972. Several aspects of the reproductive biology of *Littorina irrorata* (Gastropoda). *Nautilus*, **86**: 8–10.

BIRD, J. B. 1968. *Littorina littorea*: occurrence in a northern Newfoundland beach terrace, predating Norse settlements. *Science, New York*, **159**: 114.

BIRSTEIN, V. J. & MIKHAILOVA, N. A. 1990. On the karyology of the genus *Microphallus* and their intermediate gastropod host, *Littorina saxatilis* II. Karyological study of *Littorina saxatilis* (Gastropoda: Prosobranchia). *Genetica*, **80**: 167–170.

BJERREGAARD, P. & DEPLEDGE, M. H. 1994. Cadmium accumulation in *Littorina littorea, Mytilus edulis* and *Carcinus maenas*: the influence of salinity and calcium ion concentrations. *Marine Biology, Berlin*, **119**: 385–395.

BLAINVILLE, H. D. DE 1824. Mollusques. In: *Dictionnaire des Sciences Naturelles* (Levrault, F. G., ed.), **32**: 1–392. Strasbourg & Paris.

BLAINVILLE, H. D. DE 1828. *Turbo*. In: *Dictionnaire des Sciences Naturelles* (Levrault, F. G., ed.), **56**: 95–115. Strasbourg & Paris.

BLAINVILLE, H. D. DE 1830. Malacozoaires ou animaux mollusques. In: *Faune Française* (Vieillot, P., ed.), livr. 18, 20, 23, 28. F. G. Levrault, Paris.

BLEVGAD, H. 1915. Food and conditions of nourishment among the communities of invertebrate animals found on or in the sea bottom in Danish waters. *Report of the Danish Biological Station*, **22**: 41–78.

BOCK, C. E. & JOHNSON, R. E. 1967. The role of behavior in determining the intertidal zonation of *Littorina planaxis* Philippi, 1847, and *Littorina scutulata* Gould, 1849. *Veliger*, **10**: 42–54.

BØGGILD, O. B. 1930. The shell structure of the mollusks. *Kongelige Danske Videnskabernes Selskabs Skrifter, Naturvidensk. og mathem. Afd., 9 Raekke*, **2**: 233–326.

BORKOWSKI, T. V. 1971. Reproduction and reproductive periodicities of South Floridian Littorinidae (Gastropoda: Prosobranchia). *Bulletin of Marine Science*, **21**: 826–840.

BORKOWSKI, T. V. 1974. Growth, mortality, and productivity of south Floridian Littorinidae (Gastropoda: Prosobranchia). *Bulletin of Marine Science*, **24**: 409–438.

BOSS, K. J. 1988. References to molluscan taxa introduced by Linnaeus in the Systema Naturae (1758, 1767). *Nautilus*, **102**: 115–122.

BOULDING, E. G. 1990. Are the opposing selection pressures on exposed and protected shores sufficient to maintain genetic differentiation between gastropod populations with high intermigration rates? *Hydrobiologia*, **193**: 41–52.

BOULDING, E. G. & ALSTYNE, K. L. VAN 1993. Mechanisms of differential survival and growth of two species of *Littorina* on wave-exposed and on protected shores. *Journal of Experimental Marine Biology and Ecology*, **169**: 139–166.

BOULDING, E. G., BUCKLAND-NICKS, J. & ALSTYNE, K. L. VAN 1993. Morphological and allozyme variation in *Littorina sitkana* and related *Littorina* species from the northeastern Pacific. *Veliger*, **36**: 43–68.

BOULDING, E. G. & HAY, T. 1993. Quantitative genetics of shell form of an intertidal snail: constraints on short-term response to selection. *Evolution*, **47**: 576–592.

BOUSFIELD, E. L. 1960. *Canadian Atlantic Sea Shells*. National Museum of Canada, Ottawa.

BOUSFIELD, E. L. & THOMAS, M. L. H. 1975. Postglacial changes in distribution of littoral marine invertebrates in the Canadian Atlantic region. *Proceedings of the Nova Scotian Institute of Science. Supplement*, **3**: 47–60.

BRANDWOOD, A. 1983. Intraspecific variation and distribution of *Littorina rudis* (Maton) in the Fleet—a coastal lagoon in Dorset. *Proceedings of the Dorset Natural History and Archaeology Society*, **104**: 165–167.

BRANDWOOD, A. 1985. The effects of environment upon shell construction and strength in the rough periwinkle *Littorina rudis* Maton (Mollusca: Gastropoda). *Journal of Zoology, London. Series A*, **206**: 551–565.

BRANNER, J. C., NEWSOM, J. F. & ARNOLD, R. 1909. Description of the Santa Cruz Quadrangle, California. *United States Geological Survey Folio*, **163**: 1–11.

BRATTEGARD, T. 1966. The natural history of the Hardangerfjord. 7. Horizontal distribution of the fauna of rocky shores. *Sarsia*, **22**: 1–54.

BREBION, P. 1979. Étude biostratigraphique et paléoécologique du Quaternaire Marocain. *Annales de Paléontologie (Invertébrés)*, **65**: 1–42.

BRENCHLEY, G. A. 1982. Predation on unencapsulated larvae by adults: effects of introduced species on the gastropod *Ilyanassa obsoleta*. *Marine Ecology Progress Series*, **9**: 255–262.

BRENCHLEY, G. A. 1987. Herbivory in juvenile *Ilyanassa obsoleta* (Neogastropoda). *Veliger*, **30**: 167–172.

BRENCHLEY, G. A. & CARLTON, J. T. 1983. Competitive displacement of native mud snails by introduced periwinkles in the New England intertidal zone. *Biological Bulletin of the Marine Biological Laboratory, Woods Hole*, **165**: 543–558.

BRETON, J. LE 1970. Evolution et chute du pénis, étude de l'influence du jeûne, chez *Littorina littorea* L., mollusque, gastéropode, prosobranche. *Compte Rendu Hebdomadaire des Séances de l'Académie des Sciences, Paris, Série D*, **271**: 534–536.

BRETON, J. LE 1971. Étude experimentale de l'influence du jeûne sur le cycle de la gonade mâle et du pénis chez *Littorina littorea* L. *Haliotis*, **1** (1): 25.

BRIGGS, J. C. 1974. *Marine Zoogeography*. McGraw-Hill, New York.

BRITTON, J. C. 1990. The intertidal crevice fauna of Tolo Channel and Harbour, New Territories, Hong Kong. In: *Proceedings of the Second International Marine Biological Workshop: The Marine Flora and Fauna of Hong Kong and Southern China, Hong Kong, 1988* (Morton, B., ed.), 803–835. Hong Kong University Press, Hong Kong.

BRITTON, J. C. 1992. Evaporative water loss, behaviour during emersion, and upper thermal tolerance limits in seven species of eulittoral-fringe Littorinidae (Mollusca: Gastropoda) from Jamaica. In: *Proceedings of the Third International Symposium on Littorinid Biology* (Grahame, J., Mill, P. J. & Reid, D. G., eds), 69–83. Malacological Society of London, London.

BRITTON, J. C. 1995. The relationship between position on shore and shell ornamentation in two size-dependent morphotypes of *Littorina striata*, with an estimate of evaporative water loss in these morphotypes and in *Melarhaphe neritoides*. *Hydrobiologia*, **309**: 129–142.

BRODERIP, W. J. & SOWERBY, G. B. 1829. Observations on new or interesting Mollusca contained, for the most part, in the Museum of the Zoological Society. *Zoological Journal*, **4**: 359–379.

BRØGGER, W. C. 1901. Om de senglaciale og postglaciale nivåforandringer i Kristianiafeltet (Molluskfaunan). *Norges Geologiske Undersøgelse*, **31**: 1–731.

BROOKS, D. R. & MCLENNAN, D. A. 1991. *Phylogeny, Ecology and Behaviour. A Research Program in Comparative Biology*. University of Chicago Press, Chicago.

BROUGH, C. N. & WHITE, K. N. 1990. Localization of metals in the gastropod *Littorina saxatilis* (Prosobranchia: Littorinoidea) from a polluted site. *Acta Zoologica*, **71**: 77–88.

BROWN, K. M. & QUINN, J. F. 1988. The effect of wave action on growth in three species of intertidal gastropods. *Oecologia*, **75**: 420–425.

BROWN, T. 1827. *Illustrations of the Conchology of Great Britain and Ireland*. W. H. & D. Lizars, Edinburgh.

BROWN, T. 1837a-1844a. *Illustrations of the Recent Conchology of Great Britain and Ireland*, **2**. Smith, Elder & Co, London.

BROWN, T. 1837b-1849. *Illustrations of the Fossil Conchology of Great Britain and Ireland*. Smith, Elder & Co, London.

BRUNDIN, L. Z. 1988. Phylogenetic biogeography. In: *Analytical Biogeography: an Integrated Approach to the Study of Animal and Plant Distributions* (Myers, A. A. & Gill, P. S., eds), 343–369. Chapman & Hall, London.

BRUSINA, S. 1870. *Ipsa Chiereghini Conchylia ovvero Contribuzione Pella Malacologia Adriatica desunta dal Manoscritto...dall'Abate Stefano Chiereghini Ven. Clodiense*. Pisa.

BRUSINA, S. 1886. Appunti ed osservazioni sull'ultimo lavoro di J. Gwyn Jeffreys. *Glasnik Hrvatskoga Narodoslovnogo Drustva*, **1**: 182–221.

BRUYNE, R. H. DE 1991. *Schelpen van de Nederlanse Kust*. Jeugdbondsuitgeverij & Stichting Uitgeverij KNNV, Utrecht.

BRYAN, G. W., LANGSTON, W. J., HUMMERSTONE, L. G., BURT, G. R. & HO, Y. B. 1983. An assessment of the gastropod, *Littorina littorea*, as an indicator of heavy-metal contamination in United Kingdom estuaries. *Journal of the Marine Biological Association of the United Kingdom*, **63**: 327–345.

BUCKLAND-NICKS, J. A. 1973. The fine structure of the spermatozoon of *Littorina* (Gastropoda: Prosobranchia), with special reference to sperm motility. *Zeitschrift für Zellforschung und Mikroskopische Anatomie*, **144**: 11–29.

BUCKLAND-NICKS, J. A. & CHIA, F.-S. 1977. On the nurse cell and the spermatozeugma in *Littorina sitkana*. *Cell and Tissue Research*, **179**: 347–356.

BUCKLAND-NICKS, J. A. & CHIA, F.-S. 1984. A technique for obtaining replicas of small tubular systems by microinjection of vinyl acetate. *Stain Technology*, **59**: 311–313.

BUCKLAND-NICKS, J. A. & CHIA, F.-S. 1986. Fine structure of Sertoli cells in three marine snails with a discussion on the functional morphology of Sertoli cells in general. *Cell and Tissue Research*, **245**: 305–313.

BUCKLAND-NICKS, J. A. & CHIA, F.-S. 1990. Egg capsule formation and hatching in the marine snail *Littorina sitkana*. *Philosophical Transactions of the Royal Society of London. Series B*, **326**: 159–176.

BUCKLAND-NICKS, J. A., CHIA, F.-S. & BEHRENS, S. 1973. Oviposition and development of two intertidal snails *Littorina sitkana* and *Littorina scutulata*. *Canadian Journal of Zoology*, **51**: 359–365.

BUCKLAND-NICKS, J. A. & WORTHERN, G. T. 1992. Functional morphology of the mammiliform penial glands of *Littorina saxatilis* (Gastropoda). *Zoomorphology*, **112**: 217–225.

BULLOCK, R. C. 1995. The distribution of the molluscan fauna associated with the intertidal coralline algal turf of a partially submerged volcanic crater, the Ilhéu de Vila Franca, São Miguel, Azores. *Açoreana. Supplement*, **1995**: 9–55.

BUMPUS, H. C. 1898. The variations and mutations of the introduced *Littorina*. A third contribution to the study of variation. *Zoological Bulletin, Boston*, **1**: 247–259.

BURCH, J. Q. 1945. Family Littorinidae. *Minutes of the Conchological Club of Southern California*, **55**: 9–13, 36, 37.

BURCH, J. Q. 1946. Rediscovery of *Littorina newcombiana* (Hemphill, 1876). *Minutes of the Conchological Club of Southern California*, **64**: 10.

BURNAY, L. P. 1986. *Moluscos Testáceos Marinhos da Berlenga*. Secretaria de Estado do Ambiente e dos Recursos Naturais, Serviço Nacional de Parques, Reservas e Conservação da Natureza, Lisboa.

BURNAY, L. P. & MONTEIRO, A. A. 1977. *Seashells from Cape Verde Islands*. Lisbon.

BUSH, G. L. 1975. Modes of animal speciation. *Annual Review of Ecology and Systematics*, **6**: 339–364.

BUSH, G. L. 1993. A reaffirmation of Santa Rosalia, or why are there so many kinds of *small* animals? In: *Evolutionary Patterns and Processes. Linnean Society Symposium Series* **14** (Lees, D. R. & Edwards, D., eds), 229–249. Academic Press, London.

BUSH, G. L. 1994. Sympatric speciation in animals: new wine in old bottles. *Trends in Ecology and Evolution*, **9**: 285–288.

BUYS, J.-P. 1992. Collecting shells on the Cape Verde Islands. *Vita Marina*, **41**: 101–113.

BYERS, B. A. 1990. Shell colour polymorphism associated with substrate colour in the intertidal snail *Littorina saxatilis* Olivi (Prosobranchia: Littorinidae). *Biological Journal of the Linnean Society*, **40**: 3–10.

CAI, R. 1991. *Fauna of Zhejiang. Molluscs*. Zhejiang Science and Technology Publishing House. (Chinese).

CAJARAVILLE, M. P., MARIGOMEZ, J. A. & ANGULO, E. 1990. Ultrastructural study of the short-term toxic effects of naphthalene on the kidney of the marine prosobranch *Littorina littorea*. *Journal of Invertebrate Pathology*, **55**: 215–224.

CALEY, K. J. 1994. *The Status and Ecology of the Intertidal Gastropod Littorina neglecta Bean*. PhD thesis, University of Leeds.

CALEY, K. J., GRAHAME, J. & MILL, P. J. 1995. A geographically-based study of shell shape in small rough periwinkles. *Hydrobiologia*, **309**: 181–193.

CAMBRIDGE, P. G. 1975. Field meeting to Bramerton, near Norwich, 14–15 September 1974. *Bulletin of the Geological Society of Norfolk*, **27**: 33–46.

CAMBRIDGE, P. G. 1979. Field meeting at Bramerton Common, September 3rd 1978. *Bulletin of the Geological Society of Norfolk*, **31**: 71–77.

CANNON, J. P. & HUGHES, R. N. 1992. Resistance to environmental stressors in *Littorina saxatilis* (Olivi) and *L. neglecta* Bean. In: *Proceedings of the Third International Symposium on Littorinid Biology* (Grahame, J., Mill, P. J. & Reid, D. G., eds), 61–68. Malacological Sociey of London, London.

CARLTON, J. T. 1969. *Littorina littorea* in California (San Francisco and Trinidad Bays). *Veliger*, **11**: 283–284.

CARLTON, J. T. 1982. The historical biogeography of *Littorina littorea* on the Atlantic coast of North America, and implications for the interpretation of the structure of New England intertidal communities. *Malacological Review*, **15**: 146.

CARLTON, J. T. 1992. Introduced marine and estuarine mollusks of North America: an end-of-the-20th-century perspective. *Journal of Shellfish Research*, **11**: 489–505.

CARPENTER, P. P. 1864. Supplementary report on the present state of our knowledge with regard to the Mollusca of the west coast of North America. *Report of the British Association for the Advancement of Science*, **1863**: 517–684.

CARPENTER, P. P. 1865. Diagnoses of new forms of Mollusca from the Vancouver district. *Annals and Magazine of Natural History, Series 3*, **15**: 28–32.

CARTER, L. D., BRIGHAM-GRETTE, J. & MARINCOVICH, L. 1986. Late Cenozoic Arctic Ocean sea ice and terrestrial paleoclimate. *Geology*, **14**: 675–678.

CARVALHO, R. N. DA 1945. Catálogo da colecção de invertebrados de Portugal existentes no Museu Zoológico da Universidade de Coimbra. Mollusca 2.ª Parte: B. Classe Gastropoda Cuvier. *Memórias e Estudos do Museu Zoológico da Universidade de Coimbra*, **167**: 1–50.

CASTENHOLZ, R. W. 1961. The effects of grazing on marine littoral diatom populations. *Ecology*, **42**: 783–794.

CATALIOTTI-VALDINA, J. 1984. Littorines Quaternaires de Mediterranée. *Bulletin de l'Association Française pour l'Étude du Quaternaire*, **20**: 211–219.

CAUGANT, D. & BERGERARD, J. 1979. Variabilité de la leucine aminopeptidase chez *Littorina saxatilis* (Olivi) et *Littorina nigrolineata* (Gray) (Gastéropodes Prosobranches). *Archives de Zoologie Expérimentale et Générale*, **120**: 247–262.

CAUGANT, D. & BERGERARD, J. 1980. The sexual cycle and reproductive modality in *Littorina saxatilis* Olivi (Mollusca: Gastropoda). *Veliger*, **23**: 107–112.

CAULLERY, M. & PELSENEER, P. 1910. Sur la ponte et le développement du vignot (*Littorina littorea*). *Bulletin Scientifique de la France et de la Belgique*, **44**: 357–360.

CÉPÈDE, C. 1910. Recherches sur les infusoires astomes. *Archives de Zoologie Expérimentale et Générale*, **3**: 341–609.

Cernohorsky, W. O. 1971. Comments on the proposed designation of a type-species for *Littorina* Férussac, 1822. Z. N. (S.) 1901. *Bulletin of Zoological Nomenclature*, **27**: 210.

Chambers, M. R. 1980. Zonation, abundance and biomass of gastropods from two Hong Kong rocky shores. In: *Proceedings of the First International Workshop on the Malacofauna of Hong Kong and Southern China, Hong Kong, 1977* (Morton, B. S., ed.), 139–148. Hong Kong University Press, Hong Kong.

Chapman, A. R. O. 1989. Abundance of *Fucus spiralis* and ephemeral seaweeds in a high eulittoral zone: effects of grazers, canopy and substratum type. *Marine Biology, Berlin*, **102**: 565–572.

Chapman, M. G. & Underwood, A. J. 1994. Dispersal of the intertidal snail, *Nodilittorina pyramidalis*, in response to the topographic complexity of the substratum. *Journal of Experimental Marine Biology and Ecology*, **179**: 145–169.

Chase, M. E. & Thomas, M. L. H. 1995a. The effect of the rate and onset of temperature increase on spawning of the periwinkle, *Littorina littorea* (L.). *Journal of Experimental Marine Biology and Ecology*, **186**: 277–287.

Chase, M. E. & Thomas, M. L. 1995b. Evidence for double recruitment in the common periwinkle, *Littorina littorea* (Linnaeus, 1758), on Pendleton Island, New Brunswick, Canada. *Journal of Shellfish Research*, **14**: 153–158.

Chatwin, C. P. 1932. The Quaternary beaches of Gower. Appendix A: the marine shells. *Proceedings of the Geologists' Association, London*, **43**: 317–318.

Chemnitz, J. H. 1781. *Neues Systematisches Conchylien-Cabinet*, **5**. Nürnberg.

Chinzei, K. 1986. Faunal succession and geographic distribution of Neogene molluscan faunas in Japan. *Palaeontological Society of Japan Special Papers*, **29**: 17–32.

Choe, B. L. 1992. *Illustrated Encyclopaedia of Fauna and Flora of Korea*, **33**: *Mollusca (II)*. Ministry of Education, Republic of Korea.

Chow, V. 1975. The importance of size in the intertidal distribution of *Littorina scutulata*. *Veliger*, **18**: 69–78.

Chow, V. 1987a. Morphological classification of sibling species of *Littorina* (Gastropoda: Prosobranchia): discretionary use of discriminant analysis. *Veliger*, **29**: 359–366.

Chow, V. 1987b. Patterns of growth and energy allocation in northern California populations of *Littorina* (Gastropoda: Prosobranchia). *Journal of Experimental Marine Biology and Ecology*, **110**: 69–89.

Chow, V. 1989. Intraspecific competition in a fluctuating population of *Littorina plena* Gould (Gastropoda: Prosobranchia). *Journal of Experimental Marine Biology and Ecology*, **130**: 147–165.

Clark, B. L. 1915. Fauna of the San Pablo group of Middle California. *University of California Publications. Bulletin of the Department of Geology*, **8**: 385–572.

Clark, B. L. & Arnold, R. 1923. Fauna of the Sooke formation, Vancouver Island. *University of California Publications. Bulletin of the Department of Geological Sciences*, **14**: 123–234.

Clark, W. 1850a. Observations on the Littorinidae. *Annals and Magazine of Natural History, Series 2*, **5**: 352–364.

Clark, W. 1850b. Observations on the Lacunae. *Annals and Magazine of Natural History, Series 2*, **6**: 29–35.

Clark, W. 1855. *A History of the British Marine Testaceous Mollusca*. John van Voorst, London.

Clarke, A. 1983. Life in cold water: the physiological ecology of polar marine ecosystems. *Oceanography and Marine Biology Annual Review*, **21**: 341–453.

Clarke, A. H. 1963a. Supplementary notes on the pre-Columbian *Littorina littorea* in Nova Scotia. *Nautilus*, **77**: 8–11.

Clarke, A. H. 1963b. *Littorina littorea* as an indicator of Norse settlements. *Science, New York*, **142**: 1022.

Clarke, A. H. 1971. *Littorina littorea*, native or introduced? *The Biologist*, **53**: 160–162.

Clarke, A. H. & Erskine, J. S. 1961. Pre-Columbian *Littorina littorea* in Nova Scotia. *Science, New York*, **134**: 393–394.

Clay, E. 1967. Littorina littorea *(Linnaeus)*, Littorina littoralis *(Linnaeus)*, Littorina saxatilis *(Olivi)*, Littorina neritoides *(Linnaeus)*. *Literature Survey of the Common Fauna of Estuaries. No. 9. Revised edition*. ICI Ltd, Brixham, Devon.

Clyne, P. M. & Duffus, J. H. 1979. A preliminary study of variation in reproduction of *Littorina rudis* (Maton). *Journal of Molluscan Studies*, **45**: 178–185.

Coan, E. 1981. *James Graham Cooper. Pioneer Western Naturalist*. University Press of Idaho, Moscow, Idaho.

Coan, E. & Roth, B. 1987. The malacological taxa of Henry Hemphill. *Veliger*, **29**: 322–339.

Coates, H. 1916. Occurrence of a pearl in *Littorina littorea* Linné. *Journal of Conchology*, **15**: 10.

Coddington, J. A. 1988. Cladistic tests of adaptational hypotheses. *Cladistics*, **4**: 3–22.

Coddington, J. A. 1994. The roles of homology and convergence in studies of adaptation. In: *Phylogenetics and Ecology. Linnean Society Symposium Series* **17** (Eggleton, P. & Vane-Wright, R. I., eds), 53–78. Academic Press, London.

Coen, G. 1933. Saggio di una sylloge molluscorum Adriaticorum. *Memorie Reale Comitato Talassografico Italiano*, **192**: 1–186.

Cole, H. A. 1956. Benthos and the shellfish of commerce. In: *Sea Fisheries: Their Investigation in the United Kingdom* (Graham, M., ed.), 139–206. Edward Arnold, London.

Cole, K. R. 1991. An introductory survey of the periwinkle species found around the shores of Lundy. *Lundy Field Society Annual Report*, **41**: 67–72.

COLLIN, J. 1884. *Om Limfjordens tidligere og nuvaerende Marine Fauna, med saerligt Hensyn til Bløddyrfaunaen.* Gyldendalske Boghandels Forlag, Kjøbenhavn.

COLLINS, T. M., FRAZER, K., PALMER, A. R., VERMEIJ, G. J. & BROWN, W. M. in press. Evolutionary history of northern hemisphere *Nucella* (Gastropoda, Muricidae): molecular, morphological, ecological, and paleontological evidence. *Evolution*, in press.

COLMAN, J. 1932. A statistical analysis of the species concept in *Littorina*. *Biological Bulletin of the Marine Biological Laboratory, Woods Hole*, **62**: 223–243.

COLMAN, J. 1933. The nature of the intertidal zonation of plants and animals. *Journal of the Marine Biological Association of the United Kingdom*, **18**: 435–476.

COLMAN, J. 1940. On the faunas inhabiting intertidal seaweeds. *Journal of the Marine Biological Association of the United Kingdom*, **24**: 129–183.

COOK, L. M. 1983. Polymorphism in a mangrove snail in Papua New Guinea. *Biological Journal of the Linnean Society*, **20**: 167–173.

COOK, L. M. & GARBETT, S. D. 1992. Selection in the polymorphic mangrove snail *Littoraria pallescens*. In: *Proceedings of the Third International Symposium on Littorinid Biology* (Grahame, J., Mill, P. J. & Reid, D. G., eds), 247–253. Malacological Society of London, London.

COOKE, A. H. 1895. Molluscs. In: *The Cambridge Natural History* (Harmer, S. F. & Shipley, A. E., eds), **3**: 1–459. MacMillan, London.

COOKE, A. H. 1914. *Littorina littorea* (L.): a doubtful record. *Journal of Conchology*, **14**: 305.

COOPER, J. G. 1888. Catalogue of Californian fossils. *Seventh Annual Report of the State Mineralogist, California*, **1887**: 223–308.

COOPER, J. G. 1896. On some new Cretaceous (and Eocene?) Mollusca of California. *Proceedings of the California Academy of Sciences. Series 2*, **6**: 330–337.

COOPER, W. 1860. Report upon the Mollusca collected on the survey. In: *Reports of Explorations and Surveys, to Ascertain the Most Practicable and Economical Route for a Railroad from the Mississippi River to the Pacific Ocean* (Stevens, I. I., ed.), **12** (book 2, part 3): 369–386. U. S. Government, 36th Congress, 1st Session.

CORREA SANDOVAL, A. 1994. *Biometria y Variabilidad Genetica de* Littorina keenae *Rosewater, 1978 y* L. scutulata *Gould, 1849 (Gastropoda: Prosobranchia) en Dos Localidades de la Costa de Ensenada, Baja California, Mexico.* MSc Thesis, Centro de Investigacion Cientifica y de Educacion Superior de Ensenada, Mexico.

COSEL, R. VON 1982. Marine Mollusken der Kapverdischen Inseln Übersicht mit zoogeographischen Anmerkungen. *Courier Forschungsinstitut Senckenberg*, **52**: 35–76.

COSSMANN, M. 1915. *Essais de Paléoconchologie Comparée.* **10**. Paris.

COSTA, E. M. DA 1778. *British Conchology*. London.

CRAPP, M. 1973. The distribution and abundance of animals and plants on the rocky shores of Bantry Bay. *Irish Fisheries Investigations. Series B (Marine)*, **9**: 3–35.

CRISP, D. J. 1956. The intertidal zoology of Rockall. In: *Rockall* (Fisher, J., ed.), 177–179. G. Bles, London.

CRISP, D. J. (ed.) 1964. The effects of the severe winter of 1962–63 on marine life in Britain. *Journal of Animal Ecology*, **33**: 165–210.

CRISP, D. J. & FISCHER-PIETTE, E. 1959. Répartition des principales espèces intercotidales de la côtes atlantique française en 1954–1955. *Annales de l'Institut Océanographique*, **36**: 275–388.

CRISP, D. J. & SOUTHWARD, A. J. 1958. The distribution of intertidal organisms along the coast of the English Channel. *Journal of the Marine Biological Association of the United Kingdom*, **37**: 157–208.

CRISP, D. J. & SOUTHWARD, A. J. 1961. Different types of cirral activity of barnacles. *Philosophical Transactions of the Royal Society of London. Series B*, **243**: 271–307.

CRONIN, T. M. 1985. Speciation and stasis in marine Ostracoda: climatic modulation of evolution. *Science, New York*, **227**: 60–63.

CROSSE, H. 1862. Description d'espèces marines recueillies par M. G. Cuming dans le nord de la Chine. *Journal de Conchyliologie*, **10**: 51–57.

CROSSE, H. & DEBEAUX, O. 1863. Note sur quelques espèces nouvelles ou peu connues du littoral de l'empire Chinois. *Journal de Conchyliologie*, **11**: 253–265.

CROSSLAND, S., COATES, D., GRAHAME, J. & MILL, P. J. 1993. Use of random amplified polymorphic DNAs (RAPDs) in separating two sibling species of *Littorina*. *Marine Ecology Progress Series*, **96**: 301–305.

CROSSLAND, S., COATES, D., GRAHAME, J. & MILL, P. J. 1995. The *Littorina saxatilis* complex—interpretation using random amplified polymorphic DNAs. In: *Origin and Evolutionary Radiation of the Mollusca* (Taylor, J. D., ed.), 205–209. Oxford University Press, Oxford.

CROTHERS, J. H. 1983. Field experiments on the effects of crude oil and dispersant on the common animals and plants of rocky sea shores. *Marine Environmental Research*, **8**: 215–239.

CROTHERS, J. H. 1992. Shell size and shape variation in *Littorina littorea* (L.) from West Somerset. In: *Proceedings of the Third International Symposium on Littorinid Biology* (Grahame, J., Mill, P. J. & Reid, D. G., eds), 91–97. Malacological Society of London, London.

CURREY, J. D. & HUGHES, R. N. 1982. Strength of the dogwhelk *Nucella lapillus* and the winkle *Littorina littorea* from different habitats. *Journal of Animal Ecology*, **51**: 47–56.

DACIE, J. C. 1917. *Littorina littorea* (L.): a doubtful record. *Journal of Conchology*, **15**: 179–180.

DAGUZAN, J. 1976a. Contribution à l'écologie des Littorinidae (mollusques gastéropodes prosobranches). I. *Littorina neritoides* (L.) et *Littorina saxatilis* (Olivi). *Cahiers de Biologie Marine*, **17**: 213–236.

DAGUZAN, J. 1976b. Contribution à l'écologie des Littorinidae (mollusques gastéropodes prosobranches). II.—*Littorina littorea* (L.) et *Littorina littoralis* (L.). *Cahiers de Biologie Marine*, **17**: 275–293.

DAGUZAN, J. 1977. Analyse biometrique du dimorphisme sexuel chez quelques Littorinidae (Mollusques, Gastéropodes, Prosobranches). *Haliotis*, **6**: 17–40.

DAHL, A. L. 1964. Macroscopic algal foods of *Littorina planaxis* Philippi and *Littorina scutulata* Gould. *Veliger*, **7**: 139–143.

DALL, W. H. 1872. Preliminary descriptions of new species of mollusks from the northwest coast of America. *Proceedings of the California Academy of Sciences* (preprint from volume 4), 1–2.

DALL, W. H. 1873. Descriptions of new species of mollusks from the northwest coast of America. *Proceedings of the California Academy of Sciences*, **4**: 270–271.

DALL, W. H. 1886a. Report on the mollusks collected by L. M. Turner at Ungava Bay, North Labrador, and from the adjacent Arctic Seas. *Proceedings of the United States National Museum*, **9**: 202–208.

DALL, W. H. 1886b. Contributions to the natural history of the Commander Islands. No. 6. *Proceedings of the United States National Museum*, **9**: 209–219.

DALL, W. H. 1902. Illustrations and descriptions of new, unfigured, or imperfectly known shells, chiefly American, in the U. S. National Museum. *Proceedings of the United States National Museum*, **24**: 499–566.

DALL, W. H. 1909. Contributions to the Tertiary paleontology of the Pacific Coast. I. The Miocene of Astoria and Coos Bay, Oregon. *United States Geological Survey Professional Papers*, **59**: 1–278.

DALL, W. H. 1918. Changes in and additions to molluscan nomenclature. *Proceedings of the Biological Society of Washington*, **31**: 137–138.

DALL, W. H. 1919. The Mollusca of the Arctic coast of America collected by the Canadian Arctic Expedition west from Bathurst Inlet, with an appended report on a collection of Pleistocene fossil Mollusca. *Report of the Canadian Arctic Expedition 1913–1918*, **8**: 3A-29A.

DALL, W. H. 1920. Pliocene and Pleistocene fossils from the Arctic coast of Alaska and the auriferous beaches of Nome, Norton Sound, Alaska. *United States Geological Survey Professional Papers*, **125–C**: 23–37.

DALL, W. H. 1921. Summary of the marine shellbearing mollusks of the northwest coast of America, from San Diego, California, to the Polar Sea, mostly contained in the collection of the United States National Museum, with illustrations of hitherto unfigured species. *United States National Museum Bulletin*, **112**: 1–217.

DALL, W. H. 1923. Additions and emendations to United States National Museum Bulletin No. 112. *Proceedings of the United States National Museum*, **63**: 1–4.

DALL, W. H. 1925. Illustrations of unfigured types of shells in the collection of the United States National Museum. *Proceedings of the United States National Museum*, **66**: 1–41.

DANCE, S. P. 1967. Report on the Linnaean shell collection. *Proceedings of the Linnean Society of London*, **178**: 1–24.

DANIEL, F. 1883. Faune malacologique terrestre, fluviatile et marine des environs de Brest (Finistère). *Journal de Conchyliologie*, **31**: 330–391.

DAUTZENBERG, P. 1887. Une excursion malacologique à Saint-Lunaire (Ille-et-Vilaine) et aux environs de cette localité. *Bulletin de la Société d'Études Scientifiques de Paris, 9e année, 2e semestre*, **1887**: 1–27.

DAUTZENBERG, P. 1889. Contribution à la faune malacologique des Îles Açores. *Résultats des Campagnes Scientifiques Accomplies sur son Yacht par Albert Ier*, **1**: 1–112.

DAUTZENBERG, P. 1890. Récoltes malacologiques de M. l'abbé Culliéret aux Îles Canaries et au Sénégal. *Mémoires de la Société Zoologique de France*, **3**: 147–168.

DAUTZENBERG, P. 1893a. List de mollusques marins recueillis à Granville et à Saint-Pair. *Journal de Conchyliologie*, **41**: 16–30.

DAUTZENBERG, P. 1893b. Description d'une nouvelle espèce du genre *Littorina*, provenant des côtes de la Tunisie. *Journal de Conchyliologie*, **41**: 35–36.

DAUTZENBERG, P. 1927. Mollusques provenant des campagnes scientifiques du Prince Albert Ier de Monaco dans l'Océan Atlantique et dans le Golfe de Gascogne. *Résultats des Campagnes Scientifiques Accomplies sur son Yacht par Albert Ier Prince Souverain de Monaco*, **72**: 1–400.

DAUTZENBERG, P. & DUROUCHOUX, P. 1900. Faunule malacologique des environs de Saint-Malo. *La Feuille des Jeunes Naturalistes*, **31** (362): 39–62.

DAUTZENBERG, P. & DUROUCHOUX, P. 1906. Supplément à la faunule malacologique des environs de Saint-Malo. *La Feuille des Jeunes Naturalistes*, **36** (423): 39–45; (424): 53–60.

DAUTZENBERG, P. & DUROUCHOUX, P. 1913. Les mollusques de la Baie de Saint-Malo (suite). *La Feuille des Jeunes Naturalistes*, **43** (516, supplément): 17–24.

DAUTZENBERG, P. & FISCHER, H. 1912. Mollusques provenant des campagnes de l'Hirondelle et de la Princesse-Alice dans les Mers du Nord. *Résultats des Campagnes Scientifiques Accomplies sur son Yacht par Albert Ier Prince Souverain de Monaco*, **37**: 1–629.

Dautzenberg, P. & Fischer, H. 1915. Étude sur le *Littorina obtusata* et ses variations. *Journal de Conchyliologie*, **62**: 87–128.

Dautzenberg, P. & Fischer, P. H. 1925. Les mollusques marins du Finistère et en particulier de la région de Roscoff. *Travaux de la Station Biologique de Roscoff*, **3**: 1–180.

Davies, M. S., Hawkins, S. J. & Jones, H. D. 1992. Pedal mucus and its influence on the microbial food supply of two intertidal gastropods, *Patella vulgata* L. and *Littorina littorea* (L.). *Journal of Experimental Marine Biology and Ecology*, **161**: 57–77.

Davis, D. S. 1971. Periwinkles. *Journal of Education*, Winter, 1970–71. (Reprinted 1976 by Nova Scotia Museum, Halifax).

Davis, D. S. 1972. An unusual form of *Littorina littorea* Linné found in Nova Scotia. *Nautilus*, **86**: 87.

Davis, D. S. 1973. Variations in the northern rough periwinkle, *Littorina saxatilis* (Olivi) in Nova Scotia. *Proceedings and Transactions of the Nova Scotian Institute of Science*, **27**: 61–90.

Dawson, J. W. 1872. The post-Pliocene geology of Canada. Sub-kingdom Mollusca. *Canadian Naturalist. New Series*, **6**: 369–416.

Dawson, J. W. & Harrington, B. J. 1871. *Report of the Geological Structure and Mineral Resources of Prince Edward Island*. Government of Prince Edward Island, Montreal.

Day, J. H. 1969. *A Guide to Marine Life on South African Shores*. A. A. Balkema, Cape Town.

Deleporte, P. 1993. Characters, attributes and tests of evolutionary scenarios. *Cladistics*, **9**: 427–432.

Delessert, B. 1841. *Recueil de Coquilles Décrites par Lamarck dans Son Histoire Naturelle des Animaux sans Vertèbres et Non Encore Figurées*. Paris.

Delsman, H. C. 1913. Over de voortplanting van de Alikruik (*Littorina littorea*). *Mededelingen over Visscherij*, **20**: 53–66.

Delsman, H. C. 1914. Entwicklungsgeschichte von *Littorina obtusata*. *Tijdschrift der Nederlandsche Dierkundige Vereeniging, Serie 2*, **13**: 170–340.

Denny, M. W. 1988. *Biology and the Mechanics of the Wave-Swept Environment*. Princeton University Press, Princeton.

Denny, M. W., Daniel, T. L. & Koehl, M. A. R. 1985. Mechanical limits to size in wave-swept organisms. *Ecological Monographs*, **55**: 69–102.

Deshayes, G.-P. 1839–1853. *Traité Élémentaire de Conchyliologie. Explication des Planches*. Victor Masson, Paris.

Deshayes, G. P. & Milne Edwards, H. 1843. *Histoire Naturelle des Animaux sans Vertèbres...par J. B. P. A. de Lamarck*, **9**. Edition 2. J.-B. Baillière, Paris & London.

Deutsch, U. & Brick, M. 1993. Morphological effects of tributyltin (TBT) in vitro on the genital system of the mesogastropod *Littorina littorea* (Prosobranchia). *Helgoländer Meeresuntersuchungen*, **47**: 49–60.

Deutsch, U. & Fioroni, P. 1992. The shedding of the penis in *Littorina littorea*: some new aspects. In: *Proceedings of the Third International Symposium on Littorinid Biology* (Grahame, J., Mill, P. J. & Reid, D. G., eds), 309–311. Malacological Society of London, London.

Dexter, R. W. 1945. Zonation of the intertidal marine mollusks at Cape Ann, Massachusetts. *Nautilus*, **58**: 135–142.

Dexter, R. W. 1947. The marine communities of a tidal inlet at Cape Ann, Massachusetts: a study in bio-ecology. *Ecological Monographs*, **17**: 261–294.

Dexter, R. W. 1961. Early records of *Littorina littorea* from the coast of Massachusetts. *Nautilus*, **74**: 120–121.

Deyglun, C. 1955. *Biologie Comparée de Deux Sous-espèces de* Littorina saxatilis *Olivi*. Thèse, Diploma d'Études, Faculté des Sciences de l'Université de Paris.

Dickerson, R. E. 1917. Climate and its influence upon the Oligocene faunas of the Pacific coast, with descriptions of some new species from the *Molopophorus lincolnensis* zone. *Proceedings of the California Academy of Sciences. Series 4*, **7**: 157–192.

Dillwyn, L. W. 1817. *A Descriptive Catalogue of Recent Shells*, **2**. J. & A. Arch, London.

Dippolito, A., Sienko, J. & Smith, B. S. 1975. Relationship between mud snail and periwinkle. *Proceedings of the Malacological Society of London*, **41**: 529–532.

Dixon, D. R. & Pollard, D. 1985. Embryo abnormalities in the periwinkle, *Littorina 'saxatilis'*, as indicators of stress in polluted marine environments. *Marine Pollution Bulletin*, **16**: 29–33.

Dodge, H. 1959. A historical review of the mollusks of Linnaeus. Part 7. Certain species of the genus *Turbo* of the class Gastropoda. *Bulletin of the American Museum of Natural History*, **118**: 211–257.

Dongen, A. van 1956. The preference of *Littorina obtusata* for Fucaceae. *Archives Néerlandaises de Zoologie*, **11**: 373–386.

Donovan, E. 1800. *The Natural History of British Shells*, **1**. London.

Doughty, C. G. 1934. Additional kind of periwinkle. *Transactions of the Suffolk Naturalists' Society*, **2**: 284.

Drouët, H. 1858. *Mollusques Marins des Îles Açores*. Paris.

DUDGEON, D. & YIPP, M. W. 1990. Does competition determine the vertical zonation of high-intertidal Littorinidae (Gastropoda) in Hong Kong? In: *Proceedings of the Second International Marine Biological Workshop: The Marine Flora and Fauna of Hong Kong and Southern China, Hong Kong, 1986* (Morton, B., ed.), 795–802. Hong Kong University Press, Hong Kong.

DUDLEY, R. 1980. Crab-crushing of periwinkle shells, *Littorina littorea*, from two adjacent geographical provinces. *Nautilus*, **94**: 108–112.

DUERR, F. G. 1968. Excretion of ammonia and urea in seven species of marine prosobranch snails. *Comparative Biochemistry and Physiology*, **26**: 1051–1059.

DUFFUS, J. & JOHNSTON, C. S. 1969. Marine Mollusca from the Canary Island of Lanzarote. *Journal of Conchology*, **27**: 27–45.

DUGGAN, E. P. 1963. Report of non-indigenous marine shells collected in the state of Washington. *Veliger*, **6**: 112.

DUNKER, W. 1845. Diagnoses molluscorum quorundam novorum, quae ex itinere ad oras Africae occidentalis reportavit cl. G. Tams, Med. Dr. *Zeitschrift für Malakozoologie*, **1845**: 163–168.

DUNKER, W. 1853. Index molluscorum, quae in itinere ad Guineam inferiorem collegit Georgius Tams Med. Dr. *Novitates Conchologicae. Supplementum*, **2**: 1–74.

DUNKER, W. 1861. *Mollusca Japonica Descripta et Tabularis Tribus Iconum*. Stuttgart.

DUNKER, W. 1874. *Museum Godeffroy. Catalog V*. Hamburg.

DUNKER, W. 1882. Index molluscorum maris Japonici conscriptus et tabulis iconum XVI illustratus. *Novitates Conchologicae. Supplementum*, **7**: 1–301.

DURHAM, J. W. 1944. Megafaunal zones of the Oligocene of Washington. *University of California Publications. Bulletin of the Department of Geology*, **27**: 101–211.

DURHAM, J. W. & MACNEIL, F. S. 1967. Cenozoic migrations of marine invertebrates through the Bering Strait region. In: *The Bering Land Bridge* (Hopkins, D. M., ed.), 326–349. Stanford University Press, Stanford.

DYTHAM, C., GRAHAME, J. & MILL, P. J. 1990. Distribution, abundance and shell morphology of *Littorina saxatilis* (Olivi) and *Littorina arcana* Hannaford Ellis at Robin Hood's Bay, North Yorkshire. *Hydrobiologia*, **193**: 233–240.

DYTHAM, C., GRAHAME, J. & MILL, P. J. 1992. Esterase variation in *Littorina arcana* and *L. saxatilis* at Robin Hood's Bay, Yorkshire. In: *Proceedings of the Third International Symposium on Littorinid Biology* (Grahame, J., Mill, P. J. & Reid, D. G., eds), 39–44. Malacological Society of London, London.

DYTHAM, C., MILL, P. J., GRAHAME, J. & O'HIGGINS, P. 1992. Fourier analysis as a size-free descriptor of shape for rough periwinkle shells. In: *Proceedings of the Third International Symposium on Littorinid Biology* (Grahame, J., Mill, P. J. & Reid, D. G., eds), 127–133. Malacological Society of London, London.

EERNISSE, D. J. & KLUGE, A. G. 1993. Taxonomic congruence versus total evidence, and the phylogeny of amniotes inferred from fossils, molecules and morphology. *Molecular Biology and Evolution*, **10**: 1170–1195.

EFFINGER, W. L. 1938. The Gries Ranch fauna (Oligocene) of western Washington. *Journal of Paleontology*, **12**: 355–390.

EGOROV, R. V. 1992. Guide to Recent molluscs of northern Eurasia. 2. Fauna of gastropods of the family Littorinidae. *Ruthenica*, **2**: 169–172.

EINARSSON, T., HOPKINS, D. M. & DOELL, R. R. 1967. The stratigraphy of Tjörnes, northern Iceland and the history of the Bering Land Bridge. In: *The Bering Land Bridge* (Hopkins, D. M., ed.), 312–325. Stanford University Press, Stanford.

EKARATNE, S. U. K. & CRISP, D. J. 1982. Tidal micro-growth bands in intertidal gastropod shells, with an evaluation of band-dating techniques. *Proceedings of the Royal Society of London. Series B*, **214**: 305–323.

EKARATNE, S. U. K. & CRISP, D. J. 1983. A geometrical analysis of growth in gastropod shells, with particular reference to turbinate forms. *Journal of the Marine Biological Association of the United Kingdom*, **63**: 777–797.

EKARATNE, S. U. K. & CRISP, D. J. 1984. Seasonal growth studies of intertidal gastropods from shell micro-growth band measurements, including a comparison with alternative methods. *Journal of the Marine Biological Association of the United Kingdom*, **64**: 183–210.

EKENDAHL, A. 1994. Factors important to the distribution of colour morphs of *Littorina mariae* Sacchi & Rastelli in a non-tidal area. *Ophelia*, **40**: 1–12.

EKENDAHL, A. 1995. Microdistribution in the field and habitat choice in aquaria by colour morphs of *Littorina mariae* Sacchi & Rastelli. *Journal of Molluscan Studies*, **61**: 249–256.

ELLIOTT, W. T. 1921. Pearl in *Littorina littorea*. *Journal of Conchology*, **16**: 233.

ELLIS, D. V. 1955. Some observations on the shore fauna of Baffin Island. *Arctic*, **8**: 224–236.

ELLIS, D. V. & WILCE, R. T. 1961. Arctic and subarctic examples of intertidal zonation. *Arctic*, **14**: 224–235.

ELMHIRST, R. 1923. Notes on the breeding and growth of marine animals in the Clyde Sea area. *Annual Report of the Scottish Marine Biological Association*, **1922**: 19–43.

ELNER, R. W. & RAFFAELLI, D. G. 1980. Interactions between two marine snails, *Littorina rudis* Maton and *Littorina nigrolineata* Gray, a predator, *Carcinus maenas* (L.), and a parasite, *Microphallus similis* Jägerskiöld. *Journal of Experimental Marine Biology and Ecology*, **43**: 151–160.

EMERSON, D. N. & DUERR, F. G. 1967. Some physiological effects of starvation in the intertidal prosobranch *Littorina planaxis* (Philippi, 1847). *Comparative Biochemistry and Physiology. Series B*, **20**: 45–53.

EMLET, R. B. 1995. Developmental mode and species geographic range in regular sea urchins (Echinodermata: Echinoidea). *Evolution*, **49**: 476–489.

EMSON, R. H. & FALLER-FRITSCH, R. J. 1976. An experimental investigation into the effect of crevice availability on abundance and size-structure in a population of *Littorina rudis* (Maton): Gastropoda: Prosobranchia. *Journal of Experimental Marine Biology and Ecology*, **23**: 285–297.

ENDLER, J. A. 1977. *Geographic Variation, Speciation, and Clines. Monographs in Population Biology*, **10**. Princeton University Press, Princeton.

ENDLER, J. A. 1978. A predator's view of animal color patterns. *Evolutionary Biology*, **11**: 319–364.

ENDLER, J. A. 1981. An overview of the relationships between mimicry and crypsis. *Biological Journal of the Linnean Society*, **16**: 25–31.

ERESHEFSKY, M. 1992. Introduction to Part I: biological concepts. In: *The Units of Evolution. Essays on the Nature of Species* (Ereshefsky, M., ed.), 3–14. Massachusetts Institute of Technology, Cambridge, Massachusetts.

ERLANDSSON, J. & JOHANNESSON, K. 1994. Sexual selection on female size in a marine snail, *Littorina littorea* (L.). *Journal of Experimental Marine Biology and Ecology*, **181**: 145–157.

ERLANDSSON, J. & KOSTYLEV, V. 1995. Trail following, speed and fractal dimension of movement in a marine prosobranch, *Littorina littorea*, during a mating and a non-mating season. *Marine Biology, Berlin*, **122**: 87–94.

ETTER, R. J. 1988a. Asymmetrical developmental plasticity in an intertidal snail. *Evolution*, **42**: 322–334.

ETTER, R. J. 1988b. Physiological stress and colour polymorphism in the intertidal snail *Nucella lapillus*. *Evolution*, **42**: 660–680.

EVANS, R. G. 1947. The intertidal ecology of selected localities in the Plymouth neighbourhood. *Journal of the Marine Biological Association of the United Kingdom*, **27**: 173–218.

EVANS, R. G. 1948. The lethal temperatures of some common British littoral molluscs. *Journal of Animal Ecology*, **17**: 165–173.

EYERDAM, W. J. 1960. Mollusks and brachiopods from Afognak and Sitkalidak Islands, Kodiak Group, Alaska. *Nautilus*, **74**: 41–46, 91–95.

FABRICIUS, O. 1780. *Fauna Groenlandica*. Hafniae & Lipsiae.

FALLER-FRITSCH, R. J. 1977. Reproductive strategies of the winkle *Littorina rudis* in relation to population dynamics and size structure. In: *Biology of Benthic Organisms* (Keegan, B. F., Ceidigh, P. O. & Boaden, P. J. S., eds), 225–231. Pergamon Press, New York.

FALLER-FRITSCH, R. J. & EMSON, R. H. 1985. Causes and patterns of mortality in *Littorina 'rudis'* (Maton) in relation to intraspecific variation: a review. In: *The Ecology of Rocky Coasts* (Moore, P. G. & Seed, R., eds), 157–177. Hodder & Stoughton, London.

FAUCHEUX, O. LE 1961. Un curieux exemplaire de bigorneau, *Littorina littorea* (L.). *Penn ar Bed*, **3**: 29.

FÉRUSSAC, A. É. J. P. J. F. D'A. DE 1822. *Tableaux Systématiques des Animaux Mollusques*. A. Bertrand, Paris.

FEVOLDEN, S. E. & GARNER, S. P. 1987. Environmental stress and allozyme variation in *Littorina littorea* (Prosobranchia). *Marine Ecology Progress Series*, **39**: 129–136.

FEYLING-HANSSEN, R. W. 1955. Stratigraphy of the marine Late-Pleistocene of Billefjorden, Vestspitsbergen. *Norsk Polarinstitutt Skrifter*, **107**: 1–186.

FINET, Y. 1978. Zonation des gastéropodes sur la Côte de la Manche, près de Roscoff (France). *Informations de la Société Belge de Malacologie. Série 6*, **4**: 93–120.

FINET, Y., WÜEST, J. & MAREDA, K. 1991. *Les Radulas des Gastéropodes de la Manche (Cotentin–Baie de Seine, France)*. Museum d'Histoire Naturelle, Genève.

FINET, Y., WÜEST, J. & MAREDA, K. 1992. *Gastropods of the Channel and Atlantic Ocean: Shells and Radulas*. Mostra Mondiale Malacologia, Cupra Marittima, Italy.

FIORONI, P., DEUTSCH, U., STROBEN, E. & OEHLMANN, J. 1992. Artificially induced pseudohermaphroditism in prosobranchs and its absence in littorinids. In: *Proceedings of the Third International Symposium on Littorinid Biology* (Grahame, J., Mill, P. J. & Reid, D. G., eds), 313–315. Malacological Society of London, London.

FIORONI, P., OEHLMANN, J. & STROBEN, E. 1990. Le pseudohermaphroditisme chez les prosobranches: analyse morphologique et histologique. *Vie et Milieu*, **40**: 45–56.

FIORONI, P. & REID, D. G. (eds) 1994. Fourth International Symposium on Littorinid Biology, Roscoff, 19–25 September 1993. *Cahiers de Biologie Marine*, **35**: 237–267.

FISCHER, P. 1880–1887. *Manuel de Conchyliologie et de Paléontologie Conchyliologique*. F. Savy, Paris.

FISCHER, P.-H. 1941. Quelques cas de pigmentation chez les gastéropodes. *Journal de Conchyliologie*, **84**: 66–72.

FISCHER, P.-H. 1942. Trouvaille d'une littorine Americaine en Laponie. *Bulletin de la Société Zoologique de France*, **67**: 187.

FISCHER, P.-H. 1943. Mollusques marins des Archipels Atlantiques. Considérations sur la biogéographie littorale de l'Atlantique Nord. *Compte Rendu Sommaire des Séances de la Société de Biogéographie*, **20** (167): 3–7.

FISCHER, P.-H. 1946. Sur les mollusques littoreaux des Îles Canaries. *Mémoires de la Société de Biogéographie*, **8**: 279–294.

FISCHER, P.-H. 1953. Les mollusques et leurs milieux de vie. *Journal de Conchyliologie*, **93**: 121–151.

FISCHER, P.-H. 1967. Sur quelques espèces noduleuses du genre *Littorina*. *Journal de Conchyliologie*, **106**: 8–19.

FISCHER-PIETTE, E. 1931. Sur la pénétration des diverses espéces marines sessiles dans les estuaires et sa limitation par l'eau douce. *Annales de l'Institut Océanographique. Nouvelle Serie*, **10**: 213–243.

FISCHER-PIETTE, E. 1950. List des types décrits dans le Journal de Conchyliologie et conservés dans la collection de ce journal. *Journal de Conchyliologie*, **90**: 8–23.

FISCHER-PIETTE, E. 1953. Répartition de quelques mollusques intercotidaux communs le long des côtes septentrionales de l'Espagne. *Journal de Conchyliologie*, **93**: 39–73.

FISCHER-PIETTE, E. 1955. Répartition, le long des côtes septentrionales de l'Espagne, des principales espèces peuplement les rochers intercotidaux. *Annales de l'Institut Océanographique*, **31**: 37–124.

FISCHER-PIETTE, E. 1956. Sur des déplacements de frontières biogéographiques intercotidales, actuellement en cours en Espagne: situation en 1956. *Compte Rendu Hebdomadaire des Séances de l'Académie des Sciences, Paris*, **242**: 2782–2784.

FISCHER-PIETTE, E. 1957a. Sur des déplacements de frontières biogéographiques, observés au long des côtes Ibériques dans le domaine intercotidal. *Publicaciones del Instituto de Biologia Aplicada, Barcelona*, **26**: 35–40.

FISCHER-PIETTE, E. 1957b. Sur le progrès des espèces septentrionales dans le bios intercotidal Ibérique: situation en 1956–1957. *Compte Rendu Hebdomadaire des Séances de l'Académie des Sciences, Paris*, **245**: 373–375.

FISCHER-PIETTE, E. 1958. Sur l'écologie intercotidale Ouest-ibérique. *Compte Rendu Hebdomadaire des Séances de l'Académie des Sciences, Paris*, **246**: 1301–1303.

FISCHER-PIETTE, E. 1959. Contribution à l'écologie intercotidale du Détroit de Gibraltar. *Bulletin de l'Institut Océanographique*, **1145**: 1–32.

FISCHER-PIETTE, E. 1960. Nouvelle avance de mollusques intercotidaux septentrionaux vers le sud du Portugal. *Compte Rendu Hebdomadaire des Séances de l'Académie des Sciences, Paris*, **250**: 2288–2289.

FISCHER-PIETTE, E. & GAILLARD, J. M. 1960. Études sur les variations de *Littorina saxatilis*. II. Modifications des caractères au long des estuaires et rias. *Journal de Conchyliologie*, **100**: 10–37.

FISCHER-PIETTE, E. & GAILLARD, J. M. 1961. Études sur les variations de *Littorina saxatilis*. III. Comparison des points abrités au long des côtes françaises et ibériques. *Bulletin de la Société Zoologique de France*, **86**: 163–172.

FISCHER-PIETTE, E. & GAILLARD, J. M. 1966. Sur les variations de *Littorina saxatilis* VII. Des populations changent actuellement d'aspect. *Cahiers de Biologie Marine*, **7**: 375–382.

FISCHER-PIETTE, E. & GAILLARD, J. M. 1968. Sur les variations de *Littorina saxatilis*. 9. Exemple de distribution d'une variété donnée. *Malacological Review*, **1**: 103–118.

FISCHER-PIETTE, E. & GAILLARD, J. M. 1971. La variabilité (morphologique et physiologique) des *Littorina saxatilis* (Olivi) ibériques et ses rapports avec l'écologie. *Mémoires du Muséum National d'Histoire Naturelle, Paris. Série A, Zoologie*, **70**: 1–90.

FISCHER-PIETTE, E., GAILLARD, J. M. & DELMAS, D. 1967. Sur les variations de *Littorina saxatilis*. 8. Les populations de la Ria de Camariñas. *Bulletin du Muséum National d'Histoire Naturelle, Paris*, **38**: 874–897.

FISCHER-PIETTE, E., GAILLARD, J. M. & JAMES, B. L. 1963. Études sur les variations de *Littorina saxatilis*. V. Sur des cas de variabilité extrème. *Cahiers de Biologie Marine*, **4**: 1–22.

FISCHER-PIETTE, E., GAILLARD, J. M. & JAMES, B. L. 1964. Études sur les variations de *Littorina saxatilis*. VI. Quelques cas qui posent de difficiles problèmes. *Cahiers de Biologie Marine*, **5**: 125–171.

FISCHER-PIETTE, E., GAILLARD, J. M. & JOUIN, C. 1961. Études sur les variations de *Littorina saxatilis* IV.—comparaison de les points battus, au long des côtes européenes. A. Côtes ibériques. *Bulletin de la Société Zoologique de France*, **86**: 320–328.

FISCHER-PIETTE, E., GAILLARD, J. M., PANOUSE, J. B. & BELLON-HUMBERT, C. 1971. *Littorina saxatilis* (Moll. Gastér.) dans le sud Marocain. *Bulletin du Muséum National d'Histoire Naturelle, Paris*, **42**: 970–974.

FISCHER-PIETTE, E. & SEOANE-CAMBA, J. 1962. Sur un facteur de prospérité de *Littorina littorea* et son effet sur la frontière supérieure de l'espèce. *Bulletin du Centre d'Études et de Recherches Scientifiques, Biarritz*, **4**: 171–185.

FISH, J. D. 1972. The breeding cycle and growth of open coast and estuarine populations of *Littorina littorea*. *Journal of the Marine Biological Association of the United Kingdom*, **52**: 1011–1019.

FISH, J. D. 1979. The rhythmic spawning behaviour of *Littorina littorea* (L.). *Journal of Molluscan Studies*, **45**: 172–177.

FISH, J. D. & FISH, S. 1977. The veliger larva of *Hydrobia ulvae* with observations on the veliger of *Littorina littorea* (Mollusca: Prosobranchia). *Journal of Zoology, London*, **182**: 495–504.

FISH, J. D. & SHARP, L. 1985. The ecology of the periwinkle, *Littorina neglecta* Bean. In: *The Ecology of Rocky Coasts* (Moore, P. G. & Seed, R., eds), 143–156. Hodder & Stoughton, London.

FLETCHER, C. R. 1992. Distribution of the flat periwinkles *Littorina obtusata* (L.) and *L. mariae* Sacchi & Rastelli (Gastropoda: Prosobranchia): relationship to the slope of the shore. In: *Proceedings of the Third International Symposium on Littorinid Biology* (Grahame, J., Mill, P. J. & Reid, D. G., eds), 119–126. Malacological Society of London, London.

FLETCHER, C. R. 1995. Microgeographical variation in shell strength in the flat periwinkles *Littorina obtusata* and *Littorina mariae*. *Hydrobiologia*, **309**: 73–87.

FLOWER, B. P. & KENNETT, J. P. 1994. The middle Miocene climatic transition: East Antarctic ice sheet development, deep ocean circulation and global carbon cycling. *Palaeogeography, Palaeoclimatology, Palaeoecology*, **108**: 537–555.

FOLTZ, D. W., SHUMWAY, S. E. & CRISP, D. 1993. Genetic structure and heterozygosity-related fitness effects in the marine snail *Littorina littorea*. *American Malacological Bulletin*, **10**: 55–60.

FORBES, E. & HANLEY, S. 1850, 1852. *A History of British Mollusca and their Shells*, **3** and **4**. John van Voorst, London.

FOREY, P. L., HUMPHRIES, C. J., KITCHING, I. J., SCOTLAND, R. W., SIEBERT, D. J. & WILLIAMS, D. M. (eds) 1992. *Cladistics. A Practical Course in Systematics*. Clarendon Press, Oxford.

FORTUIN, A. W. 1979. *Systematik van* Littorina saxatilis *(Olivi) en* Littorina obtusata *(L.) in Nederland*. Eindwerk Bijzondere Opleiding MO-Biologie, Leiden.

FOSTER, M. S. 1964. Microscopic algal food of *Littorina planaxis* Philippi and *Littorina scutulata* Gould (Gastropoda: Prosobranchiata). *Veliger*, **7**: 149–152.

FOSTER, M. S., VOGELAERE, A. P. DE, OLIVER, J. S., PEARSE, J. S. & HARROLD, C. 1991. Open coast intertidal and shallow subtidal ecosystems of the northeast Pacific. In: *Ecosystems of the World* **24.** *Intertidal and Littoral Ecosystems* (Mathieson, A. C. & Nienhuis, P. H., eds), 235–272. Elsevier, Amsterdam.

FOSTER, N. R. 1981. *A Synopsis of the Marine Prosobranch Gastropod and Bivalve Mollusks in Alaskan Waters*. Institute of Marine Science, University of Alaska, Fairbanks.

FRAENKEL, G. 1960. Lethal high temperatures for three marine invertebrates: *Limulus polyphemus*, *Littorina littorea* and *Pagurus longicarpus*. *Oikos*, **11**: 171–182.

FRANK, P. 1979. Maximum size and latitude: their relation in *Littorina planaxis* and *L. scutulata*. *Western Society of Malacologists Annual Report*, **12**: 8–9.

FRETTER, V. 1980. Observations on the gross anatomy of the female genital duct of British *Littorina* spp. *Journal of Molluscan Studies*, **46**: 148–153.

FRETTER, V. & GRAHAM, A. 1962. *British Prosobranch Molluscs: Their Functional Anatomy and Ecology*. Ray Society, London.

FRETTER, V. & GRAHAM, A. 1978. The prosobranch molluscs of Britain and Denmark. Part 3—Neritacea, Viviparacea, Valvatacea, terrestrial and freshwater Littorinacea and Rissoacea. *Journal of Molluscan Studies. Supplement*, **5**: 101–152.

FRETTER, V. & GRAHAM, A. 1980. The prosobranch molluscs of Britain and Denmark. Part 5—marine Littorinacea. *Journal of Molluscan Studies. Supplement*, **7**: 242–284.

FRETTER, V. & GRAHAM, A. 1994. *British Prosobranch Molluscs: Their Functional Anatomy and Ecology*. Revised Edition. Ray Society, London.

FRETTER, V. & PILKINGTON, M. C. 1970. Prosobranchia. Veliger larvae of Taenioglossa and Stenoglossa. *Conseil International pour l'Exploration de la Mer. Fiches d'Identification. Zooplankton*, Sheets 129–132; pp. 1–26.

FRETTER, V. & SHALE, D. 1973. Seasonal changes in population density and vertical distribution of prosobranch veligers in offshore plankton at Plymouth. *Journal of the Marine Biological Association of the United Kingdom*, **53**: 471–492.

FREY, J. K. 1993. Modes of peripheral isolate formation and speciation. *Systematic Biology*, **42**: 373–381.

FRID, C. L. J. 1988. The marine fauna of the north Norfolk salt marshes, and their ecology. *Transactions of the Norfolk and Norwich Naturalists' Society*, **28**: 46–50.

FRID, C. L. J. & JAMES, R. 1988. Interactions between two species of saltmarsh gastropods, *Hydrobia ulvae* and *Littorina littorea*. *Marine Ecology Progress Series*, **43**: 173–179.

FRORIEP, L. F. 1806. *C. Dumeril's Analytische Zoologie, aus dem Französischen, mit Zusätzen*. Weimar.

FRUMHOFF, P. C. & REEVE, H. K. 1994. Using phylogenies to test hypotheses of adaptation: a critique of some current proposals. *Evolution*, **48**: 172–180.

FUJI, A. & NOMURA, H. 1990. Community structure of the rocky shore macrobenthos in southern Hokkaido, Japan. *Marine Biology, Berlin*, **107**: 471–477.

FUKUDA, H., MASHINO, K. & SUGIMURA, T. 1992. *A Review of the Molluscan Fauna of Yamaguchi Prefecture, Western Japan*. Yamaguchi Museum.

FUNNELL, B. M. 1961. The Palaeogene and Early Pleistocene of Norfolk. *Transactions of the Norfolk and Norwich Naturalists' Society*, **19**: 340–364.

FUNNELL, B. M. 1979. Palaeoenvironmental analysis of the Dobb's Plantation section, Crostwick (and comparison with type localities of the Norwich and Weybourne Crags). *Bulletin of the Geological Society of Norfolk*, **31**: 1–10.

FUNNELL, B. M., NORTON, P. E. P. & WEST, R. G. 1979. The Crag at Bramerton, near Norwich, Norfolk. *Philosophical Transactions of the Royal Society of London. Series B*, **287**: 489–534.

GABB, W. M. 1869. Section I. Tertiary invertebrate fossils. Part III. Synopsis of the Tertiary invertebrate fossils of California. *Geological Survey of California. Palaeontology*, **2**: 65–124.

GAILLARD, J. M. 1965. Aspects qualitatifs et quantitatifs de la croissance de la coquille de quelques espèces de mollusqes prosobranches en fonction de la latitude et des conditions écologiques. *Mémoires du Muséum National d'Histoire Naturelle, Paris. Série A, Zoologie*, **38**: 1–155.

GAILLARD, J. M. 1977. Le polymorphisme de *Littorina saxatilis* (Olivi) (Gastropoda, Prosobranchia). *Malacologia*, **16**: 11–14.

GALAKTIONOV, K. V. 1985. The infection of males and females of molluscs of the genus *Littorina* (Gastropoda: Prosobranchia) by trematode parasites on the coast of the Barents Sea. *Parasitologiya*, **19**: 213–219. (Russian).

GALAKTIONOV, K. V. & MARASAEV, S. F. 1986. Some patterns in the distribution of infection of the molluscan genus *Littorina* by parthenites of trematodes on the shores of the Barents and White Seas. *Vestnik Leningradskogo Gosudarstvennogo Universiteta (Biologiya)*, **1986** (1): 88–90. (Russian).

GALKIN, Y. J. & SKARLATO, O. A. 1955. Gastropoda Prosobranchia. In: *Atlas of the Invertebrates of the Far Eastern Seas of the U. S. S. R* (Pavlovskii, E. N., ed.). Akademiya Nauk S. S. S. R., Moscow. (Russian).

GALLARDO, C. S. & GÖTTING, K.-J. 1985. Reproduktionsbiologische Untersuchungen an drei *Littorina*-Arten der südlichen Nordsee. *Helgoländer Meeresuntersuchungen*, **39**: 165–186.

GANONG, W. F. 1886. Is *Littorina litorea* introduced or indigenous? *American Naturalist*, **20**: 931–940.

GANONG, W. F. 1887. Is *Littorina litorea* introduced or indigenous? *American Naturalist*, **21**: 287–288.

GEISELMAN, J. A. & MCCONNELL, O. J. 1981. Polyphenols in brown algae *Fucus vesiculosus* and *Ascophyllum nodosum*: chemical defenses against the marine herbivorous snail, *Littorina littorea*. *Journal of Chemical Ecology*, **7**: 1115–1133.

GELLER, J. B. 1991. Gastropod grazers and algal colonization on a rocky shore in northern California: the importance of the body size of grazers. *Journal of Experimental Marine Biology and Ecology*, **150**: 1–17.

GENDRON, R. P. 1977. Habitat selection and migratory behaviour of the intertidal gastropod *Littorina littorea* (L.). *Journal of Animal Ecology*, **46**: 79–92.

GEORGE, T. N. 1932. The Quaternary beaches of Gower. *Proceedings of the Geologists' Association, London*, **43**: 291–324.

GHISOTTI, F. 1972. Le conchiglie del Golfo di Gabes: parte prima. *Conchiglie*, **8**: 63–89.

GHISOTTI, F. 1978. Diagnosi originali: le specie di Giuseppe Olivi. *Conchiglie*, **14**: 23–28.

GIBSON, D. G. 1964. Mating behaviour in *Littorina planaxis* Philippi (Gastropoda: Prosobranchiata). *Veliger*, **7**: 134–139.

GILI, C. & MARTINELL, J. 1994. Relationship between species longevity and larval ecology in nassariid gastropods. *Lethaia*, **27**: 291–299.

GILKINSON, K. D. & METHVEN, D. A. 1991. Observations on the subtidal distributions of the intertidal rough periwinkle, *Littorina saxatilis*, and the common periwinkle, *L. littorea*, in a shallow embayment in eastern Newfoundland. *Canadian Field Naturalist*, **105**: 522–525.

GILL, J., RAMSAY, R. & SMITH, S. 1976. The ecology of flat periwinkles. *Journal of Biological Education*, **10**: 237–241.

GLADENKOV, Y. B. 1981. Marine Plio-Pleistocene of Iceland and problems of its correlation. *Quaternary Research, New York*, **15**: 18–23.

GLADENKOV, Y. B. 1994. Cenozoic paleogeography and climatic change in the North Pacific Ocean. *Palaeogeography, Palaeoclimatology, Palaeoecology*, **108**: 311–318.

GLADENKOV, Y. B., GLADIKOVA, V. M., KAFANOV, A. N., KONOVA, L. V., KRISHTOFOVICH, L. V., SINELNIKOVA, V. N. & POPOV, S. V. 1984. Marine molluscs. *Trudy Geologicheskogo Instituta, Leningrad*, **385**: 152–250. (Russian).

GLADENKOV, Y. B., NORTON, P. & SPAINK, G. 1980. Upper Cenozoic of Iceland. (Stratigraphy of Pliocene-Pleistocene and palaeontological assemblages). *Trudy Geologicheskii Institut, Leningrad*, **345**: 1–116. (Russian).

GLADENKOV, Y. B. & SHANTSER, A. Y. 1990. Neogene of Kamchatka: stratigraphy and correlation of geological events. In: *Pacific Neogene Events* (Tsuchi, R., ed.), 173–181. University of Tokyo Press, Tokyo.

GLADENKOV, Y. B. & SINELNIKOVA, V. N. 1990. Mollusks and climatic optima of the Miocene of Kamchatka. *Trudy Geologicheskogo Instituta, Leningrad*, **453**: 1–172. (Russian).

GLIBERT, M. 1962. Les Mesogastropoda fossiles du Cénozoïque étranger des collections de l'Institut Royal des Sciences Naturelles de Belgique. Première partie, Cyclophoridae à Stiliferidae (inclus). *Mémoires de l'Institut Royal des Sciences Naturelles de Belgique. Série 2*, **69**: 1–305.

GLYNN, P. W. 1965. Community composition, structure, and interrelationships in the marine intertidal *Endocladia muricata–Balanus glandula* association in Monterey Bay, California. *Beaufortia*, **12**: 1–198.

GMELIN, J. F. 1791. *Caroli a Linné...Systema Naturae per Regna Tria Naturae*, **1** (part 6). Edition 13. Lugduni.

GOFAS, S. 1975. Sur l'extension de *Littorina saxatilis* (Olivi) (Moll. Gaster.) dans le Détroit de Gibraltar. *Bulletin du Société des Sciences Naturelles et Physiques du Maroc*, **55**: 95–99.

GOLIKOV, A. N. 1959. Influence of environmental factors on the intraspecific variability of *Neptunea arthritica* (Bernardi) and *Littorina squalida* Broderip et Sowerby. *Zoologicheskii Zhurnal*, **38**: 1335–1343. (Russian).

GOLIKOV, A. N. 1987. Gastropoda. *Opredeliteli Fauny S. S. S. R.*, **151**: 41–149. (Russian).

GOLIKOV, A. N., DOLGOLENKO, M. A., MAXIMOVICH, N. V. & SKARLATO, O. A. 1990. Theoretical approaches to marine biogeography. *Marine Ecology Progress Series*, **63**: 289–301.

GOLIKOV, A. N. & GULBIN, V. V. 1978. Prosobranch gastropods of the Kurile Islands. I. Orders Docoglossa–Entomostoma. In: *Fauna and Vegetation of the Shelf of the Kurile Islands* (Kusakin, O. G., ed.), 159–223. (Russian).

GOLIKOV, A. N., GULBIN, V. V. & SIRENKO, B. I. 1987. Prosobranchia of Moneron Island shelf (Japan Sea). I. Orders Patelliformes–Calyptraeiformes. In: *Molluscan Fauna and Distribution: the Northern Pacific and Polar Basin* (Kafanov, A. I., ed.), 22–40. Institut Biologii Morya, Dal'nevostochnyi Nauchnyi Tsentr, Akademiya Nauk S. S. S. R., Vladivostok. (Russian).

GOLIKOV, A. N. & KUSAKIN, O. G. 1962. Fauna and ecology of Gastropoda, Prosobranchia from the intertidal zone of the Kurile Islands. *Isseldovaniya Dal'nevostochnykh Morei S. S. S. R.*, **8**: 248–376. (Russian).

GOLIKOV, A. N. & KUSAKIN, O. G. 1978. *Recent Molluscs of the Littoral Zone of the U. S. S. R.* Nauka, Leningrad. (Russian).

GOLIKOV, A. N. & SKARLATO, O. A. 1967. Molluscs of Pos'et Bay (Sea of Japan) and their ecology. *Trudy Zoologicheskogo Instituta*, **42**: 1–154. (Russian).

GOLIKOV, A. N. & SKARLATO, O. A. 1971. Some results of hydrobiological studies made by diving in Pos'et Bay (Sea of Japan). *Gidrobiologicheskii Zhurnal, Kiev*, **7** (5): 32–37. (Russian).

GOLIKOV, A. N. & SKARLATO, O. A. 1985. Shell-bearing gastropods and bivalves of the shelf of south Sakhalin and their ecology. In: *Biocoenoses and Fauna of the Shelf of South Sakhalin* (Golikov, A. N., ed.), 368–487. Nauka, Leningrad. (Russian).

GOLIKOV, A. N. & TZVETKOVA, N. L. 1972. The ecological principle of evolutionary reconstruction as illustrated by marine animals. *Marine Biology, Berlin*, **14**: 1–9.

GOODWIN, B. J. 1978. The growth and breeding cycle of *Littorina obtusata* (Gastropoda: Prosobranchiata) from Cardigan Bay. *Journal of Molluscan Studies*, **44**: 231–242.

GOODWIN, B. J. 1979. The egg mass of *Littorina obtusata* and *Lacuna pallidula* (Gastropoda: Prosobranchia). *Journal of Molluscan Studies*, **45**: 1–11.

GOODWIN, B. J. & FISH, J. D. 1977. Inter- and intraspecific variation in *Littorina obtusata* and *L. mariae* (Gastropoda: Prosobranchia). *Journal of Molluscan Studies*, **43**: 241–254.

GOULD, A. A. 1841. *Report on the Invertebrata of Massachusetts, Comprising the Mollusca, Crustacea, Annelida, and Radiata*. Commissioners on the Zoological and Botanical Survey of the State, Cambridge, Massachusetts.

GOULD, A. A. 1847. [Descriptions of species of shells from the collection of the Exploring Expedition]. *Proceedings of the Boston Society of Natural History*, **2**: 251–252.

GOULD, A. A. 1849. [Descriptions of new species of shells, brought home by the U. S. Exploring Expedition]. *Proceedings of the Boston Society of Natural History*, **3**: 83–85.

GOULD, A. A. 1852, 1856, 1861. Mollusca and shells. *United States Exploring Expedition during the Years 1839–1842 under the Command of Charles Wilkes, U. S. N.*, **12**: 1–498 (1852), 499–510 (1856), atlas (1861).

GOULD, A. A. 1870. *Report on the Invertebrata of Massachusetts Comprising Mollusca*. Edition 2. Wright & Potter, Boston.

GOULD, S. J. 1966. Allometry in Pleistocene land snails from Bermuda: the influence of size upon shape. *Journal of Paleontology*, **40**: 1131–1141.

GOULD, S. J. & VRBA, E. 1982. Exaptation—a missing term in the science of form. *Paleobiology*, **8**: 4–15.

GOWANLOCH, J. N. & HAYES, F. R. 1926. Contributions to the study of marine gastropods. I. The physical factors, behaviour and intertidal life of *Littorina*. *Contributions to Canadian Biology and Fisheries*, **3**: 135–166.

GRABAU, A. W. & KING, S. G. 1928. *Shells of Peitaiho. Peking Society of Natural History, Handbook no. 2*. Edition 2. Peking Laboratory of Natural History and Peking Leader Press, Peking.

GRAF, I. E. 1956. Die Fährten von *Littorina littorea* Linné (Gastr.) in verschiedenen Sedimenten. *Senckenbergiana Lethaea*, **37**: 305–317.

GRAHAM, A. 1988. *Molluscs: Prosobranch and Pyramidellid Gastropods: Keys and Notes for the Identification of the Species*. E. J. Brill & W. Backhuys, for the Linnean Society of London & Estuarine and Brackish-water Sciences Association. Synopses of the British Fauna (New Series) no. 2. Edition 2. Leiden.

GRAHAME, J. 1969. Shedding of the penis in *Littorina littorea*. *Nature, London*, **221**: 976.

GRAHAME, J. 1973. Assimilation efficiency of *Littorina littorea* (L.) (Gastropoda: Prosobranchiata). *Journal of Animal Ecology*, **42**: 383–389.

GRAHAME, J. 1975. Spawning in *Littorina littorea* (L.) (Gastropoda: Prosobranchia). *Journal of Experimental Marine Biology and Ecology*, **18**: 185–196.

GRAHAME, J. & BRANCH, G. M. 1985. Reproductive patterns of marine invertebrates. *Oceanography and Marine Biology Annual Reviews*, **23**: 373–398.

GRAHAME, J. & MILL, P. J. 1986. Relative size of the foot of two species of *Littorina* on a rocky shore in Wales. *Journal of Zoology, London. Series A*, **208**: 229–236.

GRAHAME, J. & MILL, P. J. 1989. Shell shape variation in *Littorina saxatilis* and *L. arcana*: a case of character displacement? *Journal of the Marine Biological Association of the United Kingdom*, **69**: 837–855.

GRAHAME, J. & MILL, P. J. 1992. Local and regional variation in shell shape of rough periwinkles in southern Britain. In: *Proceedings of the Third International Symposium on Littorinid Biology* (Grahame, J., Mill, P. J. & Reid, D. G., eds), 99–106. Malacological Society of London, London.

GRAHAME, J. & MILL, P. J. 1993. Shell shape variation in rough periwinkles: genotypic and phenotypic effects. In: *Quantified Phenotypic Responses in Morphology and Physiology* (Aldrich, J. C., ed.), 25–30. JAPAGA, Ashford.

GRAHAME, J., MILL, P. J. & BROWN, A. C. 1990. Adaptive and non-adaptive variation in two species of rough periwinkle (*Littorina*) on British shores. *Hydrobiologia*, **193**: 223–231.

GRAHAME, J., MILL, P. J., DOUBLE, M. & HULL, S. L. 1992. Patterns of variation in Aat-1 allele frequencies in rough periwinkles (*Littorina*) suggest similar selection regimes rather than conspecificity. *Journal of the Marine Biological Association of the United Kingdom*, **72**: 499–502.

GRAHAME, J., MILL, P. J., DOUBLE, M. & HULL, S. L. 1993. Reply to Johannesson & Johannesson. *Journal of the Marine Biological Association of the United Kingdom*, **73**: 250.

GRAHAME, J., MILL, P. J., HULL, S. L. & CALEY, K. J. 1995. *Littorina neglecta* Bean: ecotype or species? *Journal of Natural History*, **29**: 887–899.

GRAHAME, J., MILL, P. J. & REID, D. G. (eds) 1992. *Proceedings of the Third International Symposium on Littorinid Biology*. Malacological Society of London, London.

GRANADO, I. 1991. Feeding preferences by *Littorina striata* King and Broderip on different algal species. *Scientia Marina*, **55**: 577–581.

GRANOVITCH, A. I. 1992. The effect of trematode infection on the population structure of *Littorina saxatilis* (Olivi) in the White Sea. In: *Proceedings of the Third International Symposium on Littorinid Biology* (Grahame, J., Mill, P. J. & Reid, D. G., eds), 255–263. Malacological Society of London, London.

GRANT, U. S. & GALE, H. R. 1931. Catalogue of the marine Pliocene and Pleistocene Mollusca of California. *Memoirs of the San Diego Society of Natural History*, **1**: 1–1036.

GRATTON, P. 1969. Colour polymorphism in *Littorina mariae*. *Bios, Swansea*, **3**: 53–57.

GRAUS, R. R. 1974. Latitudinal trends in the shell characteristics of marine gastropods. *Lethaia*, **7**: 303–314.

GRAY, J. E. 1839. Molluscous animals, and their shells. In: *The Zoology of Captain Beechey's Voyage* (Beechey, F. W., ed.), 101–155. H. G. Bohn, London.

GRAY, J. E. 1840. Mollusca. In: *Synopsis of the Contents of the British Museum*. Edition 42 [A], 105–152. British Museum, London.

GRAY, J. E. 1847a. The classification of the British Mollusca. By W. E. Leach, M. D. *Annals and Magazine of Natural History. Series 1*, **20**: 267–273.

GRAY, J. E. 1847b. A list of the genera of recent Mollusca, their synonyma and types. *Proceedings of the Zoological Society of London*, **1847**: 129–219.

GRAY, J. E. 1857. *Guide to the Systematic Distribution of Mollusca in the British Museum. Part I*. British Museum, London.

GUALTIERI, N. 1742. *Index Testarum Conchyliorum*. Florentiae.

GULBIN, V. V., GOLIKOV, A. N. & SIRENKO, B. I. 1993. List of prosobranch gastropod molluscs of the Moneron Island Shelf (Sea of Japan) with reference to their distribution and biogeographical composition. *Publications of the Seto Marine Biological Laboratory*, **36**: 61–72.

HABE, T. 1951. Littorinidae in Japan (I). In: *Illustrated Catalogue of Japanese Shells* (Kuroda, T., ed.), 87–93.

HABE, T. 1958a. The fauna of Akkeshi Bay XXV. Gastropoda. *Publications of the Akkeshi Marine Biological Station*, **8**: 1–39.

HABE, T. 1958b. A study on the productivity of the Tanabe Bay (Part I) VI. Zonal arrangement of intertidal benthic animals in the Tanabe Bay. *Records of Oceanographic Works in Japan. Special Volume*, **2**: 43–49.

HABE, T. 1964. *Shells of the Western Pacific in Color*, **2**. Hoikusha, Osaka.

HABE, T. 1979. Marine molluscan species described from Hirado, Nagasaki Prefecture, as type locality. *Nagasaki-Ken Seibutsu Gakkaishi*, **18**: 1–10. (Japanese).

HABE, T. & ITO, K. 1965. *Shells of the World in Colour*. **1**: *the Northern Pacific*. Hoikusha, Osaka.

HABE, T. & KOSUGE, S. 1967. *Common Shells of Japan in Color*. Hoikusha, Osaka.

HADLOCK, R. P. 1980. Alarm response of the intertidal snail *Littorina littorea* (L.) to predation by the crab *Carcinus maenas* (L.). *Biological Bulletin of the Marine Biological Laboratory, Woods Hole*, **159**: 269–280.

HALKETT, A. 1898. List of zoological specimens, etc., collected by Dr. Wakeham and Mr. Low at various points in Hudson Bay and Strait during season of 1897. *Canadian Department of Marine and Fisheries, Session Papers*, **11B**: 80–83.

HANLEY, S. 1855. *Ipsa Linnaei Conchylia. The Shells of Linnaeus, Determined from his Manuscripts and Collection*. Williams & Norgate, London.

HANNA, G. D. 1923. Upper Miocene lacustrine mollusks from Sonoma County, California. *Proceedings of the California Academy of Sciences. Series 4*, **12**: 31–41.

HANNA, G. D. 1966. Introduced mollusks of western North America. *Occasional Papers of the California Academy of Sciences*, **48**: 1–108.

HANNA, G. D. & HERTLEIN, L. G. 1959. Marine shells of Middleton Island, Alaska. *Nautilus*, **72**: 78–84.

HANNAFORD ELLIS, C. J. 1978a. *Littorina arcana* sp. nov.: a new species of winkle (Gastropoda: Prosobranchia: Littorinidae). *Journal of Conchology*, **29**: 304.

HANNAFORD ELLIS, C. J. 1978b. Development of *Littorina rudis*. *Littorinid Tidings*, **7**: 3.

HANNAFORD ELLIS, C. J. 1979. Morphology of the oviparous rough winkle *Littorina arcana* Hannaford Ellis, 1978, with notes on the taxonomy of the *L. saxatilis* species-complex (Prosobranchia: Littorinidae). *Journal of Conchology*, **30**: 43–56.

HANNAFORD ELLIS, C. J. 1983. Patterns of reproduction in four *Littorina* species. *Journal of Molluscan Studies*, **49**: 98–106.

HANNAFORD ELLIS, C. J. 1984. Ontogenetic change of shell colour patterns in *Littorina neglecta* Bean (1844). *Journal of Conchology*, **31**: 343–347.

HANNAFORD ELLIS, C. J. 1985. The breeding migration of *Littorina arcana* Hannaford Ellis, 1978 (Prosobranchia: Littorinidae). *Zoological Journal of the Linnean Society*, **84**: 91–96.

HANSEN, J. R. & HAUGEN, I. 1989. Some observations of intertidal communities on Spitsbergen (79°N), Norwegian Arctic. *Polar Research*, **7**: 23–27.

HANSEN, T. A. 1980. Influence of larval dispersal and geographic distribution on species longevity in neogastropods. *Paleobiology*, **6**: 193–207.

HÄNTZSCHEL, W. 1938. Quer-Gliederung bei *Littorina*-Fahrten, ein Beitrag zur deutung von *Keckia annulata* Glocker. *Senckenbergiana*, **20**: 297–304.

HÄNTZSCHEL, W. 1941. Gleichlauf und Verzweigung von Fährten der Strandschnecke *Littorina*. *Natur und Volk*, **71**: 117–120.

HARDWICK-WITMAN, M. N. & MATHIESON, A. C. 1983. Intertidal macroalgae and macroinvertebrates: seasonal and spatial abundance patterns along an estuarine gradient. *Estuarine Coastal and Shelf Science*, **16**: 113–129.

HARGER, J. R. E. 1972. Competitive coexistence among intertidal invertebrates. *American Scientist*, **60**: 600–607.

HARMER, F. W. 1920–1925. *The Pliocene Mollusca of Great Britain, Being Supplementary to S. V. Wood's Monograph of the Crag Mollusca*. **2**. Palaeontographical Society, London.

HART, A. & BEGON, M. 1982. The status of general reproductive strategy theories, illustrated in winkles. *Oecologia*, **52**: 37–42.

HARVEY, P. H. & PAGEL, M. D. 1991. *The Comparative Method in Evolutionary Biology*. Oxford University Press, Oxford.

HATAI, K. & NISIYAMA, S. 1952. Check list of Japanese Tertiary marine Mollusca. *Science Reports of Tôhoku University. Series 2 (Geology). Special Volume*, **3**: 1–464.

HAWKINS, S. J. 1981. The influence of season and barnacles on the algal colonization of *Patella vulgata* exclusion areas. *Journal of the Marine Biological Association of the United Kingdom*, **61**: 1–15.

HAWKINS, S. J., BURNAY, L. P., NETO, A. I., TRISTÃO DA CUNHA, R. & FRIAS MARTINS, A. M. 1990. A description of the zonation patterns of molluscs and other important biota on the south coast of São Miguel, Azores. *Açoreana. Supplement*, **1990**: 21–38.

HAWKINS, S. J. & HARTNOLL, R. G. 1983. Grazing of intertidal algae by marine invertebrates. *Oceanography and Marine Biology Annual Review*, **21**: 195–282.

HAWKINS, S. J., HARTNOLL, R. G., KAIN, J. M. & NORTON, T. A. 1992. Plant-animal interactions on hard substrata in the north-east Atlantic. In: *Plant-Animal Interactions in the Marine Benthos* (John, D. M., Hawkins, S. J. & Price, J. H., eds), 1–32. Clarendon Press, Oxford.

HAWKINS, S. J. & HISCOCK, K. 1983. Anomalies in the abundance of common eulittoral gastropods with planktonic larvae on Lundy Island, Bristol Channel. *Journal of Molluscan Studies*, **49**: 86–88.

HAWKINS, S. J., WATSON, D. C., HILL, A. S., HARDING, S. P., KYRIAKIDES, M. A., HUTCHINSON, S. & NORTON, T. A. 1989. A comparison of feeding mechanisms in microphagous, herbivorous, intertidal, prosobranchs in relation to resource partitioning. *Journal of Molluscan Studies*, **55**: 151–165.

HAYASAKA, S. 1955. A preliminary report on the local variation of a periwinkle *Littorina brevicula* (Philippi). *Research Bulletin of the Saito Ho-on Kai Museum*, **24**: 35–37.

HAYES, F. R. 1927. The effect of environmental factors on the development and growth of *Littorina littorea*. *Proceedings and Transactions of the Nova Scotian Institute of Science*, **17**: 6–13.

HAYES, F. R. 1929. Contributions to the study of marine gastropods. 3. Development, growth and behaviour of *Littorina*. *Contributions to Canadian Biology and Fisheries*, **4**: 413–430.

HAYWARD, P. J., WIGHAM, G. D. & YONOW, N. 1990. Mollusca I: Polyplacophora, Scaphopoda, and Gastropoda. In: *The Marine Fauna of the British Isles and North-West Europe* (Hayward, P. J. & Ryland, J. S., eds), **2**: 628–730. Clarendon Press, Oxford.

HAZLETT, A. & SEED, R. 1976. A study of *Fucus spiralis* and its associated fauna in Strangford Lough, Co. Down. *Proceedings of the Royal Irish Academy. Series B*, **76**: 607–618.

HEALY, J. M. & JAMIESON, B. G. M. 1993. Euspermatozoa, paraspermatozoa and spermatozeugmata of *Littoraria (Palustorina) articulata* (Prosobranchia: Caenogastropoda) with species reference to the pseudotrich. *Acta Zoologica*, **74**: 321–330.

HEIL, K. P. & EICHELBERG, D. 1983. Studies of uric acid metabolism in *Littorina littorea* (Gastropoda). *Helgoländer Meeresuntersuchungen*, **36**: 465–472.

HELLER, J. 1975a. The taxonomy of some British *Littorina* species with notes on their reproduction (Mollusca: Prosobranchia). *Zoological Journal of the Linnean Society*, **56**: 131–151.

HELLER, J. 1975b. Visual selection of shell colour in two littoral prosobranchs. *Zoological Journal of the Linnean Society*, **56**: 153–170.

HELLER, J. 1976. The effects of exposure and predation on the shell of two British winkles. *Journal of Zoology, London*, **179**: 201–213.

HEMPHILL, H. 1877. Description of a new California mollusk. *Proceedings of the California Academy of Sciences*, **7**: 49.

HEPPELL, D. 1970. Comments on the proposed designation of a type-species for *Littorina* Férussac, 1822. Z. N. (S.) 1901. *Bulletin of Zoological Nomenclature*, **27**: 131–132.

HEPPELL, D. 1971. Further comments on *Littorina* Férussac and its type-species. Z. N. (S.) 1901. *Bulletin of Zoological Nomenclature*, **28**: 76.

HEPPELL, D. 1995. The identity of *Melanella dufresnii* Bowdich, 1822, and other species of Eulimidae described earlier. *Journal of Conchology*, **35**: 193–222.

HERMAN, Y. & HOPKINS, D. M. 1980. Arctic oceanic climate in late Cenozoic time. *Science, New York*, **209**: 557–562.

HERMANS, J. M. C. 1977. *Littorina littorea* (Linnaeus, 1758) parasité par *Balanus balanoides* (Linnaeus, 1761). *Arion*, **1977** (11–12): 1–3.

HERMANS, J. M. C. 1978. Shell deformation in European Littorinidae caused by *Balanus balanoides* (L., 1761). *Arion*, **1978** (7–8): 1.

HERRMANNSEN, A. N. 1846. *Indicis Generum Malacozoorum Primordia*, **1**. Cassell.

HERSCHBERG, J. B. 1992. A malacological encounter with northern Norway. *Vita Marina*, **41**: 85–99.

HERTLING, H. 1928. Beobachtungen und Versuche an den Eiern von *Littorina* und *Lacuna*. Bedeutung der Eihüllen. Entwicklung im naturlichen und abgeänderten Medium. *Wissenschaftliche Meeresuntersuchungen, Abteilung Helgoland*, **17** (2): 1–49.

HERTLING, H. & ANKEL, W. E. 1927. Bemerkungen über Laich und Jugendformen von *Littorina* und *Lacuna*. *Wissenschaftliche Meeresuntersuchungen, Abteilung Helgoland*, **16** (7): 1–14.

HERZENSTEIN, S. 1885. Beiträge zur Kenntniss der Fauna der Murmanküste und des Weissen Meeres. I. *Trudy Imperatorskago Sankt-Peterburgskago Obshchestva Estestvoispytatelei*, **16**: 635–814.

HEWATT, W. G. 1937. Ecological studies on selected marine intertidal communities of Monterey Bay, California. *American Midland Naturalist*, **18**: 161–206.

HEWATT, W. G. 1938. Notes on the breeding seasons of the rocky beach fauna of Monterey Bay, California. *Proceedings of the California Academy of Sciences*, **23**: 283–288.

HEWITT, C. G. 1907. Some Arthrostraca and other Invertebrata from St Kilda. *Annals of Scottish Natural History*, **1907**: 219–221.

HIDALGO, J. G. 1867. Catalogue des mollusques testacés marins des côtes de l'Espagne et des îles Baléares,—suite. *Journal de Conchyliologie*, **15**: 357–426.

HIDALGO, J. G. 1890. *Moluscos Marinos de España, Portugal y las Baleares*. Parts 18–20. Madrid.

HIDALGO, J. G. 1904–1905. *Catálogo de los Moluscos Testáceos de las Islas Filipinas, Joló y Marianas. I.—Moluscos Marinos*. Madrid.

HIDALGO, J. G. 1917. Fauna malacológica de España, Portugal y las Baleares. Moluscos testáceos marinos. *Trabajos del Museo Nacional de Ciencias Naturales. Serie Zoológica*, **30**: 1–752.

HIGHSMITH, R. C. 1985. Floating and algal rafting as potential dispersal mechanisms in brooding invertebrates. *Marine Ecology Progress Series*, **25**: 169–179.

HIGO, S. 1973. *A Catalogue of Molluscan Fauna of the Japanese Islands and the Adjacent Area*.

HIGO, S. & GOTO, Y. 1993. *A Systematic List of Molluscan Shells from the Japanese Islands and the Adjacent Area*. Eru Malacological Publishers, Yao, Osaka.

HILLIS, D. M. & MORITZ, C. (eds) 1990. *Molecular Systematics*. Sinauer Associates, Sunderland, Massachusetts.

HIRAI, E. 1963. On the breeding seasons of invertebrates in the neighbourhood of the Marine Biological Station of Asamushi. *Science Reports of Tôhoku University. Series IV (Biology)*, **29**: 369–375.

HIRASE, S. 1934. *A Collection of Japanese Shells with Illustrations in Natural Colours*. Edition 3. Matsumuru Sanshodo, Tokyo.

HIRASE, S. & TAKI, I. 1954. *An Illustrated Handbook of Shells in Natural Colors from the Japanese Islands and Adjacent Territory*. Maruzen, Tokyo.

HOEKSEMA, D. 1985. Opmerkingen over variëteiten van *Littorina saxatilis* (Olivi, 1792) in het algemeen en over de gebandeerde kleurvariëteiten van de ruwe alikruik in Nederland in het bijzonder. *Het Zeepaard*, **45**: 22–28.

HOLTHUIS, L. B. 1971. Comments on the proposed designation of a type-species for *Littorina* Férussac, 1822. Z. N. (S.) 1901. *Bulletin of Zoological Nomenclature*, **27**: 210.

HOPKINS, D. M. 1967a. Quaternary marine transgressions in Alaska. In: *The Bering Land bridge* (Hopkins, D. M., ed.), 47–90. Stanford University Press, Stanford.

HOPKINS, D. M. 1967b. The Cenozoic history of Beringia—a synthesis. In: *The Bering Land Bridge* (Hopkins, D. M., ed.), 451–484. Stanford University Press, Stanford.

HOUSSAY, F. 1884. Recherches sur l'opercule et les glands du pied des gastéropodes. *Archives de Zoologie Expérimentale et Générale. Série 2*, **2**: 171–288.

HOWE, H. V. 1922. Faunal and stratigraphic relationships of the Empire Formation, Coos Bay, Oregon. *University of California Publications. Bulletin of the Department of Geological Sciences*, **14**: 85–114.

HUBENDICK, B. & WARÉN, A. 1969–1976. *Framgälade Snäckor från Svenska Västkusten*. Göteborgs Naturhistoriska Museum, Göteborg.

HUGHES, R. N. 1978. Demography and reproductive mode in *Littorina neritoides* and the *Littorina saxatilis* species complex. *Haliotis*, **9** (2): 91–98.

HUGHES, R. N. 1979. South African populations of *Littorina rudis*. *Zoological Journal of the Linnean Society*, **65**: 119–126.

HUGHES, R. N. 1980. Population dynamics, growth and reproductive rates of *Littorina nigrolineata* Gray from a moderately sheltered locality in North Wales. *Journal of Experimental Marine Biology and Ecology*, **44**: 211–228.

HUGHES, R. N. 1995. Resource allocation, demography and the radiation of life histories in rough periwinkles (Gastropoda). *Hydrobiologia*, **309**: 1–14.

HUGHES, R. N. & ROBERTS, D. J. 1980. Reproductive effort of winkles (*Littorina* spp.) with contrasted methods of reproduction. *Oecologia*, **47**: 130–137.

HUGHES, R. N. & ROBERTS, D. J. 1981. Comparative demography of *Littorina rudis*, *L. nigrolineata* and *L. neritoides* on three contrasted shores in north Wales. *Journal of Animal Ecology*, **50**: 251–268.

HULL, S. L., GRAHAME, J. & MILL, P. J. 1996. Morphological divergence and evidence for reproductive isolation in *Littorina saxatilis* (Olivi) in northeast England. *Journal of Molluscan Studies*, **62**: 89–99.

HULL, S. L., MILL, P. J. & GRAHAME, J. 1994. Aminotransferases in *Littorina*: is there a functional story? *Cahiers de Biologie Marine*, **35**: 248–249.

HUMPHREY, R. R. & MACY, R. W. 1931. Observations on some of the probable factors controlling the size of certain tide pool snails. *Publications of the Puget Sound Biological Station*, **7**: 205–208.

HUMPHRIES, C. J. & PARENTI, L. R. 1986. *Cladistic Biogeography*. Clarendon Press, Oxford.

HUNTER, R. D. & RUSSELL-HUNTER, W. D. 1983. Bioenergetic and community changes in intertidal aufwuchs grazed by *Littorina littorea*. *Ecology, New York*, **64**: 761–769.

HUNTSMAN, A. G. 1918. The vertical distribution of certain intertidal animals. *Proceedings and Transactions of the Royal Society of Canada. Series 3*, **4**: 53–60.

HUXHAM, M., RAFFAELLI, D. G. & PIKE, A. 1993. The influence of *Cryptocotyle lingua* (Digenea: Platyhelminthes) infections on the survival and fecundity of *Littorina littorea* (Gastropoda: Prosobranchia); an ecological approach. *Journal of Experimental Marine Biology and Ecology*, **168**: 223–238.

HYLLEBERG, J. & CHRISTENSEN, J. T. 1977. Phenotypic variation and fitness of periwinkles (Gastropoda: Littorinidae) in relation to exposure. *Journal of Molluscan Studies*, **43**: 192–199.

HYLLEBERG, J. & CHRISTENSEN, J. T. 1978. Factors affecting the intra-specific competition and size distribution of the periwinkle *Littorina littorea* (L.). *Natura Jutlandica*, **20**: 193–202.

IGNATIUS, H., AXBERG, S., NIEMISTÖ, L. & WINTERHALTER, B. 1981. Quaternary geology of the Baltic Sea. In: *The Baltic Sea* (Voipio, A., ed.), 54–104. Elsevier, Amsterdam.

ILINA, A. P. 1939. Gastropod fauna from the Tertiary of the western shore of Kamchatka. *Trudy Neftyanogo Geologo-Razvedochnogo Instituta. Seriya A*, **124**. (Russian).

IMRIE, D. W. 1992. The role of pedal mucus in the feeding behaviour of *Littorina littorea* (L.). In: *Proceedings of the Third International Symposium on Littorinid Biology* (Grahame, J., Mill, P. J. & Reid, D. G., eds), 221–226. Malacological Society of London, London.

IMRIE, D. W., HAWKINS, S. J. & MCCROHAN, C. R. 1989. The olfactory-gustatory basis of food preference in the herbivorous prosobranch, *Littorina littorea* (Linnaeus). *Journal of Molluscan Studies*, **55**: 217–225.

INAGAKI, H. 1972. Sénescence dans la croissance chez le mâle de *Littorina saxatilis* var. *rudis* (Olivi), mollusque gastéropode, mise en évidence par la régression sexuelle et la diminution de l'intensité réspiratoire. *Compte Rendu Hebdomadaire des Séances de l'Académie des Sciences, Paris. Série D*, **274**: 1828–1831.

INGÓLFSSON, A. 1992. The origin of the rocky shore fauna of Iceland and the Canadian Maritimes. *Journal of Biogeography*, **19**: 705–712.

INGÓLFSSON, A. 1995. Floating clumps of seaweed around Iceland: natural microcosms and a means of dispersal for shore fauna. *Marine Biology, Berlin*, **122**: 13–21.

INTERNATIONAL COMMISSION ON ZOOLOGICAL NOMENCLATURE 1985. *International Code of Zoological Nomenclature*. International Trust for Zoological Nomenclature, London.

IRELAND, M. P. & WOOTTON, R. J. 1977. Distribution of lead, zinc, copper and manganese in the marine gastropods *Thais lapillus* and *Littorina littorea* around the coast of Wales. *Environmental Pollution*, **12**: 27–41.

IRVING, E. & MASSEY, N. W. D. 1990. Palaeomagnetism of ocean layers 2 and 3: evidence from the Metchosin Complex, Vancouver Island. *Physics of the Earth and Planetary Interiors*, **64**: 247–260.

ISARANKURA, K. & RUNHAM, N. W. 1968. Studies on the replacement of the gastropod radula. *Malacologia*, **7**: 71–91.

ITOIGAWA, J. & YAMANOI, T. 1990. Climatic optimum in the mid-Neogene of the Japanese Islands. In: *Pacific Neogene Events: Their Timing, Nature and Interrelationships* (Tsuchi, R., ed.), 3–14. University of Tokyo Press, Tokyo.

JABLONSKI, D. 1982. Evolutionary rates and modes in Late Cretaceous gastropods: role of larval ecology. *Proceedings of the Third North American Paleontological Convention*, **1**: 257–262.

JABLONSKI, D. 1986. Larval ecology and macroevolution in marine invertebrates. *Bulletin of Marine Science*, **39**: 565–587.

JABLONSKI, D. & LUTZ, R. A. 1980. Molluscan larval shell morphology: ecological and paleontological applications. In: *Skeletal Growth of Aquatic Organisms: Biological Records of Environmental Change* (Rhoads, D. C. & Lutz, R. A., eds), 323–377. Plenum Press, New York.

JABLONSKI, D. & LUTZ, R. A. 1983. Larval ecology of marine benthic invertebrates. *Biological Reviews of the Cambridge Philosophical Society*, **58**: 21–89.

JAECKEL, S. 1951. Die Mollusken der Schlei. *Archiv für Hydrobiologie*, **44**: 214–270.

JAECKEL, S. 1952. Zur Oekologie der Molluskenfauna in der westlichen Ostsee. *Schriften des Naturwissenschaftlichen Vereins für Schleswig-Holstein*, **26**: 18–50.

JAGNOW, B. & GOSSELCK, F. 1987. Bestimmungsschlüssel für die Gehäuseschnecken und Muscheln der Ostsee. *Mitteilungen aus dem Zoologischen Museum in Berlin*, **63**: 191–268.

JAMES, B. L. 1963. Subspecies of *Littorina saxatilis* around Dale. *Field Studies*, **1**: 30–31.

JAMES, B. L. 1964. Sub-speciation in *Littorina saxatilis* (Olivi). *Report of the Challenger Society*, **3** (16): 26.

JAMES, B. L. 1968a. The characters and distribution of the subspecies and varieties of *Littorina saxatilis* (Olivi, 1792) in Britain. *Cahiers de Biologie Marine*, **9**: 143–165.

JAMES, B. L. 1968b. The distribution and keys of species in the family Littorinidae and of their Digenean parasites in the region of Dale, Pembrokeshire. *Field Studies*, **2**: 615–650.

JAMES, B. L. 1968c. The occurrence of *Parvatrema homoeotecnum* James, 1964 (Trematoda: Gymnophallidae) in a population of *Littorina saxatilis tenebrosa* (Mont.). *Journal of Natural History*, **2**: 21–37.

JAMES, B. L. 1968d. Studies on the life-cycle of *Microphallus pygmaeus* (Levinsen, 1881) (Trematoda: Microphallidae). *Journal of Natural History*, **2**: 155–172.

JAMES, B. L. 1969. The larval Digenea of the intertidal prosobranch, *Littorina saxatilis* (Olivi). *Zeitschrift für Zoologische Systematik und Evolutionsforschung*, **7**: 273–316.

JANKE, K. 1986. Die Makrofauna und ihre Verteilung im Nordost-Felswatt von Helgoland. *Helgoländer Meeresuntersuchungen*, **40**: 1–55.

JANKE, K. 1990. Biological interactions and their role in the community structure in the rocky intertidal of Helgoland (German Bight, North Sea). *Helgoländer Meeresuntersuchungen*, **44**: 219–263.

JANSON, K. 1982a. Phenotypic differentiation in *Littorina saxatilis* Olivi (Mollusca, Prosobranchia) in a small area on the Swedish west coast. *Journal of Molluscan Studies*, **48**: 167–173.

JANSON, K. 1982b. Genetic and environmental effects on the growth rate of *Littorina saxatilis*. *Marine Biology, Berlin*, **69**: 73–78.

JANSON, K. 1983a. Chromosome number in two phenotypically distinct populations of *Littorina saxatilis* Olivi, and in specimens of the *Littorina obtusata* (L.) species-complex. *Journal of Molluscan Studies*, **49**: 224–227.

JANSON, K. 1983b. Selection and migration in two distinct phenotypes of *Littorina saxatilis* in Sweden. *Oecologia*, **59**: 58–61.

JANSON, K. 1985a. Variation in the occurrence of abnormal embryos in females of the intertidal gastropod *Littorina saxatilis* Olivi. *Journal of Molluscan Studies*, **51**: 64–68.

JANSON, K. 1985b. A morphological and genetic analysis of *Littorina saxatilis* (Prosobranchia) from Venice, and on the problem of *saxatilis-rudis* nomenclature. *Biological Journal of the Linnean Society*, **24**: 51–59.

JANSON, K. 1987a. Allozyme and shell variation in two marine snails (*Littorina*, Prosobranchia) with different dispersal abilities. *Biological Journal of the Linnean Society*, **30**: 245–256.

JANSON, K. 1987b. Genetic drift in small and recently founded populations of the marine snail *Littorina saxatilis*. *Heredity*, **58**: 31–37.

JANSON, K. & SUNDBERG, P. 1983. Multivariate morphometric analysis of two varieties of *Littorina saxatilis* from the Swedish west coast. *Marine Biology, Berlin*, **74**: 49–53.

JANSON, K. & WARD, R. D. 1984. Microgeographical variation in allozyme and shell characters in *Littorina saxatilis* Olivi (Prosobranchia: Littorinidae). *Biological Journal of the Linnean Society*, **22**: 289–307.

JANSON, K. & WARD, R. D. 1985. The taxonomic status of *Littorina tenebrosa* Montagu as assessed by morphological and genetic analyses. *Journal of Conchology*, **32**: 9–15.

JANSSEN, H. H. & TRIEBSKORN, R. 1987. Comparative morphology of the radulae in *Pomatias elegans* and in *Littorina littorea* (Gastropoda: Taenioglossa). *Zoologischer Anzeiger*, **219**: 73–82.

JARDINE, I. 1984. The occurrence of *Littorina* spp. in Robin Hood's Bay, North Yorkshire, with a key to British species. *Naturalist, Hull*, **109**: 3–9.

JAY, J. C. 1839. *A Catalogue of the Shells, Arranged According to the Lamarckian System; Together with Descriptions of New or Rare Species, Contained in the Collection of John C. Jay, M.D.* Edition 3. Wiley & Putnam, New York & London.

JEFFREYS, J. G. 1839. A list of marine Mollusca taken during a few days stay at Oban, in Argyleshire, in the autumn of 1838. *Malacological and Conchological Magazine*, **2**: 33–46.

JEFFREYS, J. G. 1859. Additional gleanings in British Conchology. *Annals and Magazine of Natural History. Series 3*, **4**: 189–201.

JEFFREYS, J. G. 1865, 1869. *British Conchology*. **3** and **5**. John van Voorst, London.

JEFFREYS, J. G. 1872. The Mollusca of St Helena. *Annals and Magazine of Natural History. Series 4*, **9**: 262–264.

JEFFREYS, J. G. 1874. On some species of Japanese marine shells and fishes, which inhabit also the north Atlantic. *Journal of the Linnean Society. Zoology*, **12**: 100–109.

JEFFREYS, J. G. 1877. New and peculiar Mollusca of the Patellidae and other families of Gastropoda procured in the 'Valorous' Expedition. *Annals and Magazine of Natural History. Series 4*, **19**: 231–243.

JEFFREYS, J. G. 1883. On the Mollusca procured during the 'Lightning' and 'Porcupine' Expeditions, 1868–70. (Part VI). *Proceedings of the Zoological Society of London*, **1883**: 88–115.

JELETZKY, J. A. 1973. Age and depositional environments of Tertiary rocks of Nootka Island, British Columbia (92E): molluscs versus foraminifers. *Canadian Journal of Earth Sciences*, **10**: 331–365.

JENNER, C. E. 1956. Seasonal resorption of the copulatory organ in *Nassarius trivittatus* and *Littorina littorea*. *Biological Bulletin of the Marine Biological Laboratory, Woods Hole*, **111**: 305.

JENSEN, J. T. 1981. Distribution, activity and food habits of juvenile *Tegula funebralis* and *Littorina scutulata* (Gastropoda: Prosobranchia) as they relate to resource partitioning. *Veliger*, **23**: 333–338.

JOHANNESSON, B. 1986. Shell morphology of *Littorina saxatilis* Olivi: the relative importance of physical factors and predation. *Journal of Experimental Marine Biology and Ecology*, **102**: 183–195.

JOHANNESSON, B. & JOHANNESSON, K. 1990. *Littorina neglecta* Bean, a morphological form within the variable species *Littorina saxatilis* (Olivi)? *Hydrobiologia*, **193**: 71–87.

JOHANNESSON, B. & JOHANNESSON, K. 1996. Population differences in behaviour and morphology in *Littorina saxatilis*: phenotypic plasticity or genetic differentiation? *Journal of Zoology, London*, in press.

JOHANNESSON, K. 1988. The paradox of Rockall: why is a brooding gastropod (*Littorina saxatilis*) more widespread than one having a planktonic larval dispersal stage (*L. littorea*)? *Marine Biology, Berlin*, **99**: 507–513.

JOHANNESSON, K. 1992. Genetic variability and large scale differentiation in two species of littorinid gastropods with planktotrophic development, *Littorina littorea* (L.) and *Melarhaphe (Littorina) neritoides* (L.) (Prosobranchia: Littorinacea), with notes on a mass occurrence of *M. neritoides* in Sweden. *Biological Journal of the Linnean Society*, **47**: 285–299.

JOHANNESSON, K. & JOHANNESSON, B. 1989. Differences in allele frequencies of *Aat* between high- and mid-rocky shore populations of *Littorina saxatilis* (Olivi) suggest selection in this enzyme locus. *Genetic Research, Cambridge*, **54**: 7–11.

JOHANNESSON, K. & JOHANNESSON, B. 1990. Genetic variation within *Littorina saxatilis* (Olivi) and *Littorina neglecta* Bean: is *L. neglecta* a good species? *Hydrobiologia*, **193**: 89–97.

JOHANNESSON, K. & JOHANNESSON, B. 1993. The taxonomic status of *Littorina neglecta*: a comment to Grahame, Mill, Double and Hull. *Journal of the Marine Biological Association of the United Kingdom*, **73**: 249.

JOHANNESSON, K. & JOHANNESSON, B. 1995. Dispersal and population expansion in a direct developing marine snail (*Littorina saxatilis*) following a severe population bottleneck. *Hydrobiologia*, **309**: 173–180.

JOHANNESSON, K., JOHANNESSON, B. & LUNDGREN, U. 1995. Strong natural selection causes microscale allozyme variation in a marine snail. *Proceedings of the National Academy of Sciences of the U. S. A.*, **92**: 2602–2606.

JOHANNESSON, K., JOHANNESSON, B. & ROLÁN-ALVAREZ, E. 1993. Morphological differentiation and genetic cohesiveness over a microenvironmental gradient in the marine snail *Littorina saxatilis*. *Evolution*, **47**: 1770–1787.

JOHANNESSON, K., RAFFAELLI, D. G. & HANNAFORD ELLIS, C. J. (eds) 1990. Progress in Littorinid and Muricid Biology. Proceedings of the Second European Meeting on Littorinid Biology, Tjärnö Marine Biological Laboratory, Sweden, July 4–8, 1988. *Hydrobiologia*, **193**: 1–285.

JOHANNESSON, K., ROLÁN-ALVAREZ, E. & EKENDAHL, A. 1995. Incipient reproductive isolation between two sympatric morphs of the intertidal snail *Littorina saxatilis*. *Evolution*, **49**: 1180–1190.

JOHANNESSON, K. & SUNDBERG, P. 1992. Speciation in *Littorina saxatilis* (Olivi)?—A one-dimensional selection-migration model. In: *Proceedings of the Third International Symposium on Littorinid Biology* (Grahame, J., Mill, P. J. & Reid, D. G., eds), 1–8. Malacological Society of London, London.

JOHANNESSON, K. & WARMOES, T. 1990. Rapid colonization of Belgian breakwaters by the direct developer, *Littorina saxatilis* (Olivi) (Prosobranchia, Mollusca). *Hydrobiologia*, **193**: 99–108.

JOHANSEN, A. C. 1901. On the variations observable in some northern species of *Littorina*. *Videnskabelige Meddelelser fra den Naturhistoriske Forening i Kjöbenhavn*, **53**: 295–307.

JOHANSSON, J. 1939. Anatomische Studien über die Gastropodenfamilien Rissoidae und Littorinidae. *Zoologiska Bidrag från Uppsala*, **18**: 287–396.

JOHNSON, C. W. 1915. Fauna of New England. 13. List of the Mollusca. *Occasional Papers of the Boston Society of Natural History*, **7**: 1–231.

JOHNSON, C. W. 1934. List of marine Mollusca of the Atlantic coast from Labrador to Texas. *Proceedings of the Boston Society of Natural History*, **40**: 1–204.

JOHNSON, M. S., MURRAY, J. & CLARKE, B. 1986. An electrophoretic analysis of phylogeny and evolutionary rates in the genus *Partula* from the Society Islands. *Proceedings of the Royal Society of London. Series B*, **227**: 161–177.

JOHNSON, R. I. 1964. The Recent Mollusca of Augustus Addison Gould. *Bulletin of the United States National Museum*, **239**: 1–182.

JOHNSTON, G. 1842. A descriptive catalogue of the gasteropodous Mollusca of Berwickshire. Part II. *History of the Berwickshire Naturalists' Club*, **1**: 263–276.

JONES, R. L. & KEEN, D. H. 1993. *Pleistocene Environments in the British Isles*. Chapman & Hall, London.

JÜCH, P. J. W. & BOEKSCHOTEN, G. J. 1980. Trace fossils and grazing traces produced by *Littorina* and *Lepidochitona*, Dutch Wadden Sea. *Geologie en Mijnbouw*, **59**: 33–42.

JULL, A. J. T., KUZ'MIN, Y. V., LUTAENKO, K. A., ORLOVA, L. A., POPOV, A. N., RAKOV, V. A. & SULERZHITSKII, L. D. 1994. The mid-Holocene malacofauna of the Neolithic site Boisman 2 (Primorye): composition, age, and environment. *Doklady Akademii Nauk*, **339**: 697–700. (Russian).

KAAS, P. & BROEK, A. N. C. TEN 1942. *Nederlandse Zeemollusken*. Wereldbibliotheek, Amsterdam.

KADOLSKY, D. 1970. *Littorina* Férussac, 1822 (Gastropoda): proposed designation of a type-species under the plenary powers. Z. N. (S.) 1901. *Bulletin of Zoological Nomenclature*, **27**: 51–54.

KADOLSKY, D. 1971. Nomenklatorische Bemerkungen. *Archiv für Molluskenkunde*, **101**: 191–193.

KADOLSKY, D. 1973. Die vorpliozänen Littorinidae und Lacunidae Mitteleuropas (Gastropoda: Prosobranchia). *Archiv für Molluskenkunde*, **103**: 31–62.

KALABUSHKIN, B. A. 1976. Intrapopulational variability in contemporary and mid-Holocene populations of *Littorina squalida*. *Zhurnal Obshchei Biologii, Moskva*, **37**: 369–377. (Russian).

KALABUSHKIN, B. A. & ALTUKHOV, Y. P. 1974. Stable polymorphism in an isolated population of *Littorina squalida* of the Busse Lagoon from the Holocene climatic optimum to the present. In: *Biology of Marine Molluscs and Echinoderms*, 61. Vladivostok. (Russian).

KALABUSHKIN, B. A., ZHIVOTOVSKII, L. A. & AFANAS'EV, K. I. 1985. The population structure of the mollusc *Littorina squalida* by morphophysiological and biochemical features. In: *Mikroevolyutsiya Sbornik Tezisov i Vsesoyuznoi Konferentsii po Problemam Evolyutsii*, 42–44. Moscow. (Russian).

KANWISHER, J. W. 1966. Freezing in intertidal animals. In: *Cryobiology* (Meryman, H. T., ed.), 487–494. Academic Press, London.

KARTAVTSEV, Y. F. 1976. Polymorphism of shell colour in the gastropod mollusc *Littorina brevicula* (Philippi). *Sbornik Rabot. Institut Biologii Morya, Dal'nevostochnyi Nauchnyi Tsentr*, **5**: 93–98. (Russian).

KARTAVTSEV, Y. F. & EFREMOV, V. V. 1981. Genetic similarity and variability of two littorinid species (Mollusca: Littorinidae). *Genetika*, **17**: 1029–1033. (Russian).

KAY, J. E. DE 1843. *Zoology of New-York, or the New-York Fauna. Part V. Mollusca*. Albany.

KEEN, A. M. 1970. Symposium on the rare and endangered mollusks of North America. 8. Western marine mollusks. *Malacologia*, **10**: 51–53.

KEEN, A. M. 1971. *Sea Shells of Tropical West America*. Edition 2. Stanford University Press, Palo Alto, California.

KEEN, A. M. & BENTSON, H. 1944. Check list of California Tertiary marine Mollusca. *Geological Society of America Special Papers*, **56**: 1–280.

KEEN, A. M. & DOTY, C. L. 1942. An annotated check list of the gastropods of Cape Arago, Oregon. *Oregon State Monographs. Studies in Zoology*, **3**: 1–16.

KEEP, J. 1888. *West Coast Shells*. Samuel Carson & Co, San Francisco.

KEEP, J. 1904. *West American Shells*. Whitaker & Ray, San Francisco.

KEEP, J. & BAILY, J. L. 1935. *West Coast Shells*. Stanford University Press, Stanford.

KEMP, P. & BERTNESS, M. 1984. Snail shape and growth rates: evidence for plastic shell allometry in *Littorina littorea*. *Proceedings of the National Academy of Sciences of the U. S. A.*, **81**: 811–813.

KENNARD, A. S. 1942. The *Histoire* and *Prodrome* of Férussac. *Proceedings of the Malacological Society of London*, **25**: 12–17, 105–118.

KESSEL, K. Y. & RAUKAS, A. V. 1967. *The deposits of the Ancylus Lake and Littorina Sea in Estonia*. Institute of Geology, Academy of Sciences of Estonia, Tallin. (Russian).

KILBURN, R. N. 1972. Taxonomic notes on South African marine Mollusca (2), with the description of new species and subspecies of *Conus*, *Nassarius*, *Vexillum* and *Demoulia*. *Annals of the Natal Museum*, **21**: 391–437.

KING, P. P. & BRODERIP, W. J. 1832. Description of the Cirrhipeda, Conchifera and Mollusca, in a collection formed by the officers of H.M.S. Adventure and Beagle employed between the years 1826 and 1830 in surveying the southern coasts of South America, including the Straits of Magalhaens and the coast of Tierra del Fuego. *Zoological Journal*, **5**: 332–349.

KINNE, O. 1970. Temperature: Invertebrates. In: *Marine Ecology* (Kinne, O., ed.), **1**: 407–514. Wiley-Interscience, London.

KIRA, T. 1962. *Shells of the Western Pacific in Color*, **1**. Hoikusha, Osaka.

KIZAKI, H. 1986. Ecology of *Littorina brevicula* 1.—Seasonal variation of penis length. *The Chiribotan*, **17**: 71–75. (Japanese).

KIZAKI, H. 1987. Ecology of *Littorina brevicula*—II. Seasonal variation in relation between radular length and feeding activity. *The Chiribotan*, **18**: 43–46. (Japanese).

KLEIN, J. T. 1753. *Tentamen Methodi Ostracologicae Sive Dispositio Naturalis Cochlidum et Concharum*. G. J. Wishoff, Lugduni Batavorum.

KNIGHT, A. J., HUGHES, R. N. & WARD, R. D. 1987. A striking example of the founder effect in the mollusc *Littorina saxatilis*. *Biological Journal of the Linnean Society*, **32**: 417–426.

KNIGHT, A. J. & WARD, R. D. 1986. Purine nucleoside phosphorylase polymorphism in the genus *Littorina* (Prosobranchia: Mollusca). *Biochemical Genetics*, **24**: 405–413.

KNIGHT, A. J. & WARD, R. D. 1991. The genetic relationships of three taxa in the *Littorina saxatilis* species complex (Prosobranchia: Littorinidae). *Journal of Molluscan Studies*, **57**: 81–91.

KNIGHT, J. B. 1934. A salt-marsh study. *American Journal of Science. Series 5*, **28**: 172.

KNIPOWITSCH, N. 1900. Zur Kenntnis der geologischen Geschichte der Fauna des Weissen und des Murman-Meeres (Postpliocaene Mollusken und Brachiopoden). *Verhandlungen der Russisch-Kaiserlichen Mineralogischen Gesellschaft zu St. Petersburg. Serie 2*, **38**: 1–169.

KNUDSEN, J. 1949. Geographical variation of *Littorina obtusata* (L.) in the North-Atlantic. *Videnskabelige Meddelelser fra Dansk Naturhistorisk Forening i København*, **111**: 247–255.

KOBELT, W. 1876–1878. *Illustrirtes Conchylienbuch*, **1**. Bauer & Raspe, Nürnberg.

KOBELT, W. 1888. *Prodromus Faunae Molluscorum Testaceorum Maria Europaea Inhabitantium*. Bauer & Raspe, Nürnberg.

KOBELT, W. 1906–1908. *Icongraphie der schalentragenden europäischen Meeresconchylien*, **4**. C. W. Kreidel, Wiesbaden.

KOHN, A. J. & PERRON, F. E. 1994. *Life History and Biogeography: Patterns in* Conus. Clarendon Press, Oxford.

KOJIMA, Y. 1957. On the breeding of a periwinkle, *Littorivaga brevicula* (Philippi). *Bulletin of the Biological Station of Asamushi, Tohoku University*, **8**: 59–62.

KOJIMA, Y. 1958a. On the breeding of periwinkles *Littorivaga brevicula* (Philippi) and *Littorivaga atkana* (Dall). *Venus*, **19**: 224–229.

KOJIMA, Y. 1958b. On the floating egg capsules of periwinkles, *Littorina squalida* Broderip et Sowerby and *Nodilittorina pyramidalis* (Quoy et Gaimard). *Venus*, **19**: 233–237.

KOJIMA, Y. 1958c. On the breeding of a periwinkle, *Littorivaga atkana* (Dall). *Bulletin of the Biological Station of Asamushi, Tohoku University*, **9**: 35–37.

KOJIMA, Y. 1958d. A new type of the egg capsule of a periwinkle, *Littorina squalida* Broderip et Sowerby. *Bulletin of the Biological Station of Asamushi, Tohoku University*, **9**: 39–41.

KOJIMA, Y. 1958e. On the planktonic egg capsules of *Littorivaga mandschurica* (Schrenck) and *Littoraria strigata* (Lischke). *Venus*, **20**: 81–86.

KOJIMA, Y. 1959. The relation between seasonal migration and spawning of a periwinkle, *Littorina brevicula* (Philippi). *Bulletin of the Marine Biological Station of Asamushi*, **9**: 183–186.

KOJIMA, Y. 1960. On the reproduction of periwinkles, Littorinidae. *Bulletin of the Marine Biological Station of Asamushi*, **10**: 117–120.

KONAN, A., CARDINAL, A. & LAMONTAGNE, I. 1992. Rythmes à court et à moyen termes associés au broutage de la microflore marine épilithique par *Littorina saxatilis*. *Oceanologica Acta*, **15**: 397–403.

KOZLOFF, E. N. 1987. *Marine Invertebrates of the Pacific Northwest*. University of Washington Press, Seattle.

KRAEUTER, J. N. 1974. Offshore currents, larval transport, and establishment of southern populations of *Littorina littorea* Linné along the U. S. Atlantic coast. *Thalassia Jugoslavica*, **10**: 159–170.

KURODA, T. 1931. Report on the fossil Mollusca. In: *Geology of Central Shinano* (Homma, F., ed.), **4**: 1–90.

KURODA, T. & HABE, T. 1952. *Check List and Bibliography of the Recent Marine Mollusca of Japan*. L. W. Stack, Tokyo.

KURODA, T., HABE, T. & OYAMA, K. 1971. *The Sea Shells of Sagami Bay*. Maruzen, Tokyo.

KURODA, T. & KOBA, K. 1933. Molluscan fauna of the northern Kurile Islands. *Bulletin of the Biogeographical Society of Japan*, **4**: 151–170. (Japanese).

KUSAKIN, O. G. 1958. Seasonal changes on the littoral of the South Kurile Islands. *Vestnik Leningradskogo Gosudarstvennogo Universiteta (Biologiya)*, **1958** (3): 116–130. (Russian).

KUSAKIN, O. G. 1985. The intertidal zone of Moneron Island. In: *Benthos of the Shelf of Moneron Island* (Kusakin, O. G., ed.), 5–17. Akademiya Nauk S. S. S. R., Vladivostok. (Russian).

KUSAKIN, O. G. & IVANOVA, M. B. 1978. The intertidal zone of the Bering Sea coast of Chukotka. In: *The Intertidal Zone of the Bering Sea and Southeastern Kamchatka* (Kusakin, O. G., ed.), 10–40. Nauka, Moscow. (Russian).

KÜSTER, H. C. 1853, 1856. Die Gattung *Litorina*. In: *Systematisches Conchylien-Cabinet von Martini und Chemnitz*, parts 129 (1853, pls 1, 2), 149 (1856, pls 3–5, pp. 1–24) (Küster, H. C., ed.). Bauer & Raspe, Nürnberg.

KUZNETSOV, V. V. 1947. The influence of ice cover on the morphology and population of the littoral zone. *Doklady Akademii Nauk S. S. S. R.*, **58**: 163–166. (Russian).

KUZNETSOV, V. V. 1960. *The White Sea and Biological Peculiarities of its Flora and Fauna*. Akademiya Nauk S. S. S. R., Moscow.

LAM, P. K. S. & CALOW, P. 1988. Differences in the shell shape of *Lymnaea peregra* (Müller) (Gastropoda: Pulmonata) from lotic and lentic habitats; environmental or genetic variance? *Journal of Molluscan Studies*, **54**: 197–207.

LAMARCK, J. B. P. A. DE M. DE 1822. *Histoire Naturelle des Animaux Sans Vertèbres*, **7**. Paris.

LAMBERT, T. C. & FARLEY, J. 1968. The effect of parasitism by the trematode *Cryptocotyle lingua* (Creplin) on zonation and winter migration of the common periwinkle *Littorina littorea* (L.). *Canadian Journal of Zoology*, **46**: 1139–1147.

LAMI, R. 1937. Sur des "champs de pâture" de colonies de *Littorina saxatilis* Olivi. *Bulletin du Laboratoire Maritime de Dinard*, **17**: 41–43.

LAMY, E. 1907. Liste des coquilles marines recueillies par M. Ch. Gravier à l'Île San Thome (1906). *Bulletin du Muséum National d'Histoire Naturelle, Paris*, **13**: 145–154.

LAUDER, G. V., LEROI, A. M. & ROSE, M. R. 1993. Adaptations and history. *Trends in Ecology and Evolution*, **8**: 294–297.

LAURSEN, D. 1944. Contributions to the Quaternary geology of northern west Greenland especially the raised marine deposits. *Meddelelser om Grønland*, **135** (8): 1–125.

LAURSEN, D. 1950. The stratigraphy of the marine Quaternary deposits in West Greenland. *Meddelelser om Grønland*, **151** (1): 1–142.

LEACH, W. E. 1852. *A Synopsis of the Mollusca of Great Britain* (J. E. Gray, Ed.). John van Voorst, London.

LEBOUR, M. V. 1935. The breeding of *Littorina neritoides*. *Journal of the Marine Biological Association of the United Kingdom*, **20**: 373–378.

LEBOUR, M. V. 1937. The eggs and larvae of the British prosobranchs with special reference to those living in the plankton. *Journal of the Marine Biological Association of the United Kingdom*, **22**: 105–166.

LEBOUR, M. V. 1947. Notes on the inshore plankton of Plymouth. *Journal of the Marine Biological Association of the United Kingdom*, **26**: 527–547.

LECH, W. 1878. Öfversigt öfver de af svenska expeditionerna till Novaja Semlja och Jenissej 1875 och 1876 insamlade hafs-mollusker. *Kongliga Svenska Vetenskaps-akademiens Handlingar*, **16** (2): 1–85.

LECOINTRE, G. 1950. Coquilles remarquables du Quaternaire marocain. *Journal de Conchyliologie*, **90**: 240–244.

LECOINTRE, G. 1952. Recherches sur le Néogène et le Quaternaire marins de la côte Atlantique du Maroc. Tome I: Stratigraphie. Tome II: Paléontologie. *Protectorat de la République Française au Maroc, Direction de la Production Industrielle et des Mines, Division des Mines et de la Géologie, Service Géologique, Notes et Mémoires*, **99** (I): 1–198; (II): 1–173.

LECOINTRE, G. 1963. Recherches sur le Néogène et le Quaternaire marins de la côte Atlantique du Maroc. Tome 3. Les acquisitions nouvelles durant la période de 1952 à 1962 (stratigraphie et paléontologie). *Protectorat de la République Française au Maroc, Direction de la Production Industrielle et des Mines, Division des Mines et de la Géologie, Service Géologique, Notes et Mémoires*, **174**: 1–75.

LEIN, T. E. 1980. The effects of *Littorina littorea* L. (Gastropoda) grazing on littoral green algae in the inner Oslofjord, Norway. *Sarsia*, **65**: 87–92.

LEWIS, J. B. 1960. The fauna of rocky shores of Barbados, West Indies. *Canadian Journal of Zoology*, **38**: 391–435.

LEWIS, J. R. 1964. *The Ecology of Rocky Shores*. English Universities Press, London.

LEWIS, R. I. & WILLIAMS, G. A. 1995. Extreme morphological diversity between populations of *Littorina obtusata* (L.) from Iceland and the UK. *Hydrobiologia*, **309**: 161–166.

LIEBERMAN, B. S., ALLMON, W. D. & ELDREDGE, N. 1993. Levels of selection and macroevolutionary patterns in turritellid gastropods. *Paleobiology*, **19**: 205–215.

LIMA, G. M. & LUTZ, R. A. 1990. The relationship of larval shell morphology to mode of development in marine prosobranch gastropods. *Journal of the Marine Biological Association of the United Kingdom*, **70**: 611–637.

LINK, H. F. 1807. *Beschreibung der Naturalien-Sammlung der Universität zu Rostock*. Part 3. Rostock.

LINKE, O. 1933. Morphologie und Physiologie des Genitalapparates der Nordseelittorinen. *Wissenschaftliche Meeresuntersuchungen, Abteilung Helgoland*, **19** (5): 1–60.

LINKE, O. 1934. Beiträge zur Sexualbiologie der Littorinen. *Zeitschrift für Morphologie und Ökologie der Tiere*, **28**: 170–177.

LINNAEUS, C. 1745. *Carl Linnaei...Öländska och Gothländska Resa på Riksens Högloflige Ständers Befallning Förrättad åhr 1741*. Stockholm.

LINNAEUS, C. 1746. *Fauna Svecica*. Stockholm.

LINNAEUS, C. 1747. *Carl Linnaei...Wästgöta-Resa, på Riksens Högloflige Ständers Befallning Förrättad år 1746*. Stockholm.

LINNAEUS, C. 1758. *Systema Naturae per Regna Tria Naturae, Secundum Classes, Ordines, Genera, Species, cum Characteribus, Differentiis, Synonymis, Locis*, **1**. Edition 10. Laurentii Salvii, Holmiae.

LINNAEUS, C. 1761. *Fauna Svecica. Editio Altera, Auctior*. Stockholm.

LINNAEUS, C. 1767. *Systema Naturae per Regna Tria Naturae, Secundum Classes, Ordines, Genera, Species, cum Characteribus, Differentiis, Synonymis, Locis*, **1**. Part II. Edition 12. Laurentii Salvii, Holmiae.

LISCHKE, C. E. 1869. Japanische Meeres-conchylien. *Novitates Conchologicae. Supplementum*, **4** (1): 1–192.

LISTER, M. 1678. *Historiae Animalium Angliae Tres Tractatus*. London.

LISTER, M. 1688. *Historiae Sive Synopsis Methodicae Conchyliorum*, Liber IV. London.

LITTLE, C. & SMITH, L. P. 1980. Vertical zonation on rocky shores in the Severn Estuary. *Estuarine Coastal and Marine Science*, **11**: 651–669.

LITTLE, C. & WILLIAMS, G. A. 1989. Distribution of littorinid gastropods at Lough Hyne. *Irish Naturalists' Journal*, **23**: 48–53.

LITTLER, M. M., LITTLER, D. S., MURRAY, S. N. & SEAPY, R. R. 1991. Southern California rocky intertidal ecosystems. In: *Ecosystems of the World* **24**. *Intertidal and Littoral Ecosystems* (Mathieson, A. C. & Nienhuis, P. H., eds), 273–296. Elsevier, Amsterdam.

LIVINGSTONE, D. R. 1988. Responses of microsomal NADPH-cytochrome C reductase activity and cytochrome P-450 in digestive glands of *Mytilus edulis* and *Littorina littorea* to environmental and experimental exposure to pollutants. *Marine Ecology Progress Series*, **46**: 37–43.

LOCARD, A. 1886. *Prodrome de Malacologie Française. Catalogue Générale des Mollusques Vivants de France. Mollusques Marins*. J.-B. Baillière & Fils, Paris.

LOCARD, A. 1891. Les coquilles marines des côtes de France. *Annales de la Société Linnéenne de Lyon*, **37**: 1–384.

LOCARD, A. 1897. Mollusques testacés. Tome 1. In: *Expéditions Scientifiques du Travailleur et du Talisman pendant les Années 1880, 1881, 1882, 1883* (Milne-Edwards, A., ed.). Masson & Co, Paris.

LOSSE, G. & GREVEN, H. 1993. Structure, composition and permeability of the egg covering in the viviparous prosobranch gastropod *Littorina saxatilis*. *Invertebrate Reproduction and Development*, **24**: 225–236.

LOUBERE, P. 1988. Gradual Late Pliocene onset of glaciation: a deep-sea record from the northeast Atlantic. *Palaeogeography, Palaeoclimatology, Palaeoecology*, **63**: 327–334.

LOVÉN, S. 1846. Index molluscorum litora Scandinaviae occidentalia habitantium. *Öfversigt af Kongliga Vetenskaps-Akademiens Förhandlingar, Stockholm*, **3**: 135–160, 183–204.

LOWELL, R. B., FLETCHER, C. R., GRAHAME, J. & MILL, P. J. 1994. Ontogeny of shell morphology and shell strength of the marine snails *Littorina obtusata* and *Littorina mariae*: different defence strategies in a pair of sympatric, sibling species. *Journal of Zoology, London*, **234**: 149–164.

LOWENSTAM, H. A. 1962. Goethite in radula teeth of recent marine gastropods. *Science, New York*, **137**: 279–280.

LUBCHENCO, J. 1978. Plant species diversity in a marine intertidal community: importance of herbivore food preference and algal competitive abilities. *American Naturalist*, **112**: 23–39.

LUBCHENCO, J. 1980. Algal zonation in the New England rocky intertidal community: an experimental analysis. *Ecology*, **61**: 333–344.

LUBCHENCO, J. 1982. Effects of grazers and algal competitors on fucoid colonization in tide pools. *Journal of Phycology*, **18**: 544–550.

LUBCHENCO, J. 1983. *Littorina* and *Fucus*: effects of herbivores, substratum heterogeneity, and plant escapes during succession. *Ecology*, **64**: 1116–1123.

LUBCHENCO, J. & MENGE, B. A. 1978. Community development and persistence in a low rocky intertidal zone. *Ecological Monographs*, **48**: 67–94.

LUCAS, M. 1968. *Littorina obtusata* Linné et ses variations. *Bulletin Mensuel d'Information, Les Naturalistes Belges, Section de Malacologie*, **1968** (3): 11–14; (4): 3–8.

LUCKENS, P. A. 1970. Seasonal distributional variation within a limited shore area at Asamushi. *Science Reports of Tôhoku University. Biology*, **35**: 161–170.

LYELL, C. 1835. On the proofs of a gradual rising of the land in certain parts of Sweden. *Philosophical Transactions of the Royal Society of London*, **1835** (1): 1–38.

LYNCH, J. D. 1989. The gauge of speciation: on the frequencies of modes of speciation. In: *Speciation and its Consequences* (Otte, D. & Endler, J. A., eds), 527–553. Sinauer, Sunderland, Massachusetts.

MACDONALD, K. B. 1969a. Quantitative studies of salt marsh mollusc faunas from the North American Pacific coast. *Ecological Monographs*, **39**: 33–60.

MACDONALD, K. B. 1969b. Molluscan faunas of Pacific coast salt marshes and tidal creeks. *Veliger*, **11**: 399–405.

MACGILLIVRAY, W. 1843. *A History of the Molluscous Animals of the Counties of Aberdeen, Kincardine, and Banff*. Cunningham & Mortimer, London.

MACNEIL, F. S. 1965. Evolution and distribution of the genus *Mya*, and Tertiary migrations of Mollusca. *United States Geological Survey Professional Papers*, **483**: 1–51.

MACNEIL, F. S., MERTIE, J. B. & PILSBRY, H. A. 1943. Marine invertebrate faunas of the buried beaches near Nome, Alaska. *Journal of Paleontology*, **17**: 69–96.

MACPHERSON, E. 1971. *The Marine Molluscs of Arctic Canada: Prosobranch Gastropods, Chitons and Scaphopods*. National Museum of Natural Sciences. Publications in Biological Oceanography no. 3, Ottawa.

MADDISON, W. P. & MADDISON, D. R. 1992. *MacClade. Version 3.03*. Sinauer Associates, Sunderland, Massachusetts.

MADSEN, H. 1936. Investigations on the shore fauna of east Greenland with a survey of the shores of other Arctic regions. *Meddelelser om Grønland*, **100** (8): 1–79.

MADSEN, H. 1940. A study of the littoral fauna of northwest Greenland. *Meddelelser om Grønland*, **124** (3): 1–24.

MAITLAND, R. T. 1854. Week- en schelpdieren in Nederland waargenomen. In: *Bouwstoffen voor eene Fauna van Nederland* (Herklots, J. A., ed.), **2**. E. J. Brill, Leiden.

MALATESTA, A. 1957. Terreni, faune e industrie quaternarie nell'Archipelago delle Egadi. *Quaternaria*, **4**: 165–190.

MALATESTA, A. & ZARLENGA, F. 1986. Northern guests in the Pleistocene Mediterranean Sea. *Geologica Romana*, **25**: 91–154.

MALONEY, W. E. & BURNS, R. E. 1958. A reappraisal of the tides of the Mediterranean. *Technical Reports of the Hydrographic Office, Washington*, **61**: 1–33.

MANSEL-PLEYDELL, J. C. 1898. *The Mollusca of Dorsetshire*. Dorchester.

MARCHE-MARCHAD, I. 1956. Sur une collection de coquilles marines provenant de l'Archipel du Cap-Vert. *Bulletin de l'Institut Français d'Afrique Noire. Série A*, **18**: 39–74.

MARIGOMEZ, J. A. & IRELAND, M. P. 1990. A laboratory study of cadmium exposure in *Littorina littorea* in relation to environmental cadmium and exposure time. *Science of the Total Environment*, **90**: 75–87.

MARINCOVICH, L. 1990. Molluscan evidence for early middle Miocene marine glaciation in southern Alaska. *Bulletin of the Geological Society of America*, **102**: 1591–1599.

MARION, P. VAN 1978a. Littorina saxatilis-*komplekset i Trondheimsfjordsområdet (Mollusca: Prosobranchia)*. Hovedfagsoppgave i marin zoologi ved Universitetet i Trondheim.

MARION, P. VAN. 1978b. Begroing på skallet hos vanlig strandsnegl. *Fauna, Oslo*, **31**: 271–272.

MARION, P. VAN 1981. Intrapopulation variation of the shell of *Littorina rudis* (Maton) (Mollusca: Prosobranchia). *Journal of Molluscan Studies*, **47**: 99–108.

MARSHALL, J. T. 1898. Additions to "British Conchology" (continued). *Journal of Conchology*, **9**: 120–138.

MARSHALL, J. T. 1903. *Additions to "British Conchology". Part VII*. Privately printed.

MARSHALL, J. T. 1915. Additions to "British Conchology". Part VII (cont'd). *Journal of Conchology*, **14**: 324–329.

MARSHALL, J. T. 1916. Additions to "British Conchology". Part VII (cont'd). *Journal of Conchology*, **15**: 44–47.

MARTEL, A. & CHIA, F.-S. 1991a. Foot-raising behaviour and active participation during the initial phase of post-metamorphic drifting in the gastropod *Lacuna* spp. *Marine Ecology Progress Series*, **72**: 247–254.

MARTEL, A. & CHIA, F.-S. 1991b. Drifting and dispersal of small bivalves and gastropods with direct development. *Journal of Experimental Marine Biology and Ecology*, **150**: 131–147.

MARTEL, H. 1901. Liste des coquilles marines, terrestres & d'eau douce recueillies aux environs de Cancale. Première partie—coquilles marines (suite). *La Feuille des Jeunes Naturalistes*, **31** (364): 109–114; (369): 221–222.

MASON, A. Z. 1988. The kinetics of zinc accumulation by the marine prosobranch gastropod *Littorina littorea*. *Marine Environmental Research*, **24**: 135–139.

MASTRO, E., CHOW, V. & HEDGECOCK, D. 1982. *Littorina scutulata* and *Littorina plena*, sibling species status of two prosobranch gastropod species confirmed by electrophoresis. *Veliger*, **24**: 239–246.

MASUDA, K., HANDA, M., KANNO, M., MIYASAKA, Y., NEMOTO, J., SASAKI, T., TAKAIZUMI, Y. & WAKO, K. 1990. Remarks on molluscan fauna from the Dainenji formation, Sendai, Miyagi Prefecture, Japan. *Saito Ho-on Kai Museum of Natural History, Research Bulletin*, **58**: 1–18.

MASUDA, K. & NODA, H. 1976. *Check List and Bibliography of the Tertiary and Quaternary Mollusca of Japan, 1950–1974*. Saito Ho-on Kai Museum, Sendai.

MATHIESON, A. C., PENNIMAN, C. A. & HARRIS, L. G. 1991. Northwest Atlantic rocky shore ecology. In: *Ecosystems of the World* **24**. *Intertidal and Littoral Ecosystems* (Mathieson, A. C. & Nienhuis, P. H., eds), 109–191. Elsevier, Amsterdam.

MATON, W. G. 1797. *Observations Relative Chiefly to the Natural History, Picturesque Scenery, and Antiquities, of the Western Counties of England, Made in the Years 1794 and 1796*, **1**. Salisbury.

MATON, W. G. & RACKETT, T. 1807. A descriptive catalogue of the British Testacea. *Transactions of the Linnean Society of London*, **8**: 17–250.

MATSUSHIMA, Y. & KAWAKAMI, J. 1995. Radiocarbon age of the molluscan shells from the Holocene marine deposits in the Nemuro Peninsula, eastern Hokkaido. *Bulletin of the Kanagawa Prefectural Museum (Natural Science)*, **24**: 39–46.

MATTHEWS, P. M. & MONTGOMERY, W. I. 1987. The identification and distribution of littorinid species in Northern Ireland with particular reference to *Littorina nigrolineata* in Strangford Lough. *Irish Naturalists' Journal*, **22**: 290–296.

MATTHEWS, R. S. 1978. *Natural History and Biology of the Salt-marsh Periwinkle* Littorina newcombiana. Student Report, Oregon Institute of Marine Biology, University of Oregon.

MATVEEVA, T. A. 1974. Ecology and life cycles of common species of gastropods in the Barents and White Seas. *Isseldovaniya Fauny Morei*, **13**: 65–190. (Russian).

MAYR, E. 1963. *Animal Species and Evolution*. Belknap Press, Cambridge, Massachusetts.

MCCOOK, L. J. & CHAPMAN, A. R. O. 1991. Community succession following massive ice-scour on an exposed rocky shore: effects of *Fucus* canopy algae and of mussels during late succession. *Journal of Experimental Marine Biology and Ecology*, **154**: 137–169.

MCCORMACK, S. M. D. 1982. The maintenance of shore-level size gradients in an intertidal snail (*Littorina sitkana*). *Oecologia*, **54**: 177–183.

MCGOWRAN, B. 1986. Cainozoic oceanic and climatic events: the Indo-Pacific foraminiferal biostratigraphic record. *Palaeogeography, Palaeoclimatology, Palaeoecology*, **55**: 247–265.

MCKAY, W. A. & FOX, A. A. 1991. Particulate-associated nuclides in Cumbrian winkles—implications for assessment of dose to man. *Journal of Environmental Radioactivity*, **14**: 1–21.

MCMAHON, R. F. & RUSSELL-HUNTER, W. D. 1977. Temperature relations of aerial and aquatic respiration in six littoral snails in relation to their vertical zonation. *Biological Bulletin of the Marine Biological Laboratory, Woods Hole*, **152**: 182–198.

MCMAHON, R. F., RUSSELL-HUNTER, W. D. & ALDRIDGE, D. W. 1995. Lack of metabolic temperature compensation in the intertidal gastropods, *Littorina saxatilis* (Olivi) and *L. obtusata* (L.). *Hydrobiologia*, **309**: 89–100.

McMillan, N. F. 1939. *Littorina saxatilis nigrolineata* Gray. *Journal of Conchology*, **21**: 173–174.

McMillan, N. F. 1944. Notes on *Littorina saxatilis* (Olivi). *Journal of Conchology*, **22**: 100–103.

Melville, R. V. 1980. Opinion 1159. *Littorina* (Mollusca, Gastropoda): author and date of this generic name, and type species of this nominal genus determined by use of the plenary powers. *Bulletin of Zoological Nomenclature*, **37**: 103–106.

Menke, K. T. 1828. *Synopsis Methodica Molluscorum*. H. Gelpke, Pyrmonti.

Menke, K. T. 1830. *Synopsis Methodica Molluscorum*. Edition 2. G. Uslar, Pyrmonti.

Menke, K. T. 1845. Übersicht der Mollusken der deutschen Nordsee. *Zeitschrift für Malakozoologie*, **1845**: 49–60.

Merdsoy, B. & Farley, J. 1973. Phasic activity in the digestive gland cells of the marine prosobranch gastropod, *Littorina littorea* (L.). *Proceedings of the Malacological Society of London*, **40**: 473–482.

Merklin, R. L., Zarkhidze, V. C. & Ilina, L. B. 1979. Diagnostic keys to the marine Pliocene-Pleistocene molluscs from the northeastern European portions of the U. S. S. R. *Trudy Paleontologicheskogo Instituta*, **173**: 1–96. (Russian).

Meyer, H. A. & Möbius, K. 1872. *Fauna der Kieler Bucht*. **2**: *Die Prosobranchia und Lamellibranchia*. W. Engelmann, Leipzig.

Middendorff, A. T. 1848. Vorläufige Anzeige einiger neuer Konchylien aus den Geschlechtern: *Littorina*, *Tritonium*, *Bullia*, *Natica* und *Margarita*. *Bulletin de la Classe Physico-Mathématique de l'Académie Impériale des Sciences de Saint-Pétersbourg*, **7**: columns 241–246.

Middendorff, A. T. 1849. Beiträge zu einer Malacozoologia Rossica II. Aufzählung und Beschreibung der zur Meeresfauna Russlands gehörigen Einschaler. *Mémoires de l'Académie Impériale des Sciences de Saint-Pétersbourg. Série 6. Sciences Mathématiques, Physiques et Naturelles*, **8** (2) *Sciences Naturelles*, **6**: 329–516.

Middendorff, A. T. 1851. Mollusken. In: *Reise in den Äussersten Norden und Osten Sibiriens während der Jahre 1843 und 1844* (Middendorff, A. T., ed.), **2** (1): 163–464. Kaiserlichen Akademie der Wissenschaften, St Petersburg.

Mienis, H. K. 1975. Type specimens of Littorinidae in the Zoological Museum of the Hebrew University of Jerusalem. *Littorinid Tidings*, **4**: 38–39.

Mikhailova, N. A., Granovitch, A. I. & Sergievsky, S. O. 1988. Effect of trematodes on the microbiotopical distribution of the molluscs *Littorina obtusata* and *L. saxatilis*. *Parasitologia*, **22**: 398–407. (Russian).

Mikhailova, N. A. & Naumov, A. D. 1990. The shell structure and colour polymorphism of *Littorina obtusata* (Gastropoda, Prosobranchia). *Zoologicheskii Zhurnal*, **69**: 131–134.

Mileikovsky, S. A. 1960. On the relation between temperature spawning range of a species and its zoogeographical position in marine invertebrates. *Zoologicheskii Zhurnal*, **39**: 666–669. (Russian).

Mileikovsky, S. A. 1961. The gastropod larvae in the plankton in the vicinity of the White Sea Biological Station of MGU. In: *Biology of the White Sea* (Zenkevitch, L. A., ed.). Moscow. (Russian).

Mileikovsky, S. A. 1975. Types of larval development in Littorinidae (Gastropoda: Prosobranchia) of world ocean, and ecological patterns of their distribution. *Marine Biology, Berlin*, **30**: 129–136.

Mill, P. J. & Grahame, J. 1988. Esterase variability in the gastropod *Littorina saxatilis* (Olivi) and *L. arcana* Ellis. *Journal of Molluscan Studies*, **54**: 347–353.

Mill, P. J. & Grahame, J. 1990a. The distribution of *Littorina saxatilis* (Olivi) and *Littorina arcana* Hannaford Ellis in Lyme Bay, southern England. *Journal of Molluscan Studies*, **56**: 133–135.

Mill, P. J. & Grahame, J. 1990b. Distribution of the species of rough periwinkle (*Littorina*) in Great Britain. *Hydrobiologia*, **193**: 21–27.

Mill, P. J. & Grahame, J. 1992a. Clinal changes in esterase variability in *Littorina saxatilis* (Olivi) and *L. arcana* Hannaford Ellis in southern Britain. In: *Proceedings of the Third International Symposium on Littorinid Biology* (Grahame, J., Mill, P. J. & Reid, D. G., eds), 31–37. Malacological Society of London, London.

Mill, P. J. & Grahame, J. 1992b. Distribution of the rough periwinkles in Great Britain. In: *Proceedings of the Third International Symposium on Littorinid Biology* (Grahame, J., Mill, P. J. & Reid, D. G., eds), 305–307. Malacological Society of London, London.

Mill, P. J. & Grahame, J. 1995. Shape variation in the rough periwinkle *Littorina saxatilis* on the west and south coasts of Britain. *Hydrobiologia*, **309**: 61–71.

Mill, P. J. & McQuaid, C. D. (eds) 1995. Advances in Littorinid Biology. Proceedings of the Fourth International Symposium on Littorinid Biology, held in Roscoff, France, 19–25 September 1993. *Hydrobiologia*, **309**: 1–193.

Miller, S. L. & Vadas, R. L. 1984. The population biology of *Ascophyllum nodosum*: biological and physical factors affecting survivorship of germlings. *British Phycological Journal*, **19**: 198.

Mishler, B. D. & Donoghue, M. J. 1982. Species concepts: a case for pluralism. *Systematic Zoology*, **31**: 491–503.

Miyamoto, M. M. & Fitch, W. M. 1995. Testing species phylogenies and phylogenetic methods with congruence. *Systematic Biology*, **44**: 64–76.

Mokyevsky, O. B. 1960. The littoral fauna of the northwestern coast of the Japanese Sea. *Trudy Instituta Okeanologii*, **34**: 242–328. (Russian).

Molinier, R. & Picard, J. 1954. Eléments de bionomie marine sur les côtes de Tunisie. *Bulletin de la Station Océanographique de Salammbô*, **48**: 1–47.

Møller, H. P. C. 1842. Index molluscorum Groenlandiae. *Naturhistorisk Tidsskrift. Række 1*, **4**: 76–97.

MONTAGU, G. 1803. *Testacea Britannica*. Part II. Romsey.

MONTAGU, G. 1808. *Supplement to Testacea Britannica with Additional Plates*. Exeter.

MONTEROSATO, T. A. DE 1872. *Notizie Intorno alle Conchiglie Mediterranee*. Palermo.

MONTEROSATO, T. A. DE 1878. Enumerazione e sinonimia delle conchiglie mediterranee. *Giornale Scienze Naturali ed Economiche, Palermo*, **13**: 61–115.

MONTFORT, D. DE 1810. *Conchyliologie Systématique*, **2**. Paris.

MOOLENBEEK, R. G. 1989. Does *Littorina littorea* Linné, 1758 occur on the Canary Islands? *Gloria Maris*, **28**: 102–103.

MOOR, B. VON 1973. Zur frühen Furchung des Eies von *Littorina littorea* L. (Gastropoda Prosobranchia). *Zoologische Jahrbücher Abteilung für Anatomie und Ontogenie der Tiere*, **91**: 546–573.

MOORE, H. B. 1937. The biology of *Littorina littorea* (L.). Part I. Growth of the shell and tissues, spawning, length of life and mortality. *Journal of the Marine Biological Association of the United Kingdom*, **21**: 721–742.

MOORE, H. B. 1940. The biology of *Littorina littorea*. Part II. Zonation in relation to other gastropods on stony and muddy shores. *Journal of the Marine Biological Association of the United Kingdom*, **24**: 227–237.

MOORE, P. G. 1977. Additions to the littoral fauna of Rockall, with a description of *Araelaimus penelope* sp. nov. (Nematoda: Axonolaimidae). *Journal of the Marine Biological Association of the United Kingdom*, **57**: 191–200.

MØRCH, O. A. L. 1857. Fortegnelse over Grønlands Bløddyr. In: *Grønland Geographik og Statistik Belkrevet* (Rink, H., ed.), **2**: 75–100. Kjøbenhavn.

MØRCH, O. A. L. 1864. Fortegnelse over de i Danmark forekommende Land- og Ferskvandsblöddyr. *Videnskabelige Meddelelser fra den Naturhistoriske Forening i Kjöbenhavn*, **1863**: 265–367.

MØRCH, O. A. L. 1868. Faunula molluscorum insularum Faeroënsium. *Videnskabelige Meddelelser fra den Naturhistoriske Forening i Kjöbenhavn*, **1867**: 67–111.

MØRCH, O. A. L. 1869. Faunula molluscorum Islandiae. *Videnskabelige Meddelelser fra den Naturhistoriske Forening i Kjöbenhavn*, **1868**: 185–227.

MØRCH, O. A. L. 1870. Catalogue des mollusques du Spitzberg. *Mémoires de la Société Malacologique de Belgique*, **4**: 1–32.

MØRCH, O. A. L. 1871a. Synopsis molluscorum marinorum Daniae. Fortegnelse over de i de danske Have forekommende Blöddyr. *Videnskabelige Meddelelser fra den Naturhistoriske Forening i Kjöbenhavn*, **1871**: 157–225.

MØRCH, O. A. L. 1871b. On the Mollusca of the Crag-formation of Iceland. *Geological Magazine*, **8**: 391–400.

MORETEAU, J.-C. 1976. Étude sur la croissance de *Littorina saxatilis* (Olivi) *rudis* (Maton). *Cahiers de Biologie Marine*, **17**: 463–484.

MORETEAU, J.-C. 1978. Les effets de l'agitation des eaux sur les états sexuels de *Littorina saxatilis* (Olivi) *rudis* (Maton). *Haliotis*, **7**: 56–58.

MORETEAU, J.-C. 1981. Biologie de *Littorina nigrolineata* (Gray): croissance et mortalité. *Cahiers de Biologie Marine*, **22**: 193–206.

MORETEAU, J.-C. 1983. Relations entre les variations annuelles de température et la fréquence des dates de ponte chez *Littorina rudis* (mollusque, prosobranche). *Journal of Molluscan Studies. Supplement*, **12A**: 111–115.

MORI, K., NISHIHAMA, S. & TANAKA, M. 1985. Community structure of a rocky shore in Tsuji-shima Island, Amakusa. III. The analysis of relationships between distribution of organisms and micro-topographical conditions using small quadrat. *Publications of the Amakusa Marine Biological Laboratory*, **8**: 43–63.

MORI, K., TANAKA, M. & NISHIHAMA, S. 1985. Community structure of a rocky shore in Tsuji-shima Island, Amakusa. II. Vertical distribution of dominant species and its zonation pattern. *Publications of the Amakusa Marine Biological Laboratory*, **8**: 27–41.

MORRIS, S. R. 1979. *Genetic Variation in the Genus Littorina*. PhD thesis, University of Wales.

MORSE, E. S. 1880. The gradual dispersion of certain mollusks in New England. *Bulletin of the Essex Institute*, **12**: 3–8.

MORTON, B. 1990a. The intertidal ecology of Ilhéu de Vila Franca—a drowned volcanic crater in the Azores. *Açoreana. Supplement*, **1990**: 3–20.

MORTON, B. 1990b. The rocky shore ecology of Qingdao, Shandong Province, People's Republic of China. *Asian Marine Biology*, **7**: 167–187.

MORTON, B. & MORTON, J. 1983. *The Sea Shore Ecology of Hong Kong*. Hong Kong University Press, Hong Kong.

MOTOMURA, I. 1953. On the aggregation of *Littorina*. *Ecological Review, Sendai*, **13**: 63–66.

MOYSE, J. & NELSON-SMITH, A. 1963. Zonation of animals and plants on rocky shores around Dale, Pembrokeshire. *Field Studies*, **1** (5): 1–31.

MOYSE, J., THORPE, J. P. & AL-HAMADANI, E. 1982. The status of *Littorina aestuarii* Jeffreys: an approach using morphology and biochemical genetics. *Journal of Conchology*, **31**: 7–15.

MULLER, J. E. & CAMERON, B. E. B. 1981. Geology and mineral deposits of Nootka Sound map-area (92E) Vancouver Island, British Columbia. *Geological Survey of Canada Paper*, **80** (16): 1–53.

MÜLLER, O. F. 1776. *Zoologiae Danicae Prodromus, seu Animalium Daniae et Norvegiae Indigenarum*. Havniae.

MUNTHE, H. 1931. "Litorinahavet", "Clypeushavet" och "Limnaeahavet". En orienterande översikt. *Geologiska Föreningens i Stockholm Förhandlingar*, **53**: 159–186.

MURPHY, D. J. & JOHNSON, L. C. 1980. Physical and temporal factors influencing the freezing tolerance of the marine snail *Littorina littorea* (L.). *Biological Bulletin of the Marine Biological Laboratory, Woods Hole*, **158**: 220–232.

MURRAY, T. E. 1979. Evidence for an additional *Littorina* species and a summary of the reproductive biology of *Littorina* from California. *Veliger*, **21**: 469–474.

MURRAY, T. E. 1982. Morphological characterization of the *Littorina scutulata* species complex. *Veliger*, **24**: 233–238.

MUUS, B. J. 1967. The fauna of Danish estuaries and lagoons. *Meddelelser fra Danmarks Fiskeri- og Havundersøgelser. Ny Serie*, **5**: 1–316.

NARDO, G. D. 1847. *Sinonimia Moderna delle Specie Registrate nell' Opera Intitolata: Descrizione de' Crostacei, de' Testacei e de' Pesci che Abitano le Lagune e Golfo Veneto Rappresentati in Figure, a Chiaro-scuro ed a Colori dall' Abate Stephano Chiereghini Ven. Clodiense*. Venezia.

NAVAZ Y SANZ, J. M. 1942. Estudio de los yacimientos de moluscos comestibles de la Ría de Vigo. *Trabajos. Instituto Español de Oceanografía*, **16**: 1–74.

NAYLOR, R. & BEGON, M. 1982. Variation within and between populations of *Littorina nigrolineata* Gray on Holy Island, Anglesea. *Journal of Conchology*, **31**: 17–30.

NEI, M. 1972. Genetic distance between populations. *American Naturalist*, **106**: 283–292.

NEI, M. 1978. Estimation of average heterozygosity and genetic distance from a small number of individuals. *Genetics*, **89**: 583–590.

NEI, M. 1987. *Molecular Evolutionary Genetics*. Columbia University Press, New York.

NELSON, C. M. 1978. *Neptunea* (Gastropoda: Buccinacea) in the Neogene of the North Pacific and adjacent Bering Sea. *Veliger*, **21**: 203–215.

NEUVILLE, R. & RUHLMANN, A. 1941. La place du Paléolithique Ancien dans le Quaternaire Marocain. *Collection Hesperis Institut des Hautes-Études Marocaines*, **8**: 1–156.

NEVILL, G. 1885. *Hand List of Mollusca in the Indian Museum, Calcutta*. Part 2. Indian Museum, Calcutta.

NEWCOMBE, C. F. 1893. Report on the marine shells of British Columbia. *Bulletin of the Natural History Society of British Columbia*, **1**: 31–72.

NEWELL, G. E. 1958. The behaviour of *Littorina littorea* (L.) under natural conditions and its relation to position on the shore. *Journal of the Marine Biological Association of the United Kingdom*, **37**: 229–239.

NEWELL, R. C. & PYE, V. I. 1971. Quantitative aspects of the relationship between metabolism and temperature in the winkle *Littorina littorea* (L.). *Comparative Biochemistry and Physiology*, **38** (4B): 635–650.

NEWKIRK, G. F. & DOYLE, R. W. 1975. Genetic analysis of shell-shape variations in *Littorina saxatilis* on an environmental cline. *Marine Biology, Berlin*, **30**: 227–237.

NEWKIRK, G. F. & DOYLE, R. W. 1979. Clinal variation at an esterase locus in *Littorina saxatilis* and *Littorina obtusata*. *Canadian Journal of Genetics and Cytology*, **21**: 505–514.

NICOTRI, M. E. 1977. Grazing effects of four marine intertidal herbivores on the microflora. *Ecology*, **58**: 1020–1032.

NIELSEN, C. L. 1980. On the occurrence of the prosobranchs *Littorina neritoides*, *L. mariae* and *L. obtusata* in Denmark. *Journal of Molluscan Studies*, **46**: 312–316.

NOBRE, A. 1886. Faune malacologique des bassins du Tage et du Sado (Portugal). *Journal de Conchyliologie*, **34**: 1–54.

NOBRE, A. 1898. Mollusques et brachiopodes du Portugal. *Annaes de Sciencias Naturaes*, **5**: 160–166.

NOBRE, A. 1900. Mollusques et brachiopodes du Portugal. *Annaes de Sciencias Naturaes*, **6**: 79–140.

NOBRE, A. 1930. *Materiais para o Estudo da Fauna dos Açores*. Porto.

NOBRE, A. 1931. *Moluscos Marinhos de Portugal*. Instituto de Zoologia da Universidade do Pôrto, Pôrto.

NOBRE, A. 1936. *Moluscos Marinhos de Portugal*, **2**. Companhia Editora do Minho, Barcelos.

NOBRE, A. 1937. Moluscos testáceos marinhos do Arguipélago da Madeira. *Memórias e Estudos do Museo Zoológico da Universidade de Coimbra, Série I*, **98**: 1–101.

NOMLAND, J. O. 1917. The Etchegoin Pliocene of Middle California. *University of California Publications. Bulletin of the Department of Geology*, **10**: 191–254.

NOMURA, E. 1926. The influence of coastal waves on the shell development in *Littorina sitchana* Phil. *Science Reports of Tôhoku University. Series 4 (Biology)*, **2**: 53–55.

NOMURA, S. 1938. Molluscan fossils from the Tatunokuti shell bed exposed at Gôroku Cliff in the western border of Sendai. *Science Reports of Tôhoku Imperial University. Series 2 (Geology)*, **19**: 235–275.

NOMURA, S. & HATAI, K. 1936. Fossils from the Tanagura beds in the vicinity of the town Tanagura, Hukusima-ken, northeast Honsyû, Japan. *Saito Ho-on Kai Museum Research Bulletin*, **10**: 109–155.

NOMURA, S. & ONISI, H. 1940. Neogene Mollusca from the Sennan District, Miyagi Prefecture, Japan. *Japanese Journal of Geology and Geography*, **17**: 181–194.

NORDSIECK, F. 1968. *Die europäischen Meeres-Gehäuseschnecken (Prosobranchia) vom Eismeer bis Kapverden und Mittelmeer*. Gustav Fischer, Stuttgart.

NORDSIECK, F. 1982. *Die europäischen Meeres-Gehäuseschnecken (Prosobranchia) vom Eismeer bis Kapverden, Mittelmeer und Schwarzes Meer.* **2**. *Völlig neubearbeitete und erweiterte Auflage*. Gustav Fischer Verlag, Stuttgart.

NORDSIECK, F. & GARCIA-TALAVERA, F. 1979. *Moluscos Marinos de Canarias y Madera (Gastropoda)*. Aula de Cultura, Tenerife.

NORMAN, A. M. 1888. *Museum Normanianum IV: Mollusca Marina*. Durham.

NORMAN, A. M. 1902. Notes on the natural history of East Finmark. *Annals and Magazine of Natural History, Series 7*, **10**: 341–361.

NORTH, W. J. 1954. Size distribution, erosive activities, and gross metabolic efficiency of the marine intertidal snails, *Littorina planaxis* and *L. scutulata*. *Biological Bulletin of the Marine Biological Laboratory, Woods Hole*, **106**: 185–197.

NORTON, P. E. P. 1967. Marine molluscan assemblages in the early Pleistocene of Sidestrand, Bramerton and the Royal Society borehole at Ludham, Norfolk. *Philosophical Transactions of the Royal Society of London, Series B*, **253**: 161–200.

NORTON, P. E. P. 1975. Paleoecology of the Mollusca of the Tjörnes sequence, Iceland. *Boreas*, **4**: 97–110.

NORTON, P. E. P. 1976. The problems of naming fossil *Littorina*. *Littorinid Tidings*, **1**: 4.

NORTON, T. A., HAWKINS, S. J., MANLEY, N. L., WILLIAMS, G. A. & WATSON, D. C. 1990. Scraping a living: a review of littorinid grazing. *Hydrobiologia*, **193**: 117–138.

NOY, R., LAVIE, B. & NEVO, E. 1987. The niche-width variation hypothesis revisited: genetic diversity in the marine gastropods *Littorina punctata* (Gmelin) and *L. neritoides* (L.). *Journal of Experimental Marine Biology and Ecology*, **109**: 109–116.

NUNN, J. & SMITH, S. M. 1987. Collecting marine Mollusca on the north coast of Ireland. *Conchologists' Newsletter*, **102**: 33–39.

NYBAKKEN, J. 1967. Habitat preferences of *Littorina sitkana* on two shores of differing exposure in Alaska. *Veliger*, **9**: 347–348.

O'CLAIR, C. E. 1977. Marine invertebrates in rocky intertidal communities. In: *The Environment of Amchitka Island, Alaska*, 395–449. Technical Information Center, Energy Research and Development Administration, Alaska.

O'CLAIR, C. E. & CHEW, K. K. 1971. Transect studies of littoral macrofauna, Amchitka Island, Alaska. *Bioscience*, **21**: 661–665.

ODHNER, N. H. 1915. Die Molluskenfauna des Eisfjordes. Zoologische Ergebnisse der Schwedischen Expedition nach Spitzbergen 1908 unter Leitung von Prof. G. de Geer. Teil II. 1. *Kungliga Svenska Vetenskapsakademiens Handlingar*, **54** (1): 1–274.

Ó FOIGHIL, D. 1989. Planktotrophic larval development is associated with a restricted geographic range in *Lasaea*, a genus of brooding, hermaphrodite bivalves. *Marine Biology, Berlin*, **103**: 349–358.

OGASAWARA, K. 1994. Neogene paleogeography and marine climate of the Japanese Islands based on shallow-marine molluscs. *Palaeogeography, Palaeoclimatology, Palaeoecology*, **108**: 335–351.

OHGAKI, S. 1981. Spawning activity in *Nodilittorina exigua* and *Peasiella roepstorffiana* (Littorinidae, Gastropoda). *Publications of the Seto Marine Biological Laboratory*, **26**: 437–446.

OHGAKI, S. 1983. Distribution of the family Littorinidae (Gastropoda) in Hokkaido, with special emphasis on the distribution in Akkeshi Bay. *Nankiseibutu*, **25**: 173–180.

OHGAKI, S. 1985. Distribution of the family Littorinidae (Gastropoda) on Hong Kong rocky shores. In: *The Malacofauna of Hong Kong and Southern China. II.* (Morton, B. & Dudgeon, D., ed.), **2**: 457–464. Hong Kong University Press, Hong Kong.

OHGUSHI, R. 1956. On the so-called "habitat segregation" phenomena seen between two species of Japanese periwinkles. *Japanese Journal of Ecology*, **6**: 9–12.

OHTSUKA, S. & YOSHIOKA, E. 1985. A preliminary note on the pelagic eggs of marine invertebrates. *Special Publications of the Mukaishima Marine Biological Station*, **1985**: 225–235.

OKUTANI, T. 1986. *Coloured Illustrations of Living Things: Molluscs*. Sekaibunkasha Co., Tokyo.

OLDROYD, I. S. 1924. Marine shells of Puget Sound and vicinity. *Publications of the Puget Sound Biological Station*, **4**: 1–272.

OLDROYD, I. S. 1927. *The Marine Shells of the West Coast of North America*, **2** (3). Stanford University Publication Series, Geological Sciences, Stanford, California.

OLIVI, G. 1792. *Zoologia Adriatica*. Bassano.

ORBIGNY, A. D' 1839. Mollusques, échinodermes, foraminifères et polypiers, recueillis aux Îles Canaries par MM Webb et Berthelot. In: *Histoire Naturelle des Îles Canaries* (Barker-Webb, P. & Berthelot, S., eds), **2** (2). Béthune, Paris.

OSKARSSON, I. 1973. Nokkrar fágætar skeldyrategundir vid Island. *Náttúrufrædingurinn*, **43**: 154–161.

OUGHTON, J. 1940. A visit to the Arctic of eastern Canada. *Nautilus*, **54**: 1–6.

OYAMA, K. 1953. Palaeoecological studies on Littorinidae. *Miscellaneous Reports of the Research Institute for Natural Resources, Tokyo*, **30**: 26–33.

OYAMA, K. 1973. Revision of Matajiro Yokoyama's type Mollusca from the Tertiary and Quaternary of the Kanto area. *Palaeontological Society of Japan Special Papers*, **17**: 1–148.

OYAMA, K. & TAKEMURA, Y. 1963. *The Molluscan Shells*, **6**. Resources Exploitation Institute, Tokyo.

OZAWA, T. & OKAMOTO, K. 1993. Integrated palaeontological and molecular phylogenetic approaches to the study of phylogeny: a case study of *Umbonium* (Gastropoda). *Chikyu Monthly*, **15**: 589–595. (Japanese).

PACKARD, A. S. 1867. Observations on the glacial phenomena of Labrador and Maine, with a view of the recent invertebrate fauna of Labrador. *Memoirs of the Boston Society of Natural History*, **1**: 210–303.

PACKARD, E. L. 1918. Molluscan fauna from San Francisco Bay. *University of California Publications in Zoology*, **14**: 199–452.

PADILLA, D. K. 1985. Structural resistance of algae to herbivores: a biomechanical approach. *Marine Biology, Berlin*, **90**: 103–109.

PADILLA, D. K. 1989. Algal structural defenses: form and calcification in resistance to tropical limpets. *Ecology, New York*, **70**: 835–842.

PALMER, A. R. 1981. Do carbonate skeletons limit the rate of body growth? *Nature, London*, **292**: 150–152.

PALMER, A. R. 1985. Quantum changes in gastropod shell morphology need not reflect speciation. *Evolution*, **39**: 699–705.

PALMER, A. R. 1990. Effect of crab effluent and scent of damaged conspecifics on feeding, growth, and shell morphology of the Atlantic dogwhelk *Nucella lapillus* (L.). *Hydrobiologia*, **193**: 155–182.

PALMER, A. R. & STRATHMANN, R. R. 1981. Scale of dispersal in varying environments and its implications for life histories of marine invertebrates. *Oecologia*, **48**: 308–318.

PALMER, K. VAN W. 1958. Type specimens of marine Mollusca described by P. P. Carpenter from the West Coast (San Diego to British Columbia). *Geological Society of America Memoir*, **76**: 1–376.

PALMER, P. 1989. A case for *Littorivaga* Dall 1918. *Conchologists' Newsletter*, **110**: 200–202.

PALUMBI, S. R. 1994. Genetic divergence, reproductive isolation, and marine speciation. *Annual Review of Ecology and Systematics*, **25**: 547–572.

PATERSON, H. E. H. 1985. The recognition concept of species. In: *Species and Speciation. Transvaal Museum Monograph*, **4** (Vrba, E., ed.), 21–29. Transvaal Museum, Johannesburg.

PAVIOUR, S., MILL, P. J. & GRAHAME, J. 1989. Fine structure of spermatozoa in periwinkles (*Littorina*). In: *Reproduction, Genetics and Distributions of Marine Organisms* (Ryland, J. S. & Tyler, P. A., eds), 161–165. Olsen & Olsen, Fredensborg, Denmark.

PAYNE, C. M. 1983. *Littorina obtusata* and *L. mariae* (Gastropoda: Prosobranchia) from western Sørøya, N. Norway. *Expedition Field Studies Reports, Brathay Exploration Group*, **39**: 102–112.

PAYRAUDEAU, B. C. 1826. *Catalogue Descriptif et Méthodique des Annélides et des Mollusques de l'Île de Corse*. Paris.

PELLETIER, B. R., WAGNER, F. J. E. & GRANT, A. C. 1968. Marine geology. In: *Science, History and Hudson Bay* (Beals, C. S., ed.), **2**: 557–613. Department of Energy, Mines and Resources, Ottawa.

PELSENEER, P. 1911. Recherches sur l'embryologie des gastropodes. *Mémoires de l'Académie Royale de Belgique, 4°. Série 2*, **3** (6): 1–167.

PELSENEER, P. 1923. Variations dans les mollusques. *Annales de la Société Royale Zoologique de Belgique*, **54**: 68–78.

PELSENEER, P. 1926. La proportion relative des sexes chez les animaux et particulièrement chez les mollusques. *Mémoires de l'Académie Royale de Belgique. Classe des Sciences. Collection en 4°*, **8**: 1–258.

PENNANT, T. 1777. *British Zoology*, **4**. Edition 4. B. White, London.

PENNANT, T. 1812. *British Zoology*, **4**. Edition 5. Wilkie & Robinson etc., London.

PÉREZ SÁNCHEZ, J. M. & MORENO BATET, E. 1991. *Invertebrados Marinos de Canarias*. Ediciones del Abildo Insular de Gran Canaria, Las Palmas de Gran Canaria.

PETERSEN, G. H. 1962. The distribution of *Balanus balanoides* (L.) and *Littorina saxatilis* (Olivi) var. *groenlandica* Menke in northern West Greenland, with remarks on some causative factors. *Meddelelser om Grønland*, **159** (9): 1–42.

PETIT DE LA SAUSSAYE, S. 1852. Suite et fin du catalogue des coquilles marines des côtes de France. *Journal de Conchyliologie*, **3**: 176–207.

PETIT DE LA SAUSSAYE, S. 1869. *Catalogue des Mollusques Testacés des Mers d'Europe*. F. Savy, Paris.

PETIVER, J. 1695–1703. *Musei Petiveriani Centuria Prima (Secunda-Decima)*. London.

PETPIROON, S. & MORGAN, E. 1983. Observations on the tidal activity rhythm of the periwinkle *Littorina nigrolineata* (Gray). *Marine Behaviour and Physiology*, **9**: 171–192.

PETRAITIS, P. S. 1982. Occurrence of random and directional movements in the periwinkle, *Littorina littorea* (L.). *Journal of Experimental Marine Biology and Ecology*, **59**: 207–217.

PETRAITIS, P. S. 1983. Grazing patterns of the periwinkle and their effect on sessile intertidal organisms. *Ecology*, **64**: 522–533.

PETRAITIS, P. S. 1987. Factors organizing rocky intertidal communities of New England: herbivory and predation in sheltered bays. *Journal of Experimental Marine Biology and Ecology*, **109**: 117–136.

PETRAITIS, P. S. & SAYIGH, L. 1987. *In situ* measurement of radular movements of three species of *Littorina* (Gastropoda: Littorinidae). *Veliger*, **29**: 384–387.

PETTITT, C. W. 1973a. A proposed new method of scoring the colour morphs of *Littorina saxatilis* (Olivi, 1792) (Gastropoda: Prosobranchia). *Proceedings of the Malacological Society of London*, **40**: 531–538.

Pettitt, C. W. 1973b. An examination of the distribution of shell pattern in *Littorina saxatilis* (Olivi) with particular regard to the possibility of visual selection in this species. *Malacologia*, **14**: 339–343.

Pettitt, C. W. 1973c. Possible shedding of the penis by *Littorina saxatilis* (Olivi, 1792). *New York Shell Club Notes*, **195**: 7.

Pettitt, C. W. 1974a. An indexed bibliography of the family Littorinidae [Gastropoda: Mollusca] 1758–1973. *Manchester Museum Publications. New Series*, **4.74**: 1–36.

Pettitt, C. W. 1974b. An indexed bibliography of the family Littorinidae 1758–1973. First supplement: corrections and additions. *Manchester Museum Publications. New Series. Supplement*, 1–9.

Pettitt, C. W. 1975. A review of the predators of *Littorina*, especially those of *L. saxatilis* (Olivi) (Gastropoda: Prosobranchia). *Journal of Conchology*, **28**: 343–357.

Pettitt, C. W. 1979. Bibliography of Littorinidae (supplement). *Manchester Museum Computer Produced Publications*, **4**: 1–21.

Pfeiffer, L. 1839. Beschreibung einer neuen *Litorina*, nebst Bemerkungen über die Konchylien des Ostseestrandes bei Travemünde. *Archiv für Naturgeschichte*, **5** (1): 81–84.

Philippi, R. A. 1836. *Enumeratio Molluscorum Siciliae*, **1**. S. Schroppii, Berolini.

Philippi, R. A. 1844a. *Enumeratio Molluscorum Siciliae*, **2**. E. Anton, Halis Saxonum.

Philippi, R. A. 1844b. Descriptiones testaceorum quorundam novorum, maxime chinensium. *Zeitschrift für Malakozoologie*, **1844**: 161–167.

Philippi, R. A. 1845–1850. *Abbildungen und Beschreibungen neuer oder wenig gekannter Conchylien*, **2** and **3**. Theodor Fischer, Cassel.

Philippi, R. A. 1846b. Descriptions of a new species of *Trochus* and of eighteen new species of *Littorina*, in the collection of H. Cuming, Esq. *Proceedings of the Zoological Society of London*, **1845**: 138–143.

Picton, B. E. 1978. The marine fauna of Lundy, Prosobranchia. *Lundy Field Society Annual Report*, **29**: 38–45.

Pilsbry, H. A. 1895. *Catalogue of the Marine Mollusca of Japan*. Detroit.

Poppe, G. T. & Goto, Y. 1991. *European Seashells*, **1** *(Polyplacophora, Caudofoveata, Solenogastra, Gastropoda)*. Verlag Christa Hemmen, Wiesbaden.

Posselt, H. I. & Jensen, A. S. 1898. Grønlands Brachiopoder og Bløddyr. *Meddelelser om Grønland*, **23**: 1–298.

Pourreau, C. N. 1979. Succession of five common salt marsh detritivores on *Spartina alterniflora* detritus of decreasing particle size and increasing age. *Biological Bulletin of the Marine Biological Laboratory, Woods Hole*, **157**: 390.

Provancher, L. 1891. *Les Mollusques de la Province de Québec*. Québec.

Pulteney, R. 1799. *Catalogues of the Birds, Shells, and Some of the More Rare Plants, of Dorsetshire*. London.

Pulteney, R. 1813. *Catalogues of the Birds, Shells, and Some of the More Rare Plants, of Dorsetshire*. Edition 2. Nichols, Son & Bentley, London.

Raeymaekers, D. 1889. Note sur la variété "unicarinata", nobis, observée chez le "*Littorina littorea*" (male). *Annales de la Société Royale Malacologique de Belgique*, **24**: xxxvi-xxxviii.

Raffaelli, D. G. 1977. Observations on the copulatory behaviour of *Littorina rudis* (Maton) and *Littorina nigrolineata* (Gray). *Veliger*, **20**: 75–77.

Raffaelli, D. G. 1978a. Factors affecting the population structure of *Littorina neglecta* Bean. *Journal of Molluscan Studies*, **44**: 223–230.

Raffaelli, D. G. 1978b. The relationship between shell injuries, shell thickness and habitat characteristics of the intertidal snail *Littorina rudis* Maton. *Journal of Molluscan Studies*, **44**: 166–170.

Raffaelli, D. G. 1979a. The taxonomy of the *Littorina saxatilis* species-complex, with particular regard to the systematic position of *Littorina patula* Jeffreys. *Zoological Journal of the Linnean Society*, **65**: 219–232.

Raffaelli, D. G. 1979b. Colour polymorphism in the intertidal snail *Littorina rudis* Maton. *Zoological Journal of the Linnean Society*, **67**: 65–73.

Raffaelli, D. G. 1982. Recent ecological research on some European species of *Littorina*. *Journal of Molluscan Studies*, **48**: 342–354.

Raffaelli, D. G. & Hughes, R. N. 1978. The effects of crevice size and availability on populations of *Littorina rudis* and *Littorina neritoides*. *Journal of Animal Ecology*, **47**: 71–83.

Rang, P. K. S. L. 1829. *Manuel de l'Histoire Naturelle des Mollusques et de leur Coquilles*. Paris.

Rasmussen, E. 1973. Systematics and ecology of the Isefjord marine fauna (Denmark), with a survey of the eelgrass (*Zostera*) vegetation and its communities. *Ophelia*, **11**: 1–495.

Ray, C. 1960. The application of Bergmann's and Allen's rules to the poikilotherms. *Journal of Morphology*, **106**: 85–108.

Read, K. R. H. 1965. *Littorina littorea* without shell. *Proceedings of the Malacological Society of London*, **36**: 307.

Read, K. R. H. 1966. *Littorina littorea* without shells, a further note. *Proceedings of the Malacological Society of London*, **37**: 127–128.

Récluz, C. 1869. Mélanges malacologiques. III. Quelques observations sur les littorines et catalogue des espèces de nos côtes. *Actes de la Société Linnéenne de Bordeaux*, **27**: 42–47.

Reed, E. H. 1982. Malformed mollusk shells. *New York Shell Club Notes*, **285**: 6–7.

REES, W. J. 1965. The aerial dispersal of Mollusca. *Proceedings of the Malacological Society of London*, **36**: 269–282.

REEVE, L. A. 1842. *Conchologia Systematica*, **2**. Longman, Brown, Green & Longmans, London.

REEVE, L. A. 1857–1858. Monograph of the genus *Littorina*. In: *Conchologia Iconica*, pls 1–16 (1857), 17–18 (1858). Lovell Reeve, London.

REID, D. G. 1986a. *The Littorinid Molluscs of Mangrove Forests in the Indo-Pacific Region: the Genus Littoraria*. British Museum (Natural History), London.

REID, D. G. 1986b. *Mainwaringia* Nevill, 1885, a littorinid genus from Asiatic mangrove forests, and a case of protandrous hermaphroditism. *Journal of Molluscan Studies*, **52**: 225–242.

REID, D. G. 1987. Natural selection for apostasy and crypsis acting on the shell colour polymorphism of a mangrove snail, *Littoraria filosa* (Sowerby) (Gastropoda: Littorinidae). *Biological Journal of the Linnean Society*, **30**: 1–24.

REID, D. G. 1989a. The comparative morphology, phylogeny and evolution of the gastropod family Littorinidae. *Philosophical Transactions of the Royal Society of London. Series B*, **324**: 1–110.

REID, D. G. 1989b. Systematic revision of the Recent species of *Peasiella* Nevill, 1885 (Gastropoda: Littorinidae), with notes on the fossil species. *Nautilus*, **103**: 43–69.

REID, D. G. 1990a. Note on the discrimination of females of *Littorina mariae* Sacchi & Rastelli and *L. obtusata* (Linnaeus). *Journal of Molluscan Studies*, **56**: 113–114.

REID, D. G. 1990b. A cladistic phylogeny of the genus *Littorina* (Gastropoda): implications for evolution of reproductive strategies and for classification. *Hydrobiologia*, **193**: 1–19.

REID, D. G. 1990c. Trans-Arctic migration and speciation induced by climatic change: the biogeography of *Littorina* (Mollusca: Gastropoda). *Bulletin of Marine Science*, **47**: 35–49.

REID, D. G. 1992. The gastropod family Littorinidae in Hong Kong. In: *Proceedings of the Fourth International Marine Biological Workshop: The Marine Flora and Fauna of Hong Kong and Southern China III* (Morton, B., ed.), 187–210. Hong Kong University Press, Hong Kong.

REID, D. G. 1993. Barnacle-dwelling ecotypes of three British *Littorina* species and the status of *Littorina neglecta* Bean. *Journal of Molluscan Studies*, **59**: 51–62.

REID, D. G. 1994. Species and speciation in Littorinidae. *Cahiers de Biologie Marine*, **35**: 258–259.

REID, D. G. & GOLIKOV, A. N. 1991. *Littorina naticoides*, new species, with notes on the other smooth-shelled *Littorina* species from the northwestern Pacific. *Nautilus*, **105**: 7–15.

REID, D. G., RUMBAK, E. & THOMAS, R. H. 1996. DNA, morphology and fossils: phylogeny and evolutionary rates of the gastropod genus *Littorina*. *Philosophical Transactions of the Royal Society. Series B*, **351**: in press.

REID, D. G., ZASLAVSKAYA, N. I. & SERGIEVSKY, S. O. 1991. *Littorina kasatka*, a new species from the Kurile Islands and Okhotsk Sea. *Nautilus*, **105**: 1–6.

REIMCHEN, T. E. 1979. Substrate heterogeneity, crypsis, and colour polymorphism in an intertidal snail (*Littorina mariae*). *Canadian Journal of Zoology*, **57**: 1070–1085.

REIMCHEN, T. E. 1981. Microgeographical variation in *Littorina mariae* Sacchi & Rastelli and a taxonomic consideration. *Journal of Conchology*, **30**: 341–350.

REIMCHEN, T. E. 1982. Shell size divergence in *Littorina mariae* and *L. obtusata* and predation by crabs. *Canadian Journal of Zoology*, **60**: 687–695.

REIMCHEN, T. E. 1989. Shell colour ontogeny and tubeworm mimicry in a marine gastropod *Littorina mariae*. *Biological Journal of the Linnean Society*, **36**: 97–109.

RENDALL, R. 1956. Mollusca Orcadensia. *Proceedings of the Royal Society of Edinburgh. Section B (Biology)*, **66**: 131–201.

REPENNING, C. A., BROUWERS, E. M., CARTER, L. D., MARINCOVICH, L. & AGER, T. A. 1987. The Beringian ancestry of *Phenacomys* (Rodentia: Cricetidae) and the beginning of the modern Arctic Ocean borderland biota. *United States Geological Survey Bulletin*, **1687**: 1–31.

REQUIN, E. 1848. *Catalogue des Coquilles de l'Île de Corse*. Fr. Seguin Ainé, Avignon.

RICHARD, J. & NEUVILLE, H. 1897. Sur l'histoire naturelle de l'Île d'Alboran. *Mémoires de la Société Zoologique de France*, **10**: 75–87.

RICHARDS, H. G. 1929. Rare mollusks from New Jersey. *Nautilus*, **43**: 63–65.

RICHARDS, H. G. 1936. Recent and Pleistocene marine shells of James Bay. *American Midland Naturalist*, **17**: 528–545.

RICHARDS, H. G. 1940. Marine Pleistocene fossils from Newfoundland. *Bulletin of the Geological Society of America*, **51**: 1781–1788.

RICHARDS, H. G. 1941. Post-Wisconsin fossils from the west coast of Hudson Bay. *Notulae Naturae*, **84**: 1–10.

RICHARDS, H. G. 1962. Studies on the marine Pleistocene: Part II. The marine Pleistocene mollusks of eastern North America. *Transactions of the American Philosophical Society*, **52** (3): 42–141.

RICHTER, G. 1962. Die schnecken "zung" als Werkzeug. *Natur und Museum*, **92**: 391–406.

RICKETTS, E. F. & CALVIN, J. 1939. *Between Pacific Tides*. Stanford University Press, Stanford.

RICKETTS, E. F., CALVIN, J. & HEDGPETH, J. W. 1952. *Between Pacific Tides*. Edition 3. Stanford University Press, Stanford.

RICKETTS, E. F., CALVIN, J., HEDGPETH, J. W. & PHILLIPS, D. W. 1985. *Between Pacific Tides*. Edition 5. Stanford University Press, Stanford.

RIGG, G. B. & MILLER, R. C. 1949. Intertidal plant and animal zonation in the vicinity of Neah Bay, Washington. *Proceedings of the California Academy of Sciences*, **26**: 323–351.

ROBERTS, A. E. 1958. Geology and coal resources of the Toledo-Castle Rock District, Cowlitz and Lewis Counties, Washington. *Bulletin of the United States Geological Survey*, **1062**: 1–71.

ROBERTS, D. J. & HUGHES, R. N. 1980. Growth and reproductive rates of *Littorina rudis* from three contrasted shores in North Wales, UK. *Marine Biology, Berlin*, **58**: 47–54.

ROBERTS, M. B., STRINGER, C. B. & PARFITT, S. A. 1994. A hominid tibia from Middle Pleistocene sediments at Boxgrove, U. K. *Nature, London*, **369**: 311–313.

ROBERTSON, A. 1992. The oystercatcher, *Haematopus ostralegus*, as a selective agent on littoral gastropods. In: *Proceedings of the Third International Symposium on Littorinid Biology* (Grahame, J., Mill, P. J. & Reid, D. G., eds), 153–161. Malacological Society of London, London.

ROBERTSON, A. I. & MANN, K. H. 1982. Population dynamics and life history adaptations of *Littorina neglecta* Bean in an eelgrass meadow (*Zostera marina* L.) in Nova Scotia. *Journal of Experimental Marine Biology and Ecology*, **63**: 151–172.

ROBSON, E. M. & WILLIAMS, I. C. 1971. Relationships of some species of Digenea with the marine prosobranch *Littorina littorea* (L.) II. The effect of larval Digenea on the reproductive biology of *L. littorea*. *Journal of Helminthology*, **45**: 145–159.

RÖDING, P. F. 1798. *Museum Boltenianum*. Hamburg.

ROFF, D. A. 1992. *The Evolution of Life Histories: Theory and Analysis*. Chapman & Hall, New York.

ROLÁN, E. 1984. Moluscos de la Ría de Vigo. I. Gasteropodos. *Thalassas*, **1**: 1–383.

ROLÁN, E. & TEMPLADO, J. 1987. Consideraciones sobre el complejo *Littorina obtusata–mariae* (Mollusca, Gastropoda, Littorinidae) en el noroeste de la peninsula iberica. *Thalassas*, **5**: 71–85.

ROLÁN-ALVAREZ, E., JOHANNESSON, K. & EKENDAHL, A. 1995. Frequency- and density-dependent sexual selection in natural populations of Galician *Littorina saxatilis* Olivi. *Hydrobiologia*, **309**: 167–172.

ROLÁN-ALVAREZ, E. & ROLÁN, E. 1995. Applicability of the biological species concept in taxonomy: the example of *Littorina saxatilis* (Gastropoda: Prosobranchia). *Argonauta*, **9**: 1–8.

ROLÁN-ALVAREZ, E., ROLÁN, E. & JOHANNESSON, K. 1996. Differentiation in radular and embryonic characters, and further comments on gene flow, between two sympatric morphs of *Littorina saxatilis* (Olivi). *Ophelia*,

ROLÁN-ALVAREZ, E., ZAPATA, C. & ALVAREZ, G. 1995a. Distinct genetic subdivision in sympatric and sibling species of the genus *Littorina* (Gastropoda: Littorinidae). *Heredity*, **74**: 1–9.

ROLÁN-ALVAREZ, E., ZAPATA, C. & ALVAREZ, G. 1995b. Multilocus heterozygosity and sexual selection in a natural population of the marine snail *Littorina mariae* (Gastropoda: Prosobranchia). *Heredity*, **75**: 17–25.

ROQUE, A. LA 1953. Catalogue of the Recent Mollusca of Canada. *Bulletin of the National Museum of Canada*, **129**: 1–406.

ROSEN, B. R. 1988. Biogeographic patterns: a perceptual overview. In: *Analytical Biogeography: an Integrated Approach to the Study of Animal and Plant Distributions* (Myers, A. A. & Giller, P. S., eds), 23–55. Chapman & Hall, New York.

ROSEWATER, J. 1963. Problems of species analogues in world Littorinidae. *Report of the American Malacological Union and AMU Pacific Division*, **30**: 5–6.

ROSEWATER, J. 1970. The family Littorinidae in the Indo-Pacific. Part I. The subfamily Littorininae. *Indo-Pacific Mollusca*, **2**: 417–506.

ROSEWATER, J. 1978. A case of double primary homonymy in Eastern Pacific Littorinidae. *Nautilus*, **92**: 123–125.

ROSEWATER, J. 1980. A close look at *Littorina* radulae. *Bulletin of the American Malacological Union*, **1979**: 5–8.

ROSEWATER, J. 1981. The family Littorinidae in tropical West Africa. *Atlantide Report*, **13**: 7–48.

ROSKELL, F. 1962. Barnacles epizoic on *Littorina* shells. *Journal of Animal Ecology*, **31**: 263–271.

ROSS, B. & BERRY, A. J. 1991. Annual and lunar reproductive cycles in *Littorina saxatilis* (Olivi) and differences between breeding in the marine Firth of Forth and the Forth estuary. *Journal of Molluscan Studies*, **57**: 347–358.

ROTHE, P. 1968. Mesozoische Flysch-Ablagerungen auf der Kanareninsel Fuerteventura. *Geologische Rundschau*, **58**: 314–332.

ROTHPLETZ, A. & SIMONELLI, V. 1890. Die marinen Ablagerungen auf Gran Canaria. *Zeitschrift der Deutschen Geologischen Gesellschaft*, **42**: 677–736.

RÓZYCKI, O. 1991. Mollusca. In: *Atlas of the Marine Fauna of Southern Spitzbergen* (Klekowski, R. Z. & Weslawski, J. M., ed.), **2**: 358–541. Institute of Oceanology, Polish Academy of Sciences, Gdansk.

RUMBAK, E., REID, D. G. & THOMAS, R. H. 1994. Reconstruction of phylogeny of 11 species of *Littorina* (Gastropoda: Littorinidae) using mitochondrial DNA sequence data. *Nautilus. Supplement*, **2**: 91–97.

RUMSEY, T. J. 1973. Some aspects of osmotic and ionic regulation in *Littorina littorea* (L.) (Gastropoda, Prosobranchia). *Comparative Biochemistry and Physiology. Series A*, **45**: 327–344.

RUSSELL, G. 1991. Vertical distribution. In: *Ecosystems of the World* **24**. *Intertidal and Littoral Ecosystems* (Mathieson, A. C. & Nienhuis, P. H., eds), 43–65. Elsevier, Amsterdam.

RUSSELL HUNTER, W. 1953. Notes on the Mollusca of the Garvelloch Islands. *Journal of Conchology*, **23**: 379–386.

RUSSELL HUNTER, W., MAITLAND, P. S. & YEOH, P. K. H. 1964. *Potamopyrgus jenkinsi* in the Loch Lomond area, and an authentic case of passive dispersal. *Proceedings of the Malacological Society of London*, **36**: 27–32.

RUSSELL-HUNTER, W. D. & MCMAHON, R. F. 1974. Patterns of aerial and aquatic respiration in relation to vertical zonation in four littoral snails. *Biological Bulletin of the Marine Biological Laboratory, Woods Hole*, **147**: 496–497.

SABELLI, B., GIANUZZI-SAVELLI, R. & BEDULLI, D. 1990, 1992. *Catalogo Annotato dei Molluschi Marini del Mediterraneo*, **1** and **2**. Edizioni Libreria Naturalistica Bolognese, Bologna.

SACCHI, C. F. 1961. Contribution à l'étude des rapports écologie-polychromatisme chez un prosobranche intercotidal *Littorina obtusata* (L.) à Roscoff. *Cahiers de Biologie Marine*, **2**: 271–290.

SACCHI, C. F. 1962. Relazioni ecologia-policromatismo nel prosobrancho intercotidale *Littorina obtusata* (L.) a Roscoff. II. Ricerche biometriche. *Bollettino di Zoologia*, **28**: 517–528.

SACCHI, C. F. 1963. Contribution à l'étude des rapports écologie/polychromatisme chez un prosobranche intercotidal, *Littorina obtusata* (L.). III.—Données expérimentales et diverses. *Cahiers de Biologie Marine*, **4**: 299–313.

SACCHI, C. F. 1964. Relazioni tra ecologia e policromatismo nel prosobrancho intercotidale *Littorina obtusata* (L.). IV—Studio sulla Ria di Vigo. *Archivio Zoologico Italiano*, **49**: 93–156.

SACCHI, C. F. 1966a. Sur le dimorphisme du pénis chez *Littorina obtusata* (L.) (Gastr. Pros.). *Compte Rendu Hebdomadaire des Séances de l'Académie des Sciences, Paris. Série D*, **262**: 2370–2372.

SACCHI, C. F. 1966b. Contribution à l'étude des rapports entre variabilité et écologie chez un prosobranche intertidal *Littorina obtusata* (L.). V.—Recherches dans le Plymouth Sound (Angleterre S.-W.). *Cahiers de Biologie Marine*, **7**: 281–294.

SACCHI, C. F. 1966c. *Littorina obtusata* (L.) (Gastropoda, Prosobranchia): a problem of variability and its relation to ecology. *Symposia Genetica et Biologica Italica*, **13**: 521–541.

SACCHI, C. F. 1967a. Sur les variations du rapport des sexes au cours de l'année chez *Littorina obtusata* (L.) (Gastropoda: Prosobranchia). *Compte Rendu Hebdomadaire des Séances de l'Académie des Sciences, Paris*, **264**: 1504–1506.

SACCHI, C. F. 1967b. Variabilità ed ambiente nella coppia di specie intertidali *Littorina obtusata* (L.) e *Littorina mariae* Sacchi e Rastelli (Gastropoda, Prosobranchia) a Concarneau (Bretagna meridionale). *Studia Ghisleriana*, **2**: 339–355.

SACCHI, C. F. 1968a. Une population exceptionelle de *Littorina mariae* (Gastropoda, Prosobranchia) à Brest (Nord-Finistère). *Natura, Milan*, **59**: 213–219.

SACCHI, C. F. 1968b. Sur le dimorphisme sexuel de *Littorina mariae* Sacchi et Rast. (Gastr. Prosobranchia). *Compte Rendu Hebdomadaire des Séances de l'Académie des Sciences, Paris. Série D*, **266**: 2483–2485.

SACCHI, C. F. 1969a. Recherches sur l'écologie comparée de *Littorina obtusata* (L.) et de *Littorina mariae* Sacchi et Rast. (Gastropoda, Prosobranchia) en Galice et en Bretagne. *Investigacion Pesquera*, **33**: 381–413.

SACCHI, C. F. 1969b. Ricerche biometriche sul dimorphismo sessuale della coppia di specie intertidali *Littorina obtusata* (L.) e *L. mariae* Sacchi et Rast. (Gastropoda, Prosobranchia) in Bretagna ed in Galizia. *Bollettino di Pesca, di Piscicoltura e di Idrobiologia*, **24**: 33–47.

SACCHI, C. F. 1970. Les epibiontes animaux de *Littorina obtusata* (L.) et de *Littorina mariae* Sacchi et Rast. (Gastropoda, Prosobranchia). *Cahiers de Biologie Marine*, **11**: 43–56.

SACCHI, C. F. 1971. Considérations écologiques sur les peuplements de marée: aspects généraux et problèmes méditerranéens. *Pubblicazioni della Stazioni Zoologica di Napoli. Supplemento*, **37**: 73–123.

SACCHI, C. F. 1972a. Recherches sur l'écologie comparée de *Littorina obtusata* (L.) et de *L. mariae* Sacchi & Rast. (Gastropoda, Prosobranchia). II—Recherches sur la valence thermique. *Bollettino di Pesca, di Piscicoltura e di Idrobiologia*, **27**: 105–137.

SACCHI, C. F. 1972b. Recherches sur la valence thermique du couple d'espèces intertidales *Littorina obtusata* (L.) et *L. mariae* Sacchi & Rast. (Gastropoda, Prosobranchia). *Fifth European Marine Biology Symposium*, 204–215.

SACCHI, C. F. 1974. Le polychromatisme des littorines (Gastropodes, Prosobranches) d'Europe: points du vue d'un écologiste. *Mémoires de la Société Zoologique de France*, **37**: 61–101.

SACCHI, C. F. 1975. *Littorina nigrolineata* Gray (Gastropoda, Prosobranchia). *Cahiers de Biologie Marine*, **16**: 111–120.

SACCHI, C. F. 1979. Frequenze di fenotipi cromatici in *Littorina saxatilis* (Olivi) della Ria di Vigo (Spagna Nordoccidentale) e loro significato ecologico. *Natura, Milan*, **70**: 7–21.

SACCHI, C. F. 1981. Ricerche sulle variazioni di mole in *Littorina saxatilis* (Olivi) e sul loro significato ecologico. *Bollettino del Museo Civico di Storia Naturale di Venezia*, **31**: 51–67.

SACCHI, C. F. 1984. Recherches sur l'écologie comparée de *Littorina obtusata* (L.) et de *L. mariae* Sacchi et Rast., dans la Ria de Arosa, Espagne Nord-Occidentale. I. Caractères des peuplements, polychromatisme, rapports des sexes. (Gastropoda Prosobranchia). *Atti della Società Italiana di Scienze Naturali e del Museo Civico di Storia Naturale di Milano*, **125**: 209–232.

SACCHI, C. F. 1985. Recherches sur l'écologie comparée de *Littorina obtusata* (L.) et de *L. mariae* Sacchi & Rast. dans la Ria de Arosa, Espagne nord-occidentale. II. Les epibiontes animaux. *Atti della Società Italiana di Scienze Naturali e del Museo Civico di Storia Naturale di Milano*, **126**: 212–236.

SACCHI, C. F., AMBROGI, A. O. & VOLTOLINA, D. 1981. Recherches sur le spectre trophique comparé de *Littorina saxatilis* (Olivi) et de *L. nigrolineata* (Gray) (Gastropoda, Prosobranchia) sur la grève de Roscoff. II.—cas de populations vivant au milieu d'algues macroscopiques. *Cahiers de Biologie Marine*, **22**: 83–88.

SACCHI, C. F. & MALCEVSCHI, S. 1983. Polimorphismo ed ambiente nel prosobranco intermareale *Littorina saxatilis* (Olivi) dei dintorni di Vigo (Galizia, Spagna nordoccidentale). *Bollettino del Museo Civico di Storia Naturale di Venezia*, **33**: 7–21.

SACCHI, C. F. & RASTELLI, M. 1966. *Littorina mariae*, nov. sp.: les differences morphologiques et écologiques entre "nains" et "normaux" chez l'«espèce» *L. obtusata* (L.) (Gastr. Prosobr.) et leur signification adaptive et évolutive. *Atti della Società Italiana di Scienze Naturali e del Museo Civico di Storia Naturale di Milano*, **105**: 351–369.

SACCHI, C. F. & SCONFIETTI, R. 1988. Recherches sur l'écologie comparée de *Littorina obtusata* (L.) et de *L. mariae* Sacchi & Rast. dans la Ría de Arosa, Éspagne nord-occidentale III. Dimorphisme sexuel et gradients de taille en *Littorina obtusata* (Gastropoda Prosobranchia). *Atti della Società Italiana di Scienze Naturali e del Museo Civico di Storia Naturale di Milano*, **129**: 285–304.

SACCHI, C. F. & SCONFIETTI, R. 1992. Recherches sur l'écologie comparée de *Littorina obtusata* (L.) et de *L. mariae* Sacchi & Rast. dans la Ría de Arosa, Éspagne nord-occidentale. IV. Dimorphisme sexuel et gradients de taille en *Littorina mariae* (Gastr. Prosobranchia). *Atti della Società Italiana di Scienze Naturali e del Museo Civico di Storia Naturale di Milano*, **132**: 105–118.

SACCHI, C. F., TESTARD, P. & VOLTOLINA, D. 1977. Recherches sur le spectre trophique comparé de *Littorina saxatilis* (Olivi) et de *L. nigrolineata* (Gray) (Gastropoda, Prosobranchia) sur la grève de Roscoff. *Cahiers de Biologie Marine*, **18**: 499–505.

SACCHI, C. F. & TORELLI, A. R. 1973. Présence, variabilité et cycle biotique de *Littorina saxatilis* (Olivi) (Gastropoda, Prosobranchia) dans la lagune de Venise. *Atti della Società Peloritana di Scienze, Fisiche, Matematiche e Naturali*, **19**: 181–188.

SACCHI, C. F. & TORELLI, A. R. 1979. La variabilité de *Littorina saxatilis* (Olivi) (Gastr. Prosobr.) dans la lagune de Venise: conclusions. *Rapport et Procès-Verbaux des Réunions de la Commission Internationale pour l'Exploration Scientifique de la Mer Méditerranée*, **25/26** (3): 125–126.

SACCHI, C. F., TORELLI, A. R. & VOLTOLINA, D. 1977. Contribution à l'étude du régime alimentaire chez *Littorina saxatilis* (Olivi) (Gastropoda, Prosobranchia) dans la lagune de Venise. *Malacologia*, **16**: 241–242.

SACCHI, C. F. & VOLTOLINA, D. 1987. Recherches sur l'écologie comparée des littorines (Gastropoda Prosobranchia) dans l'Île de Vancouver (British Columbia, Canada). *Atti della Società Italiana di Scienze Naturali e del Museo Civico di Storia Naturale di Milano*, **128**: 209–234.

SALIS-MARSCHLINS, C. U. VON 1793. *Reisen in Verschiedene Provinzen des Königreichs Neapel*, **1**. Zurich and Leipzig.

SANNIA, A. & JAMES, B. L. 1977. The Digenea in marine molluscs from Eyjafjördur, north Iceland. *Ophelia*, **16**: 97–109.

SARS, G. O. 1878. *Mollusca Regionis Arcticae Norvegiae*. Christiania University, Christiania.

SARS, M. 1851. Beretning om en i sommeren 1849 foretagen zoologisk reise i Lofoten og Finmarken. *Nyt Magazin for Naturvidenskaberne*, **6**: 121–211.

SAUR, M. 1990. Mate discrimination in *Littorina littorea* (L.) and *L. saxatilis* (Olivi) (Mollusca: Prosobranchia). *Hydrobiologia*, **193**: 261–270.

SAUVAGE, H. E. 1873. Note sur l'accouplement des *Littorina rudis* et *L. littorea*. *Journal de Conchyliologie*, **21**: 122–123.

SAVIN, S. M., DOUGLAS, R. G., KELLER, G., KILLINGLEY, J. S., SHAUGHNESSY, L., SOMMER, M. A., VINCENT, E. & WOODRUFF, F. 1981. Miocene benthic foraminiferal isotope records: a synthesis. *Marine Micropaleontology*, **6**: 423–450.

SAWADA, Y. 1962. The geology and paleontology of the Setana and Kuromatsunai areas in southwest Hokkaido, Japan. *Memoirs of the Muroran Institute of Technology*, **4**: 1–110.

SAY, T. 1822. An account of some of the marine shells of the United States. *Journal of the Academy of Natural Sciences of Philadelphia*, **2**: 221–248.

SCHALKE, H. J. W. G. 1973. The Upper Quaternary of the Cape Flats area (Cape Province, South Africa). *Scripta Geologica*, **15**: 1–57.

SCHELTEMA, R. S. 1971. Larval dispersal as a means of genetic exchange between geographically separated populations of shallow-water benthic marine gastropods. *Biological Bulletin of the Marine Biological Laboratory, Woods Hole*, **140**: 284–322.

SCHELTEMA, R. S. 1989. Planktonic and non-planktonic development among prosobranch gastropods and its relationship to the geographic range of species. In: *Reproduction, Genetics and Distributions of Marine Organisms* (Ryland, J. S. & Tyler, P. A., eds), 183–188. Olsen & Olsen, Fredensborg.

SCHELTEMA, R. S. 1995. The relevance of passive dispersal for the biogeography of Caribbean mollusks. *American Malacological Bulletin*, **11**: 99–115.

SCHIØTTE, T. & WARÉN, A. 1992. An annotated and illustrated list of the types of Mollusca described by H. P. C. Møller from West Greenland. *Meddelelser om Grønland, Bioscience*, **35**: 1–34.

SCHLESCH, H. 1916. Faunula Littorinidae Islandiae Borealis. *Naturalist, Hull*, **1916** (716): 279–280.

SCHLESCH, H. 1924. Zur Kenntnis der pliocänen Cragformation von Hallbjarnarstadur, Tjörnes, Nordisland und ihrer Molluskenfauna. *Abhandlungen des Archiv für Molluskenkunde*, **1**: 309–370.

SCHLESCH, H. 1930. Kleine Mitteilungen VI. *Archiv für Molluskenkunde*, **62**: 210–222.

SCHLESCH, H. 1931. Kleine Mitteilungen VII. *Archiv für Molluskenkunde*, **63**: 133–155.

SCHLESCH, H. 1937. Bemerkungen über die Verbreitung der Süsswasser- und Meeresmollusken im östlichen Ostseegebiete. *Tartu Ülikooli Juures Oleva Loodusuurijate Seltsi Aruanded*, **43**: 37–64.

SCHMITT, R. J. 1979. Mechanics and timing of egg capsule release by the littoral fringe periwinkle *Littorina planaxis* (Gastropoda: Prosobranchia). *Marine Biology, Berlin*, **50**: 359–366.

SCHOLL, D. W., BUFFINGTON, E. C. & MARLOW, M. S. 1975. Plate tectonics and the structural evolution of the Aleutian-Bering Sea region. *Special Papers of the Geological Society of America*, **151**: 1–31.

SCHRENCK, L. VON 1861. Vorläufige Diagnosen einiger neuer Molluskenarten aus der Meerenge der Tartarei und dem Nordjapanischen Meere. *Bulletin de l'Académie Impériale des Sciences de Saint-Pétersbourg*, **4**: 408–413.

SCHRENCK, L. VON 1867. Mollusken des Amur-landes und des nordjapanischen Meeres. In: *Reisen und Forschungen im Amur-lande in den Jahren 1854–1856* (Schrenck, L. von, ed.), **2**. Kaiserlichen Akademie der Wissenschaften, St Petersburg.

SCHRÖTER, J. S. 1779. *Die Geschichte der Flussconchylien*. J. J. Gebauer, Halle.

SCHRÖTER, J. S. 1784. *Einleitung in die Conchylienkenntnis nach Linné*, **2**. J. J. Gebauer, Halle.

SEAWARD, D. R. 1982. *Sea Area Atlas of the Marine Molluscs of Britain and Ireland*. Nature Conservancy Council, Peterborough.

SEAWARD, D. R. 1990. *Distribution of the Marine Molluscs of North West Europe*. Nature Conservancy Council, Peterborough.

SEED, R. 1977. A curious association—periwinkles and saddle-oysters. *Irish Naturalists' Journal*, **19**: 8–9.

SEELEY, R. H. 1982a. Periwinkle puzzles: the New England *Littorina* species. *Hawaiian Shell News*, **30**: 1, 6.

SEELEY, R. H. 1982b. Association of *Littorina obtusata* and fucoid algae: effects of competition with *Littorina littorea*. *Malacological Review*, **15**: 150.

SEELEY, R. H. 1983. Periwinkle puzzles: the seaweed specialists. *Hawaiian Shell News*, **31**: 12.

SEELEY, R. H. 1985. PhD thesis, Yale University, New Haven, Connecticut.

SEELEY, R. H. 1986. Intense natural selection caused a rapid morphological transition in a living marine snail. *Proceedings of the National Academy of Sciences of the U. S. A.*, **83**: 6897–6901.

SELIN, N. I. 1994. Distribution, morphology and production characteristics of the gastropod *Littorina kurila* in the areas of volcanic activity (the Kuril Islands). *Biologiya Morya*, **20**: 126–136. (Russian).

SELIN, N. I. 1995. The effect of hydrochemical conditions upon shell morphology of the gastropod mollusc *Littorina kurila*. *Ekologiya*, **1995**: 90–91. (Russian. Translated in *Russian Journal of Ecology*, **26**: 78–79).

SERGIEVSKY, S. O. 1982a. Ecological studies on the polymorphism of the littoral mollusc *Littorina obtusata* (L.) in the White Sea. In: *Povyshennaya Produktivnost' i Ratsional'noe Ispol'zovanie Biologicheskikh Resursov Belogo Morya. Materialy i Koordinatsionnogo Soveshchaniya, Leningrad, May, 1982*, 78–79. Leningrad. (Russian).

SERGIEVSKY, S. O. 1982b. Differential invasion of the shell-colour forms of the periwinkle *Littorina obtusata* (L.) by trematode parthenitae. *Doklady Akademii Nauk S. S. S. R*, **267**: 246–249. (Russian).

SERGIEVSKY, S. O. 1983a. Shell-colour polymorphism of molluscs: parametric system. In: *Molluski: Sistematika, Ekologiya i Raspredelenie*, 52–54. Leningrad. (Russian).

SERGIEVSKY, S. O. 1983b. The mating system in the periwinkle *Littorina obtusata* (L.) (Gastropoda: Prosobranchia). *Vestnik Leningradskogo Gosudarstvennogo Universiteta (Biologiya)*, **1983** (4): 22–27. (Russian).

SERGIEVSKY, S. O. 1983c. Parapatric differentiation of populations. *Zhurnal Obshchei Biologii*, **44**: 491–500. (Russian).

SERGIEVSKY, S. O. 1984. Analysis of phenotypic structure of one abnormal population of *Littorina obtusata* (Gastropoda: Prosobranchia). *Vestnik Leningradskogo Gosudarstvennogo Universiteta (Biologiya)*, **1984** (1): 96–102. (Russian).

SERGIEVSKY, S. O. 1985a. Population approach to the analysis of the periwinkle *Littorina obtusata* (L.) invasions with the trematode parthenitae. *Helminthologia*, **22**: 5–14.

SERGIEVSKY, S. O. 1985b. Analysis of infection of White Sea populations of the periwinkle *Littorina obtusata* by trematode parthenitae. *Parazitologicheskii Sbornik*, **33**: 99–124. (Russian).

SERGIEVSKY, S. O. 1985c. The phenotypic population structure of the prosobranch gastropod *Littorina obtusata* in the estuary of the Keret River (White Sea). *Biologiya Morya*, **1985** (3): 44–53. (Russian).

SERGIEVSKY, S. O. 1986. Different rates of infection of phenotypes of the littoral mollusc *Littorina obtusata* by trematode parthenitae. *Parazitologiya*, **20**: 61–66. (Russian).

SERGIEVSKY, S. O. 1987. Complex analysis of polymorphism adaptivity (*Littorina obtusata* L. gastropod taken as an example). *Zhurnal Obshchei Biologii*, **48**: 675–686. (Russian).

SERGIEVSKY, S. O. 1989. Population biology of invertebrates. *Soviet Scientific Reviews. Section F. Physiological and General Biology*, **3**: 153–200.

SERGIEVSKY, S. O. 1992. A review of ecophysiological studies of the colour polymorphism of *Littorina obtusata* (L.) and *L. saxatilis* (Olivi) in the White Sea. In: *Proceedings of the Third International Symposium on Littorinid Biology* (Grahame, J., Mill, P. J. & Reid, D. G., eds), 235–245. Malacological Society of London, London.

SERGIEVSKY, S. O. & BERGER, V. YA. 1984. Physiological differences between the major shell colour phenotypes of the periwinkle *Littorina obtusata* (L.). *Biologiya Morya*, **1984** (2): 36–44. (Russian).

SERGIEVSKY, S. O. & GRANOVITCH, A. I. 1987. Population studies of sibling species *Littorina obtusata* and *L. mariae* in the White Sea. In: *Problemy i Izucheniya, Ratsional'nogo Ispol'zovaniya i Okhrany Prirodnykh Resursov Belogo Morya*, 235–238. Kandalaksha. (Russian).

SERGIEVSKY, S. O. & GRANOVITCH, A. I. 1989. The spatial mosaicism in structure of *Littorina saxatilis* (Olivi) populations. *Trudy Zoologicheskogo Instituta*, **203**: 146–168. (Russian).

SERGIEVSKY, S. O. & GRANOVITCH, A. I. 1990. Mosaic phenotypic structure of populations of *Littorina saxatilis* (Gastropoda, Prosobranchia). *Zoologicheskii Zhurnal*, **69**: 5–16. (Russian).

SERGIEVSKY, S. O., GRANOVITCH, A. I. & MIKHAILOVA, N. A. 1991. The age structure of White Sea populations of *Littorina obtusata* and *L. saxatilis*. *Trudy Zoologicheskogo Instituta*, **233**: 79–126. (Russian).

SESHAPPA, G. 1947. Oviparity in *Littorina saxatilis* (Olivi). *Nature, London*, **160**: 335–336.

SESHAPPA, G. 1948. Nomenclature of the British Littorinidae. *Nature, London*, **162**: 702–703.

SESHAPPA, G. 1976. Reproduction in *Littorina saxatilis rudissima* (Bean) and two other subspecies of *Littorina saxatilis* (Olivi) at Cullercoats, England. *Indian Journal of Marine Science*, **5**: 222–229.

SESHAPPA, G. 1988. Notes on *Littorina saxatilis* and allied shells examined at the British Museum of Natural History, London. *Journal of the Marine Biological Association of India*, **30**: 218–223.

SEURAT, L. G. 1924. Observations sur les limites, les faciès et les associations animales de l'étage intercotidal de la petite Syrte (Golfe de Gabès). *Bulletin de la Station Océanographique de Salammbô*, **3**: 1–72.

SEURAT, L. G. 1929. Observations nouvelles sur les faciès et les associations animales de l'étage intercotidal de la Petite Syrte (Golfe de Gabès). *Bulletin de la Station Océanographique de Salammbô*, **12**: 1–59.

SEURAT, L. G. 1934. Formations littorales et estuaires de la Syrte Mineure (Golfe de Gabès). *Bulletin de la Station Océanographique de Salammbô*, **32**: 1–65.

SHACKLETON, N. J., BACKMAN, J., ZIMMERMAN, H., KENT, D. V., HALL, M. A., ROBERTS, D. G., SCHNITKER, D., BALDAUF, J. G., DESPRAIRIES, A., HOMRIGHAUSEN, R., HUDDLESTUN, P., KEENE, J. B., KALTENBACK, A. J., KRUMSIEK, K. A. O., MORTON, A. C., MURRAY, J. W. & WESTBERG-SMITH, J. 1984. Oxygen isotope calibration of the onset of ice-rafting and history of glaciation in the North Atlantic region. *Nature, London*, **307**: 620–623.

SHUMWAY, S. E., LESSER, M. P. & CRISP, D. J. 1993. Specific dynamic action demonstrated in the herbivorous marine periwinkles, *Littorina littorea* L. and *Littorina obtusata* L. (Mollusca, Gastropoda). *Comparative Biochemistry and Physiology. Part A*, **106**: 391–395.

SIGURDSSON, J. B. 1978. *Littorina mariae* from Iceland. *Littorinid Tidings*, **7**: 73.

SIMON, C. 1991. Molecular systematics at the species boundary: exploiting conserved and variable regions of the mitochondrial genome of animals via direct sequencing from amplified DNA. In: *Molecular Techniques in Taxonomy* (Hewitt, G. M., Johnston, A. W. B. & Young, J. W. P., eds), 33–71. Springer Verlag, Berlin.

SKELTON, P. W. 1993. Adaptive radiation: definition and diagnostic tests. In: *Evolutionary Patterns and Processes. Linnean Society Symposium Series* **14** (Lees, D. & Edwards, D., eds), 45–58. Academic Press, London.

SMART, R. W. J. & COOKE, A. H. 1885. The marine shells of Scilly. *Journal of Conchology*, **4**: 285–303.

SMIDT, E. 1944. Biological studies of the invertebrate fauna of the harbour of Copenhagen. *Videnskabelige Meddelelser fra Dansk Naturhistorisk Forening i Kjøbenhavn*, **107**: 235–316.

SMITH, A. B. & LITTLEWOOD, D. T. J. 1994. Paleontological data and molecular phylogenetic analysis. *Paleobiology*, **20**: 259–273.

SMITH, D. A. S. 1976. Disruptive selection and morph-ratio clines in the polymorphic snail *Littorina obtusata* (L.) (Gastropoda: Prosobranchia). *Journal of Molluscan Studies*, **42**: 114–135.

SMITH, E. A. 1890. Report on the marine molluscan fauna of the Island of St Helena. *Proceedings of the Zoological Society of London*, **1890**: 247–317.

SMITH, J. 1839. On the last changes in the relative levels of the land and sea in the British Islands. *Memoirs of the Wernerian Natural History Society*, **8**: 49–107.

SMITH, J. E. 1981. The natural history and taxonomy of shell variation in the periwinkles *Littorina saxatilis* and *Littorina rudis*. *Journal of the Marine Biological Association of the United Kingdom*, **61**: 215–241.

SMITH, J. E. & NEWELL, G. E. 1955. The dynamics of the zonation of the common periwinkle, *Littorina littorea* (L.) on a stony beach. *Journal of Animal Ecology*, **24**: 35–56.

SMITH, L. P. & LITTLE, C. 1980. Intertidal communities on rocky shores in the Severn Estuary. *Proceedings of the Bristol Naturalists' Society*, **38**: 61–67.

SMITH, S. 1860. On the Mollusca of Peconic and Gardiner's Bays, Long Island, New York. *Annals of the Lyceum of Natural History of New York*, **7**: 147–168.

SMITH, S. & PRIME, T. 1870. Report on the Mollusca of Long Island, N. Y., and of its dependencies. *Annals of the Lyceum of Natural History of New York*, **9**: 377–407.

SMITH, S. M. 1978. Winkles in the Uists. *Hebridean Naturalist*, **1**: 45–49.

SMITH, S. M. 1979a. *Littorina rudis* var. *scotia* and its adaptation to the extreme environment of Rockall (Mollusca: Gastropoda). *Porcupine Newsletter*, **1**: 138–139.

SMITH, S. M. 1979b. Littorinas beware—or *cave Littorinae*. *Littorinid Tidings*, **8**: 1–2.

SMITH, S. M. 1981. *Littoral Mollusca of West Sutherland and Coigach*. Nature Conservancy Council, Huntingdon.

SMITH, S. M. 1982. A review of the genus *Littorina* in British and Atlantic waters (Gastropoda: Prosobranchia). *Malacologia*, **22**: 535–539.

SMITH, S. M. & HEPPELL, D. 1991. Checklist of British marine Mollusca. *National Museums of Scotland Information Series*, **11**: 1–114.

SNELI, J.-A. & MARION, P. VAN 1979. Nye strandsnegler i norsk fauna. *Fauna, Oslo*, **32**: 4–8.

SNELI, J.-A. & STEINNES, A. 1975. Marine Mollusca of Jan Mayen Island. *Astarte*, **8**: 7–16.

SNYDER, T. P. & GOOCH, J. L. 1973. Genetic differentiation in *Littorina saxatilis* (Gastropoda). *Marine Biology, Berlin*, **22**: 177–182.

SOKOLOVA, I. M. 1995. Embryonic abnormalities in populations of *Littorina saxatilis* (Olivi) (Gastropoda: Prosobranchia) in the White Sea. *Journal of Molluscan Studies*, **61**: 303–311.

SOULEYET, F. L. A. 1852. Mollusques. In: *Voyage autour du Monde Exécuté pendant les Années 1836 et 1837 sur la Corvette La Bonite. Zoologie* (Eydoux, J. F. T. & Souleyet, F. L. A., eds), **2**: 7–633. Paris.

SOUTHWARD, A. J. & CRISP, D. J. 1954. The distribution of certain intertidal animals around the Irish coast. *Proceedings of the Royal Irish Academy*, **57B**: 1–29.

SOWERBY, G. B. 1839. *A Conchological Manual*. G. B. Sowerby, London.

SOWERBY, G. B. 1844. Description of fossil shells. In: *Geological Observations on the Volcanic Islands, Visited During the Voyage of H.M.S. Beagle* (Darwin, C., ed.), 153–154. Smith, Elder & Co, London.

SOWERBY, G. B. 1859. *Illustrated Index of British Shells*. Simpkin, Marshall & Co, London.

SOWERBY, G. B. 1887. *Illustrated Index of British Shells*. Edition 2. Sowerby, London.

SOWERBY, J. 1812–1815. *The Mineral Conchology of Great Britain*, **1**. London.

SOWERBY, J. 1832. *The Genera of Recent and Fossil Shells*. Part 37. London.

SPAINK, G. 1974. Een merkwaardige deformatie bij gastropoden. *De Kreukel*, **10**: 45–48.

SPÄRCK, R. & THORSON, G. 1933. Marine Gastropoda Prosobranchiata. *Zoology of the Faroes*, **52**: 1–56.

SPARRE SCHNEIDER, J. 1886. Undersøgelser af dyrelivet i de arktiske fjorde. III. Tromsøsundets molluskfauna. *Tromsø Museums Aarshefter*, **9**: 1–50.

SPASSKY, N. N. 1961. Littoral of the southeastern shore of Kamchatka. *Issledovaniya Dal'nevostochnykh Morei S. S. S. R.*, **7**: 261–311. (Russian).

SPIGHT, T. M. 1973. Ontogeny, environment and shape of a marine snail *Thais lamellosa* Gmelin. *Journal of Experimental Marine Biology and Ecology*, **13**: 215–228.

SPIGHT, T. M. 1976. Colors and patterns of an intertidal snail, *Thais lamellosa*. *Researches on Population Ecology, Kyoto University*, **17**: 176–190.

SPIGHT, T. M. 1981. Latitude and prosobranch larvae whose veligers are found in tropical waters. *Ecosynthesis*, **1**: 29–52.

SPJELDNAES, N. & HENNINGSMOEN, K. E. 1963a. *Littorina littorea*: an indicator of Norse settlement in North America? *Science, New York*, **141**: 275–276.

SPJELDNAES, N. & HENNINGSMOEN, K. E. 1963b. *Littorina littorea* as an indicator of Norse settlements. *Science, New York*, **142**: 1022.

SPOONER, G. M. & MOORE, H. B. 1940. The ecology of the Tamar Estuary VI. An account of the macrofauna of the intertidal mud. *Journal of the Marine Biological Association of the United Kingdom*, **24**: 283–330.

STENECK, R. S. & WATLING, L. 1982. Feeding capabilities and limitation of herbivorous molluscs: a functional group approach. *Marine Biology, Berlin*, **68**: 299–319.

STEPHENSEN, K. 1939. Crustacea Decapoda. *Zoology of Iceland*, **3** (25): 1–31.

STEPHENSON, T. A. & STEPHENSON, A. 1949. The universal features of zonation between tide marks on rocky coasts. *Journal of Ecology*, **37**: 289–305.

STEPHENSON, T. A. & STEPHENSON, A. 1952. Life between tide-marks in North America II. Northern Florida and the Carolinas. *Journal of Ecology*, **40**: 1–49.

STEPHENSON, T. A. & STEPHENSON, A. 1954a. Life between tide-marks in North America. IIIA. Nova Scotia and Prince Edward Island: description of the region. *Journal of Ecology*, **42**: 14–45.

STEPHENSON, T. A. & STEPHENSON, A. 1954b. Life between tide-marks in North America. IIIA. Nova Scotia and Prince Edward Island: the geographical features of the region. *Journal of Ecology*, **42**: 46–70.

STEPHENSON, T. A. & STEPHENSON, A. 1961a. Life between tide-marks in North America. IVA. Vancouver Island, I. *Journal of Ecology*, **49**: 1–29.

STEPHENSON, T. A. & STEPHENSON, A. 1961b. Life between tide-marks in North America. IVB. Vancouver Island, II. *Journal of Ecology*, **49**: 227–243.

STEPHENSON, T. A. & STEPHENSON, A. 1972. *Life Between Tidemarks on Rocky Shores*. W. H. Freeman & Co, San Francisco.

STEWART, R. B. 1927. Gabb's California fossil type gastropods. *Proceedings of the Academy of Natural Sciences of Philadelphia*, **78**: 287–447.

STIMPSON, W. 1851. *Shells of New England*. Phillips, Sampson & Co, Boston.

STRATHMANN, M. F. 1987. *Reproduction and Development of Marine Invertebrates of the Northern Pacific Coast: Data and Methods for the Study of Eggs, Embryos, and Larvae*. University of Washington Press, Seattle.

STRATHMANN, R. R. 1978. The evolution and loss of feeding larval stages of marine invertebrates. *Evolution*, **32**: 894–906.

STRATHMANN, R. R. 1985. Feeding and nonfeeding larval development and life-history evolution in marine invertebrates. *Annual Review of Ecology and Systematics*, **16**: 339–361.

STRAUCH, F. 1972. Phylogenese, Adaptation und Migration einiger nordischer mariner Molluskengenera (*Neptunea*, *Panomya*, *Cyrtodaria* und *Mya*). *Abhandlungen der Senckenbergischen Naturforschenden Gesellschaft*, **531**: 1–211.

STRAUGHAN, D. & HADLEY, D. 1978. Experiments with *Littorina* species to determine the relevancy of oil spill data from southern California to the Gulf of Alaska. *Marine Environmental Research*, **1**: 135–163.

STRELNIKOV, I. D. 1929. La faune de la mer de Kara et ses conditions écologiques. *Compte Rendu Hebdomadaire des Séances de l'Académie des Sciences, Paris*, **188**: 1008–1010.

STRESEMANN, E. 1957. *Exkursionsfauna von Deutschland*. Wirbellose I. Volk & Wissen, Berlin.

STROUD, D. A. & KNUDSEN, J. 1982. The demography and reproduction of *Littorina rudis* Maton, 1797, from Greenland. In: *Report of the 1979 Greenland White-fronted Goose Study Expedition to Equalungmiut Nunât, West Greenland* (Fox, A. D. & Stroud, D. A., eds), 257–267. Aberystwyth.

STRUHSAKER, J. W. 1968. Selection mechanisms associated with intraspecific shell variation in *Littorina picta* (Prosobranchia: Mesogastropoda). *Evolution*, **22**: 459–480.

STRUHSAKER, J. W. & COSTLOW, J. D. 1968. Larval development of *Littorina picta* (Prosobranchia, Mesogastropoda) reared in the laboratory. *Proceedings of the Malacological Society of London*, **38**: 153–160.

STUNKARD, H. W. 1930. The life history of *Cryptocotyle lingua* (Creplin), with notes on the physiology of the metacercariae. *Journal of Morphology and Physiology*, **1**: 143–191.

STUNKARD, H. W. 1957. The morphology and life history of the digenetic trematode, *Microphallus similis* (Jägerskiöld, 1900), Baer, 1943. *Biological Bulletin of the Marine Biological Laboratory, Woods Hole*, **112**: 145–164.

SUMNER, F. B., OSBURN, C. & COLE, L. J. 1913. A biological survey of the waters of Woods Hole and vicinity. Section III.—A catalogue of the marine fauna. *Bulletin of the Bureau of Fisheries*, **31**: 549–794.

SUNDBERG, P. 1988. Microgeographic variation in shell characters of *Littorina saxatilis* Olivi—a question mainly of size? *Biological Journal of the Linnean Society*, **35**: 169–184.

SUNDBERG, P., KNIGHT, A. J., WARD, R. D. & JOHANNESSON, K. 1990. Estimating the phylogeny in mollusc *Littorina saxatilis* (Olivi) from enzyme data: methodological considerations. *Hydrobiologia*, **193**: 29–40.

SUNDELL, K. 1985. Adaptability of two phenotypes of *Littorina saxatilis* (Olivi) to different salinities. *Journal of Experimental Marine Biology and Ecology*, **92**: 115–123.

SWAINSON, W. 1840. *A Treatise on Malacology*. Longman, Orme, Brown, Green & Longmans, London.

SWOFFORD, D. L. 1993. *PAUP: Phylogenetic Analysis Using Parsimony*. Illinois Natural History Survey, Champaign.

SWOFFORD, D. L. & MADDISON, W. P. 1987. Reconstructing ancestral character states under Wagner parsimony. *Mathematical Biosciences*, **87**: 199–229.

SYKES, E. R. 1892. On some monstrosities of *Littorina rudis*, Maton. *Proceedings of the Dorset Natural History and Antiquarian Field Club*, **13**: 191–198.

TAKADA, Y. 1992. The migration and growth of *Littorina brevicula* on a boulder shore in Amakusa, Japan. In: *Proceedings of the Third International Symposium on Littorinid Biology* (Grahame, J., Mill, P. J. & Reid, D. G., eds), 277–279. Malacological Society of London, London.

TAKADA, Y. 1995. Seasonal migration promoting assortative mating in *Littorina brevicula* on a boulder shore in Japan. *Hydrobiologia*, **309**: 151–159.

TAKADA, Y. & KIKUCHI, T. 1990. Mobile molluscan communities in boulder shores and the comparison with other intertidal habitats at Amakusa. *Publications of the Amakusa Marine Biological Laboratory*, **10**: 145–168.

TAKADA, Y. & KIKUCHI, T. 1991. Seasonal and vertical variation of the boulder shore fauna in Amakusa. *Publications of the Amakusa Marine Biological Laboratory*, **11**: 1–17.

TAKI, I. & OYAMA, K. 1954. Matajiro Yokoyama's the Pliocene and later faunas from the Kwanto region in Japan. *Palaeontological Society of Japan. Special Papers*, **2**: 1–68.

TALMADGE, R. R. 1962. Random notes on *Littorina newcombiana* Hemphill. *Annual Reports of the American Malacological Union*, **29**: 29.

TATARENKOV, A. N. 1992. Allozyme variation in *Littorina brevicula* (Philippi) from Peter the Great Bay (Sea of

Japan). In: *Proceedings of the Third International Symposium on Littorinid Biology* (Grahame, J., Mill, P. J. & Reid, D. G., eds), 25–30. Malacological Society of London, London.

TATARENKOV, A. N. 1995a. Genetic divergence between sibling species *Littorina mariae* Sacchi & Rastelli and *L. obtusata* (L.) (Mollusca: Gastropoda) from the White Sea. *Ophelia*, **40**: 207–218.

TATARENKOV, A. N. 1995b. Genetic heterogeneity in populations of *Littorina brevicula* (Philippi) (Mollusca: Gastropoda) in the northern part of Peter the Great Bay (Sea of Japan). *Veliger*, **38**: 85–91.

TATARENKOV, A. N. & JOHANNESSON, K. 1994. Habitat related allozyme variation on a microgeographic scale in the marine snail *Littorina mariae* (Prosobranchia: Littorinacea). *Biological Journal of the Linnean Society*, **53**: 105–125.

TATTERSALL, W. M. 1908a. The development of *Littorina*. *Irish Naturalist*, **17**: 238.

TATTERSALL, W. M. 1908b. The development of *Littorina*. *Nature, London*, **78**: 649.

TATTERSALL, W. M. 1920. Notes on breeding habits and life history of the periwinkle. *Scientific Investigations. Fisheries Branch, Department of Agriculture for Ireland*, **1920** (1): 1–11.

TAYLOR, D. W. 1981. Freshwater mollusks of California: a distributional checklist. *California Fish and Game*, **67**: 140–163.

TAYLOR, G. W. 1895. Preliminary catalogue of the marine Mollusca of the Pacific coast of Canada, with notes upon their distribution. *Transactions of the Royal Society of Canada. Series 2*, **1**: 17–100.

TAYLOR, J. D. & REID, D. G. 1990. Shell microstructure and mineralogy of the Littorinidae: ecological and evolutionary significance. *Hydrobiologia*, **193**: 199–215.

TAYLOR, P. M. & ANDREWS, E. B. 1987. Glucose reabsorption by the prosobranch gastropod *Littorina littorea* (L.). *Journal of Experimental Marine Biology and Ecology*, **108**: 99–111.

TAYLOR, P. M. & ANDREWS, E. B. 1988. Osmoregulation in the intertidal gastropod *Littorina littorea*. *Journal of Experimental Marine Biology and Ecology*, **122**: 35–46.

TEILMAN-FRIIS, A. 1898. Smaa Bidrag til Artsbestemmelserne indenfor Slaegterne *Littorina* og *Hydrobia*. *Videnskabelige Meddelelser fra den Naturhistoriske Forening i Kjøbenhavn*, **1897**: 211–215.

TEILMAN-FRIIS, A. 1899. Malakologisk Smaating. *Videnskabelige Meddelelser fra den Naturhistoriske Forening i Kjøbenhavn*, **51**: 429–443.

TEMPLETON, A. R. 1989. The meaning of species and speciation: a genetic perspective. In: *Speciation and its Consequences* (Otte, D. & Endler, J. A., eds), 3–27. Sinauer Associates, Sunderland, Massachusetts.

THAMDRUP, H. M. 1935. Beiträge zur Ökologie der Wattenfauna auf experimenteller Grundlage. *Meddelelser fra Kommissionen for Danmarks Fiskeri- og Havundersøgelser, Serie Fiskeri*, **10** (2): 1–125.

THIELE, J. 1929. *Handbuch der Systematischen Weichtierkunde*. Part 1. Gustav Fischer Verlag, Jena.

THOMPSON, W. 1852. On *Littorina palliata*. *Annals and Magazine of Natural History. Series 2*, **10**: 75–76.

THORODDSEN, T. 1892. Postglaciale marine Aflejringer, Kystterrasser og Strandlinjer i Island. *Geografisk Tidskrift, Kjøbenhavn*, **11**: 209–225.

THORPE, C. 1844. *British Marine Conchology*. Edward Lumley, London.

THORSON, G. 1941. Marine Gastropoda Prosobranchiata. *Zoology of Iceland*, **4** (60): 1–150.

THORSON, G. 1944. The Zoology of East Greenland. Marine Gastropoda Prosobranchiata. *Meddelelser om Grønland*, **121** (13): 1–181.

THORSON, G. 1946. Reproduction and larval development of Danish marine bottom invertebrates, with special reference to the planktonic larvae in the Sound (Øresund). *Meddelelser fra Kommissionen for Danmarks Fiskeri- og Havundersøgelser. Serie Plankton*, **4** (1): 1–523.

THORSON, G. 1950. Reproduction and larval ecology of marine bottom invertebrates. *Biological Reviews of the Cambridge Philosophical Society*, **25**: 1–45.

THORSON, G. 1951. The Godthaab Expedition 1928. Scaphopoda, Placophora, Solenogastres, Gastropoda Prosobranchiata, Lamellibranchiata. *Meddelelser om Grønland*, **81**: 1–117.

THORSON, G. 1965. The distribution of benthic marine Mollusca along the N. E. Atlantic shelf from Gibraltar to Murmansk. *Proceedings of the First European Malacological Congress (1962)*, 5–25.

TITOVA, L. V. 1994. Cenozoic history of Turritelloidea and Buccinoidea (Mollusca: Gastropoda) in the North Pacific. *Palaeogeography, Palaeoclimatology, Palaeoecology*, **108**: 319–334.

TORELLI, A. R. 1974. Bionomia di *Littorina saxatilis* (Olivi) (Gastropoda Prosobranchia) nella laguna di Venezia. *Bollettino del Museo Civico di Storia Naturale di Venezia*, **25**: 37–47.

TORELLI, A. R. 1977. Distribuzione di *Littorina saxatilis* (Olivi) nella Laguna di Venezia. *Atti della Società Italiana di Scienze Naturali e del Museo Civico di Storia Naturale di Milano*, **118**: 125–135.

TORELLI, A. R. 1978. Ricerche sui rapporti tra ecologia e policromatismo in *Littorina saxatilis* (Olivi) (Gastropoda Prosobranchia) nella laguna di Venezia (1). *Cahiers de Biologie Marine*, **19**: 91–98.

TORELLI, A. R. 1980. Notizie preliminari sulla presenza di *Littorina saxatilis* (Olivi) (Gastropoda, Prosobranchia) nelle laguna di Grado-Marano. *Bollettino del Museo Civico di Storia Naturale di Venezia*, **31**: 79–83.

TORELLI, A. R. 1982a. *Littorina saxatilis* (Olivi) (Gastropoda Prosobranchia) nella ria di Arosa (Spagna nord-occidentale). 1—Frequenze dei fenotipi cromatici. *Lavori della Società Malacologica Italiana*, **20**: 133–141.

TORELLI, A. R. 1982b. *Guide per il Riconoscimento delle Specie Animali delle Acque Lagunari e Costiere Italiane*, **8**. *Gasteropodi Conchigliati*. Consiglio Nazionale delle Ricerche, Genoa.

TORELLI, A. R. 1984. *Littorina saxatilis* (Olivi) nella laguna di Venezia: sue variazioni e ciclo biologica in relazione all'ambiente. *Atti della Società Italiana di Scienze Naturali e del Museo Civico di Storia Naturale di Milano*, **125**: 112–120.

TRESIERRA-AGUILAR, A. 1985. Between site variation in shell and body measures of *Littorina littorea*. *Biological Bulletin of the Marine Biological Laboratory, Woods Hole*, **169**: 537–538.

TROSCHEL, F. H. 1856–1863. *Das Gebiss der Schnecken*, **1**. Nicolaische Verlagsbuchhandlung, Berlin.

TRUSSELL, G. C., JOHNSON, A. S., RUDOLPH, S. G. & GILFILLAN, E. S. 1993. Resistance to dislodgement: habitat and size-specific differences in morphology and tenacity in an intertidal snail. *Marine Ecology Progress Series*, **100**: 135–144.

TRYON, G. W. 1873–1874. *American Marine Conchology*. Philadelphia.

TRYON, G. W. 1883. *Structural and Systematic Conchology*, **2**. Philadelphia.

TRYON, G. W. 1887. *Manual of Conchology*, **9**. Philadelphia.

TSUCHIYA, M. 1979. Quantitative survey of intertidal organisms on rocky shores in Mutsu Bay, with special reference to the influence of wave action. *Bulletin of the Marine Biological Station of Asamushi, Tohoku University*, **16**: 69–86.

TSUCHIYA, M. 1984. Effect of barnacle cover on the distributions of the limpet *Collisella (Conoidacmea) heroldi* (Dunker) and the periwinkle *Littorivaga brevicula* (Philippi) on the higher eulittoral zone. *Bulletin of the Marine Biological Station of Asamushi, Tohoku University*, **17**: 177–189.

TURESSON, G. 1922. The species and the variety as ecological units. *Hereditas*, **3**: 100–113.

TURTON, W. 1819. *A Conchological Dictionary of the British Islands*. J. Booth, London.

TURTON, W. 1825. Description of some new British shells; accompanied by figures from the original specimens. *Zoological Journal*, **2**: 361–367.

UNDERWOOD, A. J. 1972. Tide-model analysis of the zonation of intertidal prosobranchs. I. Four species of *Littorina* (L.). *Journal of Experimental Marine Biology and Ecology*, **9**: 239–255.

UNDERWOOD, A. J. 1973. Studies on the zonation of intertidal prosobranchs (Gastropoda: Prosobranchia) in the region of Heybrook Bay, Plymouth. *Journal of Animal Ecology*, **42**: 353–372.

URBAN, E. K. 1962. Remarks on the taxonomy and intertidal distribution of *Littorina* in the San Juan Archipelago, Washington. *Ecology*, **43**: 320–323.

USHAKOV, P. V. 1953. *Fauna of the Okhotsk Sea and Conditions of its Existence*. Akademiya Nauk S. S. S. R. (Russian).

VADAS, R. L. 1992. Littorinid grazing and algal patch dynamics. In: *Proceedings of the Third International Symposium on Littorinid Biology* (Grahame, J., Mill, P. J. & Reid, D. G., eds), 197–209. Malacological Society of London, London.

VADAS, R. L. & ELNER, R. W. 1992. Plant-animal interactions in the north-west Atlantic. In: *Plant-Animal Interactions in the Marine Benthos. Systematics Association Special Volume* **46** (John, D. M., Hawkins, S. J. & Price, J. H., eds), 33–60. Clarendon Press, Oxford.

VADER, W. 1979. Strandkrabben *Carcinus maenas* i Finnmark. *Fauna, Oslo*, **32**: 27–29.

VALEN, L. VAN 1971. Adaptive zones and the orders of mammals. *Evolution*, **25**: 420–428.

VALENTINE, J. W. 1984. Climate and evolution in the shallow sea. In: *Fossils and Climate* (Brenchley, P., ed.), 265–277. John Wiley & Sons, New York.

VALENTINE, J. W. & JABLONSKI, D. 1983. Speciation in the shallow sea: general patterns and biogeographic controls. In: *Evolution, Time and Space: the Emergence of the Biosphere. Systematics Association Special Volume* **23** (Sims, R. W., Price, J. H. & Whalley, P. E. S., eds), 201–226. Academic Press, London.

VALENTINE, J. W. & MEADE, R. F. 1960. Isotopic and zoogeographic paleotemperatures of Californian Pleistocene Mollusca. *Science, New York*, **132**: 810–811.

VEDEL, G. & DEPLEDGE, M. 1995. Temperature tolerance and selected trace metal concentration in some Azorean gastropod molluscs. *Açoreana. Supplement*, **1995**: 113–124.

VERMEIJ, G. J. 1971. Temperature relationships of some tropical Pacific intertidal gastropods. *Marine Biology, Berlin*, **10**: 308–314.

VERMEIJ, G. J. 1973. Morphological patterns in high intertidal gastropods: adaptive strategies and their limitations. *Marine Biology, Berlin*, **20**: 319–346.

VERMEIJ, G. J. 1978. *Biogeography and Adaptation: Patterns of Marine Life*. Harvard University Press, Cambridge, Massachusetts.

VERMEIJ, G. J. 1980. Gastropod shell growth rate, allometry, and adult size: environmental implications. In: *Skeletal Growth of Aquatic Organisms: Biological Records of Environmental Change* (Rhoads, D. C. & Lutz, R. A., eds), 379–394. Plenum Press, New York.

VERMEIJ, G. J. 1982. Environmental change and the evolutionary history of the periwinkle (*Littorina littorea*) in North America. *Evolution*, **36**: 561–580.

VERMEIJ, G. J. 1989. Geographical restriction as a guide to the causes of extinction: the case of the cold northern oceans during the Neogene. *Paleobiology*, **15**: 335–356.

VERMEIJ, G. J. 1991. Anatomy of an invasion: the trans-Arctic interchange. *Paleobiology*, **17**: 281–307.

VERMEIJ, G. J. 1992a. Time of origin and biogeographical history of specialized relationships between northern marine plants and herbivorous molluscs. *Evolution*, **46**: 657–664.

VERMEIJ, G. J. 1992b. Repaired shell breakage and shell thickness in gastropods of the genera *Littorina* and *Nucella* in the Aleutian Islands, Alaska. In: *Proceedings of the Third International Symposium on Littorinid Biology* (Grahame, J., Mill, P. J. & Reid, D. G., eds), 135–139. Malacological Society of London, London.

VERMEIJ, G. J., PALMER, A. R. & LINDBERG, D. R. 1990. Range limits and dispersal of mollusks in the Aleutian Islands, Alaska. *Veliger*, **33**: 346–354.

VERRILL, A. E. 1874. Brief contributions to zoology from the Museum of Yale College. No. XXVII.—Results of recent dredging expeditions on the coast of New England. No. 5. *American Journal of Science and Arts. Series 3*, **7**: 131–138.

VERRILL, A. E. 1880. Rapid diffusion of *Littorina littorea* on the New England coast. *American Journal of Science. Series 3*, **20**: 251.

VITTURI, R., LIBERTINI, A., PANOZZO, M. & MEZZAPELLE, G. 1995. Karyotype analysis and genome size in three Mediterranean species of periwinkles (Prosobranchia: Mesogastropoda). *Malacologia*, **37**: 123–132.

VOLOVA, G. N., GOLIKOV, A. N. & KUSAKIN, O. G. 1979. *Shelled Gastropod Molluscs of Peter the Great Bay*. Institut Biologii Morya, Dal'nevostochnyi Nauchnyi Tsentr, Akademiya Nauk S. S. S. R., Vladivostok. (Russian).

VOLTOLINA, D. & SACCHI, C. F. 1990. Field observations on the feeding habits of *Littorina scutulata* Gould and *L. sitkana* Philippi (Gastropoda, Prosobranchia) of southern Vancouver Island (British Columbia, Canada). *Hydrobiologia*, **193**: 147–154.

VUILLEUMIER, F. & MATTEO, M. B. 1972. Esterase polymorphisms in European and American populations of the periwinkle, *Littorina littorea* (Gastropoda). *Experientia*, **28**: 1241–1242.

WAGNER, F. J. E. 1977. Paleoecology of marine Pleistocene Mollusca, Nova Scotia. *Canadian Journal of Earth Sciences*, **14**: 1305–1323.

WAHL, M. & SÖNNICHSEN, H. 1992. Marine epibiosis. IV. The periwinkle *Littorina littorea* lacks typical antifouling defenses—why are some populations so little fouled? *Marine Ecology Progress Series*, **88**: 225–235.

WALLENTINUS, I. 1991. The Baltic Sea gradient. In: *Ecosystems of the World* **24**. *Intertidal and Littoral Ecosystems* (Mathieson, A. C. & Nienhuis, P. H., eds), 83–108. Elsevier, Amsterdam.

WALLER, T. R. 1991. Evolutionary relationships among commercial scallops (Mollusca: Bivalvia: Pectinidae). In: *Scallops: Biology, Ecology and Aquaculture* (Shumway, S. E., ed.), 1–73. Elsevier, Amsterdam.

WALRECHT, B. J. J. R. 1950. Afwijking in schelpvorm bij een alikruik (*Littorina littorea* L.). *De Levende Natuur*, **53**: 96–98.

WANG, H. 1982. The Cenozoic gastropods from Hainan Island and Leizhou Peninsula of Guandong Province. *Memoirs of the Nanjing Institute of Geology and Paleontology*, **17**: 117–196.

WARD, R. D. 1990. Biochemical genetic variation in the genus *Littorina* (Prosobranchia: Mollusca). *Hydrobiologia*, **193**: 53–69.

WARD, R. D. & JANSON, K. 1985. A genetic analysis of sympatric subpopulations of the sibling species *Littorina saxatilis* (Olivi) and *Littorina arcana* Hannaford Ellis. *Journal of Molluscan Studies*, **51**: 86–94.

WARD, R. D. & WARWICK, T. 1980. Genetic differentiation in the molluscan species *Littorina rudis* and *Littorina arcana* (Prosobranchia: Littorinidae). *Biological Journal of the Linnean Society*, **14**: 417–428.

WARD, R. D., WARWICK, T. & KNIGHT, A. J. 1986. Genetic analysis of ten polymorphic enzyme loci in *Littorina saxatilis* (Prosobranchia: Mollusca). *Heredity*, **57**: 233–241.

WARÉN, A. 1980. Marine Mollusca described by John Gwyn Jeffreys, with the location of the type material. *Conchological Society of Great Britain and Ireland Special Publication*, **1**: 1–60.

WARÉN, A. 1983. Marine Mollusca described by W. Turton and W. Clark. *Journal of Conchology*, **31**: 161–171.

WARMOES, T. 1986. *Een Inleidende Systematische en Taxonomische Studie van het Genus* Littorina *(Gastropoda, Prosobranchia)*. Licentiaatsthesis, Universitaire Instelling Antwerpen.

WARMOES, T. 1989. Trends in littorinid research in Belgium (Mollusca, Gastropoda, Prosobranchia). In: *Verhandelingen van het Symposium 'Invertebraten van Belgie'* (Wouters, K. & Baert, L., eds), 125–128. Koninlijk Belgisch Instituut voor Natuurwetenschappen, Brussels.

WARMOES, T., BACKELJAU, T. & BRUYN, L. DE 1988. The littorinid fauna of the Belgian coast (Mollusca, Gastropoda). *Bulletin de l'Institut Royal des Sciences Naturelles de Belgique. Biologie*, **58**: 51–70.

WARMOES, T., DUMOULIN, E. & REID, D. G. 1992. The distribution of *Littorina* and *Melarhaphe* along the continental coast of the English Channel. In: *Proceedings of the Third International Symposium on Littorinid Biology* (Grahame, J., Mill, P. J. & Reid, D. G., eds), 293–303. Malacological Society of London, London.

WARWICK, T. 1982. A method of maintaining and breeding members of the *Littorina saxatilis* (Olivi) species complex. *Journal of Molluscan Studies*, **48**: 368–370.

WARWICK, T., KNIGHT, A. J. & WARD, R. D. 1990. Hybridisation in the *Littorina saxatilis* species complex (Prosobranchia: Mollusca). *Hydrobiologia*, **193**: 109–116.

WATANABE, M. 1985. On the feeding activity of *Littorina brevicula*. *Mitamaki*, **17**: 22–27.

WATERSTON, J. & TAYLOR, J. W. 1906. Land and freshwater molluscs of St Kilda. *Annals of Scottish Natural History*, **57**: 21–24.

WATSON, D. C. & NORTON, T. A. 1985a. Dietary preferences of the common periwinkle, *Littorina littorea* (L.). *Journal of Experimental Marine Biology and Ecology*, **88**: 193–211.

WATSON, D. C. & NORTON, T. A. 1985b. The physical characteristics of seaweed thalli as deterrents to littorine grazers. *Botanica Marina*, **28**: 383–387.

WATSON, D. C. & NORTON, T. A. 1987. The habitat and feeding preferences of *Littorina obtusata* (L.) and *L. mariae* Sacchi et Rastelli. *Journal of Experimental Marine Biology and Ecology*, **112**: 61–72.

WATSON, R. B. 1886. Report on the Scaphopoda and Gasteropoda collected by H. M. S. Challenger during the years 1873–76. *Report on the Scientific Results of the Voyage of H. M. S. Challenger during the Years 1873–76. Zoology*, **15**: 1–756.

WATSON, R. B. 1897. On the marine Mollusca of Madeira. *Journal of the Linnean Society. Zoology*, **26**: 233–329.

WEAVER, C. E. 1909. Stratigraphy and palaeontology of the San Pablo formation in Middle California. *University of California Publications. Bulletin of the Department of Geology*, **5**: 243–269.

WEAVER, C. E. 1942. Paleontology of the marine Tertiary formations of Oregon and Washington. Parts II and III. *University of Washington Publications in Geology*, **5**: 275–562, 563–790.

WEAVER, C. E. & PALMER, K. VAN W. 1922. Fauna from the Eocene of Washington. *University of Washington Publications in Geology*, **1** (3): 1–56.

WEINKAUFF, H. C. 1868. *Die Conchylien des Mittelmeeres*, **2**. T. Fischer, Cassel.

WEINKAUFF, H. C. 1878, 1882. Die Gattung *Litorina*. Angefaugen von Dr. Küster, durchgesehen, ergänzt und vollendet von H. C. Weinkauff. In: *Systematisches Conchylien-Cabinet von Martini und Chemnitz* (Küster, H. C., Kobelt, W. & Weinkauff, H. C., eds), parts 269 (1878, pp. 25–40), 315 (1882, pls 6–11, pp. 41–72), 318 (1882, pls 12–14, pp. 73–114). Bauer & Raspe, Nürnberg.

WEINKAUFF, H. C. 1883. Catalog der Gattung *Littorina* Férussac. *Jahrbücher der Deutschen Malakozoologischen Gesellschaft nebst Nachrichtsblatt*, **10**: 213–227.

WELLS, H. W. 1965. Maryland records of the gastropod *Littorina littorea*, with a discussion of factors controlling its southern distribution. *Chesapeake Science*, **6**: 38–42.

WENZ, W. 1938. Gastropoda. Teil I: Allgemeiner Teil und Prosobranchia. In: *Handbuch der Paläozoologie* (Schindewolf, O. H., ed.), **6**. Gebrüder Borntraeger, Berlin.

WESTERLUND, C. A. 1871, 1873. *Fauna Molluscorum Terrestrium et Fluviatilium Sveciae, Norvegiae et Daniae*. A. Bonnier, Stockholm.

WHITE, E. 1959. *British Caenozoic Fossils (Tertiary and Quaternary)*. British Museum (Natural History), London.

WHITEAVES, J. F. 1901. *Catalogue of the Marine Invertebrata of Eastern Canada*. Geological Survey of Canada,

WIESER, W. 1952. Investigations on the microfauna inhabiting seaweeds on rocky coasts. IV. Studies on the vertical distribution of the fauna inhabiting seaweeds below the Plymouth Laboratory. *Journal of the Marine Biological Association of the United Kingdom*, **31**: 145–174.

WILEY, E. O. 1981. *Phylogenetics: the Theory and Practice of Phylogenetic Systematics*. John Wiley, New York.

WILHELMSEN, U. & REISE, K. 1994. Grazing on green algae by the periwinkle *Littorina littorea* in the Wadden Sea. *Helgoländer Meeresuntersuchungen*, **48**: 233–242.

WILKINS, N. P. & O'REGAN, D. 1980. Genetic variation in sympatric sibling species of *Littorina*. *Veliger*, **22**: 355–359.

WILKINS, N. P., O'REGAN, D. & MOYNIHAN, E. 1978. Electrophoretic variability and temperature sensitivity of phosphoglucose isomerase and phosphoglucomutase in littorinids and other marine molluscs. In: *Marine Organisms: Genetics, Ecology and Evolution* (Battaglia, B. & Beardmore, J. A., eds), 141–155. Plenum Press, New York.

WILLEY, C. H. & GROSS, P. R. 1957. Pigmentation in the foot of *Littorina littorea* as a means of recognition of infection with trematode larvae. *Journal of Parasitology*, **43**: 324–327.

WILLIAMS, E. E. 1964. The growth and distribution of *Littorina littorea* (L.) on a rocky shore in Wales. *Journal of Animal Ecology*, **33**: 413–432.

WILLIAMS, E. E. 1970. Seasonal variation in the biochemical composition of the edible winkle *Littorina littorea* L. *Comparative Biochemistry and Physiology*, **33**: 655–661.

WILLIAMS, G. A. 1990a. *Littorina mariae*—a factor structuring low shore communities? *Hydrobiologia*, **193**: 139–146.

WILLIAMS, G. A. 1990b. The comparative ecology of the flat periwinkles *Littorina obtusata* (L.) and *L. mariae* Sacchi et Rastelli. *Field Studies*, **7**: 469–482.

WILLIAMS, G. A. 1992a. The effect of predation on the life histories of *Littorina obtusata* and *Littorina mariae*. *Journal of the Marine Biological Association of the United Kingdom*, **72**: 403–416.

WILLIAMS, G. A. 1992b. The effects of grazing by *Littorina mariae* on diatom assemblages on natural and artificial surfaces. In: *Proceedings of the Third International Symposium on Littorinid Biology* (Grahame, J., Mill, P. J. & Reid, D. G., eds), 211–219. Malacological Society of London, London.

WILLIAMS, G. A. 1994. Variation in populations of *Littorina obtusata* and *L. mariae* (Gastropoda) in the Severn Estuary. *Biological Journal of the Linnean Society*, **51**: 189–198.

WILLIAMS, G. A. 1995. Maintenance of zonation patterns in two species of flat periwinkle, *Littorina obtusata* and *L. mariae*. *Hydrobiologia*, **309**: 143–150.

WILLIAMS, G. A. & BRAILSFORD, T. J. 1990. The relationship between flat periwinkle life histories and digenean infections. *Hydrobiologia*, **193**: 183–190.

WILLIAMS, I. C. & ELLIS, C. 1975. Movements of the common periwinkle *Littorina littorea* (L.) on the Yorkshire coast in winter and the influence of infection with larval Digenea. *Journal of Experimental Marine Biology and Ecology*, **17**: 47–58.

WILLIS, J. R. 1863. On the occurrence of *Littorina litorea* on the coast of Nova Scotia. *Transactions of the Nova Scotian Institute of Natural Science*, **1**: 88–90.

WINCKWORTH, R. 1922. Nomenclature of British Littorinidae. *Proceedings of the Malacological Society of London*, **15**: 95–97.

WINCKWORTH, R. 1932. The British marine Mollusca. *Journal of Conchology*, **19**: 211–252.

WINCKWORTH, R. 1933. Names of British marine Mollusca, I. *Journal of Conchology*, **19**: 334–338.

WINCKWORTH, R. 1951. A list of the marine Mollusca of the British Isles: additions and corrections. *Journal of Conchology*, **23**: 131–134.

WIUM-ANDERSEN, G. 1970. Haemoglobin and protein variation in three species of *Littorina*. *Ophelia*, **8**: 267–273.

WOHLENBERG, E. 1937. Die Wattenmeer-Lebenagemeinschaften im Königshafen von Sylt. *Helgoländer Wissenschaftliche Meeresuntersuchungen*, **1**: 1–92.

WOLF, H. DE, DONGEN, S. VAN, BRITO, C. & BACKELJAU, T. 1994. Population genetics of *Littorina striata* at a microgeographical scale. *Cahiers de Biologie Marine*, **35**: 243–244.

WOOD, S. V. 1842. A catalogue of shells from the Crag. *Annals and Magazine of Natural History. Series 1*, **9**: 455–462, 527–544.

WOOD, S. V. 1848. *A Monograph of the Crag Mollusca. Part I. Univalves*. Palaeontographical Society, London.

WOOD, S. V. 1872. *Supplement to the Crag Mollusca. Part I. Univalves*. Palaeontographical Society, London.

WOOD, W. 1818. *Index Testaceologicus; or a Catalogue of Shells, British and Foreign*. London.

WOOD, W. 1828. *Index Testaceologicus; or a Catalogue of Shells, British and Foreign*. Edition 2. London.

WOODRING, W. P. & BRAMLETTE, M. N. 1946. Geology and paleontology of Palos Verdes Hills, California. *United States Geological Survey Professional Papers*, **207**: 1–145.

WOODRING, W. P. & BRAMLETTE, M. N. 1951. Geology and paleontology of the Santa Maria district, California. *United States Geological Survey Professional Papers*, **222**: 1–185.

WOODRING, W. P., STEWART, R. & RICHARDS, R. W. 1940. Geology of the Kettleman Hills oil field, California. Stratigraphy, paleontology, and structure. *United States Geological Survey Professional Papers*, **195**: 1–170.

WOODS, A. J. C. & SAUL, L. R. 1986. New Neritidae from southwestern North America. *Journal of Paleontology*, **60**: 636–655.

WOODWARD, B. B. 1909. Darwinism and malacology. *Proceedings of the Malacological Society of London*, **8**: 272–286.

WOODWARD, S. 1833. *An Outline of the Geology of Norfolk*. J. Stacy, Norwich.

WOODWARD, S. P. 1856. *A Manual of the Mollusca*. Part 3. John Weale, London.

WRIGHT, F. S. 1936. Report on the Malden (Essex) periwinkle fishery together with observations on the natural history of the common periwinkle *Littorina littorea* Linnaeus and suggestions in regard to conservation. *Fishery Investigations. Series 2*, **14** (6): 1–37.

YAMAGUCHI, M. 1967. Egg capsules of a periwinkle, *Littorina brevicula*, in plankton samples. *Venus*, **25**: 73–76.

YAMAMOTO, G. & HABE, T. 1962. Fauna of shell bearing mollusks in Mutsu Bay. *Bulletin of the Biological Station of Asamushi, Tohoku University*, **11**: 1–20.

YEN, T.-C. 1936. The marine gastropods of Shantung Peninsula. *Contributions from the Institute of Zoology, National Academy of Peiping*, **3**: 165–255.

YEN, T.-C. 1942. A review of Chinese gastropods in the British Museum. *Proceedings of the Malacological Society of London*, **24**: 170–289.

YI, J. & LI, F. 1990. Distribution and number change of the littorinid (Gastropoda) on intertidal hard bed in the Jiulong Estuary, Fujian Province. *Acta Oceanologica Sinica. English Edition*, **9**: 293–302.

YOKOYAMA, M. 1920. Fossils from the Miura Peninsula and its immediate north. *Journal of the College of Science, Imperial University of Tokyo*, **39** (6): 1–193.

YOKOYAMA, M. 1925. Tertiary Mollusca from Shinano and Echigo. *Journal of the Faculty of Science, Imperial University of Tokyo. Section 2*, **1** (1): 1–23.

YOU, Z.-J. 1990. Ecological studies on Littorinidae on the Zhe Jiang coast. *Chinese Journal of Zoology*, **25** (4): 1–6. (Chinese).

YOUNG, C. M. & CHIA, F.-S. 1988. Abundance and distribution of pelagic larvae as influenced by predation, behaviour, and hydrographic factors. *Reproduction of Marine Invertebrates*, **9**: 385–463.

ZALASIEWICZ, J. A. & GIBBARD, P. L. 1988. The Pliocene to early Middle Pleistocene of east Anglia: an overview. In:

The Pliocene-Middle Pleistocene of East Anglia. Field Guide (Gibbard, P. L. & Zalasiewicz, J. A., eds), 1–31. Quaternary Research Association, Cambridge.

ZAOUALI, J. 1978. Les peuplements malacologiques de la mer de Bou-Grara. *Bulletin de l'Office Nationale de la Pêche, Tunisie*, **2**: 199–209.

ZASLAVSKAYA, N. I. 1989. Genetic variability of four Pacific species of *Littorina* (Mollusca Littorinidae). *Genetika*, **25**: 1636–1644. (Russian).

ZASLAVSKAYA, N. I. 1995. Allozyme comparison of four littorinid species morphologically similar to *Littorina sitkana*. *Hydrobiologia*, **309**: 123–128.

ZASLAVSKAYA, N. I., SERGIEVSKY, S. O. & TATARENKOV, A. N. 1992a. Allozyme similarity of Atlantic and Pacific species of *Littorina* (Gastropoda: Littorinidae). *Journal of Molluscan Studies*, **58**: 377–384.

ZASLAVSKAYA, N. I., SERGIEVSKY, S. O. & TATARENKOV, A. N. 1992b. Genetic biochemical composition of Atlantic and Pacific species of the genus *Littorina* (Mollusca: Gastropoda). *Genetika*, **28** (7): 89–98. (Russian).

ZENKEVITCH, L. A. 1963. *Biology of the Seas of the U. S. S. R.* George Allen & Unwin, London.

ZIEGELMEIER, E. 1966. Die Schnecken (Gastropoda Prosobranchia) der deutschen Meeresgebiete und brackigen Küstengewässer. *Wissenschaftliche Meeresuntersuchungen, Abteilung Helgoland*, **13**: 1–61.

ZIMMERMAN, S. T., HANSON, J. L., FUJIOKA, J. T., CALVIN, N. I., GHARRETT, J. A. & MACKINNON, J. S. 1977. *Intertidal Biota and Subtidal Kelp Communities of the Kodiak Island Area*. Northwest and Alaska Fisheries Center, Auke Bay Laboratory, National Marine Fisheries Service, N. O. A. A., Auke Bay, Alaska.

TAXONOMIC INDEX

The index lists all names that appear in the synonymies of *Littorina* species described in Chapter 4. Available names of generic and specific rank (i.e. those available as valid names or replacement names, and which enter into homonymy) are indicated in bold type. Unavailable infrasubspecific names and *nomina nuda* are also listed, but in plain type. Entries are in the form: specific (or subspecific or infrasubspecific) epithet, followed by author, date and original genus (and, for names below the rank of species, original species). Valid names of *Littorina* species are in capitals. For each epithet, the valid name (if different) follows in square brackets. Page references are to the main entries in the synonymies only. Those species excluded from *Littorina* that are specifically discussed in the text are also listed, but misidentifications and obviously incorrect subsequent spellings are not included.